Dioxin Perspectives

A Pilot Study on International Information Exchange on Dioxins and Related Compounds

NATO • Challenges of Modern Society

A series of edited volumes comprising multifaceted studies of contemporary problems facing our society, assembled in cooperation with NATO Committee on the Challenges of Modern Society.

Dioxin Perspectives

A Pilot Study on International Information Exchange on Dioxins and Related Compounds

Edited by
Erich W. Bretthauer
U.S. Environmental Protection Agency
Washington, D.C.

Heinrich W. Kraus
Bundesministerium für Umwelt
Bonn, Germany

and

Alessandro di Domenico
Istituto Superiore di Sanità
Rome, Italy

Coordinating Editors:

Frederick W. Kutz
U.S. Environmental Protection Agency
Washington, D.C.

David P. Bottimore
Versar Inc.
Springfield, Virginia

Otto Hutzinger
University of Bayreuth
Bayreuth, Germany

Heidelore Fiedler
University of Bayreuth
Bayreuth, Germany

and

A. Essam Radwan
University of Central Florida
Orlando, Florida

Published in cooperation with
NATO Committee on the Challenges of Modern Society

SPRINGER SCIENCE+BUSINESS MEDIA, LLC

Library of Congress Cataloging in Publication Data

Dioxin perspectives: a pilot study on international information exchange on
dioxins and related compounds / edited by Erich W. Bretthauer, Heinrich W.
Kraus, and Alessandro di Domenico.
        p.   cm.—(NATO challenges of modern society; v. 16)
    "Published in cooperation with NATO Committee on the Challenges of Modern
Society."
    Includes bibliographical references and index.
    ISBN 978-1-4613-6456-6   ISBN 978-1-4615-3308-5 (eBook)
    DOI 10.1007/978-1-4615-3308-5
    1. Dioxins—Toxicology—Congresses. 2. Dioxins—Environmental aspects—
Congresses. I. Bretthauer, Erich Walter, date. II. Kraus, Heinrich W. III. Di
Domenico, Alessandro. IV. Series.
RA1242.D55D563   1991   91-4262
615.9'5112—dc20   CIP

Proceedings of a NATO/CCMS Pilot Study on International Information
Exchange on Dioxins and Related Compounds, which began May 1985,
in Brussels, Belgium, and concluded April 1988, in Berlin, Germany

ISBN 978-1-4613-6456-6

© 1991 Springer Science+Business Media New York
Originally published by Plenum Press, New York in 1991

ACKNOWLEDGMENTS

This volume reflects over three years of effort from the participants in the NATO/CCMS Pilot Study on International Information Exchange on Dioxins and Related Compounds. The Pilot Study produced 15 reports that describe the findings and accomplishments of the project. This volume summarizes the products from the international information exchange activities in a format suitable for quick reference. The editors of the volume are listed below:

Editors: Erich W. Bretthauer
U.S. Environmental Protection Agency
Washington, D.C., USA

Heinrich W. Kraus
Bundesministerium für Umwelt, Naturschutz und
 Reaktorsicherheit
Bonn, Federal Republic of Germany

Alessandro di Domenico
Istituto Superiore di Sanita
Rome, Italy

Coordinating Editors: Frederick W. Kutz
U.S. Environmental Protection Agency
Washington, D.C., USA

David P. Bottimore
Versar Inc.
Springfield, Virginia, USA

Otto Hutzinger
University of Bayreuth
Bayreuth, Federal Republic of Germany

Heidelore Fiedler
University of Bayreuth
Bayreuth, Federal Republic of Germany

A. Essam Radwan
Arizona State University
Tempe, Arizona, USA

Principal authors for the contributions can be found at the beginning of each section.

Special thanks are extended to Wendy Grieder and Lynn Schoolfield, U.S. EPA, Office of International Activities; Thomas Thompson and Frank W. Karasek, Department of Chemistry, University of Waterloo, Ontario, Canada; Leslie C. Dickson, University of Bayreuth, FRG; Terry L. Stoddart and Jeffrey J. Short, U.S. Air Force, Tyndall Air Force Base, Florida; Horst Neidhard, Environmental Protection Agency, Berlin; and Antonio Spallino, President of "Alessandro Volta" Center.

Information contained in this book was collected through the contributions of the following lead delegates: Martin J. Boddington, and Hugh P. Dibbs, Environment Canada, Canada; Arne Grove, Kemiteknik Teknologisk Institut, Denmark; Christa Morawa, Umweltbundesamt, Federal Republic of Germany; Alessandro di Domenico, Istituto Superiore di Sanita, Italy; Job A. van Zorge, Ministerie van Volkshuisvestung, Rumtelijke Ordering en Milieubeheer, The Netherlands; Sigrid Louise Bjornstad, State Pollution Control Authority, Norway; Ernest A. Cox, HM Inspectorate of Pollution, United Kingdom; Frances Pollitt, Department of Health and Social Security, United Kingdom; and Erich W. Bretthauer, U.S. EPA Office of Research and Development, USA.

The information contained in this book was collected through the contributions of numerous people, including:

Donald G. Barnes, U.S. EPA, Office of the Administrator;
Werner Beckert, U.S. EPA, Office of Research and Development;
Jakob Enresmann, Gesellschaft zur Beseitigung van Sonderabfallen
    in Rheinland-Pfalz, Gerolsheim, FRG;
Sergio Facchetti, Commission of the European Communities;
Renate Fuchs, Chair of Ecological Chemistry and Geochemistry,
    University of Bayreuth, FRG;
Donald L. Grant, Health and Welfare, Canada
Helmut Greim, GSF Institut for Toxicology, Munich, FRG;
H. Hagenmaier, Organische Chemie, Universitat Tubingen, FRG;
Donald J. Hay, Chief, Urban Activities Division, Environmental
    Protection Servie, Environment Canada, Ottawa, Ontario;
Annette Heindl, Chair of Ecological Chemistry, University of
    Bayreuth, FRG;
Ralf Kilger, Umweltbehorde, Amt fur Altlastensanierung,
    Hansestadt Hamburg, Hamburg, FRG;
Michael McLachlan, Chair of Ecological Chemistry, University of
    Bayreuth, FRG;
Connie Moller, Ministry of the Environment, National Agency of
    Environmental Protection, Copenhagen, Denmark;
J. J. M. Post, MT-TNO, Appeldoorn, The Netherlands;
Stephen H. Safe, Texas A & M University, USA;
A. A. Sien, Rijksinstiuut voor Volksgezondheid en Milieuhygiene,
    Bilthoven, The Netherlands;
Susanne Sieners, Amt fur Umweltuntersuchungen der Umweltbehorde
    Hamburg, Hamburg, FRG;

Ludwig Stieglitz, Institut fur Heisse Chemie, Kernforschungsanlage
   Karlsruhe, FRG;
James Wilson, Monsanto, St. Louis, Missouri, USA;
Armon F. Yanders, Environmental Trace Substances Research Center,
   University of Missouri, Columbia, Missouri, USA;
Alvin L. Young, U.S. Department of Agriculture, Washington, D.C., USA.

In addition to the participants of the Pilot Study, attendees of the meetings, and authors of this book, acknowledgment also goes to numerous Versar personnel involved in the preparation of this book: Christel Ackerman, Kathy Bowles, Libba Colby, Kevin Cross, Juliet Crumrine, Steve Duda, Sally Gravely, Kammi Johannsen, Linda Mandeville, and Pat Wood.

CONTENTS

LIST OF TABLES

LIST OF FIGURES

CHAPTER 3.   TECHNOLOGY ASSESSMENT WORKING GROUP
Formation of Dioxins and Related Compounds from
Combustion and Incineration Processes

LIST OF ACRONYMS

AB	- afterburner
AHH	- aryl hydrocarbon hydroxylase
ASME	- American Society of Mechanical Engineers
BGA	- German Federal Health Office
BHout	- baghouse (outlet)
BLB	- black liquor boiler
BSOB	- Binghamton State Office Building
CA	- chimney ash
CB	- chlorobenzenes
CDDs/CDFs	- chlorinated dibenzo-p-dioxins and -dibenzofurans
CDWG	- Chlorinated Dioxins Work Group
CE	- combustion efficiency
CEC	- Commission of the European Communities
Cl_xDD	- polychlorinated dibenzo-p-dioxins
Cl_xDF	- polychlorinated dibenzofurans
CP	- chlorophenols
CRF	- carbon regeneration furnace
DBE	- dibromoethane
DBR	- drum and barrel reclamation incinerator
DCE	- dichloroethane
EPA	- U.S. Environmental Protection Agency
EPA-TEFs	- toxicity equivalency factors adopted by EPA in 1987
EPA-TEQs	- toxicity equivalents (based on EPA-TEFs)
EFW	- energy from waste
EISIM	- electron impact single ion mode (GC/MS detection)
EPS	- Environmental Protection Service (Canada)
ESP	- electrostatic precipitator
FA	- fly ash
FF	- fabric filter
FG	- flue gas
FP	- fire place
FRG	- Federal Republic of Germany
GerStaff	- German statuatory regulation for dangerous substances
HCB	- hexachlorobenzene
HCH	- hexachlorocyclohexane
HDPE	- high density polyethylene
HxCDD	- hexachlorinated dibenzo-p-dioxin
HxCDF	- hexachlorinated dibenzofuran
HpCDD	- heptachlorinated dibenzo-p-dioxin
HpCDF	- heptachlorinated dibenzofuran
ISW	- industrial solid waste incinerator
I-TEFs	- International Toxicity Equivalency Factors
I-TEQs	- International Toxicity Equivalents (based on I-TEFs)
LR	- low recovery
MSW	- municipal sewage waste
MWC	- municipal waste combustor

NATO-CCMS	-	North Atlantic Treaty Organization - Committee Challenges of Modern Society
NA	-	not available
ND	-	not detected
NITEP	-	National Incineration Testing and Evaluation Program
NS	-	not sampled
NYSDOH	-	New York State Department of Health
OA	-	open air
OCDD	-	octachlorodibenzo-*p*-dioxin
OCDF	-	octachlorodibenzofuran
OECD	-	Organization for Economic Cooperation and Development
PAH	-	polynuclear aromatic hydrocarbons
PBB	-	polybrominated biphenyls
PBDDs	-	polybrominated dibenzo-*p*-dioxins
PBDFs	-	polybrominated dibenzofurans
PCBs	-	polychlorinated biphenyls
PCDDs	-	polychlorinated dibenzo-*p*-dioxins
PCDFs	-	polychlorinated dibenzofurans
PeCDD	-	pentachlorinated dibenzo-*p*-dioxins
PeCDF	-	pentachlorinated dibenzofurans
PEI	-	Prince Edward Island
PCP	-	pentachlorophenol
ppb	-	parts per billion
ppm	-	parts per million
ppq	-	parts per quadrillion
ppt	-	parts per trillion
pu	-	polyurethane
pvc	-	polyvinyl chloride
PxDD	-	polyhalogenated dibenzo-*p*-dioxins
PxDF	-	polyhalogenated dibenzofurans
RDF	-	refused-derinsed fuel
RfD	-	reference dose
SAB	-	EPA's Science Advisory Board
SAR	-	structure-activity relationship
SCR	-	scrubber
SD	-	spray dryer
SSI	-	sewage sludge incinerator
2,4,5-T	-	trichlorophenoxyacetic acid
TCB	-	trichlorobenzene
TCDD	-	tetrachlorodibenzo-*p*-dioxin
2,3,7,8,-TCDD	-	2,3,7,8-tetrachlorodibenzo-*p*-dioxin
TCP	-	trichlorophenol
TE	-	toxic equivalents
TEF	-	toxicity equivalency factor
TEQ	-	toxicity equivalents
THC	-	total hydrocarbon
TOX	-	total organic halogen
UBA	-	German Federal Environmental Protection Office
UNEP	-	United Nations Environmental Programme
WB	-	wood burning
WFB	-	wood fired boiler
WHO	-	World Health Organization
WRI	-	wire reclamation incinerator
WS	-	wood stove

CHAPTER 1 - INTRODUCTION AND BACKGROUND INFORMATION

Erich W. Bretthauer*, Heinrich W. Kraus**, and
Alessandro di Domenico***

U.S. Environmental Protection Agency*, Bundesministerium fur Umwelt
Naturschutz, and Reaktorsicherheit**, and the Istituto Superiore di
Sanita***

Washington, D.C.*, Bonn, Federal Republic of Germany**, and Rome
Italy***

The Pilot Study on International Information Exchange on Dioxins and
Related Compounds was initiated to address issues associated with chlorinated dibenzo-p-dioxins (CDDs), dibenzofurans (CDFs), and related
compounds. The 3-year project was initiated in 1985 under the auspices
of the Committee on the Challenges of Modern Society (CCMS) of the North
Atlantic Treaty Organization (NATO). Numerous activities were carried
out to promote information exchange among the participating nations and
to foster cooperative research efforts. The Pilot Study produced 15 CCMS
documents, as well as numerous journal articles (Kutz and Bottimore, 1988;
Bottimore et al., 1989), that describe the specific accomplishments of the
project. The purpose of this volume is to describe, in detail, the participation, function, scope, and achievements of this project, which was
completed in April 1988.

The Pilot Study was conducted to apply the cooperative efforts of numerous nations to address a global dioxin problem. The project represented
recognition on the part of governments involved that CDDs and CDFs have
the potential to be major environmental problems. The major objectives of
the project were to exchange information to improve methods to determine
the hazards associated with these compounds and to identify technologies
and management techniques to prevent accidents and the release of CDDs,
CDFs, and related chemicals into the environment. Information exchange
activities were conducted to promote the resolution of many scientific
issues and uncertainties on a truly international level and to distribute

1

pertinent information so that better informed decisions can be made concerning future research programs and resource allocations. In addition to the major information exchange objective, secondary goals included the identification of knowledge voids during the planning process and the reduction of research program duplication. One of the other aims was to achieve consensus on methods of risk assessment for complex mixtures of CDDs and CDFs.

1. PARTICIPATION

The CCMS mechanism was identified as an appropriate forum for the coordination of an international effort to address the dioxin problem. Participating nations included Canada, Denmark, the Federal Republic of Germany, Italy, the Netherlands, Norway, the United Kingdom, and the United States. In addition, several international organizations participated, including the World Health Organization (WHO), the United Nations Environmental Programme (UNEP), the Commission of the European Communities (CEC), and the Organization for Economic Cooperation and Development (OECD). The governments of Austria and Sweden, although not members of NATO, requested to be kept informed of the progress of the project. Several of the participating governments requested representatives of industrial trade organizations and nongovernmental environmental public interest groups to serve as observers on their delegations. For example, involved as observers were the Chemical Manufacturers Association representing the American chemical industry, the *Verband der Chemischen Industrie* representing the chemical industry in the Federal Republic of Germany, and the Environmental Defense Fund representing the American public interest environmental groups.

This project was one of the numerous undertakings sponsored by the CCMS. Since NATO formally established the CCMS in 1969, the cooperative efforts of member nations have been applied to solving practical problems impacting society. Initially proposed to give the Alliance a greater influence in international social issues, the CCMS has expanded to become a powerful force by combining the knowledge and resources of many to work toward a single goal. CCMS-sponsored studies are designed to provide a better understanding of the effects of our technology-intensive way of life and to stimulate actions to remedy the problems. The Committee has addressed a broad spectrum of technical and social problems, but environmental issues have been central to the group's effort. Some other CCMS projects addressed topics such as inland water pollution, solar energy, nutrition and health, and drinking water.

In all cases, pilot studies are initiated by a lead country, which is responsible for developing and conducting the project. Following the approval of the proposed pilot study by the CCMS, the work is carried out by experts nominated by member countries or by CCMS Fellows. One of the more effective vehicles for promoting widespread participation in the pilot studies is provided through the CCMS Fellowship Programme. The Programme enables experienced researchers to contribute to the projects in an adjunct manner. The Fellows' work is carried out under the guidance of the respective pilot study director to ensure compatibility within the scope of the program. In this way, the knowledge and expertise of others may complement the overall efforts of the CCMS.

All member countries, whether they actively contribute or not, benefit from pilot studies conducted by the CCMS. In many of the programs, which address topics with a wide variety of related issues, several pilot countries may be responsible for planning, managing, and reporting on each component of the study. At the completion of each project, the pilot countries report on the conclusions of the study and provide recommendations for further work in the area. One unique aspect of the CCMS programs has been the development of a follow-up procedure to ensure that the results meet the needs of the group. In summary, the projects conducted under the auspices of CCMS benefit from the cooperative efforts of member countries by combining their expertise for the common good.

2. THE PILOT STUDY ORGANIZATION

The Pilot Study on International Information Exchange on Dioxins and Related Compounds had three pilot nations: the United States, the Federal Republic of Germany, and Italy. The project was divided into three basic working groups: (1) Exposure and Hazard Assessment, chaired by Mr. Erich W. Bretthauer of the U.S. Environmental Protection Agency (EPA); (2) Technology Assessment, chaired by Mr. Heinrich W. Kraus of the Bundesministerium fur Umwelt, Naturschutz und Reaktorsicherheit; and (3) Management of Accidents Involving Dioxins, chaired by Dr. Alessandro di Domenico of the Istituto Superiore di Sanita. This book is organized according to the products of the three working groups. Chapter 2 describes the accomplishments of the Exposure and Hazard Assessment Working Group, Chapter 3 presents the findings of the Technology Assessment Working Group, and Chapter 4 contains the work of the Management of Accidents Working Group.

The Exposure and Hazard Assessment Working Group was assigned the task of addressing several aspects of the dioxin problem, focusing specifically on the exchange of information on research, analytical laboratories, regulations and statutes, and methods of risk and exposure assessment. The Technology Assessment Working Group focused its attention on producing documents that reported the state of the art with respect to the formation of dioxins and related compounds from combustion and industrial sources, as well as methods of destruction and disposal. The Management of Accidents Working Group based much of its work on the emergency response and health surveillance activities carried out after the Seveso accident in Italy in 1976 and how that information could be used to prevent future accidents.

As noted previously, the Pilot Study was initiated in 1985 and was concluded in 1988. During the 3-year period, four plenary meetings were held, as well as three meetings of the individual working groups. In general, the plenary meetings were devoted to status reports of ongoing projects and negotiating agreements on the completion, publication, and distribution of reports. The final plenary meeting of the Pilot Study was held in Berlin, Federal Republic of Germany, in April 1988, and was a particularly productive meeting with three major accomplishments: (1) consensus was achieved on the International Toxicity Equivalency Factor (I-TEF) Method of Risk Assessment for Complex Mixtures of Dioxins and Related Compounds; (2) all of the final reports of the study were reviewed and approved for publication; and (3) agreement was reached on a formal follow-up activity of a CCMS-sponsored international information exchange session at the International Symposium on Chlorinated Dioxins and Related Compounds.

The follow-up activities, which continued for 2 years following the completion of the Pilot Study, were two CCMS-sponsored International Information Exchange Sessions on Future Aspects of Dioxin Research and Regulations at the 1988 and the 1989 International Symposium on Chlorinated Dioxins and Related Compounds. The CCMS-sponsored sessions were considered to be a logical extension of the goals of the Pilot Study to a larger international audience. The sessions consisted of presentations from each interested nation regarding the future plans for research and regulations. The follow-up activities were developed to extend the international information exchange activities to a wider audience in hope of impacting future research plans. Consistent with the goals of the collection, analysis, and distribution of research and regulatory information, the presentations at the special sessions emphasized the areas of future concern so that planners and managers can make more informed decisions with respect to knowledge voids and areas already under intense study. The first session was held at Dioxin '88 in Umea, Sweden, on August 25, 1988. The session consisted of presentations from all eight of the participating nations, as well as Switzerland and Sweden (CCMS, 1988a). The session, which was arranged by the U.S. Environmental Protection Agency, also included two presentations on the accomplishments of this Pilot Study in addition to an announcement of the availability of the CCMS reports. A second similar session was held at Dioxin '89 in Toronto, Canada, in September 1989 and included presentations from nine nations, including Japan and France.

3. THE NATURE OF THE PROBLEM

Chlorinated dibenzo-*p*-dioxins and dibenzofurans (CDDs and CDFs) constitute a family of 210 structurally related chemical pollutants (Figure 1) and contain some of the most toxic chemicals known to man. Concern for human health effects of these compounds has stemmed from a number of specific cases (Seveso, Agent Orange, Love Canal, and Times Beach) and the identification of the widespread low-level environmental contamination from a variety of sources including incinerators, pulp and paper mills, herbicide/pesticide production and use, and other industrial processes producing or using chlorinated phenols (USEPA, 1980; USEPA, 1986; CCMS 1988b,c). CDDs and CDFs have been identified in soil, water, plants, fish, human adipose tissue and breast milk, and other biological media that display the absorption and bioaccumulation of these compounds from exposure (ATSDR, 1987; USEPA, 1985; WHO, 1988). Numerous efforts have been carried out to assess the human health risks associated with exposures to these compounds (Kimbrough et al., 1984; Fries, 1986; USEPA, 1988).

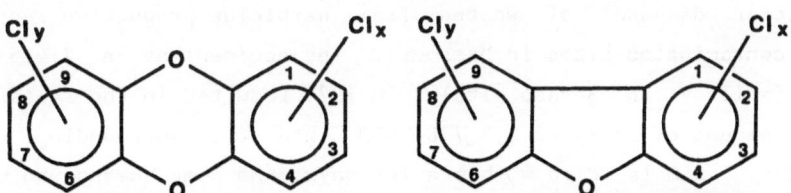

Figure 1.   Chemical Structure of CDDs and CDFs

CDDs and CDFs, unlike most other chemical pollutants, do not occur naturally, nor are they manufactured intentionally for commercial use. Rather, these compounds are generated inadvertently in a variety of chemical and combustion processes. Although there are 210 chemicals in the CDD and CDF family, attention was initially focused on 2,3,7,8-tetrachlorodibenzo-p-dioxin (2,3,7,8-TCDD). 2,3,7,8-TCDD is the most potent animal carcinogen ever tested, resulting in a concern for its presence even at extremely low levels. Humans and animals exposed to 2,3,7,8-TCDD have shown acute, chronic, and subchronic effects on the skin, liver, nervous system, and immune systems. Testing on animals such as mice, rats, guinea pigs, and mink has elucidated some of the toxic effects of the compound. The LD_{50} for 2,3,7,8-TCDD in guinea pigs is as low as 0.6 μg/kg, and is 50 million times more potent than vinyl chloride, a known carcinogen (USEPA, 1988). In addition, reproductive effects are also known to result from exposure to 2,3,7,8-TCDD. However, the mechanism of toxicity and human health effects caused by the compound are uncertainties facing scientists and regulatory agencies. Considerable research has been devoted to determining the mechanism of toxicity and determining at what concentrations these compounds are harmful to humans. Some of these efforts are described in more detail in Chapter 2 by the Exposure and Hazard Assessment Working Group.

Because of the toxicity of 2,3,7,8-TCDD, many regulatory agencies around the world have implemented regulations to limit human exposure to this compound (CCMS, 1988d). Initially, the concern stemmed from the recognition of the presence of 2,3,7,8-TCDD in chlorophenols such as 2,4,5-trichlorophenol (TCP). TCP is made in large quantities and is an intermediate in the production of several herbicides including 2,4,5-trichlorophenoxyacetic acid (2,4,5-T), hexachlorophene, and Agent Orange (a 50/50 mix of 2,4-dichlorophenoxyacetic acid (2,4-D) and 2,4,5-T). The presence of 2,3,7,8-TCDD in these chemicals became the basis for considerable controversy over their production and use. In several nations the manufacture and use of these chemicals have been regulated or prohibited, partially due to the presence of 2,3,7,8-TCDD (CCMS, 1988d). Additionally, the improper disposal of wastes from herbicide production resulted in numerous contaminated sites in Missouri. The accident at a 2,4,5-T production facility in Seveso, Italy, in 1976 resulted in the release of an estimated amount of 13 kg of 2,3,7,8-TCDD into the surrounding environment. Many animals died within a few days, and some humans who were exposed developed chloracne. As a result, hundreds of people were evacuated from the region and years of cleanup ensued (described in detail in

Chapter 4). At the time, however, many of today's analytical methods, sampling techniques, emergency response strategies, and rehabilitation strategies had not yet been developed, thereby hampering municipal, regional, and national authorities from providing rapid relief. Many of the specialized techniques that are commonplace today were developed and improved in the Seveso cleanup.

As a result of the Seveso incident and a growing body of knowledge pertaining to the presence of 2,3,7,8-TCDD in 2,4,5-T, Agent Orange, and other chemical products, more attention was devoted to dioxins. More research was conducted to identify CDDs and CDFs in the environment, and improved analytical methods were developed to detect these compounds at extremely low levels in many environmental media. Two major breakthroughs were made as a result of this work: (1) the identification of many sources of CDDs and CDFs and (2) the ability to quantify many CDD and CDF isomers in addition to 2,3,7,8-TCDD.

In the late 1970s, researchers in the Netherlands quantified the presence of CDDs and CDFs in the fly ash and flue gas from municipal waste incinerators (Olie et al., 1977). Subsequently, more combustion and industrial sources of CDDs and CDFs were identified, and concern regarding the emission of these compounds into the environment increased. Many other countries, including the United States, Canada, the United Kingdom, and the Federal Republic of Germany, initiated intense sampling and analysis programs to identify the potential risks from these sources to human health and the environment. As a result, more has been learned about the emissions of CDDs and CDFs from incinerators (municipal, hazardous, and hospital waste), pulp and paper mills, metallurgical processes, and the manufacture/use of chlorophenols such as pentachlorophenol for wood preservation. A comprehensive review of these studies to identify the sources of CDDs and CDFs was performed by the Technology Assessment Working Group and is described in Chapter 3.

Prior to the Seveso accident, there was very little guidance for dealing technically, socially, and politically with such severe cases of environmental contamination by this group of compounds. Much of the experience gained from the Seveso incident has been applied to other contaminated sites. Chapter 4 of this book describes the activities of the Management of Accidents Working Group and their experiences with Seveso. Major advances in risk assessment, toxicology, sampling and analysis, management of accidents, and hazardous waste treatment have also been made in the last 10 years because of the public concern over these toxic compounds. In addition, the cleanup of many contaminated sites (e.g., Seveso and sites in Missouri) has also promoted the improvement of the state of the art with respect to research and technology. This Pilot Study assumed a lead role in the international scientific and regulatory arena in facilitating the collection, exchange, and dissemination of information on many aspects of the problem. In addition, the project enhanced the research and regulatory approaches of several nations by serving as a vital conduit for cooperative efforts to address specific knowledge voids.

One of the most notable accomplishments of the project was the achievement of international consensus on the International Toxicity Equivalency Factor (I-TEF) Method of Risk Assessment for Complex Mixtures of Dioxins and Related Compounds (CCMS, 1988e,f). Prior to the development of the I-TEFs, at least ten slightly different methods were in use around the world. The existence of so many slightly different schemes complicated communication among scientists and regulatory agencies. As a result of this Pilot Study, there now exists one accepted TEF method by which the toxicological significance of complex mixtures of CDDs and CDFs can be determined and upon which regulations and risk assessments can be based. The I-TEFs are a prime example of the benefit of a coordinated attempt by the participating nations to solve the dioxin problem on an international scale. Subsequently, many other nations outside of the NATO group have adopted this method for assessing the risk associated with complex mixtures of CDDs and CDFs (USEPA, 1989).

Incidents involving CDDs and CDFs continue to receive high visibility in the news media. The accompanying public concern calls for solutions to problems at the local, national, and international levels. Much of the concern stems from the uncertainty associated with the health effects of these compounds and ways to prevent the release of the pollutants into the environment. Various nations have used different approaches to address the dioxin problem, and much of the inconsistency can be attributed to the uncertainty associated with the potential risks to humans. Researchers are beginning to solve some of the problems and fill the knowledge voids, and much of this progress is a result of an increased cooperation and information exchange among scientists around the world.

Several major investigations are generating data valuable in enhancing our understanding of the health effects of this group of compounds. Evaluation of these data by the international scientific community continues to fill knowledge voids. Chemical analyses of blood samples taken from exposed population after the Seveso accident show the highest human levels ever encountered (CDC, 1989). The highest blood levels (up to 56 ppb) are from people who developed chloracne. At this time, however, no threshold level for chloracne is obvious from the data because several levels from asymptomatic exposed residents overlap with levels of people having chloracne. Additionally, it will be several years until there will be data on potential cancer effects associated with this exposure. These data on human levels, in conjunction with epidemiological and health effects, may also help to answer many other questions. Other data pertaining to human exposure and latent health effects are being generated to fill the knowledge void. Results from a study of U.S. chemical workers at plants manufacturing chlorinated phenols and 2,4,5-T between 1951 to 1972 show similar blood human levels (Sweeny et al. 1989). These data are also preliminary and represent analyses from only 25 samples out of a survey of 281 workers. The levels recently found (4.6 to 717 ppt) have been extrapolated back to the time of exposure, yielding results as high as 14.7 ppb, using a half-life estimate of 7 years. These levels are from workers chronically exposed over a long period of time as compared to the acute exposure of the Seveso inhabitants. However, to date, the chronic health effects have not been analyzed.

These data, in conjunction with a growing data base on the health effects of these compounds, may prove to be indispensable in answering many questions about the human health hazards associated with CDDs and CDFs. The participants in the Pilot Study are encouraged that many of these questions may be answered as studies on the exposed people of Seveso and the U.S. chemical workers are completed and the results are published. The enhanced understanding of the risks associated with exposures to CDDs and CDFs will also be complimented with a better knowledge of risk reduction methodologies and technologies to minimize the generation of these compounds at their sources. The efforts of this NATO/CCMS Pilot Study have provided a wealth of information for those working in the future to resolve the dioxin problem, in risk assessment, regulation, analytical chemistry, emission control technologies, remediation of contaminated sites, and management of accidents.

REFERENCES

ATSDR. 1988. Agency for Toxic Substances and Disease Registry. Toxicological profile report for 2,3,7,8-TCDD. U.S. Public Health Service, Agency for Toxic Substances and Disease Registry.

Bottimore, D.P., Kutz, F.W., and E.W. Bretthauer. 1989. Accomplishments of the NATO/CCMS pilot study on international information exchange on dioxins and related compounds. J. Toxicol. Environ. Chem. 26(1-4):111-122.

CCMS. 1988a. Committee on the Challenges of Modern Society. Proceedings of symposium seminar on prospective research and regulatory issues involving dioxins and related compounds. Pilot Study on International Information Exchange on Dioxins and Related Compounds. North Atlantic Treaty Organization/Committee on the Challenges of Modern Society. Report Number 179. December 1988.

CCMS. 1988b. Committee on the Challenges of Modern Society. Emissions of dioxins and related compounds from combustion and incineration sources. Pilot Study on International Information Exchange on Dioxins and Related Compounds. North Atlantic Treaty Organization/ Committee on the Challenges of Modern Society. Report Number 172. August 1988.

CCMS. 1988c. Committee on the Challenges of Modern Society. Formation of dioxins and related compounds in industrial processes. Pilot Study on International Information Exchange on Dioxins and Related Compounds. North Atlantic Treaty Organization/Committee on the Challenges of Modern Society. Report Number 173. August 1988.

CCMS. 1988d. Committee on the Challenges of Modern Society. Inventory of regulations and statutes concerning dioxins and related compounds. Pilot Study on International Information Exchange on Dioxins and Related Compounds. North Atlantic Treaty Organization/Committee on the Challenges of Modern Society. Report Number 169. August 1988.

CCMS. 1988e. Committee on the Challenges of Modern Society. International toxicity equivalency factor (I-TEF) method of risk assessment for complex mixtures of dioxins and related compounds. Pilot Study on International Information Exchange on Dioxins and Related Compounds. North Atlantic Treaty Organization/Committee on the Challenges of Modern Society. Report Number 176. August 1988.

CCMS. 1988f. Committee on the Challenges of Modern Society. Scientific basis for the international toxicity equivalency factor (I-TEF) method of risk assessment for complex mixtures. Pilot Study on International Information Exchange on Dioxins and Related Compounds. North Atlantic Treaty Organization/Committee on the Challenges of Modern Society. Report Number 178. December 1988.

CDC. 1988. Centers for Disease Control. Preliminary Report: 2,3,7,8-TCDD exposure to humans - Seveso, Italy. Mortality and Morbidity Weekly Report 37:733-736.

Fries, G.F. 1986. Assessment of potential residues in foods derived from animals exposed to TCDD-contamianted soil. Presented at 6th International Symposium on Chlorinated Dioxins and Related Compounds; September, Fukuoka, Japan.

Kimbrough, R.; Falk, H.; Stehr, S; Fries, G. 1984. Health implications of 2,3,7,8-tetrachlorodibenzo-p-dioxin (TCDD) contamination of residential soil. J. Toxicol. Environ. Health 14:47-93.

Kutz F.W. and Bottimore D.P., eds. 1988. Proceedings of symposium seminar on prospective research and regulatory issues involving dioxins and related compounds. Chemosphere 17(11):1-67.

Olie, K., Vermeulen, P.L., and Hutzinger, O. 1977. Chlorodibenzo-p-dioxins and chlorodibenzofurans are trace components of fly ash and flue gas of some municipal incinerators in the Netherlands. Chemosphere 6:455-459.

Sweeney, M.H., Fingerhut, M.A., Connally, L.B., Halperin, W.E., Moody, W.E., and D.A. Marlow. 1989. Progress of the NIOSH cross-sectional medical study of workers occupationally exposed to chemicals contaminated with 2,3,7,8-TCDD. Chemosphere (in press).

USEPA. 1980. U.S. Environmental Protection Agency. Dioxins. Industrial Environmental Research Laboratory, Cincinnati, OH. EPA-600/2-80-197.

USEPA. 1985. U.S. Environmental Protection Agency. Health assessment document for polychlorinated dibenzo-p-dioxins. Office of Health and Environmental Assessment, Environmental Criteria and Assessment Office, Cincinnati, OH. EPA-600/8-84-014F. NTIS PB86-122546.

USEPA. 1986. U.S. Environmental Protection Agency. National dioxin study, tier 4: combustion sources. Office of Air Quality Planning and Standards. Research Triangle Park, NC. EPA-450/4-84-014g.

USEPA. 1988. U.S. Environmental Protection Agency. A cancer risk-specific dose estimate for 2,3,7,8-TCDD. EPA/600/6-88/007a,b.

USEPA. 1989. U.S. Environmental Protection Agency. Interim procedures for estimating risks associated with exposures to mixtures of chlorinated dibenzo-p-dioxins and dibenzofurans (CDDs and CDFs) and 1989 update. EPA/625/3-89/016.

WHO. 1988. World Health Organization. Assessment of health risks in infants associated with exposure to PCBs, PCDDs, and PCDFs in breast milk. Copenhagen, Denmark.

CHAPTER 2 - EXPOSURE AND HAZARD ASSESSMENT WORKING GROUP

1. INTRODUCTION

Erich W. Bretthauer, Chairman of the Exposure and Hazard Assessment
Working Group

U.S. Environmental Protection Agency
Washington, D.C.

The efforts of the Exposure and Hazard Assessment Working Group were
focused on the exchange of information on a variety of topics including
research projects, regulations/statutes, analytical laboratories, and
methods of exposure/risk assessment involving CDDs and CDFs. It was
evident to the leaders of the Working Group that several of the knowledge
voids had to be addressed on a fundamental level before expanded efforts
could be made. Several questions needed to be answered:

- Who has done research on this topic, and what do the data
  indicate?

- Who is performing research now, and what are their capabilities?

- How are other nations addressing this problem, and do they have
  legislative mandates in place?

- Is there a general consensus on the topic?

The members of the Working Group believed that these questions could
be answered by surveying the major participants in the field of interest.
Three principal survey efforts were performed by the Working Group, which
collected information on research, regulations/statutes, and analytical
laboratories from each of the participating nations. In addition to
answering these fundamental questions, these efforts also fulfilled the
major objectives of the entire Pilot Study. The collection, analysis, and
distribution of information on research projects, regulations/statutes,
and analytical laboratories were very useful efforts in helping to fill
some of the basic knowledge voids. In addition, these topics are closely

related to each other. For example, the use of analytical methods is essential to many types of research as well as to performing risk assessments and setting standards and regulations.

The following sections in Chapter 2 describe the work carried out by the Exposure and Hazard Assessment Working Group chaired by the U.S. EPA. Section 2 describes the collection, analysis, and distribution of information on research projects in the participating nations. This section represents a final report of an effort carried out over the entire 3-year project, in which two interim reports were published. The two reports, CCMS Report Numbers 160 and 166, presented the compilation of information through spring 1986 and fall 1987, respectively. Report Number 166 was distributed through CCMS channels and was also distributed to approximately 500 registrants at the Seventh International Symposium on Chlorinated Dioxins and Related Compounds (Dioxin '87). The final report (CCMS Report Number 170, published in December 1989) on research topics contains an expanded analysis of the results with special attention paid to knowledge voids and duplicative efforts. Section 3 is the final listing of analytical laboratories with expertise in the analysis of dioxins and related compounds. One interim report of the listing was also included in CCMS Report Number 166. Section 4 describes the collection of information on regulations and statutes concerning dioxins and related compounds in the participating nations. The International Toxicity Equivalency Factor (I-TEF) Method of Risk Assessment for Complex Mixtures of Dioxins and Related Compounds is presented in Section 5. This method was first published in CCMS Report 176 and presented at Dioxin '88 as part of the Special Seminar on Prospective Research and Regulatory Issues Involving Dioxins and Related Compounds, sponsored by the U.S. EPA and the Pilot Study. Subsequently, the I-TEF method has been adopted and used by numerous nations as the preferred interim procedure for estimating risks associated with exposures to CDDs and CDFs. Section 6 includes a review and update of exposure assessment methods related to 2,3,7,8-TCDD. Sections 7 and 8 describe risk assessment approaches that serve as the basis for establishing acceptable daily intake levels for 2,3,7,8-TCDD.

2. INTERNATIONAL EXCHANGE OF RESEARCH AND TECHNOLOGY INFORMATION ON DIOXINS AND RELATED COMPOUNDS

David N. McNelis*, Erich W. Bretthauer**,
Frederick W. Kutz**, and David P. Bottimore***

University of Nevada, Las Vegas*
U.S. Environmental Protection Agency**, and Versar, Inc.***

Las Vegas, Nv*, Washington, D.C.**, and
Springfield, Va***

2.1 Introduction

This section describes the international exchange of information on research addressing chlorinated dibenzo-p-dioxins (CDDs), dibenzo-furans (CDFs), and related compounds. Compilation of information on research projects concerning this group of chemicals was prepared to inform scientists, planners, regulatory, and funding organizations about areas being investigated. This information was collected in order to identify knowledge voids and duplicative efforts so that cooperative research efforts can be carried out on an international basis. In addition, the exchange of information on research projects facilitates planning future research activities and allocating resources.

2.1.1 The Exchange of Information on Research Projects

One of the primary mechanisms for fulfilling the objectives of the project was the collection, analysis, and distribution of reports containing information on research activities. The first product of this Pilot Study, CCMS Report Number 160, was a compilation of summaries (Research Formats) published in the spring of 1986. The document contains descriptions of 158 projects which were underway in seven nations and was distributed through CCMS channels. Based on the considerable interest generated by the first compilation, the participants of the Pilot Study elected to collect a second round of information on research activities. As a result of more intense solicitations, 232 Research Formats were collected. An interim version of the compilation was published in CCMS Report Number 166 and distributed through CCMS channels as well as to

15

approximately 500 registrants at the Seventh International Symposium on Chlorinated Dioxins and Related Compounds (Dioxin '87) in October 1987.

This section is a condensed version of CCMS Report Number 170, published in December 1989, which is a revised compilation of the Research Formats from the two previous reports. This updated report contains a compilation of information on dioxin-related research presently being or having been pursued in the participating nations between the years of 1985 and 1989. More detailed descriptions of the research projects can be found in CCMS Report Number 170. The number of research projects described in Reports 160, 166, and 170, are presented by participating nation in Table 1.

Table 1. Number of Formats Describing Research Activities
Submitted by Participating Nation for CCMS Reports

Participating Nation	CCMS Report 160 Spring 1986	CCMS Report 166 October 1987	CCMS Report 170 December 1989
Canada	45	46	47
Denmark	-	-	6
Federal Republic of Germany	26	26	57
Italy	17	6	33
Netherlands	12	22	22
Norway	4	-	-
Sweden[1]	-	1	1
United Kingdom	2	4	5
United States	52	127	127
WHO[1]			
Total	158	232	282

- Denotes that no information was submitted for inclusion in these publications.

[1] Sweden and the World Health Organization are involved only as observers, and are not NATO-Participating Nations.

2.1.2   Data Collection

The data presented in this section provide a qualitative indication of the research focus and resource allocation of each of the countries or agencies involved in research on CDDs and CDFs. The solicitation of input (Formats) by the chairpersons and the lead delegates from each country was committed but there is no measure of the comprehensiveness of either the canvass or the response. The Formats were distributed to and prepared by investigators and/or program managers representing the various participating agencies and laboratories and submitted through each lead delegate.

As a result, the accuracy of the information and conclusions is dependent upon the information submitted. Also, because of inconsistencies in the reporting of data for years past as well as the predictability of data for the future, data for 1986, 1987, and 1988 are considered of somewhat better quality than the data on 1985 and 1989.

Figures 1a and 1b present the two sides of the Format distributed to participants, while Figure 2 presents the accompanying set of instructions for its preparation. Project and resource information was requested for the 5-year period 1985-1989. The resources reported were to indicate the total expenditure (or expected expenditure) within each calendar year. The reverse side of the Format (Figure 1b) includes a series of key words to be selected by the authors. The second column contains primary key words that relate directly to one of the three working groups; i.e., Exposure and Hazard Assessment (Working Group A), Technology Assessment (Working Group B), and Management of Accidents (Working Group C). The third column contains secondary key words which bear a similar, albeit not as direct, relationship to one or more of the working groups. The final column contains further delimiters. The working group and the primary and secondary key words have been used to order the discussion and the graphic displays which follow. After evaluating the responses the Formats were grouped into the following 13 topical areas; 7 of which are under Exposure and Hazard Assessment, 5 under Technology Assessment, and 1 under Management of Accidents.

Working Group	Topical Area
Exposure and Hazard Assessment	Environmental Assessment
	Exposure Assessment
	Risk Assessment
	Effects/Toxicity/Assessment
	Sampling Methodology
	Properties of Dioxins
	Analytical Methods
Technology Assessment	Incineration Sources
	Other Sources
	Destruction by Incineration
	Destruction by Other Means
	Waste Management
Management of Accidents	Remedial Response

2.2 Discussion

Table 2a displays the total funds expended or projected to be expended during calendar years 1985 through 1989 for projects included under each

ON INTERNATIONAL INFORMATION EXCHANGE ON DIOXINS AND RELATED COMPOUNDS

WORKING GROUP: (circle one)   A   B   C   COUNTRY OF ORIGIN: _____

TITLE OF PROJECT OR ACTIVITY:

NAME, ADDRESS, AND TELEPHONE NUMBER OF PRINCIPAL INVESTIGATOR:

NAME, ADDRESS AND TELEPHONE NUMBER OF SUPPORTING AGENCY:(include name of Project Officer)

IDENTIFYING NUMBER: _____ STARTING DATE: ___/___/___ COMPLETION DATE: ___/___/___

GOAL/RATIONALE/SCOPE:

ESTIMATED RESOURCES (IN UNITED STATES DOLLARS TO NEAREST $1,000.)

CALENDAR YEAR	1985		1986		1987		1988		1989	
FUNDING	$	K	$	K	$	K	$	K	$	K

MAJOR OUTPUTS: (e.g., technical reports, other publications, patents)

Figure 1a.   Format For CCMS Pilot Project

18

Figure 1b. Key Words (circle appropriate key words)

Working Group	Select one only:	Select one:	Select all that apply:	
A	Assessment	Environmental Exposure Body Burden Risk Effects	Monitoring Pathways Epidemiology Transport/Mobility Pharmacological	Physiologic Genetic 2378 TCDD TCDD PCDD PCDF Other Compounds Standards QA/QC Methods Research
	Properties	Toxicity Synthesis Physicochemical Reactivity/Stability Biological Sampling	Bioaccumulation Bioavailability Biotransformation Equivalency Chemicokinetics	
	Analytical Methods		Food Animal Human Environmental Manufacturing Products Plant Cell Cultures	
B	Sources	Manufacturing Incineration Internal Combustion Engines End Products/Use Other Sources	Emissions Control Technology	
	Destruction	Incineration Detoxification Physicochemical Biological Environmental	Wastes Sludge Soil Sediment Water Air Tissue	
	Waste Management	Interim Storage Exposure Control		
C	Response	Emergency Remedial Contingency Planning	Decontamination Containment Health/Safety Surveillance/Monitoring	

19

Working Group: Circle one letter to indicate the working group under which the project most appropriately falls.
(A - Exposure and Hazard Assessment; B - Technology Assessment; C - Management of Accidents).

Country of Origin: Specify the country supporting the project.

Title of Project or Activity: Provide a clear, concise title which is descriptive of the project.

Name, Address and Telephone Number of Principal Investigator: Provide this detail on the individual(s) actually conducting the investigation.

Name, Address and Telephone Number of Supporting Agency: Provide this detail and include the name of that Agency's responsible project officer.

Identifying Number: Indicate Agency project identification code or contract/grant number uniquely associated with this task.

Starting Date: Indicate month, day and year when project was actually initiated.

Completion Date: Indicate the actual or estimated month, day and year when project will be completed.

Goal/Rationale/Scope: Describe the goal, rationale and scope of the project or activity. The key word listing on the reverse of the format provides an indication on how the projects will be categorized and sorted.

Estimated Resources: Indicate the total resources actually committed/expended within each calendar year to the nearest $1,000. If, for example, funding is provided during 1986 to cover an 18-month period, then that portion that will actually be expended during CY 1986 should be shown in that block and the balance reflected in the CY 1987 block. Each block should be completed with either zero, the actual/known level, or the estimated level.

Major Outputs: Provide a listing of the expected major outputs, such as reports, publications, patents, etc., from the project or activity. Include a listing of those already completed with their appropriate scientific literature citation.

Key Words: Key words are listed on the reverse of the formats and will be used as the basis for sorting the projects and activities and determining relationships between and among programs.

Select one word only from column two. The key words in that column represent the principal divisions of the working groups. Also circle one key word only from column three. Finally, all of the key words in column four that apply should be circled. Please note that key words listed in columns three and four are not partitioned according to Working Group and may be selected from any position(s) in the columns.

Figure 2. Instructions for the Preparation of the Format for International Information Exchange on Dioxins and Related Compounds.

of the three Working Groups; i.e., Exposure and Hazard Assessment, Technology Assessment, and Management of Accidents. Also shown is the 5-year total for each Working Group and the total for this information exchange project, approximately $142.4 million. Table 2b presents the number of project-years associated with those resources and Table 2c shows the average cost per project per year for each year and for each of the Working Groups. The value of project-year is used in this report to account for those projects which continue over longer time periods. The total funding of research within each working group for the 5-year project period is displayed graphically in Figure 3. Of the approximately $142.4 million reported in this survey, 53 percent ($75 million) is for Exposure and Hazard Assessment projects (Working Group A); 41 percent ($59 million) is for projects in Technology Assessment (Working Group B); and 6 percent ($9 million)is for Remedial Response programs (Working Group C).

Table 2a. Total Funding by Working Group (5-Year Program)
(In Thousands of Dollars)

	1985	1986	1987	1988	1989	5-YEAR TOTAL
Assessment	13,833	17,022	23,734	14,182	6,291	75,062
Technology	9,094	18,258	20,041	6,770	4,399	58,562
Remedial Response	1,828	2,828	638	2,898	558	8,750
					TOTAL	142,374

Table 2b. Total Project-Years by Working Group

Assessment	73	123	119	67	37	419
Technology	38	56	53	26	14	187
Remedial Response	6	5	4	4	3	22
					TOTAL	628

Table 2c. Average Cost/Project/Year by Working Group
(In Thousands of Dollars)

						Mean
Assessment	189	138	199	212	170	182
Technology	239	326	378	260	314	303
Remedial Response	305	566	160	725	186	388

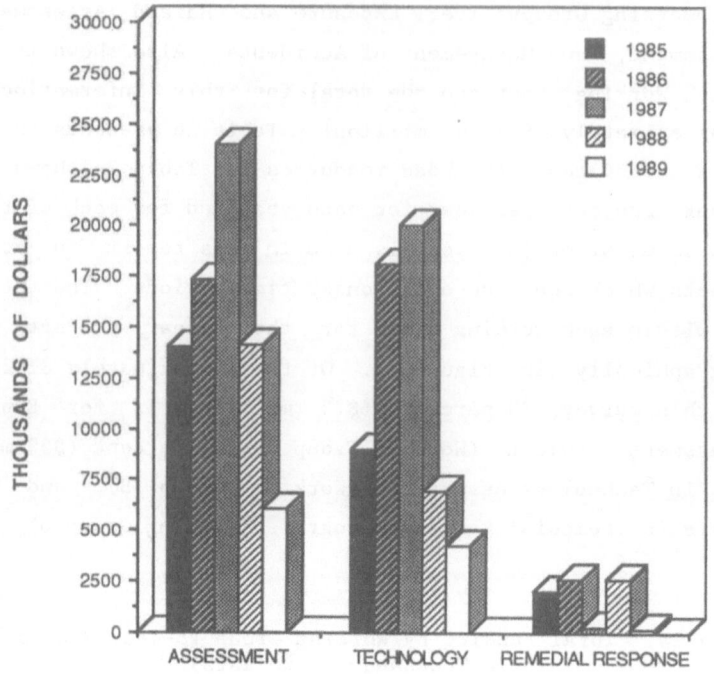

Figure 3. Total Funding (5 Year Program) by Working Group

The apparent peak in funding and in number of projects for Assessment
and Technology programs during 1987 may be indicative of an actual increase and subsequent decrease in programmatic interest in CDDs, CDFs, and
related compounds. As 1987 was the principal year for data collection, it
could also relate to differences in the thoroughness of the canvass or of
data reporting for previous and out-year data.

The average cost of projects of each year for the Exposure and Hazard
Assessment and for the Technology Assessment Working Group is approximately $180,000 and $300,000, respectively, and is relatively constant from
year to year. The average cost for Remedial Response, however, is highly
variable from year to year and ranges from $160,000 per project in 1987
to $725,000 per project in 1988. Figure 4 displays the average cost per
project year for the 13 topical areas contained within the three working
groups. The examination of the research formats also indicates a significant trend in that the funding per project in the area of assessment is
substantially lower than the waste management and incineration destruction
studies which frequently have very high costs for manpower, equipment,
and chemical analyses of samples. The analysis of secondary key words
indicate that most of these incineration studies were near completion and
that research in this area would decline in importance in the future.

22

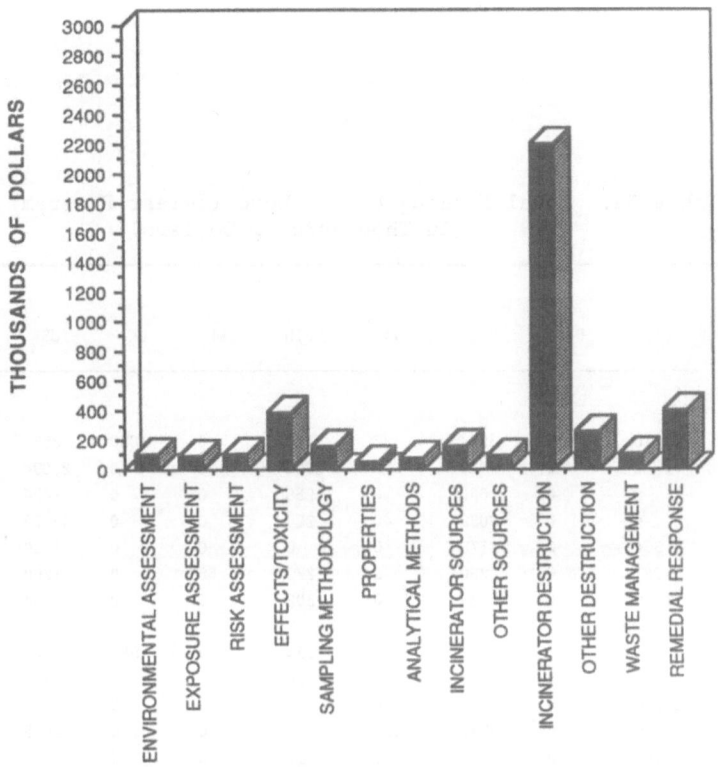

Figure 4.   Cost/Project-Year

Table 3a shows the total funds expended or projected to be expended individually by each of the participating nations over the 5-year program under each of the 13 topical areas. Table 3b shows the cumulative project-year effort by each nation. The relative participation, in terms of funding by each country and the World Health Organization, can also be seen in Figure 5.

Of note is that $15.6 million of the $22.8 million (or 68.4 percent) included for Canada is for the incineration of CDDs, CDFs, and other toxic materials. Denmark indicates that 94.5 percent of their approximately $1.2 million was, or is, to be spent on the study of incinerators as sources of dioxins and related materials. The resources of the Federal Republic of Germany are distributed over most every topical area although the emphasis is on those in the general area of Technology Assessment. Italy funded, or will fund, programs almost exclusively in the area of sources, destruction technology, and waste management. The resources of the Netherlands have been, or will be, committed to Exposure and Hazard

23

Table 3a. Total Funding by Key Word (5-Year Program) (In Thousands of Dollars)

	CAN	DEN	FRG	ITA	NETH	SWE	UK	USA	WHO	5-YEAR TOTAL
Environ Assmt.	2118	0	3924	32	100	0	1030	3531	0	10735
Exposure Assmt.	966	0	0	0	200	0	0	2,090	0	3256
Risk Assmt.	5	0	660	0	255	0	0	1474	110	2504
Effects/Toxicity	0	0	3082	40	220	0	0	43176	0	46518
Sampling Method.	364	0	2173	0	0	0	0	1250	0	3787
Properties	281	0	260	0	279	68	0	1260	0	2148
Analytical Mthds.	1825	0	933	0	302	0	0	3054	0	6114
Incin Sources	627	1115	6682	460	934	0	987	0	0	10805
Other Sources	870	40	1943	815	60	0	0	400	0	4128
Incin Destruct	15600	25	1475	0	0	0	0	13636	0	30736
Other Destruct	0	0	5637	100	0	0	0	3575	0	9312
Waste Mgmt	144	0	2234	800	18	0	0	347	38	3581
Remedial Response	0	0	1500	0	0	0	0	7250	0	8750
TOTAL	22,800	1,180	30,503	2,247	2,368	68	2,017	81,043	148	142,374

Table 3b. Total Project-Years by Key Word (5-Year Program)

	CAN	DEN	FRG	ITA	NETH	SWE	UK	USA	WHO	5-YEAR TOTAL
Environ Assmt	43	0	14	1	7	0	7	27	0	99
Exposure Assmt.	12	0	0	0	2	0	0	20	0	34
Risk Assmt.	1	0	3	0	4	0	0	12	5	25
Effects/Toxicity	0	0	18	2	5	0	0	96	0	121
Sampling Method.	9	0	8	0	0	0	0	7	0	24
Properties	11	0	3	0	6	1	0	20	0	41
Analytical Mthds.	25	0	5	0	7	0	0	36	0	73
Incin Sources	11	5	25	4	7	0	16	0	0	68
Other Sources	5	2	25	10	2	0	0	2	0	46
Incin Destruct	5	1	2	0	0	0	0	6	0	14
Other Destruct	0	0	11	5	0	0	0	21	0	37
Waste Mgmt	5	0	16	4	1	0	0	3	3	32
Remedial Response	0	0	3	0	0	0	0	19	0	22

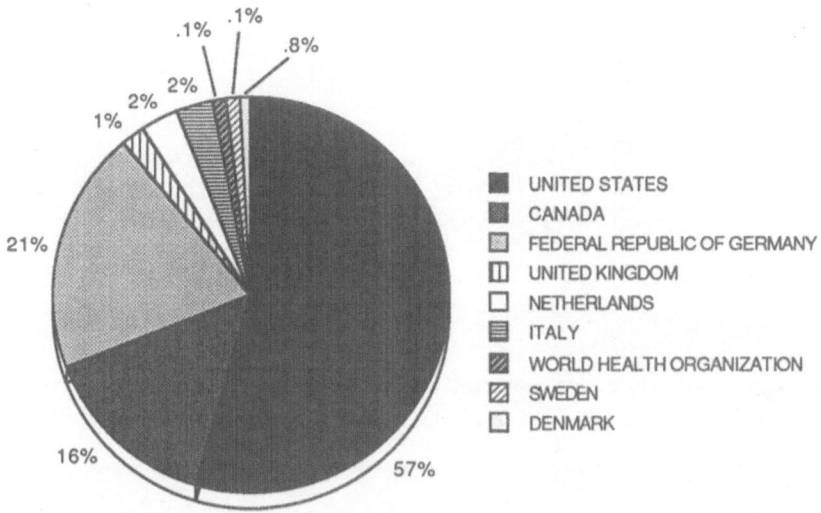

Figure 5. Total Funding by Nation ($142.4 Million)

Assessment and source investigations. The United Kingdom has funded, or
will fund, projects in environmental assessment and the study of incinerators as sources of dioxins. The resources of the United States are distributed over most every topical area with the main emphasis in the areas
associated with Exposure and Hazard Assessment.   The United States has
devoted $43.2 million to human health effects and toxicity investigations.
In fact, research devoted to health effects of CDDs and CDFs, such as
toxicology, amounted to approximately $46.5 million. The United States
accounted for 92 percent ($43 million) of that total, with the Federal
Republic of Germany, Italy, and the Netherlands filling out the balance.
The United States also contributed to research in the areas of technology
assessment, including studies on incineration/destruction, control technologies, environmental levels, and identification of sources of dioxins.
The United States expended approximately one-third ($18 million) of the
$59 million total funding in the eight nations. In the area of Remedial
Response, the United States funded 83 percent ($7.3 million) of the total
expenditure in the participating nations.

   Figure 6 shows the percent distribution of total resources among topical areas over the the 5-year program in the Exposure and Hazard Assessment Group (Working Group A).   There is a strong emphasis on health
effects and toxicity assessment investigations; i.e., approximately 62

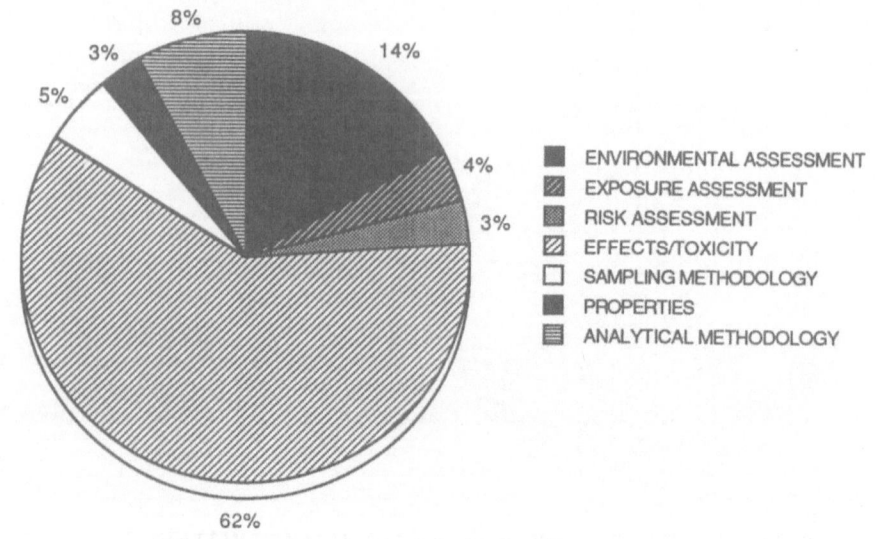

Figure 6. Working Group A - Total Funding Over 5 Years ($75.1 Million)

percent ($46.5 million) of a total of $75.1 million was committed to studies in those areas. An additional $16.7 million was spent on environmental, exposure, and risk assessment studies. A breakdown of funding by country in Working Group A is shown in Figure 7. The United States reports the major share (74 percent) of the total.

Figure 8 shows the percentage distribution of total resources over the 5-year program in Technology Assessment. Almost $55 million of the $58.6 million identified with Group B projects is committed to research programs addressing sources of CDDs and CDFs as well as destruction techniques. The balance, or $3.6 million is identified with programs for the waste management of dioxins. A breakdown of funding by country in Working Group B is shown in Figure 9. The Federal Republic of Germany, Canada, and the United States share evenly (approximately 30 percent apiece) in the total resource commitment. The remaining 9 percent is shared among Italy, Denmark, the Netherlands, United Kingdom and The World Health Organization.

Figure 10 displays the funding of remedial response projects by nation. The United States (83 percent) and the Federal Republic of Germany (17 percent) account for all of the resources dedicated over the 5-year period to Remedial Response.

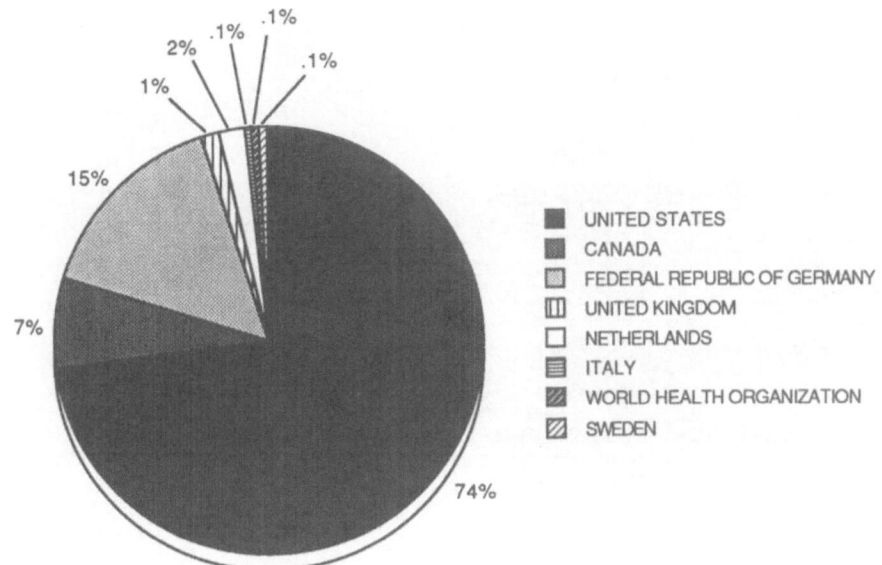

Figure 7.   Funding of Exposure and Hazard Assessment Projects by Nation
($75.1 Million)

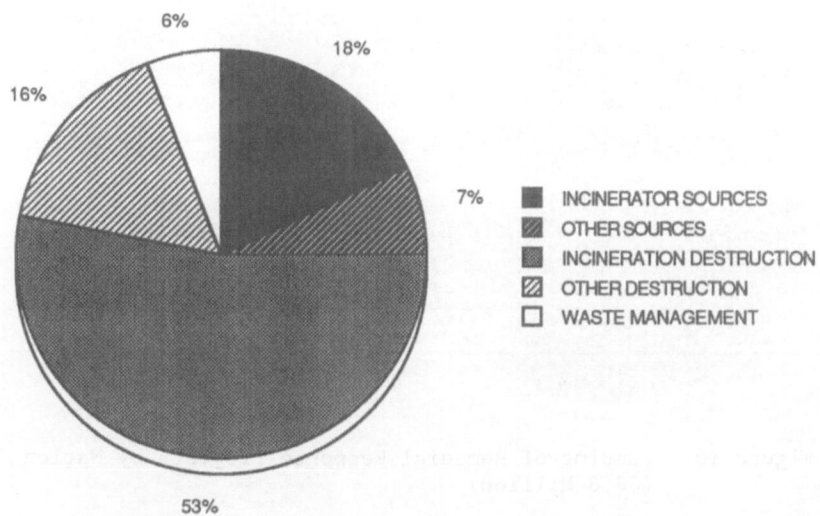

Figure 8.   Working Group B - Total Funding Over 5 Years ($58.6 Million)

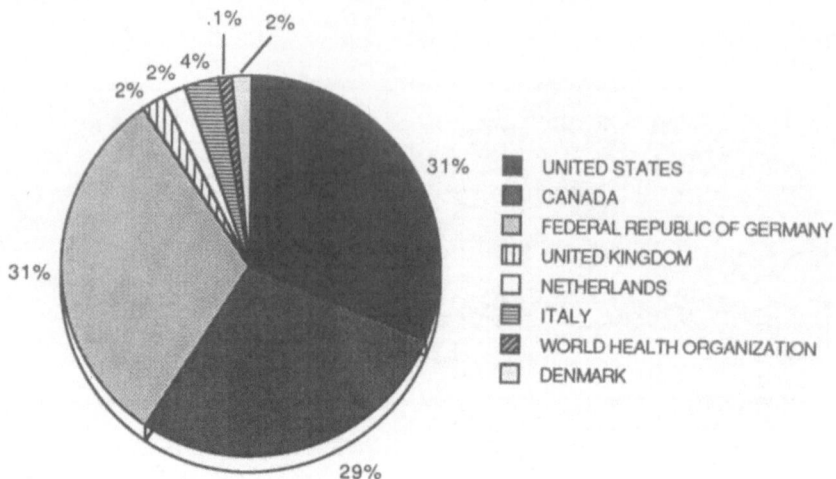

Figure 9.   Funding of Technology Assessment Projects by Nation
            ($58.6 Million)

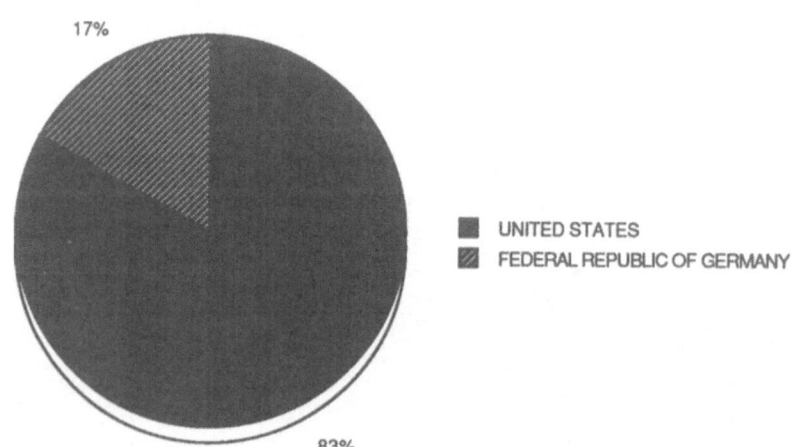

Figure 10.   Funding of Remedial Response Projects by Nation
             ($8.8 Million)

Table 4 presents a listing of the 282 research projects that are described in CCMS Report Number 170. The table includes the project title, principal investigator, country, and working group classification. Readers interested in obtaining more detailed information on individual projects are encouraged to consult CCMS Report Number 170.

2.3 Conclusions

Even in the absence of an exhaustive and quantitative canvass, the objectives of this survey have been met. The well circulated Research Formats, which include the identification of agencies, institutions, and investigators involved in both narrowly focused as well as general topical areas, have stimulated exchange and collaboration between scientific and programmatic personnel from the various participants and other interested parties. The report strongly highlights the intensity and focus of the national or institutional programs summarized. The results have been used, for example, as input into evaluations of national and international program emphasis as well as voids. They have also been used in budget discussions to demonstrate awareness of the existence or lack of potentially supportive programs.

Although each of the NATO countries, Sweden, and the World Health Organization participated in this exchange, Canada, the Federal Republic of Germany, and particularly the United States provide the preponderance of the resources involved with research on CDDs, CDFs, and related compounds.

2.3.1 Overall Trends

- Total research funding on dioxins and related compounds identified through this survey (1985-1989) = $142 million:

  - Exposure and Hazard Assessment represents 53 percent ($75 million)
  - Technology Assessment represents 41 percent ($59 million)
  - Remedial Response represents 6 percent ($9 million)

- Average cost of projects per year:

  - Exposure and Hazard Assessment ($180,000)
  - Technology Assessment ($300,000)
  - Remedial Response ($390,000)

- Exposure and Hazard Assessment funding is $75 million:

  - $47 million (62 percent) in effects/toxicity
  - $11 million (14 percent) in environmental assessment
  - $6 million (8 percent) in analytical methods

- In the area of Exposure and Hazard Assessment, approximately 75 percent of the total funding ($75 million) is from the United States ($56 million).

- Technology Assessment funding is $59 million:

  - $31 million (52 percent) in incinerator destruction
  - $11 million (18 percent) in incinerator sources
  - $9 million (6 percent) in other destruction methods

- In the Technology Assessment area, the Federal Republic of Germany, Canada, and the United States evenly share (approximately 30 percent apiece) the total resource commitment ($59 million).

- Funding of Remedial Response projects ($9 million) includes 83 percent from the United States and 17 percent from the Federal Republic of Germany.

2.3.2   National Areas of Emphasis

- Canada - (Total Funding = $23 million) 68 percent is in the area of incineration of dioxins and related compounds.

- Denmark - (Total Funding = $1.2 million) 95 percent is in identification of incinerator sources of dioxins and related compounds.

- Federal Republic of Germany - (Total Funding = $30.5 million) Funding research in almost all areas, with some emphasis on control technologies.

- Italy - (Total Funding = $2.2 million) Most funding in sources, destruction technologies, and waste management.

- Netherlands - (Total Funding = $2.4 million) Funding is primarly commited to exposure and hazard assessment and source investigations.

- United Kingdom - (Total Funding = $2 million) Emphasis on environmental assessment (background levels) and incinerator sources.

- United States - (Total Funding = $81 million) Research is devoted to every area, with emphasis ($43 milion) on human health effects and toxicity investigations.

Table 4. Listing of Research Projects

Work group	Last name	First name	Title	Country
A	ALBRO	P	DEVELOPMENT OF ANALYTICAL METHODOLOGY	USA
A	AUST	S	TOXIC & ANORECTIC EFFECTS OF INVOLVING TCDD (RATS, MICE)	USA
A	BAINBRIDGE	J	HEALTH HAZARD EVAL. & TECH ASSIST INVOLVING PCB'S, DIOXS, ETC., FOR HAZARDOUS WASTE SITES, SPILLS, ETC.	USA
A	BECK	T	PCDD/PCDF IN ENVIRONMENT, FOOD & HUMAN SAMPLES INCLUDING BREAST MILK (SAMPLING, ANALYSIS, ASSESSMENT)	FRG
A	BELTON	T	DISTRIBUTION OF 3378-TCDD AND 2378-TCDF IN FINFISH & CRUSTACEANS FROM NEWARK BAY AND THE NY BIGHT	USA
A	BELTON	T	STUDY OF DIOXIN & FURAN BIOACCUMULATION IN BLUE CRABS & AMERICAN LOBSTERS FROM THE NY BIGHT	USA
A	BIANCARDI	G	DETERMINATION OF TCDD IN ENVIRONMENTAL SAMPLES	ITA
A	BIRNBAUM	L	TCDD TERATOGENICITY--MODULATION IN MIXTURES	USA
A	BIRNBAUM	L	TCDF: METABOLITE IDENTIFICATION	USA
A	BIRNBAUM	L	DISPOSITION OF HALOGENATED DIBENZOFURANS	USA
A	BIRNBAUM	L	STUDIES OF THE CHEMICAL DISPOSITION & METABOLISM OF OCDD	USA
A	BIRNBAUM	L	DIOXIN SKIN TOXICITY-HAIRLESS MICE	USA
A	BOPP	R	SEDIMENT SAMPLNG & RADIONUCLIDE & CHLORINATED HYDROCARBON ANALYSIS IN NEWARK BAY AND RIVERS	USA
A	BOWMAN	R	PHARMACOKINETICS OF 2,3,7,8-TCDD IN MONKEYS	USA
A	BRINKMANN	F	AD HOC INVESTIGATIONS IN BEHALF OF THE INSPECTORATE OF PUBLIC HEALTH AND ENVIRONMENTAL HYGIENE	NET
A	BUDDE	W	DEVELOP COMPUTER SOFTWARE TO INTERPRET MASS SPEC DATA FROM SAMPLES CONTAINING DIOXINS & DIBENZOFURANS	USA
A	BURKE	M	OCCURRENCE OF DIOXIN IN PUBLIC WATER SUPPLIES	USA
A	BURT	V	A CASE CONTROL STUDY OF NON-HODGKIN'S LYMPHOMA	USA
A	CALLAHAN	M	ESTIMATING EXPOSURES TO 2,3,7,8-TCDD	USA
A	CARUSO	J	ATOMIC EMISSION SPECTROMETRY FOR DIOXIN TRACE ANALYSIS/EMPHASIS PLASMA SOURCE MASS SPECTROMETRY	USA
A	CHITTIM	B	SYNTHESIS OF PCDD, PCDF, BROMO DD & DF AND MIXED BR/CL DD & DF	CAN
A	CHOUDHRY	G	PHOTOINCORPORATION OF CHLORINATED AROMATIC POLLUTANTS (CAPS) INTO AQUATIC & SOIL HUMIC MATERIALS	CAN
A	CHOUDHRY	G	ENVIRONMENTAL AQUATIC PHOTOCHEMISTRY OF CHLORINATED AROMATIC POLLUTANTS (CAPS)	CAN
A	CHOUDHRY	G	ENVIRONMENTAL PHOTOCHEMISTRY OF POLYCHLORINATED DIPHENYL ETHERS IN AQUATIC SYSTEMS	CAN
A	CLEMENT	R	STORAGE STABILITY STUDY OF CDDs/CDFs IN FISH	CAN
A	COATES	J	PLANT UPTAKE & METABOLISM OF PCDD ISOMERS	USA
A	COOK	P	LAKE ONTARIO TCDD BIOACCUMULATION STUDY	USA
A	COOPER	K	BIOAVAILABILITY & PHYSIOLOGICAL EFFECTS OF DIOXINS/FURANS ON BIVALVE MOLLUSCS, CRUSTACEANS, & FINFISH	USA
A	CREASER	C	DETERMINE RANGE OF BACKGROUND LEVELS OF PCDDS & PCDFS IN SOILS	UK
A	CREASER	C	HMIP DIOXIN SURVEY	UK

Table 4. (continued)

Work group	Last name	First name	Title	Country
A	DAMM	T	AMBIENT AIR MEASUREMENT OF PARTICULATE & GAS PHASE DIOXIN & FURAN	CAN
A	DEROOS	F	EVALUATION OF AN EPA HIGH-VOLUME AIR SAMPLER FOR PCDDs/PCDFs	USA
A	DIEBOLD	G	OPTOACOUSTIC DETECTION OF CARCINOGENS	USA
A	DiDOMENICO	A	ROLE OF PLANTS IN ENVIRONMENTAL FATE OF CHLORINATED AROMATICS AND HETEROAROMATIC COMPOUNDS	ITA
A	ENDE	M	DETERMINATION OF PCDDs/PCDFs IN HUMAN MILK & FOODS	FRG
A	EPA REG 7		DIOXIN SAMPLING IN MISSOURI	USA
A	FINGERHUT	M	NIOSH DIOXIN REGISTRY & COHORT MORTALITY STUDY	USA
A	FRANCKE		IMPROVING ANALYTICAL METHODS FOR PCDDs/PCDFs	FRG
A	FREEMAN	R	MODELLING CLEARANCE OF NONMETABOLIZED CHLOROCARBONS (HEXACHLOROBENZENE)	USA
A	GALLO	M	COMPARATIVE BIOAVAILABILITY OF CHLORINATED DIOXIN IN NEW JERSEY & MISSOURI SOILS	USA
A	GASIEWICZ	T	CDDs, MECHANISMS OF TOXICITY (RATS, MICE, GUINEA PIGS)	USA
A	GASIEWICZ	T	CDDs, MECHANISMS OF TOXICITY (RATS, GUINEA PIGS)	USA
A	GASIEWICZ	T	MOLECULAR TOXICOLOGY OF TCDD (RATS, MICE, GUINEA PIGS, HAMSTERS)	USA
A	GIERTHY	J	DIOXIN-EPITHALIAL CELL INTERACTIONS - MECHANISM & ASSAY	USA
A	GOBAS	F	RELATIONSHIP BETWEEN PHYSICO-CHEMICAL PROPERTIES & BIOACCUMULATION OF PCDDS IN AQUATIC ORGANISMS	CAN
A	GRANT	D	RE-EVALUATION OF PCDDS IN FOODS	CAN
A	GROSS	M	ANALYTICAL DETERMINATION OF 2378-TCDD, 2378-TCDF, & THEIR CONGENERS IN RIVER & BAY SEDIMENTS	USA
A	GUSTAFSSON	J	STRUCTURE & FUNCTIONS OF THE TCDD RECEPTORS (RATS)	SWE
A	HAGENMAIER	H	STUDIES ON PHARMACOKINETICS AND TOXICOLOGY OF A PCDD/PCDF MIXTURE IN RODENTS & PRIMATES	FRG
A	HAGENMAIER	H	CONTENT OF PCDD/PCDF IN SELECTED SOILS, SEDIMENTS & PLANTS	FRG
A	HAGENMAIER	H	STUDY ON PCDD AND PCDF IN THE ENVIRONMENT	FRG
A	HAGENMEIER	H	STUDY OF DIOXINS/FURANS IN SELECTED SOILS, PLANTS & SEDIMENTS	FRG
A	HAILE	C	VALIDATION ON EMISSION TEST METHOD FOR DIOXINS	USA
A	HAWLEY	J	DEVELOPMENT OF HEALTH CRITERION DOCUMENT	USA
A	HELDER	T	COMPARISON OF THE TOXICITY OF SELECTED DDs & DFs & DETERMINATION OF TOXICITY OF MIXTURES	NET
A	HELDER	T	TOXICITY OF FLY-ASH AND FLUE GAS	NET
A	HELGE	H	CHANGES OF THE IMMUNOSYSTEM AND INDUCTION OF MONOOXYGENASES UPON EXPOSURE TO DIOXIN	FRG
A	HILEMAN	F	BACKGROUND POPULATION LEVELS OF SELECTED PCDDS/PCDFS	USA
A	HITES	R	ATMOSPHERIC EXPOSURES OF MAN TO COMBUSTION GENERATED CDDs/CDFs	USA
A	HOEL	D	INTERNATIONAL REGISTER OF PERSONS EXPOSED TO PHENOXY ACID HERBICIDES & CONTAMINANTS, PHASE II	USA

Table 4. (continued)

Work group	Last name	First name	Title	Country
A	HOLLEBONE	B	DEVELOPMENT OF AUTOMATED LARGE-VOLUME WATER SAMPLING METHODS USING SORBENT CARTRIDGES	CAN
A	HOLSAPPLE	M	IMMUNOTOXICOLOGY BY CHLORINATED DIBENZO-P-DIOXINS (MICE)	USA
A	HUNSINGER	R	ROUTINE & EMERGENCY MONITORING OF DRINKING WATER SUPPLIES FOR CDDs/CDFs	CAN
A	HUSKOL	R	HALF-LIFE OF CHEMICALS IN SOIL	USA
A	HUTZINGER	O	DEVELOPMENT OF NEW ANALYTICAL METHODS FOR ANALYSIS OF A COMPLEX MATRIX (PCDD/PCDF)	FRG
A	HUTZINGER	O	PHARMACOKINETICS & TOXICOLOGY OF PBDD AND PBDF IN RODENTS-ANALYSIS OF ORGANS	FRG
A	JANSEN	E	DEVELOPMENT OF A CHEMOLUMINESCENT IMMUNOASSAY FOR CDDs/CDFs	NET
A	JOHNSON	A	FISH CONTAMINANTS PROGRAM	CAN
A	JONES	D	UPTAKE OF 2,3,7,8-TCDD BY DAIRY COWS	USA
A	KAHN	P	ROLE OF ACTIVATED OXYGEN IN THE TOXIC DAMAGE PRODUCED BY 2,3,7,8-TCDD	USA
A	KANG	H	VIETNAM VETERANS MORTALITY STUDY	USA
A	KANG	H	WOMEN VIETNAM VETERANS HEALTH STUDY	USA
A	KANG	H	VA/AFIP CASE CONTROL STUDY OF SOFT TISSUE SARCOMA	USA
A	KANG	H	SOFT TISSUE SARCOMA & MILITARY SERVICE IN VIETNAM	USA
A	KANG	H	RETROSPECTIVE STUDY OF DIOXINS & FURANS IN ADIPOSE TISSUE	USA
A	KARASEK	F	DEVELOPMENT OF EFFECTIVE/RAPID CLEANUP PROCEDURES FOR ANALYSIS OF PCDD/PCDF IN FISH & OTHER BIOTA	CAN
A	KARASEK	F	METHODOLOGY FOR COMPLETE ANALYSIS OF COMPLEX MIXTURES	CAN
A	KARASEK	F	PROCEDURE FOR THE 2,3,7,8-SUBSTITUTED ANALYSIS OF PCDD/PCDF, ETC IN ENVIRONMENTAL SAMPLES	CAN
A	KERKVLIET	N	ALTERATIONS IN CELL SURFACE MARKER EXPRESSION BY DIOXINS (MICE)	USA
A	KONIETZKO		EPIDEMIOLOGICAL STUDY	FRG
A	KONIG	W	IMMUNE-TOXICOLOGICAL STUDY OF EXPOSED PERSONS FOR EFFECTS OF HALOGENATED DIOXINS, ETC. - ALLERGLOGIC STDY	FRG
A	KORTE	F	THERMOLYSIS OF PRODUCTS CONTAINING HALOGENATED ORGANIC COMPOUNDS	FRG
A	LAO	R	DEVELOPMENT OF ANALYTICAL METHODS FOR THE MEASUREMENT OF PCDD/PCDF IN ENVIRONMENTAL SAMPLES	CAN
A	LAO	R	DETERMINATION OF PHYSICO-CHEMICAL PROPERTIES OF PCDD/PCDF	CAN
A	LINTAS	C	CONTAMINANT RESIDUES IN MARINE ORGANISMS AND THEIR IMPACT ON THE FOOD CHAIN	ITA
A	LOCKWOOD	K	ANALYSIS OF 2378-TCDD, 2378-TCDF, PCB'S, & PAH'S IN FISH AND LOBSTER LIVER TISSUE	USA
A	LODGE	K	DIOXIN BIOAVAILABILITY - FOOD CHAIN	USA
A	LUCIER	G	RECEPTOR ACTION AND LIVER TUMOR PROMOTION	USA
A	LUETZKE	K	TESTING OF SAMPLE TECHNIQUES FOR EMISSION MEASUREMENTS OF PCDD/PCDF	FRG
A	LUSIS	M	STUDIES OF TOXIC DEPOSITION	CAN

33

Table 4. (continued)

Work group	Last name	First name	Title	Country
A	LUSTER	M	MECHANISM OF IMMUNO-SUPPRESSION	USA
A	MACMILLAN	J	MONOCLONAL ANTIBODIES FOR DETECTION OF DIOXINS	USA
A	MATSUMURA	F	EPIDERMAL GROWTH FACTOR AS A MODULATION OF MEDIATOR OF DIOXIN TOXICITY	USA
A	MATSUMURA	F	ALTERATION OF CELL-SURFACE MEMBRANE FOR DIOXIN TOXICITY (RODENTS, RABBITS)	USA
A	MCFARLANE	C	PLANT UPTAKE OF DIOXIN	USA
A	MEANS	J	POTENTIAL FOR 2378-TCDD TRANSPORT IN SOILS USING BOTH STATIC & DYNAMIC SYSTEMS	USA
A	MEHRLE	P	TOXICITY & BIOACCUMULATION OF 2,3,7,8-TCDD & 2,3,7,8-TCDF IN RAINBOW TROUT	USA
A	MEREDITH	M	FATE OF TOXIC CHEMICALS IN HEPATOCYTE AND HEPATOMA CELLS (RATS, HUMAN)	USA
A	MILL	T	TCDD VAPOR-PHASE PHOTOLYSIS	USA
A	MILLER	J	STUDIES ON ELECTROPHILIC ULTIMATE CARCINOGENS (MICE, RATS)	USA
A	MILLER	G	DIOXIN PHOTOLYSIS: SOIL SURFACES	USA
A	MITCHUM	R	PREP OF STDS & INVESTIGATION OF PHYS PROPERTIES & ENV CONTAMINATION BY BCDD/BCDFs	USA
A	MUIR	D	BIOAVAILABILITY OF DIOXINS	CAN
A	MUKERJEE	D	HEALTH ASSESSMENT OF PCDFS	USA
A	MULLER	W	IMMUNOTOX STUDIES OF POLLUTANT-LOADED PERSONS FOR EFFECTS OF DIOXINS, ETC. - IMMUNOBIOLOGY	FRG
A	McBRIDE	A	U.S. EPA NATIONAL DIOXIN STUDY	USA
A	NEUBERT	D	PHARMACOKINETICS AND TOXICOLOGY OF PCDD/PCDF MIXTURE, AND PBrs ON RODENTS & PRIMATES	FRG
A	NIEDING	G	PCDDs/PCDFs IN THE ENVIRONMENT & IN FOOD & HUMANS - SAMPLING, ANALYSIS, ASSESSMENT	FRG
A	NORSTROM	R	LEVEL AND DYNAMICS OF DIOXINS, FURANS & RELATED COMPOUNDS IN WILDLIFE	CAN
A	OKEY	A	STRUCTURE & FUNCTION OF THE Ah RECEPTOR	CAN
A	OLSON	J	MECHANISMS FOR TOXICITY OF CDDs (RODENTS, HUMAN)	USA
A	PATTERSON	D	SYNTHESIS OF CDDs & RELATED COMPOUNDS	USA
A	PETERSON	R	PCDD/PCDF-PERSISTENCE & TOXICITY IN FRESHWATER FISH	USA
A	PETERSON	R	ENVIRONMENTAL POLLUTANTS & TOXICOLOGY OF THE LIVER (RATS, MONKEYS)	USA
A	PINKERTON	M	COMBUSTION PRODUCTS FROM HIGH IMPACT POLYSTYRENE CONTAINING DECABROMODIPHENYL ETHER AS FLAME RETARDENT	USA
A	PIPER	W	TOXICANT DEREGULATION OF ENDOCRINE HEME BIOSYNTHESIS (RATS, HAMSTERS, GUINEA PIGS)	USA
A	PIRKLE	J	DETERMINATION OF HALF-LIFE OF 2,3,7,8-TCDD IN HUMANS	USA
A	POLAND	A	STUDIES ON THE HALOGENATED AROMATIC HYDROCARBONS (MICE, HAMSTERS, RATS)	USA
A	POLITZER	P	CAUSAL STRUCTURE-ACTIVITY METHODS APPLIED TO THE ASSESSMENT OF TOXICITY OF DD/DFs - BIOLOGICAL ACTIVITY	USA
A	POLS	H	PCDD IN INFLUENT, EFFLUENT & SLUDGE OF POWT-PLANTS	NET

Table 4. (continued)

Work group	Last name	First name	Title	Country
A	POMPA	G	PCBs TOXIC EFFECTS IN PREGNANT & VIRGIN RABBITS AND TRANSFER VIA MILK TO SUCKLING OFFSPRING	ITA
A	PUHVEL	M	MECHANISMS OF TOXICITY OF DIOXIN (MICE)	USA
A	RICE	R	KERATINOCYTE ENVELOPES--PHYSICAL & TOXIC MECHANISMS (RATS, RABBITS, HUMANS)	USA
A	RIFKIND	A	ARACHIDONATE PRODUCTS IN DIOXIN & PCB TOXICITY (CHICK EMBRYOS)	USA
A	RILEY	R	VAPOR PRESSURE & PARTITIONING BEHAVIOR OF 2,3,7,8-SUBSTITUTED DIOXINS & FURANS	USA
A	RIS	C	RISK ASSESSMENT APPROACH FOR 2,3,7,8-TCDD	USA
A	ROHLEDER	H	ENVIRONMENTAL HAZARD POTENTIALS OF CHEMICALS	FRG
A	ROZMAN	K	MECHANISMS OF SPECIES DIFFERENCES IN TCDD TOXICITY	FRG
A	RUSSELL	D	MECHANISMS OF TCDD TOXICITY (RATS)	USA
A	RYAN	J	PCDD/PCDF: METHOD DEVELOPMENT SURVEY OF FOODS & BIOLOGICAL SAMPLES TO ESTIMATE EXPOSURE & SOURCES	CAN
A	SAFE	S	PCDDs AND PCDFs	USA
A	SAFE	S	ASSAY OF AH RECEPTOR USING MONOCLONAL ANTIBODIES	USA
A	SAFE	S	IN VITRO BIOASSAY PROCEDURES FOR PCDD & PCDF COMPOUNDS	USA
A	SAFE	S	MECHANISM OF DIOXIN TOXICITY (MICE, RATS, GUINEA PIGS, HAMSTERS, AVIAN)	USA
A	SAFE	S	2,3,7,8-TCDD & RELATED COMPOUNDS & MECHANISMS OF CARCINOGENICITY	USA
A	SAFE	S	TCDD--EFFECTS OF RECEPTOR MODULATORS/ANTAGONISTS (MICE, HAMSTERS, GUINEA PIGS)	USA
A	SCHAUM	J	EXPOSURE ASSESSMENT METHODS FOR 2,3,7,8-TCDD	USA
A	SCHILLER	C	REGULATION OF INTESTINAL METABOLISM	USA
A	SCHROY	J	ENVIRONMENTAL FATE & TRANSPORT OF 2,3,7,8-TCDD	USA
A	SELENKA	F	IMMUNOTOXICOLOGIC STUDIES OF EXPOSED PERSONS FOR EFFECTS OF DIOXINS, ETC.- CHEMICAL - ANALYTICAL SURVEY	FRG
A	SHIVERICK	K	STEROID & XENOBIOTIC METABOLISM IN THE PLACENTA (RATS)	USA
A	SHIVERICK	K	TCDD EFFECTS ON STEROID HORMONE SYNTHESIS IN PREGNANCY (RATS)	USA
A	SHUSHAN	B	APPLICATION OF TAGA 6000E GC/MS/MS SYSTEM-RAPID SCREEN MUNICIPAL WASTE INCIN COMBUST PRODUCTS - CDDs/CDFs	CAN
A	SKINNER	L	LAKE ONTARIO CONTAMINANT TREND ANALYSIS	USA
A	SMITH	E	IMPACT OF DISASTER ON CHILDREN - DIOXIN & FLOOD	USA
A	SOVOCOOL	G	SUPERFUND/SARA CONTAINMENT FACILITY	USA
A	SRINIVASAN	V	DEGRADATION OF PERSISTENT ENVIRONMENTAL POLLUTANTS BY LIGNINOLYTIC MICROORGANISMS	USA
A	STARK	A	EPIDEMIOLOGIC RESEARCH ON THE HEALTH EFFECTS OF DIOXIN EXPOSURE	USA
A	STARTIN	J	DETERMINATION OF PCDDS & PCDFS IN HERBAGE & ANIMAL TISSUES	UK
A	STEVENSON	E	NEW JERSEY PHASE I & PHASE II DIOXIN SITE INVESTIGATIONS	USA

Table 4. (continued)

Work group	Last name	First name	Title	Country
A	SUNS	K	INVESTIGATION OF BIOLOGICAL MONITORS OF CDD/CDF CONTAMINATION IN THE ENVIRONMENT	CAN
A	SWEENEY	G	STUDIES ON HEPATOXICITY OF TCDD AND RELATED SUBSTANCES	CAN
A	SWEENEY	M	PERSISTENT HEALTH EFFECTS IN CHEMICAL HERBICIDE WORKERS & IN COMMUNITY RESIDENTS	USA
A	SZAKOLCAI	A	METHOD FOR DETERMINATION OF PARTICULATE & VAPOR-PHASE CDDs/CDFs IN AMBIENT AIR	CAN
A	TAGUCHI	V	IMPROVED INSTRUMENTAL TECHNIQUES FOR ANALYSIS OF CDDs/CDFs	CAN
A	TASHIRO	C	IMPROVED CLEANUP METHODS FOR CDDs/CDFs IN COMPLEX SAMPLES	CAN
A	TASHIRO	C	TOTAL TOXIC CONGENER CDD/CDF ANALYSIS OF FISH AND OTHER MATRICES	CAN
A	THEELEN	R	TOXICITY OF 2,3,7,8-TCDD & ISOSTEROMERS IN HUMAN ADIPOSE TISSUE	NET
A	THORPE	B	INVESTIGATION OF CDDs/CDFs/OTHER ORGANICS IN TYPICAL ONTARIO FOOD SUPPLIES: FOOD BASKET SURVEY	CAN
A	TONDEUR	Y	METHOD 8290: ANAL PROTOCOL FOR THE MULTIMEDIA CHARACTERIZATION OF PCDD/PCDF BY HIGH RESOLUTION GC/MS	USA
A	TURKSTRA	E	DIOXINS AND FURANS IN SEDIMENT AND FISH	NET
A	TUSCHALL	J	US EPA'S STANDARDS REPOSITORY	USA
A	VAN DE WERKEN	G	QA/QC OF THE DETERMINATION OF 2,3,7,8-TCDD IN SOIL, SEDIMENT & WATER	NET
A	VAN DEN BERG	M	TOXICOKINETICS OF PCDD AND PCDF	NET
A	VAN DER HEIJDEN	C	TOXICOLOGICAL EVALUATION IN LAB ANIMALS OF 2378-TCDD & ISOSTERIC COMPOUNDS IN FLY ASH	NET
A	VAN OTTERLOO	R	DIOXINS IN SEDIMENT & FISH IN A ROTTERDAM HARBOUR	NET
A	VANDERLAAN	M	DEVELOPMENT OF AN IMMUNOASSAY FOR CDDs BASED ON A MONOCLONAL ANTIBODY & ELISA	USA
A	VOS	J	EFFECT OF 2,3,7,8-TCDD ON T! ` EPITHELIUM	NET
A	WEBSTER	G	BIOAVAILABILITY & FATE OF SE...IED PCDDS IN AN EXPERIMENTAL AQUATIC ECOSYSTEM	CAN
A	WEBSTER	G	AVAILABILITY, BIOCONCENTRATION, & TRANSFORMATION OF PCDDS IN SEDIMENT & AQUATIC SYSTEMS	CAN
A	WEGMAN	R	DEVELOPMENT/OPTIMIZATION OF METHODS FOR THE DETERMINATION OF PCDD & PCDF IN ENVIRONMENTAL SAMPLES	NET
A	WEGMAN	R	OPTIMIZATION OF A QUANTITATIVE METHOD FOR DETERMINATION OF PCDD/PCDF CONGENERS IN HUMAN MILK & FAT TISSUE	NET
A	WEGMAN	R	PCDD AND PCDF IN HUMAN MILK	NET
A	WHITLOCK, JR	J	CONTROL OF GENE EXPRESSION BY DIOXIN (MICE)	USA
A	WHITTLE	D	GREAT LAKES FISH CONTAMINANTS MONITORING PROGRAM	CAN
A	WILLIAMS	D	HUMAN TISSUE ANALYSIS	CAN
A	WOLF	K	NEW PROCESSES/METHODS FOR REMEDIAL MEASURES OF ABANDONED SITES	FRG
A	WOLFE	W	AIR FORCE HEALTH STUDY- AN EPIDEM INVEST OF HEALTH EFFECTS IN A.F. PERSON EXPOSED TO HERBICIDES	USA
A	WONG	T	PLACENTAL MARKERS OF EXPOSURE TO TOXIC HALOGENATED AROMATICS	USA
A	YRJANHEIKKI	E	ANALYTICAL STUDIES ON LEVELS OF PCBS, PCDDS & PCDFS IN HUMAN MILK IN DIFFERENT GEOGRAPHIC AREAS	WHO

Table 4. (continued)

Work group	Last name	First name	Title	Country
A	YRJANHEIKKI	E	EPIDEM STUDIES ON HEALTH EFFECTS IN INFANTS ASSOC WITH EXPOSURE TO PCBs, PCDDs, & PCDFs IN HUMAN MILK	WHO
A	YRJANHEIKKI	E	ASSESSMENT OF HEALTH RISKS IN INFANTS ASSOCIATED WITH EXPOSURE TO PCBS, PCDDs & PCDFs IN HUMAN MILK	WHO
A	ZIMMERMAN	E	TCDD ALTERATION IN GENE EXPRESSION & CLEFT PALATE (MICE)	USA
A	ZITKO	V	ORGANIC CONTAMINANTS IN NORTH ATLANTIC FISH	CAN
B	AUST	S	BIODEGRADATION OF ENVIRONMENTAL POLLUTANTS	USA
B	BAKKER	V	SCREENING SOURCES ON DIOXINS	NET
B	BALLSCHMITER	K	CONTRIBUTION OF MOTOR VEHICLE EMISSIONS TO ENVIRONMENT LOADING BY PCDD/PCDF	FRG
B	BALLSCHMITER	K	ANALYSIS OF CAUSES & REDUCTION OF PCDD/PCDF IN A WASTE INCINERATOR IN HAMBURG	FRG
B	BARTHA	R	BIOTECHNOLOGY APPROACHES TO DETOXIFICATION OF 2,3,7,8-TCDD CONTAMINATED SOILS	USA
B	BINECCHIO	L	CHARACTERIZATION OF ORGANIC COMPOUNDS IN FLUE GAS EMITTED BY THERMOELECTRICAL POWER STATIONS	ITA
B	BOECKHOUT	C	PREVENTION OF EXPOSURE OF WORKERS TO FLY ASH	NET
B	BRIDLE	T	CHARACTERIZATION & MANAGEMENT OF MSW INCINERATION ASH	CAN
B	BROKER	G	DIOXIN-EMISSIONS FROM MUNICIPAL WASTE INCINERATORS	FRG
B	BROKER	G	EMISSIONS OF DIOXINS AT WASTE INCINERATION PLANTS	FRG
B	CAMICI	G	DETERMINATION OF PCDD, PCDF, IPA IN INCINERATOR EMISSIONS	ITA
B	CHRISTENSEN	R	SEWAGE SLUDGE INCINERATOR EMISSIONS TESTING	CAN
B	CICCIOLI	P	IDENTIFICATION OF TOXIC COMPONENTS IN STATIONARY/MOBILE SOURCES (PAH, NITRO-PAH, PCDDs & PCDFs)	ITA
B	CREASER	C	DETERMINATION OF LEVELS OF PCDDS/PCDDS IN THE VICINITY OF VARIOUS TYPES OF INCINERATORS	UK
B	DAS	G	DEVELOPMENT OF CODE OF GOOD PRACTICE-CANADIAN WOOD PRESERVATION INDUSTRY	CAN
B	DES ROSIERS	P	SUPERFUND DIOXIN SITE ACTIVITIES	USA
B	DiDOMENICO	A	OCCURRENCE OF PCDD/PCDF IN PCBs USED AS DIELECTRIC FLUIDS	ITA
B	DiDOMENICO	A	INACTIVATION OF CHLORINATED AROMATICS IN AQUEOUS MEDIA BY PHOTOCHEMICAL TECHNIQUES	ITA
B	EHRESMANN	J	REMEDIATION AT ABANDONED SITES - SPREADING WITHIN LANDFILL	FRG
B	EHRESMANN	J	REMEDIATION AT ABANDONED SITES - RECORDING & DECONTAMINATION	FRG
B	EPA REG 7		EPA MOBILE INCINERATOR	USA
B	EPLING	G	ENHANCED PHOTODEHALOGENATION OF HALOAROMATIC POLLUTANTS	USA
B	ESSERS	U	PCDD/PCDF EMISSIONS FROM INTERNAL COMBUSTION ENGINES WITH COMMERCIAL POWER FUELS (PART A)	FRG
B	FACCHETTI	S	CHEMICAL WASTE - MANAGEMENT SYSTEM	ITA
B	FANELLI	R	ENVIRONMENTAL IMPACT OF MUNICIPAL WASTE INCINERATOR EMISSIONS	ITA
B	FOGLER	H	REMOVAL OF DIOXINS FROM INDUSTRIAL WASTE WATER BY SORPTION	USA

37

Table 4. (continued)

Work group	Last name	First name	Title	Country
B	FORSTNER		TESTING & DEVELOPMENT OF VARIOUS SOLIDIFICATION TECHNIQUES FOR PCDD/PCDF CONTAMINATED LIQUID IN LANDFILLS	FRG
B	HAGENMAIER	H	PCDD/PCDF EMISSIONS FROM INTERNAL COMBUSTION ENGINES WITH COMMERCIAL POWER FUELS (PART C)	FRG
B	HAGENMAIER	H	DETERMINATION OF PCDD/PCDF & OTHER CHLORINATED HYDROCARBONS IN SEWAGE SLUDGE	FRG
B	HAY	D	NATIONAL INCINERATOR TESTING & EVALUATION PROGRAM	CAN
B	HO	A	SURVEY OF ORGANIC COMPOUNDS, INCLUDING CDDs/CDFs, IN ONTARIO SEWAGE TREATMENT PLANT EFFLUENT	CAN
B	HORCH		COMBUSTION EXPERIMENTS FOR THE MINIMIZATION OF PCDD/PCDF -EMISSIONS	FRG
B	HUTZINGER	O	PCDD/PCDF ANALYSIS OF MUNICIPAL & SPECIAL INCINERATORS	FRG
B	HUTZINGER	O	POSSIBLE RELEASES OF BROMINATED/CHLORINATED DIOXINS AND FURANES	FRG
B	HUTZINGER	O	PCDD/PCDF EMISSIONS FROM INTERNAL COMBUSTION ENGINES WITH COMMERCIAL POWER FUELS (PART B)	FRG
B	HUTZINGER	O	STUDIES ON THE EXPOSITION OF PCDD/PCDF FROM INDUSTRIAL PROCESSES	FRG
B	HUTZINGER	O	SEARCH FOR PCDD/PCDFS IN PROCESSES OR ORGANIC CHLORINE CHEMISTRY	FRG
B	HUTZINGER	O	POSSIBLE RELEASE OF PBDD/PBDF FROM FIRE PROTECTED PLASTICS DURING A FIRE	FRG
B	JACKSON	D	SORPTION/DESORPTION OF 2,3,7,8-TCDD IN CONTAMINATED SOILS	USA
B	JACKSON	D	EFFECTS OF OTHER NON-POLAR ORGANICS ON MOVEMENT OF DIOXINS & FURANS IN SOILS	USA
B	JAMES	L	TIMES BEACH RESEARCH FACILITY	USA
B	KAMINSKY	D	POLLUTANTS IN PYROLYSIS OIL WITH A COMPARISON TO OTHER TECHNICAL USED OILS	FRG
B	KARASEK	F	STUDY OF THERMAL REACTIONS OF PCDDs ON FLYASH PARTICLES UNDER INCINERATOR CONDITIONS	CAN
B	LAPERRIERE	F	RECOVERY OF SINKING CHEMICALS FROM BOTTOM OF BODY OF WATER WITH SORBENTS	CAN
B	LEONHARD	T	TREATMENT OF HALOGENATED ORGANIC LIQUIDS FROM REFUSE TIPS (SEPARATION & TREATMENT OF DIOXIN)	FRG
B	LOUW	R	FORMATION OF PCDD & PCDF IN MUNICIPAL INCINERATORS	NET
B	MANGOLD	K	ENVIRONMENTAL NEUTRAL TREATMENT & PURIFICATION OF DIOXIN LEACHATES AT THE LANDFILL - (MALSCH)	FRG
B	MCBRIDE	A	INVESTIGATION OF POTENTIAL SOURCES OF DIOXIN IN PULP & PAPER MLLS	USA
B	MILJOSTYRELSEN		DANISH INCINERATOR TESTING PROGRAM	DEN
B	MILJOSTYRELSEN		PILOT EMISSION TESTS - AMAGER INCINERATOR PLANT (COPENHAGEN)	DEN
B	MORTENSEN	H	TRIAL BURN & FIELD OPERATION OF EPA MOBILE INCINERATOR SYSTEM ON DIOXIN-CONTAMINATED LIQUIDS & SOLIDS	USA
B	OZVACIC	V	INVESTIGATION OF EMISSIONS OF CDDS/CDFS & OTHER ORGANICS FROM MEDICAL WASTE INCINERATION	CAN
B	PELIZZETTI	E	PHOTOCATALYTIC DEGRADATION OF CHLOROAROMATIC COMPOUNDS	ITA
B	PITEA	D	COMBUSTION OF MUNICIPAL WASTES & PCDD & PCDF EMISSIONS	ITA
B	ROGERS	C	CHEMICAL DECONTAMINATION OF CDD & RELATED COMPOUNDS IN CONTAMINATED LIQUIDS & SOLIDS	USA
B	ROTH	R	DESTRUCTION TECHNOLOGIES	FRG

38

Table 4. (continued)

Work group	Last name	First name	Title	Country
B	SCHAFFNER	B	ORIGIN, REDUCTION & HEALTH EFFECTS OF PCDD/PCDF EMISSIONS FROM MUNICIPAL WASTE INCINERATORS	FRG
B	SCHOLZ	M	BEHAVIOUR OF TCDD-ISOMERS DURING OZONATION	FRG
B	SEIN	A	EMISSIONS OF MUNICIPAL INCINERATORS	NET
B	SHEFFIELD	A	SALES OF DIOXIN-CONTAINING CHEMICALS	CAN
B	SHORT	J	SITE DEMO: ENVIRONMENTAL RESTORATION TECHNOLOGY-PILOT SCALE THERMAL DESORPTION/UV PHOTOLYSIS FIELD TRIAL	USA
B	SIERIG	J	PRACTICAL STRATEGIES FOR THE REDUCTION OF PCDD/PCDF FORMATON IN MUNICIPAL WASTE INCINERATORS	FRG
B	SINGH	J	TO DETERMINE LEVELS OF PCDDs/PCDFs IN AGRICULTURAL CHEMICALS, i.e., MCPA, 214-D, PCP, & DICAMBA	CAN
B	SOLWENNICKE	A	FEASIBILITY STUDY OF AN INTERIM STORE FOR MEDIUM-TERM STORAGE OF HIGHLY CONTAMINATED LANDFILL DUST	FRG
B	STEVENS	D	QA PROGRAM FOR CDD/CDF & OTHER ORGANICS/INORGANICS IN EMISSIONS FROM MUNICIPAL INCINERATORS	CAN
B	STODDART	T	SITE DEMO: ENVIRONMENTAL RESTORATION TECHNIQUE-FULL SCALE INCINERATION FIELD TRIAL	USA
B	STRABERGER		MONITORING PROGRAMME OF HOSPITAL WASTE INCINERATORS	FRG
B	TECH INSTITUTE		DIOXIN EMISSIONS FROM STRAW COMBUSTION	DEN
B	TECH INSTITUTE		WASTE OIL INCINERATOR TEST	DEN
B	TECH INSTITUTE		STATIONARY DIESEL ENGINE PERFORMANCE TEST	DEN
B	TECH INSTITUTE		WASTE OIL INCINERATOR TEST	DEN
B	ULLRICH	R	CAUSES & REDUCTION OF PCDD/PCDF EMISSIONS AT AN INCINERATOR IN HAMBURG	FRG
B	VISALLI	J	CORRELATION OF INCINERATOR CONDITIONS WITH CDD/CDF EMISSIONS	CAN
B	VROOMAN	W	INVESTIGATION OF CDDs/CDFs IN EMISSIONS FROM PULP & PAPER INDUSTRY	CAN
B	WAHL		ORIGIN & REDUCTION OF PCDD/PCDF-EMISSIONS FROM MUNICIPAL WASTE INCINERATORS	FRG
B	WAHL	H	CAUSES & REMOVAL OF PCDD/PCDF EMISSIONS FROM 3 INCINERATION PLANTS IN SCHLESWIG - HOLSTEIN	FRG
B	WEGMAN	R	DETERMINATION OF SOIL CONTAMINATION DUE TO OPEN AIR COMBUSTION OF WASTE ELECTRICAL PRODUCTS & CARS	NET
B	WEGMAN	R	DIOXINS IN PESTICIDES & RELATED COMPOUNDS	NET
B	WENTRUP	G	MEASUREMENT OF PCDD/PCDF IN WASTE OIL & SECONDARY RAFFINATES & EMISSIONS AT A WASTE OIL REFINERY	FRG
B	WILDERER		BIOLOGICAL TREATMENT OF DIOXIN CONTAMINATED LEACHATE FROM LANDFILL - HAMBURG	FRG
B	WOLF	K	NEW PROCESSES FOR REMEDIATION AT ABANDONED SITES - INCINERATION AT GEORGSWERDER LANDFILL	FRG
B	WOLF	K	NEW REMEDIATION PROCESS FOR ABANDONED SITES-SAFEGUARDING/STORAGE AT GEORGSWERDER LANDFILL	FRG
B	WOODFIELD	M	ISOLATION OF SOURCES	UK
B	YRJANHEIKKI	E	PCBs, PCDDs & PCDFs - PREVENTION & CONTROL OF ACCIDENTAL EXPOSURES	WHO
B	ZARTH		OCCUPATIONAL HEALTH & SAFETY REGULATIONS & ANALYTICAL SCREENING PROGRAMS FOR PCDD/PCDF CLEAN-UP	FRG
C	EPA REG 7		REMOVAL OF DIOXIN CONTAMINATED SOIL AT CASTLEWOOD	USA

Table 4. (continued)

Work group	Last name	First name	Title	Country
C	EPA REG 7		REMOVAL OF DIOXIN CONTAMINATED SOIL AT MINKER/STOUT/ROMAINE CREEK	USA
C	EPA REG 7		REMOVAL OF DIOXIN CONTAMINATED SOIL AT QUAIL RUN MOBIL MANNER	USA
C	EPA REG 7		CAPPING OF DIOXIN CONTAMINATED MATERIAL AT FRONTENAC	USA
C	EPA REG 7		CAPPING OF DIOXIN CONTAMINATED MATERIAL AT EAST NORTH STREET	USA
C	EPA REG 7		CAPPING OF DIOXIN CONTAMINATED MATERIAL AT COMM. CHRIST. CHURCH	USA
C	JAMES	L	DIOXIN SITE MANAGEMENT	USA
C	JAMES	L	TIMES BEACH BUYOUT/GRANT	USA
C	JAMES	L	CASTLEWOOD RELOCATION	USA
C	JAMES	L	MINKER/STOUT BUYOUT/MONITORING	USA
C	ROTH	R	DISMANTLING & DECONTAMINATION OF A CHEMICAL PLANT	FRG
C	TABASARAN	O	SITE CLEAN-UP	FRG
C	VAN DEN BERG	M	DIOXIN ANALYSIS AT LOW PPT LEVEL	NET

3. LISTING OF LABORATORIES WITH EXPERTISE IN THE ANALYSIS OF DIOXINS AND RELATED COMPOUNDS

David P. Bottimore*, Frederick W. Kutz**, and Erich W. Bretthauer**

Versar Inc.* and U.S. Environmental Protection Agency**
Springfield, Va.*, and Washington, D.C.**

This section presents a listing of laboratories with expertise in the analysis of dioxins and related compounds. The listing was compiled to facilitate the identification of analytical centers in the participating NATO nations and describe their capabilities for such analyses. Also listed are laboratories located in participating countries, or as components of international organizations, active as observers in this project. In addition, the names and locations of the laboratories, the matrices analyzed, principal analysts, and descriptions of the quality assurance schemes used are included. A series of graphic representations is also included to summarize the information in the listing.

The objective of this listing is to identify laboratories in participating countries having experience in the analysis of chlorinated dibenzo-p-dioxins (CDDs), dibenzofurans (CDFs), and related compounds. This listing was compiled as a contribution toward international information exchange on the analysis of these important compounds. It should facilitate the exchange of information among those interested in this phase of dioxin research and promote the development of improved analytical methods and standards among the participating nations. Examination of this listing of 108 laboratories in 9 nations also illuminates some of the trends with regard to laboratory affiliation, matrices analyzed, and quality assurance schemes used. Many of the laboratories listed have participated in externally moderated interlaboratory quality assurance programs to validate and compare analytical results.

Accurate analysis of dioxins in environmental matrices at extremely low detection levels is fundamental to many areas of research and regulation pertaining to these compounds. The development of precise and repeatable methods for dioxin analysis is required for assessing the risks posed by these compounds to human health and the environment even at extremely low concentrations. For example, the effectiveness of destruction technologies is measured by the ability of treatment systems to reduce the concentration of these compounds to extremely low levels. In addition, the success of a remedial action or a human effects monitoring program depends heavily on the analytical capabilities available.

Ultimately, the control of dioxins and related compounds is directly based on the ability to analyze them in a variety of environmental matrices. As a result, the state of the art in dioxin analysis directly impacts many other forms of research. In recent years, there has been a greater emphasis on the risks involved with the 209 CDDs and CDFs other than 2,3,7,8-TCDD. Because of the need to analyze and quantify these compounds at very low concentrations (ppt or ppq) in various environmental matrices, special analytical methods and equipment have been developed. Currently, a limited number of laboratories have the capabilities to analyze these compounds at the detection limits necessary for research and regulation.

3.1   Method of Compilation

This section presents the methodology used to collect and compile the information contained in the listing. It also outlines the organizations contacted for information and the criteria used to determine whether a lab should be listed. At the time that this Pilot Study was adopted by the Committee on the Challenges of Modern Society, each NATO member nation interested in participating appointed a lead delegate to coordinate and administer the project for his or her country. To compile this inventory, requests were sent to the lead delegates of the participating nations and to representatives of observer organizations. These requests asked for information on analytical centers with expertise in dioxin analysis within their organizational or political areas of representation. Information requested included: (1) name and address of the laboratory, (2) name of the principal analysts, (3) matrices analyzed, and (4) a brief summary of the quality assurance schemes in use at each laboratory. Recommended criteria for inclusion on the list were (1) actual experience with the analyses for CDDs, CDFs, and related compounds; (2) procedures utilizing combined high-resolution gas chromatography-mass spectrometry (GC/MS); and (3) demonstrated ability to perform analyses at the parts-per-trillion (or lower) sensitivity level on environmental matrices.

Several of the laboratories listed have participated in studies sponsored by international organizations. The CEC Joint Research Centre, Ispra Establishment, has been the moderating laboratory for CEC-sponsored studies and helped to identify numerous laboratories included in this listing. The laboratories affiliated with the World Health Organization (WHO) study on levels of dioxins and related compounds in human milk were identified from the report of their most recent meeting in February 1988. The report, Summary of Consultation on Results on Analytical Field Studies on Levels of PCBs, PCDDs, and PCDFs in Human Milk, contained a listing of the participating laboratories and a description of the interlaboratory quality assurance program used to provide consistent analytical results.

Laboratories in the United States were identified through several mechanisms. In addition to EPA's Contract Laboratory Program (CLP), the EPA has several laboratories that provide dioxin analyses. As part of the EPA's National Dioxin Study, which was completed in 1986, the analytical methods development and quality assurance aspects of the project were provided by the Troika. The Troika consisted of three EPA laboratories and two supporting university laboratories, and their function was to provide immediate and valid analysis for the National Dioxin Study and to serve as the nucleus for the development of the Contract Laboratory Program after private laboratories had demonstrated their analytical capabilities. Information on the Troika laboratories and quality assurance was collected from the report, Analytical Procedures and Quality Assurance Plan for the Analysis of 2,3,7,8-TCDD in Tier 3-7 Samples of the U.S. EPA National Dioxin Study.

In the United States, in addition to listing laboratories involved in government programs, a survey was made of members of the American Council of Independent Laboratories. Laboratories that were members of this council and had combined gas chromatography-mass spectrometry capabilities as reported to the Council's headquarters in Washington, D.C., were contacted by telephone and asked if their capabilities matched the inclusion criteria. This process added several laboratories to the list and confirmed the analytical capabilities of several laboratories already listed.

3.2   Description of Laboratories Listed

The listing of laboratories presents the names and addresses of the laboratories and the names of principal analysts. The matrices analyzed and quality assurance schemes employed are also included as part of the listing. This information is included in the listing to indicate the type of analyses performed and to promote interaction among labs in the participating nations. Increased awareness of the type of analyses being conducted, especially when dealing with difficult matrices, can result in greater cooperation among labs in the forms of sample exchanges and other activities in order to develop improved methods.

3.2.1 Matrices Analyzed

When the information on laboratories with dioxin analysis capabilities was compiled from the participating nations, one of the specific data elements requested was the matrices analyzed. Because of the variety of environmental matrices analyzed and the varying detail of the available information reported on each analytical center, summary groupings of matrices were developed. For example, surface waters, ground waters, and drinking waters were grouped together and presented in the listing as water. Other groupings, for which similarities were not as immediately obvious, were combined for the listing based on chemical/physical properties into categories such as hazardous waste/chemicals. In contrast to many environmental samples where the matrices are generally different from the constituents of concern, hazardous wastes and chemicals may contain constituents very similar in structure and concentration. As a result, many of the extraction and cleanup steps may be similar.

In general, the matrix groupings were summarized for this listing with respect to the type and concentrations of potentially interfering constituents, the extraction and cleanup methods required, and the overall difficulty in analyzing these matrices. It is recognized that there are potential differences in these matrices, and that, in some cases, they may require entirely different extraction or cleanup methods and/or may result in varying detection limits achievable because of interferences from other constituents. Readers with questions related to a particular laboratory and the types of analyses carried out should contact that laboratory directly.

The groupings are generally as follows:

Water	= drinking water, surface water, ground water, etc.
Soil	= soils, sediments, sludges
Hazardous Waste/Chemicals	= industrial effluents, solid waste, chemical products, oil, pesticides
Combustion Sources	= incinerator emissions, fly ash, air, stack gas, automobile emissions
Tissue	= animal and human tissue, blood, milk
Food	= food from animal sources
Vegetation	= vegetation, food from plant sources

Figure 1 displays the types of sample matrices analyzed at the laboratories providing this information. Although the information on matrices analyzed is incomplete (only 75 of the 108 labs listed reported the matrices analyzed), most of these laboratories have expertise in the analysis of soil, water, and tissue. Analysis of combustion sources, chemicals, and hazardous wastes is generally a common practice at these laboratories. Analysis of CDDs and CDFs in food and vegetation is relatively scarce, and only a few selected laboratories reported these capabilities.

3.2.2 Quality Assurance Schemes

At the time that this compilation was initiated, the lead delegates from each country were requested to determine and describe quality assurance schemes employed at each analytical center. The quality assurance schemes employed at the labs listed vary considerably. All of the labs provide elements of internal quality assurance; in addition, some reported participation in externally moderated programs. Internal quality assurance denotes that the analytical center uses procedures within the control of the individual laboratory. This is the usual case in research studies that are oriented toward analytical methods development. Externally moderated quality assurance denotes that the laboratory participates in a program under the technical direction of an independent organization outside the laboratory's direct control. This type of program is usually employed when a large number of samples are analyzed, as in organized exposure monitoring studies. It is usually resource intensive, with some type of technical assistance available from the moderating laboratory. In general, the interaction among labs in externally moderated quality assurance programs has improved the state of the art of dioxin analysis by validating methods and analytical results through interlaboratory sample exchanges. As a result, all labs, including those that use internally developed quality assurance schemes, benefit from these activities. Many labs follow the quality assurance procedures developed from these externally moderated programs, including the frequency of instrument calibration, matrix spike analyses, and method blanks.

Both types of quality assurance programs are statistically based, and their goals are to provide data on the variability, accuracy, and precision of the analytical methodology. There are no value judgments placed by listing the internal/external quality assurance guidelines in terms of

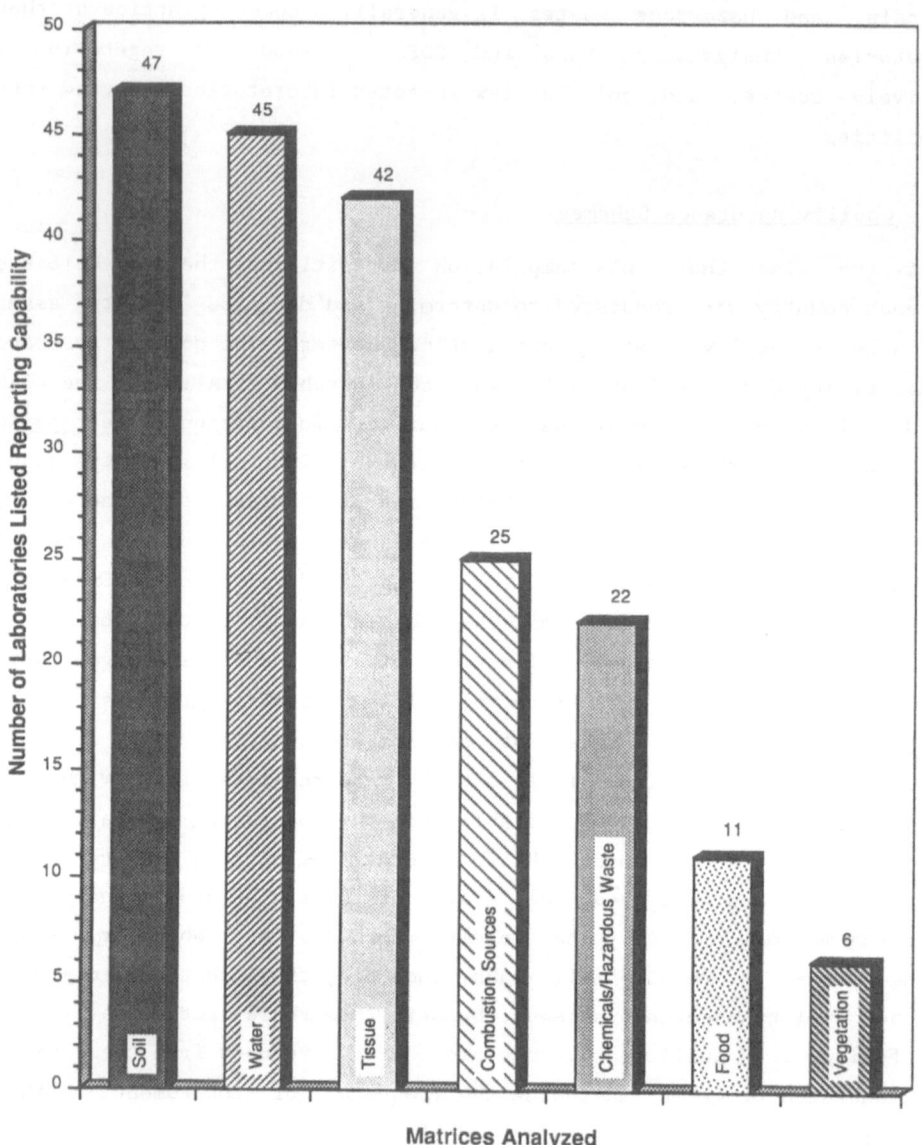

Figure 1. Matrices Analyzed at Listed Laboratories

the accuracy of the analyses performed. In many cases, the internal quality assurance standards are as rigorous as those followed by the externally moderated laboratories. This information is included to supplement the information exchange aspects of the project because the laboratories participating in externally moderated quality assurance programs are already utilizing information exchange to their benefit. Below are some descriptions of the types of quality assurance (both internal and external) being used by the laboratories. Figure 2 presents the quality assurance schemes employed at the laboratories listed. Over 30 percent of the laboratories listed reported that they participate in externally moderated quality assurance programs. Specifically, 18 laboratories are associated with the CEC Community Bureau of Reference, 16 participated in the WHO interlaboratory quality control program, 12 participated in the U.S. EPA Contract Laboratory Program, and 5 laboratories listed were part of the Troika. (All of these externally moderated quality assurance programs are described below.)

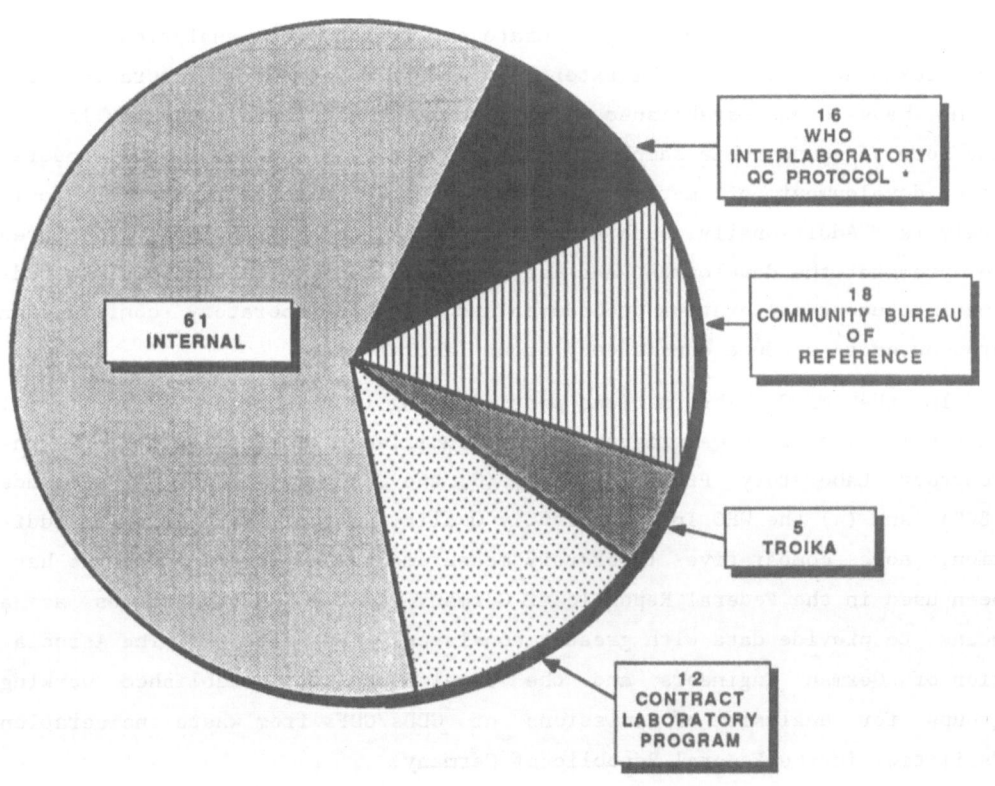

TOTAL NUMBER OF LABORATORIES LISTED = 108

Figure 2. Quality Assurance Schemes Employed at Listed Laboratories
(*5 laboratories participate in both the BCR and WHO programs)

3.2.3 Internal Quality Assurance

Laboratories reporting internal quality assurance schemes use procedures controlled within the management structure of the laboratory. Internal procedures comprise a discrete series of operations that vary from
simple to complex and usually cover the entire analytical procedure.
Operations include, but are not limited to, calibration of scales and
measuring equipment, confirmation of the decontamination of glassware and
other equipment, and more complex procedures such as recovery calculations
and instrument calibration. The underlying objective is the determination
of the accuracy, precision, and overall variability of the procedure.
Samples are analyzed in specific sets with both positive and negative controls associated with each analytical series. The exact procedures
employed at each laboratory vary, but records are kept to document procedures. The proper evaluation of a laboratory always includes a review of
the internal quality assurance data base.

3.2.4 Externally Moderated Quality Assurance

In the effort to develop accurate and reproducible analytical methods
for dioxin analysis, several externally moderated quality assurance programs have been established. In general, the interaction among labs in
the form of round-robin sample exchanges, blind QA samples, and cooperative development of methods has improved the state of the art of dioxin
analysis. Additionally, such externally moderated programs are structured
to promote the development and use of precise analytical methods, including instrument calibration procedures and multiple-laboratory confirmation
procedures to produce consistency among laboratories.

In the NATO member nations participating in this project, four formal
externally moderated QA programs were identified: (1) the Troika, (2) the
Contract Laboratory Program (CLP), (3) the Community Bureau of Reference
(BCR), and (4) the WHO Interlaboratory Quality Control Protocol. In addition, some comparative laboratory measurements and sample exchanges have
been used in the Federal Republic of Germany and the United States as a
means to provide data with greater consistency and validity. The Association of German Engineers and the Umweltbundesamt established working
groups for analysis of emissions of CDDs/CDFs from waste incineration
facilities in the Federal Republic of Germany.

A Lake Ontario TCDD bioaccumulation study conducted by the U.S. EPA,
New York State Department of Health, and Occidental Chemical Corporation
utilized interlaboratory comparisons using well-documented analytical

protocols. Although it was not developed as a formal externally moderated quality assurance program, the study utilized the analytical expertise of the Environmental Research Laboratory-Duluth, the New York State Department of Health, and the Ontario (Canada) Ministry of the Environment Laboratories in round-robin comparisons to confirm analytical results for use in bioaccumulation studies of TCDD in fish and sediment from the Hyde Park landfill to Lake Ontario.

Included below are descriptions of the four formal externally moderated quality assurance schemes that were identified.

(1) <u>The Troika</u>. The Troika quality assurance program was initiated in 1983 to provide a consistent standard for dioxin analysis in the United States. The quantification of 2,3,7,8-TCDD and related compounds was performed by a limited number of labs in the U.S. To satisfy the demand for sensitive analyses at the parts-per-trillion level of detection, the EPA delineated special measures to validate the accuracy of analytical procedures for 2,3,7,8-TCDD to be used in the conduct of analysis for the National Dioxin Study. Because the analytical portion of the National Dioxin Study was completed, the Troika program was terminated at the end of 1986.

The four basic principles of the program were (1) studies to validate methods, (2) "blind" QA samples, (3) multiple-laboratory participation, and (4) definition of analytical criteria for confirmation of 2,3,7,8-TCDD. The Troika consisted of the Environmental Research Laboratory-Duluth (ERL-D), the Environmental Monitoring Systems Laboratory-Research Triangle Park (EMSL-RTP), and the Environmental Chemistry Laboratory-Bay St. Louis (ECL). Also working with the Troika under cooperative agreements were Dr. Mike Gross of the University of Nebraska, Lincoln, and Dr. Thomas Tiernan of Wright State University.

The Troika program was structured to facilitate the exchange of QA samples, test samples, and analytical results. The validation capabilities inherent in these sample exchanges and "blind" QA samples, along with detailed sample preparation and analytical methods, provided the means for the generation of scientifically sound data. The interaction among labs not only verified the accuracy of analysis, but also enabled the labs to develop new methods or make modifications required to analyze troublesome samples. Overall, the Troika program established reference standards for analytical procedures to provide accurate and reproducible methods, and enhanced credibility and validity for the data from participating labs.

(2) <u>The Contract Laboratory Program (CLP)</u>. The Contract Laboratory Program (CLP) was established by the U.S. EPA to provide analytical services in support of EPA's Superfund program for remediating contaminated hazardous waste sites. Because of the concern for CDDs and CDFs in such contaminated sites and the need for accurate analytical data for these compounds, the Agency established detailed quality assurance criteria and validation procedures for a network of commercial analytical laboratories with capabilities for CDD/CDF analysis. The CLP provides the Agency with consistent and accurate analytical services and is closely controlled by the use of specific methods, procedures, and reporting requirements. The CLP quality assurance program is moderated by the U.S. EPA Environmental Monitoring Systems Laboratory-Las Vegas.

The QA/QC program associated with dioxin analysis conducted under the CLP includes:

- Detailed requirements for initial and periodic calibration and instrument performance checks, and specified actions that must be taken when the specified criteria are not met;

- Specifics on frequency of and criteria for method blank and matrix spike analyses;

- Field blanks with every batch of samples; and

- A blind performance evaluation sample with every batch of samples. (The acceptability of the analytical results depends on the performance evaluation sample results.)

In addition, the EPA performs complete and detailed data audits of approximately 20 percent of the results submitted by the contract laboratories. A large and increasing data base is maintained by the EPA, which facilitates establishing realistic requirements for QA/QC and for precision and accuracy. The EPA also maintains a standards repository for use by the laboratories participating in the CLP. These QA/QC measures are complemented by periodic onsite laboratory evaluations conducted by EPA teams of qualified scientists.

(3) <u>The Community Bureau of Reference (BCR)</u>. The Commission of European Communities (CEC) formed a group to improve the analytical methodologies for the determination of CDDs and CDFs. The Community Bureau of Reference (BCR) program was conducted with numerous analytical laboratories in several nations. Various isomers, in addition to 2,3,7,8-TCDD, were chosen as the subjects of the study. The analysis of CDDs and CDFs was studied in a series of collaborative exercises to improve and

standardize the analytical methods used. The participants evaluated several analytical steps for the study including sample preparation, extraction, cleanup, GC injection and separation, and MS quantification and identification.

Identification and quantification at parts-per-trillion levels have been studied using solutions of pure CDD/CDF congeners that are labeled with C^{13} isotopes. By using statistical techniques for collaborative tests and examining the experimental conditions, the sources of error in the various steps of the entire analytical procedure can be determined. As a result, the entire methodology can be improved, and through such collaborative efforts the sources of error can be minimized. After having improved their methodology, the participants disseminate their information in the form of certified materials. The overall program was moderated through the CEC Joint Research Centre, Ispra. Of the 22 analytical centers that participated in the BCR in addition to the Ispra establishment, 17 are located in NATO nations that participated in the CCMS project. These analytical centers have also been included in this listing. (Four are located in France and one is in Belgium.)

(4) The WHO Interlaboratory Quality Control Protocol. To provide consistent analytical results for a World Health Organization (WHO) study on CDDs, CDFs, and PCBs in human milk, a group of 18 laboratories participated in an interlaboratory quality assurance program. This study was initiated in 1985 to produce more data on the levels of, and routes of exposure to, CDDs, CDFs, and related compounds in various countries and geographic areas. Samples of milk were collected according to a well-defined scheme, and analyses for CDDs, CDFs, and PCBs were completed at 18 selected laboratories, 16 of which participated in the CCMS Pilot Study (and 5 of these also participated in the BCR).

The interlaboratory analytical quality assurance program was implemented to ensure the validity and comparability of the analytical results in order to produce reliable data for subsequent risk assessments. The analytical protocol was developed and moderated by Dr. Rappe of the University of Umea, Sweden. Two separate pools of human milk, along with the C^{13}-labeled standards, were distributed from the moderating laboratory to the participating laboratories. Analyses for CDDs and CDFs were received from 12 labs, and PCB results were received from 6 labs. In general, high-resolution GC/MS was used for CDD/CDF analysis, while PCB analysis was performed using gas chromatography equipped with EC detection or a similar technique.

3.3   Analysis of the Listing of Laboratories

The following tables and figures illustrate the information contained in the listing. These graphic representations are intended to illuminate some of the more notable trends pertaining to the laboratory location and affiliation. Table 1 presents the number of laboratories in each of the nations that participated in this project. There are 108 laboratories listed from 9 nations.

Table 1. Number of Laboratories Listed by Geographic Location

Country	Number of Labs Listed
Canada	13
Denmark	1
Federal Republic of Germany	23
Italy[a]	9
Netherlands	5
Norway	1
Sweden[b]	1
United Kingdom	7
United States	48
Total	108

[a] Includes CEC Joint Research Centre, Ispra Establishment, Italy.
[b] Non-NATO member observer nation.

Figure 3 presents the breakdown of laboratory affiliation (government, university, or commercial) for all the labs listed. Of the 108 laboratories listed, 40 percent are commercial labs, 37 percent are affiliated with government agencies, and 23 percent are university laboratories.

Figure 4 displays the affiliation of the laboratories in each nation. In Canada, Italy, and the United Kingdom, the majority of the labs are affiliated with government agencies; in the United States and the Federal Republic of Germany, more labs are commercially operated. It should be noted, however, that a considerable portion of the CDD/CDF analysis performed by commercial labs is in support of government programs.

Table 2 presents the listing of laboratories with expertise in the analysis of CDDs, CDFs, and related compounds.

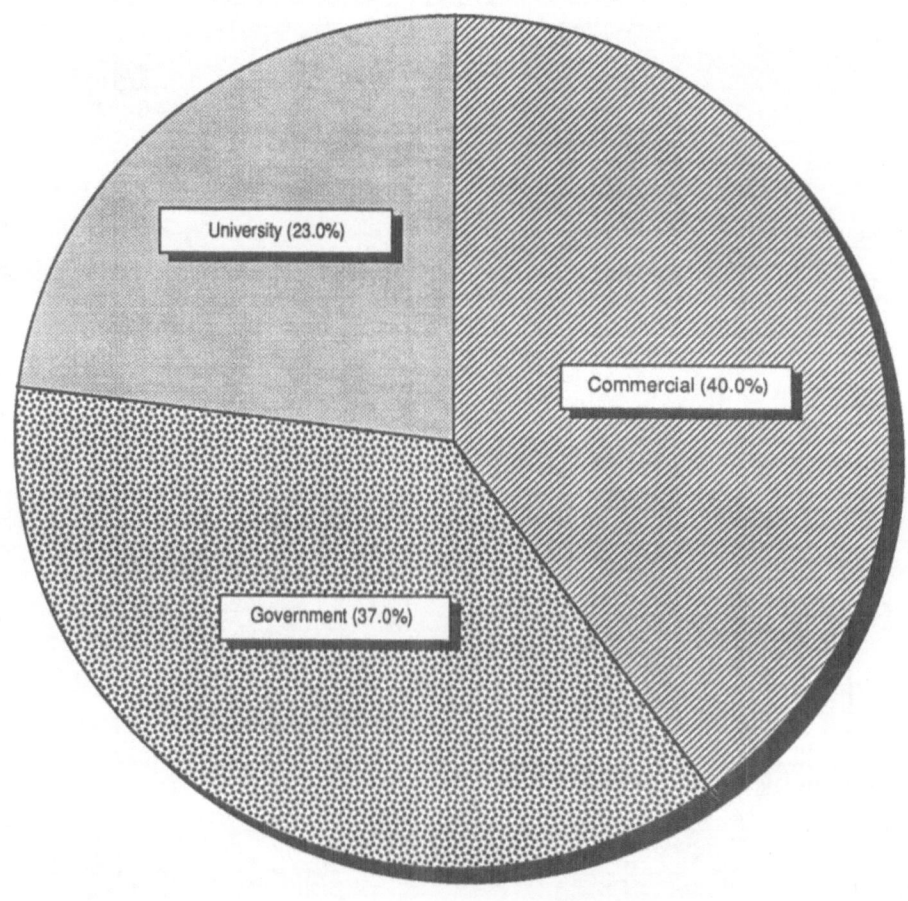

University (23.0%)

Commercial (40.0%)

Government (37.0%)

Figure 3.   Laboratory Affiliation

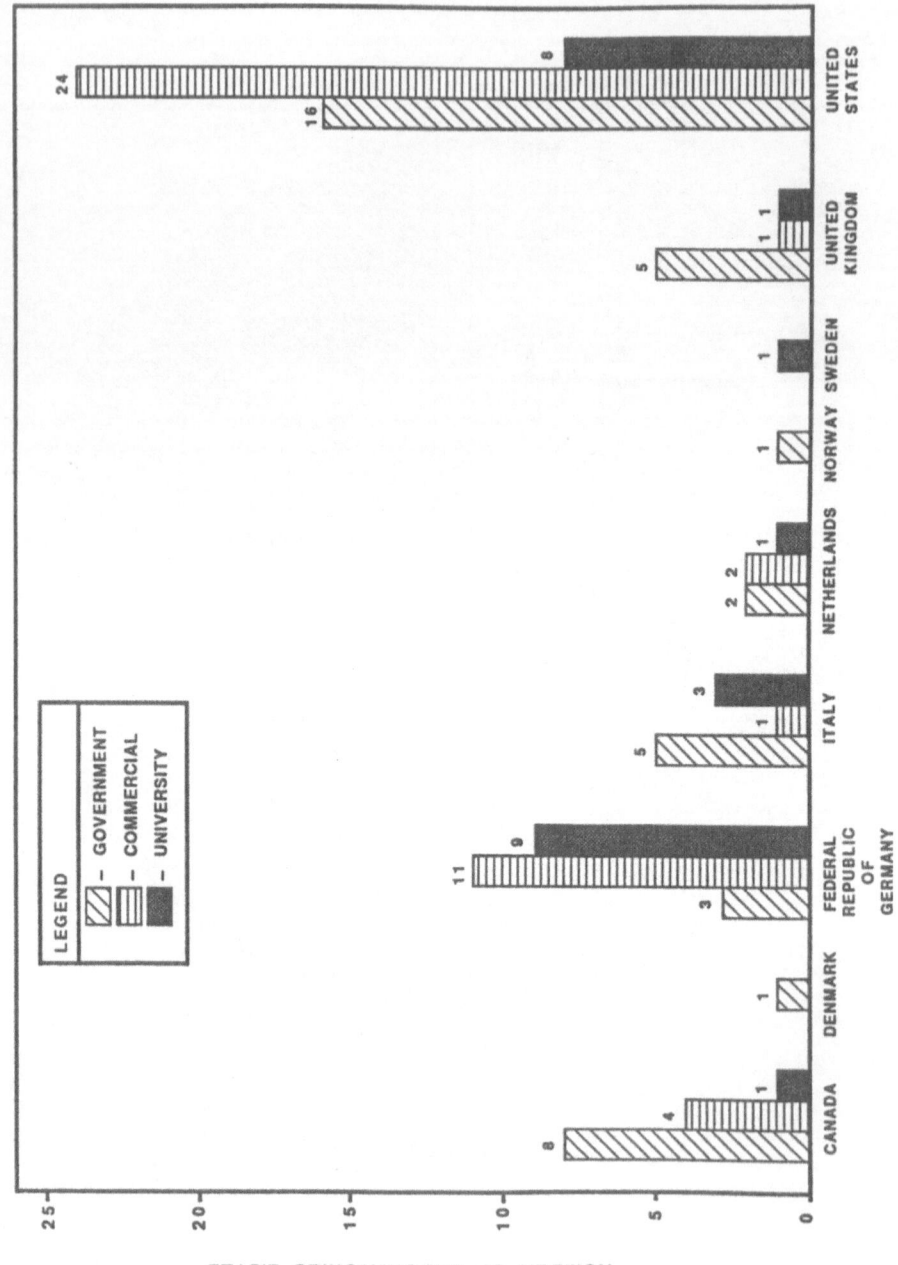

Figure 4. Laboratory Affiliation in Each Nation

54

Table 2. Listing of Laboratories with Expertise in the Analysis of Dioxins and Related Compounds

Country	Location	Principal Analysts	Matrices Analyzed	Quality Assurance Scheme
Canada	Bureau of Chemical Hazardous Environmental Health Directorate Health and Welfare Canada Ottawa, Ontario K1A 0L2 (613) 957-3126	Dr. D. Williams	Water	Internal
	Bureau Chemical Safety and Food Directorate Food Research Division Health and Welfare Canada Ottawa, Ontario K1A 0L2 (613) 959-0976	Dr. J. Ryan	Food Tissue	WHO
	Dioxin Laboratory Laboratory Services Ontario Ministry of the Environment P.O. Box 213 125 Resources Road Rexdale, Ontario M9W 5L1 (416) 235-5896	Mr. B. Bobbie Dr. R. Clement	Water Soil Tissue Combustion Sources	Internal
	Laboratory Services Division Food Inspection Directorate Agriculture Canada CEF Carling Avenue Ottawa, Ontario K1A 0C5 (613) 995-4907	Dr. J. Singh	Chemicals/ Hazardous Waste	Internal

Table 2. (continued)

Country	Location	Principal Analysts	Matrices Analyzed	Quality Assurance Scheme
Canada (continued)	Mann Testing Laboratories, Ltd. 5550 McAdam Road Mississauga, Ontario L4Z 1P1 (416) 890-2555	Mr. Tim Munshaw Mr. Pierre Beaumier	Water Tissue Soil Vegetation Combustion Sources	Internal
	National Water Quality Laboratory Inland Water Directorate, CCIW Environment Canada 867 Lakeshore Road, P.O. Box 5050 Burlington, Ontario L7R 4A6 (416) 336-44661	Dr. B.K. Afghan	Water Tissue Soil	Internal
	National Wildlife Research Institute Environment Canada 100 Gamelin Boulevard Hull, Quebec	Dr. R. Norstrom	Tissue	Internal
	Novalab Ltd. 9420 Cote de Liesse Lachine, Quebec H8T 1A6 (514) 636-6219	Dr. J.D. Fenwick	Water Soil Tissue Combustion Sources	Internal

Table 2. (continued)

Country	Location	Principal Analysts	Matrices Analyzed	Quality Assurance Scheme
Canada (continued)	River Road Environmental Technology Centre Technology Development and Technological Services Conservation and Protection Environment Canada Ottawa, Ontario K1A 0E7 (613) 998-3677	Dr. R. Lao	Water Soil Chemicals/ Hazardous Waste Combustion Sources	Internal
	Ultratrace Laboratory Bayfield Institute Great Lakes Laboratory for Fisheries and Aquatic Sciences Fisheries and Oceans Canada Burlington, Ontario (416) 336-4863	Mr. D.B. Sergeant	Tissue Water Soil	Internal
	University of Waterloo Department of Chemistry Waterloo, Ontario N2L 3G1	Prof. Frank W. Karasek	----	Internal
	Wellington Laboratories P.O. Box 1261 291 Woodlawn Road West, Unit 2 Guelph, Ontario N1H 6N6 (519) 822-2436	Mr. Brock Chittim Ms. Jocelyn Madge	Water Tissue Soil Chemicals/ Hazardous Waste Combustion Sources	Internal

57

Table 2. (continued)

Country	Location	Principal Analysts	Matrices Analyzed	Quality Assurance Scheme
Canada (continued)	Zenon Environmental Inc. 845 Harrington Court Burlington, Ontario L7N 3P3 (516) 631-6320	Dr. Glenys Foster Mr. John Coburn	Water Tissue Soil Combustion Sources	Internal
Denmark	National Agency of Environmental Protection Department of Analytical Chemistry Morkhoj Bygade 26 Dk-2860 Seborg Telephone 1-697088	Dr. Jorgen Carle	Tissue	WHO
Federal Republic of Germany	B.A.M Berlin	Dr. T. Win	----	BCR
	Bayer AG Zentrale Analytik Leverkusen/OAL Gebaude 013 D-5090 Leverkusen-Bayerwerk	Dr. Heinz Weis	----	BCR
	BASF AG Bereich Umweltschutz und Arbeitssicherheit, ZHU - E 210 Carl-Bosch-Straße 38 6700 Ludwigshafen (0621)60-0(Vermittlung)	Dr. K.S. Brenner	----	BCR

Table 2. (continued)

Country	Location	Principal Analysts	Matrices Analyzed	Quality Assurance Scheme
Federal Republic of Germany (continued)	Biochemisches Institut fur Umweltcarcinogene Sieker Landstraße 19 2070 Großhansdorf	Prof. Dr. G. Grimmer	----	Internal
	Biocontrol GmbH Hamburger Straße 1 6507 Ingelheim	Dr. Schlesing	----	Internal
	Chemisches Landesuntersuchungsamt Nordrhein-Westfalen Sperlichstrasse 19 D-4400 Munster Telephone 251-7793200	Dr. P. Furst	Tissue	WHO
	Deutsche BP AG Institut fur Forschung and Entwicklung Postfach 369 2000 Wedel/Holstein Tel.: 04103/701-310	Dr. W. Garbe Dr. Kaschani	----	Internal
	ERGO-Forschung Luruper Chaussee 145 2000 Hamburg 50 Tel.: 040/8969-2322	Dr. Ball	----	Internal
	Gesellschaft fur Arbeitsplatz und Umweltanalytik Nottulner Landweg 102 4400 Munster-Roxel	Dr. Funke	----	Internal

Table 2. (continued)

Country	Location	Principal Analysts	Matrices Analyzed	Quality Assurance Scheme
Federal Republic of Germany (continued)	Hoechst AG Analytisches labor Postfach 80 30 20 6230 Frankfurt 80	Dr. Hoffmann	----	Internal
	Ingenieur-Gemeinschaft Technischer Umweltschutz Ansbacherstraße 5 1000 Berlin 30 Tel.: 030/211 70 93 u. 7095	Dr. Joh. Jager	----	Internal
	Institut Fresenius Im Maisel 14 6204 Taunusstein/Neuhof Tel.: 06128/74 43 30	Dr. Scholz Prof. Fresenius M.N. Palaucheck	----	BCR
	Institut fur Naturwissen- Schaftliche Dienste GmbH (NATEC) Postfach 1568 Behringstraße 158 2000 Hamburg 50 Tel.: 040/8827	Dr. Eckert	----	Internal
	Institut fur Okologische Chemie und Geochemie der Universitat Bayreuth Postfach 3008 8580 Bayreuth Tel.: 0921/55 22 54	Prof. O. Hutzinger Dr. Thoma	----	BCR

Table 2. (continued)

Country	Location	Principal Analysts	Matrices Analyzed	Quality Assurance Scheme
Federal Republic of Germany (continued)	Institut fur Organische Chemie der Universitat Tubingen Auf der Morgenstelle 18 7400 Tubingen Tel.: 07071/29-2099	Prof. Dr. Hagenmeier	---	Internal
	Institut fur Ruckstandsanalytik St. Anscharplatz 10 2000 Hamburg 36	Dr. Winkelmann	---	Internal
	Institut fur Umweltanalytik Luitpoldstraße 190 6700 Ludwigshafen Tel.: 0621/69 10 51	Dr. F. Kuhlmann	---	BCR
	Institut fur Wasser-, Boden- und Lufthygiene des Bundesgesundheitsamt 1000 Berlin 33 Tel.: 030/8308-2344/2405/2290	Dr. Rotard	---	Internal
	Kernforschungszentrum Karlsruhe Institut fur Heiße Chemie Postfach 3640 7500 Karlsruhe	Dr. Stieglitz	---	Internal

Table 2. (continued)

Country	Location	Principal Analysts	Matrices Analyzed	Quality Assurance Scheme
Federal Republic of Germany (continued)	Max von Pettenkofer - Institut des Bundesgesundheitsamtes Postfach 330013 Thielallee 88-92 1000 Berlin 33 Telephone (030)83082665 Telefax (030)8308741	Dr. Beck Dr. Mathar Dr. Eckart	Tissue	BCR, WHO
	TÜV Norddeutschland Große Bahnstraße 31 2000 Hamburg 54 Tel.: 040/8957-457/454	Dr. Ullrish Dr. Schnabel	---	Internal
	Universität Ulm Analytische Chemie Oberer Eselsberg 026 Postfach 4066 7900 Ulm Tel.: 0731/176-2181	Prof. Dr. K. Ballschmiter Dr. Buchert Dr. M. Swerev	---	BCR
	Wacker Chemie Postfach 8263 Burghausen	Dr. Bienert	---	Internal

Table 2. (continued)

Country	Location	Principal Analysts	Matrices Analyzed	Quality Assurance Scheme
Italy	Commission of the European Communities* Joint Research Centre Ispra Establishment 210210 Ispra (Varese) Phone: (Italy-Ispra-) 332-789969 Telex No.: 380-042 EUR I or 380-058 EUR I Telefax No.: (Italy-Ispra-) 789-001	Dr. Sergio Facchetti	Tissue	WHO, BCR - Moderating Laboratory
	Consiglio Nazionale delle Ricerche Istituto di Inquinamento Atmosferico Via Salaria, km 29.300 - C.P. 10 00016 Monterotondo Stazione (Roma) Phone: (Italy-Rome-) 90020-652 Telex No.: 624-809 CNR MLI	Dr. Paolo Ciccioli Mr. Enzo Brancaleoni Dr. Angelo Cecinato	----	BCR
	ENEL (DCO) Laboratorio Centrale Via Bixio, 39 29100 Piacenza Phone: (Italy-Piacenza-) 791-224 Telex No.: 530-541 Telefax No.: (Italy-Piacenza-) 791-219	Dr. Luisa Binecchio	----	Internal

* Although geographically located in Italy, the CEC Joint Research Centre is the moderating lab for numerous laboratories in the participating nations.

Table 2. (continued)

Country	Location	Principal Analysts	Matrices Analyzed	Quality Assurance Scheme
Italy (continued)	Istituto di Ricerche Farmacologiche "Mario Negri" Viale Eritrea, 62 20157 Milano Phone: (Italy-Milan-) (02) 3554-546 Telex No.: 331-268 NEGRI I	Dr. Roberto Fanelli	Tissue	BCR, WHO
	Istituto Superiore di Sanita Laboratorio di Tossicologia Comparata ed Ecotossicologia Viale Regina Elena, 29 00161 Roma Phone: (Italy-Rome-) 4990, ext. 995 or 653 Telex No.: 610-071-ISTSAN I Telefax No.: (Italy-Rome-) 4957-621	Dr. Allessandro di Domenico Dr. Franco Merli	Tissue	WHO
	Servizio Multizonale di Prevenzione - USL 2 Via del Patriota, 2 54100 Massa Phone: (Italy-Massa-) 40676	Dr. Gino Camici Dr. Gino Biancardi	Soil	Internal
	Servizio Multizonale di Prevenzione - USL 10/A Via Ponte alle Mosse, 211 50144 Firenze Phone: (Italy-Florence-) 2758-4360	Dr. Moreno Berlincioni	Tissue	WHO

Table 2. (continued)

Country	Location	Principal Analysts	Matrices Analyzed	Quality Assurance Scheme
Italy (continued)	Unita Socio-Sanitaria Locale 75/11 Unita Operativa Chimica Via Juvara, 22 20129 Milano Phone: (Italy-Milan) 717-713	Prof. Aldo Cavallaro Dr. Alfredo Gorni	Tissue	WHO
	Universita di Milano Istituto di Scienze Farmacologiche Facolta di Farmacia Cattedra di Biochimica Applicata Via Balzaretti, 9 20133 Milano Phone: (Italy-Milan-) 209-817	Prof. Giovanni Galli Prof. Flaminio Cattabeni	----	Internal
The Netherlands	University of Amsterdam Laboratory of Environmental and Toxicological Chemistry Nieuwe Achtergracht 166 1018 WV Amsterdam (020) 52 56 504	Dr. K. Olie	Water Soil Tissue Combustion Sources Chemicals/ Hazardous waste Food	BCR, WHO

Table 2. (continued)

Country	Location	Principal Analysts	Matrices Analyzed	Quality Assurance Scheme
The Netherlands (continued)	National Institute of Public Health and Environmental Hygiene Laboratory for Organic-Analytical Chemistry P.O. Box 1 3720 BA Bilthoven Tel: 030-742871 Telex: 47215 rivm bl Telefax: 030-742971	Drs. Ronald C.C. Wegman (head) Drs. A.K. Djien Liem George S. Groenemeijer Gert A.L. de Korte Ir. Ad P. de Jong Drs. Gerrit van de Werken Arie C. den Boer Evert Evers G. Jan ten Hove Marjo J. Vredenbregt	Water Soil Tissue Combustion Sources Chemicals/ Hazardous waste Vegetation Food	WHO
	The Netherlands Organization for Applied Scientific Research TNO Division of Technology for Society Department of Analytical Chemistry P.O. Box 217 2500 AE Delft (031)15569330	Dr. J.W.J. Gielen P. Verhoeve	Combustion Sources Chemicals/ Hazardous Waste	BCR
	Duphar NV V. Houtenlaan 36 Weeps Telephone 02940-79794	Dr. H.A.M. DeKuik	Combustion Sources Chemicals/ Hazardous Waste	BCR

Table 2. (continued)

Country	Location	Principal Analysts	Matrices Analyzed	Quality Assurance Scheme
The Netherlands (continued)	TAUW Infra Consult B.V. Environmental Laboratory P.O. Box 479 7400 AL Deventer Tel: (05700)99911 Telex: 49545 Telefax: (05700)99270	Dr. G.H.W. Baalhuis	Water Soil	BCR
Norway	Norwegian Institute for Air Research P.O. Box 64, N-2001 Lillestrom, Norway Tel: 06/81 41 70. Telex: 74854 Telefax: 06/81 92 47	Dr. Michael Oehme Dr. Stein Mano Dr. A. Mikalsen	Combustion Sources Water Tissue	WHO
Sweden	University of Umea Department of Organic Chemistry S-901 87 Umea, Sweden	Dr. Christoffer Rappe	Tissue Combustion Sources Water Soil	WHO - Moderating Laboratory
United Kingdom	Analytical Services Centre (PCMU) Harwell Laboratory United Kingdom Atomic Energy Authority Didcot Oxfordshire OX11 ORE	Dr. Ian Stenhouse Dr. Peter Ambridge Dr. Dick Atkins	Water Soil Tissue Chemicals/ Hazardous Waste Vegetation Combustion Sources	Internal

67

Table 2. (continued)

Country	Location	Principal Analysts	Matrices Analyzed	Quality Assurance Scheme
United Kingdom (continued)	Laboratory of the Government Chemist Cornwall House Waterloo Road London SE1 8XY	Mr. R.E. Lawn Mr. D. Carter Dr. R.A. Hoodless Dr. K. Webb	Soil Tissue Chemicals/ Hazardous Waste Combustion Sources	Internal
	Ministry of Agriculture, Fisheries and Food Food Science Laboratory Haldin House Old Bank of England Court Queen Street Norwich NR2 4SX 603-611712	Dr. J.R. Startin	Tissue Vegetation	BCR, WHO
	SCHERING Agrochemical Limited Chesterford Park Research Station Saffron Walden Essex CB10 1XL (0799) 30123	Dr. D.J. Martin Dr. P.L. Carter	----	BCR
	School of Chemical Sciences University of East Anglia United Kingdom	Dr. C.S. Creaser	Soil	Internal

Table 2. (continued)

Country	Location	Principal Analysts	Matrices Analyzed	Quality Assurance Scheme
United Kingdom (continued)	Strathclyde Regional Council Regional Chemist Department 8 Elliot Place Clydeway Glasgow G3 8EJ	Dr. B.I. Brookes	Combustion Sources Soil Tissue	Internal
	Warren Springs Laboratory Gunnels Wood Road Stevenage, Herts SG1 2BX 438-741122	Dr. M.G. Kibblewhite	----	BCR
United States	American Analytical and Technical Services 10926 E. 55th Place Tulsa, OK 74146 (918) 665-2069	Mr. Robert Harris	Water Soil	CLP
	Argonne National Laboratory (U.S. DOE) Organic Analysis Group 9700 S. Cass Avenue Argonne, IL 60439 (312) 992-7625	Dr. Ron Wingender	Combustion Sources Soil	Internal
	Battelle-Columbus Division 505 King Avenue Columbus, Ohio 43201 (614) 424-6424	Dr. Judy Gephart Dr. Marcus Cook Dr. Jean Czuczwa	----	Internal

Table 2. (continued)

Country	Location	Principal Analysts	Matrices Analyzed	Quality Assurance Scheme
United States (continued)	California Analytical Labs - ENSECO 2544 Industrial Blvd. W. Sacramento, CA 95691 (916) 372-1393	Mr. Mike Miille	Water Soil	CLP
	Hazardous Material Laboratory 2151 Berkeley Way Berkeley, CA 94704 (415) 540-3003	Dr. Robert D. Stephens	Water Chemicals/ Hazardous Waste Food	
	Division of Environmental Health Laboratory Sciences Center for Environmental Health Center for Disease Control Atlanta, Georgia 30333 (404) 329-4176 (FTS) 236-4176	Dr. L.L. Needham Dr. Donald G. Patterson Jr.	Tissue	WHO
	CompuChem Labs 3308 Chapel Hill/Nelson Hwy P.O. Box 12652 RTP, NC 27709 (919) 549-8263	Ms. Diana Schamell	Water Soil	CLP
	Dow Chemical 2020 Dow Center Midland, MI 48640 (517) 636-1000	Dr. Terry Nestrick Dr. Les Lamparski	Soil Water Chemicals/ Hazardous Waste Combustion Sources Vegetation	Internal

70

Table 2. (continued)

Country	Location	Principal Analysts	Matrices Analyzed	Quality Assurance Scheme
United States (continued)	Envirodyne Engineers 12161 Lackland Road St. Louis, MO 63146 (314) 434-6960	Mr. Shaaban Ben-Poorat Ms. Margaret Winter	Water Soil	CLP
	Environmental Chemistry Laboratory Office of Pesticides and Toxic Substances U.S. Environmental Protection Agency National Space Technology Lab Bay St. Louis, Mississippi 39529 (601) 688-3212 (FTS) 494-3212	Dr. Aubry Dupuy	Water Tissue Soil Hazardous Waste/ Chemicals Combustion Sources Food	Troika
	Environmental Monitoring Systems Laboratory Office of Research and Development U.S. Environmental Protection Agency Las Vegas, Nevada 89114 (702) 798-2103 (FTS) 545-2103	Dr. Ronald K. Mitchum	Soil Water Hazardous Waste/ Chemicals Combustion Sources Tissue Food	CLP
	Environmental Monitoring Systems Laboratory Office of Research and Development U.S. Environmental Protection Agency Research Triangle Park, North Carolina 27711 (919) 541-2248 (FTS) 629-2248	Mr. Robert L. Harless	Water Tissue Soil Hazardous Waste/ Chemicals Combustion Sources Food	Troika

71

Table 2. (continued)

Country	Location	Principal Analysts	Matrices Analyzed	Quality Assurance Scheme
United States (continued)	Environmental Research Laboratory Office of Research and Development U.S. Environmental Protection Agency 6201 Congdon Boulevard Duluth, Minnesota 55804 (218) 720-5558 (FTS) 780-5558	Mr. D. W. Kuehl	Water Soil Tissue Hazardous Waste/ Chemicals Combustion Sources Food	Troika
	Environmental Services Laboratory U.S. Environmental Protection Agency Region 7 25 Funston Road Kansas City, Missouri 66115 (913) 236-3881 (FTS) 757-3881	Mr. Robert Greenall	Water Soil Hazardous Waste/ Chemicals Combustion Sources Tissue	Internal
	Environmental Testing & Certification P.O. Box 7808 Edison, NJ 08818 (201) 225-6782	Mr. Jack Farrell	Soil Water	CLP
	Florida State University Tallahasse, Florida 32306	Dr. Ralph Dougherty	Tissue	Internal
	U.S. Food and Drug Administration 200 C Street, SW Washington, DC 20204 (202) 245-1381 (FTS) 245-1381	Dr. David Firestone	Food	Internal

Table 2. (continued)

Country	Location	Principal Analysts	Matrices Analyzed	Quality Assurance Scheme
United States (continued)	U.S. Food and Drug Administration Chicago District Laboratory Facility IITRA 10 West 35th Street Chicago, IL 60616 (312) 353-9764 (FTS) 353-9764	Dr. J.C. Brucciani	Food	Internal
	U.S. Food and Drug Administration Detroit District Laboratory 1560 East Jefferson Avenue Detroit, MI 48207 (313) 226-7658 (FTS) 226-7658	Dr. L.F. Schneider	Food	Internal
	U.S. Food and Drug Administration National Center for Toxicology Research Jefferson, Arkansas 72079 (501) 541-4288 (FTS) 790-4288	Dr. Walter A. Korfmancher	Inactive	Internal
	Geochem Research, Inc. 16920 Park Row Houston, TX 77084 (713) 492-2510	Mr. Charles Bohnstedt	Water Soil	CLP
	Harvard University Cambridge, Massachusetts 02138	Dr. M. Meselson	Inactive	Internal

73

Table 2. (continued)

Country	Location	Principal Analysts	Matrices Analyzed	Quality Assurance Scheme
United States (continued)	Hazleton Labs 3301 Kinsman Blvd. P.O. Box 7545 Madison, WI 53707 (608) 241-4477	Mr. David Hills	Water Soil	CLP
	Idaho National Engineering Lab EG&G Idaho, Inc. Chemical Sciences P.O. Box 1625 Idaho Falls, ID 83415 (208) 586-1292 (FTS) 583-1292	Dr. Clyde Frank	Water Soil Hazardous Waste/ Chemicals	Internal
	U.S. Department of the Interior Columbia National Fisheries Research Laboratory Fish and Wildlife Service Columbia, Missouri 65201 (314) 875-5399 (FTS) 276-5399	Dr. David L. Stallings Mr. L.M. Smith	Tissue	Internal
	International Technology Corporation 304 Directors Drive Knoxville, TN 37923 (615) 690-3211	Ms. Helen Chandler Mr. Tom Adams Ms. Carol Pudelek Mr. Barry Hall	Water Soil Hazardous Waste/ Chemicals	Internal

74

Table 2. (continued)

Country	Location	Principal Analysts	Matrices Analyzed	Quality Assurance Scheme
United States (continued)	International Testing Laboratories, Inc. 578-582 Market Street Newark, NJ 07105 (201) 589-4772	Dr. M.M. Sackoff	Water Soil	Internal
	Laucks Testing Labs, Inc. 940 S. Harney St. Seattle, WA 98108 (206) 767-5060	Mr. Michael Nelson	Water Soil	CLP
	Midwest Research Institute 425 Volker Boulevard Kansas City, Missouri 64110 (816) 753-7600	Dr. John S. Stanley	Tissue Combustion Sources	Internal
	Mississippi Forest Products Laboratory P.O. Box FP Mississippi State, MS 39762 (601) 325-3101	Dr. Gary McGinnis	Vegetation	Internal
	University of Missouri Environmental Trace Substances Research Center Route No. 3, Sinclair Road Columbia, Missouri 65203 (314) 882-2151	Dr. Armon F. Yanders	Soil Water	Internal
	Monsanto Company Environmental Sciences Center St. Louis, MO 63167	Dr. Fred Hileman	Chemicals/ Hazardous Waste Tissue	Internal

Table 2. (continued)

Country	Location	Principal Analysts	Matrices Analyzed	Quality Assurance Scheme
United States (continued)	National Institute of Environmental Health Sciences Laboratory of Molecular Biophysics P.O. Box 12233 Research Triangle Park, North Carolina 27709 (919) 541-1966 (FTS) 629-1966	Dr. Kenneth Tomer	----	Internal
	University of Nebraska at Lincoln Department of Chemistry Hamilton Hall Lincoln, Nebraska 68588-0304 (402) 472-3501	Dr. M.L. Gross	Water Tissue Soil	Troika
	Wadsworth Center for Laboratories and Research New York State Department of Health Albany, NY 12201	Dr. P. O'Keefe	Tissue Soil	Internal
	University of Nevada, Las Vegas Environmental Research Center 4505 Maryland Parkway Las Vegas, NV 89154 (702) 739-3382	Dr. David McNelis	Water Soil Tissue Food Hazardous Waste/ Chemicals Combustion Sources	CLP Referee Lab

Table 2. (continued)

Country	Location	Principal Analysts	Matrices Analyzed	Quality Assurance Scheme
United States (continued)	O'Brien & Gere Engineers 1304 Buckley Road Syracuse, NY 13221 (315) 451-4700	Mr. Antonio LoSurdo	Combustion Sources Water Soil	CLP
	Occidental Chemical Corporation Hooker Chemical Center 360 Rainbow Boulevard - South Box 728 Niagara Falls, NY 14302 (716) 286-3000	Mr. John Nicther	Hazardous Waste/ Chemicals Soil Water	Internal
	Raba-Kistner Consultants, Inc. 12821 West Golden Lane P.O. Box 690287 San Antonio, TX 78269 (512) 699-9090	Dr. Wong	Water Soil Tissue	Internal
	Radian Corporation P.O. Box 9948 Austin, TX	Dr. Lawrence H. Keith	Hazardous Waste/ Chemicals Water	Internal
	Rocky Mountain Analytical Lab 4955 Yarrow Street Arvada, CO 80002 (303) 421-6611	Mr. Ken Faust Mr. Jeff Lowry	Water Soil	CLP

Table 2. (continued)

Country	Location	Principal Analysts	Matrices Analyzed	Quality Assurance Scheme
United States (continued)	Savannah Laboratories and Environmental Services, Inc. P.O. Box 13842 Savannah, GA 31416 (912) 354-7858	Dr. James W. Andrews	Water	Internal
	Triangle Labs 4915 F Prospectus Drive Research Triangle Park, NC 27713	Mr. Ron Haas	----	Internal
	Twin City Testing Corporation 662 Cromwell Avenue St. Paul, MN 55114 (612) 641-9485	Ms. Barbara Larka	Water Soil Tissue Hazardous Waste/ Chemicals	Internal
	University of Utah Salt Lake City, Utah 84112	Mr. Uegene Futrell	----	Internal
	United States Testing Company, Inc. 1415 Park Avenue Hoboken, NJ 07030 (201) 792-2400 x459	Dr. Seyed Dastgheyb Ms. Jane Dunn	Water Soil	Internal
	Versar, Inc. 6850 Versar Center Springfield, VA 22151 (703) 750-3000	Dr. Charles Carter	Water Soil	CLP

Table 2. (continued)

Country	Location	Principal Analysts	Matrices Analyzed	Quality Assurance Scheme
United States (continued)	Wright State University 3640 Colonel Glenn Highway Dayton, Ohio 45435	Dr. T.O. Tiernan	Tissue Water Soil	Troika

4. INVENTORY OF REGULATIONS/STATUTES CONCERNING DIOXINS AND RELATED COMPOUNDS

David P. Bottimore*, Frederick W. Kutz**, and Erich W. Bretthauer**

Versar Inc.* and U.S. Environmental Protection Agency**
Springfield, Va.*, and Washington, D.C.**

This section describes the compilation and analysis of regulations and statutes in participating nations that may be used to control CDDs, CDFs, and related compounds. Many participating countries have national regulations and statutes, as well as some regulatory mechanisms at lesser political jurisdictions, to control the release of, or exposure to, these compounds. This inventory contains numerous regulations, which vary in scope from specific concentration limits to general guidelines for the control of these compounds in the environment. A series of graphic representations that illustrate and summarize the information on the regulations/statutes is included. In addition, a table summarizing the regulations in each nation is also presented to facilitate the identification of specific regulations/statutes. Readers interested in obtaining more information on specific regulations and statutes are encouraged to consult CCMS Report Number 169, which includes descriptions of each individual regulation and statute.

The objective of compiling and analyzing regulations was to demonstrate the importance that has been placed on these chemicals in the participating nations and to promote the exchange of information regarding their regulation. In numerous instances, several nations have promulgated very similar laws, regulations, or statutes pertaining to CDDs, CDFs, and related compounds through specific exposure routes. Two of the objectives of this inventory were to demonstrate how different nations approach the regulatory process for the control of these compounds and to facilitate the examination of the science-policy decisions that lead to regulatory action. One step that has been taken to provide consistency in regulation and communication among the nations was the development of the International Toxicity Equivalency Factor (I-TEF) Method of Risk Assessment for Complex Mixtures of Dioxins and Related Compounds. Numerous regulations and statutes have set concentration limits based on toxicity equivalents. Prior to the achievement of international consensus on the I-TEF method, the existence of numerous slightly different TEF schemes complicated the communication of analytical results and the discussion of the toxicological significance of complex mixtures of CDDs and CDFs. The I-TEFs provide a common basis upon which risk assessments and regulations concerning complex mixtures of CDDs and CDFs can be based, and they facilitate communication among scientists and regulatory agencies. A more detailed description of the I-TEF method can be found in Section 5 of this chapter.

International exchange of information on regulations also highlighted future areas of concern and future research activities. For example, many regulations require the development or use of specific analytical methods for enforcement purposes. Additionally, research activities relating to the risk of these compounds and their control through various technologies are stimulated in areas associated with the regulations. As a result, the identification of regulatory activities planned for the future may provide insight into future resource allocations for research.

4.1   Method of Compilation

During the first plenary meeting in Bayreuth, Federal Republic of Germany, in 1985, it was decided that conducting a survey of the existing regulations and statutes in the participating nations that address CDDs/CDFs would be a valuable exercise and would contribute to the international information exchange objectives of the project. To compile the inventory, requests for information were sent to the lead delegates of the participating nations. The requests asked for information on regulations or statutes that may involve the control of CDDs, CDFs, and related compounds. Each lead delegate was asked to complete the regulatory formats provided. In addition, an instruction sheet was included to facilitate the accurate and complete transfer of information on the laws, regulations, and statutes addressing CDDs, CDFs, and related compounds. The information requested included: (1) the title of the regulation, (2) the citation to record, (3) the effective date or status, (4) a brief summary of the scope, and (5) the contact person or lead agency responsible for the regulation or statute. Generally, the lead delegates from each nation were responsible for completing the regulatory formats on the federal regulations/statutes as well as those on smaller political jurisdictions.

In the United States, an intense solicitation was carried out to contact federal agencies and state governments regarding the regulation of these compounds. To identify all of the federal regulations that address CDDs, CDFs, and related compounds, a comprehensive survey of the major federal agencies and departments was undertaken. Although the U.S. EPA has primary responsibility for the control of many chemicals in the environment, numerous other federal agencies have mandates to address these substances in a variety of ways. With the aid of several key figures in the regulatory arena, a list of the federal agencies that may be directly involved with promulgating regulations and setting standards was developed. As a result, the following agencies and groups were contacted to determine if they have any regulations or statutes applicable to CDDs, CDFs, or related compounds:

- Department of Defense
- Department of Energy
- Department of Transportation
- Department of Agriculture
- Department of Health and Human Services
  - Food and Drug Administration
  - Centers for Disease Control
  - National Institute of Environmental Health Sciences
  - National Institute for Occupational Safety and Health
- Veterans Administration
- Environmental Protection Agency
- Department of Labor
  - Occupational Safety and Health Administration
- Consumer Product Safety and Health Administration
- Consumer Product Safety Commission
- Office of Science and Technology Policy
- Office of Management and Budget
- Office of Technology Assessment
- Interstate Commerce Commission
- Postal Service

These organizations were asked to complete the regulatory formats for regulations and statutes that specifically or generally address CDDs, CDFs, and related compounds. Also, a literature search was conducted to identify any other compilations of this type that might have been performed previously.

In addition to the survey of federal agencies, all of the 50 states were contacted to inquire about regulations and statutes that are used to control CDDs, CDFs, and related compounds on the state or local level. These contacts were made with the assistance of EPA's Regional Offices. Other efforts to identify state regulations and statutes included the review of the Environment Reporter.

4.2   <u>Descriptions of Regulations and Statutes Reported</u>

This section provides a summary of the types of regulations and statutes on CDDs, CDFs, and related compounds for which regulatory formats were completed. The summary also includes a brief description of the scope of the regulation or statute addressing these compounds. For more detailed descriptions of each regulation or statute, the reader should consult CCMS Report Number 169. The number of regulations and statutes reported by geographic location is presented in Table 1. Table 2 presents a summary of the description of the reported regulations and statutes in the participating nations.

Table 1. Number of Regulations/Statutes Reported by Geographic Location

Country	Number of Regulations/Statutes
Canada	10
Denmark	1
Federal Republic of Germany	15
Italy	4
Netherlands	3
Norway	1
United Kingdom	5
United States	81
Total	120

Some of the trends identified from the regulatory formats submitted indicate that many of the nations have promulgated similar regulations to control CDDs, CDFs, and related compounds. In addition, the review of regulatory formats from several of the NATO nations that are also members of the Commission of European Communities (CEC), reveal that many regulations were enacted on the national level as a result of CEC Directives. Specifically, the Federal Republic of Germany, United Kingdom, and Italy have promulgated national regulations on such topics as the marketing of PCBs, emergency management planning, and air quality directly as a result of the Directives. Perhaps the most well-known Directive pertaining to CDDs/CDFs was the CEC "Seveso" Directive 82/501, June 1982, to be adopted by manufacturers to prevent and/or contain the effects of accidents involving toxic chemicals. Figure 1 presents the number of regulatory formats received from each country, subdivided on national, state/provincial, and county jurisdictional levels. Figure 2 is a matrix that presents the scope of the regulations and statutes reported in each of the nations.

84

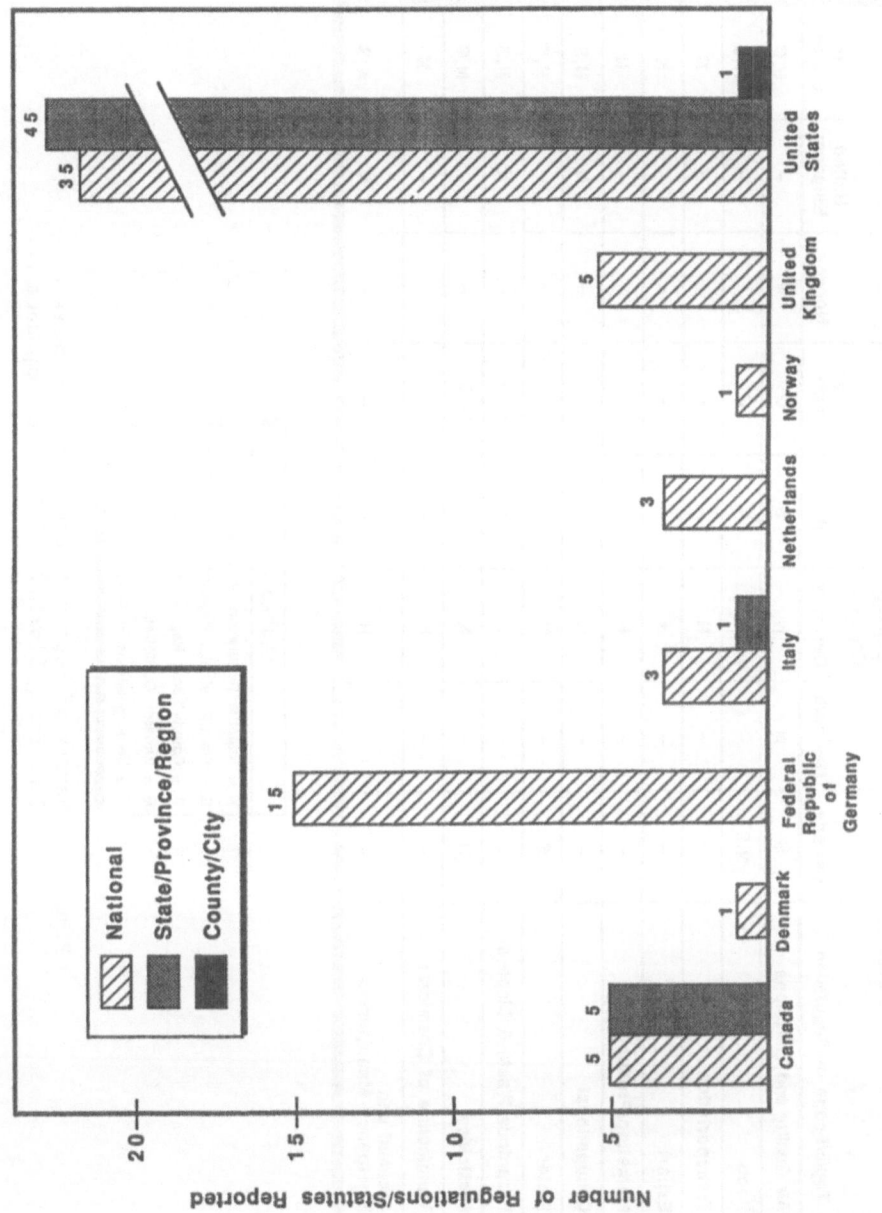

Figure 1. Number of Regulations/Statutes Reported in
Participating Nations by Political Jurisdiction

Topic/Scope of Regulation	Canada	Denmark	Federal Republic of Germany	Italy	Nether-lands	Norway	United Kingdom	United States
Air Quality and Emissions	S	N	N	S	G	N	-	S,C
Food	N,S	-	N	-	G	-	-	N,S
Transportation	-	-	N	-	-	-	-	N
Export	-	-	N	-	-	-	-	N
Marketing/Products	-	-	N	-	N	-	-	N
Occupational	-	-	N	-	-	-	N	N,S
Water	S	-	N	N	-	-	-	N,S
Hazardous Waste & Disposal	S	-	N	N	N,G	-	N	N,S
Pesticides	N	-	N	-	N,G	-	-	N,S
Manufacture of Chemicals	-	-	N	-	-	-	-	N
Accident and Emergency Management	-	-	N	N	-	-	N	N,S

LEGEND

N = National Regulation or Statute

S = State/Provincial Regulation, or Statute

C = County/Local Regulation or Statute

G = General Guidelines

- = No Regulations or Statutes Reported

Figure 2. Matrix of Reported Regulations/Statutes Concerning Dioxins and Related Compounds

4.2.1 Canada

Ten regulatory formats were received from Canada. These formats included five national regulations addressing the presence of these compounds in foods, fish, and pesticides and five provincial regulations and guidelines on drinking water, ambient air quality, fish, and hazardous waste. The national regulations set maximum levels for CDDs/CDFs in food products, fish, and pesticides/herbicides (2,4-D, 2,4,5-T, fenoprop, pentachlorophenol, and tetrachlorophenol).

4.2.2 Denmark

The Danish Environment Ministry and EPA have a comprehensive guidance manual for the operation of municipal waste incinerators to minimize the formation and emissions of CDDs and CDFs. The guideline specifies operational and control procedures but will not set limits for 2,3,7,8-TCDD (equivalents by Eadon method) until June 1991.

4.2.3 Federal Republic of Germany

Regulatory formats describing 15 regulations and ordinances were submitted from the Federal Republic of Germany. These formats presented a variety of regulations and ordinances addressing the presence of CDDs and CDFs in the water, air emissions, and marketable products and the potential for exposure as a result of the transport, occupational use, or disposal of hazardous wastes containing these compounds. Several regulations address the permitting of facilities that may handle CDDs/CDFs and provisions for emergency response.

4.2.4 Italy

Italy completed four regulatory formats that describe three national regulations and one regional guideline. The national regulations address drinking water standards, hazardous waste management, and emergency response measures for accidents; the regional guidance addresses air emissions from incinerators. All of the regulations specify acceptable concentration limits.

4.2.5 Netherlands

Three regulatory formats were received from the Netherlands. These regulations and guidelines are used to limit the CDD/CDF contamination and the use of pentachlorophenol, fuels, and used oil containing PCBs or halogenated hydrocarbons. The Netherlands also has several general guidelines that have not been promulgated as regulations. These guidelines address air emissions, pesticides, food, and hazardous waste.

4.2.6 Norway

One regulatory format was received on the establishment of an air emission standard for CDDs and CDFs from municipal waste incinerators.

4.2.7 United Kingdom

Five regulatory formats from the United Kingdom were submitted. These regulations primarily address the prevention of accidents through notification requirements for permitting of facilities that handle hazardous substances. In addition, a waste oil is defined in part by the PCB content and is prohibited from trade.

4.2.8 United States

Regulatory formats were collected from the United States representing regulations on the national, state, and county jurisdictional levels. Thirty-five of the regulatory formats were completed for national regulations administered by 10 federal agencies. The majority of the regulations for which formats were received were promulgated by the EPA, although other agencies have general or specific laws and regulations controlling the presence of CDDs and CDFs in foods, products, or other articles. For example, the Department of Defense has occupational health standards for CDDs/CDFs for reentry into workplaces following fires in electrical equipment containing PCBs.

The EPA is mandated, through several laws, to develop regulations that address the threat of chemicals to human health and the environment. The EPA has established numerous regulations to control the release of dioxins. Under authority of the Clean Water Act, the Agency has established ambient water quality criteria. Under the Federal Insecticide, Fungicide, and Rodenticide Act (FIFRA), the presence of CDDs and CDFs has resulted in the cancellation of the registration of several pesticides and herbicides including, 2,4,5-T, silvex, and some pentachlorophenol products. Hazardous wastes are listed and restricted from land disposal by regulations promulgated under the Resource Conservation and Recovery Act (RCRA). The use and destruction of PCBs are regulated under the Toxic Substances Control Act (TSCA) in part because of the formation of dioxins and furans in fire-related incidents. The cleanup of contaminated hazardous waste sites is required under the Comprehensive Environmental Response, Compensation, and Liability Act (CERCLA or Superfund). Several sites contaminated with CDDs and CDFs, including Love Canal, Hyde Park, Vertac, and the Missouri sites, are being remediated under Superfund.

Regulatory formats were also received from several state environmental agencies with the authority to establish regulations for their jurisdictions. Typically, these regulations are general in nature and follow the examples of national regulations. Some of the areas covered in these state regulations and statutes include ambient air and water standards; foods; drinking water standards; pesticides; permitting of PCB incinerators; and hazardous waste treatment, storage, and disposal facilities. In some instances, state regulations are specific in the control of these chemicals and set limits for CDDs and CDFs. For example, Montana has a regulation to control petroleum products because of the potential for the presence of hazardous substances such as dioxins. Missouri, which has had several contaminated sites remediated under CERCLA or Superfund, has a state hazardous waste regulation specifically listing 2,3,7,8-TCDD. The State of New York has set levels for dioxins in ground water and drinking water. One county-level regulation was identified for the control of CDDs and CDFs. The county of Philadelphia, Pennsylvania, has ambient air quality guidelines to limit total TCDDs to concentrations below 35 pg/m^3.

Table 2. Index of Regulations/Statutes Submitted for NATO/CCMS Inventory

Country/State, Province	Description	Concentrations specified*
Canada	Food (exception fish) adulterated with dioxins	--
Canada	Fish adulterated with dioxins (2,3,7,8-TCDD)	20 ppt
Canada	2,4-D contaminated with any isomer of dioxin	10 ppb
Canada	Limits 2,3,7,8-TCDD in 2,4,5-T and fenoprop	100 ppb
Canada	Limits total hexachlorodibenzodioxin in pentachlorophenol or tetrachlorophenol	5 ppm
Canada, Ontario	Drinking water limits for 2,3,7,8-TCDD and other isomers (TEF)	15 pg/l
Canada, Ontario	Ambient air standard for dioxins and furans	(formula provided)
Canada, Ontario	Point of impingement for ambient air for dioxins and furans	(formula provided)
Canada, Ontario	Definition of hazardous wastes including dioxin content	1 ppm
Canada, Ontario	Ban on consumption of fish contaminated with 2,3,7,8-TCDD	20 ppt
Denmark	Control of pollution from waste incineration plants (2,3,7,8-TCDD by Eadon)	1 ng/Nm^3 (into effect 1991)
Federal Republic of Germany	Provisions for marketing of PCBs, PCT, and VC products	--
Federal Republic of Germany	Protection of humans, flora, fauna, and materials from pollution	--
Federal Republic of Germany	Listing of installations that are likely to affect environment	--
Federal Republic of Germany	Emergency Response provisions for plants generating 1,2,3,7,8,9-HCDD	--
Federal Republic of Germany	Air quality and emissions as basic principles for licensing installations	--
Federal Republic of Germany	Marine pollution of dioxins	--
Federal Republic of Germany	Goods containing 2,3,7,8-TCDD banned from transport	0.002 ppm (2,3,7,8-TCDD)
Federal Republic of Germany	Transportation requirements for numerous isomers of dioxins/furans	0.1 ppm (various isomers) 0.01 ppm (2,3,7,8-TCDD)
Federal Republic of Germany	Transportation of PCBs by road	--
Federal Republic of Germany	Transportation of PCBs by rail	--
Federal Republic of Germany	Waste disposal regulations	--
Federal Republic of Germany	Ban on marketing substances containing dioxins	0.005 ppm (total) 0.002 ppm (2,3,7,3-TCDD) 0.1 ppm (total) 0.01 ppm (2,3,7,8-TCDD)
Federal Republic of Germany	Notification of handling substances and/or disposal	

Table 2. (continued)

Country/State, Province	Description	Concentrations specified*
Federal Republic of Germany	Occupational protection for dioxins and furans	5 ppm (total)
Federal Republic of Germany	Ban on processes that generate dioxins	1 ppm (2,3,7,8-TCDD)
Italy	Drinking water quality limits for organohalogenated compounds	0.003 ppm
Italy	Emergency response provisions for plants generating dioxins	100 kg (1,2,3,7,8,9-HCDD)
		1 kg (2,3,7,8-TCDD)
Italy	Classification and disposal of hazardous wastes	1 ug/kg (specific isomers)
		500 ug/kg (total)
Italy, Lombardy Region	Air emission limits for waste incineration plants	PCDDs + PCDFs = 0.01 mg/Nm^3
		TCDDs + TCDFs = 0.05 mg/Nm^3
		PCTs + PCNs = 0.1 mg/Nm^3
Netherlands	Prohibition to use, sell, or store fuels containing PCBs, halogenated hydrocarbons	2 ppm (PCBs)
		500 ppm (halogenated hydrocarbons)
Netherlands	Definition of used oil as hazardous waste	2 ppm (PCBs)
		1,000 ppm (others)
Netherlands	Contaminants in pentachlorophenol	1,000 (hexachlorobenzene)
		10 ppm (HCDDs)
		100 ppm (HpCDDs)
		1,000 ppm (trichlorophenol)
		500 ppm (PCDFs)
Netherlands	General Guidelines (non-regulations)	--
	Air emissions (2,3,7,8-TCDD equivalents)	10 ng/Nm^3
	Pesticides containing 2,3,7,8-TCDD	--
	Contaminated soil (2,3,7,8-TCDD equivalents)	1 ppb
	Human acceptable daily intake (2,3,7,8-TCDD equivalents)	4 pg/kg
Norway	Emission standard for municipal waste incinerators (by TCDD equivalents)	2 ng/Nm^3 (Eadon)

Table 2. (continued)

Country/State, Province	Description	Concentrations specified*
United Kingdom	Regulations controlling permitting of plants, pipelines, and ports	--
United Kingdom	Notification requirements of handling of hazardous substances	--
United Kingdom	Hazardous substance definition for PCBs, PCTs in oils	0.01%
United Kingdom	Hazardous waste (oils and solvents) management and disposal	--
United Kingdom	Emergency response notifications and plans for dioxins (2,3,7,8-TCDD)	1 kg
United States	Department of Agriculture - contaminants in food	--
United States	Department of Commerce - export regulations for dioxins and furans	--
United States	Department of Commerce - export regulations for hazardous substances	--
United States	Consumer Products Safety Commission - general statute to regulate chemicals	--
United States	Consumer Products Safety Commission - general statute to regulate chemicals	--
United States	Department of Defense - occupational re-entry levels for PCBs, PCDFs following fires (2,3,7,8-TCDD equivalents)	10 pg/m^3 (air) 3 ng/m^2 (surface <8 ft) 10 ng/m^2 (surface >8 ft)
United States	USEPA - Clean Air Act	--
United States	USEPA - Clean Water Act	--
United States	Ambient water quality criteria for 2,3,7,8-TCDD	--
United States	USEPA - CERCLA/Superfund	--
United States	USEPA - Federal Insecticide, Fungicide and Rodenticide Act (FIFRA)	--
United States	Cancels non-wood preservative uses of pentachlorophenol and sets limits on HxCDD level	6 ppm (HxCDD in product) 3 ppm (avg. HxCDD for one month)
United States	Requires testing of certain pesticides for HDDs and HDFs	--
United States	Cancels 2,4,5-T for all uses	--
United States	Maximum level of HxCDD in pentachlorophenol	--
United States	Cancels pesticidal uses of silvex	--
United States	Cancels certain uses of 2,4,5-T and silvex	--
United States	Cancels all uses of 2,4,5-T and silvex	--
United States	USEPA - Resource Conservation and Recovery Act (RCRA)	--
United States	Lists as hazardous waste containing CDDs, CDFs, chlorinated phenols	--
United States	Prohibits land disposal of dioxin containing wastes	1 ppb

Table 2. (continued)

Country/State, Province	Description	Concentrations specified*
United States	USEPA - Toxic Substances Control Act (TSCA)	--
United States	Requirement of analytical testing of chemicals for HDD/HDF	--
United States	Restriction of use of PCBs in electrical transformers	--
United States	Suspends all pesticidal uses of hexachlorophene	--
United States	Federal Emergency Management Agency - emergency response for dioxins	--
United States	Federal Emergency Management Agency - relocation activities	--
United States	Federal Emergency Management Agency - temporary relocation assistance	--
United States	Federal Emergency Management Agency - permanent relocation assistance	--
United States	FDA - regulation of dioxins and related chemicals in food	--
United States	Interstate Commerce Commission - transportation requirements for hazardous wastes	--
United States	U.S. Postal Service - packaging, labelling, and disclosure regulation	--
United States	U.S. Postal Service - unmailable poisons prohibition	--
United States	U.S. Postal Service - definitions of poisonous (unmailable) substances	--
United States	U.S. Veterans Administration - claims of exposure to herbicides	--
United States, California	Cancellation of 2,4,5-T registrations	--
United States, California	Right to know and drinking water regulations for dioxins	--
United States, California	Definition of hazardous wastes	0.001 mg/kg (TCDD)
United States, Colorado	Definition and control of hazardous wastes	0.01 mg/kg (total)
United States, Connecticut	Definition and control of hazardous wastes including dioxins/furans	--
United States, Delaware	General laws for control of air, water, and hazards waste pollution	--
United States, District of Columbia	Definition and control of hazardous wastes	--
United States, District of Columbia	General air pollution regulation	--
United States, Florida	Drinking water monitoring and analyses for 2,3,7,8-TCDD	--
United States, Florida	Ground-water protection standards	--
United States, Georgia	Definition and control of hazardous wastes and CERCLA sites	--
United States, Idaho	Air quality permitting for PCB incinerators	--

Table 2. (continued)

Country/State, Province	Description	Concentrations specified*
United States, Kentucky	Control of air emissions of 2,3,7,8-TCDD from existing facilities	--
United States, Kentucky	Control of air emissions of 2,3,7,8-TCDD from newer modified facilities	--
United States, Kentucky	Definition of hazardous wastes	--
United States, Kentucky	Hazardous waste management requests	--
United States, Kentucky	Hazardous waste management requests for interim/closure/post-closure	--
United States, Maine	Land application/disposal of sludges and residuals containing PCDDs and PCDFs	--
United States, Maryland	General air pollution regulations	--
United States, Maryland	General hazardous waste regulations	--
United States, Massachusetts	General environmental quality regulations	--
United States, Mississippi	Hazardous waste (listed wastes under RCRA)	--
United States, Missouri	Listing of hazardous wastes, contaminated soils and debris for 2,3,7,8-TCDD	1 gram
United States, Missouri	Water quality standards for PCBs, pesticides, 2,3,7,8-TCDD	--
United States, Montana	Definition and control of hazardous wastes	--
United States, Montana	Regulation of petroleum products and substances	--
United States, New Hampshire	Definition and control of hazardous wastes containing dioxins, furans and phenols	--
United States, New Jersey	Discharge of 2,3,7,8-TCDD into groundwater	--
United States, New Jersey	Hazardous waste definition including TCDD, TCDF, and 2,3,7,8-TCDD	--
United States, New York	Ambient water quality standard (2,3,7,8-TCDD)	0.000001 $\mu g/l$
United States, New York	Groundwater quality standard (2,3,7,8-TCDD)	0.000035 $\mu g/l$
United States, Pennsylvania	Air pollution regulations and permitting requirements (using TEF for 2,3,7,8-TCDD)	--
United States, County of Philadelphia	Ambient air quality guidelines for toxic air contaminants (total TCDD)	35 pg/m^3
United States, Rhode Island	Definition and control of hazardous wastes	--
United States, Tennessee	Definition and control of hazardous wastes	--
United States, Vermont	Definition and control of hazardous wastes	--
United States, Vermont	Occupational health	--
United States, Vermont	Air pollution regulations	--
United States, Virginia	General hazardous waste disposal regulation	--
United States, Virginia	General air pollution control regulation	--
United States, Washington	Definition and control of hazardous wastes	--

Table 2. (continued)

Country/State, Province	Description	Concentrations specified*
United States, West Virginia	General hazardous waste disposal regulation	--
United States, West Virginia	General air pollution control	--
United States, Wisconsin	Air emissions from incinerators (total TCDD)	0.0001 lb/yr
United States, Wisconsin	Groundwater quality standards (TCDD)	0.00000022 ppb
United States, Wyoming	Disposal of hazardous substances	--

* Please consult the regulatory format for the specific requirements regarding regulatory standards or levels.

5. INTERNATIONAL TOXICITY EQUIVALENCY FACTOR (I-TEF) METHOD OF RISK

ASSESSMENT FOR COMPLEX MIXTURES OF DIOXINS AND RELATED COMPOUNDS

Donald G. Barnes[1], Frederick W. Kutz[1], David P. Bottimore[2],
Donald L. Grant[3], Helmut Greim[4], James Wilson[5], and
Erich W. Bretthauer[1]

U.S. Environmental Protection Agency[1], Versar Inc.[2], Health and
Welfare Canada[3], GSF Institute for Toxicology[4], and Monsanto Chemical
Co.[5]
Washington, D.C.[1], Springfield, Va.[2], Ottawa, Canada[3], Neuherberg
Federal Republic of Gremany[4], and St. Louis, Mo.[5]

As one of the activities of this Pilot Study, the Exposure and Hazard
Assessment Working Group desired to achieve international consensus on
methods to estimate risks associated with exposures to complex mixtures of
CDDs and CDFs. Prior to the establishment of the Pilot Study, the communication among the international scientific and regulatory communities of
the toxicological significance of complex mixtures had been complicated
by the existence of numerous slightly different approaches using toxicity
equivalency factors (TEFs). The objective of the Working Group was to
resolve the discrepancies among the existing TEF schemes used throughout
the participating nations and to develop an updated TEF scheme that would
provide a consistent basis for risk assessment and regulation.

A special subgroup (the TEF Subgroup) was established to explore the
issues associated with TEFs and to determine whether a consensus position
on the use of one method was achievable. The original charge to the TEF
Subgroup was broken down into separate tasks and assigned to the Subgroup
members. Additional insight was gained at a meeting of available Subgroup
members at Dioxin '87, held in October 1987, in Las Vegas, Nevada. These
contributions were synthesized and edited by the Chair into a draft report, which was distributed to all Subgroup members for review and comment
in February 1988. The final report reflects the comments received as a
result of that review, as well as the discussion during the concluding

plenary meeting of the Pilot Study in Berlin on April 27-28, 1988. These new TEFs, the International-TEFs (I-TEFs), were published in CCMS Report Number 176 and were presented at the Eighth International Symposium on Chlorinated Dioxins and Related Compounds in Umea, Sweden, on August 25, 1988 (CCMS, 1988a; Kutz and Bottimore, 1988). An expanded background document that presents the scientific basis for the development of TEFs, as well as the literature sources, methodology, and data used to determine the specific I-TEFs, was published as CCMS Report Number 178 (CCMS, 1988b). The I-TEF method was also described to several international audiences at conferences (Kutz et al., 1990a,b,c). Subsequently, regulatory agencies in several nations--these include the United States, the Nordic countries, the Netherlands, Canada, the United Kingdom, Ontario (Canada), and New York State--have adopted the I-TEF method for addressing complex mixtures such as incinerator flyash, contaminated soils, and biological media (USEPA, 1989). This report represents a synthesis of the two CCMS reports with some ideas developed for the EPA report which describes the adoption of the I-TEF method for Agency use in estimating risks associated with exposures to complex mixtures of CDDs and CDFs.

Interest in this topic stemmed from the fact that during the late 1970s and early 1980s analytical chemists achieved the capability of quantifying, in environmental samples, the presence of chlorinated dibenzo-p-dioxins and dibenzofurans (CDDs and CDFs), other than the long-studied 2,3,7,8-tetrachlorodibenzo-p-dioxin (2,3,7,8-TCDD). The toxicological significance of long-term exposures to these other congeners was, and is, relatively unknown; therefore, it has been difficult to rigorously ascribe a particular level of concern to mixtures of these compounds. In addition, simply summarizing and succinctly presenting analytical data on the scores of individual CDD and CDF congeners in some meaningful manner has also been difficult. For these reasons, different schemes have been proposed for estimating the toxicological significance of complex mixtures of CDDs and CDFs in terms of equivalent amounts of 2,3,7,8-TCDD.

The existence of a variety of TEF schemes has complicated communication among scientists and regulators from different countries. Since the advent of the TEF approach, at least 10 different schemes have been used. Building upon the experience gained in the application of TEF approaches in a number of different countries and jurisdictions, the TEF Subgroup examined four central questions:

- Is there scientific/regulatory consensus on the appropriateness of the use of the TEF approach?

- Is there consensus on a particular TEF approach?

- How can a TEF approach be applied to different types of analytical data sets; e.g., homologue-specific vs. congener-specific?

- What additional research and data are needed to test, refine, or replace the current I-TEF approach?

To address these issues, individual members of the TEF Subgroup drafted answers to specific questions. The drafts were circulated within the Subgroup for review and comment. Several members of the Subgroup met at The Seventh International Symposium on Chlorinated Dioxins and Related Compounds (Dioxin '87), held in Las Vegas, Nevada, in October 1987, and discussed a TEF scheme appropriate for common international use. The results of these efforts and comments received at the plenary meeting in Berlin on April 27-28, 1988, have been synthesized in this section.

The participants of the study do not suggest that these I-TEFs have been definitively derived. The I-TEFs are based on an evaluation of toxicity data for CDD and CDF congeners in a variety of *in vitro* and *in vivo* tests, and on current TEF practices of several countries. This section endorses the TEF concept as an interim measure of prudent science policy and recommends specific I-TEF values to facilitate communication and consistency among the member nations. Further, regulatory authorities are encouraged to collect congener-specific data on all CDD/CDF-containing environmental samples and to summarize the estimated combined effect of these chemicals in terms of International Toxicity Equivalents (I-TEQs). The I-TEQs are obtained by applying the I-TEFs to the congener-specific data and summing the results. Each statement of I-TEQs in a sample should also be accompanied by an indication of the percent of those I-TEQs that is contributed by 2,3,7,8-TCDD itself. The congener-specific data will be indispensable in evaluating data in terms of any modified TEF schemes that might appear in the future. In addition, such data might prove helpful in identifying the possible source(s) of contamination by applying pattern-recognition techniques to "fingerprints" of congener distributions found in environmental and source samples. At the same time, the Subgroup recommends additional research to test, refine, and eventually replace the I-TEFs.

Section 2 discusses the need for the development and use of TEFs. It includes a brief description of some of the TEF approaches developed by different countries.

Section 3 presents the International TEF scheme developed by consensus within the participating nations in the Pilot Study on International

Information Exchange on Dioxins and Related Compounds. Representatives of the member nations participating in the Pilot Study adopted this scheme during the concluding plenary meeting of the project in Berlin on April 27-28, 1988. The section closes with a discussion of how the scheme can be applied to different data sets.

Section 4 discusses some of the research (data and method development) activities that would strengthen the current I-TEF approach or enable its replacement with a more rigorously scientific approach. Section 5 presents the overall conclusions regarding the I-TEF approach.

5.1   Background

In the early 1970s, the term "dioxin" was synonymous with 2,3,7,8-TCDD, a compound that was widely thought to be present only as an impurity in a limited number of chemical products. Extensive physicochemical and toxicologic research has been conducted on 2,3,7,8-TCDD. From such studies it has been learned that the compound exhibits a broad range of toxic effects in laboratory animals, including carcinogenicity, teratogenicity, immunotoxicity, thymic atrophy, lethality, and reproductive effects. These effects are observed at exceptionally low doses--in some cases, on the order of submicrograms per kilogram of body weight per day (ug/kg-d). It is these striking toxicological properties exhibited at unprecedented low doses that have attracted the attention and concern of the scientific and regulatory communities. While much remains to be learned, sufficient information has been developed to support crude assessments of the risks posed by exposure to 2,3,7,8-TCDD (e.g., NRCC, 1981; OME, 1985; Kimbrough, 1984; USEPA, 1985, 1988; FRG, 1985).

In the late 1970s and continuing into the 1980s, additional environmental sources of 2,3,7,8-TCDD were identified. Most of these sources contain CDDs and CDFs in addition to 2,3,7,8-TCDD. These findings are testimony to the innovation, diligence, and skill of analytical chemists in different countries. The findings also pose a challenge to toxicologists and regulators who are asked to interpret and act upon the significance of the analytical data.

5.1.1   The Need

Since there is a relative dearth of information on the toxicity of CDDs and CDFs, other than 2,3,7,8-TCDD, the traditional approaches to assess the significance of exposure to these additional compounds are generally not applicable. That is, long-term toxicity data in animals exist for few mixtures of CDDs and CDFs or for individual components. There-fore, to interpret this growing body of data, a new interim approach is needed until more definitive methods can be developed.

The wealth of information gathered by analytical chemists on the 210 CDD and CDF congeners present in some environmental samples, seems to defy comprehension when the results are displayed as chromatograms or lists of numbers in tables. While this detailed information is of inesti-mable value in some instances (such as using isomer profile recognition techniques to identify the source of contamination), for other purposes requiring succinct communication, a more concise, comprehensible form of presentation is preferred. The TEF method facilitates concise communica-tion by summarizing these data into a more comprehensible form.

5.1.2 The TEF Concept

A review of the literature reveals that CDD and CDF congeners are not equally toxic. Also, it is clear that the scientific effort spent on determining the toxicity of 2,3,7,8-TCDD cannot be reasonably duplicated for each of the other congeners (AOWG, 1987; CCMS, 1987). Long-term cancer studies have been conducted only on 2,3,7,8-TCDD and a mixture of HxCDDs. Teratogenicity and reproductive effects studies have been con-ducted on a somewhat greater number of CDDs and CDFs. Although limited, this information from whole-animal studies, coupled with data that demon-strate a strong structure/activity relationship among the members of the CDD and CDF congeners, enables one to devise a means by which the relative toxicities of these compounds can be assessed for crude, but practical, purposes.

A hypothesis first developed by Poland and coworkers (1979) high-lighted the importance, with respect to biochemical activity, of chlorine atoms on the carbons at the 2, 3, 7, and 8 positions on the dibenzo-p-dioxin nucleus. This hypothesis provided the initial indication that the CDDs and CDFs could be assessed on a comparative basis. The hypothesis envisioned the binding of these molecules to a cytosolic receptor as the first step in inducing a pleiotropic response in the organism. One of the responses most frequently observed is an increase in the level of the en-zyme aryl hydrocarbon hydroxylase (AHH) (hence the designation "Ah recep-tor"). The affinity of the CDD or CDF congeners for the Ah receptor is related to the magnitude of the enzymic induction. Subsequent investiga-tions have demonstrated relationships between the ability of the CDD and CDF congeners to bind to the receptor and to induce AHH (and other micro-sonial enzymes) and manifestation of toxicity such as thymic atrophy and

101

body weight loss (Bandiera et al., 1984; Safe et al., 1985), LD_{50}, carcinogenicity, reproductive/teratogenic effects, cell transformation, and
dysfunction of the immune system (Schwetz et al., 1973; Poland et al.,
1976; Bradlaw et al., 1979; Moore et al., 1979; Murray et al., 1979;
Poland et al., 1979; Bradlaw et al., 1980; Knutson and Poland, 1980;
McKinney and McConnell, 1982; Bandiera et al., 1983; Bandiera et al.,
1984; Hassoun et al., 1984; Weber et al., 1984; Dencker et al., 1985;
Gierthy and Crane, 1985; Greenlee et al., 1985; Nagayama et al., 1985a,b;
USEPA, 1985; USEPA, 1987; Olson et al., 1988). For detailed discussions,
see the background support document (CCMS Report No. 178).

It should be noted, however, that information questions the validity
of extending the Ah receptor mechanism for TCDD toxicity to species other
than the two strains of mice (C57BL/6J and DBA/2J) in which the correlations are well established. For example, acute lethality (Pohjanvirta et
al., 1988), microsomal enzyme induction (Rozman et al., 1985a,b; Henry
and Gasiewicz, 1986), thymic involution (Gorski et al., 1988), induction
of cleft palate (Lamb et al., 1986), induction of hepatic prophyria (Greig
et al., 1984), induction of hyperkeratosis (Puhvel and Sakamoto, 1987),
and suppression of antibody response to sheep red blood cells (Pazdernik
and Rozman, 1985), as a consequence of 2,3,7,8-TCDD administrations have
been shown not to correlate with binding affinity to the Ah-high affinity
protein in other strains/species. Also, Gasiewicz and Rucci (1984) indicate that the correlation between sensitivity to 2,3,7,8-TCDD toxicity and
Ah receptor binding affinity seen in mice is not necessarily applicable
to other species. Therefore, whenever possible, TEFs are preferentially
based upon whole-animal data, rather than solely upon "Ah-receptor" binding affinities or enzyme induction.

Both *in vivo* and *in vitro* data have been used as the basis for
the relative ranking of toxicities of CDD and CDF congeners by various
regulatory authorities. Figure 1 displays the various TEF schemes that
have been developed to assess the risk associated with exposures to complex mixtures of CDDs and CDFs. Each group proposing such a scheme has
reached the conclusion that these relative rankings, expressed in the
form of "toxicity equivalency factors" (TEFs), can be used to express the
appropriate toxicity of a complex mixture of CDDs and CDFs as a "toxicological equivalent amount (TEQ)" of 2,3,7,8-TCDD.

In addition to providing a toxicological summary about the mixture
under examination, the TEQ provides a convenient means of summarizing the
often extensive amount of analytical data resulting from a detailed gas

chromatography/mass spectrometry (GC/MS) investigation from a hazard assessment point of view. However, in any published report on the analysis of environmental samples, the individual CDD and CDF data should always be featured, with the TEQ value providing a secondary summary. Each statement of TEQs should also include a percentage of those TEQs that are contributed by 2,3,7,8-TCDD itself. Analytical results on all congeners are important, even on those with TEFs presently assumed as zero; i.e., mono-, di-, tri-, octa-, and non-2,3,7,8-substituted CDDs and CDFs. Such information is valuable, for instance, in the identification of particular sources (through comparison of homologue profiles and isomer patterns) and the evaluation of environmental transport, transformation, and fate.

				TEF Scheme				
	Swiss[a]	Grant[b] Olie[c] Commoner[d]	New York State[e]	Ontario[f]	FDA[g]	CA[h]	EPA 1981[i]	EPA-TEF/87[j]
(Basis)	Enzyme		LD$_{50}$	Various Effects	Various Effects			Various Effects
Compound								
Mono thru di CDDs	0	0	0	0	0	0	0	0
Tri CDDs	0	0	0	1	0	0	0	0
2378-TCDD	1	1	1	1	1	1	1	1
Other TCDDs	0.01	1	0	0.01	0	0	1	0.01
2378-PeCDD	0.1	0.1	1	1	0	1	0	0.5
Other PeCDDs	0.1	0.1	0	0.01	0	0	0	0.005
2378-HxCDDs	0.1	0.1	0.03	1	0.02	0.03	0	0.04
Other HxCDDs	0.1	0.1	0	0.01	0.02	0	0	0.0004
2378-HpCDD	0.01	0.1	0	1	0.005	0.03	0	0.001
Other HpCDDs	0.01	0.1	0	0.01	0.005	0	0	0.00001
OCDD	0	0	0	0	<0.00001	0	0	0
2378-TCDF	0.1	0.1	0.33	0.02	0	1	0	0.1
Other TCDFs	0.1	0.1	0	0.0002	0	0	0	0.001
2378-PeCDFs	0.1	0.1	0.33	0.02	0	1	0	0.1
Other PeCDFs	0.1	0.1	0	0.0002	0	0	0	0.001
2378-HxCDFs	0.1	0.1	0.01	0.02	0	0.03	0	0.01
Other HxCDFs	0.1	0.1	0	0.0002	0	0	0	0.0001
2378-HpCDFs	0.1	0.1	0	0.02	0	0.03	0	0.001
Other HpCDFs	0	0.1	0	0.0002	0	0	0	0.00001
OCDF	0	0	0	0	0	0	0	0

[a] Swiss Government 1982. [d] Commoner et al. 1984. [g] U.S. DHHS 1983.
[b] Grant 1977. [e] Eadon et al. 1982. [h] CA Air Resources Bd. 1986.
[c] Olie et al. 1983. [f] Ontario 1982. [i] U.S. EPA 1981.
 [j] U.S. EPA 1987.

Figure 1.   Various TEF Approaches for CDDs and CDFs

5.1.3 Consensus and Accuracy

In the opinion of the Pilot Study, there is international consensus that the TEF approach is an appropriate science policy tool for use in the absence of more definitive toxicity information. It has been recognized that the TEFs for some congeners are more soundly supported by data than others, and those congeners thought to be of greater concern should be subjected to further investigation. It should also be noted that the accuracy of an assessment of the risks associated with exposure to a mixture of CDDs and CDFs depends upon a number of factors, of which the uncertainty in the TEF approach is only one (and perhaps not the most significant). Estimated intakes, bioavailability, interspecies extrapolation, safety factors or mathematical models, possible antagonistic or synergistic interactions, etc. are likely to carry as much or more uncertainty than the TEFs. As such, the TEF method is a useful interim procedure for estimating risks associated with exposures to complex mixtures of CDDs and CDFs.

5.2 International TEF Scheme

The existence of a variety of TEF schemes has complicated communication among scientists and regulators from different countries when discussing the significance of the analytical results from a particular environmental sample. It is interesting to note, however, that in many cases the TEQs derived by the different TEF schemes (Figure 1) often differ by less than an order of magnitude when applied to the same environmental mixture of CDDs and CDFs (see Table 1). It is commonly acknowledged that any TEF approach is only an interim procedure to use in the absence of more definitive data and that the procedure should be updated to reflect significant changes in the emerging scientific data base.

Table 1. Toxicity Equivalents Using Different TEF Methods

Source	EPA '81	EPA '87	Swiss	NY	CA
St. Louis (Air Particulates)	1	0.3	1	32	40
PCB Fire Soot (Isomer-Specific)	1	0.03	4	3	30
MSW ESP Dust	1	0.2	3	2	30
Lake Sediment	1	–	2	2	30
Milorganite	1	0.6	2	0.9	30
Oslo MSW Flyash	1	–	1	2	20
Ontario MSW Flyash	1	0.8	1	2	3
Japanese Plant A	1	0.3	1	2	7
Japanese Plant B	1	0.6	0.8	2	3
Albany	1	0.3	0.4	2	5
Wright-Patterson (Best)	1	0.2	2	3	20
Wright-Patterson (Worst)	1	0.4	2	2	20

[a] Calculated using the Toxicity Equivalence Factors shown in Figure 1.

5.2.1 The I-TEF Scheme

Given this situation, the members of the this Pilot Study developed a set of TEFs, the International TEFs (I-TEFs), to provide a uniform approach to summarizing analytical data and their toxicological significance. The I-TEFs adopted by the Pilot Study are displayed in Table 2. Table 2 also displays the congeners of concern in each homologue group.

Table 2. The I-TEFs

Congener of Concern	I-TEF	Congeners of Concern in a Homologous Group
2,3,7,8-TCDD	1	1 out of 22 (5%)
1,2,3,7,8-PeCDD	0.5	1 out of 14 (7%)
1,2,3,4,7,8-HxCDD 1,2,3,7,8,9-HxCDD 1,2,3,6,7,8-HxCDD	0.1	3 out of 10 (30%)
1,2,3,4,6,7,8-HpCDD	0.01	1 out of 2 (50%)
OCDD	0.001	1 out of 1 (100%)
2,3,7,8-TCDF	0.1	1 out of 38 (3%)
2,3,4,7,8-PeCDF	0.5	1 out of 28 (4%)
1,2,3,7,8-PeCDF	0.05	1 out of 28 (4%)
1,2,3,4,7,8-HxCDF 1,2,3,7,8,9-HxCDF 1,2,3,6,7,8-HxCDF 2,3,4,6,7,8-HxCDF	0.1	4 out of 16 (25%)
1,2,3,4,6,7,8-HpCDF 1,2,3,4,7,8,9-HpCDF	0.01	2 out of 4 (50%)
OCDF	0.001	1 out of 1 (100%)

In developing the I-TEFs, the Subgroup was guided by the following principles:

- The scheme should be as simple as practicable. A complex scheme suggests greater precision and sophistication than can be scientifically supported.

- The focus should be on the CDD and CDF congeners that are preferentially accumulated in mammalian tissue. These are principally the congeners that are substituted at the 2,3,7, and 8 positions and are the more toxic forms.

- The TEFs should reflect the relative toxicity exhibited by the various congeners in a variety of toxicological endpoints.

In assigning the I-TEF values, a data hiearchy was established. In general, priority was given to results from long-term, whole animal studies such as carcinogenicity. Since there is a dearth of this data for most isomers, short-term, whole animal studies (such as reproductive effects) and/or other subchronic effects data (such as thymic atrophy, reduced body weight gain, etc.) were used. Acute toxicity studies like lethality were also used in determining the I-TEFs. Among the remaining short-term *in vivo* and *in vitro* data enzyme induction data were generally used to confirm the values determined by the *in vivo* toxicological data. This is due to the fact that a good correlation has been observed between enzyme induction activity (AHH) and short-term, whole-animal results; i.e., thymic atrophy (r = 0.91), body weight loss (r = 0.84) in rats, and inhibition of body weight gain in guinea pigs (r = 0.93) (CCMS, 1988b). In addition, some structure-activity relationships and pharmacokinetic considerations were used in developing the I-TEF values.

In general the I-TEF method is similar to the other existing TEF schemes in use throughout the world. These similarities are reflected in the fact that the toxicities of the mono- through tri-substituted compounds are considered to be negligible. However, there are some distinct differences between the I-TEF scheme and some others. The principal differences between the I-TEF scheme and these earlier schemes are the following: (1) the non-2378-substituted congeners are assigned a value of zero, (2) differentiation of the two 2378-substituted PeCDFs isomers, and (3) OCDD and OCDF are assigned a non-zero value.

(1) Data generated in the past two years indicate that the 2378-substituted congeners are selectively absorbed and/or retained in higher animals; e.g., fish, humans, and other mammals. As a result, although nearly all of the 210 CDDs and CDFs can be found at low levels in the environment, the 2378-substituted congeners clearly predominate in tissues of exposed animals. This is even true when the source of the CDD/CDFs is relatively low in the concentration of 2378-substituted congeners (Kuehl et al., 1986; Van den Berg et al., 1985). Because of this tendency to preferentially bioaccumulate the 2378-substituted congeners over long periods of time, the greatest concern is placed on these congeners. However, since the presence of the non-2378-substituted congeners is effectively ignored, the somewhat higher I-TEFs for the 2378-substituted congeners tend to compensate for any small toxic contribution of these congeners.

106

(2)   For the homologous class of 2378-substituted PeCDFs, the I-TEF scheme differentiates between the 2,3,4,7,8-PeCDF and 1,2,3,7,8-PeCDF isomers.   This assignment reflects the higher potency of 2,3,4,7,8-PeCDF, as demonstrated in a variety of tests including teratogenicity (Birnbaum et al., 1987).   The 0.5 value for 2,3,4,7,8-PeCDF also gains support from the *in vivo* thymic atrophy data (0.43) and mouse immunotoxicity data (0.8).   The lower 0.05 value for 1,2,3,7,8-PeCDF is supported by *in vivo* investigations of thymic atrophy (0.05).

In addition, some structure-activity relationships can be used to explain the differences in the biological responses to these two PeCDF isomers.   When superimposed on the molecular structure of 2,3,7,8-TCDD, the C-4 of the 2,3,4,7,8-PeCDF is more stereochemically laterally positioned than the C-1 of the 1,2,3,7,8-PeCDF (see Figure 2).   As a result of the more lateral positioning of chlorines, 2,3,4,7,8-PeCDF exhibits properties more similar to 2,3,7,8-TCDD than does 1,2,3,7,8-PeCDF.

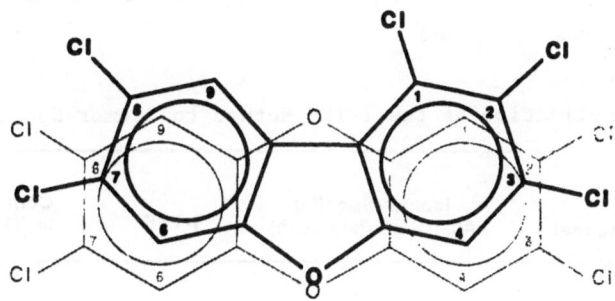

1, 2, 3, 7, 8 - PeCDF on 2, 3, 7, 8 - TCDD

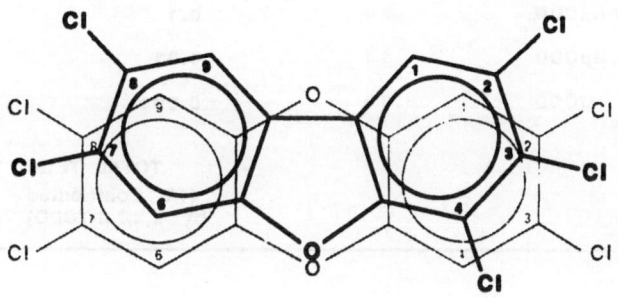

2, 3, 4, 7, 8 - PeCDF on 2, 3, 7, 8 - TCDD

Figure 2.   Structure-Activity Relationships of PeCDFs

(3) .The other most significant deviation from previous TEF schemes is the assignment of a value of 0.001 to OCDD and OCDF. This reflects recent information on the toxicity of higher chlorinated species, where male rats were exposed to OCDD for 13 weeks. At the end of the experiment, the animals had accumulated detectable levels of OCDD and were beginning to show signs of "dioxin toxicity" (Couture et al., 1988). Earlier experiments on OCDD had involved short-term, high doses, which were ineffectively absorbed and therefore elicited no toxic response. As a result of the long term low level doses and resulting toxicity, a value of 0.001 has been assigned to both OCDD and OCDF. It should be noted, however, that this assignment is a result of only one experiment and may be revised as additional data are generated.

5.2.2 Application of the I-TEF Scheme to Different Data Sets

I-TEQs are generated by summing the products of the concentrations of each of the 2,3,7,8-substituted congeners by their respective I-TEFs (see Table 3). It should be noted that when applied to complex environmental mixtures of CDDs and CDFs, the I-TEQs calculated by the I-TEF scheme are generally quite similar to those generated by several of the existing TEF schemes (see Figures 3, 4, and 5).

Table 3. Application of the I-TEF Method to Isomer-Specific Data

Congener	Isomer-Specific Analytical Data (ppb)	I-TEF	Contribution to TEQ (ppb)
2,3,7,8-TCDD	2	1	2
1,2,3,7,8-PeCDD	6	0.5	3
1,2,3,4,7,8-HxCDD	20	0.1	2
1,2,3,7,8,9-HxCDD	25	0.1	2.5
1,2,3,6,7,8-HxCDD	20	0.1	2
1,2,3,4,6,7,8-HpCDD	15	0.01	0.15
OCDD	10	0.001	0.01
		TOTAL TEQs	12
		(1/6 Contributed by 2,3,7,8-TCDD)	

Increasingly, isomer-specific data are being generated in the analytical laboratory. In other cases, various factors (e.g., limited analytical funds or difficult analytical matrices) necessitate the reporting of homologue-specific data; e.g., the total amount of TCDDs present, without

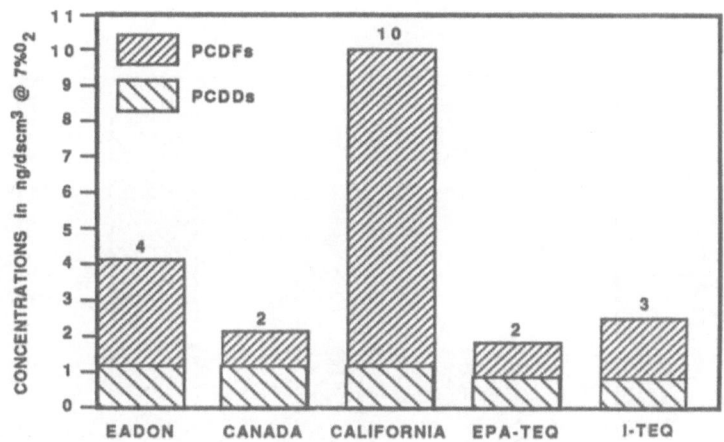

Data for Above Figure (Concentrations in ng/dscm^3 @ 7% O$_2$):

SPECIES	SOURCE DATA	TEF SCHEME				
		EADON	CANADA	CALIFORNIA	EPA-TEF	I-TEF
2378-TCDD	0.30	0.30	0.30	0.30	0.30	0.30
TCDDs (OTHER)	2.7	0	0.27	0	0.027	0
12378-PeCDD	0.79	0.79	0.39	0.79	0.39	0.39
PeCDDs (OTHER)	2.2	0	0.011	0	0.011	0
123478-HxCDD	0.16	0.0047	0.016	0.0047	0.0063	0.016
123678-HxCDD	0.39	0.012	0.039	0.012	0.015	0.039
123789-HxCDD	0.059	0.0018	0.0059	0.0018	0.0023	0.0059
HxCDDs (OTHER)	3.6	0	0.0036	0	0.0014	0
1234678-HpCDD	0	0	0	0	0	0
HpCDDs (OTHER)	5.9	0	0.00059	0	0.000059	0
OCDD	9.5	0	0.00095	0	0	0.00095
TOTAL PCDDs		**1.1**	**1.0**	**1.1**	**0.8**	**0.8**
2378-TCDF	2.3	0.76	0.23	2.3	0.23	0.23
TCDFs (OTHER)	30	0	0.03	0	0.030	0
12378-PeCDF	4.2	1.4	0.42	4.2	0.42	0.21
23478-PeCDF	2.5	0.82	0.25	2.5	0.25	1.2
PeCDFs (OTHER)	12	0	0.012	0	0.012	0
123478-HxCDF	1.6	0.016	0.079	0.047	0.016	0.16
123678-HxCDF	0	0	0	0	0	0
234678-HxCDF	0.46	0.0046	0.023	0.014	0.0046	0.046
123789-HxCDF	0.0095	0.000095	0.00048	0.00029	0.000095	0.00095
HxCDFs (OTHER)	17	0	0.0086	0	0.0017	0
1234678-HpCDF	0	0	0	0	0	0
1234789-HpCDF	0	0	0	0	0	0
HpCDFs (OTHER)	11	0	0.011	0	0.011	0
OCDF	0.41	0	0.000041	0	0	0.00041
TOTAL PCDFs		**3.0**	**1.1**	**9.1**	**1.0**	**1.8**
TOTAL TEQs		**4**	**2**	**CATEQ=10**	**EPA-TEQ=2**	**I-TEQ=3**
		7% Contributed by 2,3,7,8 - TCDD	15% Contributed by 2,3,7,8 - TCDD	3% Contributed by 2,3,7,8 - TCDD	15% Contributed by 2,3,7,8 - TCDD	10% Contributed by 2,3,7,8 - TCDD

Figure 3. Toxicity Equivalents in Emissions from a Municipal Waste Incinerator

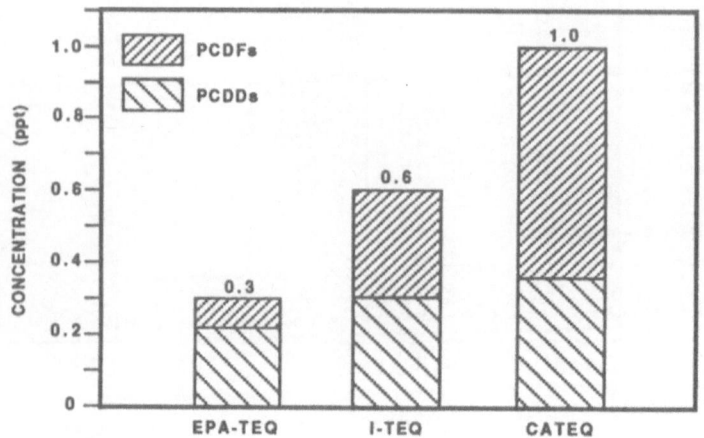

Data for Above Figure:

CONGENER	SOURCE DATA (ppt)	TEF SCHEME		
		EPA-TEF	I-TEF	CATEF
2378 - TCDD	0.11	0.11	0.11	0.11
12378 - PeCDD	0.18	0.09	0.09	0.18
123478 - HxCDD	0.08	0.0032	0.008	0.0024
123678 - HxCDD	0.73	0.029	0.073	0.022
123789 - HxCDD	0.15	0.006	0.015	0.0045
1234678 - HpCDD	1.3	0.0013	0.013	0.039
OCDD	5.7	0	0.0057	0
TOTAL PCDDs		0.24	0.31	0.36
2378 - TCDF	0.12	0.012	0.012	0.12
12378 - PeCDF	0.022	0.0022	0.0011	0.022
23478 - PeCDF	0.51	0.051	0.26	0.51
123478 - HxCDF	0.097	0.00097	0.0097	0.0029
123678 - HxCDF	0.078	0.00078	0.0078	0.0023
234678 - HxCDF	0.04	0.0004	0.004	0.0012
1234678 - HpCDF	0.19	0.00019	0.0019	0.0027
OCDF	0.052	0	0.000052	0
TOTAL PCDFs		0.068	0.30	0.66
TOTAL TEQs		EPA-TEQ=0.3	I-TEQ=0.6	CATEQ=1.0
		37% Contributed by 2,3,7,8 - TCDD	18% Contributed by 2,3,7,8 - TCDD	11% Contributed by 2,3,7,8 - TCDD

Figure 4. Toxicity Equivalents in Human Milk Sample

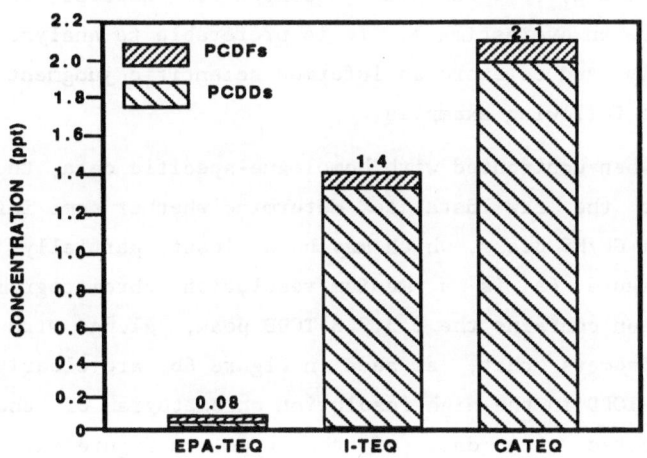

Data for Above Figure:

COMPOUND	CONCENTRATION (ppt)	EPA-TEF	I-TEF	CATEF
2378 - TCDD	-	-	-	-
Other TCDDs	-	-	-	-
2378 - PeCDD	-	-	-	-
Other PeCDDs	-	-	-	-
2378 - HxCDDs	-	-	-	-
Other HxCDDs	26	0.01	0	0
2378 - HpCDD	66.4	0.066	0.66	2.0
Other HpCDDs	393	0.0039	0	0
OCDD	678	0	0.68	0
TOTAL PCDDs		**0.079**	**1.34**	**2.0**
2378 - TCDF	-	-	-	-
Other TCDFs	-	-	-	-
12378 - PeCDF	-	-	-	-
23478 - PeCDF	-	-	-	-
Other PeCDFs	-	-	-	-
2378 - HxCDFs	-	-	-	-
Other HxCDFs	3.7	0.00037	0	0
2378 - HpCDFs	3.3	0.0033	0.033	0.1
Other HpCDFs	16	0.00016	0	0
OCDF	41.5	0	0.042	0
TOTAL PCDFs		**0.003**	**0.08**	**0.1**
TOTAL TEQs		**EPA-TEQ=0.08**	**I-TEQ=1.4**	**CATEQ=2.1**

Figure 5.   Toxicity Equivalents in a Pentachlorophenol Wood
Treatment Site

111

distinguishing between the individual isomers. In these cases, some assumptions about the isomeric composition of each homologue must be made.

As a worst case, one could assume that all of the homologue signal came from the 2,3,7,8-substituted congeners. However, in most instances, this would be an overestimate. It is preferable to analyze the situation more closely and to exercise informed scientific judgment, as illustrated in the three following examples.

First, when confronted with homologue-specific data, the risk assessor can examine the raw data to determine whether some information can be gleaned from GC/MS peaks, which may be at least partially resolved. For example, Figure 6a is a medium-resolution chromatogram of TCDDs. The shaded portion contains the 2,3,7,8-TCDD peak, along with a variety of coeluting isomers that, as seen in Figure 6b, are clearly separated from the 2,3,7,8-TCDD in the high-resolution chromatogram of the same sample. When confronted with data of the type in Figure 6a, it is prudent to assume that the entire shaded peak is the result of the presence of 2,3,7,8-TCDD. This approach will not underestimate the amount of 2,3,7,8-TCDD in the sample. At the same time, the approach will avoid the clear overestimate resulting from the assumption made in the previous paragraph that all of the TCDD homologue consists of 2,3,7,8-TCDD.

Second, over the past decade, a considerable body of information has been compiled regarding the congeneric CDD/CDF composition of several types of sources. For instance, analyses of thermal sources such as incinerators and automobile exhaust generally show that about 5 percent of the total amount of TCDDs is attributable to 2,3,7,8-TCDD. Thus, it would be a reasonable first approximation to assume that 5 percent of the total TCDDs from any combustion source sample consists of 2,3,7,8-TCDD. Adopting this assumption would result in an estimated contribution to the TEQs that is 95 percent lower than the worst case and that is more likely to reflect the actual situation. A similar procedure could be applied to other homologue groups, given the appropriate information about probable isomer distributions associated with particular sources.

It should be noted that for the larger TEFs and lower fractional contribution to their homologous class of 2,3,7,8-substituted tetra- and penta-CDDs and -CDFs (see Table 2) combine to place a premium on obtaining an accurate estimate of the relative contribution of the 2,3,7,8-substituted congeners to these classes. By contrast, in the case of HpCDD, the worst-case estimate will be no more than twofold higher than an estimate based upon the percentage of the 2,3,7,8-substituted congener

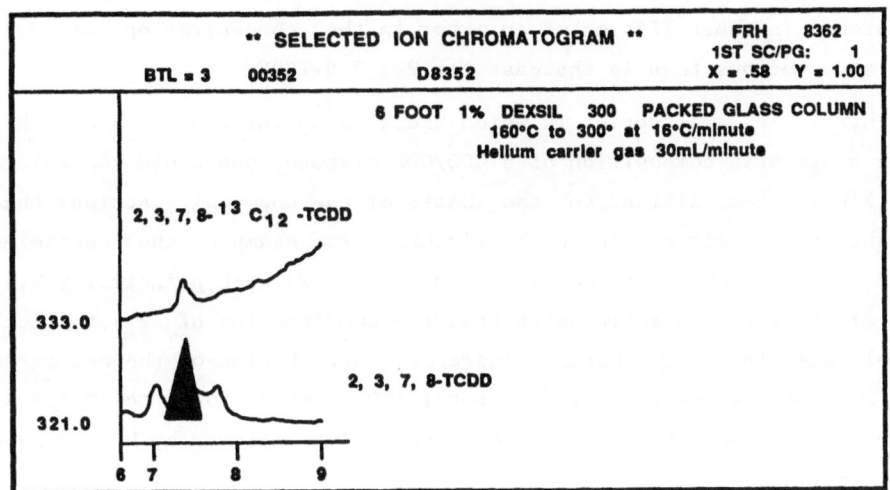

Figure 6a.   Medium Resolution GC/MS Analysis of 2,3,7,8-TCDD

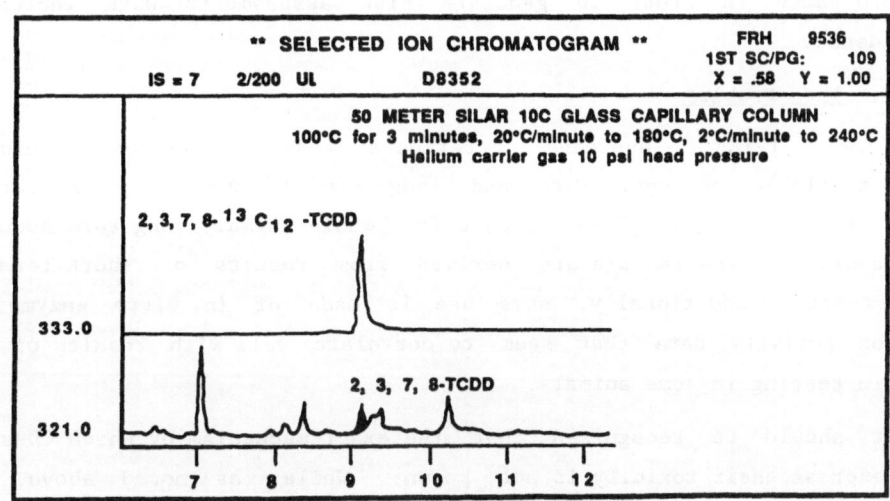

Figure 6b.   High Resolution GC/MS Analysis of 2,3,7,8-TCDD

found in most combustion emission samples, i.e., about 50 percent. Further, the I-TEF of 0.01 for 2,3,7,8-HpCDD suggests much less of an impact on the total TEQ relative to the impact of 2,3,7,8-TCDD, which has an I-TEF of 1.0. Similarly, although 2,3,7,8-TCDD and 2,3,7,8-TCDF constitute a similar fraction of the tetra-substituted congeners, the tenfold difference in their TEFs makes an error in the concentration of 2,3,7,8-TCDD more serious than is the case for 2,3,7,8-TCDF.

Third, in a situation in which there is no information about the specific congeneric composition of a CDD/CDF mixture, one could speculate on the likely composition on the basis of the chemical reactions that are thought to have given rise to the mixture. For example, the contamination of 2,4,5-trichlorophenoxyacetic acid (2,4,5-T) with principally 2,3,7,8-TCDD might have been anticipated from the condensation of 2,4,5-trichlorophenol used in its synthesis. Therefore, attribution of the entire amount of a homologue-specific analyses (total TCDDs) of 2,4,5-T to 2,3,7,8-TCDD would be appropriate (as has been generally confirmed by high-resolution GC/MS analysis).

5.3   Directions for the Future

The I-TEF approach described in this report is an interim procedure for assessing the toxicity of CDDs and CDFs. For the reasons outlined below, additional research should be conducted to test, refine, or replace the procedure in order to generate risk assessments with increasing confidence.

5.3.1   The Problem

The I-TEF approach reflects the limited information available on short-term and long-term *in vivo* experimentation. While preference is given to data from whole-animal, long-term studies, the majority of the factors are derived from results of short-term *in vivo* tests. Additionally, some use is made of *in vitro* enzyme induction activity data that seem to correlate well with results of subchronic testing in some animals.

It should be recognized that the exact mechanism by which CDDs and CDFs express their toxicity is not known. While, as noted above, the receptor binding AHH induction model can rationalize some of the whole-animal data, other toxic effects of these chemicals cannot be explained by this hypothesis alone. Alternative mechanisms have been offered, e.g., interference with the effect of thyroid hormones (Rozman et al., 1984,

1985a,b; McKinney et al., 1985a,b,c), general interference with cellular mechanisms (Rozman and Greim, 1986; Weber et al., 1987), and reduction in vitamin A storage (Thunberg et al., 1979). Some reports (e.g., Holsapple et al. 1986) have raised questions about the underlying structure/activity argument that strongly undergirds most of these hypotheses. In summary, there is no one proposed mechanism of action that can account for all of the *in vivo* experimental observations.

While the correlations cited above provide sufficient evidence for moving forward with the TEF procedure, these approaches contain inherent limitations. For example, *in vitro* studies may not represent the biological effect of chemical administered *in vivo*. Also, *in vitro* studies generally involve single high-level exposures. Such studies do not mimic the chronic low-level ambient human exposures of concern to whole animals (including humans) in the environment. Chronic exposure involves an array of pharmacokinetic and metabolic complexities that are absent in the *in vitro* situation. Possible interaction resulting from simultaneous exposure to a combination of CDDs and CDFs contributes additional uncertainties. Teratogenic studies in rats (Birnbaum et al., 1987) suggest additivity of effects from combinations of the 2,3,7,8-substituted CDDs and CDFs. Other studies indicate antagonism between the 2,3,7,8-substituted CDDs and CDFs, as well as the related chlorinated biphenyls (PCBs) (Safe, 1987).

5.3.2 Addressing the Problem

From the discussion above, it is clear that to test, refine, or replace the I-TEF approach, a vigorous program of research should be conducted to address the areas of uncertainty. Such a program should include approaches developed along the following lines:

- Investigate the models proposed for the mechanism of action of CDDs and CDFs.

- Test the structure/activity hypothesis in additional endpoints, e.g., immunotoxicity.

- Further examine the bioavailability and toxicokinetics of the various CDDs and CDFs.

- Use the findings from the first three steps to test the I-TEF's predictions in the long term using additional single compounds and complex mixtures of CDDs and CDFs.

- Investigate the basis for the wide species variability and the position of humans on this spectrum.

- Conduct any feasible epidemiological studies that have the power to critically test the risk estimates made on the basis of animal testing.

- Widen the area of similar research to include efforts with bromo- and mixed chlorobromodibenzo-p-dioxins and dibenzofurans.

It is clear that the pursuit of this line of research will be complicated, time consuming, and costly. Competing legitimate priorities limit what any one country can accomplish in this area. Therefore, it is important that the international community continue to freely exchange information on research in progress and research in planning, as has been the case with this NATO/CCMS Pilot Study on International Information Exchange on Dioxins and Related Compounds.

5.4   Conclusions

Considerable progress has been made in dealing with the issues associated with the presence of CDDs and CDFs in the environment. Scientific research has made us aware of the problems posed by these substances and has indicated some ways to resolve those problems. At the same time, research to date has not answered all of our questions or addressed all of our concerns. Unanswered questions include fundamental questions ("How toxic are CDDs and CDFs to humans and other environmental species?"), more subtle questions ("How are these CDDs and CDFs transformed and transported to humans?"), and clearly practical questions ("How can the presence of CDDs and CDFs be minimized in the environment?").

The I-TEF scheme addresses only a portion of this larger complex of questions. It rests on a rational, but not definitive, base of scientific fact. Despite all of its limitations, the I-TEF approach provides a practical interim vehicle for the scientific and regulatory community to use in communicating within itself and with the public in estimating the significance of complex mixtures of CDDs and CDFs in the environment.

REFERENCES

AOWG. 1987. Agent Orange Work Group. Report on Agent Orange-related Research in the Federal Government. Washington, DC: Department of Health and Human Services.

Bandiera, S., et al. 1983. Competitive binding of the cytosolic tetra-chlorodibenzo-*p*-dioxin receptor. Biochem. Pharmacol. 32:3803-3813.

Bandiera, S., Sawyer, T., Romkes, M., Zmudzka, B., Safe, L., Mason, G., Keys, B., and Safe, S. 1984. Polychlorinated dibenzofurans (PCDFs): Effects of structure on binding to the 2,3,7,8-TCDD cytosolic receptor protein, AHH induction and toxicity. Toxicology 32:131-144.

Birnbaum, L.S., Harris, M.W., Crawford, D.D., and Morrissey, R.E. 1987. Teratogenic effects of polychlorinated dibenzofurans in combination in C57BL/6N mice. Toxicol. Appl. Pharmacol. 91:246-255.

Boddington, M. 1988. Personal communication to Don Barnes.

Bradlaw, J.A., and Casterline, J.R. 1979. Induction of enzyme activity in cell culture: A rapid screen for detection of planar polychlorinated organic compounds. J. Assoc. Off. Anal. Chem. 65:904-916.

Bradlaw, J.A., Garthoff, L.L., Hurley, N.E., and Firestone, D. 1980. Comparative induction of aryl hydrocarbon hydroxylase activity *in vitro* by analogues of dibenzo-*p*-dioxins. Food Cosmet. Toxicol. 18:627-635.

CCMS. 1987. Committee on Challenges of Modern Society. Report on research activities and analytical centers. Pilot Study on International Information Exchange on Dioxins and Related Compounds. North Atlantic Treaty Organization/Committee on the Challenges of Modern Society. Report Number 166. October 1987.

CCMS. 1988a. Committee on the Challenges of Modern Society. International Toxicity Equivalency Factor (I-TEF) Method of Risk Assessment for Complex Mixtures of Dioxins and Related Compounds. Pilot Study on International Information Exchange on Dioxins and Related Compounds. North Atlantic Treaty Organization/Committee on the Challenges of Modern Society. Report Number 176. August 1988.

CCMS. 1988b. Committee on the Challenges of Modern Society. Scientific basis for the international toxicity equivalency factor method of risk assessment for complex mixtures. Pilot Study on International Information Exchange on Dioxins and Related Compounds. North Atlantic Treaty Organization/Committee on the Challenges of Modern Society. Report Number 178. December 1988.

Commoner, B., Shapiro, K., and Webster, T. 1984. Environmental and economic analysis of alternative municipal solid waste disposal technologies. I. An assessment of the risks due to emissions of chlorinated dioxins and dibenzofurans from proposed New York City incinerators.

117

Couture, L.A., Elwell, M.R., and Birnbaum, L.S. 1988. Dioxin-like effects observed in male rats following exposure to octachlorodibenzo-p-dioxin (OCDD) during a 13 week study. Toxicol. Appl. Pharmacol. 93:31-46.

Dencker, L., Hassoun, E., D'Aroy, R., and Alm, G. 1985. Fetal thymus organ culture as an *in vitro* model for the toxicity of 2,3,7,8-tetrachlorodibenzo-*p*-dioxin and its congeners. Mol. Pharmacol. 27:133-140.

Eadon, G., Kaminsky, L., Silkworth, J., Aldous, K., Hilker, D., O'Keefe, P., Smith, R., Gierthy, J., Hawley, J., Kim, M., and Decaprio, A. 1982. Comparisons of chemical and biological data on soot samples from the Binghamton State Office Building. Unpublished report. Department of Health, New York State.

FRG. 1985. Federal Republic of Germany. Report on dioxins. Compiled by Federal Environmental Agency and Federal Health Office.

Gasiewicz, T.A., Rucci, G. 1984. Cytosolic receptor for 2,3,7,8-tetra-chlorodibenzo-*p*-dioxin: Evidence for homologous nature among various mammalian species. Mol. Pharmacol. 26:90-98.

Gierthy, J.F. and Crane, D. 1985. *In vitro* bioassay for dioxin-like activity based on alterations in epithelial cell proliferation and morphology. Fundam. Appl. Toxicol. 5:754-759.

Gorski, et al. 1988. Elevated plasma corticosterone levels and histopathology of the adrenals and thymuses in 2,3,7,8-terachloro-dibenzo-p-dioxin (TCDD)-treated rats. Toxicology, in press.

Grant, D.L. 1977. Proceedings of the 12th annual workshop on pesticide residues analysis. Winnipeg, Canada. p. 251.

Gravitz, N., et al. 1983. Interim guidelines for acceptable exposure levels in office settings contaminated with PCB and PCB combustion products. Nov. 1, 1983. Epidemiological Studies Section, California Department of Health Services.

Greenlee, W.F., Osborne, R., Dold, K.M., Hudson, L.G., and Toxcano, Jr. WA. 1985. Toxicology of chlorinated aromatic hydrocarbons in animals and humans: *In vitro* approach to toxic mechanisms. Environ. Health Perspect. 60:69-76.

Greig, J.B., et al. 1984. Incomplete correlation of 2,3,7,8-tetrachloro-dibenzo-*p*-dioxin hepatoxicity with Ah phenotype in mice. Toxicol. Appl. Pharmacol. 74:17-25.

Hassoun, E., D'Argy, R., and Deacker, L. 1984. Teratogenicity of 2,3,7,8-tetra-chlorodibenzofuran in the mouse. J. Toxicol. Environ. Health 14:337-351.

Henry, E.C. and Gasiewicz, T.A. 1986. Effects of thyroidectomy on the Ah-receptor and enzyme inducibility by 2,3,7,8-tetrachlorodibenzo-p- dioxin in the rat liver. Chem-Biol Interactions 59:29-42.

Holsapple, M.P., et al. 1986. Immunosuppression without liver induction by subchronic exposure to 2,4-dichlorodibenzo-*p*-dioxin in adult female B6C3F1 mice. Toxicol. Appl. Pharmacol. 83:445-455.

Kimbrough, R.D., Falk, H., Stehr, P., and Fries, G. 1984. Health implications of 2,3,7,8-tetrachlorodibenzo-p-dioxin (TCDD) contamination of residential soil. J. Toxicol. Environ. Health 14:47-93.

Knutson, J. and Poland A. 1980. Keratinization of mouse teratoma call line XB produced by 2,3,7,8-tetrachlorodibenzo-p-dioxin: An *in vitro* model of toxicity. Cell 22:27-36.

Kuehl, D.W., Cook, P.M., and Batterman, A.R. 1986. Update and depuration studies of PCDDs and PCDFs in fresh water fish. Chemosphere 15:2023-2026.

Kutz, F.W. and Bottimore, D.P., eds. 1988. Proceedings of symposium seminar on prospective research and regulatory issues involving dioxins and related compounds. Chemosphere 17(11):1-67.

Kutz, F.W., Barnes, D.G., Bottimore, D.P., Greim, H., and Bretthauer, E.W. 1990a. The international toxicity equivalency factor (I-TEF) method for estimating risks associated with exposures to complex mixtures of dioxins and related compounds. J. Toxicol. Environ. Chem. 26(1-4):99-109.

Kutz, F.W., Barnes, D.G., Bottimore, D.P., Greim, H., and Bretthauer, E.W. 1990b. The international toxicity equivalency factor (I-TEF) method of risk assessment for complex mistures of dioxins and related compounds. Presented at the Ninth International Symposium on Chlorinated Dioxins and Related Compounds (Dioxin '89), in Toronto, Canada, September 1989. Chemosphere (in press).

Kutz, F.W., Barnes, D.G., Bottimore, D.P., and Bretthaver, E.W. 1990c. The toxicity equivalency factor method for estimating risks associated with exposures to complex mistures. Presented at the Total Exposure Assessment Methodology Conference, Las Vegas, Nevada, November 1989. Arch. Toxicol. Ind. Health (in press).

Lamb, J.L., Harris, M.W., McKinney, J.D., and Birnbaum, L.S. 1986. Effect of thyroid hormones on the induction of cleft palate by 2,3,7,8-tetrachlorodibenzo-p-dioxin (TCDD) in C57BL/6N mice. Toxicol. Appl. Pharmacol. 84:115-124.

Lindstrom, G. and Rappe, C. 1988. Analytical method for analysis of polychlorinated dibenzo-p-dioxins and dibenzofurans in milk. Chemosphere 17:921-935.

McKinney, J. and McConnell, J. 1982. Structural specificity and the dioxin receptor. Perg. Ser. Environ. Sci. 5:367-381.

McKinney, J.D., et al. 1985a. Structure-induction versus structure-toxicity relationships for polychlorinated biphenyls and related aromatic hydrocarbons. Environ. Health Perspect. 60:57-68.

McKinney, J.D., Fawkes, F., Chae, K., Oatley, S., Coleman, R.E., and Briner, W. 1985b. 2,3,7,8-tetrachlorodibenzo-p-dioxin (TCDD) as a potential and persistent thyroxine agonist: A mechanistic model for toxicity based on molecular reactivity. Environ. Health Perspect. 61:41-53.

McKinney, J.D., et al. 1985c. Molecular binding of toxic chlorinated dibenzo-*p*-dioxins and dibenzofurans with thyroxine binding prealbumin. J. Med. Chem. 28:375-381.

Moore, J.A., McConnell, E.E., Dalgard, D.W., and Harris, M.W. 1979. Comparative toxicity of three halogenated dibenzofurans in guinea pigs, mice, and rhesus monkeys. Ann. N.Y. Acad. Sci. 320:151-163.

Murray, F.J., Smith, F.A., Nitschke, K.D., Humiston, G.G., Kociba, R.J., and Schwetz, B.A. 1979. Three-generation reproduction study of rats given 2,3,7,8-tetrachlorodibenzo-*p*-dioxin in the diet. Toxicol. Appl. Pharmacol. 50:241-252.

NRCC. 1981. National Research Council of Canada. Polychlorinated dibenzo-*p*-dioxins: Criteria for their effects on man and his environment. NRCC/CNRC Associate Committee on Scientific Criteria for Environmental Quality, Ottawa, Canada. Publ. No. NRCC 18574, ISSN 0316-0114.

Nagayama, J., Kiyohara, C., Masuda, Y., and Kuratsune, M. 1985a. Inducing potency of aryl hydrocarbon hydroxylase in human lymphoblastoid cells and mice by polychlorinated dibenzofurans. Environ. Health Perspect. 59:107-112.

Nagayama, J., Kiyohara, C., Masuda, Y., and Kuratsune, M. 1985b. Genetically mediated induction of aryl hydrocarbon hydroxylase activity in human lymphoblastoid cells by polychlorinated dibenzofuran isomers and 2,3,7,8-tetrachlorodibenzo-*p*-dioxin. Arch. Toxicol. 56:230-235.

OME. 1985. Ontario Ministry of the Environment. Scientific Criteria Document for Standard Development. Polychlorinated dibenzo-*p*-dioxins (PCDDs) and polychlorinated dibenzofurans (PDCFs). Ministry of the Environment, Toronto, Canada. No. 4-84.

Olie, K., Van den Berg, M., and Hutzinger, O. 1983. Formation and fate of PCDD and PCDF from combustion processes. Chemosphere 12:627-737.

Olson, J.R., Bellin, J.S., and Barnes, D.G. 1989. Re-examination of data used for establishing toxicity equivalence factors (TEFs) for chlorinated dibenzo-*p*-dioxins and dibenzofurans (CDDs and CDFs). Chemosphere 18(1-6):371-381.

Ontario Government. 1982. Chlorinated dioxins and chlorinated dibenzo-furans. Ambient air guidelines. Dec. 16, 1982. Health Studies Service, Ministry of Labour.

Pazdernik, T. and Rozman, K. 1985. Effect of thyroidectomy and thyroxine on 2,3,7,8-TCDD-tetrachlorodibenzo-*p*-dioxin-induced immunotoxicity. Life Sci. 36:695-703.

Pohjanvirta, R., et al. 1988. Hepatic Ah-receptor levels and the effect of 2,3,7,8-tetrachlorodibenzo-*p*-dioxin (TCDD) on hepatic mono-oxygenase activities in a TCDD-susceptible and -resistant rat strain. Toxicol. App. Pharmacol. 92:131-140.

Poland, A., et al. 1976. 3,4,3'4'-tetrachloroazoxybenzene and azobenzene: Potent inducers of aryl hydrocarbon hydroxylase. Science 194:627-630.

Poland, A., Greenlee, W.F., and Kende, A.S. 1979. Studies on the mechanisms of action of the chlorinated dibenzo-*p*-dioxins and related compounds. Ann. N.Y. Acad. Sci. 320:214-230.

Poland, A., et al. 1985. Studies on the mechanism of action of halogenated aromatic hydrocarbons. Clin. Physiol. Biochem. 3:147-155.

Puhvel, S.M. and Sakamoto, M. 1987. Response of murine epidermal keratinocyte cultures to 2,3,7,8-tetrachlorodibenzo-*p*-dioxin (TCDD): Comparison of haired and hairless genotypes. Toxicol. Appl. Pharmacol. 89:29-36.

Rozman, K. and Greim, H. 1986. Metabolism of palmitic acid in TCDD-treated rats. Toxicologist 6:207.

Rozman, K., et al. 1984. Effect of thyroidectomy and thyroxine on 2,3,7,8-tetrachlorodibenzo-*p*-dioxin (TCDD)-induced toxicity. Appl. Pharmacol. 72:372-376.

Rozman, K., Rozman, T., Scheufler, E., Pazdernik, T., and Greim, H. 1985a. Thyroid hormones modulate the toxicity of 2,3,7,8-tetrachloro-dibenzo-*p*-dioxin. J. Toxicol. Environ. Health 16:481-491.

Rozman, K. 1985b. Effect of thyroid hormones on liver microsomal enzyme induction in rats exposed to 2,3,7,8-tetrachlorodibenzo-*p*-dioxin. Toxicology 37:51-63.

Safe, S. 1987. Hazard identification - a plenary lecture. Presented at the Seventh International Symposium on Chlorinated Dioxins and Related Compounds, Las Vegas, NV.

Safe, S. et al. 1985. Polychlorinated dibenzofurans: quantitative structure-activity relationships. Chemosphere 14:675-684.

Schwetz, B.A., Norris, J.M., and Sparschu, G.L. 1973. Toxicity of chlorinated dibenzo-*p*-dioxins. Environ. Health Perspect. 5:87-89.

Swiss Government (Bundesamt fur Umweltschutz, Bern). 1982. Environmental pollution due to dioxins and furans from chemicals rubbish incineration plants. Schriftenreighe Umweltschutz. No. 5.

Thunberg T., et al. 1979. Vitamin A (retinol) status in the rat after a single oral dose of 2,3,7,8-tetrachlorodibenzo-*p*-dioxin. Arch. Toxicol. 42:265-274.

USDHHS. 1983. U.S. Department of Health and Human Services. Levels of concern for hexa- (HCDD), hepta- (HpCDD) and octachlorodibenzo-*p*-dioxin (OCDD) in chicken and eggs. Memorandum. April 29, 1983.

USEPA. 1981. U.S. Environmental Protection Agency. Interim evaluation of human health risks associated with emissions of tetrachlorinated dioxins from municipal waste resource recovery facilities. Washington, DC: Office of the Administrator, U.S. Environmental Protection Agency.

USEPA. 1985. U.S. Environmental Protection Agency. Health assessment document for polychlorinated dibenzo-*p*-dioxins. Springfield, VA: National Technical Information Service. EPA-600/8-84-014F. PB86-122546/AS.

USEPA. 1987. U.S. Environmental Protection Agency. Interim procedures for estimating risks associated with exposures to mixtures of chlorinated dibenzo-*p*-dioxins and -dibenzofurans (CDDs and CDFs). Washington, DC: U.S. Environmental Protection Agency. EPA/625/3-87/012.

USEPA. 1988. U.S. Environmental Protection Agency. A Cancer Risk-Specific Dose Estimate for 2,3,7,8-TCDD. Washington, DC. EPA/600/6-88/007.

USEPA. 1989. U.S. Environmental Protection Agency. Interim procedures for estimating risks associated with exposures to mixtures of chlorinated dibenzo-p-dioxins and dibenzofurans (CDDs and CDFs) and 1989 update. Washington, DC: U.S. Environmental Protection Agency. EPA/625/3-89/016.

Van den Berg M, de Broom E, van Grrvenbroek M, Olie K, Hutzinger O. 1985. Bioavailability of PCDDs and PCDFs absorbed on fly ash in rat, guinea pig, and Syrian golden hamster. Chemosphere 14:865-869.

Weber, H., Lamb, J.C., Harris, M.W., and Moore, J.A. 1984. Teratogenicity of 2,3,7,8-tetrachlorodibenzofuran (TCDF) in mice. Toxicol. Lett. 20:183-188.

Weber L, et al. 1987. Metabolism and distribution of (14C) glucose in rats treated with 2,3,7,8-tetrachlorodibenzo-*p*-dioxin (TCDD). J. Toxicol. Environ. Health 22:195-206.

6.   A REVIEW AND UPDATE OF CURRENT KNOWLEDGE AND METHODS   FOR
        PERFORMING EXPOSURE ASSESSMENTS FOR 2,3,7,8-TCDD

Exposure Assessment Group - Office of Health and   Environmental
Assessment

U.S. Environmental Protection Agency, Washington, D.C.

This report is a summary of the document entitled "Estimating Exposures to 2,3,7,8-TCDD" (USEPA, 1988), which was prepared by scientists and
engineers from the Exposure Assessment Group, Office of Health and Environmental Assessment (the names of the authors are included in the acknowledgments).   The primary purpose of the document was to provide a review
of information related to exposure   to   2,3,7,8-tetrachlorodibenzo-p-
dioxin (2,3,7,8-TCDD) and to update information that has come to light
since 1984. In addition, the report included an illustration of the
application of this information in performing exposure assessments for
2,3,7,8-TCDD.

The scope of this report is to address only the update of previous
work and an analysis of key issues related to exposure assessment for
chlorinated dibenzo-p-dioxins (CDDs), with emphasis on   2,3,7,8-TCDD.
If the reader is interested in learning how the updated information is
applied in the development of exposure scenarios, then he/she should refer
to USEPA (1988).

This report is outlined as follows: (1) introduction, (2) physical/
chemical properties and general exposure parameters, (3) fate, (4) exposure, (5) post-exposure, (6) conclusions, and (7) references developed
during the preparation of the original report.

6.1   Introduction

2,3,7,8-TCDD is a substance that is of major concern to public health
because of its extreme toxicity. In the family of chemicals known as
CDDs, 2,3,7,8-TCDD has been shown to be the most toxic compound to animals
that has been isolated and tested. Humans and animals exposed to 2,3,7,8-
TCDD have shown acute, subchronic, and chronic effects. 2,3,7,8-TCDD

adversely affects the skin, the liver, the nervous system, and the immune system of humans and animals. As evidenced by its high cancer potency slope factor (q_1*) (USEPA, 1986a), 2,3,7,8-TCDD is quite potent compared to other known carcinogens, and is classified as a probable carcinogen in humans on the basis of animal carcinogenicity studies that were positive in multiple species and organs.

Human exposure is likely to result from ingestion of contaminated fish, beef, dairy products, and other foods; from ingestion of contaminated soil, especially by children with pica tendencies; from dermal contact with 2,3,7,8-TCDD-contaminated soil, dust, and sediment; and from inhalation of contaminated dust and 2,3,7,8-TCDD vapors.

This report presents an update of previous work and an analysis of key issues related to exposure assessments for CDDs, with emphasis on 2,3,7,8-TCDD. The updated information builds upon exposure assessment methods and general concepts developed as part of the Dioxin Strategy (USEPA, 1983a) and published by the Exposure Assessment Group in November of 1984 (USEPA, 1984a). The 1984 document includes standard factors, assumptions, computation methods, and nomographs for estimating exposure under various exposure scenarios involving 2,3,7,8-TCDD-contaminated soil. USEPA (1984a) addressed five pathways of exposure: dust inhalation, fish ingestion, dermal absorption from soil contact, soil ingestion, and ingestion of beef and dairy products. In the interim, results of new studies have provided information on and an understanding of scientific issues that now enable us to make more informed judgments regarding qualitative and quantitative inputs to exposure assessments for 2,3,7,8-TCDD and related compounds, including the ability to make estimates of exposure through a number of additional exposure pathways. This new information has enabled us to construct more realistic scenarios, narrow ranges of estimates for parameter input values, and broaden the scope of assessments by including municipal waste incinerators and disposal of flyash generated from the incinerators, thereby lessening the degree of uncertainty in resulting exposure assessments.

Several of the items discussed in this presentation were originally identified in the USEPA 1984a document. An expanded and updated analysis is provided relative to the physical and chemical properties of 2,3,7,8-TCDD, the behavior of 2,3,7,8-TCDD in soil and sediment, the concentrations of 2,3,7,8-TCDD in indoor versus outdoor dust and soil, and the inclusion of additional pathways. Also discussed are biologically related issues such as bioavailability from soil; body weight and respiration rates; absorption rates of soil-sorbed 2,3,7,8-TCDD through human lung, gut, and skin; consumption rates for fish, beef, and dairy products; soil ingestion rates; and bioaccumulation in fish, beef, and dairy products. Methods for considering exposure resulting from inhalation of vapor-phase 2,3,7,8-TCDD are presented; exposures caused by inhalation of dust from vehicular traffic, and loading and unloading of ash are considered; food chain contamination by direct and indirect routes is incorporated; and the analysis of pharmacokinetics for estimating exposure to 2,3,7,8-TCDD is discussed. Available information on the plant uptake of 2,3,7,8-TCDD is incorporated in estimating human exposures to make the inclusion of pertinent exposure pathways as complete as possible.

6.2   Physical/Chemical Properties and General Exposure Parameters

This section will discuss several refinements in what is known about general exposure parameters (e.g., pulmonary ventilation rates and human body weights as a function of age and sex) and physical and chemical properties including solubility enhancements when other organic materials are present. Some of these refinements have little effect on exposure assessments, but they add to our confidence in such assessments by making the input data more precise. An exception may be the volatilization potential of high molecular weight organics such as 2,3,7,8-TCDD, which often in the past has been viewed as unimportant. Implications of the increased importance of volatilization from soil and water are developed more fully later in this chapter.

6.2.1 Physical/Chemical Properties

The most essential properties of 2,3,7,8-TCDD for understanding environmental fate as well as pharmacokinetic behavior appear to be vapor pressure (P_v), octanol/water partition coefficient (K_{ow}), and water solubility (S_w). Numerous other important but less frequently investigated parameters are available by derivation or through published correlations, e.g., to soil or sediment partition coefficients and bioconcentration factors (Lyman et al., 1982, Chapters 4 and 5, respectively). The ratio of P_v to S_w (P_v/S_w) yields Henry's law constant (H_c) for low-solubility organic compounds, an index of partitioning for a compound between the atmosphere and the water phase (Mackay et al., 1982).

Brief summaries of recent scientific articles on the physical-chemical properties of 2,3,7,8-TCDD and some related compounds are provided in the following paragraphs. These have been chosen based on credibility of experimental methods and results. No discussions relative to earlier findings are included here, although such discussions usually are offered in the original papers and in reviews such as Webster et al. (1985), Schroy et al. (1985a), and Mackay et al. (1985). (The most recent comprehensive list of the physical properties of 2,3,7,8-TCDD is found in Schroy et al. (1985a).)

Podoll et al. (1986) recently measured the vapor pressure of 2,3,7,8-TCDD using ^{14}C-labeled 2,3,7,8-TCDD and a gas saturation technique followed by combustion to $^{14}CO_2$. The mean value and standard error of five determinations were $7.4 \pm 0.4 \times 10^{-10}$ mm Hg at 25°C. Henry's law constant was then calculated as 12 mm Hg M^{-1} (or 1.6×10^{-5} atm m^3 mol^{-1}), using a water solubility reported by Marple et al. (1986a). Based on this Henry's law constant, Lyman et al. (1982) offers guidelines, though not specific to 2,3,7,8-TCDD, to compare organic compounds that may or may not be volatilized from water at a significant amount, and provides the ranges of Henry's law constant at which volatilization represents "significant transfer mechanisms" and at which volatilization would be insignificant. The compounds listed as having the potential for volatilization include polycyclic aromatic hydrocarbons and other halogenated aromatics such as PCBs. Further discussion can be found in Section 6.4.1.

Marple et al. (1986a) reported the octanol/water partition coefficient of 2,3,7,8-TCDD as $(4.24 \pm 2.73) \times 10^6$ at 22 ± 1°C. Two similar experimental techniques were used, but the more expeditious and reliable one involved equilibration of octanol presaturated with water and containing the 2,3,7,8-TCDD, with water, presaturated with octanol, over 6 to 31 days.

Burkhard and Kuehl (1986) used reverse-phase HPLC and LRMS detection to determine octanol/water partition coefficients for 2,3,7,8-TCDD and a series of seven other tetrachlorinated planar molecules, including three more TCDD isomers and 2,3,7,8-tetrachlorodibenzofuran (TCDF). Log K_{ow} = 7.02 ± 0.50 was reported for 2,3,7,8-TCDD. These authors also reevaluated data on 13 chlorinated dioxins and dibenzofurans previously obtained by Sarna et al. (1984) by very similar experimental techniques. In the reevaluation, Burkhard and Kuehl (1986) used experimental rather than estimated log K_{ow} values in correlations with gas chromatographic retention time. This approach yielded log octanol-water partition coefficients ranging from about 4.0 for the nonchlorinated parent molecules to about 8.6 for the octa-chlorinated compounds, much lower than previously reported. Coefficients in this range usually mean that the substance tends to adsorb strongly to organic components in the soil.

The water solubility of 2,3,7,8-TCDD recently was reported as 19.3 $\pm$ 3.7 ppt at 22°C (or 5.99 $\pm$ 1.15 x 10^{-11} M) by Marple et al. (1986a) after equilibrating thin films of resublimed 2,3,7,8-TCDD with a small volume of water followed by gas chromatography (GC) analysis with ^{63}Ni electron capture detection. A corresponding experiment using radiolabeled 2,3,7,8-TCDD and GC plus scintillation counting was found to be less reliable due to only 80 percent purity of radiolabeled material versus 98 percent purity of nonradiolabeled material, low scintillation count rates (three to four times background), and unexplained losses after equilibration.

The very low solubility of 2,3,7,8-TCDD would imply a low level of the contaminant in leachate to ground water from the soil surface if pure water were the only transporting medium. However, some hazardous waste landfill leachates may be viewed as mixed solvent systems, with an aqueous phase containing a variety of man-made organic chemicals and possibly a second, organic phase, which could dissolve and transport significant amounts of chemicals such as 2,3,7,8-TCDD that have very low solubilities in pure water or predominantly aqueous mixed solvents (demonstrated below). (Solubility of 2,3,7,8-TCDD has been reported for benzene (570 mg/L), methanol (10 mg/L), and acetone (110 mg/L) (USEPA, 1980a).) Naturally, partition coefficients could be determined for 2,3,7,8-TCDD between water and any other solvent not miscible with water (i.e., any solvent not soluble in all proportions). More details on mixed solvent effects on solvent solubility follow shortly.

Recent work by Kapila et al. (1987) focused on used motor oil mixed with 2,4,5-T process still bottom material that was involved in a pollution incident at Times Beach, Missouri. No contact with an experimental equivalent of ground water was involved in the work of Kapila et al. (1987), although contact with an aqueous phase in an unsaturated zone was introduced through simulated rainfall. Their results suggested that 2,3,7,8-TCDD migration through soil, losses through volatilization and photolysis, or losses resulting from surface photolysis alone appear much lower than previously reported.

The effects of dissolved organic matter and very finely divided suspended solids on apparent partition coefficients have been discussed in numerous publications reviewed by Lyman and Loreti (1987). The dissolved organic matter of surface and ground water before and after pollution shows wide variability in sorption capacity. The potential effect of the presence of dissolved organic carbon must be increased transport of sorbed pollutants. Details and examples of these and other factors affecting sorption, such as pH and ionic strength, are available in Lyman and Loreti (1987).

Because little information currently is available on the frequency of occurrence of a second liquid phase in contact with ground water, description of pollutant transport in a second phase is viewed as a topic for future consideration.

The potential effect on solubility of organics in mixed water/organic solvent systems is well-known. A variety of thermodynamic approaches have been developed for estimating solute solubility in mixed solvent systems (reviewed in USEPA, 1985a). More recently, Webster et al. (1986) reported on adsorption and transport of 1,3,6,8-TCDD on dissolved and suspended organics (humic and fulvic acids) in natural waters. Investigators also have reported on the solubility and sorption properties of a number of aromatic organic compounds in methanol/water and acetone/water systems (Nkedi-Kizza et al., 1985; USEPA, 1985a; Fu and Luthy, 1986). (Because methanol and acetone are miscible with water, no second phase would form.)

Earlier work by Yalkowsky et al. (1972) related solubility of organics in water and binary solvents by showing a semilogarithmic increase in solubility with increasing volume fraction of organic solvent,

$$\ln S_m = \ln S_w + \sigma f^c \qquad \text{(Equation 1)}$$

where

S_m = solute solubility in mixed solvent (moles/L)
S_w = solute solubility in water
σ = a system-specific empirical parameter related to surface area and surface free energy of solute
f^c = volume fraction of organic cosolvent, $0 < f^c < 1$.

Nkedi-Kizza et al. (1985) noted an exponential decrease in sorption coefficient with increasing fraction of cosolvent for both methanol/water and acetone/water systems,

$$\ln (K^m/K^w) = -\alpha\sigma f^c \qquad \text{(Equation 2)}$$

where

K^m = sorption coefficient in mixed solvent systems
K^w = sorption coefficient in water
α = an empirical constant.

128

These investigators also found that s was unique to each sorbate (organic solute)/mixed solvent combination and was independent of the soil (sorbent) used in different experiments, suggesting that this might be an unusual site-independent phenomenon.

A similar exponential or semilogarithmic relationship between solubility and fraction of organic cosolvent was reported in USEPA (1985a), where it was also noted that solubility increases, upon adding a fixed amount of cosolvent, are most pronounced for very hydrophobic compounds (such as 2,3,7,8-TCDD). It was also noted that increases in solubility with increasing fraction of organic cosolvent did not result in directly proportional decreases in the sorption coefficient. (This effect is attributed to solvent swelling of the soil organic carbon material and a corresponding increase in accessibility of the latter for sorption.)

Equation 1 may be applied to the question of 2,3,7,8-TCDD solubility in methanol/water mixed solvents, since the Hildebrand solubility parameters (d) of methanol (15.15 calories$^{1/2}$ cm$^{-3/2}$) and water (23.50 calories$^{1/2}$ cm$^{-3/2}$) are more than three units larger than that of 2,3,7,8-TCDD (approximately 10 calories$^{1/2}$ cm$^{-3/2}$) (Martin et al., 1982). A technique described in USEPA (1985a, p. 107) may be used to approximate the solubility parameter, s, for 2,3,7,8-TCDD in a mixed solvent system if the solute solubility in both pure solvents is known. The solubility parameter estimate is obtained by taking the difference between the log of solute solubility in pure solvent and the log of solute solubility in pure water.

Substituting into Equation 1, the calculated solubility of 2,3,7,8-TCDD in 1 percent methanol/water (by volume) is 14 percent greater than in pure water (22.0 ng/L versus 19.3 ng/L, a 1.14-fold increase). In 10 percent methanol/water, 2,3,7,8-TCDD solubility is predicted to be 7.19 ng/L, or a 3.73-fold increase, while in the environmentally unrealistic case of 50 percent methanol/water, solubility should increase to 13,800 ng/L (a 715-fold increase). If similar calculations are performed for a saturated solution of benzene in water (1.78 g/L), 2,3,7,8-TCDD solubility is predicted to be 20.0 ng/L (1.036-fold, or less than a 4 percent increase). The application criteria for the use of Equation 1, referred to earlier, are not fully satisfied for 2,3,7,8-TCDD in benzene/ water, but generally this only poses problems at high-volume fractions.

Increases in solubility resulting from the presence of cosolvents thus are predicted as relatively small for low percentages of cosolvent (typical in landfills), as a consequence of the logarithmic variation of solubility with linear variation in volume fraction of cosolvent. Walters et al. (1987) recently reported measuring sorption isotherms for 2,3,7,8-TCDD in soil/water and soil/water/methanol systems. Isotherms were linear up to 0.5 of solubility, and logarithmic variation of solubility and partition coefficients with linear variation of fraction cosolvent was confirmed. (Further details await publication.)

In summary, prediction of increased solubility for hydrophobic solutes is feasible for binary systems of related solvents, either through UNIFAC calculations (from activity coefficients) or through existing information in the literature, which might allow calculation of s and/or a for the desired cosolvent/solute combination. UNIFAC calculations apparently have been published (Lyman et al. 1982).

Changes in the vapor pressure, pure water solubility, and octanol/water partition coefficient of the properties of 2,3,7,8-TCDD between USEPA (1984a) and the values cited above will have negligible effect on exposures calculated for any of the five routes covered in USEPA (1984a). The major influence of changes in vapor pressure and water solubility (hence, Henry's law constant) concerns volatilization from water and soils.

6.2.2 Body Weights and Pulmonary Ventilation Rates in Exposure Assessments

(1) Body weights. Traditionally, the assumption has been made that human body weight was equal to 70 kg, when calculations are made for large populations. When considering smaller numbers of individuals or subpopulations of concern (e.g., children), a different value might be more appropriate.

USEPA (1984a) presented a procedure that allowed average body weight to be considered as a function of age, in particular for individuals under 18 years old (those 18 years of age or older were considered to be 70 kg). One of the concerns in that report was to obtain representative values for small children who had a tendency to ingest soil, so that more appropriate exposure values could be calculated.

A more recent report, entitled "Development of Statistical Distributions or Ranges of Standard Factors Used in Exposure Assessments" (USEPA, 1985b), refines the treatment of body weight. It provides weight data on age-sex distributions (with percentiles) for tailoring exposure assessments to whatever level of detail is necessary.

The source of the data for the above report was the second National Health and Nutrition Examination Survey, conducted from 1976 to 1980 (National Center for Health Statistics, as referenced in USEPA, 1985b). The survey was a probability sampling of approximately 28,000 people from the ages of 6 months to 74 years. Being concerned with nutrition, the survey oversampled subpopulations thought to be at high risk of malnutrition.

(2) Pulmonary ventilation rates. One of the most critical and variable factors in considering the inhalation route in exposure assessments is the pulmonary ventilation rate. The resting ventilation rate is related to an individual's basal metabolic rate for oxygen consumption. As the activity of individuals increases, so does their metabolism and, hence, the ventilation rate. To obtain a representative description of exposure, information on the activity, its duration, and the associated ventilation rates must be integrated. A report by the USEPA (1985b) on standard factors provides tables on distributions of minute volumes and activity levels and patterns for various age-sex groups. These data are especially useful in situations when subpopulations are being considered for their particular exposure levels.

Some factors to consider in addressing ventilation rates relate to individuals in compromised states, such as emphysema and fibrosis, or those who exhibit age-related changes. The USEPA (1985b) report contains a table listing various formulas for calculating minute volume (empirically) based on variables including surface area, weight, height, age, and sex. It should be noted that little information is available for preschool children, mainly because of the problems involved in clinically studying this age group.

The estimated ventilation rates for different activity levels, which were originally presented in Table 4-3 of the USEPA (1985d) report on standard factors, are reproduced here for the convenience of the reader (Table 1).

The activity patterns, when used with the ventilation data, will provide a time-weighted average ventilation rate for use in assessment. The data on activity patterns indicate the amount of time that individuals might spend at various activities or in different microenvironments. These activity data are averaged for both sexes and all age groups.

Table 1. Estimated Minute Ventilation Associated with
Activity Level for Average Male Adult

Level of work	Watts	kg-m/min[a]	L/min	Representative activities
Light	25	150	13	Level walking at 2 mph; washing clothes
Light	50	300	19	Level walking at 3 mph; bowling; scrubbing floors
Light	75	450	25	Dancing; pushing wheelbarrow with 15-kg load; simple construction; stacking firewood
Moderate	100	600	30	Easy cycling; pushing wheelbarrow with 75-kg load; using sledge hammer
Moderate	125	750	35	Climbing stairs; playing tennis; digging with spade
Moderate	150	900	40	Cycling at 13 mph; walking on snow; digging trenches
Heavy	175	1,050	55	Cross-country skiing; rock climbing; stair climbing with load; playing squash and handball
Heavy	200	1,200	63	
Very heavy	225	1,350	72	Chopping with axe
Very heavy	250	1,500	85	Level running at 10 mph; competitive cycling
Severe	300	1,800	100+	Competitive long distance running; cross-country skiing

[a] kg-m/min = work performed each minute to move a mass of 1 kg through a vertical distance of 1 m against the force of gravity.

Source: Adapted from USEPA, 1985d.

Appendix D in the above-referenced report contains detailed information on activity patterns for 56 population subgroups.

When performing assessments for general populations, the ventilation rates (USEPA, 1985d and activity pattern data (93 percent light activity, 6 percent moderate, and 1 percent heavy (USEPA, 1985a)) can be combined as follows to provide an approximate overall daily ventilation rate:

$$\text{Ventilation rate} = [(22.4 \text{ hr/d} \times 13.8 \text{ L/min}) \quad \text{(Equation 3)}$$
$$+ (1.4 \text{ hr/d} \times 40.9 \text{ L/min})$$
$$+ (0.2 \text{ hr/d} \times 80 \text{ L/min})]/(1,000 \text{ L/m}^3 \times \text{hr/60 min}) = 23 \text{ m}^3/\text{d}.$$

This value is the same as that reported in a previous EAG report (USEPA 1984a), and is similar to values used traditionally that range from 20 to 23 m^3/d as a daily ventilation rate.

6.3   Fate

6.3.1   Fate of 2,3,7,8-TCDD in Soil

(1) Transient behavior of TCDD profile in soil layer.   2,3,7,8-TCDD, like any other organic contaminant, can undergo transport and transformation processes in soil. These processes may be chemical, biological, or physical in nature. The chemical processes may include hydrolysis, photolysis, or breakdown of chemical bonds in the molecules, resulting in the change of the contaminant to products that could be more harmful or less harmful than the original contaminant. The biological processes occur in the presence of soil microbes to enhance the breakdown of the contaminant to the other products. Finally, the physical processes can be thought of as transport processes in which the contaminant retains its chemical identity, but is transported from one location to another by diffusion or advection mechanisms, and may be transferred to different media. Examples of transport include diffusion of 2,3,7,8-TCDD vapor through soil pores and ultimate volatilization into the air, or the leaching of the contaminant into soil by precipitation or floods. The severity of the migration will be dependent on the mobility of 2,3,7,8-TCDD in the soil. As a result, the initial concentration distribution will change as the contaminant is subjected to these transport and transformation processes over a long period of time.

Freeman and Schroy (1986), USEPA (1985d), and Tung et al. (1985) simulated the concentration and thermal profile of 2,3,7,8-TCDD in soil with initially uniform contamination. As time passed, the simulated concentration profile tended to be bell-shaped, with a maximum concentration somewhere in the core of the soil column. The bell-shaped concentration profiles, calculated as a function of depth in soil, were compared with the results of analyses for 2,3,7,8-TCDD in soil core samples taken at various depths. Samples were taken from plots at Times Beach, Missouri (USEPA 1985d), and Eglin Air Force Base, Florida (Tung et al., 1985). They showed good agreement between the model simulation and the measured data in both cases. USEPA (1985d) noted that "the floods at Times Beach, Missouri, have not redistributed the TCDD over a large area," and concluded that based on a simulation of the measured concentration profile at some time periods, the volatilization process is a major mechanism by which 2,3,7,8-TCDD is depleted from the soil. USEPA (1985e) used field soils to measure the soil/water partition coefficients, which ranged from $3x10^4$ L/kg to $1.3x10^7$ L/kg, and evaluated the leachability of 2,3,7,8-TCDD from the soils. Based on these partition coefficients and the use of solute transport models, they concluded that the worst-case movement of 2,3,7,8-TCDD in leaching from soil media is so slow that leaching by water is unimportant compared with other transport mechanisms, such as volatilization and erosion. Note that in other situations concentration profiles may differ because of differences in mode of application and weather conditions.

According to USEPA (1985e), USEPA (1985c), and Freeman and Schroy (1986), the rate of movement of 2,3,7,8-TCDD in soil during the leaching process is insignificant compared to the depletion of 2,3,7,8-TCDD by volatilization. Over a long period of time, this depletion will create a new profile of 2,3,7,8-TCDD concentration in soil. This redistribution should be an extremely slow process in view of the very low vapor pressure of 2,3,7,8-TCDD.

(2) Transient profile in soil column. For assessment of exposures over extended periods, it is appropriate to obtain the average concentration of 2,3,7,8-TCDD in soil over the depth of concern and the exposure duration. For some pathways, this average concentration is important; for other pathways, surface concentrations are important. Although the surface concentration may theoretically appear to be relevant in some cases, the soil surface is not always quiescent and could be subject to disturbances resulting from construction activities, erosion, or digging. These activities will expose the subsurface soil and make these soils available for human exposure.

Tung et al. (1985) experimentally measured the variation of the soil temperature profile as a function of depth and time of a day. The temperature variation along the soil column in a typical day can affect the volatilization rate of 2,3,7,8-TCDD vapor because certain properties of 2,3,7,8-TCDD vapor influencing the volatilization rate is temperature-dependent. These properties are particularly important when dealing with low-volatility organics in a soil matrix and include vapor pressure and diffusivity. The experimental data showed that the diurnal temperature variations are noticeable at soil depths 2 and 10 cm from the surface, and diverge to an essentially constant temperature at a depth of 48 cm regardless of time of day.

The significant diurnal temperature change on the surface of soil would be an important factor in dealing with volatilization of 2,3,7,8-TCDD vapors from the shallow surface of soil. This phenomenon would be significant during the initial stage of contamination. As time passes, the bulk concentration of contaminant will remain at a depth of the soil column, as demonstrated by Freeman and Schroy (1986). Since the 2,3,7,8-TCDD vapor molecules in the soil pore in the bulk of the soil column should diffuse out to the air-soil interface, the bulk of the transport process occurs at the depth where the bulk of the contaminant is contained. The transport process will be affected by the vapor pressure and diffusivity corresponding to the temperature in the bulk of the soil column. The bulk soil temperature deep inside the soil should remain fairly constant without being affected by the diurnal temperature variation. In considering landfills with depth of 10 feet or more, the bulk phenomena would be more important that the surface phenomena when dealing with long-term exposure. The methods of estimating the average concentration along a soil depth are available for two cases in which initially the contaminant is uniformly distributed to a specific depth in soil. In one case, it is assumed that contamination occurred from the soil surface to a certain depth. In the other case, it is assumed that the contaminated surface is covered with a soil material of known thickness initially free of the contaminant. Another process that needs to be considered in estimating the average contaminant concentration in soil is the depletion of the contaminant caused by leaching. Although it may be a tedious process to correct for the time- and depth-dependent concentration variation, such a correction is particularly important for sites that were contaminated a long time ago or are being evaluated for long-term exposures. However, the use of the initial concentration in uniformly contaminated soil will provide an upper-limit value. If the site-specific concentration profile data are available, the maximum value for the bell-shaped concentration profile can approximate the initial value. Vapor-phase diffusion and leaching can occur downward and laterally as well as upward for vapor-phase diffusion.

(3) Degradation of 2,3,7,8-TCDD in soil. If the concentration of 2,3,7,8-TCDD in soil changes significantly over a period of time, the exposure evaluation should reflect this change in concentration. Such changes can be incrementally accounted for in the exposure computations requiring the summation of all exposures over the lifetime period, or, alternatively, a representative concentration of 2,3,7,8-TCDD in soil averaged over the exposure period can be used.

In addition to volatilization, leaching, and atmospheric photolysis, another possible mechanism of reducing the concentration of 2,3,7,8-TCDD in soil is biodegradation by action of microorganisms.

Many researchers claim that photodegradation is the primary means by which degradation of 2,3,7,8-TCDD occurs at the soil surface. Czuczwa and Hites (1986) studied lake sediments and concluded that little photolysis occurred during the long-range transport of atmospheric 2,3,7,8-TCDD on particulates. However, final conclusions cannot be drawn, since the sources of lake sediments are not known with certainty. Additionally, this work focused on incinerator flyash so it may not be applicable to soils. Chemical degradation via hydrolysis and oxidation in soil is very unlikely in view of the insignificant rate of these reactions in aquatic media (USEPA, 1985c). Recent evidence indicates that photolytic reactivity on flyash behaves differently from the soil surface: SRI found experimentally that photolytic degradation of 2,3,7,8-TCDD on the two types of flyash tested is negligible (Mill et al. 1987). Investigators have shown that 2,3,7,8-TCDD on the soil surface photolyzes at a rate slower than observed in organic solutions (Zepp et al. 1988).

(4) Biodegradation of 2,3,7,8-TCDD in soil. The environmental persistence of 2,3,7,8-TCDD has created concern on a national level (USEPA, 1985g). It is thought that one reason for this persistence is that soil microorganisms cannot degrade 2,3,7,8-TCDD, or that they do so very slowly (Bumpus et al., 1985). However, very few studies have been done on the biodegradation of 2,3,7,8-TCDD in soil. A biodegradation rate constant and a half-life will be derived below on the basis of the limited available information.

Young (1983) estimated the half-life of 2,3,7,8-TCDD on his test grid (grass field located at Eglin Air Force Base, Florida) to be 10 to 12 years. He cited several mechanisms to account for the disappearance of 2,3,7,8-TCDD in the herbicide applied within the area of his study. Crosby and Wong (1977) had observed that trace amounts of 2,3,7,8-TCDD in Herbicide Orange exposed to sunlight on leaves, soil, or grass, were apparently photodegraded during dissemination of the herbicide. On the basis of their data, Young calculated that less than 1 percent of the original 2,3,7,8-TCDD that was sprayed actually remained in the ecosystem of the test site, and that the majority of the remaining 1 percent was retained in the soil. He reasoned that the processes of water and wind transport of contaminated particles and biomass removal would be unimportant mechanisms in removing the 2,3,7,8-TCDD retained in the soil. Although he stated that the role of volatilization and microbial degradation in removing 2,3,7,8-TCDD from soil is not clear, he estimated the half-life as 10 to 12 years, based on observed changes in soil concentrations.

Ward and Matsumura (1978) summarized the conclusions of other investigators to the effect that microbial degradation of 2,3,7,8-TCDD was found to be very low or nonexistent. Freeman and Schroy (1986) also concluded that biodegradation of 2,3,7,8-TCDD does not occur on the basis of tests they conducted with soil from Eglin Air Force Base. Bumpus et al. (1985) conducted experiments on the degradation of 2,3,7,8-TCDD mixed in cultures containing the white rot fungus, and measured the rate of CO_2 evolution as an indication of the progression of biodegradation.

It is difficult to analyze the Bumpus et al. (1985) data in a strict kinetics sense, since the progress of the reaction was monitored by analyzing the reaction product (CO_2) being evolved, and exact stoichiometric relations between the reaction product and 2,3,7,8-TCDD are not known. For example, the CO_2 may have evolved from oxidation of compounds other than 2,3,7,8-TCDD. Additionally, the applicability of experimental conditions to actual soil conditions is rather speculative because the distribution of microorganisms in actual versus experimental soil is different and because fungus favors aerobic conditions for biodegradation. If it is assumed that aerobic conditions are poorly maintained in soil, except perhaps in surface layers, the biodegradation constant derived from these data would tend to overestimate biodegradation.

For these and other reasons, it is widely believed that the application of such experiments, particularly those based on pure cultures, to actual field conditions may be inappropriate. Accordingly, the results of this study were not applied anywhere in this report. However, for illustrative purposes only, a rate constant was derived from the Bumpus et al. (1985) data. Assuming first-order kinetics, the rate constant derived was 6.6×10^{-5} day^{-1}. This corresponds to a half-life of about 29 years. It should be noted that these data are applicable for one fungus type only. The effects of the presence of other microorganisms, including bacteria, on the rate constant cannot be evaluated at present. It appears reasonable to expect a longer half-life than that observed by Young (1983), since his observation includes the effect of photolytic degradation.

The nondegradation exposure is multiplied by this ratio to reflect the effects of degradation. The ratio of 1 assumes that biodegradation will not occur (i.e., half-life equals infinity). The lower limit of this ratio will provide the maximum degradation.

Microbial degradation reduces the concentration of 2,3,7,8-TCDD available for human exposure. Although the biodegradation rate for 2,3,7,8-TCDD has not been established, it appears that its half-life in soil is in the range of several decades. For purposes of illustration, if we assume a 30-year half-life for biodegradation in soil, approximately 80 percent of the original 2,3,7,8-TCDD would have transformed to other products at the end of a 70-year (lifetime) period. For a lifetime exposure evaluation, it is appropriate to take into account the gradually decreasing 2,3,7,8-TCDD concentration in soil from which the contaminant is released for human exposure. For a 70-year exposure period, data indicate that a 30-year half-life causes a 50 percent reduction in exposure relative to an infinite half-life (i.e., no degradation).

6.3.2 Fate of 2,3,7,8-TCDD in Sediments

(1) Aquatic sediments. Low concentrations of various CDDs have been reported to exist in sediments in the Great Lakes and other water bodies (Czuczwa and Hites, 1986). Czuczwa and Hites attribute the existence of CDDs in aquatic sediments to the long-range transport through the atmosphere of dioxin-related material in emissions of combustion products. Erosion of contaminated soil may also result in accumulation of 2,3,7,8-TCDD in sediments. Contaminated sediments will slowly release the contaminant to the water body in dissolved form or as suspended sediment. The impacts of contaminated sediments must be considered in exposure estimation.

Recent theoretical discussions (e.g., Thibodeaux et al., 1986) of the manner in which 2,3,7,8-TCDD may partition from sediments to water to fish and other aquatic biota, when compared to empirical observations (e.g., Kuehl et al. 1987a), allow some feeling for the importance of the water-to-fish route for bioaccumulation as compared to the sum of all routes (including ingestion of suspended sediment and other biota) as implied by empirical data. The following paragraphs discuss the sediment-to-water partitioning of 2,3,7,8-TCDD and the implications this partitioning may have for the water-to-biota (alone) route, as well as a way to calculate potential surface water-drinking water concentrations resulting from contaminated sediments.

(2)   Sediment-to-water transport process. In estimating the 2,3,7,8-TCDD concentration in a water body above a sediment of known contamination level, it is useful to picture a system where, initially, the water is free of 2,3,7,8-TCDD and therefore a concentration gradient is set up between sediment and water, providing a driving force for the contaminant to enter the water. In the process, dilution will occur as a result of mixing with the moving water and diffusion through it. If the 2,3,7,8-TCDD molecules reach the interface of the water body and atmosphere, volatilization will occur. In the process of transporting from bottom layer to top surface, 2,3,7,8-TCDD in water may be subject to photolysis, biodegradation, and other interrelated reactions, resulting in some disappearance of the contaminant. It is a complex process to track the movement of 2,3,7,8-TCDD from the sediments to the atmosphere across the water body.

Although the process of the transport may be inherently transient, the average concentration in the water body may approximate steady state; that is, the concentration in the bulk water remains constant by assuming that the amount of 2,3,7,8-TCDD leached from the sediments into the water is equal to that lost to the atmosphere. The steady-state (nonequilibrium) process can be modeled using steady-state transport models (Thibodeaux et al., 1986). The results of such a model should represent an average value of 2,3,7,8-TCDD concentration across the water body. Impacts on the organisms inhabiting near the sediments may be greater than the exposure estimated by this average value. For such organisms, the use of the equilibrium partitioning relationship will provide an upper-bound limit of exposure (for the sediment-water-organism route alone). The more appropriate parameter would be the sediment-to-organism transfer coefficient obtained from field data.

(3) Estimation of TCDD concentration in water body. Transient aspects of diffusional processes through the sediments and water body boundary layer will require applications of Fickian-type models. Hwang (1987a) presented a simplified model based on the consideration of two-phase resistances in the sediment and water sides, and pointed out important parameters controlling the rate of the contaminant transfer from sediment to water body. These parameters include the effect of surface winds creating mixing and moving of lake water, thermal stratification, flow-through, and lake geometry. The effects of these parameters are combined into mass transfer coefficients based on experimental data. Although the model is based on the widely applied two-phase resistance theory and uses experimentally derived mass transfer coefficients, it has not been validated via field measurement.

6.3.3 Bioaccumulation of 2,3,7,8-TCDD in Fish and Cattle

(1) Bioaccumulation in fish. 2,3,7,8-TCDD has been shown to be bioavailable to fish and other aquatic organisms from sediments and flyash (Rappe et al., 1981; Kuehl et al., 1985, 1987a,b). Also, some fish bioaccumulate 2,3,7,8-TCDD preferentially over all other 2,3,7,8-substituted dioxin congeners in the tetra through octa homologous groups (Kuehl et al., 1987a), because of a combination of elimination rate differences and kinetic effects involving decreasing amounts of each isomer with respect to 2,3,7,8-TCDD in each step along the food chain, as well as decreased uptake rates across the gills with increasing degrees of chlorination (Cook, 1987a).

The exposure assessment procedures detailed in this chapter consider bioaccumulation in fish as a function of a fish/sediment distribution factor. The principal difficulty in using recently updated bioconcentration factors (BCFs) for 2,3,7,8-TCDD (66,000 for carp, 97,000 and 159,000 for fathead minnows at two exposure concentrations (Cook, 1987b)) arises from the inability to detect environmental water concentrations with the most sensitive techniques now available. The fish/sediment distribution factor approach assumes that the fish and sediment at a given location are at some "steady state," rather than true thermodynamic equilibrium, over the exposure period. Some fish species, such as bottom feeders, will move toward steady-state conditions faster than others. Many species may never reach equilibrium since they do not spend enough time in one location, and some species will bioaccumulate more 2,3,7,8-TCDD than others because of a greater lipid content.

As a first approximation, the ratio of 2,3,7,8-TCDD in fish to 2,3,7,8-TCDD in sediment may be assumed to range from 1 to 10. For example, laboratory-derived fish/sediment ratios for 2,3,7,8-TCDD for the catfish, Ictalurus (after only 6 days of exposure), ranged from 0.2 to 2.0, with the higher ratios corresponding to the less-contaminated sediment. Fish/sediment ratios for the mosquito fish, Gambusia (after only 3 days of exposure), ranged from 0.2 to 12.0, again with higher ratios being calculated for the less-contaminated sediment (Isensee and Jones, 1975). A fish/sediment ratio of 0.44 for 2,3,7,8-TCDD can be calculated for the northern brook silverside, Laludesthes, from the laboratory data of Matsumura and Benezet (1973); exposure in this case was 4 to 7 days. A study cited by Kenaga and Norris (1983) associates fish (species unspecified) 2,3,7,8-TCDD residues of 4 to 85 ppt with sediment concentrations of 10 to 35 ppt; the means of these ranges (44.5/22.5) would indicate a fish/sediment ratio of about 2.0.

Newer data support the range of fish/sediment ratios presented above. Kuehl et al. (1987a) presented field data which indicate that carp (Cyprinus) fish (70 pg/g)/sediment (30 to 200 pg/g) ratios are in the range of 0.4 to 2.3 for 2,3,7,8-TCDD. Laboratory data presented by the same authors for 2,3,7,8-TCDD yield a fish (7.5 pg/g)/sediment (39 pg/g) ratio of 0.2 after 55 days of exposure; however, this ratio probably does not represent a steady state. In some Missouri streams, bottom fish such as sculpins appear to have 2,3,7,8-TCDD concentrations exceeding ten times the sediment concentrations (Cook, 1986).

Thus, fish/sediment ratios for 2,3,7,8-TCDD can be variable depending on a series of interdependent factors, including species, lipid content, weight, ratio of surface area to weight, organic carbon content of the sediment, food intake rate, density of suspended particulate matter, and concentration of 2,3,7,8-TCDD in the sediment. In addition, fish in a stream may approach a steady state with higher sediment concentrations upstream resulting from food drift and/or suspended sediment with a higher 2,3,7,8-TCDD concentration than the bottom sediment concentration.

Some of the variability in the values derived for fish/sediment ratios could be eliminated if they were derived based on 2,3,7,8-TCDD in the lipid of fish and organic carbon of the sediment. Lake et al. (1984) reported partitioning (preference) factors for polychlorinated biphenyls (PCBs) in field studies with aquatic invertebrates of 0.1 to 0.5 (2- to 10-fold greater concentration of PCB per gram of lipid than per gram of organic carbon of the sediment). These authors also report 3.3- to 5.9-fold greater concentrations of chlordane, DDD, and tetrachlorodiphenyl in the lipid of these organisms than in organic carbon of sediments. Kuehl et al. (1987a) also used this approach to facilitate comparison of data for laboratory (10-g carp) and field (1.5-kg carp) experiments, and to determine the congener-dependent bioavailability of select 2,3,7,8-substituted PCDDs and PCDFs.

(2) <u>Bioaccumulation in cattle</u>. Beef and dairy cattle have been shown to accumulate significant levels of 2,3,7,8-TCDD and compounds such as PCBs, DDT, and PBBs following administration in the diet or following ingestion of contaminated soil, the major pathway for transmission of 2,3,7,8-TCDD to these animals. The amount of soil ingested by grazing cattle can vary between 2 and 15 percent of dry matter intake, depending on whether vegetation is lush or sparse (Healy, 1968).

A number of studies of related compounds such as PCBs, PBBs, and DDT, correlate the resulting level in body fat or milk fat to the dietary level. Fries (1982) reported that under constant feeding these compounds reach an upper estimate, steady-state milk fat/diet ratio of approximately 5, with body fat levels being slightly lower. Jensen et al. (1981) conducted similar studies, using 2,3,7,8-TCDD, and found the beef fat/diet ratio (after 28 days of feeding) to be about 4. Jensen and Hummel (1982) developed data from dairy cows given 2,3,7,8-TCDD in their feed that indicated a cream/diet ratio of 1.6 to 2.2 (cream containing 18 percent to about 40 percent butterfat).

Fries (1985) analyzed data from a study in which cattle were kept in feed lots on four Michigan farms where soils in the confinement areas were the sole source of PBB. Under these conditions, the beef fat/soil ratio was in the range of 0.27 to 0.39 for beef cows, beef calves, and dairy heifers (never lactated), and the milk fat/soil ratio ranged from 0.02 to 0.06 for multiparous and primiparous dairy cows. Assuming that the conditions on the Michigan farms represent the typical situation on U.S. farms, a beef fat/soil bioconcentration ratio of 0.3 to 0.4 and a milk fat/soil bioconcentration ratio of 0.04 are suggested for use in the procedures described for exposure assessment in this document and in USEPA (1984a). More case-specific information should be used to adjust these values. However, in the United States lactating dairy cows are rarely pastured. Beef cattle that may have been pastured are often fattened for as long as 150 days in feed lots before slaughter, thus providing considerable opportunity for elimination and dilution of tissue residues (Fries, 1986).

6.3.4 Plant Uptake

The degree to which 2,3,7,8-TCDD can be taken up into plants has not been well established.

In one study (Wipf et al., 1982; Wipf and Schmid, 1983), investigators collected soil and vegetation samples in Seveso, Italy, over the period 1976 through 1979, following the runaway reaction incident at the ICMESA plant. They found 2,3,7,8-TCDD concentrations on the order of 1 ppm in plant material in 1976; however, the levels dropped by several orders of magnitude over the following years. The authors suspect that the contamination was due to 2,3,7,8-TCDD absorption deposited from local dusts through leaves. They also conducted greenhouse tests using carrots grown in highly contaminated soil collected from the Seveso area. The 2,3,7,8-TCDD levels in the peeled edible portions of the carrots were approximately 3 percent of the levels in the soil. On this basis, the authors concluded that plant uptake was minimal.

Cocucci et al. (1979) conducted similar tests using Seveso soils and various types of vegetation, including carrots, potatoes, and onions. In contrast to Wipf et al. (1982), they found 2,3,7,8-TCDD levels (sources unspecified) in inner parts of the vegetables were approximately equal to levels in the soil.

Young (1983) studied uptake in perennial grasses and small broadleaf plants located at a field in Eglin Air Force Base, Florida. The field was sprayed with 2,4,5-T during 1962-1970. In 1978 and 1979 the levels of TCDD (isomers unspecified) in roots (~700 ppt) were found at levels similar to those in the soil (~500 ppt). Young concluded that this result suggested a "passive" uptake process, in which soil particles are incorporated into the epidermis of the root tissue. The upper portions of the plants were found to have 10 to 75 ppt of TCDD (unspecified isomers).

A study by Facchetti et al. (1986) looked at the potential for maize and beans to absorb, translocate, and accumulate 2,3,7,8-TCDD from contaminated soils. Although poor translation makes the study somewhat difficult to interpret, the authors appear to have concluded the following: (1) roots had higher levels of 2,3,7,8-TCDD than the surrounding soil; (2) 2,3,7,8-TCDD was lost from the soils over time through volatilization; (3) 2,3,7,8-TCDD contamination of aboveground parts of plants was due primarily to volatilization from soils; (4) 2,3,7,8-TCDD levels in aboveground parts of plants did not increase significantly over time or with increasing levels of soil contamination (1 to 752 ppt); and (5) high absorption of 2,3,7,8-TCDD by the roots warrants that precaution be taken in the consumption of root vegetables such as carrots and potatoes.

Isensee and Jones (1971) studied uptake of 2,3,7,8-TCDD into oats and soybeans from treated soils. They observed a time dependency in which uptake peaked soon after planting and declined to very low levels by maturity. They concluded that accumulation of 2,3,7,8-TCDD in plants from soil uptake is highly unlikely. It should be noted, however, that these are aboveground crops, as opposed to root crops.

Sacchi et al. (1986) measured levels of 2,3,7,8-TCDD in bean and maize plants grown inside greenhouses using soil that was dosed with various concentrations of 2,3,7,8-TCDD. They found that 2,3,7,8-TCDD accumulated in the aerial parts of the plants. The accumulation levels generally increased with plant age and soil concentration levels. The ratio between levels in soil and in aerial parts ranged from about 0.3 percent to 30 percent and varied inversely with soil level. At 1 ppb, soil-plant ratio was 1 percent for beans and 3 percent for maize. Substantially lower uptake levels were found when peat was added to the soil.

Because of the contradictory nature of the current literature, it is not possible at this time to establish an equilibrium partition ratio between the plant and soil, nor the rate of plant uptake from soil or other overlying media (air or dust on leaves). If uptakes such as those observed by Cocucci actually occur in a situation where an individual obtains a significant portion of his/her root vegetable diet from a contaminated home garden, very high risks would result. The findings of Young (1983) and Sacchi et al. (1986) may indicate that ingestion of certain aboveground plant parts would pose a smaller, but still potentially significant risk.

6.4   Exposure

This section looks at some of the assumptions made in the parameters used to calculate exposure from various pathways. Specifically, the parameters discussed relate to exposure through inhalation of dust, inhalation of vapor, dermal contact, ingestion of soil, ingestion of beef and dairy products, and ingestion of fish. As discussed earlier, some of the refinements of the past few years (e.g., in treatment of indoor dust levels, soil contact rates, and consumption rates for fish, meat, and dairy products) have increased in the exposure estimates, without substantially changing the estimates themselves. The refinements in the data on soil ingestion have resulted in the reduction, by approximately a factor

144

of five, in the estimate of the high end of the range of the "normal" amount of soil ingested by children while playing. It should be noted that incidental ingestion of soil is common in children through mouthing of hands with soil on them or through ingestion of airborne soil. The estimates in this report are not for the so-called "pica child," who intentionally ingests nonfood material (e.g., "eating mud pies").

6.4.1 Inhalation--Vapors

Based on the vapor pressure consideration, Paustenbach et al. (1986) discounted the importance of 2,3,7,8-TCDD uptake via vapor inhalation in risk assessment evaluation, and assumed that the human intake via inhalation is related to the intake of airborne, respirable particulates only. Thibodeaux (1983) compared estimated environmental 2,3,7,8-TCDD exposures from vapor and dust inhalation and concluded that vapor inhalation is a significant exposure pathway. Eitzer and Hites (1986), based on a limited experimental study, found that CDD in the ambient air was present primarily in the vapor phase. (This study is discussed in more detail below.)

Despite its low vapor pressure, 2,3,7,8-TCDD can volatilize from spill and disposal sites and can be emitted into the air from a variety of other sources, including incineration and combustion processes, facilities manufacturing PCBs, paper products and pentachlorophenol, and pyrolysis of PCBs and other chlorinated benzene derivatives (Radian Corp., 1983; Freeman and Schroy, 1985; Commoner et al., 1985; Czuczwa and Hites, 1986). For example, Nash and Beall (1980) found ambient air concentrations of 2,3,7,8-TCDD when silvex spiked with 2,3,7,8-TCDD was applied to turf and field sites. These ambient concentrations were due to volatilization from the field, which the authors noted was a major pathway of 2,3,7,8-TCDD dissipation.

Because of the difficulty in measuring TCDD in air, models are very convenient tools for estimating the ambient air concentrations. The more distant the exposure location is from the area source where the emissions occur, the less concentrated the contaminant will be in the ambient air. The major cause for this dilution is the mixing of the contaminant with the winds, dispersion in the air, and possibly some degradation in the atmosphere by action of sunlight and free radicals. Since some models describing the emission rate from soil and the dispersion of volatilized CDD are complex and lengthy, the reader who is interested in details of the model derivation and different applications is encouraged to consult the relevant references (USEPA, 1986b; USEPA, 1981a; USEPA, 1979; Turner,

1970). This part of the report will concentrate on presenting the model results and pertinent applications.

(1) Emission potential. In estimating the onsite or offsite ambient air concentrations of volatilized 2,3,7,8-TCDD to which people would be exposed, the first task should be to estimate the emission rate from the contaminated area. As a result of changing concentrations of 2,3,7,8-TCDD in the soil column as the emission proceeds, no matter how small the vapor pressure, the emission rate estimation involves consideration of a non-steady-state process. Because of this, the emission rate can be presented either on an instantaneous basis or on an average basis; in the latter case, the emission rate should be averaged over the time period of interest. For evaluation of long-term exposure, it is most often appropriate to make use of the average values.

Other factors affecting the emission rate are the the moisture content of the soil, the biodegradation rate in the soil, and the existence of gases generated by decomposing material. While cover material can reduce the emission rate, once the cover material is saturated it loses the capability for adsorption, and any enhancement in retardation is due to the increased path length of vapor diffusion. Vaporization of moisture and generation of other gases from decomposing material will lead to convective transport, as well as increased diffusion rate, and thus will increase the emission rate.

When it can be assumed that 2,3,7,8-TCDD initially contaminates soil from the soil surface down to a specified depth, with a uniform concentration distribution along the soil column, the time-averaged emission rate, N_D, in grams per cm^2 per second, can be estimated from USEPA (1986b) and Hwang and Falco (1986) as follows:

$$N_D = 2D_i \ E^{4/3} \ K_{as} \ C_{so}/\sqrt{3.14\alpha T} \qquad \text{(Equation 4)}$$

The parameters in Equation 4 can be defined as follows:

D_i = the molecular diffusivity of dioxin vapor in air (= 4.7 x 10^{-2} cm^2/s)
E = the porosity of soil
T = the exposure duration (seconds)
K_{as} = the air/soil partition coefficient (mg/cm^3 air/mg/g soil)
C_{so} = the initial 2,3,7,8-TCDD concentration in soil (g/g), and
α (cm^2/s) is

$$\alpha = D_i \ E^{4/3}/[E + p_s(1 - E)/K_{as}] \qquad \text{(Equation 5)}$$

where

p_s = the true density of soil (g/cm^3).

146

For wet soils, 2,3,7,8-TCDD is partitioned among the soil, water, and air phases. Henry's law constant and the soil/water partition coefficient can be used to describe partitioning of 2,3,7,8-TCDD between the soil and air phases, or

$$K_{as} = 41\ H_c/K_d \qquad \text{(Equation 6)}$$

where

H_c = Henry's law constant (1.6×10^{-5} - 4.88×10^{-5} atm m^3/g mol) (Podoll et al., 1986; Schroy, 1985a)
K_d = the soil/water partition coefficient (mg/g soil/mg/cm^3 water)
Conversion factor 41 is needed to convert H_c/K_d in the specified units to the units of g soil/cm^3 air.

(2) <u>Dilution of emissions in ambient air</u>. The 2,3,7,8-TCDD emanating from the soil surface in vapor or particulate form is diluted by winds at the site and is transported through the atmosphere until it is ultimately reduced in concentration, possibly destroyed through photolytic reactions or deposited on land or in water bodies. The atmospheric phenomena that influence the dilution, transport, and transformation of 2,3,7,8-TCDD emissions can change extensively as a function of time and location.

Gaussian dispersion modeling cannot be used in estimating the onsite concentrations or the receptor concentrations within 100 meters of the center of the facility. Despite the importance of the contaminant dispersion phenomena at short-range or onsite locations, experimental data are lacking with which to calibrate the dispersion coefficients (Hwang, 1987b). The onsite or short-range concentrations along the center of an area source can be estimated from:

$$C = \sqrt{2/3.14}\ ((q)a/\mu\sigma_z)\ \text{EXP}\ (-1/2(z/\sigma_z)^2)\ \text{erf}\ (b/2\sqrt{2\sigma_y}) \qquad \text{(Equation 7)}$$

where

a, b   = dimensions of the site parallel and perpendicular to the wind direction, respectively
q   = emission rate per unit area
u   = average wind speed
σ_y and σ_z = horizontal and vertical dispersion coefficients in air, respectively
z   = receptor height, and it is assumed that the source is at ground level.

Alternatively, the box-model approach could be used, which considers mixing of the emitted contaminant with the winds, ignoring the dispersion effects. Based on the use of an average value for wind speeds (which actually varies logarithmically with respect to height) and for a mixing height, this approach provides an estimate of the onsite concentrations of 2,3,7,8-TCDD in ambient air, as follows:

$$\text{Onsite } C_a = Q/[(LS)V(MH)] \qquad \text{(Equation 8)}$$

where

C_a = the ambient air concentration of 2,3,7,8-TCDD at the exposure location (g/m^3)
Q = total emission rate (g/s)
LS = an equivalent side length of the site perpendicular to the direction of the winds (m)
V = the average wind speed at the inhalation height (about 2.2 m/s for 10-mph winds)
MH = the mixing height before being inhaled by an individual (m).

Two approaches are possible for estimating the ambient air concentrations of 2,3,7,8-TCDD at offsite locations. In the current state of model development, both approaches make use of virtual source approximations for area sources. The first approach, the Industrial Source Complex (ISC) model, requires extensive site-specific meteorological data. These data are used to sum the frequency distribution of wind speed for each radial/ concentric sector of exposure around the source, and to evaluate the vertical standard deviation of the plume for each observed stability class on an annual basis. The details of the equation can be found elsewhere (USEPA, 1986b).

The second approach, which is applicable for estimating concentrations at distances of 100 meters or more from the source, uses an approximation of the ISC model for a quick assessment. The simplified form of the equation is given in Turner (1970) as:

$$\text{Offsite } C_a = 2.03 \ (Q/L_v \sigma_z V) \qquad \text{(Equation 9)}$$

where

L_v = the total virtual downwind distance to the receptor (m), obtained from $L_v = L + 2.5S$
L = the distance from the center of the facility to the receptor
S = the width of the facility perpendicular to the wind direction (m)
C_a, Q, V = as defined for Equation 8
σ_z = the vertical standard deviation (m).

The values for σ_z can be found in most standard air pollution textbooks in graphic or formula form (Turner, 1970; USEPA, 1986b). In estimating the annual average values, the concentration obtained by Equation 9 is often adjusted by multiplying by a frequency factor with which the wind blows toward a particular sector of interest.

(3) <u>Exposure estimation</u>. The estimation of average lifetime exposure via vapor inhalation requires information on the contact rate, the duration of exposure, and the fraction of 2,3,7,8-TCDD vapor absorbed through the lungs upon inhalation, as well as the ambient air concentrations being inhaled. The contact rate is the air respiration rate for children and adults and should be consistent with the air respiration rate discussed for dust inhalation. The exposure lifetime period is normally assumed to be 70 years (as discussed further in this chapter).

The rate of absorption of vapor-phase 2,3,7,8-TCDD through the membranes of human lungs into the bloodstream has not been adequately addressed in the literature. Such a toxicokinetic analysis may require consideration of the diffusional process through the cell membranes of the trace amount of 2,3,7,8-TCDD vapor that entered the alveoli. In view of the data showing that 2,3,7,8-TCDD tightly bound on the soil is absorbed through the gastrointestinal (GI) tract at the rate of 20 percent to 26 percent (Poiger and Schlatter, 1980), the vapor-phase 2,3,7,8-TCDD, which is not bound on adsorptive solids, should find its way through the single-layer membrane between the air space and blood capillaries in excess of the rate reported for the GI tract. Until rigorous kinetic models or experimental data are available, it appears reasonable to assume 50 percent to 100 percent absorption of inhaled 2,3,7,8-TCDD vapor through the thin membrane layer, which at the alveoli and blood capillaries contains both hydrophilic and hydrophobic components.

(4) <u>Effect of photodegradation on exposure estimation</u>. Once emitted, 2,3,7,8-TCDD vapors or particulates can be destroyed in the atmosphere through photodegradation. This may occur by direct absorption of sunlight energy, resulting in the breakdown of 2,3,7,8-TCDD by photolysis, or through photo-induced oxidation by the reaction of 2,3,7,8-TCDD with free radicals present in the atmosphere. Photolysis requires sunlight and is facilitated by the availability of a hydrogen donor (Crosby and Wong, 1977). Also, the presence of a solvent on soil appears to make adsorbed compounds more available for photolysis.

Podoll et al. (1986) compared the half-lives of 2,3,7,8-TCDD vapors under the conditions of photolysis and hydroxyl (OH) radical oxidation in the atmosphere. In estimating the half-lives for photolysis of 2,3,7,8-TCDD vapor, the authors used the quantum yield observed in hexane and estimated an upper-bound photolysis rate of $t_{1/2} = 58$ minutes. At an average concentration of OH radical of 3×10^{-15} M in the atmosphere, they also estimated the half-life for oxidation of 2,3,7,8-TCDD vapor by

OH radicals as $t_{1/2}$ = 200 hours. The authors presented these results with reservations, suggesting the need for an accurate measurement of the vapor-phase quantum yield for 2,3,7,8-TCDD.

Crosby and Wong (1977) conducted experimental photolysis studies for 2,3,7,8-TCDD to evaluate its photolysis rates in different herbicide formulations and on different surfaces. The half-lives for loss of 2,3,7,8-TCDD from herbicide formulations on glass surfaces ranged from 2 to 6 hours. Crosby and Wong (1977) reported that thin films of 2,3,7,8-TCDD on glass plates were found to be stable in sunlight for at least 14 days. It should be noted that this test did not use herbicide formulations. They determined the half-lives on leaves as about 1 to 2 hours and those on soil as longer than 7 hours. All experiments were conducted under natural sunlight without using organic solvents. Accounting for daily and annual fractions of sunlight, Thibodeaux and Lipsky (1985) adjusted the 6-hour half-life derived by Crosby and Wong to obtain an effective photodegradation half-life of 7.2 days.

These findings appear to indicate that photolysis plays a role in the degradation of 2,3,7,8-TCDD only under certain conditions. The use of solvents, such as olive oil or hexane, can enhance the photolysis of 2,3,7,8-TCDD on solid surfaces. The surface itself, and perhaps the organic films present on that surface, are factors influencing the photolysis rate; for example, 2,3,7,8-TCDD was stable on glass surfaces without being affected by photolysis, but underwent photolytic degradation when organic solvent film was present on solid surfaces. This suggests that in the environment, 2,3,7,8-TCDD on a clean surface, such as glass, may react quite differently from 2,3,7,8-TCDD on soil or leaf surfaces with regard to photolysis.

A preliminary photolysis experiment using 2,3,7,8-TCDD adsorbed on flyash particulates suspended in recirculating air indicated that the photolysis of 2,3,7,8-TCDD in particulate form underwent virtually no photolytic reactions after 30 hours of illumination (Mill et al., 1987). Further experiments will be needed to compare the relative magnitude of the photolysis rates for 2,3,7,8-TCDD vapors and particulates. Eitzer and Hites (1986) reported that 2,3,7,8-TCDD in the atmosphere is all in vapor form. The vapor was captured by adsorption on polyurethane foam. They collected ambient air particulates using a high-volume sampler and 0.1-um pore size filters and could not detect 2,3,7,8-TCDD in the particulates. The study did not determine how much 2,3,7,8-TCDD may have been present on the particles measuring less than 0.1 um. However, if it is true that 2,3,7,8-TCDD is present primarily in the vapor phase, then the disappearance of 2,3,7,8-TCDD in the atmosphere would be controlled by the vapor-phase photolysis. Further study is needed to confirm these rather unexpected results before final conclusions can be drawn.

Photolysis is generally assumed to decrease the ambient air concentrations that are subject to human inhalation. However, higher chlorinated CDDs could degrade to 2,3,7,8-TCDD, and the degradation products of 2,3,7,8-TCDD could still be toxic. In the absence of further experimental data under sunlight conditions, it appears that a reasonable value for 2,3,7,8-TCDD vapor-phase half-life is in the range of 2 to 6 hours. This range of half-life is supported by recent experimental results (Mill et al., 1987) on the vapor phase photolysis of 2,3,7,8-TCDD under simulated sunlight conditions, where results show the vapor-phase half-life being several hours rather than the flyash particulate half-life of several hundred hours, as noted above. Even with this relatively short vapor-phase half-life, the onsite ambient air concentrations or nearby offsite ambient air concentrations will not have degraded significantly before the population breathes the air. This is because the time required for the 2,3,7,8-TCDD vapor emissions from the soil surface to reach a human or structure is short compared with the half-life.

For ambient air concentrations at many offsite locations, however, degradation by photolysis could be significant. For example, with winds at 10 miles per hour, the original amount of 2,3,7,8-TCDD vapor would have been degraded by one-half at a distance of about 20 miles from the site, if it is assumed that the half-life is 2 hours.

In estimating exposure to 2,3,7,8-TCDD vapor of the population residing at a distance from the contaminated site, the ambient air concentration estimated by Equation 9 should be corrected by the amount of photodegradation. For first-order photodegradation kinetics, this correction is exponential with respect to the travel time:

$$(C_a)_{corrected} = C_a e^{-kt}. \qquad \text{(Equation 10)}$$

The corrected concentration, as given by Equation 10, should be used in place of Equation 9 in subsequent exposure evaluations for the offsite population for which photodegradation is significant. In Equation 10, k is the first-order rate constant (= 0.12 hours^{-1}, based on a half-life of 6 hours), and t is the time-of-travel for 2,3,7,8-TCDD vapor to reach the receptor by blowing winds. The effect of photodegradation could be used in evaluating risk associated with the long-range transport of

2,3,7,8-TCDD in particulate or vapor form in the atmosphere. For the short-range transport of the contaminant in the atmosphere (e.g., within a 5-mile radius from the site), the effect of photodegradation on the results of exposure evaluation can be discounted.

6.4.2 Inhalation--Particulates

Dust emissions and the resulting TCDD emissions occur as a result of vehicular traffic, loading and unloading operations, spreading operations, transportation in trucks, and wind erosion. Such operations also occur at contaminated soil and landfill sites.

(1) Vehicular traffic. The emissions from a dump or landfill site caused by vehicular traffic can be estimated from an emission factor. This factor can be found in USEPA (1985e), and takes the form of

$$E_v = K(1.7 \ (S/12) \ (S/48) \ (W/2.7)^{0.7} \ (w/4)^{0.5} \ (365-p/365)) \quad \text{(Equation 11)}$$

where

E_v = emission factor (kg/VKT)
K   = particle size multiplier (0.8 for particulates < 30 um)
s   = silt content of road surface material (%)
S   = mean vehicle speed, km/hr
W   = mean vehicle weight, MG
w   = mean number of wheels
p   = number of days with at least 0.254 mm (0.01 in) of precipitation per year.

This emission factor is provided in units of kilogram of particulate emitted per vehicle kilometer traveled (kg/VKT). Equation 11 can be used to solve for the quantity of particulate emitted if the VKT is known and the other parameter values can be defined. USEPA (1985e) provides 4.3 to 20 percent as the range of road silt content.

(2) Loading and unloading operations. Emissions during loading and unloading of contaminated soil or flyash can be estimated using an equation given in USEPA (1985e). The emission factor equation is

$$E = k(0.0009) \left[(s/5)(U/2.2)(H/1.5)/(M/2)^2 \ (Y/4.6)^{0.33} \right] \quad \text{(Equation 12)}$$

where

E = emission factor (kg/Mg)
k = particle size multiplier (dimensionless)
s = material silt content (%)
U = mean wind speed, m/s
H = drop height, m
M = material moisture content (%)
Y = dumping device capacity, m^3.

Equation 12 provides an emission factor for kilograms of particulate emitted per megagrams (Mg) of soil unloaded or loaded. The particle size multiplier varies with aerodynamic particle size, and is given numerical values of 0.73 and 0.77 for batch drop and continuous drop operations for particle sizes less than 30 um.

(3) <u>Spreading operations</u>. It is assumed that spreading operations cause emissions on the same order as agricultural tilling. The emission factor derived from USEPA (1985e) is used to estimate emissions:

$$E_S = k(5.38)(s)^{0.6} \qquad \text{(Equation 13)}$$

where

E_S = emission factor (kg/ha)
k   = particle size multiplier
s   = silt content.

The particle size multiplier (k) varies with aerodynamic particle size, and is given as 1 for total particulate and 0.33 for particulates less than 30 um.

(4) <u>Transportation in trucks</u>. After the ash is discharged from the storage bin to trucks, emissions can be controlled by hauling the flyash in closed trucks or open trucks that use water or other wetting agents for dust control.

Currently, no emission factors are available for estimating emissions from open trucks. It is assumed that each truckload of ash can be treated as an individual aggregate pile. The emission factor for estimating emissions from open trucks is:

$$E_T = 1.9 \ (s/1.5)[(365-p)/235](f/15) \qquad \text{(Equation 14)}$$

where

E_T= total suspended particulate emission factor (kg/d/ha)
s = silt content of aggregate(%)
p = number of days with > 0.25 mm (0.01 in) of precipitation per year
f = percentage of time that the unobstructed wind speed exceeds 5.4 m/s at the mean pile height.

For wetted flyash, the value for p can be assumed to be 365 days based on the assumption that the surface of the ash is wetted prior to hauling to the landfill. Since the truck is normally moving at a speed faster than 12 mph, the value for f is 100 percent.

(5) <u>Wind erosion</u>. A method for estimating dust emissions generated by wind is described below. This method assumes that the uncrusted contaminated surface is exposed to the wind and consists of finely divided

particles. This creates a condition defined by USEPA (1985d) as an "unlimited reservoir" and results in maximum wind-caused dust emissions. Surface activities such as those described above would increase the emissions and should be computed separately if they occur.

The flux of dust particles less than 10 um from surfaces with an "unlimited reservoir" of erodible particles can be estimated as shown below (USEPA, 1985d):

$$E = 0.036 \ (1 - V) \ (Um/U_t)^3 \ F_{(z)} \qquad \text{(Equation 15)}$$

where

E = total dust flux rate of < 10 um particle ($g/m^2 \cdot hr$)
V = fraction of vegetation cover
Um = mean annual wind speed (m/s)
U_t = threshold wind speed (m/s)
$F_{(z)}$ = a function specific to this model.

Threshold wind speed (U_t) is the wind velocity at a height of 7 meters above the ground needed to initiate soil erosion. It depends on nature of surface crust, moisture content, size distribution of particles, and presence of nonerodible elements.

6.4.3 Dermal--Soil Contact Rates and Dermal Absorption

(1) Contact rates. Measurements of the amount of soil that adheres or accumulates on skin surfaces have been conducted by Lepow et al. (1975) and Roels et al. (1980). Lepow et al. (1975), using an adhesive, found an average of 11 mg of dirt per 21.5 cm^2 on the hands of children. Roels et al. (1980) obtained similar results from a study with older children (11-year-olds) using a technique involving the lead content of the dirt. In the above studies, the amount of adhered soil was 0.5 mg/cm^2 and 0.6 mg/cm^2, respectively. Hawley (1985) used the results of Lepow et al. (1975) and Roels et al. (1980) in his assessment of risk from exposure to contaminated soil and used a value of 0.51 mg/cm^2. This value was used for estimating exposure to children playing outdoors. For adults, Hawley (1985) assumed a value of 3.5 mg/cm^2 from doing yard work. USEPA (1984a), after considering Snyder (1975), Lepow et al. (1975), and Roels et al. (1980), assumed a contact range of 0.5 to 1.5 mg/cm^2; the study also assumed that this range represents an average for the entire exposed area of the human body for both adults and children.

The duration (that is, contact time per event [hr/d] times frequency of the event [d/yr]) of the exposure to the contaminated soil is an important determination. USEPA (1984a) adopted a range of 247 to 365 d/yr as the exposure frequency. Paustenbach et al. (1986), in their examination of assumptions used for risk calculations, stated that it appeared that the Centers for Disease Control (CDC) assumed exposure to soil for about 180 d/yr. Hawley (1985) assumed outdoor exposure of 5 d/wk for 6 months for young children (2 years of age), with a 12-hour contact time (since children often retain soil on their skin after coming indoors). This corresponds to about 130 d/yr. For older children, Hawley (1985) assumed an average outdoor playtime of 5 hr/d from May to September or 150 d/yr. Adults were assumed to have about 43 d/yr of outdoor soil exposure, based on 8 hours of yard work 2 d/wk for 5 months.

A similar issue is the length of time the soil is in contact with the skin surface. This is an important factor, since it will help determine the amounts absorbed. The CDC (Kimbrough et al., 1984) assumed that the contact period was sufficient to cause 1 percent absorption. This assumption was based on work by Poiger and Schlatter (1980), who found 0.05 to 2.2 percent absorption in rats from a soil and water paste after 24 hours. USEPA (1984a) assumed 24 hours of contact for the days of exposure. In contrast, Hawley (1985) assumed 12 hr/d of contact for children and 8 hr/d for adults.

Most of the above determinations and assumptions have recognized that contact with soil is dependent on various factors, including weather, age, and activity patterns of subgroups or individuals.

The skin surface area that is exposed to soil may vary because of age, type of activity, or outside conditions such as temperature. USEPA (1984a) used values developed by Sendroy and Cecchini (1954); that is, 2,940 cm^2 for an adult wearing a short-sleeved shirt and no gloves and 910 cm^2 for an adult wearing a long-sleeved shirt and gloves. For children, USEPA (1984a) computed the exposed surface area by multiplying the above values by the ratio of a child/adult total surface area. Hawley (1985) derived surface areas of various body parts from Diem and Lentner (1973) and Berkow (1924), and then made assumptions about the body parts that might be exposed. For young children, he assumed that both hands and the legs and feet would come into contact with soil, giving a surface area of 0.21 m^2. For older children, he assumed soil contact over both hands, the forearms, and the legs from the knees down (0.16 m^2). For adults, Hawley (1985) assumed contact on both hands and the forearms (0.17 m^2), estimating 3.5 mg/cm^2 of soil on the skin for adults.

Obviously, these surface area designations will change according to the attire of the individuals or assumptions made as to which body areas come in contact with soil.

So far, only outdoor exposure has been considered in this section. If the soil in the surrounding area is contaminated, then it can be assumed that soil is carried into the house and that household dust is contaminated to significant levels approaching those occurring outdoors. Hawley (1985) assumed that about 80 percent of indoor dust is identical in contaminant content to outdoor soil and that the concentration of suspended particulate matter indoors is three-quarters of that outdoors. He also assumed that the dust covering indoor surfaces was 560 mg/m^2, based on the assumption that the indoor dustfall rate was 20 percent of that outdoors, or 80 mg/m^2/d, and that cleaning took place once every 2 weeks. Therefore, according to Hawley (1985), dermal contact to children indoors was 560 mg/m^2 on a surface area of 0.05 m^2. The exposed area was taken as half that of the feet, hands, and forearms. The contact area for older children was assumed to cover both hands (0.04 m^2) and to continue for 4 hr/d. Adults were assumed to be in contact with an area equivalent to the surface area of their hands (0.09 m^2).

(2) <u>Absorption</u>. The skin constitutes a major interface between humans and the environment, and influences percutaneous absorption by physical, physiological, and biochemical means. The factors that affect transfer across this membrane, and which can modify the cutaneous penetration rate, include the intrinsic properties of the skin itself (such as age, location on the body, and any modifications from trauma or disease), in addition to the environmental conditions of exposure (temperature, humidity, concentration gradient, and duration of exposure).

Suskind (1977) states that there are two major routes of penetration: (1) the epidermis itself and (2) the hair follicles and sebaceous glands. According to Suskind, the latter are particularly important in initial or transient exposures and also in cases where chlorinated hydrocarbons and dioxins seem to impact the skin (for example, producing chloracne at those sites). Therefore, the skin may need to be considered as a target organ. Tschirley (1986) reported that in an experiment with human volunteers where 2,3,7,8-TCDD was applied in a dose of 3 to 114 ng/kg to their skin, no chloracne developed. However, at a dose of 107,000 ng/kg, 80 percent of the subjects developed chloracne.

With regard to the absorption of 2,3,7,8-TCDD, Poiger and Schlatter (1980) looked at the radiolabeled amounts of the compound in the liver of rats as an indication of its uptake. When 2,3,7,8-TCDD was dermally applied as a pure compound (in methanol) for 24 hours, they found that the highest liver content was 14.8 percent of the dose applied to the skin. When 2,3,7,8-TCDD was applied in a mixture of soil, Poiger and Schlatter (1980) found the liver content to be significantly less than half of that measured in the previous test. When 2,3,7,8-TCDD was applied in a soil-water paste, the liver content varied with dose from 0.05 to 2.2 percent. USEPA (1984a), using the information reported by Poiger and Schlatter (1980) to the effect that 70 percent of total body burden is found in the liver, modified the range for dermal absorption to 0.07 percent to 3 percent.

Hawley (1985) first considered studies by Bartek et al. (1972) and Feldman and Maibach (1970) on dermal uptake of various compounds in humans when applied as pure compounds or in acetone for 24 hours. On the basis of these studies, he assumed the percutaneous absorption rate to be 11 percent in 24 hours for adults. Therefore, for a 12-hour contact time, the rate would be about 6 percent. For children, Hawley (1985) assumed the absorption rate to be twice that of adults. He then considered the results of Poiger and Schlatter (1980) and modified this rate of 1.8 percent for children and 0.9 percent for adults on the basis of a soil matrix effect (i.e., bioavailability) of 15 percent and a contact time of 12 hours.

6.4.4 Ingestion--Soil

(1) Available studies. Binder et al. (1986) studied the ingestion of soil in children 1 to 3 years of age wearing diapers. The children studied were part of a larger study of residents living near a lead smelter in East Helena, Montana. Soiled diapers were collected over a 3-day period from 65 children (42 males and 23 females), and composited samples of soil were obtained from the children's yards. Both excreta and soil were analyzed for aluminum, silicon, and titanium, elements thought to be poorly absorbed in the gut and to have been present in diet only in limited quantities, making them reasonable to use as tracers in a mass-balance calculation. Both soil and excreta measurements were obtained for 59 children. Using a standard assumed fecal dry weight of 15 g/d, soil ingestion by each child was estimated using each of the three tracer elements (assuming no absorption or nonsoil source of these elements). The average quantity of soil ingested by the children was estimated at 121 mg/d (range 25 to 1,324) (aluminum tracer); 184 mg/d (range 31 to 799) (silicon tracer); and 1,830 mg/d (range 4 to 17,000) (titanium). The

overall soil ingestion estimate based on the minimum of the three individual element ingestion estimates for each child was 108 mg/d (range 4 to 708).

The authors were not able to explain the difference between the results for titanium and for the other two elements. The frequency distribution graph of soil ingestion estimates based on titanium shows that a group of 21 children had particularly high titanium values, > 1,000 mg/d; the remainder of the children showed titanium ingestion estimates at lower levels, with a distribution more comparable to that for the other elements. Clausing et al. (1987) conducted a soil ingestion study with Dutch children, using a tracer element methodology similar to that of Binder et al. (1986). Aluminum, titanium, and acid-insoluble residue (AIR) contents were determined for fecal samples from children, ages 2 to 4, attending a nursery school, and for samples of playground dirt at that school. Fecal samples were obtained daily over a 5-day period for the 18 children examined. Using the average soil concentrations present at the school, and assuming a standard fecal dry weight of 10 g/d, the authors calculated mass-balance estimates of soil ingestion for each material: aluminum, average 230 mg/d (range 21 to 878); AIR, average 127 mg/d (range 48 to 362); and titanium, average 1,080 mg/d (range 53 to 9,588). As in the Binder et al. study, a fraction of the children (5/19) showed titanium values of well above 1,000 mg/d, with most of the remaining children showing substantially lower values. Based on the minimum of the three chemical measurements for each child, an estimate of 100 mg/d, with a range of 21 to 362, was obtained.

In a second sample, Clausing et al. (1987) collected fecal samples for six hospitalized, bedridden children. A mass-balance calculation for these children, who presumably had very limited access to soil, yielded estimates of 46 mg/d based on aluminum. For titanium, two of the children had estimates well in excess of 1,000 mg/d, with the remaining four children in the range of 23 to 48 mg/d. The data on hospitalized children suggest a major nonsoil source of titanium for some children and may suggest a background nonsoil source of aluminum. However, conditions specific to hospitalization, e.g., medications, need to be considered. AIR measurements were not reported for the hospitalized children. Speculation on the source of titanium includes diet, the white coloring in (disposable) diapers, and several other items, but this has not as yet been resolved.

(2) <u>Evaluation</u>. The data from the tracer element studies, Binder et al. (1986) and Clausing et al. (1987), provide support for a preliminary estimate of average soil ingestion by children on the order of 100 to 200 mg/d, consistent with the "low" estimate used by USEPA (1984a). These estimates are based on findings with silicon or AIR and aluminum. Estimates based on a titanium tracer are higher by a factor of 5 to 10. This discrepancy has not been explained, but may be due to dietary and other sources. Estimates based on the minimum quantity ingested as calculated using the three tracers are not utilized here, because use of a minimum will tend to bias the estimated ingestion downwards. Hawley (1985), who estimated quantities of soil likely to be present on skin and subsequently ingested, also arrived at an estimate in the above range. It should be noted that Hawley's approach would not address children who deliberately ingest dirt or mouth soiled objects.

Binder et al. (1986) and Clausing et al. (1987) also provided some limited information on the upper range of soil ingestion in children. With the exception of the titanium data, the two studies provide evidence of an upper range of soil ingestion in children on the order of 1,000 mg/d or more. It should be noted that both studies had limited sample sizes and that neither specifically included (or excluded) children with pica (the tendency to eat nonfood materials). Again, estimates based on titanium would be substantially higher, on the order of 20 g/d.

Based on this review of the limited data now available, particularly the studies of Binder et al. (1986) and Clausing et al. (1987), the following values for soil ingestion are suggested. Average soil ingestion in the population of young children (under the age of 7) is estimated at approximately 0.1 to 0.2 g/d. For calculation purposes, an estimate of 0.2 g/d is suggested as an average value. An upper-range ingestion estimate among children with a higher tendency to ingest soil materials is 1 g/d. These estimates are based on data using silicon and aluminum as trace elements. The reason for the higher estimates for titanium is not understood, but the increase seems likely to be due to nonsoil factors.

6.4.5 <u>Ingestion--Beef and Dairy Products</u>

Lengthy accounts of the factors involved in calculating human exposure to 2,3,7,8-TCDD from beef and dairy products are available (USEPA, 1984a, 1985c; Fries, 1985, 1986). USEPA (1984a) and Fries (1985, 1986) cited studies of soil consumption by cattle and studies examining ratios of con-taminant levels in the diet to resulting levels in body fat and milk fat for chemicals similar to 2,3,7,8-TCDD, such as PCBs, PBBs, and DDT (under various production scenarios). Exposure duration effects are also discussed in USEPA (1984a).

The potential effects of "market dilution" of beef and dairy products on human exposure are discussed briefly in USEPA (1984a), at more length by Fries (1986), and at much greater length in USEPA (1985c) for the particular case of cattle production in Missouri. Aspects of the beef industry in this region specifically noted in USEPA (1985c) as important to estimating exposure were type of activity within the industry (e.g., cow-calf production, "backgrounding" (preparing calves for feed lots), feeding (for slaughter)), replacement rates as a function of activity, fractions of cattle fed to maturity outside contaminated areas before slaughter, and slaughter categories and rates relative to national figures. USEPA (1984a, 1985c) concluded that dilution will vary widely among different marketing areas. USEPA (1984a, 1985c) and Fries (1986) noted that the subpopulations most likely to receive high exposures are beef producers, dairy farmers, and their direct consumers and, further, that exposure evaluations should be very location specific.

Average consumption rates and fat content data for beef and dairy products are presented in Table 2, which has been adapted from USEPA (1984a) by addition of information from USEPA (1984b,c,d) and Fries (1986). Much greater "resolution" actually is available in USEPA (1984b,d) than is found in Table 2, since both studies (USEPA 1984b,d) are based on a U.S. Department of Agriculture (USDA) Nationwide Food Consumption Survey (NFCS) conducted in 1977-1978. The NFCS covered intake of 3,735 possible food items by 30,770 individuals characterized by age, sex, geographic location, and season of the year. A further description of the survey design is given in USEPA (1984d).

The average beef fat consumption noted in Table 2 ranges from 14.9 to 26.0 g per 70-kg person/day, with a single high consumption estimate of 30.6 g per 70-kg person/day that might be more appropriate for families of beef producers who home slaughter. Milk fat consumption from all dairy products ranges from 18.8 to 43 g per 70-kg person/day, with the lower end of this range appearing best supported at present. Considering fresh milk only, milk fat consumption is reported to average 8.9 to 10.7 g per 70-kg person/day, with a single high consumption estimate of 35 g per 70-kg person/day perhaps appropriate for dairy farm families. (Age range-specific information is available in both USEPA (1984b) and USEPA (1984c).)

160

Table 2. Rates of Ingestion of Beef and Dairy Products

Total consumption rate/std. error (g/70 kg person-d)	Percentage of fat	Fat consumption rate/std. error (g/70 kg person-d)[e]	Reference
Beef			
124	15	19	USEPA (1981h)
110.7/1.7	23	26.0/0.3	USEPA (1984b)
87.6/1.1[a]	(23)	(20.1/0.3)[b]	USEPA (1984c)
96.3	(23)	(22.1)	Berglund (1984)
66.8 (average)	22	14.9	Fries (1986)[c]
137.1 (high)	22	30.6	Fries (1986)
Dairy products			
550	7.8	43	USEPA (1981b)
308.6/5.3	(7.8)	(24.1/0.4)[d]	USEPA (1984c)
431.6/5.6	(4.4 implied)	18.8	USEPA (1984b)
Fresh milk (only)			
253.5/4.9	(3.5)	(8.9)	USEPA (1984c)
305 (average)	3.5	10.7	Fries (1986)
1000 (high)	3.5	35.0	Fries (1986)

[a] The categories established in USEPA (1984c) exclude beef in meat mixtures (e.g., meat loaf), meat by-products (e.g., wieners), and organ meats. The basic data set underlying both USEPA (1984b) and USEPA (1984c) was the USDA National Food Consumption Survey 1977-1978. The basis for the difference in total dairy products consumption rates noted for USEPA (1984b) and USEPA (1984c) has not yet been resolved.
[b] Beef fat consumption rates in parentheses are calculated using percentages of fat derived from USEPA (1984b).
[c] This and succeeding values from Fries (1986) were reportedly derived from Breidenstein (1984).
[d] Dairy fat consumption rates in parentheses are calculated using percentages of fat derived from USEPA (1981b).

Differences in beef and dairy fat consumption rates cited above from those used by USEPA (1984a), and revised bioconcentration factors, will result in a reduction in estimated exposure.

6.4.6 Ingestion--Fish Consumption Data

(1) Available studies. A variety of definitions have been used for fish consumption. Some studies examine only commercially caught fish, while others do not distinguish between marine and freshwater fish. Still others do not differentiate between fin and shellfish or fresh versus processed fish. Some data have been published that provide only nation-wide averages, while others provide data for regions or states. Conse-quently, drawing meaningful comparisons between figures derived from dif-ferent sources often is difficult.

Several recent studies of fish consumption by the U.S. population are summarized below. These studies for the most part estimate consumption of certain population subgroups and thus do not indicate an unequivocal need for changes in the average fish consumption estimates presented by the Ambient Water Quality Criteria for 2,3,7,8-TCDD (USEPA, 1984a), where an average daily consumption of 6.5 g/d per capita of freshwater and estuarine fish and shellfish was derived from analysis of a survey of fish and shellfish consumption in the United States (USEPA, 1980). The variety of results do emphasize the need to base consumption assumed in a particu-lar exposure assessment on studies involving similar populations.

Data from the Nationwide Food Consumption Survey conducted by the USDA in 1977-1978 were not available when the USEPA derived the 6.5 g/d figure for consumption of freshwater and estuarine fish (Stephan, 1980). The USDA survey obtained information on both household and individual intake of food products. Interviews were conducted to determine food consumption in households during the previous week and included a 1-day recall plus a 2-day diary of individual food intake. A national probability sample of households in the continental United States was obtained by means of approximately 15,000 interviews covering over 36,000 individuals. Supple-mental surveys of households with elderly and low-income individuals were conducted. Separate data were gathered for Alaska, Hawaii, and Puerto Rico.

Specific information was gathered, including use of specific fish species, as well as the state of processing (e.g., fresh, frozen, etc.). Analysis of the data indicated an average individual intake as 12 g/d fish and shellfish (edible weight) on a per capita basis nationwide, although geographic regional averages ranged from 9 to 14 g/d, with highest consumption in the Northeast (U.S. Department of Agriculture, 1985). Total population figures, including nonconsumers, were used in computing these averages. This survey also presents fish consumption by age group and season of the year. Other USDA publications have provided average figures for fresh commercial fish; in 1983, for example, an average of 6.4 g/d was estimated to have been consumed per person.

The most recent fish consumption data from the National Marine Fisheries Service (NMFS) report total per capita fish and shellfish consumption at 6.2 kg/yr (16.9 g/d) (USDOC, 1985). This estimate is based on the commercially landed fish and shellfish catch only.

An earlier report in the series (USDOC, 1983) gave consumption of edible weights of fresh and processed commercial marine fish and shellfish as 9.9 g/d per capita, based on yearly catches, imports, exports, and existing inventories. The recreational catch has been estimated to contribute an additional 3.7 to 5 g/d, based on information from the National Oceanographic and Atmospheric Administration (as cited in USEPA, 1986b).

The U.S. Department of Commerce (1985) also reports that 3.7 to 5.3 g/d (edible weight) marine fish and shellfish are consumed by recreational fishermen, fairly close to an SRI analysis showing consumption of 5.3 g/d of fish from recreational sources. Cordle et al. (1982) reported a 90th percentile consumption of 15.7 g/d for Great Lakes region consumers only and a 99th percentile figure of 36.8 g/d. No average figures were presented.

A National Seafood Consumption Survey was conducted by the NMFS in 1981 with a panel of 7,500 households (NMFS, undated). The households kept diaries on the amount of fish and other seafood purchased for household use, as well as the amount actually eaten both at home and away from home. Purchase data are broken out by species, nature of product (e.g., fresh, frozen, etc.), region, and a variety of demographic factors. The Longwoods Research Group (1984) analyzed some of the 1981 NMFS data based upon frequency of use rather than quantities consumed. This analysis revealed that 82 percent of all projected U.S. households eat seafood or fish.

Still earlier (1969-1970), a Market Facts Survey conducted for NMFS revealed a per capita total fish consumption figure of 16.8 g/d, of which 6.1 g/d consisted of fresh and frozen fin fish. This survey did not discuss explicitly whether portions were based on edible weight or if recreational sources were considered.

A National Purchase Diary (NPD) Fish Consumption Survey was performed for the Tuna Research Institute in 1973-1974, with the results based only on actual consumers of fish rather than total population (USEPA, 1980b). The questionnaire was administered to a total of 7,662 families (around 25,000 people) over 1 year, with 1/12 of the families responding in a given month to eliminate seasonal effects. Cordle et al. (1982) later used data from this survey as the basis for their estimate of consumption totaling 18.7 g/d. According to Conner (1984), the NPD Survey data show that 6.5 g/d of estuarine fish and shellfish and 2 g/d of freshwater fish are consumed. SRI International later reexamined the data tapes of the NPD Survey and found numerous discrepancies (USEPA, 1980b). A corrected version of the data base resulted in an average consumption figure of 14.3 g/d total fish, with a 95th percentile value of 41.7 g/d. SRI International also presents average and 95th percentile figures for each sex and different age groups (USEPA, 1980b).

Race and religion, as well as regional factors and age, may have strong impacts on fish consumption rates. The Market Facts Survey reported seafood consumption by U.S. blacks and people of Jewish faith to be approximately twice that of whites as a whole (USEPA, 1980b). However, the NPD Research Survey for the Tuna Research Institute reported only a 13 percent higher consumption rate among blacks (USEPA, 1980b). The NPD survey also found that Orientals consumed fish at a rate 47 percent above Caucasians. The 95th percentiles of fish and shellfish consumption typically were a factor of three above averages for the different population groups.

According to the USDA publication Foods Commonly Eaten by Individuals: Amounts Per Day and Per Eating Occasion, consumers of fin fish other than canned, dried, or raw, consumed an average (mean) of 54 g/d/person. It is not apparent what percentage of fin fish is recreationally caught. The following percentile distribution was given: 50th percentile - 38 g/d; 90th percentile - 96 g/d; 95th percentile - 132 g/d; 99th percentile - 221 g/d.

Puffer et al. (1983) reported the results of a survey of fishing habits and fish consumption rates among fishermen in the Los Angeles area. Interviewers obtained information from fishermen at 12 representative locations identified as frequently fished. A total of 1,059 interviews were conducted from an estimated sport fishing population of at least

31,000. Approximately one-half of the fishermen fished one or more times per week, with 14 percent of those interviewed fishing three to seven times per week. The majority of fishermen interviewed consumed the fish they caught. The median amount of fish consumed by the fishermen themselves was reported to be 37 g/d, with the 90th percentile at 225 g/d. (The report estimated that at least 100,000 family members of fishermen shared fish they caught.) These consumption rates are substantially above those for the general population. These data do not take into account consumption of fish purchased from stores.

Individuals in other areas are known to have a high intake of sport fish. A study by the Michigan Department of Public Health (Humphrey et al., 1976) examined the health status of individuals who consumed at least 30 g/d (annual average) of Great Lakes fish. The highest recorded fish consumption over the 2-year study period was 224 g/d. The New York State Department of Environmental Conservation (NYSDEC) uses a figure of 32.4 g/d in their health advisories, as the average fish consumption for recreational fishermen, based upon the 90th percentile of nationwide fish consumption figures (A. Newell, personal communication). A survey of users of the 1983 Guide to Eating Ontario Sport Fish (Ontario Ministry of the Environment, 1984) revealed that Ontario sports fishermen eat locally caught fish approximately once every 3 weeks, with an average meal size of 289 g (10.2 oz), corresponding to an average daily figure of 13.8 g/d. A substantial percentage of respondents (26 percent) ate at least a pound of fish per meal. However, as this survey is based on voluntary responses to a questionnaire, it may be subject to self-selection biases.

(2) _Evaluation_. Substantial recent information on fish consumption rates has become available through the surveys conducted by USDA and NMFS. While these surveys do not indicate the need for major revision of previous fish consumption estimates, they can provide more recent information and will allow examination of the fish consumption habits of particular population subgroups. The data from these surveys would also allow recalculation of the USEPA's estimate of human consumption of freshwater and estuarine fish and shellfish. Because current information indicates that some population groups consume fish at rates much above the national average, this work could be of significant value in determining the risks from 2,3,7,8-TCDD contamination that may be encountered by specific population groups.

Based on this review, the 6.5 g/d average consumption rate for freshwater and estuarine fish and shellfish that has been used in previous

USEPA assessments is still appropriate. To account for individuals who habitually consume larger quantities of fish, a value of 30 g/d is suggested based on the Los Angeles and Great Lakes data.

USEPA (1987) references the following values of average consumption rate that may be assumed when site-specific data are unavailable:

(1) 6.5 g/d represents an estimate of average consumption of fish and shellfish from estuarine and fresh waters by the U.S. population;

(2) 20 g/d represents an estimate of the average consumption of fish and shellfish from marine, estuarine, and fresh waters by the U.S. population;

(3) 165 g/d represents average consumption of fish and shellfish from marine, estuarine, and fresh waters by the 99.9 th percentile of the U.S. population; and

(4) 180 g/d represents a "reasonable worst case" based on the assumption that some individuals would consume fish at a rate equal to the combined consumption of red meat, poultry, fish, and shellfish in the U.S.

6.5   Post-Exposure

Within the last few years, two important areas have come under increased study with regard to the fate of 2,3,7,8-TCDD once exposure has occurred; they are bioavailability and pharmacokinetics. Bioavailability, the first of these areas, refers to an organism's ability to remove 2,3,7,8-TCDD from an ingested or inhaled particle and then to absorb it. Strictly speaking, bioavailability is a property of both the material to which an organism is exposed and the organism's capabilities and pharmacokinetic responses. However, it is useful to assume for the present that the organism's extraction ability is constant and to look at bioavailability as a property of the material. Recent research has observed that 2,3,7,8-TCDD adsorbed on various substrates can differ in bioavailability by approximately an order of magnitude. The current state of knowledge about the causes of bioavailability differences is incomplete, but early hypotheses (based on a very small data set) hold that the bioavailability of 2,3,7,8-TCDD from various materials can be related to chemical availability measured by solvent extraction. Most contaminated soils tested so far (five) show bioavailability in animal tests of about 25 to 50 percent that of 2,3,7,8-TCDD in corn oil given by gavage. Three soil samples spiked with 2,3,7,8-TCDD had bioavailabilities in the 40 to 70 percent range compared with corn oil. Based on limited data, 2,3,7,8-TCDD in flyash proved roughly 25 percent as bioavailable as 2,3,7,8-TCDD from the solvent extract of the flyash. (It should be noted that in this experiment 2,3,7,8-TCDD from both flyash and solvent extract were recovered from the rat liver in low quantities, making interpretation of the experiment difficult.) Studies with soil from one site and with activated carbon with dioxin added, however, showed gut bioavailabilities of <10 percent and <1 percent compared with 2,3,7,8-TCDD in solvents, respectively.

The implications of this early work are important, since estimated risk following intake of an environmental matrix will be proportional to bioavailability. At this point, it is unclear what the long-term implications of bioavailability differences will be to risk assessment. Some soils have shown high 2,3,7,8-TCDD bioavailability, while bioavailability in one tested soil is lower. No data on the distribution of contaminated soils by bioavailability currently exist to allow this difference to be systematically considered, nor is there an accepted protocol for measuring bioavailability from soil on a site-by-site basis.

The second important post-exposure area of study is pharmacokinetics. Theoretically, if one knows what happens to the 2,3,7,8-TCDD once it is absorbed, one can look at body burdens and back-calculate an average dose or average exposure. In practical terms, this procedure can lead to reduced uncertainty in a risk assessment by allowing calculation of exposure from two independent methods. Currently, the ability to perform these pharmacokinetics calculations is in its early stages of development. There are significant difficulties in current approaches to using body burden data to back-calculate exposure, and at this point verification of certain not-easily-verified assumptions needs to be done before such calculations can become standard tools for exposure assessment. It is safe to say that, in the future, the role of pharmacokinetics in exposure assessment will increase.

6.5.1 Absorption from Environmental Matrices (Bioavailability)

(1) General considerations. Following ingestion of a material containing 2,3,7,8-TCDD or other toxic species, the toxic effect of the material is modified by the degree of absorption, principally in the small intestine. In several experimental studies, investigators administered 2,3,7,8-TCDD-containing environmental matrices to experimental animals and measured parameters relating to bioavailability. These studies included quantitation of 2,3,7,8-TCDD in liver and other tissues following treatment; comparison of toxicities of contaminated environmental materials with pure 2,3,7,8-TCDD; and examination of enzyme induction. The results

of these different approaches, their limitations, and needs for further research are discussed below.

(2) <u>Review of data on bioavailability</u>. Umbreit et al. (1985, 1986a,b) conducted experiments in guinea pigs, administering 2,3,7,8-TCDD in corn oil, 2,3,7,8-TCDD added to chemically decontaminated soil, or soil from two industrial sites in Newark, New Jersey (a manufacturing site and a salvage site), contaminated with CDDs. 2,3,7,8-TCDD was the principal lower chlorinated isomer (dioxin or furan) present in the soil from the manufacturing site (for which a chemical analysis was presented). Soil from the manufacturing site was found to have 1,500 to 2,500 ppb 2,3,7,8-TCDD under soxhlet extraction; release under ambient temperature manual solvent extraction was much lower, reported as ">2.5 ppb." The soil from the salvage site was reported as approximately 180 ppb 2,3,7,8-TCDD under soxhlet extraction.

In this study, groups of two or four male and two or four female guinea pigs received single gavage doses of the test materials and were observed until death or sacrifice at 60 days. 2,3,7,8-TCDD in corn oil or in recontaminated soil (6 mg/kg in both) proved highly toxic, without similar toxicity being observed in animals treated with up to twice this dose of 2,3,7,8-TCDD in the soil from the manufacturing site. The limited data on 2,3,7,8-TCDD levels in the liver showed much higher levels following administration of recontaminated soil versus contaminated soil from the manufacturing site.

Umbreit et al. (1986a) thus demonstrated that gavaged 2,3,7,8-TCDD containing soil from the manufacturing site was substantially less toxic than equivalent doses of 2,3,7,8-TCDD in corn oil. However, quantitative comparison of the effective doses in this study is difficult. Approaches to a quantitative comparison are outlined below.

(1) Guinea pigs receiving 12 mg/kg 2,3,7,8-TCDD in contaminated soil experienced no deaths, while five out of eight guinea pigs receiving 6 mg/kg 2,3,7,8-TCDD in corn oil died, with no groups tested having lower doses in corn oil. Other authors have provided data on the toxic effects of 2,3,7,8-TCDD in corn oil that could aid in the comparison. McConnell et al. (1984) observed one out of six animals dying at 1 mg/kg and six out of six animals dying at 3 mg/kg. Silkworth et al. (1982) observed three out of six animals dying at 2.5 mg/kg and no deaths out of six at 0.5 mg/kg. Comparing these data directly with the Umbreit et al. 1986b results would suggest that the 2,3,7,8-TCDD in the Newark manufacturing site soil was less effective, by a factor of 10 or greater, in producing toxicity than 2,3,7,8-TCDD in corn oil.

(2) Umbreit et al. 1986b reported a "slightly reduced" weight gain in guinea pigs receiving 6 mg/kg of 2,3,7,8-TCDD in Newark manufacturing site soil, and a "greater reduction" at the 12 mg/kg dose. No other signs of toxicity were noted in these groups. The animals receiving 6 mg/kg 2,3,7,8-TCDD in corn oil, in contrast, exhibited a marked loss of body weight and showed toxicity and mortality. Silkworth et al. (1982) also provided data on weights of guinea pigs receiving 2,3,7,8-TCDD in corn oil. Those receiving 2.5 mg/kg exhibited a marked reduction in weight gain among three out of six survivors, while those receiving 0.5 mg/kg showed a weight gain comparable to vehicle controls. Comparison of these weight data with those of Umbreit et al. 1986b suggests that the 2,3,7,8-TCDD in corn oil was more than 5 times but less than 25 times as potent as 2,3,7,8-TCDD in the Newark soil. This comparison assumes that the effect of the Newark manufacturing site soil on weight gain was due to 2,3,7,8-TCDD as opposed to other compounds in the soil. Numerous other dioxin and furan compounds and other chemicals have been identified in this soil (Umbreit et al., 1987a). It has not been established that 2,3,7,8-TCDD is the sole or prime source of toxicity in the soil.

(3) Umbreit et al. 1987a presented liver concentrations of 2,3,7,8-TCDD after death or sacrifice at 60 days following gavage (see USEPA, 1988). Much lower concentrations of 2,3,7,8-TCDD were found in the livers of animals receiving soil from the manufacturing site compared with those receiving the dose in corn oil. Two factors, however, limit the conclusions that can be drawn from this comparison. First, the corn oil group experienced major toxicity and weight loss, particularly complete loss of body fat. These changes may have affected the partitioning of 2,3,7,8-TCDD within the body, leading to a higher concentration in the livers of the animals experiencing toxicity. Second, the animals gavaged with corn oil died early--half were dead by 26 days, while all of the guinea pigs treated with soil survived to 60 days (with the exception of one gavage death). The USEPA (1985a) reported a half-life for 2,3,7,8-TCDD elimination of 30 ± 6 or 22 to 43 days from two studies in guinea pigs. Additionally, the USEPA (1985a) stated that elimination in the guinea pig may follow zero-order kinetics. Differences in elimination caused by differences in periods of survival are likely to have affected the relative quantities of 2,3,7,8-TCDD found in the livers of the test groups.

Perhaps a more appropriate comparison can be made with the four animals receiving 0.32 mg/kg of 2,3,7,8-TCDD in contaminated soil from the Newark salvage site. These animals experienced no reported toxic signs (weight data not presented) and survived the full 60-day experiment. Approximately 6 percent of the gavage dose was found in the liver of these animals, while only about 0.06 percent of the gavage dose was found in the livers of guinea pigs in the 12 mg/kg group receiving the Newark manufacturing site soil (USEPA, 1988). This would suggest that the 2,3,7,8-TCDD in the manufacturing site soil was 100 times less bioavailable. However, given the different doses used and the fact that only a single pooled liver tissue sample was analyzed for 2,3,7,8-TCDD in each group, caution must be used in interpreting this comparison.

McConnell et al. (1984) treated Hartley guinea pigs (2.5 weeks old) with single gavage doses of either 2,3,7,8-TCDD or dioxin-contaminated soil from two sites in Missouri. The 2,3,7,8-TCDD concentrations from the two sites were reported at 700 and 880 ppb, respectively; total tetra-chlorodibenzofurans (TCDF) concentrations in the soil were 40 to 80 ppb, and polychlorinated biphenyl (PCB) concentrations were 3 to 4 ppm. Taking into account the relative toxicities, the authors concluded that toxicity from the other compounds was likely to be small compared with that from 2,3,7,8-TCDD. The results of the study are shown in USEPA (1988). Livers were analyzed for 2,3,7,8-TCDD at death or sacrifice at 30 days following treatment. Treatment deaths occurred between 5 and 21 days post-gavage.

Guinea pigs that died exhibited severe loss of body fat, markedly reduced thymus and testicle size, and adrenal hemorrhage. No adverse effects were noted in animals treated with decontaminated soil. For 2,3,7,8-TCDD in corn oil and for both contaminated soils, there were clear dose-responses in mortality. The calculated LD_{50} values for the two soil types were lower than the LD_{50} for 2,3,7,8-TCDD in corn oil by a factor of three to four.

There was a dose-response between the liver concentration of 2,3,7,8-TCDD and the gavage dose; the details of this relationship are complex. Animals dying during the experiment had liver concentrations a factor of 1.4 to 3.2 higher than animals in the same dose groups that survived for 30 days. This observation makes quantitation of the dose-response rela-tionships difficult (all or most of the animals in the low-dose groups survived the experiment, while all of the animals in the high-dose groups died). The authors note that the differences in liver concentrations observed in the study may reflect varying partitioning of the 2,3,7,8-TCDD among internal organs, since dying animals suffered a major loss of body weight and fat content. In addition, surviving animals would have had a greater opportunity to metabolize and excrete 2,3,7,8-TCDD because of a longer lifetime.

Therefore, liver concentrations of animals in the different dosing groups can best be compared among groups that experienced similar mortality.

(1) Animals in dose groups in which all animals died within 30 days: for 2,3,7,8-TCDD in corn oil (group 3), approximately 20 percent of the administered dose was in the liver (assuming liver makes up 4.7 percent of body weight). For the soil-treated groups (groups 6 and 9), 13 percent and 11 percent of the doses, respec-tively, were in the liver. Comparison of these data suggests

that 2,3,7,8-TCDD was approximately twice as available through corn oil as through soil.

(2) Animals surviving the 30-day experiment (in groups where at least 4 out of 6 survived): For 2,3,7,8-TCDD in corn oil (group 2), 7.5 percent of the administered dose was in the liver. For soil-treated animals (groups 4, 5, 7, and 8) < 3.6, 1.3, < 4.2, and 2.0 percent of the doses, respectively, were in the liver. Comparison here would suggest that 2,3,7,8-TCDD was approximately four times as available through corn oil as through soil.

Umbreit et al. (1986a) reported additional chemical analyses of the Times Beach soil. Soxhlet extraction of the Times Beach soil yielded a similar quantity of 2,3,7,8-TCDD to the solvent extraction reported by McConnell et al. (1984). This is in contrast to the Newark manufacturing site soil used in the Umbreit et al. (1987a) experiments, where only a small fraction of soxhlet-extractable 2,3,7,8-TCDD was extractable by the solvent extraction methodology used by McConnell et al. (1984).

McConnell et al. (1984) also reported an experiment in which groups of six Sprague-Dawley rats were given single gavage doses of 2,3,7,8-TCDD in corn oil or dioxin-contaminated soil from the Minker site. Induction of aryl hydrocarbon hydroxylase (AHH) in the rat livers was measured at sacrifice 6 days after dosing. Experimental doses ranged from 0.4 to 5.0 mg/kg 2,3,7,8-TCDD. For the five dose groups, the AHH activity for the soil group ranged from 50 percent to 110 percent of the activity in the corn oil group. Therefore, the McConnell et al. 1984 rat data indicate that the bioavailability of 2,3,7,8-TCDD from the Minker site soil was at least 50 percent of that of equivalent doses of 2,3,7,8-TCDD in corn oil.

Lucier et al. (1986) provided additional information on the induction of hepatic enzymes in rats by the 2,3,7,8-TCDD-contaminated soil from the Minker site tested by McConnell et al. (1984). AHH induction was similar for the groups of rats receiving 2,3,7,8-TCDD in corn oil and contaminated soil (within a factor of two) over a broader range of doses (0.015 mg/kg to 5 mg/kg) than reported by McConnell et al. (1984). In a second enzyme assay using the same animals, UDP glucuronyltransferase activity was found to be slightly higher in groups receiving 2,3,7,8-TCDD in corn oil than groups receiving equal doses in contaminated soil.

Liver concentrations of 2,3,7,8-TCDD for the rats were also reported by Lucier et al. (1986). For the corn oil vehicle, the liver concentrations were 40.8 $\pm$ 6.5 ppb at the 5 mg/kg dose and 7.6 $\pm$ 2.5 ppb at the 1 mg/kg dose. Assuming that the liver comprises 4.0 percent of body weight, the retention rates for the 5 and 1 mg/kg doses were 33 percent

and 30 percent, respectively. In rats receiving 2,3,7,8-TCDD in contaminated soil, the 5.5 mg/kg group had liver concentrations of 20.3 $\pm$
12.9 ppb, and the 1.1 mg/kg group had concentrations of 1.8 $\pm$ 0.3. Thus,
retention rates for the 5.5 and 1.1 mg/kg groups are estimated at 14 percent and 7 percent, respectively. These data indicate that liver
retention in the soil group was 20 to 40 percent of that in the corn oil
vehicle groups.

Umbreit et al. (1986b) report additional studies of mortality in
guinea pigs treated with soil containing 2,3,7,8-TCDD from Newark (manufacturing site) and Missouri (Times Beach) previously tested by Umbreit et
al. (1985, 1986a) and McConnell (1984), respectively. Guinea pigs
received a single gavage dose of a soil suspension and were observed for
60 days. After autopsy, deaths were classified according to whether or
not they appeared to be due to TCDD toxicity. Substantial mortality
(25 percent overall) from conditions not attributed to TCDD was observed
across all groups.

The data for both the Newark and Missouri sites are similar in trend
for the previous data on these sites, and clearly indicate the greater
toxicity of the Newark soil for given equal administered doses of 2,3,7,8-
TCDD. A toxicity-related death was observed in both the 5 and 10 mg/kg
dose groups of 14 and 10 animals, respectively (excluding non-TCDD-related
deaths), for Newark soil, while no deaths were observed in corresponding
dose groups (7 animals at both 6 and 12 ug/kg) in Umbreit et al. (1986a).

Comparing groups within this study, similar mortality (one or two
deaths in 10 to 16 animals) was seen in both the 5 and 10 mg/kg Newark
groups and the 1 and 3 mg/kg Missouri groups. These results suggest that
the toxicity of these materials differs by an order of magnitude or less.
As noted above, the degree to which toxicity from these soils can be
attributed to 2,3,7,8-TCDD in the presence of numerous other related toxic
compounds is not known. 2,3,7,8-TCDD tissue concentrations were not reported in this work.

In another comparative study, Umbreit et al. (1987b) compared the
Newark manufacturing site and Times Beach soils in the induction of aryl
hydrocarbon hydroxylase (AHH) in rats. While the use of only single dose
levels prevents detailed analysis, the two soils proved quite similar in
their ability to induce AHH. The explanation for the difference in this
finding from those observed in the toxicity studies discussed above is

not clear, but may relate to the presence of other toxic and/or AHH-inducing compounds.

Umbreit et al. (1987a) report a reproductive toxicity study with soils from the Newark manufacturing site and salvage yard previously studied by Umbreit (1986a). Female mice were treated thrice weekly with soil from these sites, with treatment continuing through fertilization and weaning of pups. The total doses of 2,3,7,8-TCDD received by the mice were 720 mg/kg in manufacturing site soil and 86 mg/kg in salvage yard soil. A corn oil vehicle group and a contaminated soil group received a total of 225 mg/kg.

The results of this study demonstrate that acute and reproductive effects occurred in animals receiving manufacturing site soil. These effects were of a lesser magnitude than those seen in animals treated with 2,3,7,8-TCDD in corn oil at a dose threefold lower. The authors note the presence of substantial quantities of other toxic substances in the manufacturing site soil (chemical analyses presented). No toxic effects were noted in animals treated with salvage site soil that received a much smaller 2,3,7,8-TCDD dose. The data do not allow a more quantitative evaluation of the bioavailability of 2,3,7,8-TCDD.

Kaminski et al. (1985) and Silkworth et al. (1982) reported the results of a series of studies on the toxicity of soot containing dioxin and furan compounds from a fire involving transformer fluid containing PCBs. Hartley guinea pigs (500 to 600 g) received single oral doses of soot in an aqueous vehicle, a soxhlet extract of the soot in the same vehicle, or 2,3,7,8-TCDD in either an aqueous vehicle or corn oil.

The soot was reported to contain 2.8 to 2.9 ppm 2,3,7,8-TCDD and 124 to 273 ppm 2,3,7,8-TCDF. The total polychlorinated dibenzofuran content was estimated at 5,000 ppm. Animal weights and mortality were recorded for 42 days, at which point the survivors were sacrificed and LD_{50} values were calculated.

Silkworth et al. (1982) noted that the LD_{50}'s for contaminated soot and soot extract were similar at 410 and 327 equivalent mg/kg, indicating that the matrix had only a small effect on toxicity. If expressed in terms of the content of 2,3,7,8-TCDD, the LD_{50} from soot is 2.5 mg/kg, which is a factor of seven below the LD_{50} for 2,3,7,8-TCDD in an aqueous vehicle, suggesting that other compounds contributed to the toxicity of the soot and soot extract.

The data from these experiments also demonstrate that use of an oil vehicle leads to substantially greater 2,3,7,8-TCDD toxicity than does an aqueous vehicle. Therefore, it is likely that a larger difference in toxicity would have been observed if the soot extract were in an oil vehicle.

Van den Berg et al. (1983) fed small groups of male Wistar rats flyash from a municipal incinerator (pretreated with HCl) containing dioxins and furans, a soxhlet extract of the flyash, or a purified extract of the ash that was obtained using column chromatography. 2,3,7,8-TCDD was present as 3.3 percent of the TCDD isomer group in the flyash extract. (The authors did not specify whether this reference was to crude or purified extract.) 2,3,7,8-TCDF was present as 17.9 percent of the tetra-CDF isomer group in the extract. The rats were fed 2 g/d flyash mixed with diet or the residual from 2 mL/d extract after the extract was mixed with diet and the solvent was evaporated. The animals were exposed to the treated diet for 19 days and then sacrificed, and the liver tissue was analyzed for the presence of dioxins and furans. This yielded daily doses of 490 ug, 644 ug, and 950 ug of tetra-CDD congeners in the flyash, Krude extract, and purified extract groups respectively.

Approximately 1 percent of the 2,3,7,8-TCDD dose from flyash was retained in the liver, and approximately 4 percent of the dose of this isomer from flyash extract was so retained. The corresponding percentages for 2,3,7,8-TCDF are 0.3 and 1.0. The percentages of 2,3,7,8-TCDD recovered from liver tissue in this study are low for both the flyash and flyash extract groups in comparison with other studies in which 2,3,7,8-TCDD was administered to rats. For example, Fries and Marrow (1975) fed rats diets containing 7 or 20 ppb of 2,3,7,8-TCDD for a period of up to 42 days. After 14 days of feeding, the rat livers contained an average of 32 percent of the cumulative administered dose; at 28 days, 21 percent of the dose; and at 42 days, 18 percent of the dose. Data from Kociba et al. (1976), Rose et al. (1976), and Kociba et al. (1978) lead to conclusions similar to those from the Fries data regarding the fraction of cumulative 2,3,7,8-TCDD dose retained in the rat liver.

An explanation of the low level of recovery for the animals receiving the soxhlet extract of soot is not apparent. It is possible that the presence of multiple compounds affected absorption or metabolism in the rats fed soot and soot extract.

An approach to the van den Berg et al. (1983) data is to compare the ratios of liver concentrations for dioxins in flyash-treated animals to

174

the concentrations in extract-treated animals. These ratios, based on measurements in small numbers of animals, indicate a substantial bioavailability of dioxin and furan compounds from the tested flyash. This availability varied among the different isomers with the value of 0.3 for 2,3,7,8-TCDD, indicating that this isomer was three times as available from flyash extract as from flyash.

Van den Berg (1985) fed flyash (pretreated with HCl) to Wistar rats, guinea pigs, and Syrian golden hamsters. Flyash was mixed with standard laboratory diet at 2.5 percent by weight, and animals were allowed to eat ad libitum. The amount of flyash consumed by each group of five rodents was determined by the authors. For each species there were three groups of animals, each fed flyash for approximately 32 days (group I), 60 days (group II), or 94 days (group III). Concentrations of dioxin and furan isomer groups in the food were presented and include 1.4 ng/g TCDD compounds and 2.1 ng/g TCDF compounds.

The authors presented calculated recovery percentages for the cumulative dose of specific isomers in the rodent liver. For 2,3,7,8-TCDD in guinea pigs, 3.7 percent, 0.9 percent, and 1.4 percent of the administered dose were recovered in the liver in groups I, II, and III, respectively. The 32-day (group I) recovery percentage is comparable to that seen in the lower dose groups receiving 2,3,7,8-TCDD-contaminated soil in McConnell et al. (1984). The value in hamsters was approximately 2 percent (only reported for group II), and analytical problems prevented this determination in rats. No other TCDD compounds were quantitated. Similarly, for 2,3,7,8-TCDF, guinea pigs showed retention of 4.7 percent, 2.2 percent, and 2.5 percent of the administered dose in groups I, II, and III, respectively. For both 2,3,7,8-TCDD and 2,3,7,8-TCDF, the recovery percentages in guinea pigs at 32 days were approximately a factor of 4 to 15 higher than that observed in the van den Berg et al. (1983) study in rats.

As with other experiments in which the retention of dioxins in the liver has been determined, these percentages place a lower bound on the bioavailability of the dioxins, but, because not all dioxin is localized in the liver, do not permit bioavailability to be estimated without knowledge of the elimination of the administered dose over time and the quantity of dioxins in the remainder of the organism. No positive control group receiving 2,3,7,8-TCDD was included for comparison.

Poiger and Schlatter (1980) conducted several experiments in Sprague-Dawley rats (180 to 220 g) in which liver concentrations of tritium label from 2,3,7,8-TCDD were determined using various doses and vehicles. All experiments consisted of a single gastric intubation of 2,3,7,8-TCDD-containing material, followed by animal sacrifice at predetermined times. The doses used were well below the LD_{50} in the rat (the maximum dose applied was 5 mg/kg), and no deaths or toxic effects were reported.

A preliminary experiment in rats treated with 14.7 ng/rat 2,3,7,8-TCDD in ethanol indicated substantial localization of 2,3,7,8-TCDD in the rat liver, with a decrease of a factor of two in the fraction of the dose in the liver between 1 and 4 days. Poiger and Schlatter (1980) conducted all further studies with sacrifice at 24 hours to maximize the recovery of 2,3,7,8-TCDD from the liver.

In a further experiment, 2,3,7,8-TCDD was administered at low dose in a series of vehicles. These results are shown in Table 3. These data demonstrate that administration of 2,3,7,8-TCDD in soil reduced the retention of the dose in the liver to 66 percent, or 44 percent of the retention seen with 2,3,7,8-TCDD in ethanol. The lower value, 44 percent, was obtained for soil that was aged for 8 days at 30 to 40°C following addition of 2,3,7,8-TCDD. The aqueous suspension of 2,3,7,8-TCDD in activated carbon showed little evidence of bioavailability. It is relevant that the authors' measurements showed that 2,3,7,8-TCDD was only slightly extractable from the activated carbon matrix by various solvents. In contrast, 58 to 70 percent of 2,3,7,8-TCDD could be recovered from soil samples by washing with hexane/acetone (4:1 v/v).

Table 3. Percentage of Tritiated 2,3,7,8-TCDD Dose in the Liver 24 Hours after Oral Administration of 0.5 ml of Various Media

Formulation	TCDD dose (ng)	# Rats	% Dose in liver
50% ethanol	14.7	7	36.7 ± 1.2
Aqueous suspension of soil after TCDD contact for:			
8-15 hr (room temp.)	12.7, 22.4	17	24.1 ± 4.8
8 d (30°C - 40°C)	21.2, 22.7	10	16 ± 2.2
Aqueous suspension of activated carbon	14.7	6	$< or = 0.07$

Source: Poiger and Schlater (1980).

Poiger and Schlatter (1980) also presented results from several skin application experiments with TCDD-containing materials using rats and rabbits (not reviewed here).

Bonaccorsi et al. (1984) reported the results of a study of gut absorption of 2,3,7,8-TCDD from soil taken from the Seveso, Italy, accident site. Soil containing 81 $\pm$ 8 ppb 2,3,7,8-TCDD from the "highly contaminated" area in Seveso was administered to albino male rabbits (2.6 $\pm$ 0.3 kg) in daily gavage doses for 7 days. Additional samples of clean soil were spiked with 2,3,7,8-TCDD in the laboratory to yield 10 and 40 ppb contamination levels and were administered to rabbits following the same protocol. For comparison, rabbits were also treated with 2,3,7,8-TCDD in solution in acetone-vegetable oil (1:6) or alcohol-water (1:1). Rabbits were sacrificed on the day after treatment stopped and liver concentrations of 2,3,7,8-TCDD were measured. The authors did not remark on the presence or absence of toxicity in the treated rabbits. USEPA (1985e) reports values for the single dose LD_{50} of 2,3,7,8-TCDD in rabbits of 115 and 275 mg/kg. The total doses received by the rabbits in this study were approximately 54, 107, and 215 mg/kg over 7 days. Based on this comparison, there is a likelihood that toxic effects occurred in the Bonaccorsi work, and, as noted above, toxicity has the potential to affect the tissue concentrations of 2,3,7,8-TCDD. For this reason, the most appropriate comparisons among these data are between groups showing similar liver concentrations of 2,3,7,8-TCDD, which may then be inferred to have experienced similar toxic effects.

Similar liver concentrations of 2,3,7,8-TCDD were seen in the 40 mg/d solvent vehicle and 80 mg/d Seveso soil groups. Comparing the percentage of liver retention in these two groups indicates absorption from Seveso soil was 40 percent of that from the solvent vehicle. Using the same approach, comparison of the 80 mg/d solvent vehicle and 160 mg/d Seveso soil groups indicates that absorption from the soil was 41 percent of that from the solvent.

The same approach can be used to compare absorption from the solvent vehicle and from the spiked soil. In this case, the 40 mg/d solvent vehicle group had the liver concentrations closest to either the 40 or 80 mg/d spiked soil groups. Comparison of the percentage of dose in the liver indicates absorption from spiked soil is 68 to 73 percent of that from the solvent vehicle. Bonaccorsi et al. (1984) conducted work with either aged or non-aged spiked soil, but do not present data to allow a comparison of these groups.

Shu et al.'s (1987, as cited by Leung and Paustenbach, 1987) study of 2,3,7,8-TCDD from the Missouri site tested by McConnell et al. (1984) also examined oral bioavailability. Their paper reports an oral bioavailability of approximately 43 percent in the rat dosed with environmentally contaminated soil from Times Beach, Missouri. This figure did not change significantly over a 500-fold dose range of 2 to 1,450 ng 2,3,7,8-TCDD per kg of body weight for soil contaminated with approximately 2, 30, or 60 ppb of 2,3,7,8-TCDD.

(3) Summary. Table 4 summarizes data that are pertinent to the bioavailability of 2,3,7,8-TCDD from environmental matrices. Studies of bioavailability, which examined soil samples, soot, and flyash, have utilized three methodologies: measuring acute toxicity, retention of 2,3,7,8-TCDD in the liver, and induction of hepatic enzymes.

Among the five samples of soil from contaminated sites that have been tested, three have shown substantial bioavailability, e.g., 25 to 50 percent, when compared with 2,3,7,8-TCDD in corn oil gavage. A fourth soil sample was compared with 2,3,7,8-TCDD administered in a solvent vehicle and fell in this range. The fifth soil, tested by Umbreit et al. (1986a,b; 1987a,b), showed bioavailability substantially less than the other soils tested. While difficult to gauge quantitatively, dioxin from this soil may be an order of magnitude less available than from the other soils.

Additionally, three samples of soil spiked with 2,3,7,8-TCDD have been tested for bioavailability, including one sample in which the 2,3,7,8-TCDD was incubated with soil at an elevated temperature. The 2,3,7,8-TCDD added to these soil samples proved to be highly available (e.g., 40 to 70 percent).

In one study, soot from a transformer fire containing dioxins and furans proved similarly toxic to a soxhlet extract of the soot in an aqueous vehicle. However, the soot extract may have proved more toxic if delivered in corn oil, as was 2,3,7,8-TCDD in the soil studies. The availability of 2,3,7,8-TCDD and other dioxins and furans from incinerator flyash has been addressed by van den Berg et al. (1985) in extended feeding studies. In these studies, liver retention of 2,3,7,8-TCDD from either flyash or flyash extract proved low, with availability from flyash being approximately 25 percent of that from the extract.

The individual studies reviewed have a variety of limitations, as dis-

Table 4. Summary of Data on the Bioavailability of 2,3,7,8-TCDD Following Ingestion of Environmental Materials

Author	Material	Species	Dosing	Observation
Umbreit et al. (1986a,b)	Soil Newark Manuf. Site	Guinea pig	Single gavage	2,3,7,8-TCDD in soil <10% as toxic as in corn oil), based on lethality and weight loss.
				2,3,7,8-TCDD in the manuf. site soil had retention in liver approx. 1% as great as with salvage site soil.
McConnell et al. (1984)	Soil Newark Salvage Site	Guinea pig	Single gavage	Liver retention similar to 2,3,7,8-TCDD in corn oil from lower dose data.
	Recontaminated soil	Guinea pig	Single gavage	Toxicity similar to equal dose of 2,3,7,8-TCDD in corn oil.
McConnell et al. (1984)	Soil Times Beach, MO	Guinea pig	Single gavage	LD$_{50}$ data indicate 2,3,7,8-TCDD in soil approx. 25% as toxic as in corn oil.
				Comparing animals dying early, liver retention of 2,3,7,8-TCDD in soil group approx. 50% of that in corn oil vehicle group.
				Comparing animals surviving experiment, liver retention of 2,3,7,8-TCDD in soil group approx. 20% of that in corn oil vehicle group.
McConnell et al. (1984)	Soil Minker Site, MO	Guinea pig	Single gavage	LD$_{50}$ data indicate soil approx. 30% as toxic as 2,3,7,8-TCDD in corn oil.
				Comparing animals dying early, liver retention approx. 50% of that in corn oil vehicle group.
				Comparing animals surviving experiment, liver retention approx. 25% of that of corn oil vehicle group.

179

Table 4. (continued)

Author	Material	Species	Dosing	Observation
McConnell et al. (1984) and Lucier et al. (1986)	Soil Minker Site, MO	Rat	Single gavage	Introduction of AHH and UDP glucuronyltransferase activity > 50% of that in groups receiving 2,3,7,8-TCDD in corn oil.
				Liver retention 20-40% of that in rats receiving an equal dose of 2,3,7,8-TCDD in corn oil.
Poiger and Schlatter (1980)	Soil with 2,3,7,8-TCDD added			Liver retention approx. 40-70% of that in ethanol vehicle groups.
Bonaccorsi et al. (1984)	Soil from Seveso Accident Site	Rabbit		2,3,7,8-TCDD 40% as bioavailable from soil as from solvent vehicle, based on liver concentrations.
	Soil with 2,3,7,8-TCDD added			2,3,7,8-TCDD 70% as bioavailable from spiked soil as from solvent, based on liver concentrations.
Kaminski et al. (1985) and Silkworth et al. (1982)	Soot from fire	Guinea pig	Single gavage	LD_{50} data indicate soot containing dioxins and furans approx. equal in toxicity to soxhlet extract of soot in aqueous vehicle.
van den Berg et al. (1983)	Incinerator fly ash	Rat	19-day feeding	Liver retention of 2,3,7,8-TCDD from ash and ash extract 1% and 4% respectively, indicating 2,3,7,8-TCDD approx. 25% as avail from ash as from extract. Both ash and extract retentions are low compared with other feeding and gavage studies.
van den Berg et al. (1985)	Incinerator fly ash	Guinea pig	Feeding	4%, 1%, and 1% retention of total dose in liver following feeding for 32, 60, and 94 days, respectively.
van den Berg et al. (1983)	Incinerator fly ash	Hamster	60-day feeding	2% of total dose retained in liver following feeding.
Poiger and Schlatter (1980)	2,3,7,8-TCDD on activated carbon	Rat	Single gavage	<0.1% retention in liver.

cussed in the preceding text. A notable limitation was that some experiments were conducted, using highly toxic doses of 2,3,7,8-TCDD, so that determination of bioavailability was complicated by wasting and early death of the test animals. It should also be noted that while the relative retention of 2,3,7,8-TCDD in the liver can serve as an appropriate indication of differences in bioavailability between samples, the percentage of dose found in the liver only places a lower bound on absorption. This is particularly relevant to experiments where animals have been maintained for many weeks after dosing and an undetermined quantity of 2,3,7,8-TCDD has been metabolized and/or excreted. Finally, toxicity data for mixtures for which both toxicity and bioavailability of individual compounds may vary are difficult to interpret quantitatively in terms of bioavailability.

As presented in USEPA (1985a), Rose et al. (1976) determined gut absorption of 2,3,7,8-TCDD in a 1:25 mixture of acetone to corn oil (by volume) in the rat. In both single dose and multiple dose experiments, measured absorption was approximately 85 percent. Assuming that absorption from pure corn oil is similar to that from this mixture, and assuming that absorption in other species for which data are not available is similar, the 85 percent factor can be applied to the data presented here to obtain an approximate range for typical 2,3,7,8-TCDD absorption from soil. Using this factor and the range of 25 percent to 50 percent for the typical relative bioavailability of 2,3,7,8-TCDD from soil compared with corn oil, gut absorption can be estimated at 20 to 40 percent of ingested 2,3,7,8-TCDD in soil.

Recognizing these limitations, the weight of evidence indicates that 2,3,7,8-TCDD is often highly available from environmental materials. In one tested soil sample, however, the compound was substantially less bioavailable. While the data are too sparse to allow a prediction as to whether a particular environmental sample will prove more or less bioavailable, one important suggestion has emerged. In the two samples that have proved least bioavailable (the Umbreit et al. (1986a) manufacturing site soil sample and 2,3,7,8-TCDD on activated carbon tested by Poiger and Schlatter (1980)), the 2,3,7,8-TCDD was largely resistant to solvent extraction. This was not the case for more bioavailable materials.

Further research, using short-term experiments in which animals are handled under identical conditions and are fed dioxins in different media, is needed for an improved comparison of absorption between different environmental samples. Acutely toxic doses should be avoided to ensure that tissue concentrations are directly interpretable. Experiments studying both tissue retention and enzyme induction should prove valuable for this research. Whole-body levels of 2,3,7,8-TCDD need to be related to liver concentrations, and the effects of metabolism must be addressed. The vehicle of administration has been shown to affect acute 2,3,7,8-TCDD toxicity, and vehicle effects should be considered in designing experiments.

6.5.2 Pharmacokinetics and Body Burden of Dioxins

The pharmacokinetic profiles of CDDs and other related compounds, such as the polychlorinated dibenzofurans, are quite complex. However, a thorough analysis and understanding of these pharmacokinetic data could prove very helpful in ensuring that exposure assessments for CDDs are reliable. In addition, such profiles may be useful in providing information to help understand and apply the data from animal studies to human exposure and risk assessments.

Wroblewski and Olson (1985) reported differing levels of responses to CDDs by various species. It may be that some of these differences can be explained by examining and quantifying the species differences in disposition and metabolism (Wroblewski and Olson, 1985; King et al., 1983). The varying responses to various isomers may also be explained by differences in the disposition and metabolism (Burant and Hsia, 1984). It may be well to note that the definition of "disposition" may have to be extended to include suborgan or even subcellular sites in order to more fully describe some of the noted differences. As will be discussed in greater detail later in this section, CDDs have been implicated to bind with very specific loci within cells. The structure and concentration of these intracellular receptors appear to be under genetic control and thus may exhibit considerable inter- and intra-species variation.

Pharmacokinetic analysis may also allow for predicting the time required for eliminating the body burden after exposure ceases. With sufficient data and proper understanding, these analyses can account for various exposure and physiologic conditions.

The redistribution of a CDD among the various tissues and organs, which may occur during elimination, can be accounted for and tracked. Effects on disposition, which may result from altered physiology such as from sudden weight loss or from lactation, can be incorporated and thus can be adequately considered in exposure and risk assessments. Lactation

(Astrila, et al., 1981) and pregnancy (Nau and Bass, 1981) are known to accelerate the removal of CDDs from the body. This increased elimination may be at the expense of accumulation by the embryo, fetus, or offspring.

One potentially powerful and practical application of pharmacokinetic analysis is to estimate exposure levels from body burden data. The goal here would be to use as much human data as are available in order to avoid the uncertainties that arise when making extrapolations from animal data.

(1) Physiologically based pharmacokinetic modeling. Physiologically based pharmacokinetic modeling is a method for describing and predicting disposition within the body. These models take into account physiologic and biochemical processes such as blood flows, metabolism, and renal clearance, and describe the body according to its normal anatomy. Physiologically based pharmacokinetic models can, given adequate data, predict disposition from one exposure scenario to another and even from species to species. One such model was developed for 2,3,7,8-TCDF (King et al., 1983) and is described here with some modification for the dioxins. First, the anatomic regions of the body are blood, liver, fat, skin, and muscle. The other organs are all lumped together as the "carcass." Input may be by a variety of routes, but for the purposes of this discussion, input is considered to occur through the gastrointestinal system by continuous or chronic dosing. See USEPA (1988) for the pertinent equations and derivations.

Some assumptions are made to simplify the model for use in estimating intake from fat (adipose) tissue concentrations determined in actual human samples: Assuming fat tissue to be in steady state, and the other various organs to be in steady state, then the equation for intake can be derived as:

$$I = Mk_a F_o = (C_{F,ss} k_m)/R_F \qquad \text{(Equation 16)}$$

where

$C_{F,ss}$ = steady-state fat concentration
k_m = rate constant for metabolic clearance
RF = partitioning ratio of TCDD between fat and blood.

Thus, it can readily be seen that three important factors that govern the bioaccumulation of these compounds are bioavailability, lipid-to-blood partitioning, and rate of elimination.

With the steady-state assumptions, Equation 16 may be used to estimate the fat concentration at steady state, given an average intake over time, for example, per day. Alternatively, given the concentration in the fat, an individual's daily intake may be estimated. In doing so, total body clearance is substituted for metabolic clearance. To use Equation 16 to estimate intake, several pieces of information are needed. Obviously, the concentration of the CDD in the fat is needed. These data are generally available for humans from various "adipose tissue" banks. For animals, these data may be determined when conducting experimental studies. Another parameter needed is the fat-blood partition ratio. This may be difficult to obtain for humans, but may be estimated from animal data. There is evidence that for similar compounds this parameter is nearly equal in all tested species (King et al., 1983). Total body clearance is the most difficult parameter to estimate for humans. One way to obtain this parameter would be to directly monitor humans for the disappearance of the chemical from the body after exposure was known to have stopped. From these data, the elimination rate constant can be determined and thus some estimate of clearance can be made. Elimination rate constants and clearance may be related according to:

$$C_L = (k_e \, V_d) \qquad \text{(Equation 17)}$$

where

C_L = clearance
k_e = elimination rate constant
V_d = volume of distribution.

For a one-compartment model,

$$k_e = \ln 2/t_{1/2} \qquad \text{(Equation 18)}$$

where

$t_{1/2}$ = half-life of the chemical in the body.

The volume of distribution may be estimated as in King et al. (1983) by summing the product of organ volume times tissue-blood partition ratio for each of the individual organs.

From these equations and laboratory data, some approximation of clearance in the human may be made, as follows:

- From animal experiments, determine elimination rate constant and partition ratios. If data in humans exist, the elimination rate constant need not be extrapolated from animals. The elimination constant used to calculate daily intake is taken from the published data.

- From known organ volumes and partition ratios, calculate the volume of distribution for 2,3,7,8-TCDD.

• From the volume of distribution and the elimination rate constant, determine the clearance.

Commoner et al. (1985) reported that a survey of the available literature revealed an average human half-life for 2,3,7,8-TCDD of 4.95 years; Poiger and Schlatter (1986) found it to be 5.8 years. In addition, extrapolation from clearance in nonhuman primates results in estimates of a human half-life of between 3 and 7 years. A survey of human adipose tissues from the human adipose tissue bank reveals a wide range of concentration of 2,3,7,8-TCDD. Data were gathered from a variety of geographical locations across the United States and categorized according to three age groups. As an example, the data used here were reported for persons over 45 years of age. The 45-year-old age group could be approximated to be in steady state if exposures over that period were fairly constant, regardless of living conditions and geographical moves during the span.

From certain equations it was determined that it would require approximately four half-lives, or 20 years, to reach 90 percent of the steady-state concentration, and seven half-lives, or 35 years, to reach 99 percent of the steady-state concentration.

(2) <u>Calculation of daily intake</u>. As discussed, a growing body of evidence may imply that there exists an overall body burden of CDDs, including 2,3,7,8-TCDD, in the general population of the United States. If true, a number of important issues arise. First, sources of such a widespread body burden need to be identified. Second, if such background exposures exist, then a calculation of the carcinogenic risks based on such background levels must be performed.

To review, some basic questions require resolution in order to determine daily intake and subsequent risk estimates from such background exposure. First, is it reasonable to assume, after examining the available data, that a body burden of 2,3,7,8-TCDD exists in the general population of the United States? Second, if such a body burden exists, what are the average daily intake levels that result in such background levels? The following sections address these questions, attempt to identify and estimate the uncertainties associated with such assumptions and calculations, and try to estimate ranges of risk from such putative intake levels.

(a) Data. Because of their high lipid solubilities, CDDs tend to preferentially partition into adipose tissues and reside there for long periods of time. Depending upon the exact elimination rates, they can then remain in human adipose tissue for well over 25 years. Thus, the presence of CDDs in adipose tissue is good evidence of previous exposure.

Further, if steady-state conditions are assumed, the adipose tissue levels may be used to calculate average daily intake, as described earlier in the report. As a result, most investigators have taken fat biopsies to determine the body burden of dioxin in humans.

(b)   Data use.   Review of the literature reveals many studies with results that show various levels of 2,3,7,8-TCDD in human adipose tissues. Several of the samples were taken from persons, such as Vietnam war veterans, with known exposures to substances containing 2,3,7,8-TCDD. Others were obtained from persons with no "known" specific exposure that might result from activities such as spraying of herbicides or pesticides, industrial sources, or living or working in areas where Agent Orange was sprayed. An important issue needing resolution regards the suitability of using such selected samples to generalize for the entire U.S. population. These "controls" were apparently not randomized and sample sizes were usually quite small. Also, some were based on biopsy, while others were samples taken at autopsy. Details and findings are discussed in the subsequent section.

(c)   Findings.   EPA staff and staff from contractors supporting EAG have reviewed several studies reporting the levels of 2,3,7,8-TCDD in the adipose tissues of individuals with a known or believed exposure. Thirteen persons in Sweden with a history of exposure to phenoxy herbicides were reported to have adipose tissue concentrations of 2 ppt (Rappe et al., 1986). Gross (1982) reported that three American Vietnam war veterans with known exposure to Agent Orange had an adipose tissue concentration of 99 ppt 2,3,7,8-TCDD.

In the same study discussed above, Rappe et al. (1981) reported that 18 nonexposed persons had a concentration of 3 ppt in their adipose tissue. The fact that both "exposed" and "nonexposed controls" showed essentially the same levels could indicate some discrepancy in the exposure history of either group. Also, the level of exposure in those individuals having an exposure history could have been low, as the exposure could have been sporadic and could have occurred long before the monitoring. Thus, the levels of 2,3,7,8-TCDD resulting from phenoxy herbicide exposure, in those persons, could have had little impact on their total concentration observed at the time of monitoring. In other words, the concentrations in these persons might have returned to background levels.

186

Ono (1986) reports 9 ppt as an average concentration in the adipose tissues of 13 Japanese people with no known exposure to 2,3,7,8-TCDD-containing substances. Persons in rural areas of Georgia and Utah with no known exposure had 7.1 ppt in their adipose tissues (Patterson et al., 1986). Graham et al. (1985) reported similar levels for samples taken at autopsy.

Results of the National Human Adipose Tissue Survey (NHATS) showed 6.2 + 3.3 ppt in the composited adipose tissue samples taken at random from throughout the United States at surgery or autopsy.

(d) Calculation, assumptions, uncertainties, and actual parameter values. The method chosen to calculate daily intake from adipose tissue concentrations is developed and described in USEPA (1988). Two major assumptions go into the simplification of the model for purposes of calculations from the available data. First, steady-state conditions are assumed. Given that the elimination of the compound from the body would probably require at least 5 years, it could take well over 15 years to reach such steady-state conditions. Thus, the assumption of steady state could be considered reasonable only if background environmental concentrations are relatively similar throughout the nation. Under those conditions, even if people move from one geographic area to another, exposure concentrations would be relatively constant. Also implicit in the scenario of steady state is an assumption that the bioavailability of 2,3,7,8-TCDD is virtually the same throughout the United States. However, even if environmental concentrations are the same for a person moving from one area of the country to another, the amount absorbed into the body could vary significantly. Hence, steady-state conditions may not be constantly maintained. The result of this is that the adipose tissue concentrations measured at any one time may not reflect actual steady-state concentrations. Alternatively, concentrations measured at one time, although at steady state, may not reflect concentrations of a steady state reached with a previous exposure to a form of 2,3,7,8-TCDD with different bioavailability. Errors resulting from this assumption could either overpredict or underpredict daily intake. Without knowing the bioavailability of the 2,3,7,8-TCDD at the various locations to which an individual has been exposed and the length of time that he or she may have been exposed at these locations, it is not possible to measure the amount of uncertainty associated with this assumption.

The second major assumption is that 2,3,7,8-TCDD is eliminated from the body by monophasic kinetics. Data gathered within only a few years of exposure might not reveal the existence of any second or even third elimination phases. Such an error in assumption could result in an underestimation of the half-life. As a consequence, the daily intake levels required to reach the steady-state levels in the adipose tissues would be overestimated. The impact of these miscalculations can be more easily minimized. Intake values can be calculated using a range of half-lives. Choosing a range with sufficiently long half-lives assures that possible slower phase elimination kinetic constants would also be included.

Also, the elimination kinetics are assumed to be constant over the entire life of the individual. This may not be accurate in many cases. Variation in renal function, metabolic capabilities caused by disease, and exposure to other compounds would alter the kinetics of elimination over the lifetime. For the purposes of these calculations, because of a lack of relevant data, it is assumed that the elimination rate constant is relatively stable with time. Again, by choosing half-lives far greater than those estimated in the literature, one is assured of including those individuals with reduced elimination rates.

(e) Parameters chosen. The total fat volume for a 70-kg adult person was assumed to range between 5 and 14 liters. A common assumption is 10 liters. Based on the discussion above, a concentration of 2,3,7,8-TCDD in adipose tissue at steady state of 6.72 ppt was chosen. Based on several reports (Commoner et al., 1985; Poiger and Schlatter, 1980) and extrapolation from elimination data in nonhuman primates, a half-life of approximately 5 to 8 years is assumed. Because of the possible underestimation of this value as discussed earlier, however, a wider range of 5 to 30 years was chosen. As noted previously, the value of the partition coefficient between adipose and lean body tissues was set at 100.

(f) Daily intakes calculated. The smallest fat volume (5 liters), an adipose tissue concentration of 6.72 ppt, and the longest half-life in the range (30 years) were used for calculating the lowest reasonable daily intake. The largest fat volume (10 liters), the same concentration in the adipose tissue (6.72 ppt), and the shortest half-life in the range (5 years) were used for calculating the highest reasonable daily intake. A fat volume of 7 liters, the same adipose tissue concentration of 6.72 ppt, and a half-life of 10 years were arbitrarily chosen as a reasonable expectation between extremes of the range. Table 5 shows the results.

Table 5.   Calculated Average Daily Intake

$t_{1/2}$ (yr)	Fat vol. (L)	Vol. of dist. (L x 100)	Conc. in fat (pg/gm)	Daily intake (pg/kg)
5	10	14	6.72	0.51
5	7	10	6.72	0.36
10	7	10	6.72	0.18
20	7	10	6.72	0.09
30	7	10	6.72	0.06
30	5	7	6.72	0.04

In summary, these calculations resulted in estimates of daily intake of 2,3,7,8-TCDD between 0.04 picogram per kilogram body weight per day (pg/kg) and 0.51 pg/kg. The "reasonable expectation" value is 0.18 pg/kg. This latter value is in reasonable agreement with those reported by others (Geyer et al., 1986; Graham et al., 1985). Also, the values calculated above are similar to those calculated using formulas discussed in a recent paper by Leung and Paustenbach (1987).

(g) Impact of daily background on risk. Calculations are then performed as follows to estimate the upper bound risk that could result from the background intake levels calculated in the preceding section. A potency based on animal studies of 1.5×10^{-4} is multiplied by the average daily intake (adjusted for fraction absorbed) to calculate the estimated upper limit of risk. Because the average daily intake was determined from steady-state conditions as described above, the exposure duration is assumed to be over the entire lifetime. For illustrative purposes, the highest calculated daily intake was used for estimation.

These upper limit risks are then compared with incidence of specific tumors and all tumors for the general population. Table 6 summarizes the results of these calculations and comparisons. As may be observed from examining Table 6 if 2,3,7,8-TCDD were assumed to cause only human soft tissue sarcomas, the background intake levels presently calculated would account for, at most, about 10 percent of the soft tissue sarcomas observed in the general population. Similarly, the background levels would account, at most, for 1 percent of all non-Hodgkins lymphomas and less than 0.1 percent of all cancers in the general population.

(h) Recommendations for future activities. TCDDs have interesting and important pharmacokinetic characteristics. There appear to be significant differences between species in several pharmacokinetic properties

Table 6. Risks Associated with Background Daily Intake of 2,3,7,8-TCDD Compared with Annual Cancer Incidence in U.S. Population

Upper limit of animal potency $(pg/kg-day)^{-1}$	Daily dose[a] (pg/kg)	Incremental cancer risk resulting from daily intake	Lifetime upper limit annual cancers resulting from daily intake[b]	Background probability cancer in U.S.	Annual background cancer[b]
1.5×10^{-4}	0.96	1.47×10^{-4}	504	1.9×10^{-3} (Soft Tissue Sarcoma)	6,500
1.5×10^{-4}	0.96	1.47×10^{-4}	504	9.2×10^{-3} (Non-Hodgkins Lymphoma)	32,000
1.5×10^{-4}	0.96	1.47×10^{-4}	504	(All Cancers)	965,000

[a] Daily intake (absorbed) converted to applied dose to be consistent with animal-derived potency, which is based on applied dose.

[b] Based on a U.S. population of 240,000,000.

and parameters. For example, rodents and primates show very different elimination kinetics. The impact of this difference and even the effect of the larger fat compartment of primates on risk estimates could be elucidated by the use of physiologically based pharmacokinetics. The role of the lymphatic system on absorption and transport of dioxin within the body remains an unexplained process. A well-formulated and validated physiologically based pharmacokinetic model will accurately assess the dose received by infants from lactating mothers with a body burden of TCDDs.

EPA's Exposure Assessment Group (EAG) is developing a physiologically based pharmacokinetic model for TCDDs that with properly gathered data for formulation and validation, will:

- More accurately account for simultaneous exposure by more than one route and not depend on the implicit assumption that absorption fractions are the same over all concentrations and times and for all species;

- Not be restricted to steady-state conditions;

- Account for elimination by other than monophasic kinetics and assess the potential impact of the change in elimination kinetics that may occur throughout life with changing exposure and physiologic conditions;

- Account for elimination by more than one simultaneous process, including lactational shedding;

- Realistically represent the role that absorption of TCDD by lacteals and its transport by lymphatics may have;

- Be extended to food-producing animals using proper data, and thus will be used to more fully evaluate human exposure by this route; and

- Help explain and account for obvious pharmacokinetic differences (e.g., metabolism, lipid sequestering, etc.) between species. This is especially important regarding the differences between rodents and humans.

With regard to risk assessments, there are several potential ways in which pharmacokinetics may be applied in exposure assessments. In a conventional risk assessment for carcinogens, some dose-response function is generated, and from that function human risk is calculated at various exposure concentrations. Usually such a process involves extrapolation from animal high-dose experiments to calculated risk for animals at low doses and then further extrapolation from animals to humans. First, the pharmacokinetic model enables the risk assessor to utilize some internal body concentration of the parent compound of the metabolite (depending upon the mechanism of action) as the dose for the dose-response curve. The dose-response function is then calculated from the model-generated target concentrations. To do this, something must be known about the mechanism of action. At the very least, it should be known whether the parent compound or the metabolite is the carcinogen. An example of such a process was performed for tetrachloroethylene (Chen and Blancato, 1987), where the physiological pharmacokinetic model described total metabolite formation. The amount of metabolite was then used as the dose in the two-stage cancer model to define the incremental cancer risk for mice and the compound's potency factor. The next step involves adjusting the parameters so that the pharmacokinetic model describes the target concentration of the carcinogenic species in human tissues after exposure. This calculated dose is then used with the potency factor to estimate the incremental risk for humans (Chen and Blancato, 1987). As more detail becomes known about the mechanism of action, the pharmacokinetic model is further refined to give detailed concentrations of the toxin at very specific target sites.

6.6   Conclusions

The authors are in consensus on the following conclusions. Where risk is discussed, it refers to the upper bound incremental lifetime cancer risk only.

(1) Recent literature is divided and seemingly contradictory on the issue of whether, and how much, 2,3,7,8-TCDD is taken up into plants from contaminated soil. The authors of this report conclude that there is evidence that 2,3,7,8-TCDD is taken up by plants growing in contaminated soils, but the amount taken up, or subsequent transport within the plant itself (say, to edible portions), is very uncertain. The worst-case calculations (using the highest plant-to-soil ratio from the literature) result in very high exposures, at least as high as other pathways. On the other hand, using other values from the literature would result in exposures of little concern.

(2) The properties of 2,3,7,8-TCDD make widespread ground-water contamination from landfills unlikely, provided uncontaminated water filters through the capped landfill. Preliminary calculations of the effects of codisposed solvents indicate a slight increase in solubility of 2,3,7,8-TCDD with one-phase solutions (i.e., saturated solutions or moderate mixtures of miscible solvents with water), but the effect of this solubility increase on mobility has not been fully investigated. For systems where two distinct liquid phases exist (water and a relatively nonpolar organic solvent), the authors believe much greater mobility of 2,3,7,8-TCDD is possible. For these cases, and in cases where physical transport of soil particles to ground water can occur, there may be an associated threat to ground water.

(3) The weight of evidence indicates that 2,3,7,8-TCDD is often bio-available from contaminated soils, although certain soils may bind 2,3,7,8-TCDD very tightly, decreasing the bioavailability by an order of magnitude or more. The reasons for this difference in bioavailability from one material to another are not well understood at this time. The data base upon which this conclusion is drawn is very slim. The implications of this conclusion are that after additional data are collected, sufficient to draw more firm conclusions, bioavailability may be an important factor in site-specific assessments.

(4) Pharmacokinetics have been considered in order to calculate, from body burden data, "background" daily intake levels in the U.S. population. While the data do not allow an estimation of an average body burden in the U.S. population from which to calculate an average daily intake, an upper limit of 6.72 ppt in the adipose tissue has been estimated. From this upper bound estimate of body burden, the upper bound daily intake ranges from 0.04 to 0.51 pg/kg. The upper limit of risk that results from such estimates is then compared with the annual cancer incidence in the U.S. Further aims, future research, and application goals regarding physiologically based pharmacokinetic models are discussed.

(5) The authors believe that a significant amount of uncertainty in the exposure assessment could be reduced by a focused, limited research program addressing the areas where critical information is needed.

REFERENCES

Astrila AV, Reggiani G, Sonvani TE, Raisaneu S, Wipf HK. 1981. Elimination of 2,3,7,8-TCDD in goats milk. Toxicol. Lett. 9(3):215.

Bartek MJ, LaBudde JA, Maibach HI. 1972. Skin permeability in vitro: comparison in rat, rabbit, pig and man. J. Invest. Dermatol. 58:114-123 (as cited in Hawley (1985)).

Berkow SG. 1924. A method of estimating the extensiveness of lesions (burns and scalds) based on surface area proportions. Arch. Surg. 8:138-148.

Binder S, Sokal D, Maughn D. 1986. The use of tracer elements in estimating the amount of soil ingested by young children. Arch. Environ. Health 41:341-345.

Bonaccorsi A, di Domenico A, Fanelli R, Merli F, Motta R, Vanzati R, Zapponi GA. 1984. The influence of soil particles adsorption on 2,3,7,8-tetra-chlorodibenzo-p-dioxin biological uptake in the rabbit. Arch. Toxicol. Suppl. 7:431-434.

Breidenstein BC. 1984. Contribution of red meat to the U.S. diet. National Livestock and Meat Board, Chicago, IL.

Bumpus JA, Tien M, Wright D, Aust SD. 1985. Oxidation of persistent environmental pollutants by a white rot fungus. Science 228:1434-1436.

Burant CF, Hsia MT. 1984. Excretion and distribution of two occupational toxicants, tetrachloroazobenzene and tetrachloroazoxybenzene in the rat. Toxicology 29:(3)243.

Burkhard LP, Kuehl DW. 1986. n-Octanol/water partition coefficient by reverse-phase liquid chromatography/mass spectrometry for eight tetrachlorinated planar molecules. Chemosphere 15(2):163-167.

Chen C, Blancato JN. 1987. Role of pharmacokinetic modeling in risk assessment: Perchloroethylene (PCE) as an example. Pharmacokinetics in Risk Assessment - Drinking and Health, National Academy Press, Washington, DC. p. 369-390.

Clausing P, Brunekreff B, Van Wijen JH. 1987. A method for estimating soil ingestion by children. Int. Arch. Occup. Environ. Health 59:73-82.

Cocucci S, DiGerolamo F, Verderio A, Cavallaro A, Colli G, Gorni A, Invernizzi G, Luciani L. 1979. Absorption and translocation of tetrachlorodibenzo-p-dioxin by plants from polluted soil. Experientia 35(4):482-484.

Commoner B, Webster T, Shapiro K. 1985. Environmental levels and health effects of PCDDs and PCDFs. Presented at the 5th International Symposium on Chlorinated Dioxins and Related Compounds, September, Bayreuth, FRG.

Commoner B, Webster T, Shapiro K. 1986. Environmental levels and health effects of chlorinated dioxins and furans. Presented at the AAAS Annual Meeting, May, Philadelphia, PA.

Connor MS. 1984. Comparison of the carcinogenic risks from fish vs. groundwater contamination by organic compounds. Environ. Technol. (18):628-631.

Cook PM. 1986. Memorandum dated September 2, 1986, to F.W. Kutz of the Office of Environmental Processes and Effects Research, Washington, DC, from Philip Cook. U.S. Environmental Protection Agency, Office of Environmental Processes and Effects Research, Duluth, MN.

Cook PM. 1987a. Memorandum titled 2,3,7,8-TCDD in aquatic environments, dated February 4, 1987, to J. Cummings of the Office of Solid Waste and Emergency Response. U.S. Environmental Protection Agency, from Philip Cook, Office of Environmental Processes and Effects Research, Duluth, MN.

Cook PM. 1987b. Memorandum with attachment titled Bioavailability of polychlorinated dibenzo-p-dioxins and dibenzofurans from contaminated Wisconsin River sediment to carp, to Jim Cummings of the Office of Solid Waste and Emergency Response, from Philip Cook, U.S. Environmental Protection Agency, Office of Environmental Processes and Effects Research, Duluth, MN.

Cordle F, Licke R, Springer J. 1982. Risk assessment in a federal regulatory agency: an assessment of risk associated with the human consumption of some species of fish contaminated with polychlorinated biphenyls (PCBs). Environ. Health Perspect. 45:171.

Crosby DG, Wong AA. 1977. Environmental degradation of 2,3,7,8-tetra-chlorodibenzo-p-dioxin (TCDD). Science 195:1337-1338.

Czuczwa JM, Hites RA. 1986. Airborne dioxins and dibenzofurans: sources and fates. Environ. Sci. Technol. 20(2):195-200.

Diem K, Letner C. 1973. Documenta Geigy (Scientific tables. Ciba-Geigy, Ltd., Basle) p. 528.

Eitzer BD, Hites RA. 1986. Concentrations of dioxins and dibenzofurans in the atmosphere. Int. J. Environ. Anal. Chem. 27:215-230.

Facchetti S, Balasso A, Fichtner C, Frare G, Leoni A, Mauri C, Vascvo M. 1986. Studies on the absorption of TCDD by some plant species. Chemosphere (15) 9-12, pp. 1387-1388.

Feldman RJ, Maibach HI. 1970. Absorption of some organic compounds through the skin in man. J. Invest. Dermatol. 54:399-404 (as cited in Hawley (1985)).

Freeman RA, Schroy JM. 1985. Environmental mobility of dioxins. In: Bahner RC, Hansen DJ, eds. Aquatic toxicology and hazard assessment: eighth symposium. Philadelphia, PA: American Society for Testing and Materials.

Freeman RA, Schroy JM. 1986. Modeling the transport of 2,3,7,8-TCDD and other low volatility chemicals in soils. Environ. Prog. 5(1):28-33.

Fries GF. 1982. Potential polychlorinated biphenyl residues in animal products from application of contaminated sewage sludge to agricultural land. J. Environ. Qual. 11:14-20.

Fries GF. 1985. Bioavailability of soil-borne polybrominated biphenyls ingested by farm animals. J. Toxicol. Environ. Health 16:565-579.

Fries GF. 1986. Assessment of potential residues in foods derived from animals exposed to TCDD-contaminated soil. Presented at 6th International Symposium on Chlorinated Dioxins and Related Compounds, September, Fukuoka, Japan.

Fries GF, Marrow GS. 1975. Retention and excretion of 2,3,7,8-tetra-chlorodibenzo-p-dioxin (TCDD) by rats. J. Agric. Food Chem. 23:265-269.

Fu JK, Luthy RG. 1986. Effect of organic solvent on sorption of aromatic solutes onto soils. J. Environ. Sci. 112:346-366.

Geyer H, Scheunert, Korte F. 1986. Bioconcentration potential of organic environmental chemicals in humans. Regul. Toxicol. Pharmacol. (6):313-347.

Graham M, Hileman F, Kirk D, Wnedling J, Wilson J. 1985. Background human exposure to 2,3,7,8-TCDD. Chemosphere 14(6/7):925-928.

Gross ML. 1982. Application of mass spectrometric methods to analysis of xenobiotics in biological systems. IARC Sci. Publ. 39:443-462, Lyon, France.

Hawley JK. 1985. Assessment of health risk from exposure to contaminated soil. Risk Analysis 5(4):289-302.

Healy WB. 1968. Ingestion of soil by dairy cows. New Zealand J. Agric. Res. 11:487-499.

Humphrey HEB, Price HA, Budd ML. 1976. Evaluation of changes of the level of polychlorinated biphenyls (PCBs) in human tissue. Final report of FDA Contract no. 223-73-2209, 1976 (as cited in Cordle, et al. (1982)).

Hwang ST. 1987a. Multimedia approach to risk assessment for contaminated sediments in a marine environment. Superfund 87, Proceeding of the 8th National Conference, Washington, DC.

Hwang ST. 1987b. Methods for estimating on-site ambient air concentrations at disposal sites. Nuclear and Chemical Waste Management, 7:95-98.

Hwang ST, Falco J. 1986. Estimation of multimedia exposures related to hazardous waste facilities. In: Cohen Y, ed. Pollutants in a multimedia environment. New York, NY: Plenum Publishing Co.

Isensee AR, Jones GE. 1971. Absorption and translocation of root and foliage applied 2,4-dichlorophenol, 2,7-dichorodibenzo-p-dioxin and 2,3,7,8-tetrachlorodibenzo-p-dioxin. J. Agric. Food Chem. 19(6):1210-1214.

Isensee AR, Jones GE. 1975. Distribution of 2,3,7,8-tetrachloro-dibenzo-p-dioxin (TCDD) in aquatic model ecosystems. Environ. Sci. Technol. 9:668-672.

Jensen DJ, Hummel RA, Mahle NH, Kocher CW, Higgings HS. 1981. A residue study on beef cattle consuming 2,3,7,8-tetrachlorodibenzo-p-dioxin. J. Agric. Food Chem. 29:265-268.

Jensen DJ, Hummel RA. 1982. Secretion of TCDD in milk and cream following the feeding of TCDD to lactating dairy cows. Bull. Environ. Contam. Toxicol. 19:440-446.

Kaminski LS, DeCapiro AP, Gierthy JF, Silkworth JB, Tumasonis C. 1985. The rule of environmental matrices and experimental vehicles in chlorinated dibenzodioxin and dibenzofuran toxicity. Chemosphere 14:685-695.

Kapila S, Yanders AF, Orazio C, Meadows J, Malhorta RK, Cerlesi S. 1987. Field and laboratory studies on the movement of and fate of tetrachloro-dibenzo-p-dioxins in soil. Chemosphere (in press).

Kenaga EE, Norris LA 1983. Environmental toxicology of TCDD. In: Tucker RE, Young AL, Gray AP, eds. Human and environmental risks of chlorinated dioxins and related compounds. New York, NY: Plenum Press, pp. 277-299.

Kimbrough R, Falk H, Stehr S, Fries G. 1984. Health implications of 2,3,7,8-tetrachlorodibenzo-p-dioxin (TCDD) contamination of residential soil. J. Toxicol. Environ. Health 14:47-93.

King FG, Dedrick RL, Collins JM, Matthews HB, Birnbaum LS. 1983. Physiological model for the pharmacokinetics of 2,3,7,8-tetrachlorodi-benzofuran in several species. Toxicol. Appl. Pharmacol. 67:390.

Kociba RJ, Keeler PA, Park CN, Gehring PJ. 1976. 2,3,7,8-Tetrachloro-dibenzo-p-dioxin (TCDD): results of a 13-week oral toxicity study in rats. Toxicol. Appl. Pharmacol. 35:553-573.

Kociba RJ, Keyes DG, Beyer JE, Carreon RM, Wade EE, Dittenber DA, Kalning RP, Frauson LF, Park DN, Barnard SD, Hummel RA, Humiston CG. 1978. Results of a two-year chronic toxicity and oncogenicity study of 2,3,7,8-tetrachlorodibenzo-p-dioxin in rats. Toxicol. Appl. Pharmacol. 46(2):279-303.

Kuehl DW, Cook PM, Batterman AR, Lothenback DB, Butterworth BC, Johnson DL. 1985. Bioavailability of 2,3,7,8-tetrachlorodibenzo-p-dioxin from municipal incinerator fly ash to freshwater fish. Chemosphere 14:427-437.

Kuehl DW, Cook PM, Batterman AR, Lothenback D, Butterworth BC. 1987a. Bioavailability of polychlorinated dibenzo-p-dioxins and dibenzofurans from contaminated Wisconsin River sediment to carp. Chemosphere 16(4):667-676.

Kuehl DW, Cook PM, Batterman AR, Lothenback DB, Butterworth BC. 1987b. Isomer dependent bioavailability of polychlorinated dibenzo-p-dioxins and dibenzofurans from municipal incinerator fly ash to carp. Chemosphere 16(4):657-666.

Lake JL, Rubinstein N, Pavignano S. 1984. Predicting bioaccumulation: development of a simple partitioning model for use as a screening tool for regulating ocean disposal of waste. Presented at the Sixth Pillston Workshop, August 12-17, Florissant, CO. In press: Fate and effects of sediment-bound chemicals in aquatic systems, Dickson KL, Maki AW, Brungs W, eds.

Lepow ML, Bruckman L, Rubino RA, Markowitz S, Gillette M, Kapish J. 1975. Investigations into sources of lead in the environment of urban children. Environ. Res. 10:415.

Leung H, Paustenbach J. 1987. A proposed occupational exposure limit for 2,3,7,8- tetrachlorodibenzo-p-dioxin. Environmental Health and Safety, Syntex U.S.A. Inc., Palo Alto, CA, J. Amer. Indust. Hygiene Assoc. (submitted).

Longwoods Research Group, Limited. 1984. A usage segmentation analysis of the 1981 U.S. seafood consumption study (final report). Prepared for the Fisheries Council of Canada (as cited in Kleiman CG (1985)).

Lucier GW, Rumbaugh RC, McCoy Z, Hass R, Harvan D, Albro P. 1986. Ingestion of soil contaminated with 2,3,7,8-tetrachlorodibenzo-p-dioxin (TCDD) alters hepatic enzyme activities in rats. Fundam. Appl. Toxicol. 36:364-371.

Lyman WJ, Reehl WF, Rosenblatt DH. 1982. Handbook of chemical property estimation methods. New York, NY: McGraw-Hill.

Lyman WJ, Loreti CP. 1987. Prediction of soil and sediment sorption for organic compounds. Final report submitted by Arthur D. Little, Inc. to the U.S. Environmental Protection Agency, Office of Water Regulations and Standards under EPA Contract no. 68-01-6951, June 1987.

Mackay D, Bobra A, Chan DW, Shiu WY. 1982. Vapor pressure correlations for low-volatility environmental chemicals. Environ. Sci. Technol. 16(10):645-649.

Mackay D, Paterson S, Cheung B. 1985. Evaluating the environmental fate of chemicals: the fugacity level III approach as applied to 2,3,7,8-TCDD. Chemosphere 14(6/7):859-863.

Marple L, Brunck R, Throop L. 1986a. Water solubility of 2,3,7,8-tetra-chlorodibenzo-p-dioxin. Environ. Sci. Technol. 20(2):180-182.

Marple L, Berridge B, Throop L. 1986b. Measurement of the water-octanol partition coefficient of 2,3,7,8-tetrachlorodibenzo-p-dioxin. Environ. Sci. Technol. 20(4):397-399.

Martin A, Wu PL, Adjei A, Lindstrom RE, Elworthy PH. 1982. Extended Hildebrand solubility approach and the log-linear solubility equation. J. Pharmacol. Sci. 71(8):849-856.

Matsumura F, Benezet HJ. 1973. Studies on the bioaccumulation and microbial degradation of 2,3,7,8-tetrachlorodibenzo-p-dioxin. Environ. Health Perspect. 5:253-258.

McConnell EE, Lucier GW, Rumbaugh RC, Albro PW, Harvan DJ, Hass JR, Harris MW. 1984. Dioxin in soil: bioavailability after ingestion by rats and guinea pigs. Science 223:1077-1079.

Mill T, Rossi M, McMillen D, Coville M, Leung D, Spang J. 1987. Photolysis of tetrachlorodioxin and PCBs under atmospheric conditions. Internal report prepared by SRI, International for the U.S. Environmental Protection Agency, Office of Health and Environmental Assessment, Washington, DC.

Nash RG, Beall ML. 1980. Distribution of silvex, 2,4-D, and TCDD applied to turf in chambers and field plots. J. Agric. Food Chem. 28:614-623.

National Marine Fisheries Service (NMFS). 1983. Fisheries of the U.S. 1982. Current Fisheries Statistics No. 8300., U.S. Department of Commerce.

National Marine Fisheries Service (NMFS). 1984. Fisheries of the U.S. 1983. Current Fisheries Statistics No. 8320., U.S. Deptartment of Commerce.

Nau H, Bass R. 1981. Transfer of 2,3,7,8-tetrachlorodibenzo-p-dioxin (TCDD) to the mouse embryo and fetus. Toxicology 20(4):299.

Nkedi-Kizza P, Rao PSC, Hornsby AG. 1985. Influence of organic cosolvents on sorption of hydrophobic organic chemicals by soils. Environ. Sci. Technol. 19:975-979.

Ono M, Wakimoto T, Tatsukawa R, Masuda Y. 1986. Polychlorinated dibenzo-p-dioxin and dibenzofurans in human adipose tissue of Japan. Chemosphere 15(9-12):1629-1634.

Patterson DG, Holler JS, Smith SJ, Liddle JA, Sampson EJ, Needham LL. 1986. Human adipose data for 2,3,7,8-tetra-chlorodibenzo-p-dioxin in certain U.S. samples. Chemosphere 15(9-12):2055-2060.

Paustenbach DJ, Shu HP, Murray FJ. 1986. A critical examination of assumptions used in risk assessments of dioxin contaminated soil. Regul. Toxicol. Pharmacol. 6:284-307.

Podoll RT, Jaber HM, Mill T. 1986. Tetrachlorodibenzodioxin: rates of volatilization and photolysis in the environment. Environ. Sci. Technol. 20(5):490-492.

Poiger H, Schlatter C. 1980. Influence of solvents and adsorbents on dermal and intestinal adsorption of TCDD. Food Cosmet. Toxicol. 18:477-481.

Puffer HW, Zen SPA, Duda MJ, Young DR. 1983. Consumption rates of potentially hazardous marine fish caught in the metropolitan Los Angeles area. Project summary. U.S. Environmental Protection Agency, Environmental Research Laboratory, Corvallis, OR. EPA-600/53-82-070.

Radian Corporation. 1983. Review and development of chlorinated dioxins and furans emissions data. Prepared for the Office of Air Quality Planning and Standards, U.S. Environmental Protection Agency, Research Triangle Park, NC.

Rappe C, Buser HR, Stalling DL, Smith LM, Dougherty RC. 1981. Identification of polychlorinated dibenzofurans in environmental samples. Nature 292:524-526.

Rappe C, Marklund S, Kjeller LO, Tysklind M. 1986. PCDDs and PCDFs in emissions from various incinerators. Chemosphere 15(9-12):1213-1217.

Roels H, Buchet J, Lauwerys RR, Bruaux P, Thoreau FC, Lafontaine A, Verduyn G. 1980. Exposure to lead by the oral and pulmonary routes of children living in the vicinity of a primary lead smelter. Environ. Res. 22:81-94.

Rose JQ, Ramsey JC, Wentzler TH. 1976. The fate of 2,3,7,8-tetrachloro-dibenzo-p-dioxin following single and repeated oral doses to the rat. Toxicol. Appl. Pharmacol. 36:209-226.

Sacchi GA, Vigano P, Fortunati G, Cocucci SM. 1986. Accumulation of 2,3,7,8-TCDD from soil and nutrient solution by bean and maize plants. Experientia 42:586-588.

Sarna LP, Hodge PE, Webster GRB. 1984. Octanol-water partition coefficients of chlorinated dioxins and dibenzofurans by reversed-phase HPLC using several C18 columns. Chemosphere 13(9):975-983.

Schroy JM, Hileman FD, Cheng SC. 1985a. Physical/chemical properties of 2,3,7,8-tetrachlorodibenzo-p-dioxin. In: Bahner RC, Hansen DJ, eds. Aquatic toxicology and hazard assessment: eighth symposium. Philadelphia, PA: American Society for Testing and Materials, pp. 409-421.

Schroy JM, Hileman FD, Cheng SC. 1985b. Physical/chemical properties of 2,3,7,8-TCDD. Chemosphere 14(6/7):877-880.

Sendroy J, Cecchini LP. 1954. Determination of human body surface area from height and weight. J. Appl. Physiol. 7(1):1-12.

Sheffield A. 1985. Sources and releases of PCDD's and PCDF's to the Canadian environment. Chemosphere 14(6/7):811-814.

Silkworth J, McMartin D, DeCaprio A, Rej R, O'Keefe P, Kaminsky L. 1982. Acute toxicity in guinea pigs and rabbits of soot from a polychlorinated biphenyl-containing transformer fire. Toxicol. Appl. Pharmacol. 65:425-439.

Snyder WS. 1975. Report of the task group on reference man. International Commission of Radiological Protection, No. 23, New York, NY: Pergamon Press.

Stephan CE. 1980. Memorandum titled Per capita consumption of nonmarine fish and shellfish dated July 3, 1980, to J. Stara, U.S. EPA Environmental Criteria and Assessment Office, Cincinnati, OH, from Charles E. Stephan, Environmental Research Laboratory, Duluth, MN.

Suskind RR. 1977. Environment and the skin. Environ. Health Perspect. 20:27-37.

Thibodeaux LJ. 1983. Offsite transport of 2,3,7,8-tetrachlorodibenzo-p-dioxin from a production disposal facility. In: Choudhary, G, ed. Chlorinated dioxins and dibenzofurans in the total environment. Chapter 5, Boston, MA: Butterworth.

Thibodeaux LJ, Lipsky D. 1985. A fate and transport model for 2,3,7,8-tetrachlorodibenzo-p-dioxin in fly ash on soil and urban surfaces. Hazardous Waste and Hazardous Materials 2:225-235.

Thibodeaux LJ, Reible DD, Fang CS. 1986. Transport of chemical contaminants in the marine environment originating from offshore drilling bottom deposits--a vignette model. In: Cohen Y, ed. Pollutants in a multimedia environment. New York, NY: Plenum Publishing Corp.

Tschirley FH. 1986. Dioxin. Scientific American 254(2):29-35.

Tung LS, Freeman RA, Schroy JM. 1985. Prediction of soil temperature profiles: a concern in the assessment of transport of low volatility chemicals in the soil column. Presented at the 1985 Summer National Meeting of the American Institute of Chemical Engineers (AIChE); August, Seattle, WA.

Turner DB. 1970. Workbook of atmospheric dispersion estimates. PHS Publication No. 999-AP-26 (NTIS PB 191482), U.S. Environmental Protection Agency, Research Triangle Park, NC.

USDA. 1985. U.S. Department of Agriculture. Food and nutrient intakes: individuals in four regions, 1977-1978. Report I-3. Nationwide Food Consumption Survey 1977-1978.

USDOC. 1985. U.S. Department of Commerce. Fisheries of the United States, April 1985. Current fisheries statistics No. 8360 (as cited in U.S. EPA (1986b)).

USEPA. 1979. U.S. Environmental Protection Agency. Industrial source complex (ISC) dispersion model user's guide. Office of Air Quality Planning and Standards, Research Triangle Park, NC. EPA-450/4-79-030.

USEPA. 1980a. U.S. Environmental Protection Agency. Dioxins. Industrial Environmental Research Laboratory, Cincinnati, OH. EPA-600/2-80-197.

USEPA. 1980b. U.S. Environmental Protection Agency. Seafood consumption data analysis. Prepared for the Office of Water Regulations and Standards, Washington, DC, by SRI International under EPA Contract No. 68-01-3887.

USEPA. 1981a. U.S. Environmental Protection Agency. The potential atmospheric impact of chemicals released to the environment. Office of Toxic Substances, Washington, DC. EPA-560/5-80-001.

USEPA. 1981b. U.S. Environmental Protection Agency. Risk assessment on (2,4,5-trichlorophenoxy)acetic acid (2,4,5-T), (2,4,5-trichloro-phenoxy) propionic acid (silvex), and 2,3,7,8-tetrachlorodibenzo-p-dioxin (TCDD). Office of Health and Environmental Assessment, Washington, DC. EPA-600/6-81-003. NTIS PB81-234825.

USEPA. 1983a. U.S. Environmental Protection Agency. Dioxin strategy. Internal report dated November 28, 1983. Office of Water Regulations and Standards, Office of Solid Waste and Emergency Response in conjunction with Dioxin Strategy Task Force, Washington, DC.

USEPA. 1984a. U.S. Environmental Protection Agency. Risk analysis of TCDD contaminated soil. Office of Health and Environmental Assessment, Washington, DC. EPA-600/8-84-031.

USEPA. 1984b. U.S. Environmental Protection Agency. Personal communication, September 4, 1986, Barbara Peterson, Peterson and Associates, Inc., Bethesda, MD, regarding Office of Pesticide Programs Tolerance Assessment System: Crop to Food Map, Draft Report, August 1984. (Data analyzed were compiled in the USDA Nationwide Food Consumption Survey, 1977-78.)

USEPA. 1984c. U.S. Environmental Protection Agency. An estimation of the daily average food intake by age and sex for use in assessing the radionuclide intake of individuals in the general population. Office of Radiation Programs, Washington, DC. EPA-520/1-84-021.

USEPA. 1984d. U.S. Environmental Protection Agency. An estimation of the daily food intake based on data from the 1977-1978 USDA Nationwide Food Consumption Survey. Office of Radiation Programs, Washington, DC. EPA-520/1-84-015.

USEPA. 1985a. U.S. Environmental Protection Agency. Pollutant sorption to soils and sediments in organic/aqueous solvent systems. Office of Environmental Processes and Effects Research, Athens, GA. EPA-600/3-85-050. NTIS PB85-242535.

USEPA. 1985b. U.S. Environmental Protection Agency. Health assessment document for polychlorinated dibenzo-p-dioxins. Office of Health and Environmental Assessment, Environmental Criteria and Assessment Office, Cincinnati, OH. EPA-600/8-84-014F. NTIS PB86-122546.

USEPA. 1985c. U.S. Environmental Protection Agency. Development of statistical distributions or ranges of standard factors used in exposure assessments. Office of Health and Environmental Assessment, Washington, DC. EPA-600/8-85-010. NTIS PB85-242667/AS.

USEPA. 1985d. U.S. Environmental Protection Agency. Dioxin transport from contaminated sites to exposure locations: a methodology for calculating conversion factors. Office of Health and Environmental Assessment, Washington, DC. EPA-600/8-85-012. NTIS PB85-214310.

USEPA. 1985e. U.S. Environmental Protection Agency. Rapid assessment of exposure to particulate emission from surface contamination sites. Office of Health and Environmental Assessment, Washington, DC. EPA-600/8-85-002. NTIS PB85-192219/AS.

USEPA. 1985f. U.S. Environmental Protection Agency. Compilation of air pollutant emission factors, Vol. 1. Office of Air Quality Planning and Standards, Research Triangle Park, NC.

USEPA. 1985g. U.S. Environmental Protection Agency. Proceeding of the Eleventh Annual Research Symposium, Leaching potential of 2,3,7,8-TCDD in contaminated soils. U.S. EPA Hazardous Waste Engineering Research Laboratory, Cincinnati, OH. EPA-600/9-85-013.

USEPA. 1986a. U.S. Environmental Protection Agency. Guidelines for carcinogen risk assessment. Federal Register 51(185):33992-34003.

USEPA. 1986b. U.S. Environmental Protection Agency. Development of advisory levels for polychlorinated biphenyls (PCBs) cleanup. Office of Health and Environmental Assessment, Washington, DC. EPA-600/6-86-002. NTIS PB86-232774/AS.

USEPA. 1988. U.S. Environmental Protection Agency. Estimating exposures to 2,3,7,8-TCDD, Draft document, Office of Health and Environmental Assessment, Washington, DC, EPA/600/6-88/005A, March 1988.

Umbreit TH, Patel D, Gallo MA. 1985. Acute toxicity of TCDD contaminated soil from an industrial site. Chemosphere 14:945-947.

Umbreit TH, Hesse EJ, Gallo MS. 1986a. Bioavailability of dioxin in soil from a 2,4,5-T manufacturing site. Science 232:497-499.

Umbreit TH, Hesse EJ, Gallo MS. 1986b. Comparative toxicity of TCDD contaminated soil from Times Beach, Missouri, and Newark, New Jersey. Chemosphere, 15(9-12):2121-2124.

Umbreit TH, Hesse EJ, Gallo MS. 1987a. Reproductive toxicity of female mice of dioxin-contaminated soils from a 2,4,5-trichlorophenoxyacetic acid manufacturing site. Arch. Environmental Contamination & Toxicology 16, 461-466.

Umbreit TH, Hesse EJ, Gallo MS. 1987b. Differential bioavailability of 2,3,7,8-tetrachlorodibenzo-p-dioxin from contaminated soils. American Chemical Society Symposium Series No. 338, Solving hazardous waste problems: Learning from dioxin, American Chemical Society, Washington, DC.

van den Berg M, Olie K, Hutzinger O. 1983 Uptake and selective retention in rats of orally administered chlorinated dioxins and dibenzofurans from fly-ash and fly-ash extract. Chemosphere 12:537-544.

van den Berg M, Vroom A, van Greevenbroek M, Olie K, Hutzinger O. 1985. Bioavailability of PCDDs and PCDFs adsorbed on fly ash in the rat, guinea pig, and Syrian golden hamster. Chemosphere 14:865-869.

Walters RW, Guiseppi-Elie A, Yousefi Z, Means JC. 1987. Sorption of dioxin to soils. Chemosphere (in press).

Ward CT, Matsumura F. 1978. Fate of 2,3,7,8-tetrachlorodibenzo-p-dioxin (TCDD) in a model aquatic environment. Arch. Environ. Contam. Toxicol. 7:349-357.

Webster GRB, Friesen KJ, Sarna LP, Muir DCG. 1985. Environmental fate modelling of chlorodioxins: determination of physical constants. Chemosphere 14(6/7):609-622.

Webster GRB, Muldrew DH, Graham JJ, Sarna LP, Muir DCG. 1986. Dissolved organic matter mediated aquatic transport of chlorinated dioxins. Chemosphere 15(9-12):1379-1386.

Wipf HK, Homberger E, Neuner N, Ranalder UB, Vetter W, Vuilleumier JP. 1982. TCDD levels in soil and plant samples from Seveso area. In: Hutzinger O, ed. Chlorinated dioxins and related compounds: impact on the environment. Vol. 5, Series on Environmental Science. New York, NY: Pergamon Press, pp. 115-126.

Wipf HK, Schmid J. 1983. Seveso - an environmental assessment. In: Tucker RE, Young AL, Gray AP, eds. Human and environmental risks of chlorinated dioxins and related compounds. New York, NY: Plenum Press.

Wroblewski VJ, Olson JR. 1985. Hepatic metabolism of 2,3,7,8-tetra-chlorodibenzo-p-dioxin in the rat and guinea pig. Toxicol. Appl. Pharmacol. 81:(2)231.

Yalkowsky SH, Flynn GL, Amidon GL. 1972. Solubility of nonelectrolytes in polar solvents. J. Pharm. Sci. 61(6):983-984.

Young AL. 1983. Long term studies on the persistence and movement of TCDD in a national ecosystem. In: Tucker A, ed. Human and environmental risks of chlorinated dioxins and related compounds. New York, NY: Plenum Press.

Zepp RG, Miller GC, Herbert VR, Mille MJ, Mitzel R. 1988. Photolysis of octachlorodibenzo-p-dioxin on soils; production of 2,3,7,8-TCDD. Chemosphere (in press).

7.   QUANTITATIVE IMPLICATIONS OF THE USE OF DIFFERENT EXTRAPOLATION
      PROCEDURES FOR LOW-DOSE CANCER RISK ESTIMATES FROM EXPOSURE TO
      2,3,7,8-TCDD

Steven P. Bayard

U.S. Environmental Protection Agency
Washington, D.C. 20460

7.1   Introduction and Definition of Terms

7.1.1 Introduction

2,3,7,8-Tetrachlorodibenzo-*p*-dioxin   (2,3,7,8-TCDD) is   the   most
potent animal carcinogen ever tested. It is 50 times more potent than
aflatoxin B1 on a per mole basis and 50 million times more potent than
vinyl chloride.   In addition to its carcinogenic potency, 2,3,7,8-TCDD is
also the most potent animal teratogen known, and it causes other reproductive and immune system effects at extremely low doses as well.

Because of these severe toxicities, many U.S. federal and state agencies,   as   well   as   foreign, regulatory, and health agencies, have proposed
or implemented regulations or advisories based on levels of concern for
2,3,7,8-TCDD.   Among these agencies, the U.S. Environmental Protection
Agency (EPA) was, to this writer's knowledge, the first to actually produce an upper-limit estimate of cancer risk for 2,3,7,8-TCDD exposure to
humans (USEPA, 1980). This estimate was based on a methodology that
extrapolated from cancer responses at doses of 1, 10, and 100 ng/kg-d in
an animal lifetime feeding study (Kociba et al., 1978; Table 1) to humans
at still lower levels.   Initially, this extrapolation was based on a
simple linear extrapolation from the lowest dose to show a significant
elevated response (10 ng/kg-d caused a statistically significant increase
in liver tumors--hyperplastic nodules--versus control).   Shortly after-
ward, however, the methodology was modified to include all dose-response
points in the extrapolation procedure. The new methodology was based on
a multistage model for carcinogenesis as the result of a specific time

205

Table 1. 2,3,7,8-TCDD 2-Year Oral Rat Study (1978) Using Kociba's Histopathology Analysis (Female Sprague-Dawley Rats – Spartan Sub-
strain[a]) and Eliminating Animals That Died During the First
Year

	Dose level (ug/kg-day)			
Tissue and diagnosis	0 (Control)	0.001	0.01	0.1

Kociba analysis

1. Lung
   Keratinizing squamous cell carcinoma

	0/85 (0%)	0/48 (0%)	0/48 (0%)	7/40 (18%) $(p = 2.3 \times 10^{-4})$[b]

2. Nasal turbinates/hard palate
   Stratified squamous cell carcinoma
   (revised diagnosis 2/19/79)

	1/54 (2%)	0/30 (0%)	1/27 (4%)	5/24 (21%) $p = 9.6 \times 10^{-3})$[h]

3. Liver
   Hepatocellular hyperplastic nodules/
   hepatocellular carcinoma

	9/85 (11%)	3/48 (6%)	18/48 (38%) (two had both) $(p = 3.1 \times 10^{-4})$[b]	34/40 (82%) $(p < 10^{-8})$

Total combined (1, 2, or 3 above)
(each rat had at least one tumor)

	9/85 (11%)	3/48 (6%)	18/48 (38%) $(p = 3.1 \times 10^{-8})$[b]	34/40 (82%) $(p = <10^{-8})$[b]

Squire's review

1. Lung
   Squamous cell carcinoma

	0.85 (0%)	0/48 (0%)	0/48 (0%)	8/40 (20%) $(p = 6.5 \times 10^{-5})$[b]

2. Nasal turbinates/hard palate
   Squamous cell carcinoma

	0/54 (0%)	0/30 (0%)	1/27 (4%)	5/22 (23%) $(p = 1.4 \times 10^{-3})$[b]

3. Liver
   Neoplastic nodules/hepatocellular
   carcinoma

	16/85 (19%)	8/48 (16%)	27/48 (56%) $(p = 1.3 \times 10^{-5})$[b]	33/40 (82%) $(p = 10^{-8})$[b]

Total combined (1, 2, or 3 above)
(each rat had at least one tumor)

	16/85 (19%)	8/48 (16%)	27/48 (56%) $(p = 1.3 \times 10^{-5})$[b]	34/40 (82%) $(p < 10^{-8})$[h]

[a] Average body weight of a female rat is 450 g.
[h] Fisher Exact Test (one-tailed).

206

ordering of changes, first proposed by Armitage and Doll (1954) and later modified by Crump et al. (1977, 1979) to include dose-response for extrapolation purposes. Often called the linearized multistage model, the Crump model is distinguished by its approach of providing upper-limit estimates of risk consistent with nonthreshold low-dose linearity. This model is presented in Section 7.2.1.

Following EPA's efforts, two other U.S. agencies, the Food and Drug Administration (FDA, 1983) and the Centers for Disease Control (CDC) (Kimbrough et al., 1984), produced cancer risk estimates based on slight modifications of EPA's methods but with similar resulting estimates. In another minor variation, the state of California (1984) used the Crump linearized multistage model to extrapolate the cancer response to humans from a mouse gavage study performed by the National Cancer Institute (NTP, 1982). All efforts produced results within a factor of 10. These are discussed in Section 7.2.2.

Within the framework of the linearized multistage model and the Kociba et al. rat feeding study, two other efforts appearing in the literature are noteworthy. First, Longstreth and Hushon (1983) applied several mathematical nonthreshold, nonlinear models (Logit, Probit, Weibull, and multihit) to the cancer response in the Kociba et al. study and compared the extrapolated results with those of the linearized multistage model. Second, Sielken (1987) fit the Kociba data with the multistage model (but not the Crump linearized version applying upper limits) and also with a modified version that allowed the input of actual observation times. He then compared actual estimates derived from the multistage model with those of the EPA, which used the upper limits based on the Crump version. The Sielken paper is discussed further in Section 7.2.4.

In contrast to all of the above attempts at extrapolating from animal data to humans by nonthreshold models, several U.S. nonregulatory agencies have applied safety or uncertainty factors, not models. The uncertainty factors of between 100 and 1,000 are applied to doses that have shown no adverse effects in animal cancer or other studies, and the resulting numbers are presumed safe for humans. This methodology has been used by EPA and many other agencies for animal-to-human extrapolations for toxic effects other than cancer, but EPA has never used this methodology to estimate cancer risk. The estimates of cancer risk provided by the uncertainty factor approach are in the range of 150 to 1,500 times lower than the estimates provided by EPA's use of Crump's multistage model. These estimates are presented in Section 7.3.1.

The differences in the magnitude of the estimates provided by these two approaches require a closer look at the methodologies involved in each and at the reasoning as to which is the proper one to use for cancer risk extrapolation of 2,3,7,8-TCDD. The argument focuses on the model for complete carcinogens versus promoters. Complete carcinogens have both initiating and promoting ability, and it is the mechanism leading to the initiating part of carcinogenesis, the attachment of the carcinogen to the DNA, that can be modeled on either a linear or multistage basis. Those in favor of modeling 2,3,7,8-TCDD as a complete carcinogen argue that 2,3,7,8-TCDD causes rare cancers of the hard palate and nasal turbinates and of the tongue (in male rats), and a rare form of lung cancer, and that such rare tumors would be unlikely to be initiated except by the 2,3,7,8-TCDD in the experiment. Conversely, those in favor of the uncertainty factor approach point to the strong evidence for the promoting effects of 2,3,7,8-TCDD in the liver, where most of the tumors are occurring. They argue that promotion is effectively a toxic reaction with a threshold and that all promoters show not only thresholds but also reversibility upon cessation of dosing. Treating 2,3,7,8-TCDD as a promoter and using an uncertainty factor approach has the further advantage of allowing a comparison of its cancer effects with its other toxic effects using the same methodology.

A third approach is also possible. This is an approach that models for the cancer effects of 2,3,7,8-TCDD through its known mechanism of binding to a receptor. (This actual modeling and results are presented in Section 7.4.2.) While the basic model, the Moolgavkar-Venzon-Knudson (M-V-K) two-stage model, with a promotion phase, has been used in the literature to explain many cancers and has been found to predict well the tumor promotion in mouse skin (Chu et al., 1987), the approach is new in that modeling for promoters has never been done before by regulatory agencies.

The purpose of the presentations that follow is to compare quantitatively the estimates derived from each of the separate approaches and then to compare them with each other. To do this, common terms must first be introduced.

7.1.2 <u>Definition of Terms</u>

Terms associated with modeling (defined here as an estimation of the incremental cancer risk to humans arrived at by fitting a mathematical function to animal response data) include the following:

Maximum likelihood estimate (MLE)--The statistical procedure by which the parameters of the model are estimated. The MLE has many properties, in a statistical sense, that allow it to be referred to as a "statistical average" or "best" estimate. In risk terms it might be thought of as a term that, if the assumed model is true, provides overestimates and underestimates of the true risk 50 percent of the time.

Parameter--A constant in the model, associated with either the control response or the time or dose variable inputs. For example, in the Crump linearized multistage model, the parameter associated with the linear dose variable is denoted as q_1 and is defined as the increase in cancer risk associated with an incremental increase per unit of dose. For this reason, q_1 is expressed in units of reciprocal dose such as $(ng/kg-d)^{-1}$.

Upper-confidence limit (UCL) estimates--The estimates resulting from a statistical procedure in which the upper-limit values of the parameters still consistent with the data are estimated. In the linearized multistage model (Crump), the upper-limit estimate associated with the linear term is designated q_1. In a statistical sense it is the 95 percent upper-limit estimate of the linear term associated with the fitting of the linearized multistage model to the animal data. In making cross-species extrapolations to humans, however, the "95 percent" label is dropped since the uncertainty associated with cross-species extrapolations is considered far greater than the statistical uncertainty associated with the model-fitting procedure. Also, because the linearized multistage model becomes linear at low doses, the UCL on q_1 is the same as the UCL on the incremental risk. This is not true of the nonlinear models such as the Logit, Probit, and Weibull models discussed in Section 7.2. Upper-limit incremental risk estimates, however, are comparable, and the ratio of these estimates from two different models can be expressed as the relative potency.

Risk-specific dose (RsD)--A dose associated with a specified cancer risk. For example, assume a linearized multistage model is fit to the data and the parameter estimates are $q_1 = 3.0 \times 10^{-3}$ $(ng/kg-d)^{-1}$, $q_1 = 0$ for all $i \neq 1$, and $q_1 = 7.5 \times 10^{-3}$ $(ng/kg-d)^{-1}$. Then for an incremental risk of 1 in 1,000,000, the dose would be the solution to $10^{-6} = 1 - \exp(-3.0 \times 10^{-3}$ d), and RsD $= 3.3 \times 10^{-4}$ ng/kg-d would be called the risk-specific dose. Likewise, the RsD could be defined in terms of the lower limit of the dose corresponding to a risk of 10^{-6}.

In this case q_1 would be substituted for q_1 and the solution would be RsD = 1.3 x 10^{-4} ng/kg-d.

A ratio of two RsDs can also be thought of as a measure of relative potency, but in this case the higher the RsD, the lower the potency. The RsD thus becomes a common unit to discuss relative potency between different approaches and different types of toxicity.

Virtually safe dose (VSD)--A dose associated with a very small risk. The general reasoning in discussing RsDs and VSDs is identical. The only difference is in one's definition of a "very small risk."

Lowest-observed-adverse-effect-level (LOAEL)--The lowest dose in an experiment at which there is a statistically significant increase over the control group in the proportion of animals for which adverse effects are observed. The *no-observed-adverse-effect-level* (NOAEL) and *no-observed-effect-level* (NOEL) are straightforward negations (Crockett and Crump, 1986). The uncertainty or safety factor is an arbitrary factor applied to these levels for the purpose of establishing concern or no-concern levels for humans.

7.2   EPA's Use of the Linearized Multistage Model for Carcinogen Risk Extrapolation and Comparison with Other Models

7.2.1 Description of the Multistage and Linearized Multistage Models

EPA's reasons for using the linearized multistage model for risk extrapolation, in general, are discussed in the Guidelines for Carcinogen Risk Assessment (USEPA, 1986). For the 2,3,7,8-TCDD risk assessment, additional discussions are presented in the Health Assessment Document (HAD) for Polychlorinated-Dibenzo-p-Dioxins (USEPA, 1985), as well as in the document on the cancer risk-specific dose estimate for 2,3,7,8-TCDD. Therefore, only an abbreviated review of its development will be presented here. Basically, its genesis came from Armitage and Doll (1954), who proposed a theory that a cancer cell was generated from a series of several heritable mutations in a specific order, the end result of each change being termed a stage. The transition rate from one stage to the next was hypothesized as being related to a probability of occurrence. The time rate of occurrence of the ith event is $a_i + b_i d_i$, i=1, ..., k, where a_i, $b_i \geq 0$ and d=dose. This, along with some other assumptions, leads to a dichotomous response probability of the form

$$P(d) = 1 - \exp\left[-c \prod_{i=1}^{k} (a_i + b_i d) \right] \qquad a_i b_i \geq 0$$

where a_i is the background transition rate for a progression to stage i, b_i is the incremental increase in that rate per unit of dose, c is a function of exposure duration, and P(d) is the probability of a ·tumor by some fixed age t for a dose d. This model achieved some popularity mainly because of its success in predicting many of the human epithelial cancers and because the model presented the probability of a tumor as a function of dose. In addition, the reparameterization of the individual transition rates leads to

$$P(d) = 1\text{-}\exp\text{-}[q_0 + q_1 d + q_2 d^2 + \ldots + q_k d^k)$$

where P(d) is the lifetime probability of cancer at dose d. Since q_0 is the parameter associated with the background rate, an assumption of independent background leads to

$$P_t(d) = 1\text{-}\exp\text{-}[q_1 d + \ldots + q_k d^k) \quad \text{all } q_i \geq 0$$

where $P_t(d)$ is the incremental (often called the extra) risk associated with dose d. In the linearized form of this model, an upper-limit estimate of the linear term, q_1, consistent with the data, is calculated. At low doses, this upper-limit linear term predominates, forcing the model to low-dose linearity.

7.2.2 Use of the Linearized Multistage Model for Risk Extrapolation of 2,3,7,8-TCDD: Comparison of Four U.S. Agencies

Besides the choice of the linearized multistage model for animal-to-human risk extrapolation, the final risk estimates are dependent on a choice of several other factors. Specifically, Table 2 presents a summary of 2,3,7,8-TCDD cancer risk extrapolation by four U.S. agencies, all of which used the linearized multistage model. Even with the use of the same model, however, the results varied over 10-fold because of a selection of different factors relating both to the animal data and to the procedure. Such factors are:

- Choice of animal bioassay;

- Adjustment made for differential nontumor mortality among the animal treatment groups;

- Selection of tumor types for modeling;

- Animal-to-human dose equivalence; and

- Dose used for curve fit.

Table 2.   Factors Used by Various Agencies in Calculating Their Upper-Limit Risk Estimates for 2,3,7,8-TCDD Using the Linearized Multistage Model

Factor	EPA	FDA	CDC	State of California	Effect of difference on upper-limit unit risk estimate
Animal study used	Kociba female rat feeding study	Kociba	Kociba	NTP male mouse mouse gavage study	Based on dose/surface area, dose conversion difference is less than 10%
Pathologist (Kociba or Squire)	Both	Kociba	Squire	NA	Less than 10%
Adjustment for early mortality in high-dose group	Yes	No	No	NA	Adjustment changes estimate by a factor of +1.7 or $1/2.6$[a]
Selection of tumor types	Liver, lung hard palate/nasal turbinates	Liver only	Liver only	Liver only	Less than 10%
Animal-to-man dose equivalence	$\dfrac{\text{dose}}{\text{surface area}}$	$\dfrac{\text{dose}}{\text{body weight}}$	Liver concentration	$\dfrac{\text{dose}}{\text{surface area}}$	Dose/surface increases estimate by a factor of 5.38 over the other two for the rat; by a factor of 13 for the mouse
Dose used for curve fit	Administered	Administered	Liver concentration at terminal sacrifice	Administered	Using administered dose decreases estimate by a factor of 2
Upper-limit incremental unit $(\text{fg/kg-day})^{-1}$	1.56×10^{-7}	1.75×10^{-8}	3.5×10^{-8} (when reconverted to administered dose)	1.5×10^{-5}	
Reference dose for upper-limit risk of 10^{-6}, units of fg/kg-day	6.4	57.2	27.6	6.7	

[a] Adjustment increases by a factor of about 1.7 compared with unadjusted Kociba analysis; adjustment decreases estimate by a factor of 2.6 compared with unadjusted Squire analysis where high dose is excluded because of poor fit.

212

As seen in Table 2, EPA, FDA, and CDC all used the cancer response data from the female Sprague-Dawley rat in the 2-year feeding study conducted by Kociba et al. (1977, 1978). In the EPA HAD for Polychlorinated Dibenzo-p-Dioxins, the choice of the Kociba study was based on the female rat's providing the largest slope factor, q_1, of all the available data sets. However, there were other, unstated (but probably better) reasons for the selection, such as (1) the high quality of the study, (2) response at multiple sites, (3) more applicable route of exposure to the human experience than the gavage study, and (4) less controversial tumor sites than the mouse liver. The state of California used the liver tumor response from the male mouse in the National Toxicology Program (1982) gavage study and estimated an upper-limit incremental unit risk estimate of $q_1 = 1.5 \times 10^{-7}$ $(fg/kg-d)^{-1}$, nearly the same as that of EPA $(q_1 = 1.56 \times 10^{-7})$. Also, in its analysis EPA contracted with an independent pathologist, Dr. Robert Squire, to provide a second examination of the liver slides in the Kociba study. Even though Squire's analysis found more liver tumors in the control and low- and mid-dose groups, the estimates of an incremental increase in cancer risk differed by less than 10 percent.

Another factor of concern in the extrapolation procedure was nontumor toxicity among the female rats. To correct for high early mortality in the high-dose rats, EPA's analysis eliminated all animals that died during the first year of the study (before the appearance of the first tumor). The elimination of nine animals in the high-dose group, one in the controls, and one in each of the other dose groups changed the upper-limit estimates by a factor of either +1.7 or 1/2.5 depending on which pathologist's analysis was used. Since EPA was the only agency to make the adjustment, its estimate incorporating both pathologists' analyses actually decreased by a factor of 2.7 compared with the unadjusted analysis.

In selection of tumor types, all the agencies modeled the liver tumor response. EPA also included the cancer response in the lung and hard palate/nasal turbinates, but this led to only a minor increase in the final estimate since the liver produced the major response.

Animal-to-man dose equivalence factors are discussed in the HAD. Both EPA and the state of California used dose/surface area equivalences between animals and humans. The FDA used dose/body weight, which reduces human risk estimates compared to surface area by a factor of 5.4 for rat-to-human extrapolation. The CDC used liver concentration at terminal sacrifice, a measure that would be preferable if human tissue distribution were also known. In the present case, however, the known rat liver concentration measures of dose equivalence had to be equated back to the rat administered dose without a comparable known relationship in humans. The dose used for the curve fit by EPA, FDA, and the state of California was the dose actually administered to the animals. The CDC's use of liver concentrations at terminal sacrifice resulted in the risk estimate's being increased by a factor of 2. Species conversion factors are discussed further in Section 7.2.3.

The end result of all of these factors in the risk extrapolation procedure was a maximum difference of a factor of 9, that being between the EPA and the FDA. This can be seen by comparing either the upper-limit incremental unit risk estimates or the risk-specific doses (RsDs), which are just the reciprocals x 10^{-6}.

7.2.3 Allometric and Body Burden Considerations

The dose metric or allometric equivalence for rat-to-human extrapolation has a potentially large quantitative impact on 2,3,7,8-TCDD risk estimation because of the major differences in half-lives in the rat and human. However, what little attention this topic has received from regulatory agencies until now has taken into account only the standard dose metrics. As shown in Table 2, both EPA and the state of California used the administered dose/surface area conversion, the FDA applied an administered dose/body weight, and the CDC applied the actual rat liver concentration at terminal sacrifice. When extrapolating from rat to human, use of the dose/surface area allometry increases the risk estimate by a factor of 5.4 versus either dose/body weight or liver concentration. When extrapolating from the smaller mouse to human, the corresponding use of dose/surface area allometry results in a factor of 13 greater risk estimate.

The EPA has used the dose/surface area metric, often called a species extrapolation correction factor, as a conservative, prudent policy. It is based on the observations that among different mammalian species many physiological rates--especially ventilation, basal metabolic, and clearance rates--tend to scale in proportion to a fractional power of body weight. It has also been found to hold for the acute therapeutic effects of anticancer agents. That fractional power, often between 0.6 and 0.8, is very close to the fractional power of 2/3 relating the surface area of cylindrical or round objects to their volume. Since the density of most mammalian bodies is about the same, mass or body weight can be used

214

instead of volume, and hence the term surface area or (body weight)$^{2/3}$ correction. In simplified terms, the allometry of basal metabolism is often explained by the observation that the number of calories a warm-blooded animal will consume is enough to maintain body temperature and that loss of heat is related to surface area and not mass. However, the allometry of species conversion for carcinogen risk assessment is far more complicated than that for simple basal metabolism. Even assuming that the basic mechanism of the carcinogen stays the same from high to low doses, there are often large species differences both in tissue distribution and in metabolic pathways to form the active carcinogen. Often, it is not known whether the parent compound or one (or more) of its metabolites is the active carcinogen. Almost never is there a good understanding of the mechanism.

It is because of these many unknowns that regulatory agencies have been forced to adopt a general default position of a surface area or body weight or parts per million (ppm) in an air species conversion factor. EPA most often uses surface area, but sometimes uses ppm in air as a species dose equivalence, based on the known cross-species allometry for O_2 consumption. The FDA position is to use dose/surface area allometry when the active carcinogen is a metabolite of the administered compound and to use dose/body weight when the active carcinogen is thought to be the administered compound itself. Their reasoning for the latter case, of which they consider 2,3,7,8-TCDD an example, is that if the administered compound does not have to be metabolized to be carcinogenic, then strict dose/body weight considerations should apply. The CDC apparently agrees. EPA counters, however, that even if the parent compound is the active carcinogen, its activity is related to its time in the body, which in turn is related to clearance time and hence to dose/surface area allometry.

At this point, it may be instructive to derive the dose/surface area allometry in a slightly more rigorous manner. The rat-to-human species correction factor of 5.4 means that if the human were to receive the dose/body weight of a compound 5.4 times that of the rat, the different elimi-nation capacities of the two species would cause both species' concentra-tion x time exposures to the compound to be equal. In terms of first-order elimination kinetics for a single dose of a nonmetabolized compound, a rat given concentration $C_o/5.4$ with an elimination constant k_e would have a total area under the concentration-time curve of C_o/k_e. A human given a concentration of $C_o/5.4$ would eliminate the material, if allometric considerations hold, at a rate of $k_e/5.4$, so that his/her total area under the concentration-time curve would be equal to that of the rat.

For continuous daily exposure, the total area under the concentration-time curve is (C/k^2) $(TK - 1 + e^{-kT})$, where C is the daily dose/body weight and T and k are units of days and reciprocal days, respectively. If T is large, for example 730 days for a 2-year rat study, and k is not very small, then this area becomes approximately CT/k. Thus, for the total areas to be the same for a 2-year rat and 70-year human dosing period, the human would have to be given a concentration C/(5.4x4x35) or 1/189 that of the rat. EPA's position in this metric is that one rat year is equivalent to 35 human years in the cancer development process and that the cancer age distributions for rats and humans are alike when T is viewed as representing a lifetime. Therefore, over a lifetime a human should be allowed 35 times the C/k that of the rat as an equivalent dose.

Since rats and humans seem to follow closely enough for most compounds, the clearance rate dose allometry discussed above, adjusting the species correction factor for actual clearance rates, is not often done in risk extrapolation. Furthermore, consideration of all the unknowns of actual mechanism and metabolism makes it apparent that the species correction factor is only a rough approximation, meant to somehow convey the concept of increased sensitivity of the human compared to the smaller animals. Still, the appropriateness of the use of the surface area correction factor is not clear in the case of 2,3,7,8-TCDD because of its extremely long half-life in humans as compared to rats. The potential effect of this difference on quantitative risk estimation is examined below.

Rose et al. (1976) examined the fate of 2,3,7,8-TCDD following single and repeated oral doses to Sprague-Dawley rats. For a single oral dose of 1.0 ug TCDD/kg body weight, they found a half-life, assuming a one-compartment open model, of 31 $\pm$ 6 days. For repeated oral doses of 0.01, 0.1, or 1.0 ug TCDD/kg-d, 5 days a week for 7 weeks, they found a half-life of 23.7 days. For the single dose, after 22 days nearly all of the compound had concentrated in either the fat or the liver, with equal concentrations in each. For the rats administered the repeated oral doses, the compound again concentrated mostly in the liver and fat, with the liver concentration being three to five times as high as that of the fat at the end of 7 weeks. This observation is consistent with that of Kociba et al. (1978) who, in their 2-year feeding study, found liver concentrations three to five times as high as those in adipose tissue when the daily dose was at least 0.01 ug/kg-d and about the same as the adipose tissue concentration when the daily intake was 0.001 ug/kg-d. Rose et al. estimated the elimination constants for liver, fat, and whole body as all about equal, 0.026 days^{-1}, 0.029 days^{-1}, and 0.029 days^{-1}, respectively, corresponding to half-lives of 24 to 27 days. The relationship between half-life and clearance times for first-order kinetics is $t_{1/2} = \ln_2/k_e$.

In humans the half-life of 2,3,7,8-TCDD in the body has been variously estimated as 3 to 5 years, 6 years, 10 years, and (if 2,3,7,8-TCDD acts according to two-compartment kinetics with the fat acting as a "deep" second compartment) up to 30 years (USEPA, 1988). The CDC (1980) estimates a half-life of 6 to 10 years; this estimate will be used here.

An additional complication is that nonhuman primates, unlike rats, apparently accumulate a higher concentration of 2,3,7,8-TCDD in the adipose tissue than in the liver, with ratios ranging from 10:1 to 67:1. The very limited human data also suggest an adipose tissue-to-liver concentration ratio of 10:1, with minor deposition in other organs. One experiment exposing rats and both infant and adult monkeys to a single intraperitoneal injection (400 ug TCDD/kg body weight) found that after 7 days the rat had concentrated 43 percent of the administered dose in its liver versus only 10 percent and 4.5 percent for adult and infant monkeys. In monkeys, the larger percentages were found in adipose tissue (USEPA, 1985). Thus, if the liver is the organ of primary concern, for tissue distributions alone a human would have to be given anywhere from 10 to 50 times the dose on a mg/kg-body weight basis to have the same liver concentrations as the rat. If one is not concerned with the liver alone but with total body burden, it is not these figures but the relative body half-lives that would apply.

Estimates of the ratio of half-lives in the human versus the rat show that for a human half-life of 2,190 to 3,650 days (6 to 10 years) and a rat half-life of approximately 25 days, the ratios are 88:1 to 146:1, far higher than the 5.4:1 correction used by EPA. If liver is the focus and comparative liver tissue distributions are factored in, however, the rat-to-human correction factor ranges from 1.8(88/50) to 36.5(146/4).

The quantitative risk implications of these adjustments for tissue distribution and half-life differences between the rat and the human are presented in Table 3. As can be seen, if only total body burden (area under the time-concentration curve) is considered, the very long half-life in the human leads to risk estimates between 16 and 27 times that in EPA's

HAD (1985). If liver concentrations are considered, however, the relative risks range from 0.3 to 6.8 times that of EPA's estimate. All estimates are higher than the FDA's. The limited evidence suggests that if liver tissue concentration-time species equivalence is correct, then the EPA HAD (1985) might underpredict the upper-limit risk by a factor of 1.6 to 6.8.

Table 3. Risk Extrapolations for 2,3,7,8-TCDD Using the Linearized Multi-stage Model and Various Estimates of Rat and Human Half-Lives and Tissue Distributions

Half-life (days)		Relative liver:fat tissue concentration at low doses		Calculated rat to human correction factor	Incremental risk estimates q_1^* (pg/kg-day)$^{-1}$	Potency estimates relative to EPA dose metric
Human	Rat	Human	Rat			
2190	25	Total body burden		87.6	2.5×10^{-3}	16.2
2190	25	1:4	1:1	21.9	6.4×10^{-4}	4.1[a]
2190	25	1:10	1:1	8.8	2.5×10^{-4}	1.6[a]
2190	25	1:4	5:1	4.4	1.2×10^{-4}	0.8
2190	25	1:10	5:1	1.8	4.7×10^{-5}	0.3
3650	25	Total body burden		146.0	4.2×10^{-4}	27.0
3650	25	1:4	1:1	36.5	1.1×10^{-3}	6.8[a]
3650	25	1:10	1:1	14.6	4.2×10^{-4}	2.7[a]
3650	25	1:4	5:1	7.3	2.2×10^{-4}	1.4
3650	25	1:10	5:1	2.9	7.8×10^{-5}	0.4
	EPA:	surface area correction		5.4	1.56×10^{-4}	1.0
	FDA:	mg/kg bw-day equivalence		1.0	1.75×10^{-5}	0.2

[a] Considered more likely scenarios.

7.2.4 Other Extrapolation Models

(1) Longstreth and Hushon (1983). Alternative models have been used for extrapolating to low-dose risk. Three of these models, the one-hit, the Probit, and the Weibull, have been discussed and modeled in Appendix C of the HAD (USEPA, 1985). The latter two, plus two others, the Logit and the multihit, have been modeled by Longstreth and Hushon (1984), but they have not adjusted their data for high early nontumor-related mortality. This has been done in Tables 4 and 5 for the Kociba and Squire histopathology analyses, respectively. The resulting MLE and upper-limit risk estimates for several low-dose levels are consistent for both data sets.

Table 4. Estimates of Low-Dose Risk to Humans Exposed to 2,3,7,8-TCDD Based on Female Sprague-Dawley Rats from the Dow Chemical Co. Feeding Study Derived from Four Different Models (Data: Kociba Analysis, Adjusting for Early Mortality)

Dose (mg/kg-day)	Maximum likelihood estimates of extra risks				Upper confidence limit of additional risks			
	Multistage model/One-hit model [a]	Weibull model [b]	Log-probit model	Logit model	Multistage model/One-hit model	Weibull model	Log-probit model	Logit model
10^{-5}	1.1×10^{-6}	1.8×10^{-5}	0	3.1×10^{-7}	1.5×10^{-6}	9.7×10^{-5}	7.7×10^{-18}	2.0×10^{-6}
10^{-4}	1.1×10^{-5}	1.1×10^{-4}	1.2×10^{-13}	4.5×10^{-6}	1.5×10^{-5}	5.3×10^{-4}	3.0×10^{-12}	2.5×10^{-5}
10^{-3}	1.1×10^{-4}	7.1×10^{-4}	4.9×10^{-9}	6.5×10^{-5}	1.5×10^{-4}	2.9×10^{-3}	7.5×10^{-8}	3.0×10^{-4}
10^{-2}	1.1×10^{-3}	4.5×10^{-3}	1.7×10^{-5}	9.4×10^{-4}	1.5×10^{-3}	1.5×10^{-2}	1.5×10^{-4}	3.4×10^{-3}
10^{-1}	1.1×10^{-2}	2.9×10^{-2}	5.2×10^{-3}	1.3×10^{-2}	1.5×10^{-2}	7.2×10^{-2}	2.3×10^{-2}	3.6×10^{-2}
1	1.1×10^{-1}	1.7×10^{-1}	1.7×10^{-1}	1.7×10^{-1}	1.5×10^{-1}	3.0×10^{-1}	3.1×10^{-1}	2.8×10^{-1}

[a] Both models gave identical results.
[b] The value of k, determined by best fit to the data, is <1.

Human equivalent dose (ug/kg-day):	0	0.186	1.86	18.6
Animal tumors/number examined	9/85	3/48	18/48	34/40

Human equivalence conversion: 1 ug/kg-day(oral) = 25.0 ug/m^3 in air.

Table 5. Estimates of Low-Dose Risk to Humans Exposed to 2,3,7,8-TCDD Based on Female Sprague-Dawley Rats from the Dow Chemical Co. Feeding Study Derived from Four Different Models (Data: Squire Analysis, Adjusting for Early Mortality)

Dose (mg/kg-day)	Maximum likelihood estimates of extra risks				Upper confidence limit of additional risks			
	Multistage model/ One-hit model[a]	Weibull model[b]	Log-probit model	Logit model	Multistage model/ One-hit model	Weibull model	Log-probit model	Logit model
10^{-5}	1.1×10^{-6}	3.0×10^{-4}	2.2×10^{-12}	1.0×10^{-5}	1.6×10^{-6}	1.3×10^{-3}	4.4×10^{-10}	5.8×10^{-5}
10^{-4}	1.1×10^{-5}	1.2×10^{-3}	7.6×10^{-9}	8.3×10^{-5}	1.6×10^{-5}	4.4×10^{-3}	1.1×10^{-7}	4.1×10^{-4}
10^{-3}	1.1×10^{-4}	4.9×10^{-3}	5.8×10^{-6}	6.9×10^{-4}	1.6×10^{-4}	1.5×10^{-2}	5.3×10^{-5}	2.9×10^{-3}
10^{-2}	1.1×10^{-3}	2.0×10^{-2}	9.2×10^{-4}	5.7×10^{-3}	1.6×10^{-3}	5.1×10^{-2}	5.8×10^{-3}	1.9×10^{-2}
10^{-1}	1.1×10^{-2}	7.8×10^{-2}	3.3×10^{-2}	4.6×10^{-2}	1.6×10^{-2}	1.6×10^{-1}	1.1×10^{-1}	1.1×10^{-1}
1	1.1×10^{-1}	2.8×10^{-1}	2.9×10^{-1}	2.9×10^{-1}	1.6×10^{-1}	4.3×10^{-1}	4.8×10^{-1}	4.8×10^{-1}

a Both models gave identical results.
b The value of k, determined by best fit to the data, is <1.

Human equivalent dose (ug/kg-day):	0	0.186	1.86	18.6
Animal tumors/number examined	9/85	3/48	18/48	34/40

Human equivalence conversion: 1 ug/kg-day(oral) = 25.0 ug/m^3 in air.

For both data sets the linearized multistage and one-hit models yielded identical results. Also, for both data sets the upper-limit estimates based on the multistage model were consistent, while those based on the other three models varied considerably. For example, at a dose level of 10^{-5} ng/kg-d the UCLs for the multistage model were 1.5×10^{-6} and 1.6×10^{-6} for the Kociba and Squire pathology analyses, respectively. However, at the same dose level of 10^{-5} ng/kg-d, the UCLs for the Logit model varied by a factor of 29, from 2.0×10^{-6} for the Kociba pathology to 5.8×10^{-5} for the Squire pathology. Also, at the same dose level, the UCL from the Weibull model varied by a factor of 13, while those based on the log Probit model varied by a factor of over 50 million. In general the UCL risk estimates based on these models exhibit the relationship:

Weibull > Logit > linearized multistage = one-hit >> log Probit.

All of these models fit the data satisfactorily in the animal experimental range, yet as can be seen from Tables 4 and 5, the estimates can vary over several orders of magnitude at lower environmental doses. The choice of the model must rely on factors other than goodness of fit.

(2) <u>Sielken (1987)</u>. A second risk modeling analysis of the female rat liver response in the Kociba et al. 2-year feeding study has been published by Sielken (1987), whose arguments have also been reproduced by Paustenbauch et al. (1986). Sielken fits the multistage model to the data but focuses on the large difference in low-dose behavior depending on whether or not the high experimental animal dose of 0.1 ug/kg-d is included. Sielken's measure of risk is the VSD, which he defines in terms of 10^{-6} lifetime incremental risk. The VSD is derived from an extrapolation model of the MLE of the multistage model and not from the upper-limit estimate. Sielken thus uses Crump's reparameterization of the multistage model but not Crump's linearized form.

Sielken's analysis is based on the argument that use of the multistage model with the 0.1 ug/kg-d dose-response included forces the estimate of the linear term to be positive non-zero and distorts the true shape of the dose-response curve.

In particular, he argues, even though the multistage model fits the data with the high-dose point included, "the (resulting) fitted models do NOT [his emphasis] reflect the observed behavior at the lower experimental doses." A proper low-dose shape, he claims, is seen when the highest dose is removed. In this case, the linear term of the multistage becomes zero and the quadratic term becomes positive. Extrapolation with the quadratic curve goes rapidly to zero, compared to low-dose extrapolation with a linear model.

An examination of the results derived from the type of analysis suggested by Sielken can be seen in Table 6. In this table, both the MLE and 95 percent lowerbound estimates of the VSD are calculated using different permutations of the Kociba female rat data (animals that died before the appearance of the first tumor have been eliminated). In every case, the model with estimated parameters fit the data satisfactorily. The first row contains the observed liver tumors for all doses and parameter estimates, while the second and lower rows omit the high-dose group. The third through the seventh rows permutate the low-dose data by increments of one tumor-bearing animal. The proportion of 6/48 represents the 95 percent upper limit on the observed proportion of 3 tumor-bearing animals out of 48.

Table 6. MLEs and 95% Upper-Limit Estimates of Parameters and VSDs for Various Permutations[a] of the Female Rat Liver Data Using the Multistage Model. Kociba Histopathology, Eliminating Animals that Died Prior to First Tumor

| | | | | MLE-parameter estimates | | VSD (ug/kg-day) | | |
0	0.001	0.01	0.1	Linear $(ug/kg\text{-}day)^{-1}$	Quadratic $(ug/kg\text{-}day)^{-1}$	MLE	95% Lower limit	Ratio
9/85	3/48	18/48	34/40	$2.1 \times 10^{+1}$	0	4.8×10^{-8}	3.6×10^{-8}	1.3
9/85	3/48	18/48	Omitted	0	$3.7 \times 10^{+3}$	1.6×10^{-5}	2.0×10^{-8}	800
9/85	4/48	18/48	Omitted	0	$3.7 \times 10^{+3}$	1.7×10^{-5}	1.9×10^{-8}	900
9/85	5/48	18/48	Omitted	0	$3.6 \times 10^{+3}$	1.7×10^{-5}	1.8×10^{-8}	950
9/85	6/48[b]	18/48	Omitted	$2.0 \times 10^{+1}$	$1.6 \times 10^{+3}$	5.0×10^{-8}	1.8×10^{-8}	2.8
9/85	7/48[c]	18/48	Omitted	$3.6 \times 10^{+1}$	0	2.8×10^{-8}	1.7×10^{-8}	1.6

[a] All models fit the data sets satisfactorily.

[b] 6/48 corresponds to the 95% upper limit of the observed response of 3/48.

[c] 7/48 corresponds to the 99% upper limit of the observed response of 3/48.

An examination of Table 6 shows the instability of the MLEs under this model. As was pointed out above, omitting the high-dose group changes the form of the model from linear to quadratic, with a corresponding 330-fold increase in the VSD from 4.8 to 10^{-8} to 1.6×10^{-5}. The form of the model remains quadratic until the number of tumor-bearing animals is incremented by 2. However, when the increment becomes 3, to total 6 out of 48, which is the 95 percent upper limit of the actual observed response, the picture again changes. When this 95 percent upper limit of the observed low-dose response is fit, the model again incorporates a linear MLE, and the MLE of the VSD returns to 5.0×10^{-8}.

In contrast to the instability of the MLEs, the 95 percent lower limits of the VSDs remain quite stable over the range of permutations, varying by a factor of 2. The ratios of the VSDs to their lower limits reflect the instability of the MLEs for low-dose extrapolation with the multistage model.

Sielken's results are consistent with the findings of Portier and Hoel (1983), who concluded that for a multistage model fit of the data, the MLE of the linear term is usually multimodel, the number of modes being equal to the degree of the chosen polynomial. This instability of the MLEs from the multistage family of models has also been confirmed by several other authors; it is further seen in Table 6.

Crump (1988) provided a critical review of Sielken's analysis and observed that even if "the true shape of the dose response is a straight line connecting the background response and the response at the mid-dose," the probability of an MLE estimate of zero for the linear term in the multistage model is about 1/3. He concluded that while the data are consistent with Sielken's interpretation of a higher RsD, they are also consistent with those much lower, as displayed in the confidence limits.

It is based on these types of analyses and a consideration of the typical animal bioassay data to be fit that Crump had originally advised that an upper-limit estimate of the low-dose risk be used for his model. Under his reparameterization of the multistage model, an MLE of zero for the linear term becomes possible, and this can cause great instability in low-dose risk estimation. However, under the original development of the model, it is not possible to have higher degree polynomials without having a positive linear or first-stage term. Crump's reparameterization is necessary with quantal data in order to limit the number of parameters. In doing this, however, the MLEs can become unstable, as is seen above.

7.2.5 Time-to-Tumor Analyses

An extension of the multistage model analysis can be conducted when time to observation of the tumors is known. This extension is modeled into the formulation of the multistage model:

$$P(d,t) = 1 - \exp - (q_0 + q_1 d + q_2 d^2 + q_3 d^3 t^k)$$

where q_0, q_1, q_2, q_3, and k are parameters estimated from the data and are all constrained to be nonnegative. This model is often called the Weibull model, but it is more appropriately described as the multistage Weibull, since it is multistage in dose but Weibull in time. Its generalization over the multistage model allows an estimation of the probability of cancer by a fixed age in the absence of any competing risks. Its superiority over the quantal or multistage form is in its ability to adjust for treatment group differences in nontumor-related mortality, as is seen in the Kociba study. However, the analysis requires a pathology decision as to whether the tumors of interest were fatal or incidental, a condition not available in the Kociba study pathology.

The results of a multistage Weibull model are presented in Table 7 and are compared with two quantal analyses using the linearized multistage model. In the first quantal approach, no adjustment is made for the high early mortality of the high-dose group. In the second approach, all animals dying before the appearance of the first liver tumor were dropped from the analysis (see Section 7.2.2). These two analyses are compared with the multistage Weibull under the extremes of either all fatal or all incidental tumors.

Table 7. Results of Various Time-to-Tumor Adjustments for Female Rat Liver (Neoplastic Nodules or Carcinomas) and the Multistage Model[a] Human Equivalent Dosage (Kociba Pathology)

Model	Linear coefficient q_1 $(pg/kg\text{-}day)^{-1}$ of multistage model		Power of time
	MLE	UCL	
Multistage model-quantal			
1. All animals, no adjustment	6.6×10^{-5}	9.0×10^{-5}	NA
2. Only animals surviving to time of first tumor	1.1×10^{-4}	1.5×10^{-4}	NA
Multistage Weibull model			
1. Incidental tumor[b]	2.1×10^{-4}	3.1×10^{-4}	11.56
2. Fatal tumor analysis[b]	6.6×10^{-4}	7.6×10^{-4}	7.07

[a] Three-stage model used for all analyses.
[b] Values shown for q_1 and q_1^* were derived results from WEIBULL82 by multiplying by t_k.

NA = not applicable.

As can be seen from Table 7, the largest difference in risk estimates is between the unadjusted analysis and the fatal tumor for this analysis; 8.4 (for the MLE term) and 1Q (for the upper limit). However, both these extremes are thought to be somewhat misrepresentative of the data, and either of the middle two approaches is considered superior. While the present EPA analysis (see Section 7.2.2) yields the lower of these estimates by a factor of 2, this difference is considered minor.

7.3 Treatment as a Promoter

Evidence for the promoting action of 2,3,7,8-TECC in the rat liver has been well documented in the HAD (USEPA, 1985). This section compares the quantitative cancer risk estimates under the assumption that 2,3,7,8-TECC's action on the liver is one of promotion. It is shown that even when treated solely as a promoter the estimates of risk can vary greatly according to assumptions that one is prepared to make. In Section 7.3.1, the treatment of 2,3,7,8-TCDD as a classical toxicant or promoter with a threshold is presented, along with the results of Canadian and several European regulatory agencies who estimate "virtually safe" levels by applying uncertainty factors to dose levels at which no adverse effects are observed. In Section 7.3.2, the results of a new approach to modeling for promoters are discussed in which cancer response is modeled as a function of liver cell proliferation of initiated cells.

7.3.1 Uncertainty Factor Approach

Several countries and the state of New York have estimated RsDs for 2,3,7,8-TCDD's potency by the application of the uncertainty factor to NOELs, NOAELs, or LOAELs. The general approach is to use uncertainty factors ranging from 10 to 1,000 based on a rule-of-thumb approach (Cook and Page, 1986).

- A factor of 10 where adequate chronic human toxicity data as well as adequate chronic oral toxicity data in more than one animal species are available;

- A factor of 100 where adequate chronic animal toxicity data are available in more than one species, but human toxicity data are lacking; and

- A factor of 1,000 where limited chronic animal toxicity data are available (in only one species) or inconclusive results in more than one species.

The National Academy of Sciences Safe Drinking Water Committee was more explicit with regard to carcinogenicity studies (NAS, 1977).

1. Valid experimental results from studies on prolonged ingestion by man with no indication of carcinogenicity.

Uncertainty Factor = 0

2. Experimental results of studies of human ingestion not available or scanty (e.g., acute exposure only). Valid results of long-term feeding studies on experimental animals or in the absence of human studies, valid animal studies on one or more species. No indication of carcinogenicity.

Uncertainty Factor = 100

3. No long-term or acute human data. Scanty results on experimental animals. No indication of carcinogenicity.

Uncertainty Factor = 1,000"

Others have sought to break down uncertainty factors into components. Weil (1972) interpreted the application of the uncertainty factor of 100 as a product of a factor of 10 to account for differential human sensitivities and the second factor of 10 to extrapolate results from animals to humans. When cancer became the end point of concern, others included a third factor of 10 to raise the total to 1,000. The uncertainty factors, the toxic end points, and the RsDs used by several agencies for the evaluation of 2,3,7,8-TCDD are presented in Table 8. The table includes results from the agencies that have used the linearized multistage model as a comparison with those agencies using the uncertainty factor approach. As can be seen, the RsDs generally segregate into two groups, with those in the lower potency groups all using the NOEL approach. The state of New York, as well as the countries of Canada (plus the Province of Ontario), Switzerland, and Germany, consider the cancer study of Kociba to establish a NOEL at 1 ng/kg-d. The Netherlands also uses the Kociba study as its pertinent cancer study, but instead considers the 1 g/kg-d dose as non-noxious because of the slight increase in liver cell changes in the female rats at this level. All the agencies and all countries except Switzerland, chose the uncertainty factor approach to establish an RsD. Switzerland actually estimated inhalation and oral exposure and divided those exposures into the 1 ng/kg-d NOEL to determine a "safety factor" attributable to those exposures.

As can be seen in Table 8, potency ratios seem to gather in factors of 10. The RsDs for the five agencies above the dotted line span one order of magnitude, as do those for the agencies below the dotted line. Separating the higher- and lower-potency groups is a factor of 15.6 (10 +

Table 8. Comparison of RsDs and Potency Ratio Estimates for 2,3,7,8-TCDD of Various U.S. and Foreign Regulatory Agencies

Agency	Risk-specific dose RsD (pg/kg-day)	Potency ratio	Effect	Approach[a]	Animal study
EPA	6.4×10^{-3}	100	Cancer	LMS[a]	Kociba
California	6.7×10^{-3}	96	Cancer	LMS	NCI male
CDC	2.8×10^{-2}	23	Cancer	LMS	Kociba
FDA	5.7×10^{-2}	11	Cancer	LMS	Kociba
National Research Council (Canada)	6.5×10^{-2}	10	Cancer	LMS	Kociba
Germany	1.0	0.64	Cancer/reproductive effects	Uncertainty factor (1000)	Kociba and Murray
New York State	2.0	0.32	Cancer/reproductive effects	Uncertainty factor (500)	Kociba and Murray
The Netherlands	4.0	0.16	Cancer	Uncertainty factor (250)	Kociba
Health and Welfare (Canada) (also Ontario)	10.0	0.06	Cancer/reproductive effects	Uncertainty factor (100)	Kociba and Murray
Switzerland	--[b]	--	Cancer/reproductive effects	--	Kociba and Murray

[a] LMS = the linearized multistage model.
[b] Used the Kociba study to determine a NOEL of 1 ng/kg-day, but did not provide additional uncertainty factor.

0.64). The total potency range is 1.564. All but one agency use the Kociba data.

7.3.2 Modeling as a Promoter Under the Moolgavkar, Venzon, and Knudson Two-Stage Model

A variation of the multistage model has been developed by Moolgavkar, Venzon, and Knudson (M-K-V, 1979, 1981 which models cancer as a two-stage process with a promotion phase. This model has been shown to predict very well tumor promotion in the mouse skin (Chu et al., 1987). A variation of this model has been developed by Dr. Todd Thorslund to extrapolate from animals to provide human risk estimates of liver cancer deaths. The model considers that the carcinogenic action of 2,3,7,8-TCDD is through its dose-response relationship on the proliferation of initiated liver cells. By including what is known about the receptor-mediated mechanism involved, cell proliferation is itself considered a function of 2,3,7,8-TCDD's binding capacity, which can be shown to follow linear but saturable kinetics. When these parameters are factored into the model, the cumulative probability function, P(x) for dose x at a fixed time t, becomes

$$P(x) = 1-exp-M \ [expG(x)t-1-G(x)t]/g^2(x)$$

$$G(x) = G(0) + [G(\infty)-G(0)][1-exp-Vx]$$

where

P(x) = the probability of a tumor with dose
M = the background mutation rate proportional to background rate
G(x) = the liver cell proliferation rate of initiated cells associated with dose x
G(0) = the background liver cell proliferation rate
G(∞) = the maximum cell proliferation rate possible
V = the parameter associated with the liver saturable kinetics of 2,3,7,8-TCDD-receptor binding.

Even though this model is termed a two-stage model, conceptually it is radically different from the multistage model in several respects. The multistage model, as computed by EPA, is a basic curve-fitting model with all parameters estimated from the data. The two-stage model with promotion as constructed can actually be fit without the estimation of any parameter from the cancer bioassay dose-response data. For example, M and G(0) can be estimated entirely from the control data, V can be estimated from a separate experiment measuring 2,3,7,8-TCDD uptake by the receptor, and G(∞) can be estimated by the incorporation of thymidine into the nuclei following administration of a saturation level of 2,3,7,8-TCDD.

In theory, then, with the exception of the use of control animals for the estimation of background rates, all parameters would be estimated separately and a goodness-of-fit test would be a good measure of how well the model fits the actual data.

An extension of the above distinction between the multistage model and the M-V-K two-stage model is that the parameters derived from a bioassay based on one rat strain can be used to predict the response from a second strain, with only the background rate M having to be estimated from the control group data of the second strain. This procedure was extended from animal-to-human extrapolation where human liver cancer death rates in the United States from 1980 were used to estimate the background human rates for M and G(0), and the values of the other parameters estimated from the rat data were used to provide human risk estimates.

In addition to modeling liver cell proliferation as a nearly linear function of 2,3,7,8-TCDD receptor binding (called here the negative exponential form of the M-V-K model), a second form of the model results when the 2,3,7,8-TCDD-induced cell proliferation rate is assumed to be proportional to the product of cellular 2,3,7,8-TCDD levels and the number of 2,3,7,8-TCDD receptors. This results in a model for the induced cell proliferation rate that is log-logistic in form.

Both models are used to extrapolate from the animal to human cancer response. The results are presented in Table 9, which is reproduced from the Appendix. While both forms of the M-V-K model fit the observed data quite well and both can be justified on theoretical and some experimental grounds, their use for low-dose extrapolation leads to a wide variation in risk estimates. For a lifetime daily dose of 0.1 ng/kg-d, or one order of magnitude below the animal experimental level, the estimates vary by a factor of more than 10,000. Furthermore, even though the negative exponential form of the model results in a linear low dose-response relationship, the risk estimates are two orders of magnitude below the upper confidence limit obtained by EPA, which modeled 2,3,7,8-TCDD as a complete carcinogen. This difference between the estimates with the linearized multistage model and the negative exponential form of the promoter model is due primarily to the low human background liver cancer rates compared to those of the female rat. However, these estimates pertain only to human liver cancer. In its present form, the model is target organ-specific from animals to humans.

Table 9. Estimates of Low-Dose Incremental Risk to Humans
Exposed to 2,3,7,8-TCDD Based on Female Sprague-
Dawley Rats[a] Comparison of the Two Forms of the
Two-Stage Model with Promotion with EPA's Risk
Extrapolation Using the Linearized Multistage
Model[a]

| Dose (ng/kg-day) | Promotion[b] | | Linearized multistage model EPA (Upper confidence limit) |
	Negative exponential	Log-logistic	
10^{-5}	1.7×10^{-8}	--	1.6×10^{-6}
10^{-4}	1.7×10^{-7}	--	1.6×10^{-5}
10^{-3}	1.7×10^{-6}	$<10^{-13}$	1.6×10^{-4}
10^{-2}	1.7×10^{-5}	8.8×10^{-10}	1.6×10^{-3}
10^{-1}	1.8×10^{-4}	8.8×10^{-7}	1.6×10^{-2}
1	2.4×10^{-3}	9.3×10^{-4}	1.6×10^{-1}

[a] Taken from Table 11 of the Appendix.
[b] Squire's pathology analysis adjusting for early mortality.

Estimates with the negative exponential form of this promoter model
might be considered as providing a conservative, on the high side, estimate of induced liver cancer risk, since cell proliferation is modeled as
a linear function of dose. However, not enough research has been done on
the low-dose properties of these models to characterize the variability
of the low-dose risk estimates. While the upper-limit estimates should
certainly be below those of the linearized multistage model, further
research needs to be done before these models can be used for regulatory
decision making.

7.4   Comparison of Animal Prediction with Actual Human Data

With the exception of one study on 2,3,7,8-TCDD in Holmesburg,
Pennsylvania, in 1967, all human exposure data are derived from accidental
exposures of unknown quantity. In the Holmesburg study, 2,3,7,8-TCDD was
topically applied to volunteer prisoners in total doses ranging from
0.4 ug to 7,500 ug. According to testimony and exhibits in EPA's 2,4,5-T
cancellation hearings (Rowe, 1980), doses below 16 ug did not elicit a
chloracne response, while a dose of 7,500 ug did cause chloracne in 8 out
of 10 subjects. This high dose of 0.05 mL of a 1 percent 2,3,7,8-TCDD
solution in 50/50 alcohol chloroform solvent was applied to one square
inch of the backs every other day for a month and was covered by a non-
occlusive patch. If we can assume that the subjects' average age was 35,
a 25 percent absorption rate, an infinite half-life, and lifetime
(70 years) average daily dose (LADD) proportionality, one can estimate an

internal dose of 2,3,7,8-TCDD that caused chloracne to be in the 2 to 1,000 pg/kg body weight-day range. If we assume either a 5- or 10-year half-life with first-order elimination kinetics and with the other assumptions the same, then the chloracne causing the LADD range is unchanged at one significant digit. These figures correspond to an external oral dose of 4 to 2,000 pg/kg body weight-day.

Human cancer risk estimates based on the Kociba female rat feeding study with 55 percent absorption yield an upper-limit risk estimate of 1.56×10^{-4} $(pg/kg-d)^{-1}$. Multiplying this upper-limit estimate by 4 to 2,000 pg/kg-d and adjusting for the different absorption fractions yields an upper-limit lifetime incremental cancer risk of between 6×10^{-4} and 3×10^{-1} for humans developing chloracne. Only 10 of the Holmesburg prisoners were exposed to the highest dose, and follow-up is unclear. Nevertheless, even if their lifetime projected incremental cancer risks were as great as 0.3, and even if they were observed for their full remaining lifetimes, less than three additional cancers would be expected. Put another way, with three additional cancers expected, an observation of no additional cancers would not be highly unusual (p = 0.055). Clearly, unless specific types of cancer were to appear in these tested prisoners, no conclusions could be drawn.

On the other hand, Tschirley (1986), in his review of the 2,3,7,8-TCDD literature, displays nine cohorts of some 599 subjects who developed chloracne following 2,3,7,8-TCDD and phenoxy herbicide exposure. This is reproduced as Table 10. Of those cohorts, only the study of the 1949 accident at the Monsanto plant in Nitro, West Virginia (Zack and Suskind, 1980), and the 1953 accident at the BASF plant in Germany (Thiess et al., 1982) have a sufficient latent period and other information to allow a comparison to be made between observed and predicted risks. These involved 177 chloracne cases and either a 26- or 30-year latency period. Quantitative cancer risk estimates based on these cohort experiences are calculated below. However, the large uncertainty in the exposure estimates should be emphasized, and the assumptions used to derive these exposure estimates are necessary simplifications.

7.4.1 <u>Thiess et al. (1982)</u>

(1) <u>Description</u>. At a factory in Ludwigshafen, Federal Republic of Germany, in 1953, during the hydrolization of 1,2,4,5-tetrachlorobenzol to 2,4,5-Trichlorophenol, an accident happened exposing at least 70 persons. These 70 as well as 4 additional persons who were only exposed "for a short time" during 1954 and 1955 were included in the cohort. Of the

Table 10.   Exposure to 2,3,7,8-TCDD from Industrial Accidents

Date	Workers exposed	Location of accident	Remarks
1949	250	Monsanto plant in Nitro, WV	122 cases of chloracne being studied; 32 deaths vs. 46.4 expected; no excess deaths from malignant neoplasms or circulatory disease
1953	75	BASF plant in Ludwigshafen	55 cases of chloracne, 42 severe; 17 deaths vs. 11 to 25 expected (four gastrointestinal cancers and two oat-cell lung cancers); most common injuries were impaired senses and liver damage
1956	?	Rhone-Poulenc plant in Grenoble	17 cases of chloracne; also elevated lipid and cholesterol levels in the blood
1963	106	NV Philips plant in Amsterdam	44 chloracne cases (42 severe) of whom 21 also had internal damage or central nervous system disturbances; eight deaths (six possible myocardial infarctions); some symptoms of fatigue
1964	61	Dow Chemical plant in Midland, MI	49 cases of chloracne; 4 vs. 7.8 expected deaths; 3 cancer deaths vs. 1.5 expected; one a soft tissue sarcoma
1965-69	78	Continuing leaks in Spolana plant near Prague	78 cases of chloracne; five deaths; many of the 50 workers studied for more than 10 years have hypertension, elevated blood levels of lipid and cholesterol, prediabetes; significant amounts of severe liver and neurologic damage

Source: Tschirley, 1986.

74 persons, 66 suffered chloracne or severe dermatitis. All 74 persons were successfully traced through 1979. Of the 74 persons in the cohort, 21 had died during the 26 years of observation, just slightly more than were expected in any of five different control groups. However, seven cancer deaths were observed versus 4.03 to 4.35 that were expected in these control groups. Of these seven cancers, three were stomach (ICD 151), one was colon (ICD 153), and three were lung (ICD 162). All seven occurred at least 10 years after the accident. A 10-year latent period will be assumed. The results are presented in Table 11. The control group represents the expected deaths based on the mortality rates of Rhinehessia-Palatinate 1970-1975. The mortality for stomach cancer was statistically significant ($p = 0.016$), while that for lung cancer was marginally significant ($p = 0.09$).

Table 11. Observed and Expected Deaths from Stomach, Colon, and Lung Cancer in the 74 Cases with Chloracne or Severe Dermatitis from the 1953 BASF Plant in Ludwigshafen (26-Year Follow-Up with at Least 10 Years' Latency)

Cause of death	ICD No.	Observed	Expected	SMR	p-value	95% Upper confidence limit based on Poisson distribution
Stomach cancer	151	3	0.52	5.76	0.016	1.15 - 16.8
Colon cancer	153	1	0.24	4.17	0.21	0.05 - 23.2
Lung cancer	162	3	1.05	2.86	0.09	0.57 - 8.3

Person-years at risk = 972.6.

(2) Exposure estimates. [Contributions to this section were made by Drs. Jerry Blancato and Lorenz Rhomberg of the Office of Health and Environmental Assessment.] Although there were no concurrent estimates or measurements of 2,3,7,8-TCDD exposure, Rappe et al. (1987) reported a mean level of 100 pg/g in adipose tissue of four exposed BASF workers some 30 years after exposure. If one is willing to make several simplifying assumptions, then one can estimate a range of exposure. The assumptions are:

- The body is treated as a one-compartment open model.

- A range of half-lives is assumed to be between 5 and 30 years.

- Adipose tissue in a 70-kg human is about 20 percent or 14 kg.

- Absorption into the body is assumed to be between 5 and 25 percent. This figure is arbitrary and is chosen because there is less than 55 percent absorption in the rat feeding studies.

Under these assumptions, the pertinent equation is:

$$C(t) = C_o e^{-k\,t}$$

where

C(t) = the concentration in the fat immediately following absorption in 1953

k = the total body elimination rate constant, assumed to be the sum of the elimination rate constants by various physiological processes

t = the time since absorption was completed, assumed to be 30 years.

Further, it can be noted:

$$k = 0.693/t_{0.5}$$

where

$t_{0.5}$ = half-life of elimination from the body, and

$$\text{Exposure} = C_o V/r$$

where

V = weight of the fat compartment (14 kg)

r = the absorption traction.

In this case, the weight of the fat compartment is used to calculate the dose. The other organs may be neglected because of the propensity of 2,3,7,8-TCDD to partition into the fat. For example, the concentration in the fat is about 100 times that of the blood.

Based on the above assumptions and estimates, the range of estimates of 2,3,7,8-TCDD exposure in the BASF plant accident is between 10 and 1,800 ug. Based on a LADD, these estimates range from 1.5 to 50 pg/kg body weight-day. The calculations are shown in Table 12. Clearly, this factor of 180 in the range of exposures and a factor of 30 in the range of LADDs creates significant uncertainty in the risk estimate calculations. To narrow the range, however, it can be noted that in the Holmesburg study exposures below 16 ug caused no chloracne, while exposures of 7,500 ug caused 80 percent chloracne. Considering that greater absorption is probable in the Holmesburg study, the exposure estimates in the top three rows of Table 12, showing the range of 220 to 1,800 ug, seem the most likely. Therefore, LADDs in the range of 6 to 50 pg/kg body weight-day. The risk estimates below will be calculated on the LADD of 50 pg/kg/body weight-day; risk estimates based on the lower end of the range would be about eight times higher.

234

Table 12. Estimates of 2,3,7,8-TCDD Exposure to Workers at the BASF
Plant in Ludwigshafen Based on Mean Adipose Levels in Four
Subjects 30 Years Later and Various Assumptions

Mean adipose concentration at T=30 years[a] (pg/g)	Half-life[b] $T_{0.5}$ (yr)	Absorption fraction r(0.05-0.25)	Elimination constant $k(yr^{-1})$[c]	Adipose concentration at T=0 years[d] C_o (pg/g)	Exposure (ug)[e]	Lifetime average daily dose[f] (pg/kg bw-day)
100	5	0.25	0.139	6471	1800	50
100	5	0.25	0.139	6471	360	50
100	10	0.05	0.069	799	220	6
100	15	0.05	0.046	397	110	3
100	20	0.05	0.035	286	80	2
100	30	0.05	0.023	199	55	1
100	30	0.25	0.023	199	10	1

[a] Measured by Rappe (1987).
[b] Different half-lives are assumed.
[c] $k = 0.693/T_{0.5}$.
[d] $C(t) = C_o \exp(-30k)$.
[e] Exposure = C_o x 14,000 g/r.
[f] LADD = dose per kg bw. Assumes area under time-concentration curve dose equivalence, average age at
accident = 35 years. Formula is (see Section II.C.): LADD = $[C_o \bullet (1-\exp(-35K))K/70(365)K-1]$ 14/70.

(3) Risk Estimates. Cancer risk estimates are calculated for the
stomach cancer and lung cancer mortality presented in Table 11. These
estimates are presented in Table 13. Two models are considered, the additive and relative risk models, which have been used in several EPA risk
assessments. They are developed and more fully explained in the recent
EPA update on dichloromethane (USEPA, 1987). Both models require estimates of LADDs, and these have been estimated above as between 6 and
50 pg/kg body weight-day. The results show that incremental cancer risk
estimates based on these human data are higher than those based on ·the
animal data in every case. If the lower end of the range of LADDs had
been used, the human estimates would have been greater still. The conclusion based on the above analysis is that the upper-limit incremental unit
risk estimates based on the animal studies do not overestimate the human
risk from 2,3,7,8-TCDD if either the lung or stomach cancer mortality
response in humans is bona fide.

7.4.2 Zack and Suskind (1980)

(1). Description. In Nitro, West Virginia, on March 8, 1949,
excessive temperatures in an autoclave involved in the production of

Table 13. Comparison of Incremental Cancer Risk Estimates for Lifetime Exposure to 2,3,7,8-TCDD Based on the Kociba Female Rat Liver, Lung, and Hard Palate/Nasal Turbinates (HP/NT) Response with Estimates Based on Lung and Stomach Cancer Mortality in 74 Workers at the BASF Plant in Ludwigshafen (Assume LADD = 50 pg/kg bw-day)[a]

Species/ sex	Model used	Cancer type	Parameter estimates[b] Δ	Asymptotic variance of estimates	Lifetime incremental cancer risk per 1 ng/kg-day continuous exposure		
					Lower limit	MLE[c]	95% Upper limit[d]
Humans/males	Additive	Stomach	2.6×10^{-2}	3.2×10^{-4}	0	1.8	3.8
		Lung	2.0×10^{-2}	3.2×10^{-4}	0	1.4	3.4
	Multiplicative	Stomach	47.7 $P_o = 0.013$	1.1×10^{3}	0	6.2×10^{-1}	1.3
		Lung	19.2 $P_o = 0.038$	2.6×10^{2}	0	6.5×10^{-1}	1.7
Rats/females	Linearized multistage	Lung, liver, HP/NT			0	1.15×10^{-1}	1.56×10^{-1}

[a] A LADD of 50 pk/kg bw-day via skin and oral route with estimated 5% absorption is equivalent to an administered dose of 100 pg/kg bw-day when exposure is via ingestion with estimated 50% absorption. All the above estimates are standardized to the administered dose via ingestion.

[b] For the additive model, the parameter estimates are in units of kg-day/ng-year. For the multiplicative model, the estimates are in units of kg-day/ng.

[c] MLE = maximum likelihood estimate. For the additive model, MLE = 70 x Δ; for the multiplicative model, MLE = $P_o \times \Delta$.

[d] For humans, the upper limits are based on the asymptotic variances. This produces lower upper-limit values.

2,4,5-trichlorophenoxyacetic acid caused a relief valve to open, allowing fumes and residues to escape into the atmosphere and into the interior of the building. A total of 121 white males were identified as having developed chloracne following this incident, and these were included in the subsequent follow-up study, with vital status ascertained through the last day of 1978, nearly 30 years later.

There was 100 percent follow-up of subjects; 89 were still alive and "32 were verified deceased by death certificate." With 46.41 expected deaths, the standardized mortality ratio (SMR) for all causes, 69, was significantly ($p < 0.05$) lower than expected. The only cause of death that displayed normal rates was cancer, which had 9 observed and 9.04 expected, SMR = 99.6. Of those cancer deaths, five were from lung (2.85 expected, SMR = 175), three were from lymphatic or hematopoietic tissue (0.88 expected, SMR = 341), and one was a soft tissue sarcoma (STS) (0.15 expected). Only one of the cancer deaths, a lung cancer, was a non-smoker.

(2) Risk estimates. To make any kind of quantitative risk estimation of the potency of 2,3,7,8-TCDD, several assumptions must be made. These assumptions are:

(a) Exposure. As discussed above for the Holmesburg study, it is assumed that the LADD necessary to cause chloracne was in the 2 to 1,000 pg/kg-d range. An estimate in the middle, or 500 pg/kg-d, appears to be a reasonable starting point, as does a value of 150 pg/kg-d, which is close to the geometric mean. The value of 500 pg/kg-d will be chosen as the LADD dose for those who developed chloracne, but the uncertainty of this estimate must be stressed. This dose estimate is 10 times greater than that estimated for the BASF study; the incremental unit risk cancer estimates will be correspondingly lower.

(b) Expected deaths from cancer. Table 14 presents the observed and expected cancer deaths for the 29.8-year latency. However, all the cancer deaths appeared after at least a 10-year latency, and it seems reasonable that 10 years is an appropriate latent period for any 2,3,7,8-TCDD-related cancer to express itself. Therefore, the expected deaths presented by Zack and Suskind must be adjusted by subtracting the first 10 years of experience. An examination of vital statistics rates for lung cancer and STS suggests that for lung cancer deaths approximately 20 percent of a 30-year death experience happens in the first 10 years; for STS the figure is approximately 30 percent. Based on these adjustments, the expected

Table 14.   Follow-Up of 121 Chloracne Cases at the Monsanto Company
(Nitro, West Virginia) Used to Derive Quantitative
Cancer Risk Estimates

| | Lung cancer | | STS | |
	Unadjusted	Adjusted	Unadjusted	Adjusted
Deaths				
Observed	5	5	1	1
Expected	2.85	2.3	0.15	0.10
Person-years at risk	--	2082	--	2082
SMRs				
Observed	175	217	667	1000
95% Confidence limits based on Poisson distribution	57-410	70-508	9-3707	13-5560

deaths for this exposed cohort become 2.3 and 0.10 for lung and STS cancers, respectively.

(c)   Person-years at risk.   A figure needed for the additive risk
model but not the relative risk model is the person-years at risk.   Since
the first 10 years are considered to be a latent or risk-free period,
only the last 19.8 years are counted as person-years at risk. The follow-
up was complete and there were 32 deaths. It can be assumed that the
average time until death was 20 years from first exposure (for the nine
cancer deaths the average time from the accident until death was
22 years). Therefore, the total person-years at risk can be estimated as

$$P\text{-}Y = 19.8 \times 89 + 10 \times 32 = 2,082.$$

The results of the analysis using both the additive and multiplicative models are presented in Table 15. Based on the assumptions discussed
above, both models yield similar results.   For extrapolation based on
lung cancer deaths, the MLEs for incremental risks are 9.1 x 10^{-2} and
4.5×10^{-2} $(\text{mg/kg-day})^{-1}$ for the additive and multiplicative models,
respectively. For extrapolation based on the one STS death, the MLEs are
lower than are those for lung, 3.0×10^{-2} and 8.6×10^{-3} for the additive and multiplicative models, respectively. The 95 percent upper-limit
estimates presented in Table 15 are those based on asymptotic normality
theory and are, therefore, lower than would be produced by applying the
upper confidence limits on the SMRs from Table 14, which are based on the
Poisson distribution.   Those Poisson upper-limit adjusted SMRs, for
example, would yield 95 percent upper-limit cancer unit risk estimates of

Table 15. Comparison of Incremental Cancer Risk Estimates for Lifetime Exposure to 2,3,7,8-TCDD Based on the Kociba Male Rat Liver, Lung, and Hard Palate/Nasal Turbinates (HP/NT) Response with Estimates Based on Sarcoma Mortality in 121 Monsanto Employees (Assume LADD = 50pg/kg bw-day)[a]

Species/ sex	Model used	Cancer type	Parameter estimates[b] Δ	Asymptotic variance of estimates	Lifetime incremental cancer risk per 1 ng/kg-day continuous exposure		
					Lower limit	MLE[c]	95% Upper limit[d]
Humans/males	Additive	STS	4.3×10^{-4}	2.3×10^{-7}	0	3.0×10^{-2}	8.5×10^{-2}
		Lung	1.3×10^{-3}	1.2×10^{-6}	0	9.1×10^{-2}	2.2×10^{-1}
	Multiplicative	STS $P_o = 0.00095$	9.0	10.0	0	8.6×10^{-3}	1.4×10^{-2}
		Lung $P_o = 0.038$	1.2	0.94	0	4.5×10^{-2}	1.1×10^{-1}
Rats/females	Linearized multistage	Lung, liver, HP/NT			0	1.15×10^{-1}	1.56×10^{-1}

[a] A LADD of 50 pk/kg bw-day via skin and oral route with estimated 5% absorption is equivalent to an administered dose of 100 pg/kg bw-day when exposure is via ingestion with estimated 50% absorption. All the above estimates are standardized to the administered dose via ingestion.

[b] For the additive model, the parameter estimates are in units of kg-day/ng-year. For the multiplicative model, the estimates are in units of kg-day/ng.

[c] MLE = maximum likelihood estimate. For the additive model, MLE = 70 × Δ; for the multiplicative model, MLE = $P_o \times \Delta$.

[d] For humans, the upper limits are based on the asymptotic variances. This produces lower upper-limit values.

239

5.2 x 10^{-2} (ng/kg-day)$^{-1}$ for STS and 1.6 x 10^{-1} for lung cancer deaths for the multiplicative model. These values are significantly greater than those presented in Table 15 (1.4 x 10^{-2} and 1.1x10^{-1} (ng/kg-d)$^{-1}$, respectively) and reflect the high degree of variability resulting from the small sample size.

Also presented in Table 15 is the quantitative cancer risk estimate derived from the Kociba feeding study (USEPA, 1985). While the MLEs based on the animal data are slightly higher than those based on the Monsanto data, the differences are not greater than a factor of 2.4 for lung cancer and 13.4 for STS. All the 95 percent upper-limit estimates based on the human data are within a factor of 2 of the upper-limit estimate based on the rat data.

The conclusion based on the above analysis is similar to that derived from the BASF analysis--the available human cancer data on 2,3,7,8-TCDD do not provide any evidence that the unit risk estimate based on rat data overpredicts the human experience. The information on human exposure is just too uncertain to allow for a more definitive statement.

7.5   Discussion and Summary

This report has presented the effects on the human cancer risk estimates from 2,3,7,8-TCDD exposure under varying assumptions involved in the animal-to-human extrapolation procedure. It has compared EPA's risk estimates with those of other agencies, both U.S. and foreign, has discussed the rationale used in each, and has shown the effects of slightly different assumptions on the estimates. In general, the risk extrapolations can be roughly divided into two groups--those agencies using the linearized multistage family of models for extrapolation and those using an uncertainty factor approach. The agencies using the linearized multistage model all produce RsDs within a factor of 10; similarly, the agencies using the uncertainty factor approach are also produce RSds within a factor of 10. The two groups, however, separated by a factor of 16. Thus, the lowest RsD, that of EPA (which used the linearized multistage model), is 1,600 times lower than that of Health and Welfare Canada, which used an uncertainty factor of 100.

While EPA's cancer risk estimates were the highest of the agencies presented, other methodologies consistent with the data have yielded still higher estimates. For example, fitting the data with both the Logit and the Weibull models would have produced significantly higher estimates-- the Logit by roughly one order of magnitude and the Weibull by two orders

240

of magnitude. In addition, even extending the multistage model to the Weibull-in-time model under a time-to-tumor analysis would have increased the upper-limit estimates by as much as a factor of 5. Of even greater uncertainty is the extremely long half-life of 2,3,7,8-TCDD in the human compared to the rat. If half-life is related to species sensitivity, as implied by the cross-species extrapolation factor, then recent estimates of human half-life of 6 to 10 years imply that rat-to-human extrapolation estimates should be significantly higher, probably by a factor of 2 to 7.

Finally, a new methodology has been presented that models 2,3,7,8-TCDD as a carcinogen promoter only. The methodology models the promotion (liver cell proliferation) phase of the M-V-K two-stage model as a linear saturable function of 2,3,7,8-TCDD-receptor binding. The risk estimates of induced human liver cancer are two orders of magnitude less than those using EPA's current methodology. However, research on this new methodology is ongoing, and not enough is known about the lowdose estimation properties to make definitive statements.

Which of these is the "correct answer"? The answer is probably "none of the above." What the above analyses show are that all the answers are consistent with the observed data, and all have some credence depending upon the believability of the assumptions used. The most pertinent fact is that 2,3,7,8-TCDD causes liver, tongue, hard palate/nasal turbinates, and lung tumors in rats at doses and conditions to which humans would never be exposed. As such, even with animal bioassays as well-conducted as were those for 2,3,7,8-TCDD, the information they contain for low-dose extrapolation is very limited. Within a 100-fold decrease from experimental dose levels, the range of estimates predicted by models that fit the data well, rapidly diverge to a "pay your money, take your choice" level of three orders of magnitude. Below that, divergence is even more rapid (see Tables 4 and 9). Furthermore, when extrapolation is made from animals to humans, the uncertainty about the effect of the extremely long half-life in humans raises concern about the conservativeness of the upper limit based on the surface area correction for extrapolation.

Use of human data for risk assessment purposes is also impossible with 2,3,7,8-TCDD. First, the evidence for human carcinogenicity of 2,3,7,8-TCDD alone is judged inadequate. Second, the studies providing positive evidence for carcinogenicity are of a case-control design; these lack both a population base and exposure estimates. The few cohort studies available lack sensitivity because of a combination of factors including exceptionally small cohorts, insufficient latency, and, in some instances,

little evidence of exposure. Only with one study was there even an estimate of 2,3,7,8-TCDD exposure, and that study lacked the power to discredit any of the predictions provided by the animal data.

The next question is whether any of these estimates can be considered
superior to the others. To answer this question, one must first presume
that mathematical models can be used for prediction and then decide
whether to model 2,3,7,8-TCDD as a complete carcinogen or as a promoter
only. When modeling as a complete carcinogen, only the one-hit and multi-
state models have a theoretical backing in carcinogenesis; the other
models presented are merely well-known tolerance distribution models. The
EPA position is that use of the linearized multistage model for extrapolating upper limits of incremental risk is prudent and protective. When
both animal and human data have been available for risk estimation, the
linearized multistage model's use with animal data has provided estimates
comparable to those derived from human data. Furthermore, low-dose supra-
linearity is rarely seen, so that using the linearized multistage model
for extrapolating from experimental levels probably represents a protective level for humans.

When one models 2,3,7,8-TCDD as a promoter only, many additional
uncertainties arise. The two most important for modeling are those associated with reversibility and threshold. Classical promoters are known
to show reversibility of lesions when the promoter is no longer administered and cleared from the system. Furthermore, large doses are typically
required for promotion, indicative of a toxicity effect either leading
directly to cell damage and regeneration, or overwhelming the cell's ability to prevent the promoter from reaching its site of action. In the case
of 2,3,7,8-TCDD, the evidence for liver cancer in animals points to a
mechanism of promotion via receptor binding, yielding a very strongly
bound complex. Even if this complex is broken down, the persistence of
2,3,7,8-TCDD allows it to bind with another receptor. In terms of mathematical modeling, this information translates to a dose-response function
for cell proliferation that excludes threshold and reversibility, but
otherwise can be defined by Michaelis-Menten-type kinetics. When this
function is substituted into the promoter form of the M-V-K model, the
results yield low-dose estimates of liver cancer for humans that are two
orders of magnitude lower than the upper-limit estimates provided by the
one-hit and linearized multistage models.

242

While the upper-limit estimates can be considered upper limits for total human 2,3,7,8-TCDD-induced cancer, the predictions with the promoter form of the M-V-K model require further examination. First, they apply solely to human liver cancer, a condition reported only in the cohort exposed to dibenzofuran-contaminated PCBs in Yusho, Japan (Amand et al., 1984). They do not include estimates for STS or non-Hodgkins lymphomas. Second, they are considered "best" estimates compared with the upper-limit estimates calculated from the linearized multistage model. Third, the form of the M-V-K model used predicts carcinogenic response only on the basis of promotion. If a 2,3,7,8-TCDD-induced initiation stage had been incorporated (as is suggested by Holder and Rosenthal, 1987), the low-dose cancer predictions would have been higher.

For these reasons, the estimates provided by the promoter form of the M-V-K model might be considered prudent lower bounds on cancer risk, while those upper-limit estimates provided by EPA's current methodology are to be considered upper bounds. While any number of assumptions could produce either lower or higher risk predictions, the biological facts incorporated into both models provide what should be considered reasonable working limits. In addition, 2,3,7,8-TCDD's many other toxicities should also be a consideration in setting a lower limit. Any criteria level lower than that provided by the M-V-K model would be at a level where these concerns would prevail. A final caveat remains, however, on the impact of the extremely long half-life of 2,3,7,8-TCDD in man. If 2,3,7,8-TCDD is not sequestered in the fat but is bioavailable, the quantitative risk estimates would be considerably larger.

REFERENCES

Amand M, Yagi K, Nakajima H, Takehora R, Sakai H, Umed G, Labor, K. 1984. Statistical observations about the causes of the death of patients with oil poisoning. Japan Hygiene 39(1):(translation)

Armitage P, Doll R. 1954. The age distribution of cancer and a multistage theory of carcinogenesis. Br. J. Cancer 8:1-12.

Barnes D. 1987. Holmesburg prison-based assessment. Unpublished notes. March 10, 1987.

California Department of Health Services. 1985. Health effects of 2,3,7,8-TCDD and related compounds (draft). Berkeley, California. Unpublished. April 19, 1985.

CDC. 1980. Centers for Disease Control. Serum dioxin in Vietnam-era veterans. Preliminary report. Morbidity and Mortality Weekly Report 36 (28):470-475.

Chu KC, Brown CC, Tarone RE, Tan WY. 1987. Differentiating among proposed mechanism for tumor promotion in mouse skin with the use of the multievent model for cancer. J. Natl. Cancer Inst. 79(4)789-796.

Cook BT, Page NP. 1986. Comparison of carcinogenicity risk assessment of 2,3,7,8-TCDD (draft). EPA Contract no. 68-02-4131, Work Order No. 3. February 3, 1986.

Crockett PW, Crump KS. 1986. Methods for assessment of non-cancer health risks. Prepared by K.S. Crump and Co. for the Electric Power Research Institute. Contract no. RP 1826-17.

Crump KS. 1983. How to handle animal survival data in quantitative risk assessment: a discussion paper. Prepared for Research Triangle Institute, EPA Contract no. 68-01-6826, Delivery Order 28.

Crump KS. 1988. A critical evaluation of a dose-response assessment for TCDD. Food. Chem. Toxicol. In press.

Crump KS, Guess HA, Deal LL. 1977. Confidence intervals and test of hypothesis concerning dose-response relations inferred from animal carcinogenicity data. Biometrics 33:437-451.

Crump KS, Watson WW. 1979. GLOBAL 79: A FORTRAN program to extrapolate dichotomous animal carcinogenicity data to low dose. Prepared for the National Institute of Environmental Health Sciences. Contract no. I-ES-2123.

Environmental Canada. 1984. Chlorophenols and their impurities in the Canadian environment. 1983 Supplement. Economic and technical review report. EPS3-EP-84-3. Environmental Protection Programs, Directorate, Canada.

FDA. 1983. Food and Drug Administration. Statement by S.A. Miller, Director, Bureau of Foods, FDA, before the Subcommittee on Natural Resources, Agriculture Research and Environment, U.S. House of Representatives. June 30, 1983.

Germany. 1984. Report on dioxins. Update to November 1984. Federal Environmental Agency.

Holder JM, Rosenthal S. 1987. Issue paper concerning the mechanism of 2,3,7,8-TCDD carcinogenicity. Office of Health and Environmental Assessment, U.S. Environmental Protection Agency, Washington, DC. Unpublished draft report. February 13, 1987.

Howe RB, Crump KS. 1982. WEIBULL82: a FORTRAN program for low-dose extrapolation using time to tumor data. K.S. Crump and Company, Inc., Ruston, LA.

Kimbrough RD, Falk H, Stehr P, Fries G. 1984. Health implications of 2,3,7,8-tetrachlorodibenzo-p-dioxin (2,3,7,8-TCDD) contamination of residential soil. J. Toxicol. Environ. Health 14:47-93.

Kociba RJ, Keyes DG, Beyer JE, et al. 1977. Results of a two-year chronic toxicity and oncogenicity study of 2,3,7,8-tetrachlorodibenzo-p-dioxin in rats. Submitted to the U.S. Environmental Protection Agency. Unpublished.

Kociba RJ, Keyes DG, Beyer JE, et al. 1978. Results of a two-year chronic toxicity and oncogenicity study of 2,3,7,8-tetrachlorodibenzo-p-dioxin in rats. Toxicol. Appl. Pharmacol. 46(2):279-303.

Longstreth JD, Hushon JM. 1983. Risk assessment for 2,3,7,8-TCDD. In: Tucker, RE, Young, AL, Gray, AP, eds. Human and environmental risk of chlorinated dioxins and related compounds. New York, NY: Plenum, pp. 639-664.

Moolgavkar SH, Knudson AG. 1981. Mutation and cancer: a model for human carcinogenesis. J. Natl. Cancer Inst. 66:1073-1052.

Moolgavkar SH, Venzon DJ. 1979. Two-event models for carcinogenesis. Incidence curves for childhood and adult tumors. Math Biosci. 47:55-77.

NAS. 1977. National Academy of Sciences. Drinking water and health. Safe Drinking Water Committee, National Academy of Sciences. Washington, DC.

NRCC. 1981. National Research Council of Canada. Polychlorinated dibenzo-p-dioxins: criteria for their effects on man and his environment. Publ. no. 18474, ISSN 0316-0114. NRCC/CNRC Associate Committee on Scientific Criteria for Environmental Quality, Ottawa, Canada. 251 pp.

NTP. 1982. National Toxicology Program. Carcinogenesis bioassay of 2,3,7,8-tetrachlorodibenzo-p-dioxin in Osborne-Mendel rats and B6C3F1 mice. Technical Report no. 209. Research Triangle Park, NC.

Ontario, Canada, Ministry of the Environment. 1985. Scientific criteria document for standard development. No. 4-84. PCDDs and PCDFs.

Paustenbauch DJ, Shu HP, Murray FJ. 1986. A critical examination of assumptions used in risk assessment of dioxin contaminated soil. Regul. Toxicol. Pharmacol. 6:284-307.

Portier C, Hoel D. 1983. Low-dose-rate extrapolation using the multistage model. Biometrics 39:897-906.

Rappe C, Aldersson R, Bergquist P. 1987. Sources and relative importance of PCDD and PCDF emissions. Presented at the ISWA/WHO-EURO/ DAKOFA Specialized Seminar on Emissions of Trace Organics from Municipal Solid Waste Incinerators, Copenhagen, Denmark.

Rose JO, Ramsey JC, Wentzlin TH, Hummel RA, Gehring PJ. 1976. The fate of 2,3,7,8-TCDD following single and repeated oral doses to the rat. Toxicol. Appl. Pharmacol. 36:209-226.

Rowe VK. 1980. Direct testimony at U.S. EPA's cancellation hearings of 2,4,5-T. Exhibit 865. FIFRA Docket Nos. 415 et al.

Sielken RL, Carlborg FW, Paustenbauch DJ, Shu HP, Murray FJ. 1986. Alternative approaches to mathematically analyzing the bioassay data for 2,3,7,8-TCDD (Abstract 1133.) Presented at the 25th Annual Meeting of the Society of Toxicology, New Orleans, LA.

Sielken RL. 1987. Quantitative cancer risk assessments for TCDD. Food Chem. Toxicol. 25(3):257-267.

Squire RA. 1980. Pathologic evaluations of selected tissues from the Dow Chemical TCDD and 2,4,5-T rat studies. Prepared for the Carcinogen Assessment Group, U.S. Environmental Protection Agency, Contract no. 68-01-5092.

Switzerland. 1982. Environmental pollution due to dioxins and furans from communal rubbish incineration plants. Federal Office of Environmental Protection, Bern.

Thiess AM, Frentzel-Beyme R, Link R. 1982. Mortality study of persons exposed to dioxin in a trichlorophenol-process accident that occurred in the BASF on November 17, 1983. Am. J. Ind. Med. 3:179-189.

Tschirley F. 1986. Dioxin. Scientific American 25(2):29-35.

USEPA. 1980. U.S. Environmental Protection Agency. Risk assessment on (2,4,5-trichlorophenoxy)acetic acid (2,4,5-T), (2,4,5-trichloro-phenoxy) propionic acid, and 2,3,7,8-tetrachlorodibenzo-p-dioxin (TCDD). Office of Health and Environmental Assessment, Washington, DC. EPA 600/6-81-003. NTIS PB81-234825.

USEPA. 1985. U.S. Environmental Protection Agency. Health assessment document for polychlorinated dibenzo-p-dioxins. Office of health and Environmental Assessment, Washington, DC. EPA/600/8-84/014F. NTIS PB86-122546/AS.

USEPA. 1986. U.S. Environmental Protection Agency. Guidelines for carcinogen risk assessment. Federal Register 51(185)33992-34003.

USEPA. 1987. U.S. Environmental Protection Agency. 1987. Epidemiology: recent Kodak study. Chapter 6. In: Technical analysis of new methods and data regarding dichloromethane hazard assessments. Office of Health and Environmental Assessment, Washington, DC. External Review Draft. EPA/600/8-87/029A.

USEPA. 1988. U.S. Environmental Protection Agency. Estimating exposures to 2,3,7,8-TCDD. Exposure Assessment Group, Office of Health and Environmental Assessment, Washington, DC. External Review Draft. EPA/600/6-88/005A.

Weil ES. 1972. Statistics vs. safety factors and scientific judgment in evaluation of safety for man. Toxicol. Appl. Pharmacol. 21:254-463.

Zack J, Suskind R. 1980. The mortality experience of workers exposed to TCDD in a trichlorophenol process accident. J. Occup. Med. 22(1):11-14.

8.   RATIONALE FOR A HORMONE-LIKE MECHANISM OF 2,3,7,8-TCDD FOR USE
       IN RISK ASSESSMENT

Michael A. Gallo

University of Medicine and Dentistry of New Jersey
Robert Wood Johnson Medical School
Piscataway, N.J. 08854

The   mechanisms   of   action   of   2,3,7,8-tetrachlorodibenzo-p-dioxin
(2,3,7,8-TCDD) have been intensely investigated since the pioneering work
of Kimmig and Schultz (1957) that elucidated the chloracnegen in the
chlorophenol processes. The most complete, thought-provoking treatise on
the subject is that of Poland and Knutson (1982). These authors, along
with others in Poland's laboratory, drew upon their own work and that of
others to present a unified hypothesis to account for the varied responses
in animals exposed to 2,3,7,8-TCDD. In essence, this hypothesis, which is
partly based on the classic receptor theories invoked for steroid action,
suggests that there is a cytosolic receptor for arylhydrocarbons (the Ah
receptor) that binds several compounds and then translocates to the
nucleus. Moreover, there is a second stage to the toxic reaction(s) that
is related to, but not congruent with, the induction of cytochrome
P_1-450. Activation of the cytosolic receptor leads to a cascade of
reactions, culminating with the association of the receptor-2,3,7,8-TCDD
complex with nuclear DNA. This association leads to the synthesis of a
specific microsomal protein, designated cytochrome P_1-450. To date,
the chemical that binds this putative receptor with the greatest avidity
is 2,3,7,8-TCDD.

Many other halogenated and nonhalogenated compounds also bind to this
cytosolic protein (Nebert et al., 1972). The xenobiotics with the greatest affinity are those that are planar, with two phenyl rings, and those
that contain substitutions in the lateral positions.   Several investiga-
tive teams have examined the structure activity relationships (SARs) among

the polyhalogenated biphenyls (PCBs, PBBs), polychlorinated dibenzo-*p*-dioxins (PCDDs), and polychlorinated dibenzofurans (PCDFs) (Knutson and Poland, 1980). In several studies, Safe and his coworkers have synthesized several of the highly active PCBs and have demonstrated remarkable SARs for several biological end points (Mason et al., 1987). Excellent reviews on the SAR for PCDDs, PCDFs, and PCBs can be found in recent issues of the Annual Reviews in Pharmacology and Toxicology (Vols. 22 (Poland and Knutson, 1982) and 26 (Safe, 1986), respectively).

The hypothesis also states that "... there is a second stage to the toxic reactions of 2,3,7,8-TCDD that are related to, but not congruent with, the induction of cytochrome P_1-450." This portion of the hypothesis is supported by the reports of Poland and his coworkers with XB cells in culture (Knutson and Poland, 1980); Safe and coworkers' findings of PCB inhibition of 2,3,7,8-TCDD toxicity (Haake et al., 1987); and the Umbreit et al. report of apparent maximal induction of arylhydrocarbon hydroxylase (the major system affected by cytochrome P_1-450) (Umbreit et al., 1987) by complex mixtures of polyaromatic hydrocarbons and very low bioavailability of PCDDs/PCDFs as measured by tissue levels and signs of toxicity.

At this time, it seems clear that 2,3,7,8-TCDD is working through the proposed receptor mechanism for the first phase of its activity (i.e., binding to a putative cytosolic receptor with subsequent induction of P-450). However, all of the biological effects of this molecule cannot be explained by simple receptor binding and induction of cytochrome P_1-450. Recent evidence from several laboratories has expanded on the initial studies of Poland and Knutson (1982), Neal et al. (1982), and Barsotti et al. (1979) to show quite dramatically that 2,3,7,8-TCDD markedly affects the interaction of steroids with their respective receptors (Romkes et al., 1987; Gallo et al., 1986), and 2,3,7,8-TCDD alters the number of Epidermal Growth Factor (EGF) receptors in susceptible cell lins (Matsumura et al., 1984). Molloy et al. (1987) recently reported the alteration of specific epidermal keratins in the HRJ/S strain of mice after treatment with 2,3,7,8-TCDD. This finding is especially relevant to the data base since it was in this strain of mice that Poland and Knutson (1982) reported the model for chloracne. Matsumura et al. (1984) studied the role of 2,3,7,8-TCDD and EGF receptors, while several laboratories (Safe, 1986; Gierthy et al., 1987; Umbreit and Gallo, 1988; Goldstein et al., 1987) have been pursuing the interactions of 2,3,7,8-TCDD with steroids, primarily estrogen-sensitive tissues.

Interaction with glucocorticoids has been studied extensively by several laboratories (Luster et al., 1984; Sunihara et al., 1987). In general, the response can be summarized as a decrease in the number of available cytosolic receptors for EGF or the steroids without a decrease in the affinity for the respective ligand. This phenomenon is termed "down-regulation" of the cytosolic receptor. The measurement of receptor binding is biochemical. The estimate of affinity and binding site number is by extrapolation of the response curve(s) by Scatchard analysis (1949). The strengths of this analysis are obvious, but the weaknesses are diffi-cult to reconcile. The major weakness is the lack of direct binding in-formation; this does not allow segregation of the receptors by tertiary structure, nor does it completely account for nonspecific binding. The second weakness of ligand binding experiments is the inability of the analysis to shed any light on the reason for the down-regulation. It must be emphasized at this point that none of the steroid receptor research has demonstrated an antagonism between 2,3,7,8-TCDD and the endogenous steroid for the respective steroid receptor, nor has any competitive binding of steroids by the Ah receptor been demonstrated. However, the steroid receptors are products of a supergene family that is responsible for the protein synthesis of all these receptors (Nebert et al., 1972), and the Ah receptor has many of the structural and functional characteristics of the steroid receptors (Poellinger et al., 1986).

To better understand the role of 2,3,7,8-TCDD in cellular function (or dysfunction), one must look to the results of the laboratories working on the mechanisms of action of 2,3,7,8-TCDD at the molecular level. The major groups involved in this research are Poland, Nebert's group at the National Institutes of Health, and Whitlock's laboratory at Stanford Uni-versity. As stated above, Poland established the role of the Ah receptor in some of the actions of 2,3,7,8-TCDD. Whitlock (1987) summarized his data and that of other investigators regarding the regulation of the cytochrome P-450 gene family, along with the data supporting the hypothe-sis that the Ah locus is part of a supergene family responsible for the metabolism of xenobiotics and endogenous compounds. Recent work in this laboratory has also elucidated a region on DNA that is sensitive to the 2,3,7,8-TCDD-cytosolic receptor complex, upstream from the cytochrome P_1-450 gene (Neuhold et al., 1986). These findings are critical in light of the findings of the 2,3,7,8-TCDD-responsive gene expression en-hancer system (region) described by Whitlock (Jones et al., 1986). Hence, the two laboratories have defined the regulatory mechanisms by which

2,3,7,8-TCDD controls gene expression of the cytochrome P_1-450 (Whitlock, 1987).

The significance of these findings for 2,3,7,8-TCDD risk analysis is the congruency between gene regulation for the Ah receptor, the glucocorticoid receptor, and the estrogen receptor (Becker et al., 1986). The importance of these findings cannot be underestimated. There is a direct analogy with the steroid receptor mechanisms and the control of the steroid receptor messenger RNA (Yamamoto, 1985). The role of the estrogen receptor (ER) and other steroid receptors is understood to a greater extent than the Ah receptor, probably because of the greater emphasis on the physiology of steroids. The analogy between the receptor complexes and DNA leads to the obvious comparison of effects of 2,3,7,8-TCDD and steroids. Many of the changes seen in animals after dosing with 2,3,7,8-TCDD mimic estrogen or antiestrogen effects. Umbreit and Gallo (1988) reviewed these findings, which are presented in Table 1. Kociba et al. (1978) demonstrated the hepatocarcinogenic effect of orally administered 2,3,7,8-TCDD, but in the same study there was a marked dose-dependent decrease in tumors of the mammary glands and uteri, which are estrogen-sensitive organs. These highly significant observations have been pursued by some laboratories. If 2,3,7,8-TCDD is acting through hormonal (estrogen) mechanisms, then alteration of ovarian function, exogenous estrogens, or antiestrogens should modify the response(s) to 2,3,7,8-TCDD.

Recent results have shown that 2,3,7,8-TCDD effects can be overridden by exogenous estradiol (Gallo et al., 1986), and the down-regulation of the estrogen receptor is also antagonized by estradiol (Romkes et al., 1987). The significance of these findings is amplified if one couples the reports of regulation of the EGF receptor by estrogens (Mukku and Stancel, 1985; Madhukar et al., 1984) with the consistent observation that the lowest doses in the lifetime bioassays of 2,3,7,8-TCDD decrease tumor yield in rodent livers but do not affect the background levels of breast or uterine tumors (Kociba et al., 1978). 2,3,7,8-TCDD inhibition of tumor growth at low doses and enhancement at higher levels (in the bioassays) is supported by the recent report of a marked decrease in tumorigenesis in the two-stage liver model at the lowest dose of 2,3,7,8-TCDD after diethylnitrosamine (DEN) initiation (Pitot et al., 1987). These findings are consistent with the hypothesis that 2,3,7,8-TCDD may be working through an endocrine-sensitive mechanism to yield its toxic effects. If one accepts this premise, then it is reasonable to assume that the actions of 2,3,7,8-TCDD can be explained using a physiologically based model. The

Table 1. Association of Estrogens with 2,3,7,8-TCDD Toxicity

Many of the toxic effects of 2,3,7,8-TCDD are similar to effects of estrogens in non-2,3,7,8-TCDD-treated animals.

1. Some effects of 2,3,7,8-TCDD resemble effects of elevated estrogens.

2. Other effects of 2,3,7,8-TCDD resemble antiestrogenic effects (most antiestrogens have estrogenic effects at different doses).

3. Some 2,3,7,8-TCDD effects are not straightforward estrogenic or antiestrogenic effects. For some of these, an influence of estrogen on the effect is known.

4. Other signs of 2,3,7,8-TCDD toxicity may be related to cholesterol mobilization for estrogen synthesis.

Fat loss 1,4

Wasting 1,4

Changes in serum lipids: 1,3,4
   increased cholesterol
   increased LDL, VLDL

Anorexia 1

Hypophagia 1

Hypoinsulinemia 1

Altered serum fatty acids 1,4

Hypoglycemia 1,4

Lowered O_2 consumption

Membrane damage 1,4

Stimulates differentiation 1
   in certain epithelial cells

Lower serum testosterone 3

Uterine suppression 2

Reproductive failure 1,2,3

Immunosuppression 1

Thymic involution 1

Decreased thymic cellularity

Hirsutism 1

Chloracne 3

Skin keratinization 1

Lowered T4 in serum 3

Increased:
   thyroid weight 3
   serum TSH
   T4 excretion as glucuronide

Table 1.   (continued)

Blockage of E2 uterotrophism 1

Terata 1,2,3

Lowered serum corticoids 3,4

Blockage of ACTH stimulation of corticosteroid synthesis 3,4

Down-regulates:

  EGF receptor 1
  Prolactin receptor 1
  Glucocorticoid receptor 1
  LDL receptor 1
  Estrogen receptor 1

Ascites 1,4

Hepatocyte membrane damage 1,4

Hepatocyte membrane cAMP reduced

Enzyme inductions 3

  AHH (EROD, P-448, P-450c, and d)
  Ornithine decarboxylase
  UDP-GTs
  ALA synthetase
  Others

Anemia

Porphyria cutanea tarda 1

Iron accumulation in liver

Altered iron transport in gut

physiological implications of an endocrine mechanism may explain many of the responses seen in animals after exposure to 2,3,7,8-TCDD (see Table 1), since these responses are similar to hyper- and hypo-hormonal states (O'Malley and Buller, 1977; Potter et al., 1986; Jones et al., 1986; Gustafsson et al., 1987). As stated above, the analogy to the estrogen system is arguably the strongest to 2,3,7,8-TCDD effects (both hyperplastic and dysplastic responses in endocrine sensitive organs), and the similarity between the cytosolic receptors and their stabilization by molybdate (Denison et al., 1986), activation of a nuclear site, and the anti-2,3,7,8-TCDD effect of estradiol strengthens the analogy.

The effects of estrogens are widespread throughout the body. Some of these effects may not be receptor-mediated, but the majority of the effects are directly attributable to receptor binding. The toxic effects of estrogens have recently been summarized (Umbreit and Gallo, 1988) and include thymic involution and decreased response to septic challenge (Luster et al., 1984; Grossman, 1984), a wasting syndrome characterized by weight loss, hirsutism, and epidermal lesions. It should be noted that there are recent reports of estrogens enhancing or causing cachexia and wasting, which are the major effects of 2,3,7,8-TCDD seen in intoxicated animals.

The role of estrogens as immunosuppressives is not well understood, but it is hypothesized that the putative suppressant is either an excess of circulating estradiol or perhaps an excess of trophic hormones. Estrogens also play a role in the action of other hormones and trophic factors such as EGF (Kirkland et al., 1981; Gardner et al., 1978; Mukku, 1984; Mukku and Stancel, 1985; Hsueh et al., 1981; Gonzales et al., 1984; Dickson and Lippman, 1987). These findings lead to the conclusion that the multiple effects of TCDD could be mediated through an endocrine mechanism. The weakness of this assumption is that 2,3,7,8-TCDD causes effects that appear similar to both hyperestrogenemia and hypoestrogenemia. It has been hypothesized that this apparent paradox is the result of 2,3,7,8-TCDD or the ligand complex preventing the endogenous substrate from interacting "correctly" with both the active site and a secondary binding site (Umbreit and Gallo, 1988; Umbreit et al., 1988).

Pleiotropism is not an uncommon finding with molecules such as hormones or, in this case, 2,3,7,8-TCDD. One has only to review the early experiments on multistage mouse skin carcinogenesis of 2,3,7,8-TCDD to see that in some cases it inhibited tumor formation by PAH initiators (DiGiovanni et al., 1977; Berry et al., 1978). It must be emphasized that the responses in multistage models are dependent on time, sequence of administration, dose, and species. Hence, inhibition under some conditions might have been predictable. This is juxtaposed to the two-stage liver model (Pitot et al., 1980) in which it has been shown that orally administered 2,3,7,8-TCDD enhances the tumorigenic action of DEN. However, in subsequent experiments at lower doses of 2,3,7,8-TCDD, a parabolic dose-response curve has been reported in the DEN/2,3,7,8-TCDD initiation-promotion protocol (Pitot et al., 1987). This paradoxical effect is not well understood, but it does not appear to be solely the function of enhanced metabolism or Ah receptor binding (Mason et al., 1987). Perhaps it is the result of alteration of EGF receptors at low doses (Madhukar et al., 1984), which displays a commonality with several steroid hormone receptors.

The importance of these findings to the approximation of human and animal health risks from exposure to PCDDs and related molecules cannot be overstated. Mathematical modeling of physiological phenomena, especially those related to receptor function, is conducted using the Michaelis-Menten equation (1913) as modified by Clark for the "classical" receptor model (1933). The weight of evidence for the most prevalent 2,3,7,8-TCDD effects falls into the category of the receptor model (Poland and Knutson, 1982). The recent findings that the hepatocarcinogenesis is related to estrogen levels or to the presence of functional ovaries (Goldstein et al., 1987) and that DEN hepatocarcinogenesis in partially hepatectomized rats is first inhibited, then promoted by 2,3,7,8-TCDD (Pitot et al., 1980; 1987) indicate that 2,3,7,8-TCDD is not causing its myriad of effects in liver by a simple one-step event such as binding to the Ah receptor and subsequent induction of cytochrome P_1-450. Operationally, however, 2,3,7,8-TCDD is a potent hepatocarcinogen in some species and strains of rodents.

Risk modeling for carcinogenic zenobiotics has recently been segregated into three classes or types of models: physiologically based pharmacokinetic (PBPK) models in which the body is considered to be a small group of physiological compartments (Hoel et al., 1983; Krewski et al., 1986; Bischoff, 1987); biologically motivated models of carcinogenesis (BMMC) in which the carcinogenic process is considered to occur through a series of linked reactions that result from two or more molecular events followed by cellular amplification by "promoter" molecules (Moolgavkar, 1986; Thorslund et al., 1987; Krewski et al., 1987); and the linearized multistage model (LMS) of Armitage-Doll as modified by Crump and Howe (1984) in which it is assumed that a sequence of mutational events occur within a single cell leading to the neoplastic change (Armitage, 1985).

The model that appears to accommodate most of the critical components from the biological data base on 2,3,7,8-TCDD is the BMMC model, which is generally referred to as the Moolgavkar-Venzon-Knudson (M-V-K) model (Moolgavkar and Venzon, 1979; Moolgavkar and Knudson, 1981). This model allows for several of the concepts of initiation-promotion-progression, along with the growth-stimulating role of endogenous substrates such as hormones (Moolgavkar, 1986). Incorporation of some of the factors necessary for the PBPK model can also be done using the M-V-K model as modified, or, more correctly, expanded by Thorslund et al. (1987). These expansions of the M-V-K model give the risk assessor a powerful tool for looking at cancer risk mechanistically.

This option is not available with the LMS model as originally proposed. The use of the LMS model may not be appropriate for the 2,3,7,8-TCDD data set since this model assumes an initiating event, such as a point mutation, to start the process. However, the LMS model can be accommodated if one hypothesizes that (1) the initiating event is the result of an indirect action of 2,3,7,8-TCDD through modification of exogenous or endogenous compounds; (2) a population of initiated cells exists; or (3) 2,3,7,8-TCDD leads to focal necrosis, which serves as a mitogenic stimulus.

Recent reports have shown that 2,3,7,8-TCDD and other promoters in liver enhance stimulation of DNA synthesis in situ and stimulate repair of O-6-methylguanine in liver DNA (Busser and Lutz, 1987; Den Engelse et al., 1986). Lutz et al. (1984) presented a scheme for promoter potency based on stimulation of DNA synthesis and the assumption that cell division is a prerequisite for several stages in the carcinogenesis process. These reports indicate that 2,3,7,8-TCDD can act as a complete indirect-carcinogen, including promoter activity, despite the lack of DNA binding or direct mutagenesis. The sum of all these findings, along with the myriad of toxic responses, suggests a model for 2,3,7,8-TCDD carcinogenesis in rodent liver as shown in Figure 1. This model may account for the dose-response data in the bioassays and the multistage promotion experiments, as well as allow for incorporation into existing risk models. The scheme is not incongruent with the reports of decreased tumor formation in some tissues. If pathway (A) can be verified by demonstration of reactive intermediates after 2,3,7,8-TCDD treatment, then the LMS model, slightly

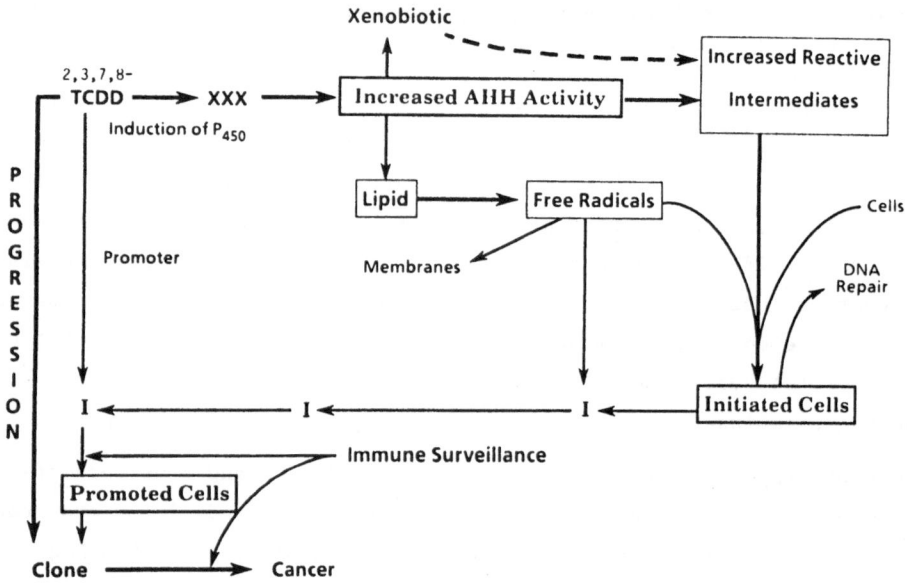

Figure 1.   Potential Secondary Mechanisms of Carcinogenic
Activity of 2,3,7,8-TCDD

modified, can be used. The preponderance of evidence at the moment supports a mechanistic model(s) that is at variance with the LMS model. However, Figure 1 presents possibilities that are not mutually exclusive for the existing models. The scheme also presents several testable hypotheses that should be examined.

REFERENCES

Armitage P. 1985. Environ. Health Perspect. 63:195-201.

Barsotti DA, Abrahamson LJ, Allen JR. 1979. Bull. Environ. Contam. Toxicol. 21:463-469.

Becker PB, Gloss B, Schmid W, Strahle U, Schultz G. 1986. Nature (London) 324:686-688.

Berry DL, DiGiovanni J, Juchau MR, Bracken WM, Gleason GL, Slaga TJ. 1978. Res. Commun. Chem. Pathol. Pharmacol. 20:101-107.

Bischoff KB. 1987. In: Drinking water and health 8:36-64. Washington, DC: National Academy Press.

Busser MT, Lutz WK. 1987. Carcinogenesis 8:1433-1437.

Clark AJ. 1933. The mode of action of drugs on cells. Baltimore, MD: Williams and Wilkins Press.

Crump KS, Howe RB. 1984. Risk Analysis 4:163-176.

Den Engelse L, Floot BGJ, Menkveld GJ, Tates AD. 1986. Carcinogenesis 7:1941-1947.

Denison MS, Vella LM, Okey AB. 1986. J. Biol. Chem. 261:3987-3995.

DiGiovanni J, Viaje A, Berry DL, Slaga TJ, Juchau MR. 1987. Bull. Environ. Contam. Toxicol. 18:552-557.

Dickson RB, Lippman ME. 1987. Endocr. Rev. 8:29-43.

Gallo MA, Hesse EJ, Macdonald GJ, Umbreit TH. 1986. Toxicol. Lett. 32:123-132.

Gardner RM, Kirkland JL, Ireland JS, Stancel GM. 1978. Endocrinology 103:1164-1172.

Gierthy JF, Dickerman HW, Seeger JL, Lincoln DW, Martirez H, Kumar SA. 1987. Toxicologist 7:159.

Goldstein JA, Graham JM, Sloop T, Maronpot R, Goodrow T, Lucier GW. 1987. Dioxin 87, Abstract #MA-07.

Gonzales F, Lakshman S, Hoath S, Fisher DA. 1984. Acta Endocr. 105:425-428.

Grossman CJ. 1984. Endocr. Rev. 5:435-455.

Gustafsson JA, Carlstedt-Duke J, Poellinger L, Okert S, Wikstrom AC, Bronnegard M, Gillner M, Dong Y, Fuxe K, Harfstrand A, Agnati L. 1987. Endocr. Rev. 8:185-234.

Haake JF, Magura F, Phillips T, Safe S. 1987. Toxicologist 7:124.

Hoel DG, Kaplan NL, Anderson MW. 1983. Science 219:1032-1037.

Hsueh AJW, Welsh TH, Jones PBC. 1981. Endocrinology 108:2002-2004.

Jones PBC, Durrin LK, Galeazzi DR, Whitlock JP Jr. 1986. Proc. Natl. Acad. Sci. USA 83:2802-2806.

Kimmig J, Schultz KH. 1957. Dermatologica 115:540-546.

Kirkland JL, Gardner RM, Mukku VR, Akhtar M, Stancel GM. 1981. Endocrinology 108:2346-2351.

Knutson JC, Poland A. 1980. Cell 30:225-234.

Kociba RJ, Keyes DG, Beyer JE, Carreon RM, Wade CE, Henber DA, Klanins RP, Frauson LE, Park CN, Barnard SD, Hummel RA, Humiston CG. 1978. Toxicol. Appl. Pharmacol. 46:279-303.

Krewski D, Murdoch DJ, Dawanje A. 1986. In: Chronic disease epidemiology, New York, NY: Wiley.

Krewski D, Murdoch DJ, Withey JR. 1987. In: Drinking water and health 8:441-467. Washington, DC: National Academy Press.

Luster MI, Hayes HT, Korach K, Tucker AN, Dean JH, Greenlee WF, Boorman GA. 1984. J. Immunol. 133:110-116.

Lutz WK, Busser MT, Sagelsdorff P. 1984. Toxicol. Pathol. 12:106-111.

Madhukar BV, Brewster DW, Matsumura F. 1984. Proc. Natl. Acad. Sci USA 81:7407-7411.

Mason G, Zacherewski T, Denomme MA, Safe L, Safe S. 1987. Toxicologist 7:161.

Matsumura F, Madhukar BV, Bombick DW, Brewster DW. 1984. Banbury Report 18:267-287, Cold Spring Harbor Laboratory, New York.

Michaelis L, Menten ML. 1913. Biochem. Z. 49:333.

Molloy CJ, Gallo MA, Laskin JD. 1987. Carcinogenesis 8:1193-1199.

Moolgavkar SH, Knudson AG Jr. 1981. J. Natl. Cancer Inst. 66:1037-1042.

Moolgavkar SH, Venzon DJ. 1979. Mathematical Biosci. 47:55-77.

Moolgavkar SH. 1986. Annu. Rev. Public Health 7:151-169.

Mukku VR, Stancel GM. 1985. J. Biol. Chem. 260:9820-9824.

Mukku VR. 1984. J. Biol Chem. 259:6543-6547.

Neal RA, Olson JR, Gasiewicz, TA, Geiger LE. 1982. Drug Metabl. Rev. 13:355-385.

Nebert DW, Goujon FM, Gielen JE. 1972. Nature New Biol. 236:107-110.

Neuhold LA, Gonzales FJ, Jaiswal AK, Nebert DW. 1986. DNA 5:403-411.

O'Malley BW, Buller RE. 1977. Invest. Dermatol. 68:1-4.

Pitot HC, Goldsworthy T, Campbell HA, Bland A. 1980. Cancer Res. 40:3616-3620.

261

Pitot HC, Goldsworthy TL, Moran S, Keenan W, Glauert HP, Maronpot R, Campbell HA. 1987. Carcinogenesis 8:1491-1499.

Poellinger L, Lund J, Soederkvist P, Gustafsson JA. 1986. Chemosphere 15:1649-1656.

Poland A, Knutson JC. 1982. Annu. Rev. Pharmacol. Toxicol. 22:517524.

Potter CL, Moore RW, Inhorn SL, Hagen TC, Peterson RE. 1986. Toxicol. Appl. Pharmacol. 84:45-55.

Romkes M, Piskorska-Pleszczynska J, Safe S. 1987. Toxicol. Appl. Pharmacol. 87:305-314.

Safe S. 1986. Annu. Rev. Pharmacol. Toxicol. 26:371-399.

Scatchard G. 1949. Ann. N.Y. Acad. Sci. 51:660-672.

Sunihara GI, McCoy Z, Bresnick EH, Sanchez ER, Nelson KG, Lucier GW. 1987. Dioxin 87, Abstract # MA-14.

Thorslund TW, Brown CC, Charnley G. 1987. Risk Analysis 7:109-119.

Umbreit TH, Gallo MA. 1988. Toxicol. Lett. In press.

Umbreit TH, Hesse EJ, Gallo MA. 1987. Arch. Environ. Contam. Toxicol. 16:461-466.

Umbreit TH, Hesse EJ, Macdonald GJ, Gallo MA. 1988. Toxicol. Lett. 40:1-9.

Whitlock JP Jr. 1987. Pharmacol. Rev. 39:147-161.

Yamamoto KR. 1985. Annu. Rev. Genet. 19:209-252.

CHAPTER 3 - TECHNOLOGY ASSESSMENT WORKING GROUP

1. INTRODUCTION TO WORKING GROUP B ACTIVITIES

1.1   Introduction

The Technology Assessment Working Group focused its attention on producing documents that reported the state of the art with respect to the formation of dioxins and related compounds from combustion and industrial sources, as well as methods of destruction and disposal.

Chlorinated dibenzo-p-dioxins and dibenzofurons (CDDs and CDFs) do not occur naturally, nor are they manufactured intentionally for commercial use. Rather, these compounds are generated inadvertently in a variety of chemical and combustion processes. Humans and animals exposed to CDDs and CDFs have shown acute, chronic, and subchronic effects on the skin, liver, nervous system, and immune systems. Because of these toxic effects, the manufacture and use of many chemicals such as some herbicides, have been regulated or prohibited. Due to intense sampling and analysis programs initiated by several countries, more has been learned about the emissions of CDDs and CDFs from incinerators, pulp and paper mills, metallurgical processes, and the manufacture/use of chlorophenols for wood preservation.

The following summaries and sections identify the chemical processes that generate CDDs and CDFs and present state-of-the-art destruction technologies.

1.2   Summary: Formation   of   Dioxins   and   Related   Compounds   from Combustion and Incineration Processes

The purpose of this report is to summarize information about PCDD/PCDF formation from combustion sources. Although progress is being made in this field, not all answers are at hand.

Polychlorinated dibenzodioxins and dibenzofurans are considered ubiquituous in technicologically advanced countries and combustion processes are considered one main source. PCDD/PCDF and related compounds were found in the emissions of various combustion processes: municipal and hazardous waste incinerators, fossil fuels power plants, automobile exhaust, private heating, wood and forest fires, cigarette smoking as well as accidents like PVC or PCB fires.

Even though several theoretical and experimental investigations have been carried out to elucidate the formation mechanism of PCDD/PCDF, many questions still remain unanswered; e.g., the point when chlorine or oxygen are introduced into the molecule. Furthermore, there are uncertainties, whether all combustion processes lead to PCDD/PCDF and what type of catalytic effects are important. It is likely that several mechanisms will lead to the formation of PCDD/PCDF or their precursors. The exact mechanisms involved in dioxin formation remain unclear. It is generally accepted, that thermal reactions involving direct precursors like PCP, PCB, and polychlorinated benezenes and diphenylethers yield PCDDs/PCDFs.

The presence of PCDD/PCDF in flue gas and fly ash of municipal incinerators is universal. International comprehensive tests of municipal waste incinerators have established the following facts:

- All waste incinerators, independent of incinerator type or garbage composition, produce all of the possible 75 PCDD and 135 PCDF isomers and congeners as well as 400 other organic compounds. These compounds are detected in the emitted stack gases and are partially adsorbed to fly ash particles (ppb-ppm range)

- The exact reaction mechanism for dioxin formation from combustion processes is unknown.

- The total and relative amounts of PCDD/PCDF and other compounds vary daily and somehow depend on the incinerator operation conditions.

264

Because tons of plastics and discarded each year with garbage, it has long been debated whether PVC serves as a chlorine donor during municipal waste incineration. During random tests of various waste incinerators, no significant increases in fly ash dioxin concentrations were registered even when 300% more PVC than normal was added to the refuse. Nevertheless, PCDD and PCDF congeners were formed during laboratory pyrolysis of PVC. However, negligible amounts of dioxins were detected from a warehouse fire in Sweden involving 200 tons of pure PVC as well as 500 tons of plastics and other materials.

Well managed large systems for combustion of hazardous waste such as PCBs and chlorinated benzenes present no problems. $2,3,7,8\text{-}Cl_4DD$ could not be detected in the gaseous emissions even when PCBs (chlorine content 56-65%) are burnt; concentrations of other PCDD/PCDF were found in the low ppt-range, in smaller amounts than all threshold requirements.

The arcing and subsequent burning of an electrical transformer in a Binghamton State office building on February 5, 1981, produced the first recognized large scale case in which toxic products, PCDDs and PCDFs, were produced outside of the laboratory from the combustion of chlorinated aromatic compounds in the transformer fluid. Since the accident of Binghamton some major PCB transformer and capacitor fires have occurred within the U.S. and Europe. Of all these fires, the Binghamton transformer fire is the only one in which PCDDs were clearly identified. The presence of chlorinated benzenes in the transformer fluid and the intense heat of the fire provided conditions similar to those in earlier laboratory studies which produced PCDDs from the pyrolysis of chlorinated benzenes.

Polychlorinated dioxins and furans can likewise occur by burning untreated wood. Recent studies have shown fairly high levels of dioxins in the soot from home furnaces heated with wood and/or coal. There is conflicting and limited data regarding PCDD/PCDF formation by the burning of fossil fuels, mostly of coal. In some investigations polychlorinated dioxins were detected in ppb-amounts. Small home heating units using oil and coal, however, consistently produce moderately high levels of PCDD/PCDF.

Polychlorinated as well as polybrominated dibenzodioxins and -furans have been identified in the exhaust gases of cars and trucks fueled with leaded gasoline or diesel. Dibromoethane (DBE) and dichloroethane (DCE), which are added as scavengers to leaded gasoline to prevent lead deposits in engines, are likely involved in combustion mechanisms leading to dioxin formation.

PVC as a chlorine donor in municipal waste incineration has been discussed for a long time. New studies and samples of fly ash from municipal waste incinerators fueled with high contents of PVC could not confirm PVC as source for the formation of high concentrations of PCDD/PCDF. Copper smelters and plants for metal recovery are to be considered separately, where cables covered with PVC are thermolyzed in the presence of, e.g., copper. The levels of PCDD/PCDF (in 100 ppb to ppm-range) found in the emissions may be due to catalytic inference of the metals present. One should note, that the maximum and minimum values deviate by several orders of magnitude. Furthermore, the average fly ash concentrations of dioxins from clinic waste incineration are greater than those from municipal waste incineration. However, it is unclear if these variations depend on technical conditions or on the higher proportions of plastics typically found in hospital wastes. Initial tests regarding soil contamination in the emission area of metal recovery plants revealed that all PCDDs/PCDFs were detected with the isomer distribution characteristic for dioxins and furans from thermal sources.

Based on the results of the Tier 4 study and information extracted from the literature, it can be concluded that total organic haolgen (TOX) levels showed the strongest association with the total level of PCDDs and PCDFs emitted while sites with high percentages of plastics in the feed were found to yield high PCDD emissions in the Tier 4 sites studied. Afterburners operating above 800°C do not necessarily destroy all PCDDs and PCDFs since factors such as residence time and mixing are also significant.

1.3 Summary: Formation of Dioxins and Related Compounds in Industrial Processes

Because they are potentially dangerous compounds polychlorinated and/or polybrominated dibenzo-p-dioxins (PCDD/PBDD) and dibenzofurans (PCDF/PBDF) have in recent years aroused public interest. In the search of dioxin and furan sources the processes of the chemical industry have been shown to be potential sources for the formation of PXDD and PXDF. The following processes have been investigated: manufacture of chlorophenols, chlorobenzenes and its derivatives, aliphatic chlorine compounds, methods involving chlorine-containing intermediates and processes of the inorganic chlorine chemistry and those applying chlorinated intermediates and solvents. Furthermore, processes of the pulp and paper industry, metallurgical processes, reactivation of granular carbon and the manufacture of brominated flame retardants have been identified as new sources within the last years.

Because PCDD/PCDF are often found in chlorophenols and their derivatives, production processes involving these chemicals rank first on the priority list. A short summary about PCDD/PCDF formation in these processes will be given. Considerably fewer analytic data have been published about other chloroaromatics and non-aromatic chlorine compounds. Mechanistic studies in the formation of PCDD/PCDF from short chain chloroaliphatic compounds indicate dichloroacetylene to be a key intermediate in the production of chlorinated aromatics. Metal chlorides ($FeCl_3$, $AlCl_3$) contain Cl_8DF and Cl_7DF in ppb- and hexachlorobenzene in ppm-range; $CuCl_2$ and $CuCl$ contain lower concentrations of Cl_8DF, Cl_7DF and additionally Cl_8DD, Cl_7DD and hexachlorobenzene in ppb/ppt-range. The origin of the chlorinated aromatics in these inorganic products could be chlorination of oils on the waste metals which are used in the manufacture of such metal chlorides at high temperatures.

In fresh motor oils no PCDD/PCDF could be detected, whereas waste oils and recycled oils contained detectable amounts of PCDD/PCDF (hexa- through octachlorinated dibenzodioxins and hepta- through octa dibenzofurans). PCDD/PCDF have been found in commercial products of PCP and PCP-Na.

PCDF have been found as contaminants of polychlorinated bipenyls chlorinated phenols, and as combustion products in fires involving PCB-containing transformers and capacitors according to formation mechanisms discussed for PCDDs but they can also be formed by intramolecular conversion of PCBs.

To date, work has started to identify the by-products produced during wood conversion to bleached pulp. In the effluents and sludges from pulp and paper mills sub-ppt levels of PCDD/PCDF have been identified as well as in consumers' paper products.

The aim of the Dioxin Study of Products by the German Chemical Industry was to investigate products of chlorine chemistry with respect to their content of selected isomers of 2,3,7,8-substituted PCDD/PCDF. The results of this study indicate that in no case the concentrations exceeded the limiting values of 2 ppb for 2,3,7,8-Cl_4DD, and 5 ppb for the total of the isomers.

1.4   Summary: Methods for Degradation, Destruction, and Detoxification of Dioxins and Related Compounds

Appropriate methods of degradation, destruction, detoxification, and disposal have to be applied if dioxins and related compounds exceed levels of concern in the environment which should be based on realistic health risk assessments. Over the last 10 years, a wide variety of technologies has been developed to be used for this task. This report characterizes a variety of methods under consideration by technical description, demonstrated field, or experimental, performance, and outlook for future application.

Biological degradation of dioxins has not as yet had sufficiently satisfactory results to call this method applicable for field use. However, for future decontamination of widespread surface/soil contamination, particularly with wastes associated with CDDs/CDFs, this appears to be a promising technology worthwhile to pursue testing with high priority.

Chemical dechlorination processes have passed the status of pilot-scale studies and are now applicable to real-life remediation tasks. Disposal/fixation of mainly inert wastes in suited landfills, engineered according to the degree of contamination, presents an acceptable solution from environmental, as well as economical perspectives. Decomposition of soils/fly ashes with low levels of dioxins in landfills represents the area of low-risk wastes. Controlled disposal in underground salt mines appears a promising method for wastes with medium levels of contamination or in surface mines with soils with lower contamination levels, if local regulations permit.

Other methods, aimed at complete molecular destruction of dioxins and related compounds, have resulted in the most widespread development of destruction methods. At present, fully field-tested methods such as high temperature hazardous waste incineration, together with an impressive number of new and emerging technologies for special application cases are available. A number of test reports yielding satisfactory destruction and removal efficiencies are included in this report.

In general, it can clearly be stated that the current status of methods for degradation, destruction, detoxification, and disposal of dioxins, together with reported and future-planned full-scale field testing, is well in accordance with requirements to protect the environment from these unwanted chemical species.

1.5   Summary: Waste Disposal Sites Contaminated with Dioxins and Related Compounds

Landfill and surface dumping has historically been cheap but within the last 10-15 years it has become obvious that this advantage was lost. Largely unregulated in the past, the practice of landfilling and dumping has created many locations where ground and surface water, soils, and air have been contaminated or are threatened with contamination by many undesirable and often toxic substances. During the search for the lost barrels of contaminated soil from the Seveso accident in April 1983, questions arose concerning the potential danger for man and the environment caused by polychlorinated dioxins and furans. Contaminated industrial areas as well as industrial production residues and fly ash have been disposed in mixed disposal waste sites. Residues from pesticides production often act as point sources of known location. Fly ash from municipal waste incinerators with a concentration of PCDD/PCDFs of total >1 ppm has often been disposed on mixed and solid waste dumps. Transport of organic pollutants may be possible via water leaking and via air by volatilization of decomposition gases.

This report presents case studies of waste sites and sites of production and/or application of pesticides and other chemicals contaminated with PCDD/PCDF. It must be emphasized that information is limited and difficult to obtain; therefore, this report has to remain incomplete.

In Germany municipal and hazardous waste dumps PCDD/PCDF have been detected mostly in oily layers of aqueous leachates from the waste body. Outside the landfill area no PCDD/PCDFs have been detected.

To reveal possible sources of PCDD/PCDF contamination the distribution of the isomers may reflect the pattern found by thermal processes, indicating fly ash as source of PCDD/PCDF. Where $2,3,7,8-Cl_4DD$ is found as the single tetra-isomer and also relatively high concentration among the sum of all PCDD/PCDF, specific contamination due by dumping of organic residues from chemical production is indicated.

Because contamination of the environment around the dumping areas cannot be excluded and possible threats may arise in future it is necessary to start remedial action to minimize and eliminate possible risks. At the moment no proven technology exists for either clean-up of large dumps or destruction of highly contaminated material. Clean-up could result in a large contamination of the surrounding area: for transport, the hazardous waste has to be removed and packed. Both may pose severe threats. Often decontamination by thermal methods on the waste site often is not possible. Top-layers to protect the dump from infiltration of rainwater will be constructed to reduce the pollution.

As part of its Dioxin Strategy, the U.S. Environmental Protection Agency (EPA) conducted the National Dioxin Study, an investigation to determine the extent of 2,3,7,8-tetrachlorodibenzo-p-dioxin (2,3,7,8-Cl_4DD) contamination in the environment. Selected sites of facilities and associated waste disposal sites where 2,4,5-T and its derivatives were formulated into pesticidal products have been analyzed. At contaminated sites, the extent of contamination was usually limited to one or two soil samples with concentrations of 2,3,7,8-Cl_4DD above 1 ppb. In areas where 2,4,5-trichlorophenol (TCP) and pesticides derived from it were used on a commercial basis contamination was found at several sites sampled and in various media. However, the levels were generally very low. Sludges from paper mills on the Petenwell Flowage were found to have levels of 2,3,7,8-Cl_4DD over 100 ppt, even though chlorophenol-based slimicides are not longer used. Organic chemical and pesticide manufacturing facilities were investigated where improper quality control may have caused products or waste streams to be contaminated with 2,3,7,8-Cl_4DD.

PCDD/PCDF analyses from samples of sewage sludge are presented in the report. Potential hazards resulting from application of sewage sludge as amendments on farm land will be discussed.

2. FORMATION OF DIOXINS AND RELATED COMPOUNDS FROM COMBUSTION AND INCINERATION PROCESSES

2.1 GENERAL ASPECTS

2.1.1 Introduction

This report describes combustion and other thermal processes which lead to the formation and emission of polychlorinated dibenzo-p-dioxins (PCDD) and polychlorinated dibenzofurans (PCDF). It is not the goal of this report to give a detailed historical review of the research and studies which have been carried out in studying these processes. Instead the purpose is to update the state-of-the-art knowledge on combustion and thermal processes which lead to the formation and emission of dioxins and furans. The report is divided into two sections, the first of which is devoted to the emissions from incineration of waste material. The second section deals with the production of PCDD and PCDF from other combustion and thermal processes such as transformer fires, wood burning, and car engines.

2.1.2 Sources of Dioxins and Furans

Numerous combustion and other thermal processes lead to the formation and emission of PCDD, PCDF and various related compounds. Examples include the incineration of municipal and industrial wastes, fossil fuel combustion, the burning of natural materials, such as wood or straw and accidental fires involving chlorine-containing dielectric fluids (Ontario Ministry, 1985; Hutzinger et al., 1985; des Rosiers, 1986). In the United States alone millions of combustion sources produce PCDD and PCDF (Brenner, 1986). Fuels such as oil, gas, coal, and wood are burned in residential areas to produce heat. Larger commercial, institutional, and utility boilers burn fossil fuels to produce heat, steam or electricity. Incineration of municipal and industrial wastes is commonly used to reduce the volume of refuse, yield a hygienic end product and, sometimes, to recover energy. Open fires such as agricultural burning and unintentional blazes involving residential or commercial buildings as well as forest fires are further combustion sources.

273

2.2 INCINERATION OF SOLID WASTE

2.2.1 <u>General and Historical</u>

Urban waste is generally divided into two broad categories: municipal solid waste and hazardous, industrial or chemical waste. Incineration of municipal refuse has become an attractive alternative to the common practice of depositing garbage in landfill sites. The trend towards incineration rather than landfilling is due largely to the lack of suitable landfill sites and also to growing concern for groundwater contamination (Hay et al., 1987). The concept of producing energy by burning waste has led to the design of energy-from-waste (EFW) incinerators (Cook, 1987). Canada burns 1.5 to 2 million tons of refuse per year. The U.S.A. burns about 28 million tons, only about seven percent of the total waste produced, which is about 150 million tons of municipal waste and 250 million tons of hazardous (industrial) waste. This figure which represents about 1.5 tons/year of waste per capita is the highest in the world. The practice of incineration is more widespread and older in Europe and Japan where landfill sites are scarce and where there is a greater need for energy produced by incineration. The city of Paris, France alone incinerates over 3 million tons per year. In France about 0.8 ton/year waste per capita is generated. Germany incinerates about 8 million tons of solid wastes per year in 47 incinerators. Only 16-18 million tons per year are landfilled. Japan incinerates about 50 million tons/year in 2,000 incinerators. Sweden burns about 7 million tons/year (0.7 tons/year per capita) in 25 plants. Incinerating garbage offers the advantages of reduction in the volume of solid wastes, the possibility of energy recovery, and in the case of units close to urban areas, lower transportation costs. However, the incineration of refuse, and combustion processes in general, may emit potentially hazardous substances into the environment. Public concerns over the safety of these plants and the cost of providing adequate pollution controls have limited the growth of incineration as a waste disposal technology in North America (Cook, 1987).

The first incinerator for combustion of municipal waste was built in 1876 in Great Britain. The cities of Hamburg and New York soon followed in 1896 and 1903. England had 200 units in place by 1914. North America has generally lagged behind Europe in the use of advanced incineration technology. In 1979 there were only 23 units installed in the United States, while 180 were operating in western Europe. Over 50 percent of the installed units in western Europe were designed for energy recovery as compared to the present two percent in the United States. The extensive use of refuse incineration in Europe is due to several reasons: the high cost of fossil fuels, a shortage of landfill sites, and a high level of capital funding available for post-World War II reconstruction. In the United States its acceptance was hampered by cheap landfill, stringent air pollution controls, the low cost of fossil fuels, and a negative public reaction.

There are many different incinerator designs and large variations in the types of materials burned at such facilities. Incinerators generally consist of fuel feed mechanisms, a grate, and combustion chamber, a boiler (heat exchanger), particulate collectors, gas scrubbers and a stack. Some plants incorporate a two-stage design in which the first combustion chamber operates at a lower temperature and under fuel-rich conditions to provide partial burning and fuel-gasification. Here large organic molecules are decomposed and small molecules become oxidized. The second afterburning chamber operates with excess air to burn the remaining uncombusted organic gases. Often an auxiliary burner using natural gas or oil is used to maintain temperatures above 900 °C. Most incinerators can be considered to be one of the broadly categorized types described below.

1. Mass burn
   - used for the combustion of unsorted municipal solid waste
   - usually equipped with heat recovery units (e.g. water wall)

2. Refuse derived fuel
   - accept waste in processed form
   - refuse is mechanically pre-sorted to remove non-combustibles
   - refuse may be shredded or made into pellets
   - usually equipped with heat recovery units

3. Rotary kiln
   - used for solid, slurried, pasty and liquid wastes
   - liquid wastes often used as fuel to incinerate other wastes
   - frequently equipped with heat recovery units

4. Liquid injection
   - used for the combustion of liquid wastes which can be pumped
   - sometimes equipped with heat recovery units

5. Fume incinerator
   - used to burn by-product fumes in industrial processes
   - sometimes equipped with heat recovery units.

6. Pyrolysis

The first two types of incinerators are commonly used for the combustion of municipal waste while the latter three are usually employed for hazardous or industrial waste incineration.

Municipal refuse is not an ideal fuel for incineration because its composition is neither constant nor homogeneous. The composition depends upon the location of the incinerator. The levels of plastic materials and paper products is higher in urban waste as compared to rural waste. In addition, waste composition varies on a seasonal and weekly basis (e.g. grass cuttings in summer). The fuel can be raw refuse burned without sorting or pre-treatment (mass burning) , or fuel that has been sorted for non-combustibles or treated in some way (refuse-derived fuel).

2.2.2 Emissions and By-products of Municipal Refuse Incineration

The emissions from municipal refuse incinerators include carbon dioxide, water vapor, particulate matter in the form of fly ash and bottom ash, and process water. A large number of different metals have been found in the residues and effluents from refuse incinerators. Metals such as silver, cadmium, chromium, manganese, lead, tin, and zinc are from non-combustible refuse material. In addition, organic compounds may be found in the fly and bottom ashes as well as in the process water and stack-emitted flue gases.

Since 1964, there have been reports of organic compounds identified in incinerator residues and effluents. Compounds identified include aliphatic and polynuclear aromatic hydrocarbons (PAH), substituted benzenes, carbonyl, sulfur and nitrogen compounds, and polychlorinated aromatic compounds such as dioxins and furans. Dioxins and furans were first detected in samples of stack gases and the precipitated fly ash from three incinerators in Arnhem, Amsterdam, and Alkmaar (The Netherlands) (Olie et al., 1977). By 1980 dioxins and furans had been found in many incinerators in Europe and North America. These compounds have since been detected in every municipal incinerator tested, despite their differences in design, operation, and nature of the refuse burned. Some studies have also shown that the emissions contain related chlorinated aromatics such as polychlorinated biphenyls (PCBs) and chlorinated benzenes and phenols.

(1) Mechanism of Dioxin and Furan Formation in Municipal Waste Incinerators

The process by which dioxins and furans are formed during incineration are not well understood nor agreed upon. Three possibilities have been proposed to explain the presence of dioxins and furans in incinerator emissions (Hutzinger et al., 1985):

1. Dioxins and furans are already present in the incoming waste and are incompletely destroyed or transformed during combustion.
2. Dioxins and furans are produced from related chlorinated precursors such as PCBs, chlorinated phenols and chlorinated benzenes.
3. Dioxins and furans are formed via de novo synthesis. This is, they are formed from the pyrolysis of chemically unrelated compounds such as polyvinyl chloride (PVC) or other chlorocarbons, and/or the burning of non-chlorinated organic matter such as polystyrene, cellulose, lignin, coal, and particulate carbon in the presence of chlorine-donors.

(2) Survival of Dioxins and Furans in Refuse

Municipal refuse may contain herbicide formulations, wood treated with chlorinated phenol preservatives, or products containing PCBs. These minor constituents could provide a source for low level dioxin and furan contamination. However, several studies have shown that dioxin and furan levels in municipal refuse are very low to undetectable (Lustenhouwer et al., 1980; Tosine et al., 1985). The levels of these compounds in incinerator emissions are much higher than can be accounted for by the dioxin and furan content of the refuse. Also, the isomer distribution of the dioxins and furans in emissions are very different from the isomer distribution in the refuse, indicating that the dioxins and furans emitted by the incinerator were formed in the combustion process.

(3) Thermal and Catalytic Reactions of Precursor Compounds

Our present understanding of the role of precursor compounds in the formation of dioxins come from two lines of investigation. One line is the study of the pyrolysis of pure compounds such as chlorinated phenols, benzenes, diphenylethers, 2-phenoxyphenols and PCBs, and studies of the emissions of combustion processes. The second line of investigation is the study of the role of inorganic compounds and fly ash in catalyzing gas-solid phase reactions of precursor compounds.

(4) Pyrolysis and Combustion Studies

Dioxins can be easily formed by heating chlorophenols or alkali metal chlorophenates at temperatures over 250 °C. A major byproduct of the reaction is hexachlorobenzene, formed by the dehydration of two penta-chlorophenols (Sandermann et al., 1957).

Nestrick et al. heated potassium salts of di-, tri-, and tetrachloro-phenols under controlled flow pyrolysis conditions at 300 °C (Nestrick et al., 1979). Twenty-two discrete tetrachlorodioxin isomers were isolated by a combined chromatographic system for use as standards. When potassium chlorophenates are pyrolyzed at 320 °C, the chlorine substitution pattern of the reactants was visible in the products. At 440 °C the substitution patterns were lost (Zoller and Ballschmiter, 1986).

(5) Formation of Dioxins via Catalytic Reactions of Precursor Compounds on Fly Ash

Incinerator fly ash is composed of the inorganic constituents of refuse and adsorbed particulate carbon and organic compounds. The basic matrix of fly ash, based on the elemental composition, is that of an alumino-silicate. A typical fly ash could be composed of: 40 percent oxygen; 20 percent carbon; with 5 to 10 percent concentrations of silicon, chlorine, sulfur, and potassium; 1-5 percent concentrations of sodium, calcium, aluminum, and zinc; and less than 1 percent concentrations of lead, tin, and copper. Chlorine would be in the form of inorganic chlorides. The ratio of particulate carbon to adsorbed organic compounds is not known. The particles have specific surface areas of 2 to 4 m^2/g and average pore diameters of 7 to 15 nm (Dickson, 1987).

Fly ash is not an inert substance, rather is promotes gas-solid phase reactions on its surface (Rghei and Eiceman, 1985a). Dibenzodioxins adsorbed on the surface of fly ash readily react with HCl and NO_2 to form chlorinated and nitrated dioxins (Eiceman and Rghei, 1984; Rghei and Eiceman, 1985a; Rghei and Eiceman, 1985b). Selective adsorption of some organic compounds has been observed. Irreversible adsorption of dioxins by fly ash may be due to bond rupture and decomposition to undetected products (Rghei and Eiceman, 1985a).

The concept that surface reactions catalytically induced on the surface of fly ash produce dioxins and furans from a host of precursors would explain most of the observed and established information about dioxins and furans in municipal incinerators. Some simple experiments by Dickson confirmed that this is a basic mechanism involved in the formation of dioxins (Dickson, 1987; Dickson and Karasek, 1987; Karasek and Dickson, 1987).

The ability of fly ash from municipal refuse incinerators located in Ontario and Machida, Japan and from a copper smelter in Noranda, Quebec to catalyze the formation of dioxins and furans was probed using ^{13}C-isotopically labelled pentachlorophenol. The use of a stable isotope labelled precursor eliminated interferences from residual levels of unlabelled dioxins co-extracted from the fly ash. In separate experiments the labelled pentachlorophenol was passed through flow-tube reactors which contained one of the fly ashes at 300 °C and in which a flow of 10 ml/min of nitrogen was maintained for one hour.

GC-MS analyses of the fly ash extracts showed that $^{13}C_{12}$-chlorinated dioxins were produced from the pentachlorophenol precursor. No chlorinated furans were detected in the reaction products of any of the experiments conducted using chlorinated phenol precursors. The experiment using Ontario fly ash produced about seven times the amount of dioxins as the experiment with Machida fly ash. The amount of dioxins produced using the Noranda fly ash was less than one percent of that produced using Ontario fly ash. It was interesting to note that the distribution of congeners produced using Ontario and Machida fly ashes are very similar, despite the large difference in the total amounts produced. Only very small amounts of octachlorodioxin was detected in the experiments using ground firebrick as a control.

Table 1.   Amounts of Dioxins Produced on Different Fly Ashes Using 100 μg of [$^{13}C_6$]-labelled Pentachlorophenol; Amounts in [ng]; ND = Not detected (Dickson, 1987)

Fly Ash	Cl_4DD	Cl_5DD	Cl_6DD	Cl_7DD	Cl_8DD
Ontario	331	803	936	1,005	185
Machida	54	115	121	132	40
Noranda	ND	ND	ND	11	18

The results indicated that the Ontario fly ash, and to a lesser extent the Machida fly ash, promoted the production of dioxins from pentachlorophenol. Since all the fly ash and firebrick samples had similar physical characteristics, differences in surface elemental composition of the fly ashes were likely responsible for the differences in the catalytic activity. The chlorine content of Ontario and Machida fly ashes was in the range of 6.2 to 6.9 percent while it was absent in the Noranda fly ash. The surface concentration of aluminum in Ontario fly ash was greater than the other fly ashes. The surface sulfur content of the Noranda fly ash was the highest of all.

Other dioxin precursor compounds have been studied in pyrolysis experiments. Chlorinated phenol formulations often contain dimeric impurities called 2-phenoxyphenols. They are formed when the hydroxyl group of one phenol reacts with the ortho-chlorine of another chlorinated phenol to eliminate a molecule of HCl. These "pre-dioxins" can readily undergo intramolecular cyclization and ring closure to form a dioxin (Nielsson et al., 1974). Dioxins and furans have been reported from the pyrolysis and photolysis of chlorinated diphenyl ethers (Lindahl et al., 1980; Norstrom et al., 1977). The dioxins form via intramolecular cyclization through the loss of ortho-Cl_2, ortho-HCl, and ortho-H_2 (Lindahl et al., 1980).

For many years PCBs were used in electrical transformers and capacitors because of their excellent thermal stability and insulating properties. Furans have been found in the residues from fires involving PCBs (Jansson and Sundström, 1982).

Dioxins at parts-per-billion (ppb) levels have been found in the residues and emissions of residential fireplaces (Crummett, 1982; Nestrick and Lamparski, 1982; Lao et al., 1983; Clement et al., 1985). Recently chimney soot samples from home heating furnaces in Bayreuth, Germany were analyzed for dioxins and furans (Thoma, 1988b). The furnaces burned wood, coal, wood/coal, or oil. Chlorinated dioxins and furans were found in all samples. In another study, no dioxins were detected when virgin wood was burned at 750 °C in air (Sheffield, 1985). However, when HCl-saturated air was used, dioxins were produced.

The combustion of vegetable matter has also produced dioxins. Liberti burned chestnut, mimosa, fruit and vegetables and detected phenolic compounds along with dioxins (Liberti et al., 1983). The phenolics may arise from tannins which pyrolyze to give polyhydroxyphenols and phenols (Liberti and Brocco, 1982). As may be expected, dioxins were produced when leaves and wood wool impregnated with chlorophenates were burned (Rappe et al., 1978). The products from the combustion of several industrial polymers were found to contain benzene, PAH, and chlorinated alkyl benzenes as the major compounds from the combustion of PVC (Hawley-Fedder, 1984). No dioxins or dioxin precursors were found.

The pentachlorophenol experiments were repeated with $^{13}C_6$-phenol to determine if inorganic chlorides on the fly ash would chlorinate the aromatic ring and promote the formation of dioxins. The phenol did react with the fly ash surface to form mostly labelled tetrachlorodioxins and a trace amount of labelled pentachlorodioxin. These results tend to confirm that inorganic chlorides on the surface are involved in the reaction mechanism.

The effect of temperature on the amount and distribution of dioxins produced from pentachlorophenol on Ontario fly ash was investigated. At 150 °C only octachlorodioxin was detected, however it was present at a level too low for accurate quantitation. At 250 °C the octachlorodioxin isomer predominated. With a temperature increase of only 50 °C, large amounts of tetra-, penta-, and hexachlorinated dioxins were formed. At 340 °C the levels of dioxins decreased to approximately 25 percent of the levels at 300 °C.

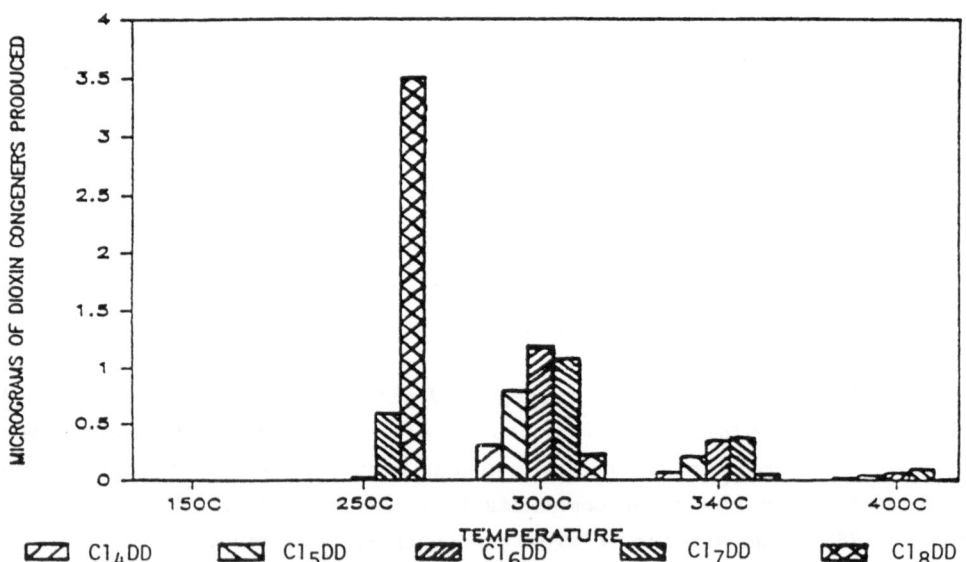

Fig. 1. Formation of Dioxins from Pentachlorophenol on Fly Ash: Effect of Temperature

A change in residence time, which is dependent upon the nitrogen flow rate, changed the amounts and distribution of dioxins produced from pentachlorophenol. A shorter residence time of 9 seconds shifted the distribution of dioxins towards the higher chlorinated congeners as compared to the usual residence time of 90 seconds. The shorter residence time was insufficient to produce detectable quantities of tetra- and pentachlorodioxins.

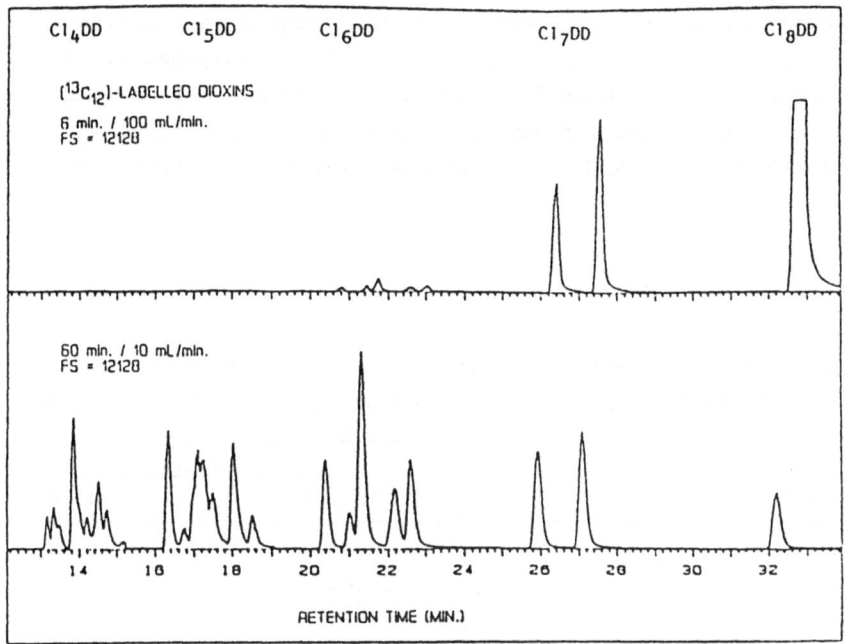

Fig. 2. Chromatograms from GC/MS-EISIM Analyses of Labelled Dioxins Produced from Labelled Pentachlorophenol on Ontario Fly Ash at Two Different Flow Rates

Chlorinated dioxins were also produced from 2,4,5- and 3,4,5-trichlorophenol precursors. The dioxin isomer distributions were different from that produced from pentachlorophenol. The experiment with 2,4,5-trichlorophenol produced primarily hexachlorodioxins, whereas octachlorodioxin and heptachlorodioxin were the major products of the experiment with 3,4,5-trichlorophenol.

Aliquots of two solutions, one containing either selected PVC pyrolysis products (cinnamyl chloride, 2-chloro-p-xylene, and naphthalene), the other containing products of combustion of a commercial PVC formulation were passed through a reaction tube containing Ontario fly ash at 300 °C in separate experiments. In both cases dioxins were detected in the reaction products but were not detected in the starting solutions.

Hagenmaier and co-workers have studied the catalytic effects of copper powder and fly ash on the dechlorination/hydrogenation of dioxins and furans (Hagenmaier et al., 1987a).

When 100 μg of a mixture of octachlorodioxin and octachlorofuran were heated in an open tube at 280 °C for five minutes, tetra- through heptachlorinated dioxins and furans were produced. The amount of chlorinated congener produced increased with decreasing degree of chlorination. Over 99 percent of the starting material was either decomposed or converted to the lower congeners. Only 0.15 percent of the octachlorodioxin and 0.75 percent of octachlorofuran was recovered after five minutes. After 15 minutes all PCDD and PCDF levels were below detection limits (Table 2).

Table 2. Catalytic Dechlorination/Hydrogenation of a Mixture of 200 μg of Cl_8DD and 4 μg of Cl_8DF with 1 g of Copper at 280 °C

PCDD/PCDF	Heat Treatment	
	5 min [ng]	15 min [ng]
2,3,7,8-Cl_4DD	750	<0.1
Cl_4DDs	12,000	<0.1
Cl_5DDs	1,930	<0.1
Cl_6DDs	360	<0.1
Cl_7DDs	290	<0.1
Cl_8DDs	290	<0.1
Σ PCDD	14,870	
Cl_4DFs	480	<0.1
Cl_5DFs	290	<0.1
Cl_6DFs	200	<0.1
Cl_7DFs	180	<0.1
Cl_8DFs	30	<0.1
Σ PCDF	1,180	

The reactions also proceed at lower temperatures. At 150 °C tetra-through hexachlorinated dioxins and furans are found after a 5 minute heating period. After 30 minutes 80 percent of the octachlorodioxin and 60 percent of the octachlorofuran were dechlorinated to yield the lower chlorinated dioxins and furans. The dioxin was found to be slightly less stable than the furan. Octachlorodioxin could be dechlorinated in the presence of copper at a temperature as low as 120 °C.

When octachlorodioxin was heated in the presence of copper and deuterated water, deuterated tetra- through heptachlorodioxins were observed, indicating that the moisture is the likely source of the hydrogen. No interconversion of dioxins to furans or furans to dioxins was observed.

Refuse incinerator fly ash was also shown to catalyze dechlorination/hydrogenation reactions of dioxins and furans (Hagenmaier et al., 1987b). Aliquots of fly ash taken from three different refuse incinerators were heated in flasks at 280 °C for two hours.

Table 3. Fly Ash Samples from Two Municipal Waste Incinerators (Fly Ash A and C) and a Waste Incinerator Operated by a University (Fly Ash B) were Heated at 280 °C for 2 Hours

	Fly Ash A [ng/g]		Fly Ash B [ng/g]		Fly Ash C [ng/g]	
	untreat.	280°C,2h	untreat.	280°C,2h	untreat.	280°C,2h
$2,3,7,8-Cl_4DD$	0.5	0.9	2.0	3.2	0.6	1.6
Cl_4DDs	24	21	24	59	12	30
Cl_5DDs	116	19	116	88	59	29
Cl_6DDs	185	6	233	50	110	18
Cl_7DDs	159	2	1,067	22	200	15
Cl_8DDs	88	1	6,204	11	458	13
Σ PCDD	572	49	7,644	230	839	105
Cl_4DFs	111	14	139	41	62	38
Cl_5DFs	188	12	393	69	250	41
Cl_6DFs	123	4	415	34	377	30
Cl_7DFs	35	1	844	9	292	16
Cl_8DFs	26	nd*	1,368	4	400	7
Σ PCDF	483	31	3,159	157	1,381	132

* The PCDD/PCDF concentrations obtained are compared with those of the untreated samples.

nd = not determined.

In all cases the concentrations of all furans and penta- through octachlorodioxins decreased while the amount of tetrachlorodioxins increased. The total amounts of dioxins and furans decreased to less than 12 percent of the original levels. Similar results were obtained using sealed tubes. When a mixture of octachlorodioxin and octachlorofuran was added to a refuse incinerator fly ash sample and heated at 280 °C for 15 minutes, approximately 99 percent of the added octachlorodioxin disappeared and high concentrations of lower chlorinated dioxins were formed (Table 4). Mono- through heptachlorodioxins accounted for 70 percent of the dechlorinated octachlorodioxin. After 2 hours most of the dioxins had disappeared. Fly ash treated with hydrochloric acid to remove the heavy metals present in the ash did not catalyze the dechlorination of octachlorodioxin or furan, indicating that metals present in the fly ash may be the catalytic agents for dechlorination/hydrogenation.

Table 4. Cl_8DD (2 mg), Containing about 40 μg of Cl_8DF as a Contaminant, was Added to 5 g Fly Ash and Treated at 280 °C for the Time Specified [*]

	280 °C for 15 min, [ng/g]	280 °C for 2 h, [ng/g]
Cl_4DDs	52,000	12,900
Cl_5DDs	82,500	2,600
Cl_6DDs	23,700	470
Cl_7DDs	9,300	210
Cl_8DDs	3,700	170
Σ PCDD	174,100	17,850
Cl_4DFs	1,400	620
Cl_5DFs	3,100	150
Cl_6DFs	840	<1
Cl_7DFs	230	<1
Cl_8DFs	23	<1
Σ PCDF	5,593	770

[*] The PCDD/PCDF concentrations are expressed as [ng/g] of fly ash.

Fly ash taken from a hospital incinerator was not as effective a dechlorination catalyst as fly ash taken from municipal refuse incinerators. However, the concentrations of dioxins and furans in the untreated fly ash from the hospital incinerator were considerably higher than those observed for the municipal refuse incinerator fly ash. The researchers suggest that this difference may be due to the higher percentage of plastic and lower percentage of metals in the waste feed.

When fly ash from a municipal incinerator was heated in a stream of air the levels of dioxins and furans increased. The levels decreased when the fly ash was heated in a stream of nitrogen or in a sealed tube. The addition of HCl to the air stream increased the levels of dioxins and furans over that of air alone, and favoured the production of the higher chlorinated isomers. This suggests that chlorine formation may be involved in the production of dioxins and furans on fly ash. Fly ash heated in a stream of air and subsequently in a sealed tube was found to produce a significant quantity of dioxins and furans. The catalytic activity of the fly ash was suggested to have decreased during the heating in air resulting in a slower rate of dechlorination. The experiment appeared to favour the production of lower chlorinated dioxins and furans.

In order to investigate the role of precursors in the formation of dioxins and furans during the heating of fly ash, a simple experiment was performed. Fly ash was heated in a sealed tube in order to eliminate dioxins and furans as well as precursors through dechlorination. After treatment the fly ash was heated in a stream of air. Increased levels of dioxins and furans were found in the treated fly ash which implies that they were formed via de novo synthesis and not from precursor originally present in the fly ash. This de novo synthesis may involve metal chlorides present in the fly ash matrix.

(7) De Novo Synthesis of Dioxins and Furans during Incineration

Some studies have provided evidence for the possibility for the formation of dioxins and furans from reactions of chemically dissimilar compounds and sources of chlorine. Bituminous coal, when oxidized at high temperature in the presence of a source of chlorine, does produce dioxins. The best yields of dioxins were obtained using HCl or Cl_2 in air (Mahle and Wittig, 1980). Phenol and HCl reacted to produce dioxins when heated in a sealed quartz tube at high temperature (Eklund et al., 1986). The reaction was pH dependent, with the yield of dioxins dependent upon HCl concentration. Chlorinated benzenes are also important intermediates which can thermally react to form PCBs and chlorinated furans (Choudhry and Hutzinger, 1983).

Some of these chlorinated compounds can be produced from very simple molecules. Burning methane with dichloromethane produced chlorinated and non-chlorinated PAH. When fuel was enriched with HCl, no chlorinated organics were detected (Krishnan and Hites, 1980). Natural processes such as forest fires can produce chlorinated organics in the presence of chlorine-containing minerals (Palmer, 1976).

The de novo synthesis of dioxins formed the basis of a study conducted in the late 1970s which reached the conclusion that dioxins are ubituitous in the environment (Bumb et al., 1980; Crummett and Townsend, 1984). The theory advanced as the "trace chemistry of fire" was used to explain the presence of dioxins and related compounds in refuse incinerators, fossil fuel power-plants gasoline and diesel-powered vehicles, fireplaces, charcoal grills, and cigarettes. This hypothesis stated that dioxins could be formed as a result of reactions of compound at very low concentrations or from very low yield reactions of more concentrated compounds. The reactions involved were not well defined.

(8) De Novo Synthesis from Particulate Carbon

Vogg and Stieglitz have established that particulate carbon can react with oxygen and inorganic chlorides, with copper(II) as a possible catalyst, to form organochlorine compounds, including dioxins and furans (Vogg and Stieglitz, 1986; Stieglitz and Vogg, 1987; Vogg et al., 1987; Stieglitz et al., 1987). The researchers have concluded that this de novo synthesis of organohalogen compounds is the primary source of dioxins and furans produced during MSW incineration (Stieglitz et al., 1987).

The researchers noted that when fly ash was heated in a stream of air the dioxin and furan content increased (Vogg and Stieglitz, 1986). The maximum increase was at temperatures in the range of 250 °C to 350 °C. With increasing temperature the concentration maximum shifted to lower chlorinated congeners. At 500 °C decomposition occured, and nearly complete degradation was noted at 600 °C after several hours.

Table 5.   Thermal Behaviour of PCDD on Fly Ash [ng/g fly ash]

	Non treated	Temperature [°C]					
		120	200	300	400	500	600
A. Fly Ash							
Cl_8DD	120	120	90	640	6	0.2	0.1
Cl_7DD	125	90	100	1,000	15	0.2	0.1
Cl_6DD	85	65	65	1,640	22	1.	0.1
Cl_5DD	45	35	40	570	35	1.	0.1
Cl_4DD	20	15	15	65	13	2.	0.1
Sum (solid)	395	325	310	3,915	91	4.4	0.1
B. Gas Phase							
Cl_8DD	-	-	-	0.2	9	0.3	0.1
Cl_7DD	-	-	-	0.4	45	0.8	0.1
Cl_6DD	-	-	-	0.8	88	2.	0.1
Cl_5DD	-	-	-	0.15	100	3.	0.1
Cl_4DD	-	-	-	0.1	37	3.	0.1
Sum (vaporized)	-	-	-	1.6	280	9.	0.1
Total Sum (A+B)	395	325	310	3,916	370	13.5	0.1

Table 6.   Thermal Behaviour of PCDF in Fly Ash [ng/g fly ash]

	Non treated	Temperature [°C]					
		120	200	300	400	500	600
A. Fly Ash							
Cl_8DD	12	11	12	218	4	0.1	0.04
Cl_7DD	48	42	48	1,030	37	0.4	0.1
Cl_6DD	56	51	61	1,253	80	1.4	0.1
Cl_5DD	129	119	129	1,570	187	8.	0.2
Cl_4DD	113	95	122	506	140	8.	0.1
Sum (solid)	358	318	372	4,577	448	18	0.4
B. Gas Phase							
Cl_8DD	-	-	-	0.2	8	1.5	0.1
Cl_7DD	-	-	-	0.4	75	15	1.
Cl_6DD	-	-	-	0.8	180	34.	2.
Cl_5DD	-	-	-	0.15	500	93.	9.
Cl_4DD	-	-	-	0.1	390	95.	15.
Sum (vaporized)	-	-	-	4.	1,151	238.	27.
Total Sum (A+B)	358	318	372	4,581	1,601	256.5	27.5

Further studies showed that the presence of oxygen and ions of transition or heavy metals were necessary for a concentration increase to take place (Stieglitz and Vogg, 1987). Decreases in the carbon content of the fly ashes upon heating led to the assumption that particulate organic carbon may be involved in dioxin and furan formation (Stieglitz et al., 1987).

Table 7. Formation of PCDD/PCDF (System: Fly ash with 1% active charcoal, 1% potassium chloride and 0.4% copper(II); 2 hours at 300 °C)

Cl_4DD	215	[ng/g]	Cl_4DF	715	[ng/g]
Cl_5DD	879	[ng/g]	Cl_5DF	3,876	[ng/g]
Cl_6DD	2,406	[ng/g]	Cl_6DF	2,958	[ng/g]
Cl_7DD	2,440	[ng/g]	Cl_7CF	2,904	[ng/g]
Cl_8DD	1,435	[ng/g]	Cl_8DF	1,204	[ng/g]
Total PCDD	7,375	[ng/g]	Total PCDF	11,660	[ng/g]

Experiments with model mixtures of carbon-free fly ash or of magnesium-aluminum-silicate with particulate carbon, potassium chloride and metal chlorides showed that at 300 °C in an air stream considerable amount of dioxins and furans can be produced (Stieglitz et al., 1987) (Table 8).

Table 8. Influence of Temperature on the Formation of PCDD/PCDF

	250 °C	300 °C	350 °C
PCDD [ng/g]	18.6	1,060	15.5
PCDF [ng/g]	65	5,337	126

The yields depended upon the reaction time, concentration and nature of the particulate carbon, and the presence of water vapor. A minimum of two hours was necessary to produce detectable levels of dioxins and furans (Table 9).

Table 9. Concentration of PCDD Congeners after 2 Hours of Annealing
          Time at 300 °C (System: Mg-Al-silicate, 4% Charcoal, 3%
          Chloride, 0.4% Cu(II); 300 °C)

Cl_8DD	10,000	[ng/g]
Cl_7DD	7,300	[ng/g]
Cl_6DD	2,000	[ng/g]
Cl_5DD	80	[ng/g]
Cl_4DD	50	[ng/g]

Of a large number of metal chlorides tested, only copper(II)chloride
resulted in dioxin and furan formation. Yields were proportional to copper concentration. The chemistry is not clearly understood, but two
processes are likely involved: the production of elemental chlorine from
reactions between inorganic chlorides and oxygen (the Deacon reaction)
and copper-catalyzed radical formation.

Table 10. Influence of Copper(II) Concentration on the Formation of
           PCDD/PCDF (System: Mg-Al-silicate, 1% charcoal, 1% KCl, hea-
           ted at 300 °C, 2 hours) in Air Stream; (150 mg H_2O/l)

Congener	PCDD/PCDF Concentrations [ng/g] Percent Cu^{2+} added			
	0	0.08	0.24	0.4
Cl_4DD	1.3	8	20	13
Cl_5DD	1.0	20	80	65
Cl_6DD	0.9	37	240	400
Cl_7DD	0.3	24	230	860
Cl_8DD	1.0	12	200	110
PCDD	4.50	101	770	1,448
Cl_4DF	7	100	310	260
Cl_5DF	11	310	1,290	1,550
Cl_6DF	3	230	1,150	3,100
Cl_7DF	1.6	100	690	2,730
Cl_8DF	0.07	20	200	840
PCDF	22.6	760	3,640	8,480

Only negligible levels of dioxins and furans were produced using graphite, while large amounts of dioxins and furans were produced using sugarcoal, charcoal, domestic soot. The authors speculated that the crystalline lattice of graphite resisted attack by chlorine and oxygen much more than the other types of carbon which are composed of degenerated graphitic structures.

Table 11. Role of Particulate Carbon on the Formation of PCDD/PCDF

Origin of carbon	PCDD [ng/g]	PCDF [ng/g]
Soot [*]	212	
Sugar coal [**]	944	5,050
Charcoal [***]	2,439	7,750
Graphite [****]	0.3-0.5	0.3-0.5

[*] from a domestic oil burner, purified by extraction with methylene chloride and thermal treatment at 500 °C (5 hrs.) in vacuum

[**] obtained by pyrolysis of glucose and conditioning with CO_2 at 800 °C

[***] active charcoal, purified by extraction with methylene chloride (15 hrs.) and vacuum drying at 500 °C (2 hrs.)

[****] purified as described for soot.

The authors postulated that the reactions found in the laboratory experiments occur in a similar way also in the incinerator and are a pathway for dioxin and furan formation. This hypothesis might explain the simultaneous production of a variety of different chlorinated compounds which have been identified in the emission of municipal solid waste incinerators.

Table 12.   Formation of PCDD/PCDF on Fly Ash from Various Municipal Waste
            Incinerators
            (300 °C, 2 hrs., carrier gas: air)

	# Chlorine	Concentrations in [ng/g] Dioxins		Furans	
		Untreated	After annealing	Untreated	After annealing
Plant A	Cl_8	120	640	12	218
	Cl_7	125	1,000	48	1,030
	Cl_6	85	1,640	50	1,250
	Cl_5	45	570	130	1,570
	Cl_4	20	65	110	500
Plant B	Cl_8	160	280	40	70
	Cl_7	50	360	60	230
	Cl_6	24	310	30	250
	Cl_5	10	170	20	240
	Cl_4	3	50	3	180
Plant C	Cl_8	23	330	3	240
	Cl_7	18	530	11	520
	Cl_6	13	360	14	450
	Cl_5	6	180	30	500
	Cl_4	5	30	14	220

(9) Relative Importance of De Novo Synthesis and Precursor Reactions to Dioxin Formation during Incineration

A recent study by Dickson investigated the relative importance of
dioxin and furan formation via de novo synthesis and formation via reactions of precursors on fly ash (Dickson et al., 1988).   Model fly ash
mixtures containing silica gel, activated charcoal, copper(II)chloride,
and ^{13}C-labelled pentachlorophenol were heated at 300 °C in flow-tube
reactors at four combinations of heating time and air flow rate.   The
levels of dioxins produced from pentachlorophenol were very much higher
than those produced via de novo synthesis from carbon.   Yields from both
processes were strongly influenced by flow rate and heating time, however
the two processes were affected differently.   The levels of dioxins formed from pentachlorophenol were about 400,000 times higher than those
formed from carbon at an air flow rate of 3.2 ml/minute for a heating
time of 10 minutes.   At 19 ml/minute and 60 minutes heating time the
levels of dioxins from pentachlorophenol were 10,000 times higher than
those produced from carbon.

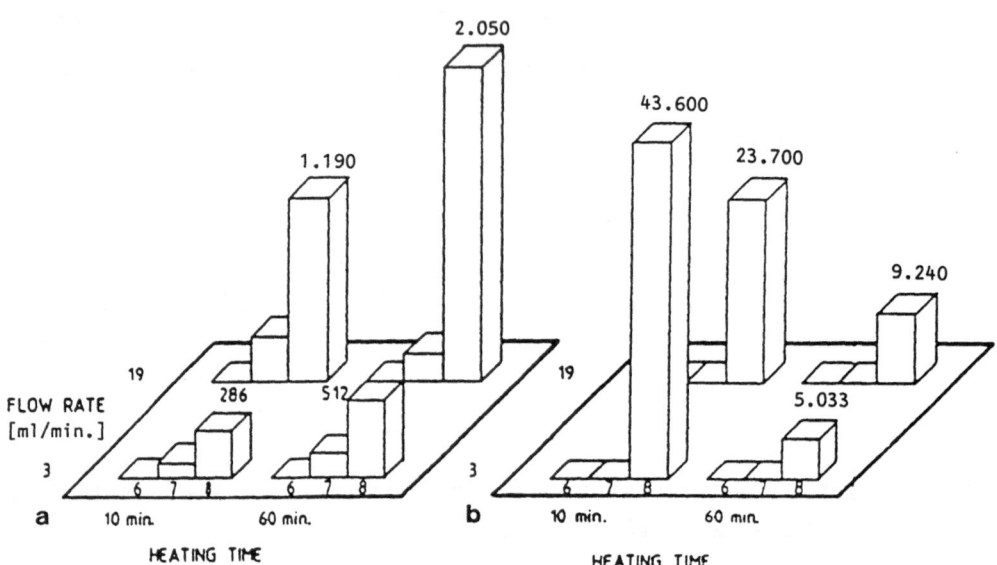

Fig. 3   Amount of Dioxins [ng] Formed from Carbon (a) and Pentachlorophenol (b) on a Model Fly Ash Consisting of 2.1% Cop-
per(II)chloride, 2.3% C and Silica Gel at 300 °C in a Flow of
Dry Air at Different Combinations of Air Flow and Heating Time

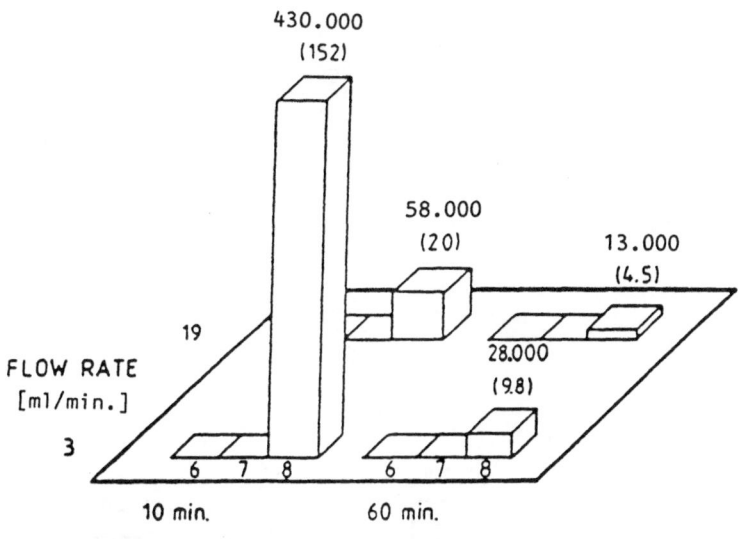

FLOW RATE
[ml/min.]

430.000
(152)

58.000
(20)

13.000
(4.5)

28.000
(9.8)

19

3

6 7 8   6 7 8

10 min.   60 min.

HEATING TIME

Fig. 4 .   Ratios of Amounts of Dioxins Formed from Pentachlorophenol Divided by the Amounts Formed on Carbon on a Model Fly Ash Consisting of 2.1% Copper(II)chloride, 2.3% Carbon and Silica Gel
at 300 °C in a Flow of Dry Air at Different Combinations of
Air Flow and Heating Time.
(The measured data are indicated in brackets, the values above
are normalized for the molar ratio of carbon to pentachlorophenol)

Dioxins were also formed in large amounts from pentachlorophenol on a
mixture of carbon and silica gel which did not contain cop-
per(II)chloride. The authors concluded that the carbon acted as a catalyst for the formation of dioxins from pentachlorophenol. The function
of copper(II)chloride therefore may be to catalyze the dechlorination of
dioxins, rather than catalyzing the initial formation of dioxins.

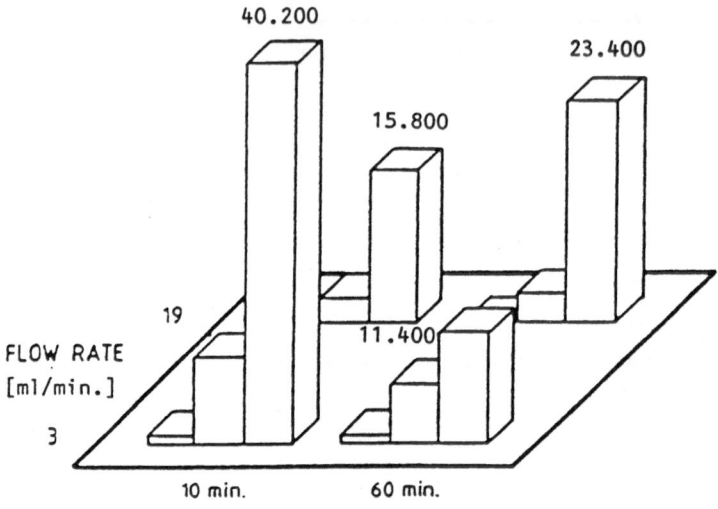

Fig. 5 .   Amounts of Dioxins [ng] Formed from Pentachlorophenol on a
Model Fly Ash Consisting of 2.3% Carbon and Silica Gel at
300 °C in a Flow of Dry Air at Different Combinations of Air
Flow and Heating Time

2.2.3 Incinerator Studies

(1) Waste Incineration in Canada

Environment Canada established the National Incinerator Testing and Evaluation Program (NITEP) (EPS Canada, 1985; 1986; 1988) in order to:

1. Find optimum design and operating parameters for municipal incinerators to minimize emissions of substances of environmental concern;
2. Relate operating conditions to the emissions observed for each incinerator design;
3. Investigate alternative combustion and emission control technologies for the determination of the best practical incinerator operation;
4. Provide design and operating guidelines for future incinerators;
5. Develop internationally recognized protocols for the sampling and analysis of PCDD, PCDF, and other emission pollutants;
6. Determine if surrogates exist for PCDD which can be readily and inexpensively monitored as indicators for PCDD emission levels.

NITEP planned to study three generic types of municipal incinerators:

1. Two-stage combustion system
2. Mass burning system
3. Refuse derived system.

The results of a comprehensive study of the first incinerator design were published in a report released in 1985 (EPS Canada, 1985).

The two-stage incinerator is illustrated and briefly described in Figure 6. The test program carried out on this facility was one of the most extensive ever performed. The first phase characterized the key operational variables for the incinerator. The characteristics studied in the performance testing are summarized in Figure 7. Based on the initial studies, four specific test conditions were chosen for the extensive performance assessment. The rationale behind the choice of the particular test conditions is summarized in Table 13. These tests were performed in triplicate. Individual control readings taken for triplicate test runs showed that consistent operational control was achieved. The key operating parameters for the four test conditions are listed in Tables 14 and 15.

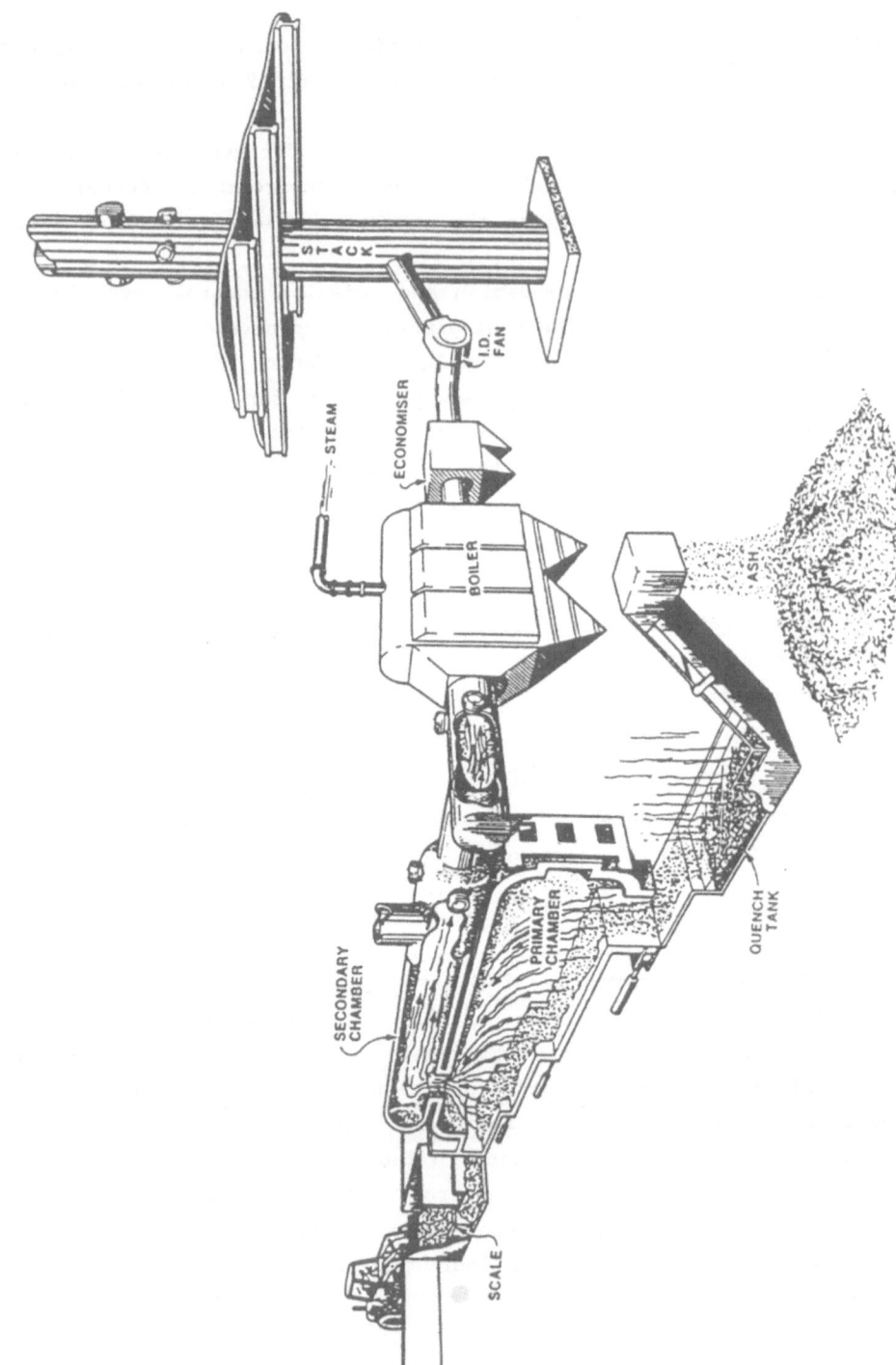

STACK

STEAM

ECONOMISER

I.D. FAN

BOILER

ASH

SECONDARY CHAMBER

PRIMARY CHAMBER

QUENCH TANK

SCALE

Fig. 6. Process Schematic of P.E.I. Two-Stage Combustion System

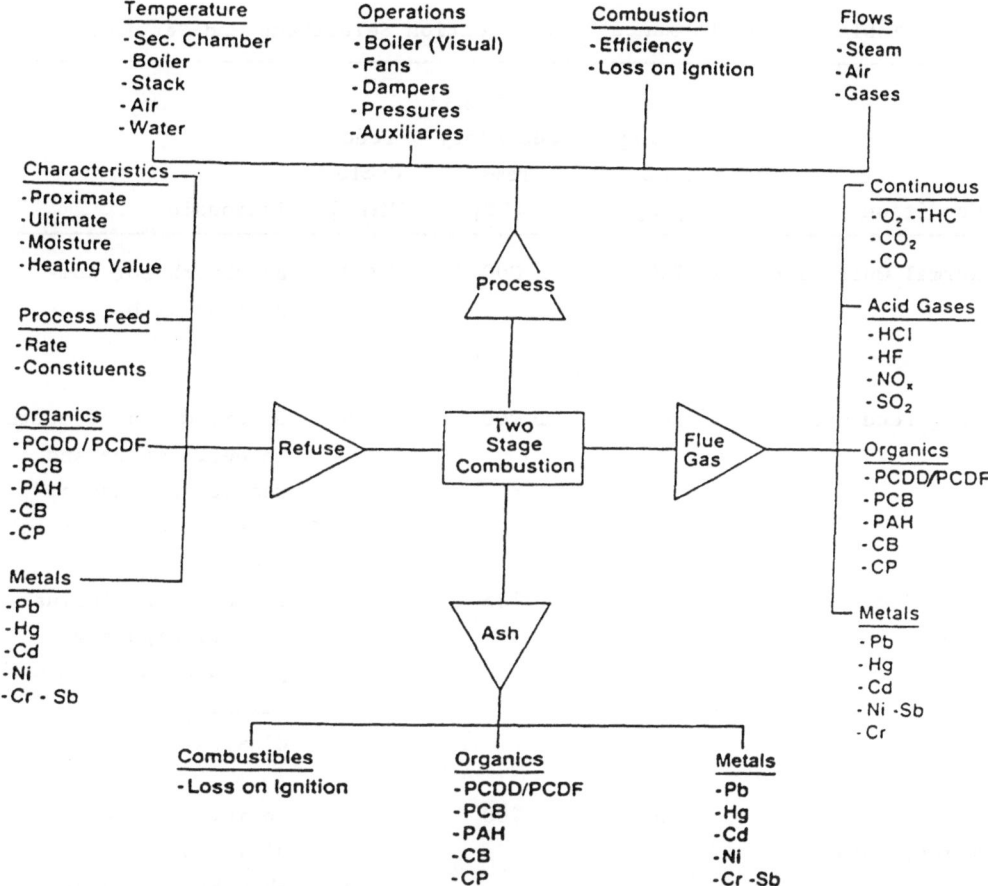

Temperature
- Sec. Chamber
- Boiler
- Stack
- Air
- Water

Operations
- Boiler (Visual)
- Fans
- Dampers
- Pressures
- Auxiliaries

Combustion
- Efficiency
- Loss on Ignition

Flows
- Steam
- Air
- Gases

Characteristics
- Proximate
- Ultimate
- Moisture
- Heating Value

Process Feed
- Rate
- Constituents

Organics
- PCDD / PCDF
- PCB
- PAH
- CB
- CP

Metals
- Pb
- Hg
- Cd
- Ni
- Cr - Sb

Process

Refuse

Two
Stage
Combustion

Flue
Gas

Ash

Continuous
- O_2 - THC
- CO_2
- CO

Acid Gases
- HCl
- HF
- NO_x
- SO_2

Organics
- PCDD/PCDF
- PCB
- PAH
- CB
- CP

Metals
- Pb
- Hg
- Cd
- Ni - Sb
- Cr

Combustibles
- Loss on Ignition

Organics
- PCDD/PCDF
- PCB
- PAH
- CB
- CP

Metals
- Pb
- Hg
- Cd
- Ni
- Cr - Sb

Fig. 7.   Performing Testing Summary;   P.E.I. Two Stage Combustion
          System.

Table 13. Performance Test Condition Selections and Settings

Condition	Primary Temp. [°C]	Settings Secondary Temp. [°C]	Feed Cycle [Min.]	Rationale
Normal Operation	700	1,000	8.7	Benchmark for comparison with other tests
Long Feed Cycle	700	1,000	12.5	Observed improvements in performance and reduced demands on plant loader operator
High Secondary Temperature	700	1,135	8.7	To determine influence of high temperature on organic emissions such as dioxins and furans
Low Secondary Temperature	700	900	8.7	To investigate a condition which was thought to create higher organic emissions

Table 14. Summary of Key Operating Parameters I

Operating Conditions	Normal Condition	Long Cycle	High Secondary Temperature	Low Secondary Temperature
Primary Temp. [°C]	695	690	703	676
Secondary Temp. [°C]	910	890	1,080	780
Steam Rate [kg/h]	4,360	4,020	4,470	3,960
Refuse - Rate [kg/h wet]	1,590	1,600	1,700	1,520
- Calorific Value [kJ/kg as fired]	10,530	10,220	10,430	10,770
- Moisture [%]	35	35	32	35
- Non-combustibles [% wet basis]	14.0	13.9	15.0	13.0
Energy Input [kJ/h]*10^6	16.8	16.2	17.6	16.0
Efficiency -Input/Output	59.6	56.6	58.4	55.9
-Heat Loss	52.0	50.9	58.1	50.7

Table 15. Summary of Key Operating Parameters II

Operating Description	% CO_2	% O_2	% Excess Air
Normal Condition	8	12	140
Long Cycle	8	12	145
High Secondary Temperature	10	10	85
Low Secondary Temperature	7	13	185

The stack emissions (corrected to 12% CO_2) averaged for the triplicate tests of each condition are shown in Table 16. From this data it is apparent that organic emission rate is generally higher at the lower secondary chamber temperature. The emission rate for metals, however, appears to be affected in just the opposite manner (i.e. decreases with decreasing temperature). Therefore there is a conflict of environmental impact interests. The amount of PCDD emitted appears to increase with decreasing secondary chamber temperature whereas the PCDF emission is lower at both the higher and lower operating temperatures than at the normal conditions. The latter result is particularly puzzling and no explanation was proposed.

Table 16. Comparison of Emission Data
* Based on average of two tests only
ND — Not Detected

| Contaminant | Stack concentration (at 12% CO_2) Dry | | | |
	Normal	Long	High	Low
TSP [mg/Nm^3]	208	230	247	167
HCl [mg/Nm^3]	1,085 *	1,070	1,165	950
Cl$_5$DD [ng/Nm^3]	107	107	62	123
Cl$_5$DF [ng/Nm^3]	153	156	95	98
PCB [ng/Nm^3]	801	58	ND	126 *
PAH [ng/Nm^3]	7,005	8,010	6,653	12,490 *
Chlorophenol [ng/Nm^3]	4,346	3,773	2,706	6,591 *
Chlorobenzene [ng/Nm^3]	4,321	3,161	3,968	4,884 *
Cadmium [mg/Nm^3]	0.9	0.8	0.8	0.6
Lead [mg/Nm^3]	13.5	15.2	15.2	8.4
Chromium [mg/Nm^3]	0.04	0.03	0.1	0.03
Nickel [mg/Nm^3]	0.2	0.3	0.5	0.5
Mercury [mg/Nm^3]	0.7	0.5	0.9	0.5
Antimony [mg/Nm^3]	0.6	2.6	0.5	0.5

Table 16. Continued

Contaminant	Stack Emissions per ton of Feed (as fired)			
	Normal	Long	High	Low
TSP [g/ton]	843	874	977	682
HCl [g/ton]	4,400	4,130	4,480	3,930
Cl_5DD [μg/ton]	428	400	228	516
Cl_5DF [μg/ton]	570	574	340	411
PCB [μg/ton]	3,413	245	ND	574[*]
PAH [μg/ton]	29,305	33,201	26,956	54,514[*]
Chlorophenol [μg/ton]	18,403	15,042	10,814	28,973[*]
Chlorobenzene [μg/ton]	18,014	12,807	16,061	22,045[*]
Cadmium [g/ton]	3.8	3.0	3.2	2.6
Lead [g/ton]	54.8	57.8	60.0	34.2
Chromium [g/ton]	0.2	0.1	0.4	0.1
Nickel [g/ton]	1.0	1.0	2.2	1.9
Mercury [g/ton]	2.8	2.0	3.6	2.2
Antimony [g/ton]	2.3	9.6	2.1	1.9

The distributions of PCDD and PCDF homologue congener groups in the stack emissions for the four test conditions are illustrated in Figures 8 and 9. The amount of PCDD homologue was found to increase with increasing degree of chlorination for all four conditions. The distribution of the PCDF homologues took on a bell-shaped form centered about the hexa-CDFs. The homologue distributions for the PCDD and PCDF do not vary significantly between the various test conditions.

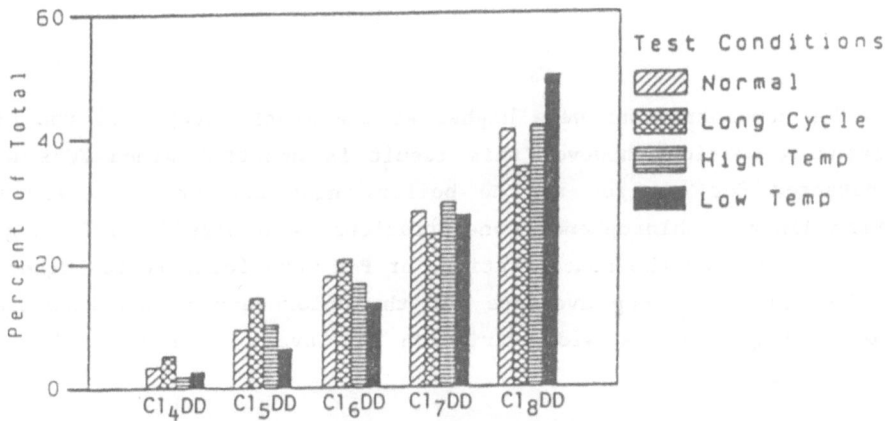

Fig. 8.   Stack Dioxin Homologue Distribution by Test Condition (P.E.I. Incinerator)

The concentrations of the various organic compound classes were compared at the boiler inlet and at the stack (see Table 17). The PCDD and PCDF concentrations measured at the boiler inlet were low or not detected for all three conditions differing from the normal. The concentrations at the stack were in all cases higher than at the boiler inlet.

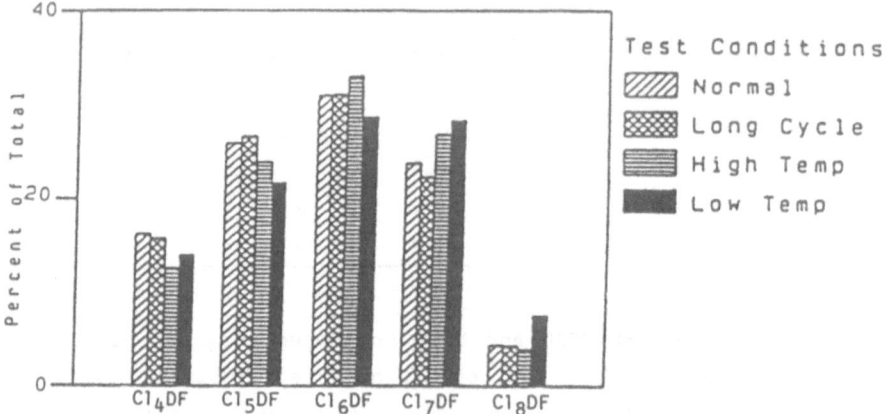

Fig. 9.   Stack Furan Homologue Distribution by Test Condition (P.E.I. Incinerator)

The PCB concentrations were higher at the stack except for the high temperature condition, however this result is doubtful since PCBs were only detected for one run at the boiler inlet and at very near the detection limit. Chlorophenol concentrations were significantly higher at the stack whereas the concentrations of PAH were found to be higher at the boiler inlet. Group averages for the chlorobenzene concentrations were misleading since a wide variation of levels were found between individual runs.

The possibility of chemical transformations occurring upon cooling of the flue gas was suggested as an explanation for the higher levels of some compounds in the stack as compared to the boiler inlet.

Table 17. Concentrations of Organic Compounds at Boiler Inlet in Comparison to Stack

| | Concentration [ng/Nm^3 @ 12% CO_2] | | | | | |
| | Long Cycle | | High Temperature | | Low Temperature | |
Compound	Boiler Inlet	Stack	Boiler Inlet	Stack	Boiler Inlet	Stack
PCDD	ND	107	1	62	14	123
PCDF	ND	156	ND	95	10	98
PCB	11 *	58	8 *	ND	61	126 **
PAH	10,475	8,010	12,226	6,653	29,481	12,490 **
Chlorophenol	486	3,773	479	2,706	2,941	6,591 **
Chlorobenzene	6,595	3,161	4,895	3,968	8,906	4,884 **

* — only one non-zero value (two readings were below detection limits)
** — based on average of two tests only.

The mass flow rates of PCDD and PCDF entering with the refuse and the mass flow rates of the stack emissions are compared in Figure 10. These results show that although the amounts of PCDD and PCDF in the input garbage are quite different, the emissions for all four test conditions are very similar. In fact these results imply that the operating parameters do not have a significant effect on the levels of the PCDD and PCDF emitted.

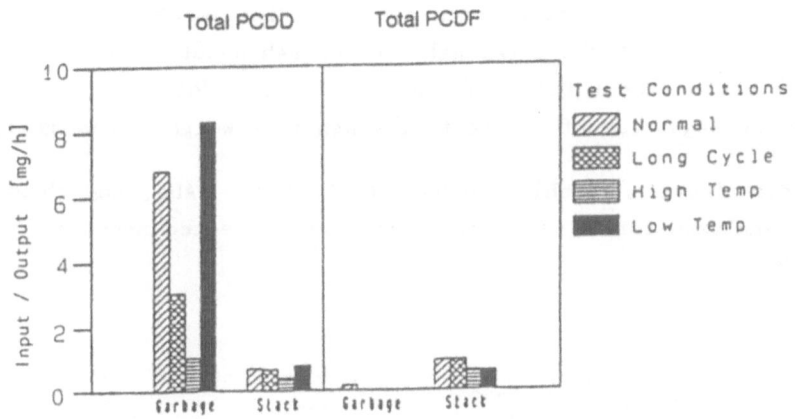

Fig. 10. Total Dioxin/Furan Input/Output

The concentrations of PCDD and PCDF in the incoming refuse and the ashes collected from three different sites within the incinerator were also compared. Figure 11 illustrates the differences in the incoming levels with the levels found in the different ashes (averaged for 12 test runs, i.e. triplicate runs of the four test conditions). No dioxins or furans were detected in the bottom ash samples. The PCDD were approximately equally distributed between the boiler and economizer ashes while the PCDF distribution appears to be slightly biased towards the economizer ash (however, the variation between the runs is not reported and therefore this difference may not be significant).

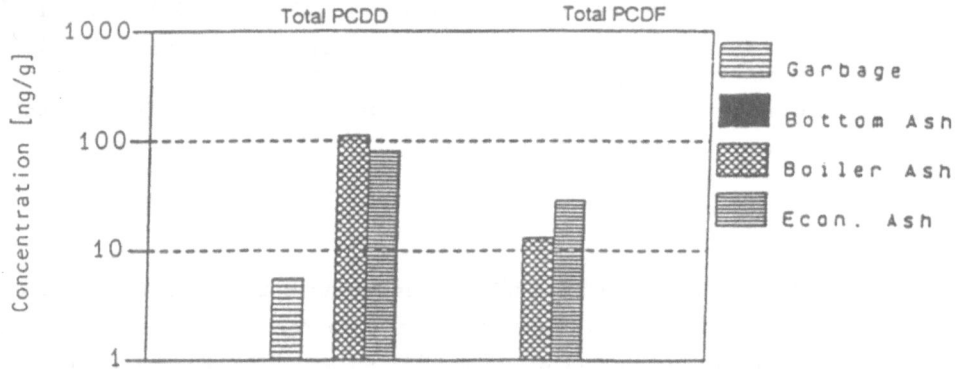

Fig. 11. Distribution of Dioxins/Furans in Solids (as analysed)

The significance of these results is misleading if one does not consider the relative amounts of each ash. The incinerator bottom ash comprises about 98.5% of the total ash, boiler ash about 1.25%, and economizer ash makes up about 0.13% of the total. In addition, one must also consider the reduction of refuse to fly ash on a weight-to-weight basis.

The distribution of chlorophenols (CP), PCBs, PAHs, and chlorobenzenes (CB) analyzed for in the garbage and the collected ashes are shown in Figure 12.

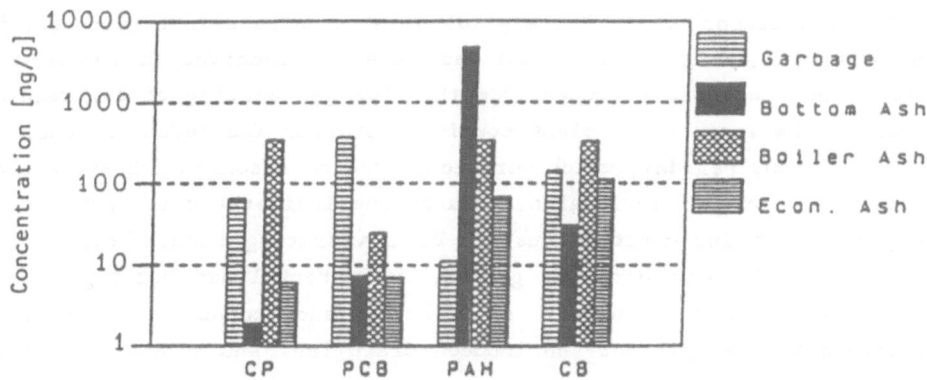

Fig. 12.   Distribution of Organics in Solids (as analysed)

The range of operating conditions for which the two-stage incinerator was tested do not reveal any significant differences with respect to the PCDD/PCDF emission levels.   The system efficiently removes PCDD and PCDF from the stack emissions regardless of the which particular set of operating conditions were used.   Good design of flue gas scrubbing systems and controlled operation of the incinerator appear to minimize the emissions of concern.

Most of the modern municipal waste incinerators tend to be operated under much more careful control than their older counterparts.   Increasingly stringent regulations and technological improvements in the design of flue gas cleaning systems have resulted in lower emissions in the modern incinerator.   NITEP undertook this study (EPS Canada, 1986) because the role of emission control technology was not fully evaluated by conducting full-scale or large-scale pilot tests, previously.

Extensive testing was performed on two pilot-scale air pollution control systems:

   1.   Wet-dry scrubber (sprayer-dryer) plus fabric filter
   2.   Dry scrubber plus fabric filter

These designs were tested for the efficiencies in reducing the emission of organic compounds, including PCDD and PCDF, as well as heavy metals and acid gases such as hydrogen chloride and sulphur dioxide.   The emissions of these contaminants were studied as function of the operating conditions in order to determine optimal conditions for incinerator operation.   The study involved the determination of the compounds of concern in the emissions and the ashes resulting from the incineration process.

The incinerator on which the pilot plant studies were carried out was a mass burning design. The refuse was burned as received in a water-wall furnace and steam was produced from the flue gas heating of the recovery boilers. The incinerator plant consisted of four incinerators, each rated at 227 tons per day, which were connected to a common ash storage pit and stack. A cross-sectional diagram of one incinerator is shown in Figure 13. Each incinerator consists of a vibrating feeder-hopper, feed chute, drying/burning/burn out grates, refractory-lined burning zone, a waste heat recovery boiler with super heater and economizer, a two-field electrostatic precipitator, an induced draft fan, and a wet ash quench/removal system.

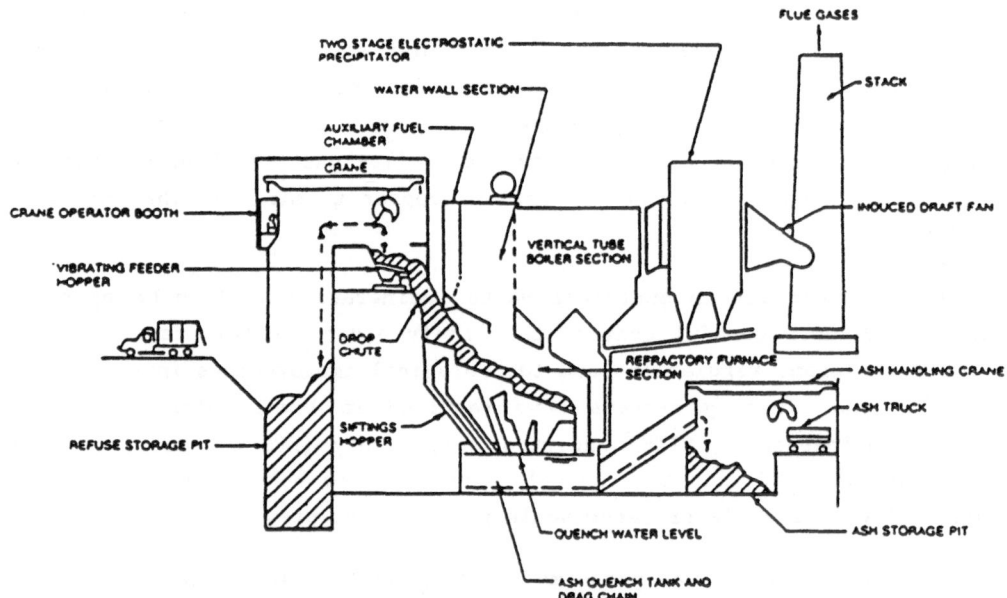

Fig. 13. Quebec Incinerator Schematic Cross-Section

The system was modified for the pilot-scale studies by installing a flue gas take-off duct at the inlet to one of the electrostatic precipitators, as shown in Figure 14. The study was set up to evaluate the performance of the pilot plant in both the dry and wet-dry modes of operation.

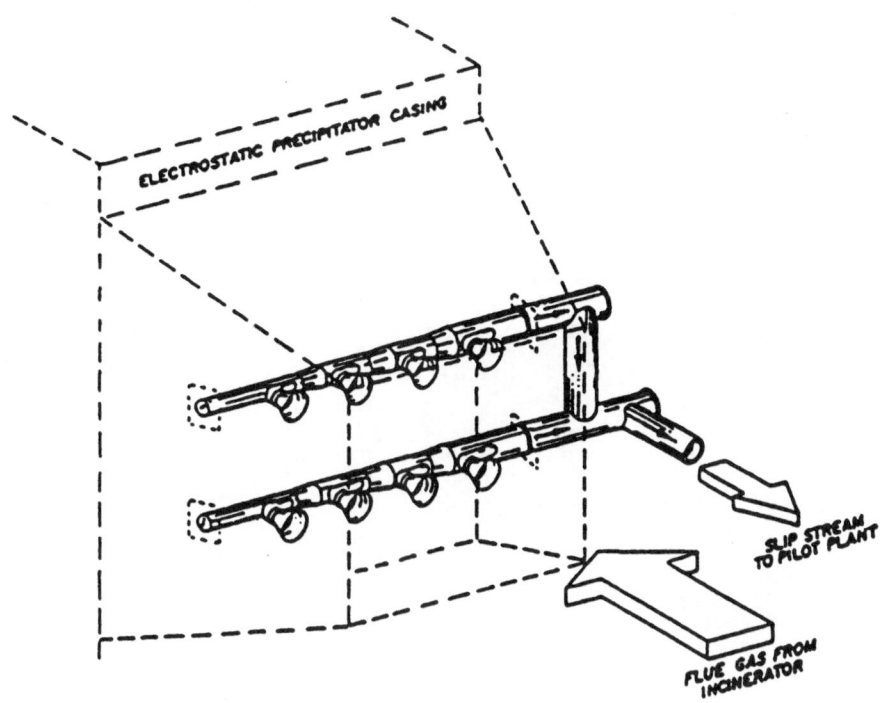

Fig. 14.   Incinerator Gas Slip Stream Take-off Arrangement

The dry system, as shown in Figure 15, consists of a single dry lime injection nozzle and an internal cyclone. The hot flue gas first enters the wet-dry scrubber which in this mode of operation is used only to cool the gas. The flue gas enters this chamber at a temperature of 255-270 °C and is cooled to 110-140 °C by a water spray. The cooled flue gas enters the bottom of the dry scrubber tangentially with respect to the internal cyclone where large particles of fly ash are removed. A dry, hydrated lime powder is injected in a stream of compressed down upon the gas stream. The chemical treatment of the flue gas results in a dry stream of particles and lime which continues on towards the fabric filter dust collector. The fabric filter is a single compartment baghouse containing a 6 x 12 array of 72 high-temperature Teflon bags with Gortex scrim (air-to-cloth ratio of 4.4). The bags are cleaned using low pressure air pulses directed downward from discharge nozzles located above the bags. The two major advantages of dry scrubbing versus wet scrubbing are the absence of a liquid effluent and the reduced corrosion of equipment.

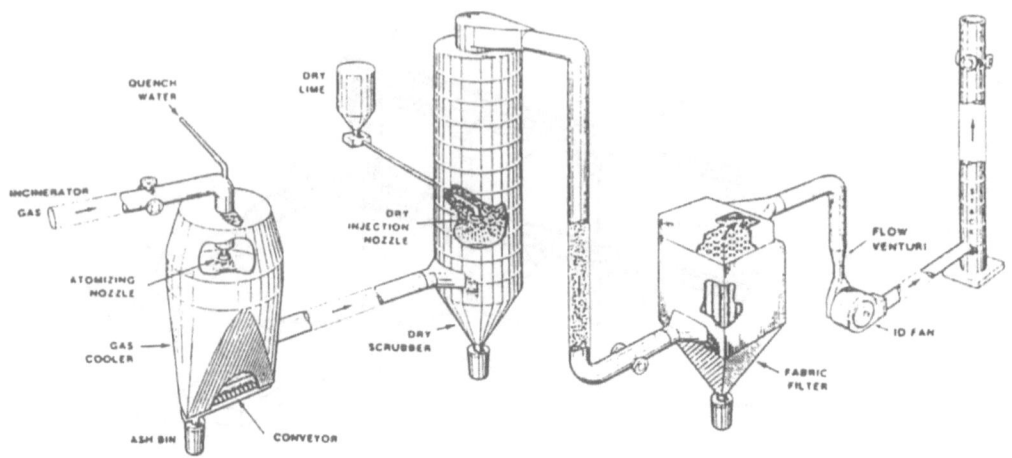

Fig. 15.   FLAKT's Dry-System

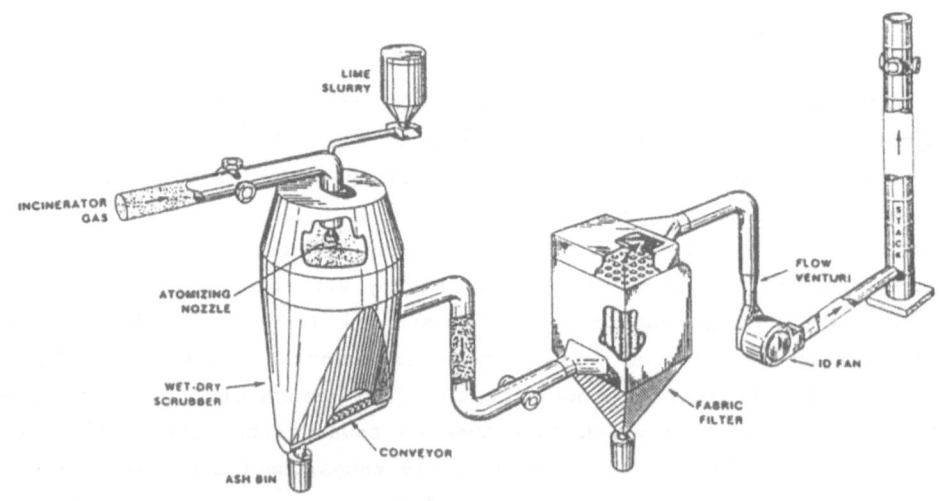

Fig. 16.   FLAKT's Wet-dry System

In the wet-dry mode of operation, the scrubber receives hot flue gas from the top of the vessel, as shown in Figure 16. The flue gas is thoroughly mixed with a spray of finely atomized lime slurry. The slurry is carefully mixed in proportion to the flue gas so that all the water is evaporated, leaving the lime as suspended fine particulates in the gas. The wet-dry scrubber creates a cool, dry mixture of particles (110-140°C) similar to that produced in the dry scrubbing process. The entrained particulates are passed on towards the fabric filter system previously described.

Both the dry and wet-dry systems can be operated with recycling of the fabric filter ash by mixing the ash, which contains some unreacted lime, with fresh lime in either the dry or slurried form.

As was done for the previous NITEP study on the two-stage incinerator, characterization tests were done on a wide variety of operating conditions to determine the parameters to be used for the intensive performance evaluation. The parameters measured during the characterization testing phase are outlined in Figure 17. Both scrubber systems were tested over a range of operating temperatures and hydrated lime feed rates since these were the two most important independent variables. The characterization test conditions studied are summarized in Table 18.

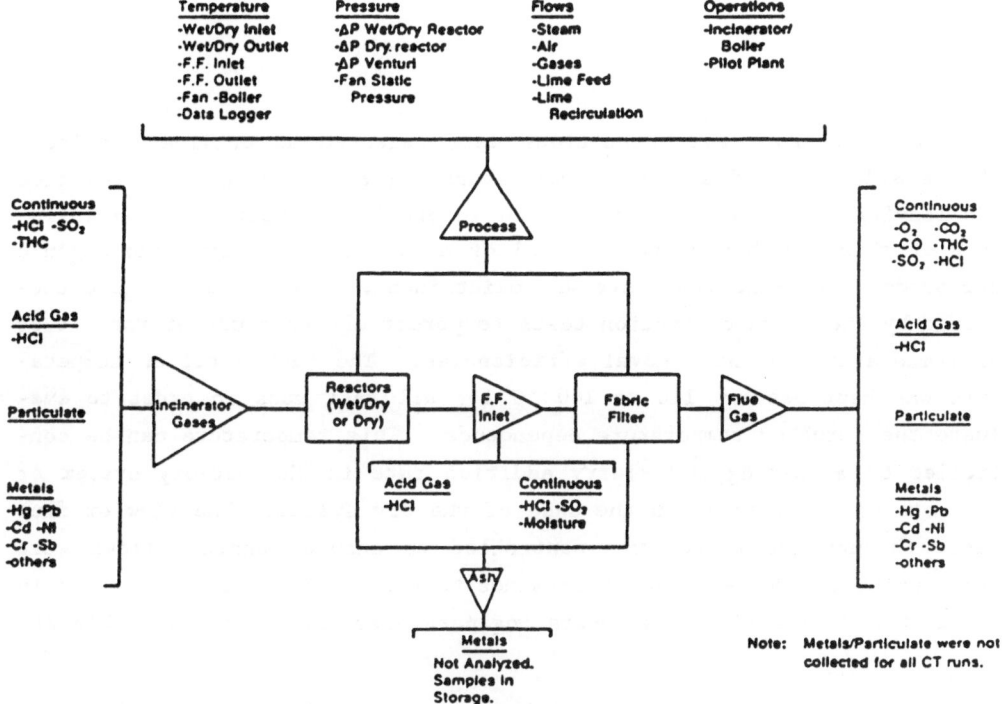

Fig. 17.   Parameters Measured during Characterization Testing.

311

Table 18. Summary of Characterization Test Conditions

Operating Mode	Temperature After Cooling: Target °C	Actual °C	Lime Feed Rate
Dry System:			
- Normal Temp. Operation	140	145	Normal
	140	151	High
	140	152	High*
	140	151	Very high
- High Temp. Operation	>200	222	Very high
- Low Temp. Operation	125	133	Normal*
Wet-Dry-System:			
-Normal Temp. Operation	150	156	Normal
	150	158	High
	150	158	High*
	150	156	Very high
- Low Temp. Operation	135	141	Normal
	135	144	Normal*

* - + Recycle

Six performance test conditions were selected as outlined in Table 19. A wide range of scrubber temperatures (i.e. temperature of the flue gas leaving the scrubber) were chosen in order to investigate the influence of this variable on the removal of acid gases, mercury, and organic compounds. The lime feed rate was maintained at normal levels as established by the characterization tests to permit closer study of the effect of temperature on the removal efficiencies. The fabric filter temperature was kept between 110 to 140 °C for all test runs in order to evaluate the scrubber temperature dependence. This temperature can be controlled by adjusting the slurry addition rate in the wet-dry system or the rate of water spray in the case of the dry system. The flow of flue gases through the system was controlled to ensure constant flows were used during the tests. The performance test conditions are summarized in Table 20 and some of the key operating parameters are listed in Table 21.

Table 19. Selection of Performance Test Conditions

Operating Mode	Target Temp. Before Fabric Filter (°C)	Rationale
Dry System	>200	To collect data with no gas cooling for comparison to the three other temperatures.
	140	Benchmark for comparison with other tests. Considered normal for full-scale.
	125	To observe potential improvements in organic removal at lower than normal temperatures.
	110	To push the limits of low temperature operation, particularly for impact on organics removal.
Wet-Dry-System	140	Benchmark for comparison with other tests.
(+ Recycle)	140	To investigate whether lime recycle enhances the removal efficiency of organics and acid gases.

Table 20. Summary of Performance Test Conditions

	Temperature before Fabric Filter	
	Target °C	Actual °C[*]
Dry System	> 200	209
	140	141
	125	125
	110	113
Wet-Dry System	140	140
(+ Recycle)	140	141

[*]   Thermocouple Data from Pilot Plant Control System

Table 21. Summary of Key Operating Parameters

	Dry System				Wet-Dry-System	
	110°C	125°C	140°C	>200°C	140°C	140°C + Recycle
Incinerator:						
-Steam Flow [kg/h $*10^3$]	32.8	33.4	33.2	33.0	32.1	31.3
-Gas Temp.-Boiler Outlet [°C]	303	287	291	287	279	293
Fabric Filter:						
-Pressure Drop [cm water gauge]	15.7	14.4	14.9	14.7	15.2	15.9
Lime:						
-Flow [kg/h]	3.7	3.6	3.7	3.6	3.5	3.5
Temperature [°C]:						
-Inlet to Pilot Plant	267	258	261	253	254	263
-Inlet to Fabric Filter	113	125	142	209	140	141

Sampling and data collection were done for duplicate runs of each test condition. Incinerator operations were found to be quite stable throughout the test program and pair averages were therefore used to represent the results of each test condition. Although in general pair group averages were concluded to provide a relative indication of the levels of emission, seemingly significant differences were sometimes seen in results between a pair of runs for the same condition.

To evaluate the performance of the two pollution control systems, the emissions were sampled at:

1. The pilot plant inlet (i.e. between the incinerator and pilot plant) to characterize the untreated raw gas coming from the incinerator

2. The midpoint between the scrubber system and the fabric filter to evaluate the scrubber performance

3. The stack to evaluate the performance of the pollution control system as a whole.

The ashes collected in the hoppers (ash bins) of the scrubber and fabric filter vessels were also subjected to analysis. The results for the emissions of PCDD, PCDF, and other chlorinated organics are discussed in the ensuing sections. Before treating the results, it is necessary to mention the detection limits for the analytical methodology employed in this study (entries of "ND" in the tables implies that the levels were below the detection limit). For the analysis of PCDD and PCDF in flue gas samples, the limits of detection per analyte peak were:

- 0.2 ng for Cl_4DD, Cl_4DF, Cl_5DF, Cl_6DF
- 0.4 ng for Cl_5DD, Cl_6DD, Cl_7DF, Cl_8DF
- 0.6 ng for Cl_7DD, Cl_8DD.

The concentrations of PCDD and PCDF in the flue at the three sampling locations for each test condition are given in Tables 22 and 23. The same results are given in Figure 18 in order to more readily illustrate the efficiency of the scrubbers and fabric filter in removing PCDD and PCDF. The removal efficiencies were calculated and are given in Tables 22 and 23. The overall removal of PCDD was greater than 99.9% for all tests except for the high temperature condition where the efficiency was slightly lower (99.4%). The stack emissions all had very low PCDD contents (all below 6 ng/Nm^3) and were not even detectable for most of the test runs. The overall removal efficiencies for PCDF ranged from 99.3 to greater than 99.9%. Clearly, both systems under all test conditions are very efficient in removing PCDD and PCDF from the flue gas.

Drawing conclusions regarding the removal efficiencies for the different scrubbers under the conditions studied should perhaps be avoided since the most important consideration should be the overall performance of the air pollution control equipment. In Table 23, one can see that for the wet-dry system at both operating conditions the efficiency for PCDF removal at the midpoint (i.e. for the wet-dry scrubber) is negative. This implies the formation of PCDF somewhere after the inlet to the pilot plant. The overall efficiencies of these two operating conditions, however, were amongst the highest observed for the entire test program. Therefore the only significant conclusion to be drawn from these results is that there does not appear to be any advantage (with respect to emission of PCDD and PCDF) to using one scrubbing design or set of operating conditions over the other as long as a fabric filter is used following the scrubber.

Table 22. PCDD Concentrations [ng/Nm^3 @ 8% O$_2$] in Flue Gas and Efficiency of Removal

| | Dry System | | | | Wet-Dry System | |
	110°C	125°C	140°C *	200°C	140°C	140°C + Recycle
Inlet [ng/Nm^3]	580	1,400	1,300	1,030	1,100	1,300
Midpoint [ng/Nm^3]	310	570	540	1,140	840	1270
Outlet [ng/Nm^3]	0.2	N.D.	N.D.	6.1	N.D.	0.4
Efficiency [%]						
- Inlet/Midpoint	47	60	57	- 11	24	2
- Overall	>99.9	>99.9	>99.9	99.4	>99.9	>99.9

* based on one test

Table 23. PCDF Concentrations [ng/Nm^3 @ 8% O$_2$] in Flue Gas and Efficiency of Removal

| | Dry System | | | | Wet-Dry System | |
	110°C	125°C	140°C *	200°C	140°C	140°C + Recycle
Inlet [ng/Nm^3]	300	940	1,000	560	660	850
Midpoint [ng/Nm^3]	270	440	630	490	690	1,030
Outlet [ng/Nm^3]	2.3	N.D.	N.D.	6.1	N.D.	0.9
Efficiency [%]						
- Inlet/Midpoint	11	54	37	13	- 4	- 21
- Overall	99.3	>99.9	99.9	99.8	>99.9	99.9

* based on one test

It is interesting to note that the distribution for PCDD homologues takes on a crude bell-shape centered about the hepta-CDDs while the concentration of PCDF homologues appears to decrease within increasing degree of chlorination. These distribution patterns are somewhat different than those reported by other workers.

The overall removal efficiencies for other chlorinated organic compounds are given in Table 24. Only the high-temperature dry scrubber/fabric filter combination shows a generally poor efficiency in removing organic material. The best system appears to be the wet-dry system operated with a continuously fresh feed of lime slurry.

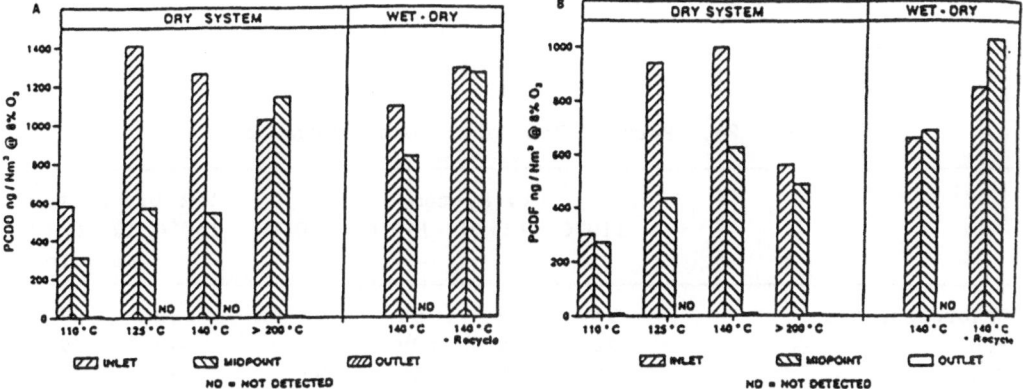

Fig. 18. PCDD (left) and PCDF (right) Concentrations in Flue Gas

In addition to sampling the flue gas at different points in the incinerator system, hopper ashes were collected in each vessel of the pilot plant. As the flue gas moves throughout the system, some of the fly ash falls out and is collected in the scrubbers while the remainder is largely collected by the fabric filter. The approximate proportion of the ash collected in each vessel is given in Table 25. The concentrations [ng/g] of PCDD in the various ashes for the different test conditions are listed in Tables 26 and 27 and illustrated in Figure 19. The proportion of PCDD found in each hopper does not match the proportion of fly ash distributed amongst the hoppers. Presumably the distribution of fly ash particles of different sizes is also not proportional to the relative amounts of the ash collected in each hopper. One would expect the smaller particles to travel further along the system. No mention was made regarding the distribution of ash particles nor was any correlation made between particle sizes and the level of PCDD present. The highest concentration of PCDD were found in the fabric filter ash for all of the test conditions studied. The ash in this vessel should consist largely of the smaller particles. The resulting surface area would be very high (for a given weight of ash, the smaller the particle, the larger the total surface area) and thus more PCDD would be expected to be present due to the greater degree of adsorption per unit weight of fly ash. The same sort of trends are apparent for the concentrations of PCDF found in the various ashes collected (see Table 26 and Figure 19). The highest concentrations of PCDF were found in the fabric filter ashes for all six conditions.

Table 24. Percent Removal of Other Organics

| | Dry System | | | | Wet-Dry System | |
	110°C	125°C	140°C *	200°C	140°C	140°C + Recycle
Chlorobenzenes	95	98	98	62	>99	99
PCB	72	>99	>99	54	>99	>99
PAH	84	82	84	98	>99	79
Chlorophenols	97	99	99	56	99	96

*   based on one test

Table 25. Proportion of Ash Collected in Each Vessel
          (Note: Excludes incinerator bottom ash and boiler ash)

Source of Ash	Dry System	Wet-Dry System
Wet-Dry Scrubber (or Conditioning Tower)	30 %	33 %
Dry Scrubber	40 %	-----
Fabric Filter	30 %	67 %
	100 %	100 %

Table 26.   PCDD Concentrations in Ash [ng/g]

Operating Condition	Dry System				Wet-Dry System	
	110°C	125°C	140°C *	200°C	140°C	140°C + Recycle
Wet-Dry Scrubber Ash	13	14	6	6	6	12
Dry Scrubber Ash	160	64	94	31	N/A	N/A
Fabric Filter Ash	280	570	570	740	160	230

N/A   Not Applicable

Table 27.   PCDF Concentrations in Ash [ng/g]

Operating Condition	Dry System				Wet-Dry System	
	110°C	125°C	140°C *	200°C	140°C	140°C + Recycle
Wet-Dry Scrubber Ash	10	12	4	5	5	82
Dry Scrubber Ash	87	36	68	22	N/A	N/A
Fabric Filter Ash	160	320	380	320	130	170

N/A   Not Applicable

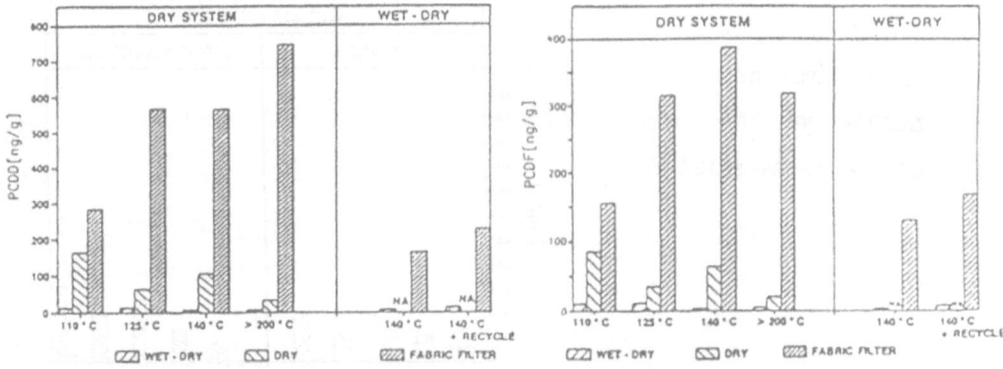

Fig. 19.   PCDD (left) and PCDF (right) Concentrations [ng/g] in Hopper Ashes

The distribution of the PCDD and PCDF homologues in the hopper ashes
are shown in Figures 20 and 21, respectively. The homologue distributions of the PCDD and the PCDF in the fabric filter ash are very similar
to their corresponding homologue distributions for the inlet flue gas.
This should not be too surprising considering that the bulk of the
dioxins and furans are trapped at the fabric filter. In fact, for the
dry scrubber system, approximately 80% of the PCDD and 75% of the PCDF
are found in the fabric filter ash. For the wet-dry system, about 98% of
the PCDD and 98% of the PCDF are present in the fabric filter ash. The
PCDD and PCDF concentrations in the other hopper ashes are too low to
permit any reasonable conclusions from being drawn.

The concentrations of organic compounds, other than PCDD and PCDF,
are shown in Figure 20. The highest concentrations, as for the PCDD and
PCDF, were found in the fabric filter flyash. This also implies that the
larger surface area per unit weight of ash will result in higher concentrations per gram of ash for the smaller particles, which would be expected to collect mainly in the fabric filter.

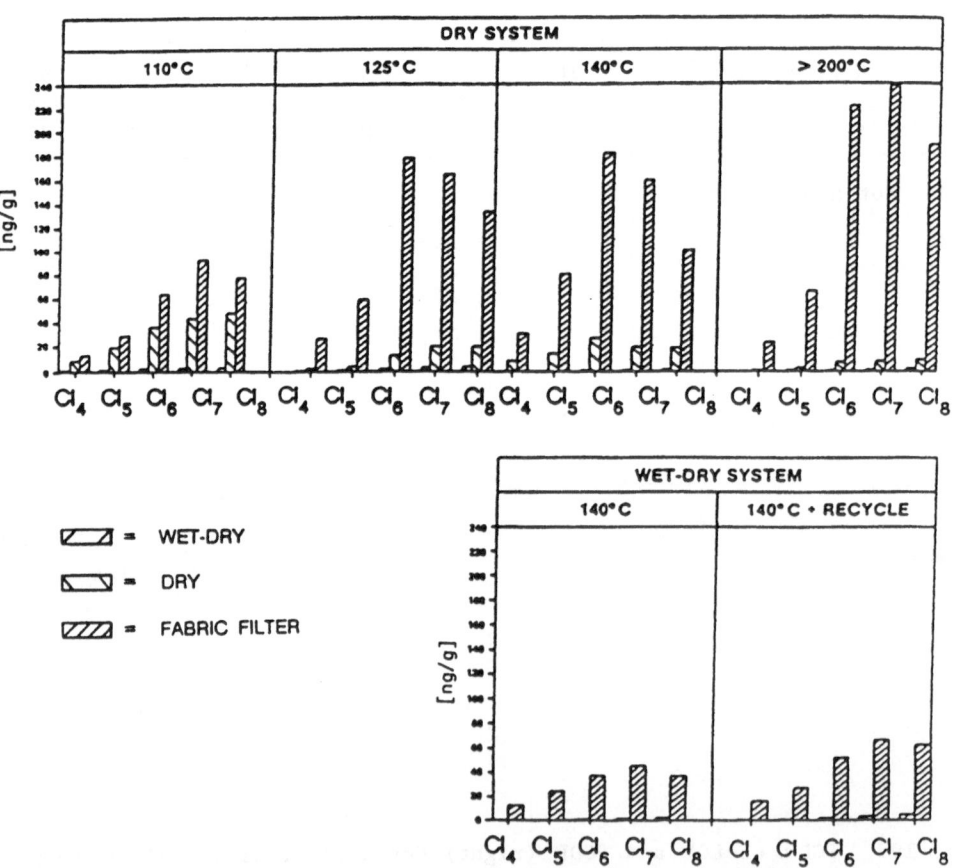

Fig. 20. PCDD Homologue Distribution in Hopper Ashes [ng/g]

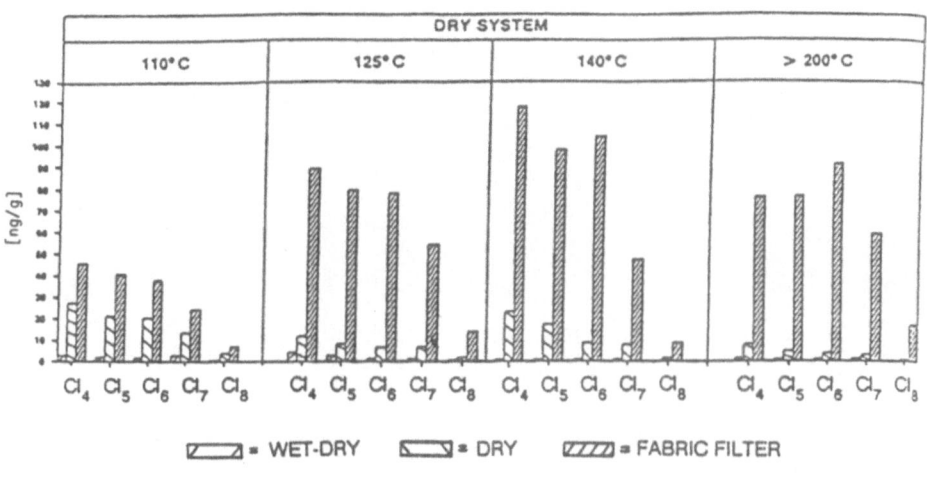

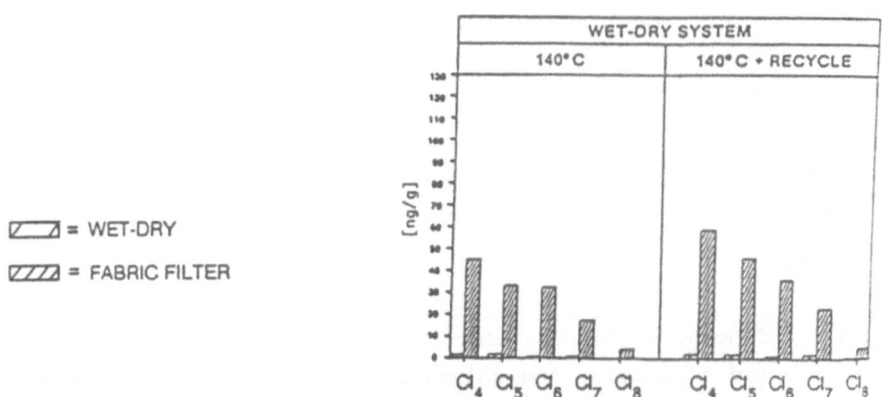

Fig. 21.   PCDF Homologue Distribution in Hopper Ashes   [ng/g]

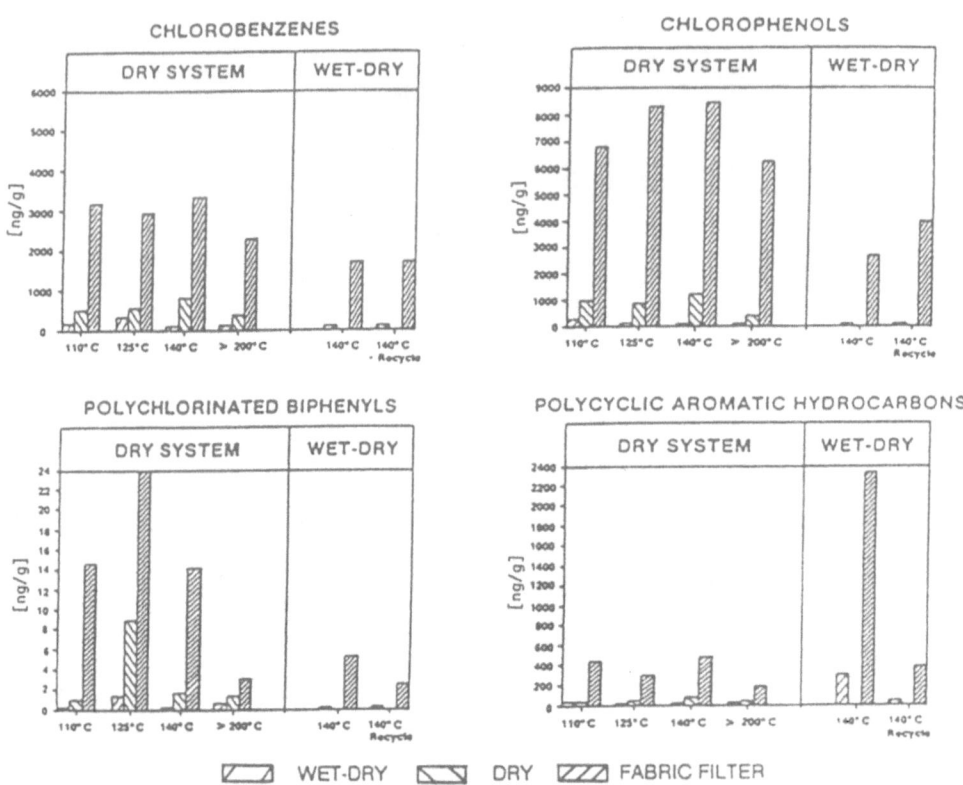

Fig. 22.   Concentrations of Chlorobenzenes, Chlorophenols, Polychlorinated Biphenyls, and Polycyclic Aromatic Hydrocarbons in Hopper
Ashes

The conclusions of this study are:

1. Both dry and wet-dry systems, when used in conjunction with a fabric filter, were very effective for removing pollutants. Neither system showed a significant difference in removal efficiency.
2. Cooling of the flue gas temperature below 200 °C was a key parameter in both systems for the removal of hydrogen chloride, sulphur dioxide, and mercury.
3. Removal efficiencies for PCDD and PCDF were very high (>99% for all tests) and for most runs, the concentrations of PCDD and PCDF emitted after the air pollution control system approached the detection limits of the analytical methodology used in the study.
4. Highest concentrations of PCDD and PCDF were found in the fabric filter ash (expected since the fabric filter exhibited high removal efficiency for PCDD and PCDF).
5. Other trace organic compounds were efficiently removed (80-99%) by both scrubber/fabric filter systems operated under cooled flue gas conditions.
6. Concentrations of the other trace organic compounds in the various hopper ashes followed a similar pattern as observed for PCDD and PCDF.
7. Heavy metal collection efficiencies generally exceeded 99.9% for both systems, except for mercury for which high removal was obtained only for cooled flue gas conditions.

The results of the tests with a mass burning system in Quebec (second incinerator design) have been published recently (EPS, 1988). As the facility was designed more than one decade ago a design modernization study was added to identify modernizations required to upgrade the plant to a modern state-of-the-art design. The improvements made on the furnace design resulted in a significant reduction in dioxin emissions. By employing good operation practices, dioxin concentrations were reduced to between 40 and 100 times below the 1984 test results (i.e. on average, 1,985 ng/m^3 PCDD and 476 ng/m^3 PCDF); under poor operating conditions, the reduction was an order of magnitude. The reduction in particulates was approximately the same.

The total dioxin and furan emissions for the new design were in the range of:

19 to 298 ng/m^3	for PCDD
44 to 306 ng/m^3	for PCDF.

The homologue distribution pattern for dioxins and all test conditions indicates that the lowest concentrations occurred for the tetra homologues and that concentrations increased progressively towards the octa group. On the other hand the furan homologue pattern showed highest concentrations for the tetra or penta group, decreasing towards the octa group. These results are consistent with the other NITEP test programs. The results for three different test conditions (low, design and high feed) and for good and poor burning conditions are given in Table 28.

Table 28. PCDD/PCDF Homologues per Performance Test Mode (Mass burn at Quebec City)

Homologue	Good conditions/ Burning rate			Poor conditions/ Design burning rate	
	Low	Design	High	Low Temp.	Poor Temp.
Cl_4DD	3.9	0.6	3.8	15.8	1.5
Cl_5DD	5.4	1.1	7.5	33.9	20.3
Cl_6DD	10.3	2.9	19.0	77.0	51.1
Cl_7DD	15.2	6.1	27.7	98.0	68.6
Cl_8DD	17.7	8.1	33.4	73.9	67.5
Σ PCDD	52.6	18.8	91.4	298.5	218.9
Cl_4DF	39.1	18.7	42.0	89.1	103.6
Cl_5DF	39.9	15.0	34.9	98.9	108.0
Cl_6DF	22.5	7.8	18.5	70.6	62.2
Cl_7DF	12.5	2.9	10.9	37.1	30.2
Cl_8DF	0.5	0.1	0.4	2.8	2.4
Σ PCDF	114.5	44.5	106.6	298.3	306.4

Statistical evaluation for PCDD/PCDF exhaust emissions can be summarized that there exist strong correlations with the the following parameters:
a) carbon monoxide emissions;
b) particulate emissions exhaust gas and primary air flow;
c) chlorobenzene and chlorophenol emissions; and
d) copper exhaust emissions.

No correlations was found between PCDD/PCDF emission and:
a) dioxin and furan concentration in the refuse; and
b) polycyclic aromatic hydrocarbon and polychlorinated biphenyl emissions.

(2) Municipal Waste Incinerator Studies in Canada, Sweden, and Germany

In a study similar to that carried out under Canada's NITEP, a pilot plant consisting of a spray dryer adsorption chamber and baghouse was set up to study its effectiveness in removing PCDD and PCDF from municipal waste incinerator flue gas (Nielsen et al., 1986). The pilot plant is illustrated in Figure 23 and its operating parameters are given in Table 27. This system was tested on a municipal incinerator which received household and light industrial waste. The incinerator had a capacity of about 10 tons/hour, a temperature of 850-900 °C, a flue gas rate of 50,000 Nm^3/h, and an outlet oxygen concentration of 10.5-11.5%. The flue gas, introduced through a gas disperser at the top of the scrubber vessel, and the hydrated lime slurry, introduced via a spray atomizer, are intimately mixed. The flue gas was sampled at two sites, before and after the scrubber/baghouse combination, SDA_{in} and BH_{out}, respectively. The sampling apparatus used in the measurements is shown in Figure 24. The heated area is kept at a temperature equal to that of the gas in the duct at the sampling point to minimize adsorption or desorption of organic vapours on the particulates. The results obtained for two test runs are given in Tables 28 and 29. The only difference between the two runs was the temperature of the gas at the outlet of the spray dryer absorption chamber. The outlet temperature in the first run was considerably higher than in the second run. It should be noted that the temperature at the SDA in sampling site (280-300 °C) is high enough to permit the formation of PCDD as it has been illustrated (Dickson, 1987). The inlet concentrations may not be realistic, however this is not critical since the main concern is the level of PCDD and PCDF emitted at the stack. Higher levels of PCDD and PCDF were observed for the test condition with the higher gas SDA outlet temperature. While it is difficult to make conclusions based on this work since only one run at each condition was reported and the reproducibility has not been demonstrated, the temperature of the gas exiting the scrubber was claimed to be as significant variable as in the NITEP study (EPS Canada, 1986).

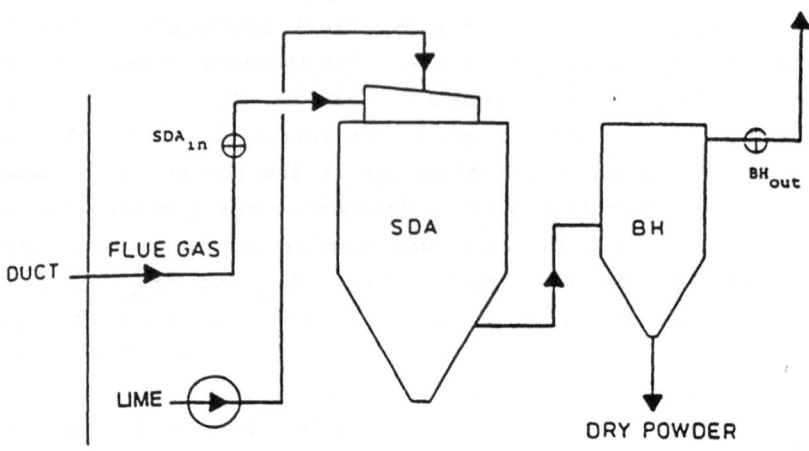

Fig. 23.   Municipal Waste Incinerator (Pilot Plant) with Spray Absorption by Rotary Atomizer

Table 29.   Average Parameters and Conditions of the Flue Gas Cleaning System

		SDA_{in}	BH_{out}
Gas Temp.	[°C]	280-300	100-160
HCl	[mg/Nm^3]	400-500	4
SO$_2$	[mg/Nm^3]	60-130	5
Particulate	[mg/Nm]^3	2,000	10
Gas Flow	[Nm^3/h]	300	

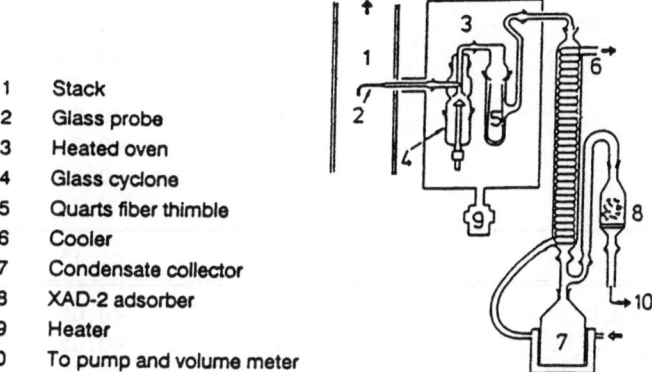

1	Stack
2	Glass probe
3	Heated oven
4	Glass cyclone
5	Quarts fiber thimble
6	Cooler
7	Condensate collector
8	XAD-2 adsorber
9	Heater
10	To pump and volume meter

Fig. 24. Flue Gas Sampling System (by Studsvik Energiteknik, Sweden)

Table 30. Gasphase PCDD and PCDF Removal, Run 1   (Inlet Samples)

Compound	SDA_{in} [ng/Nm^3]	BH_{out} [ng/Nm^3]	Removal [%]
2,3,7,8 Cl$_4$DD	0.13	<0.04	
1,2,3,7,8 Cl$_5$DD	3.5	0.4	
Σ Cl$_4$DD	2.1	<1.0	>52
Σ Cl$_5$DD	52	13	75
Σ Cl$_6$DD	67	4.9	93
Σ Cl$_7$DD	25	4.4	82
Cl$_8$DD	Not analyzed	Not analyzed	
Σ PCDD	146	23	84
2,3,7,8 Cl$_4$DF	1.0	0.3	
1,2,3,7,8 Cl$_5$DF	10	0.9	
2,3,4,7,8 Cl$_5$DF	8.5	0.8	
Σ Cl$_4$DF	37	0.9	98
Σ Cl$_5$DF	97	12	88
Σ Cl$_6$DF	48	6.8	36
Σ Cl$_7$DF	55	4.3	92
Cl$_8$DF	Not analyzed	Not analyzed	
Σ PCDF	237	24	90

Table 31. Gasphase PCDD and PCDF Removal, Run 2 (Inlet Samples)

Compound	SDA_{in} [ng/Nm^3]	BH_{out} [ng/Nm^3]	Removal [%]
2,3,7,8 Cl_4DD	1.7	<0.02	
1,2,3,7,8 Cl_5DD	4.2	<0.02	
Σ Cl_4DD	12.7	0.35	>97
Σ Cl_5DD	46	<0.2	>99.6
Σ Cl_6DD	38	<0.2	>99.5
Σ Cl_7DD	50	<0.2	>99.6
Cl_8DD	88	<0.2	>99.8
Σ PCDD	235	<1.15	>99.5
2,3,7,8 Cl_4DF	3.1	<0.02	
1,2,3,7,8 Cl_5DF	13	<0.02	
2,3,4,7,8 Cl_5DF	3.8	<0.02	
Σ Cl_4DF	61	<0.35	99.4
Σ Cl_5DF	99	<0.35	99.6
Σ Cl_6DF	122	<0.30	99.7
Σ Cl_7DF	115	<0.20	>99.8
Cl_8DF	134	<0.20	>99.8
Σ PCDF	531	<1.4	>99.7

In another study refuse burned at two Canadian incinerators was investigated. The average composition of this refuse is given in Table 32
(Tosine et al., 1985). The levels of PCDD and PCDF found in representative samples for each incinerator are listed in Table 33. In all cases
the levels of PCDD were higher than those for PCDF. In fact, PCDF were
only found in one sample of refuse. Similar findings were later reported
by NITEP (EPS Canada, 1986). For the PCDD, the higher chlorinated species, in particular Cl_8DD and Cl_7DDs, were the dominant isomers.

Table 32. Average Composition of Refuse in Two Ontario Incinerators [%]

Paper and Textile	56
Garden and Foodstuff	16
Metal and Glass	15
Plastic and Rubber	13

Table 33. Concentrations of PCDD and PCDF in Refuse Samples from Two Ontario Incinerators. Concentrations in [ng/g]

Test	Incinerator A			Incinerator B		
	1	2	3	1	2	3
Cl_4DD	ND *	ND *	ND *	1.5	ND *	ND *
Cl_5DD	ND	ND	ND	2.7	ND	ND
Cl_6DD	ND	ND	ND	26	22	ND
Cl_7DD	2.0	5.3	19	28	87	20
Cl_8DD	8.0	6.5	11	36	330	55
Total	10	12	30	94	440	75
Cl_4DF	ND	ND	ND	0.8	ND	ND
Cl_5DF	ND	ND	ND	41	ND	ND
Cl_6DF	ND	ND	ND	7.7	ND	ND
Cl_7DF	ND	ND	ND	9.0	ND	ND
Cl_8DF	ND	ND	ND	0.8	ND	ND
Total	ND	ND	ND	22	ND	ND

* None detected at the isomer detection limits ranging from 1 pg/g for tetrachlorinated

to 4 pg/g for octachlorinated species.

All data corrected for recovery of internal spike ($Cl^{37} Cl_8DD$).

In addition to determining the PCDD and PCDF contents of the refuse, the levels of PCBs, chlorophenols, and chlorobenzenes present in the incoming waste were studied. The results for these chlorinated compounds are given in Table 34. It should be noted that these values were given for the total amounts of the higher chlorinated isomers, those containing more than 2 chlorine atoms. The analytical methodology employed in this study was not suitable for the lower chlorinated isomers and therefore the levels in Table 35 represent the minimum amounts present in the garbage.

Table 34. Concentrations of Chlorinated Organics in Refuse Samples from
          Two Ontario Incinerators
          Concentrations in [ng/g]

| | Incinerator A | | | Incinerator B | | |
Test	1	2	3	1	2	3
PCB	110	1,500	200	120	120	110
Chlorobenzenes	34	41	45	10	10	5
Chlorophenols	150	240	680	420	600	570

This study illustrates that levels of PCDD and PCDF in the refuse feedstock should be considered in studies dealing with mass balances in the incineration process. The presence of potential precursors such as PCBs and the chlorinated phenols and benzenes in the initial refuse may also play a significant role in the amounts of PCDD and PCDF formed during the incineration of municipal refuse.

In the fall of 1984 and the spring of 1985, 15 experiments were carried out at the municipal solid waste incinerator located in Umea, Sweden (Rappe et al., 1986). The experiments carried out are summarized in Table 35 along with the corresponding results of the PCDD and PCDF levels given in terms of ng of "TCDD equivalents" per Nm^3 of dry gas (10% CO_2). The total levels of PCDD and PCDF as expressed in terms of ng equivalents of Cl_4DD were found to vary relatively little over the ten day period in which the experiments were conducted. The actual levels of the individual PCDD and PCDF homologues were not included in the report and thus it is difficult to get a full comprehension of this data without this information. Although it is important to get some understanding regarding the significance of results in terms of toxicity, analytical information must not be sacrificed. In future studies, results should be not be given only in terms of "TCDD equivalents".

Table 35. Experiments at the Municipal Waste Incinerator in Umea, Sweden

Experiments	No. of exp.	Temperature [°C] mean	range	Cl_4DD [ng/Nm^3]
Normal conditions (fall)	3	803	736-846	9
Normal with wood chips	2	764	737-790	10
Normal with oil burner	2	827	811-842	11
Low temperature	3	539	484-580	11
Low with oil burner	3	625	602-658	10
Start-up	1	-	20-790	54
Sart-up with oil burner	1	-	28-816	12
Normal conditions (spring)	3	784	700-850	5.6

Table 36 lists a comparison of results for a study of several incineration and related processes, again given in terms of "TCDD equivalents". In the laboratory pyrolysis of PVC (SARAN), a series of PCDF were produced while only very low amounts of PCDD were found. Based on this result, PVC may potentially act as a precursor for the formation of PCDD and PCDF in incinerators. When one compares the chromatographic profiles of penta-CDFs produced from municipal incineration, PVC pyrolysis, and hexachloroethane pyrolysis, the patterns are quite similar despite the fact that the chlorine contents of the burn materials were 1%, 57%, and 90%, respectively.

Table 36. Comparison of Emissions from Incineration and Related Processes in Sweden

Place	Type	Mean Value "TCDD equivalents" [ng/Nm^3] d g 10% CO_2	
Umea, fall	MSW cross-gate	10	
Umea, spring	MSW cross-gate	5.6	
Avesta, fall	MSW cross-gate	80	
Avesta, spring	MSW cross-gate	2.0	
Boras	MSW cross-gate	38	
Mid Sweden	MSX fluid-bed	1.8	
Rönnskär	Ind. copper melter	11	
Mid Sweden	Ind. steel melter	0.8	[ng/g]
Laboratory	SARAN pyrolysis	0.15	[ng/g]

In one study, fly ash samples from six different municipal incinerators from four different countries were compared (Tong and Karasek, 1986). Very similar isomer distributions for PCDD and PCDF were found between incinerators even though the concentration of the dioxins and furans in the ashes varied considerably. This finding implies that the PCDD and PCDF were formed through the same mechanisms, independent of the garbage composition or incinerator design. Although different incinerator designs and operating conditions have been studied and their refuse characterized, the information reported in the literature is quite scattered. Therefore the mechanism for the formation of PCDD and PCDF in incinerators still has not been conclusively proven.

Figure 25 shows the isomeric distribution patterns for the Cl_4DDs, Cl_4DFs, Cl_5DDs, and the Cl_5DFs for four of the fly ash samples studied. The distributions are very similar within a given congener group for all the fly ashes. This trend is also observed for the other congener groups.

Table 37 gives the concentrations of the tetra-, penta-, and hexachlorinated dioxins and furans for all of the samples analyzed. The Machida incinerator fly ash had the lowest levels of PCDD and PCDF. This incinerator first segregated its refuse into classes and then burned only the combustible material. Metal waste and plastics were not fed into the incinerator. The remaining incinerators studied did not sort their refuse prior to burning.

Table 37. Quantitative Comparison of Some Selected PCDD/PCDF in Fly Ash from Different Countries

	Ontario	Oslo	Paris	Kyoto	Hiroshima	Machida
Cl_4DD	436	27	18	8	29	0.2
Cl_5DD	504	77	50	17	95	0.8
Cl_6DD	668	149	142	38	149	4
Cl_4DF	294	55	81	15	90	2
Cl_5DF	508	74	136	23	92	7
Cl_6DF	420	80	192	22	85	12

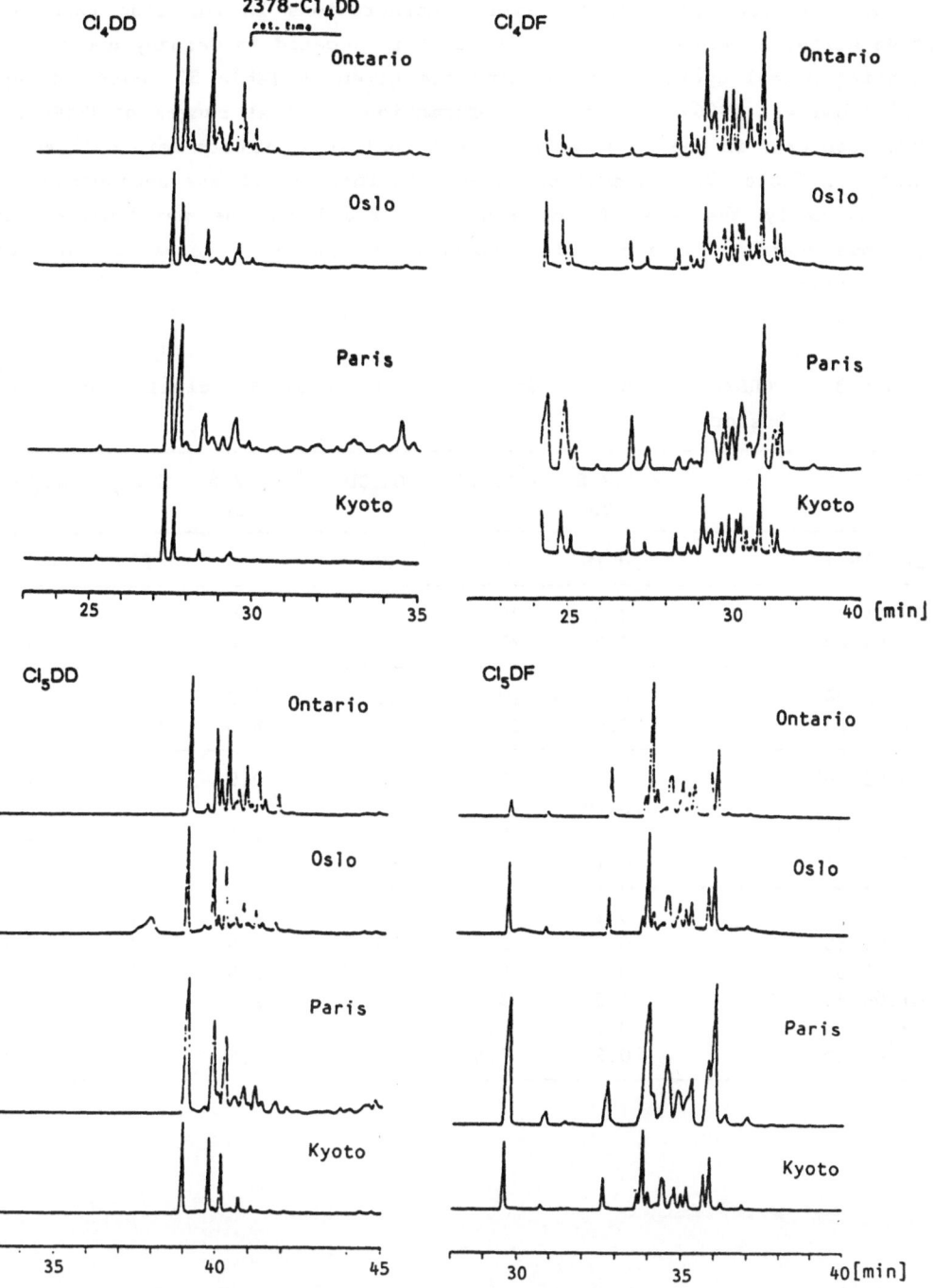

Fig. 25. GC/MS/SIM Data Showing the Isomer Distribution Pattern of Some Selected PCDD and PCDF in the Fly Ash Samples Collected from Different Countries

A comparison of PCDD and PCDF concentrations in the flue gases of five different municipal waste incinerators located in Germany and representing normal operating conditions are given in Table 38 (Nottrodt and Ballschmiter, 1986). From this information, typical ranges of PCDD and PCDF contents in the flue gases of municipal incinerators were defined as shown in Table 39. It must be noted that these stack gas concentrations are actually the sum of the PCDD/PCDF found in the particulate and gaseous phase. The particulate content of the stack gases ranged from 10-20 mg/m^3.

Table 38. PCDD/PCDF-Contents in Flue Gases from Municipal Waste Combustors

Date	2,3,7,8 Cl_4DD	Σ Cl_4DD	Cl_8DD	2,3,7,8 Cl_4DF	Σ Cl_4DF	Cl_8DF
14.-19.12.81	0.16	4	49		22	8
09.04.84	0.2	24	14	3.0	64	5
11.04.84	0.2	25	11	3.0	66	1
01.05.84	0.7	19	15	2.0	110	2
02.05.84	0.6	19	14	6.0	144	2
04.05.84	0.05 ±0.09	5	5	0.9	38	1
15.05.84	0.1	7	13	1.4	41	4
17.05.84	0.1	6	16	1.4	31	1.4
27.03.85	0.2	2.6	5	1.4	40	1.2
29.03.85	0.5	5.1	13	2.9	59	3.4
30.03.85	0.4	5.1	15	2.5	53	3.7
01.04.85	0.3	3.3	15	1.5	34	7.8
02.04.85	0.1	2.1	4	1.2	29	1.5
03.04.85	0.2	3.5	9	2.1	48	2.8

Table 39. Typical Ranges of PCDD/PCDF-Contents in Flue Gases and
Residues from Municipal Waste Combustors

Flue Gas:	$2,3,7,8-Cl_4DD$	0.05 - 0.07	$[ng/m^3]$
	$\Sigma\ Cl_4DD$	4 - 25	$[ng/m^3]$
	$\Sigma\ Cl_5DD$ (3-7)	20 - 145	$[ng/m^3]$
	Cl_8DD	4 - 49	$[ng/m^3]$
	$2,3,7,8-Cl_4DF$	0.9 - 6	$[ng/m^3]$
	$\Sigma\ Cl_4DF$	22 - 144	$[ng/m^3]$
	$\Sigma\ Cl_5DF$ (3-7)	91 - 361	$[ng/m^3]$
	Cl_8DF	1 - 8	$[ng/m^3]$
Clean Gas Dust:		0.25 - 3	$[ng/g]$
Filter Dust:		1.2 - 1.4	$[ng/g]$
Blag:		< 20	$[pg/g]$

A mass burning municipal incinerator was operated under two different
load conditions and the levels of PCDD and PCDF produced under these conditions were studied (Nottrodt and Ballschmiter, 1986). The purpose of
this study was to investigate the influence of the time at which the flue
gas flowing through the combustion chamber was held above 800 °C. Regular operating parameters are given in Table 40 along with the parameters
used during the sampling done for this study. Two 24 hour runs were used
for each of the two different load conditions.

At a charging rate of 16 Mg/h, the amount of steam produced is 52.4
Mg/h. The studied load conditions were chosen as those required to produce about 45 and 30 Mg/h steam. Thus the residence time at greater than
800 °C for the flue gas under load condition 1 should be approximately
two-thirds that of load condition 2.

Table 40. Normal Operating Parameters of the Plant

Charging Rate	16	[mg/h]
At Max. Heating Value	11,100	[kJ/kg]
Grid Area	60	[m^2]
Steam Production	52.4	[mg/h]
Steam Temperature	400	[°C]
Steam Pressure	40	[bar]
Gas Cleaning	1.	Electrost. Precipitator
	2.	Wet Scrubber

Parameters During Sampling

	Steam	Temperature		Flue Gas,Wet
		1	2	
	[g/h]	[°C]	[°C]	[ml/h]
Load 1	45	1,090	950	100,000
Load 2	30	1,050	880	70,000

Temperature 1   Combustion Chamber
Temperature 2   Top of Boiler

The PCDD and PCDF contents of the fly ash and filter dust (combined prior to sample preparation and analysis) for the duplicate tests of the load conditions are given in Tables 41 and 42. Each run was characterized by the 24 hour averages of steam production and the temperatures in the combustion chamber and at the top of the boiler. The incinerator was operated at constant conditions during the 24 hour runs. The standard deviations of one hour averages for steam production and temperature were within 1-2% of the 24 hour averages. The PCDD and PCDF levels under both conditions are quite low. The levels under load 2 generally appear to be lower than the levels under load 1, however this may not be statistically significant. The variation in the levels between the duplicate runs for the same load is sometimes quite large in comparison with the average of the two runs. Therefore care must be taken in making any conclusions.

Table 41. PCDD-Contents in Fly Ash and Filter Dust

| Load | Steam | Temperature | | ΣCl_4DD | 183 | 189 | 201 | 202 | 205 | 210 |
| | | 1 | 2 | | | | | | | |
	[mg/h]	[°C]	[°C]	ppb	ppb	ppb	ppb	ppb	ppb	ppb
1	44.9	1,088	947	1.4	0.87	0.3	0.6	1.7	0.9	80
1	44.6	1,088	950	1.2	0.1	0.5	1.2	3.9	2.1	215
2	31.0	1,052	878	1.6	0.1	0.4	0.7	1.8	1.2	81
2	31.7	1,049	882	0.9	0.1	0.4	0.7	1.6	1.0	42

Table 42. PCDF-Contents in Fly Ash and Filter Dust

| Load | Steam | Temperature | | ΣCl_4DF | 83 | 117 | 121 | 135 |
| | | 1 | 2 | | | | | |
	[mg/h]	[°C]	[°C]	ppb	ppb	ppb	ppb	ppb
1	44.9	1,088	947	14.8	0.6	1.1	1.0	10
1	44.6	1,088	950	9.7	0.4	1.0	1.1	18
2	31.0	1,052	878	8.1	0.3	0.8	0.6	10
2	31.7	1,049	882	9.5	0.4	0.8	0.6	11

83	2,3,7,8 Cl_4DF	189	1,2,3,7,8 Cl_5DD
117	2,3,4,7,8 Cl_5DF	201	1,2,3,4,7,8 Cl_6DD
121	1,2,3,6,7,8 Cl_6DF	202	1,2,3,6,7,8 Cl_6DD
135	Cl_8DF	205	1,2,3,7,8,9 Cl_6DD
183	2,3,7,8 Cl_4DD	210	Cl_8DD

Temperature 1:   Combustion Chamber
Temperature 2:   Top of Boiler

The concentrations of PCDD and PCDF in the stack flue gas were also determined and are given in Tables 43 and 44. The concentrations are again quite low and load 2 appears to have slightly lower levels than for load 1. Limitations inherent with the sampling of flue gases and variations between duplicate runs do not permit any conclusive statements.

Table 43. PCDD-Contents, Flue Gas

| Load | Steam | Temperature | | Σ Cl$_4$DD | 183 | 189 | 201 | 202 | 205 | 210 |
| | | 1 | 2 | | | | | | | |
	[mg/h]	[°C]	[°C]	ng/m^3	ng/m^3	ng/m^3	ng/m^3	ng/m^3	ng/m^3	ng/m^3
1	44.9	1,088	947	5.1	0.5	0.9	0.8	1.6	1.0	13
1	44.6	1,088	950	5.1	0.4	0.6	0.6	1.7	0.9	15
2	31.0	1,052	878	3.3	0.3	0.7	0.6	1.1	0.8	15
2	31.7	1,049	882	2.1	0.1	0.3	0.3	0.7	0.5	4

Table 44. PCDF-Contents, Flue Gas

| Load | Steam | Temperature | | Σ Cl$_4$DF | 83 | 117 | 121 | 135 |
| | | 1 | 2 | | | | | |
	[mg/h]	[°C]	[°C]	ng/m^3	ng/m^3	ng/m^3	ng/m^3	ng/m^3
1	44.9	1,088	947	59	2.9	4.7	4.1	3.4
1	44.6	1,088	950	53	2.4	4.1	5.6	3.7
2	31.8	1,052	878	34	1.5	2.0	1.0	7.8
2	31.7	1,049	882	29	1.2	2.6	3.0	1.5

83	2,3,7,8 Cl$_4$DF		189	1,2,3,7,8 Cl$_5$DD
117	2,3,4,7,8 Cl$_5$DF		201	1,2,3,4,7,8 Cl$_6$DD
121	1,2,3,6,7,8 Cl$_6$DF		202	1,2,3,6,7,8 Cl$_6$DD
135	Cl$_8$DF		205	1,2,3,7,8,9 Cl$_6$DD
183	2,3,7,8 Cl$_4$DD		210	Cl$_8$DD

Temperature 1:   Combustion Chamber
Temperature 2:   Top of Boiler

A study on the formation of PCDD and PCDF in a refuse-derived fuel (RDF) fired incinerator compared the levels formed when the boilers were operated under RDF-natural gas firing and RDF-only firing (see Tables 45 and 46) (Hahn et al., 1986). The levels of PCDD and PCDF produced were low for all for all tests performed but were typical of the levels reported in studies conducted on other incinerators. The inconsistencies observed in the operation of the incinerator do not permit any conclusions being made regarding the correlation of operating conditions with the resulting levels of PCDD and PCDF emitted.

Table 45. Comparison of PCDD and PCDF Emissions during RDF-Only and RDF-Natural Gas Firing

Test Series	Homologue Group	Average ng/Nm^3 PCDD at 12% CO_2	Average ng/Nm^3 PCDF at 12% CO_2	PCDD/PCDF Ratio
RDF-Natural Gas at 9.4% CO_2				
	Cl_4	23	56	
	Cl_5	232	55	
	Cl_6	232	13	
	Cl_7	191	3	
	Cl_8	23	3	
	Total	701	130	5.39:1
RDF-Only at 10% CO_2				
	Cl_4	19	43	
	Cl_5	125	28	
	Cl_6	68	5	
	Cl_7	91	1	
	Cl_8	13	2	
	Total	316	79	4.00:1

Table 46. Comparison of 2,3,7,8-Cl_4DD, 2,3,7,8-Cl_4DF, and Cl_4DF Emissions during RDF-Only and RDF-Natural Gas Firing

	CO_2	2,3,7,8-Cl_4DD	Σ Cl_4DD	2,3,7,8-Cl_4DD of Total Cl_4DD	2,3,7,8-Cl_4DF	Σ Cl_4DF	2,3,7,8-Cl_4DF of Total Cl_4DF
	[%]	[ng/Nm³ at 12% CO_2]			[ng/Nm³ at 12% CO_2]		
RDF-Natural Gas	9.4	0.34	27.9	1.22	2.45	44.8	5.47
	9.3	0.50	18.6	2.69	2.43	48.5	4.99
	9.5	0.70	22.1	3.17	4.04	74.3	5.44
Average	9.4	0.51	22.9	2.36	2.97	55.9	5.30
RDF-Only	10.1	0.81	20.2	4.01	2.88	41.8	6.89
	9.9	0.40	12.9	3.10	2.17	36.2	5.99
	10.0	0.52	24.2	2.15	3.00	51.2	5.86
Average	10.0	0.58	19.1	3.09	2.68	43.1	6.25

The emissions of two Belgian incinerators under routine operations were studied over a one year period by taking samples daily for two weeks in each of the four seasons (De Fré, 1986). The average concentrations of PCDD and PCDF in the total gas stream for each incinerator are summarized in Table 47. The waste burned at incinerator I is fairly homogeneous and is representative of average Belgian municipal refuse, with the addition of some waste from small industries. Incinerator II is located in a rural community. The refuse presented to it is poor in caloric value and is often made up with city waste from a different source. Neither incinerator was equipped with an energy recovery system and therefore gases are emitted at relatively high temperatures.

The standard deviations calculated for the results given in Table 47 are of the same magnitude and usually larger than the averages themselves. These averages were obtained from values which often consisted of a large number of low measurements plus fewer high measurements that tend to contribute more to the value of the average. For example, two Cl_5DD isomers (or possibly artifacts) were present at extremely high concentrations (10.3 μg/Nm³) on one day at incinerator I; if this measurement was omitted, the results in the parentheses would be obtained.

The emission factors, the number of mg of PCDD/PCDF produced per ton of refuse, are given in Table 47. It was not possible to obtain statistically significant conclusions from the small differences observed because of the irregular distribution and large standard deviations for the data collected. The furnace temperatures were operated in a range from 800 to 1,050 °C and no clear relationship to the levels of PCDD and PCDF emitted was found. By observation of incinerator II, waste could be classified into two distinct sources, urban and rural refuse. For the days on which only urban waste was burned, the average sum of the dioxin and furan concentrations was nearly twice that observed when only rural refuse was incinerated. The urban waste was found to produce higher furnace temperatures and yield higher chlorine content in the flue gases. Since the PVC content was not measured directly (it was estimated from the HCl concentrations in the gases based on the assumption that PVC is the most important source of organic chlorine and the fact that PVC gives a nearly 100% yield of HCl on combustion), no relationship between the amount of PVC and the emission of PCDD and PCDF could be determined.

Table 47. Average Emission Concentration in $[ng/Nm^3]$ (left side) and Emission Factors [mg/ton] (right side)

Average Emission	Concentration $[ng/Nm^3]$		Emission Factors [mg/ton]	
Component	Plant I	Plant II	Plant I	Plant II
Σ PCDD (4-8)	1,005	354	13.3	3.1
Cl_4DD	20	40	0.27	0.35
$2,3,7,8-Cl_4DD$	0.97	3.8	0.015	0.038
Cl_5DD	396	34	5.2	0.3
Cl_6DD	185	53	2.5	0.46
Cl_7DD	206	67	2.7	0.59
Cl_8DD	202	153	2.68	1.34
Σ PCDF (4-8)	1,185	1,216	15	6.9
Cl_4DF	116	196	1.5	1.0
$2,3,7,8-Cl_4DF$	9	36	0.14	0.22
Cl_5DF	209	188	2.7	0.97
Cl_6DF	318	372	4.3	1.9
Cl_8DF	204	433	2.6	1.5

(3) Municipal Waste Incineration in Japan

Approximately 110,000 tons of municipal refuse are produced daily in Japan of which about 68% is incinerated (as of 1983) (Hiraoka et al., 1986). The incinerators are operated in three different modes:

1. Fully continuous (24 hour operation)
2. Semi-continuous (16 hour operation)
3. Batch incineration (8 hour operation).

The presence of dioxins in the fly ash from a Japanese incinerator was first reported in 1979 (Eiceman et al., 1979). In 1984 an extensive survey of 33 municipal waste incinerators was conducted in which the flue gas, fly ash, bottom ash, and waste water were analyzed for their PCDD and PCDF contents. The results of this study are summarized below.

Fly Ash (35 samples):

- Cl_4DDs ranged from ND to 14.5 ng/g; averaged 1.6 ng/g
- Cl_5DDs ranged from 2.7 to 10,700 ng/g; averaged 523 ng/g
- Cl_5DFs ranged from 1.0 to 1,510 ng/g; averaged 150 ng/g
- $2,3,7,8-Cl_4DD$ ranged from ND to 4.7 ng/g (12 samples analysed)
ND = < 0.05 ng/g

Bottom Ash (30 samples):

- Cl_4DDs ranged from 0.3 to 7.3 ng/g; averaged 1.6 ng/g
- Cl_5DDs ranged from 5.0 to 2470 ng/g; averaged 216 ng/g
- Cl_5DFs ranged from 2.3 to 864 ng/g; averaged 81.9 ng/g
- $2,3,7,8-Cl_4DD$ ND (11 samples analyzed)
ND = < 0.05 ng/g

Flue Gas (25 samples):

- Cl_4DDs ranged from ND to 109 ng/Nm^3; averaged 54.9 ng/Nm^3
- Cl_5DDs ranged from 133 to 13,600 ng/Nm^3; averaged 1,510 ng/Nm^3
- Cl_5DFs ranged from 436 to 10,000 ng/Nm^3; averaged 2,520 ng/Nm^3
- $2,3,7,8-Cl_4DD$ ND (1 sample analyzed)
ND = < 2 ng/Nm^3

Discharged Waste Water (17 samples):

- Cl₄DDs ranged from ND to 9.7 ng/L; 12 samples were ND
- Cl₅DDs ranged from ND to 46.7 ng/L; 8 samples were ND
- Cl₅DFs ranged from ND to 285 ng/L; 8 samples were ND

ND not defined for water samples

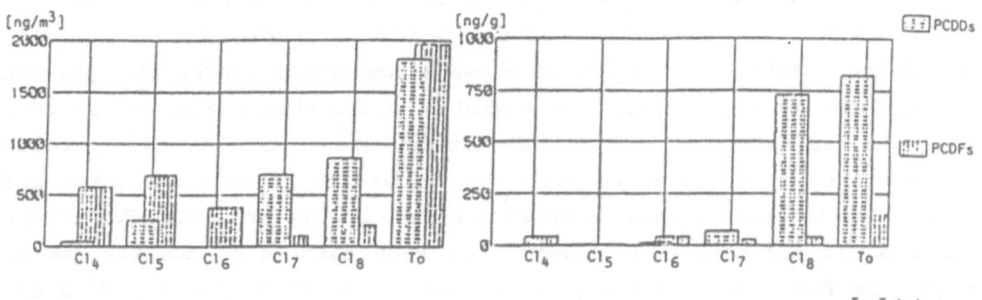

Fig. 26.   PCDD/PCDF in Exhaust Gas (left) and in Fly Ashes (right)

The average concentrations of PCDD and PCDF in the flue gas of 15 fully-continuous incinerators are shown in Figure 26 (left). The more chlorinated dioxins dominate the PCDD homologue distribution while the lower chlorinated furans, in particular the Cl₅DFs, dominate the PCDF distribution. These results are similar to those observed in other studies.

The average concentrations of PCDD and PCDF in the fly ashes of 19 fully-continuous incinerators are given in Figure 26 (right). The total amount of PCDD produced on average is about four times the amount of PCDF produced with Cl₈DD being by far the dominant species.

The incinerators studied were chosen from different districts in Japan. They were representative of the different types of incinerators and consequently were operated under various conditions. No conclusions could be drawn regarding the dependence of PCDD/PCDF formation upon operating conditions.

In 1985 two of the incinerators studied in the earlier survey were chosen for further investigation, primarily because of the high concentrations of PCDD which were detected in their flue gases (Hiroaka et al., 1987). Incinerators A and B have the same capacity, 600 tons/day, but incinerator A uses a dry scrubber and electrostatic precipitator for pollution control whereas B uses a wet scrubber. The operating conditions for the two incinerators are given in Table 48. Flue gas, fly ash and discharged water were sampled, the results of the analyses are given in Tables 49-51. The composition of the refuse as studied in 1984 and 1985 are shown in Table 52. Figure 26 shows a comparison of the flue gas concentrations for both years. It should be noted that packed column GC/MS was used for determinations in the earlier study while capillary column GC/MS was used in the later analyses. Thus the first study gives much higher levels because interferences are undoubtedly present. For incinerator A, it appeared that whenever the PCDD content of the fly ash was high, the flue gas levels were low. Similarly, high flue gas PCDD content usually corresponded to low levels present in the fly ash. The concentration of PCDD in the flue gas exiting the wet scrubber appeared to higher than for the dry scrubber, however the latter system had higher amounts of PCDD in its fly ash. The effects of refuse composition could not be determined from the data collected.

Table 48. Operating Conditions of Municipal Waste Incinerators A and B

	Incinerator A			Incinerator B		
	2/4	2/5	2/6	2/18	2/19	2/20
Incinerated Amounts:						
- per hour	6.82	6.53	6.91	8.88	9.13	8.79
- per day	163.7	156.8	156.8	213.2	219.2	211.0
Temperature of Incinerator [°C]:						
- Average	690	683	725	890	880	900
- Max.	811	812	794	1,110	1,040	1,080
- Min.	607	503	611	740	660	700
Inlet Gas Temp. of Ep [°C]:						
-Average	297	295	309	-	-	-
- Max.	330	338	339	-	-	-
- Min.	254	254	276	-	-	-
Exhaust Gas Volume [KNm^3/h]:						
- Average	43.3	42.4	42.8	52	54	55
- Max.	57.3	54.1	58.0	72	68	68
- Min.	36.0	33.6	35.2	43	43	44
HCl concentration [ppm]:						
- Average	313	334	343	0.5	0.8	0.4
- Max.	497	636	572	5	20	7
- Min.	176	146	203	0	0	0
O_2 concentration [%]:						
- Average	14.1	14.1	13.7	9.8	9.3	11.1
- Max.	15.7	17.2	15.6	12.8	13.2	15.0
- Min.	12.6	11.8	11.9	6.8	7.5	9.2

Table 49. Analyses of Flue Gas (Exhaust Gas) [ng/Nm^3]

	Date	Samples	Cl_4DD	Cl_5DD	Cl_6DD	Cl_7DD	Cl_8DD	PCDD	$2,3,7,8-Cl_4DD$
Incinerator A:	2/4	1	12	70	225	570	1,000	1,900	<0.69
		2	13	89	290	420	760	1,600	<1.08
		3	<3	31	580	1,100	1,700	3,400	-
	2/5	1	75	67	450	850	2,200	3,600	<0.54
		2	7	5	11	28	62	110	<1.07
		3	83	310	600	720	1,200	2,900	-
	2/6	1	96	740	2,800	4,500	6,600	14,700	3.7/2.0*
		2	24	140	580	1,500	1,300,	3,500	-
		3	19	100	1,000	2,400	4,400	7,900	-
Incinerator B:	2/18	1	8.6	49	200	470	920	1,600	<0.17
		2	4.2	62	17	340	1,500	1,900	1.5
		3	5.8	19	11	140	400	580	-
	2/19	1	5.3	<2	<3	130	460	590	<0.19/<0.10**
		2	<1	<2	<4	56	190	250	-
		3	8.3	12	12	100	250	380	-
	2/20	1	<1	18	<3	25	93	140	<0.69
		2	8.0	<2	13	37	100	160	<0.52
		3	1.5	<3	<3	57	160	220	-

* — Particulate matter

** — Particulate matter Gas

Table 50. Analyses of Fly Ash [ng/g]

Date	Samples	Cl_4DD	Cl_5DD	Cl_6DD	Cl_7DD	Cl_8DD	PCDD	2,3,7,8-Cl_4DD
Incinerator A:								
2/4	1	2.8	9.8	25	22	31	91	0.77
	2	3.1	11	14	11	43	82	0.32
	3	1.6	1.7	7.9	8.9	7.2	27	-
2/5	1	3.1	14	28	22	16	83	0.53
	2	12	45	91	64	37	250	0.56
	3	5.8	1.9	29	24	18	79	-
2/6	1	4.3	16	37	20	37	110	1.06
	2	7.8	50	66	55	33	220	-
	3	12	44	71	60	41	230	-
Incinerator B:								
2/18	1	1.5	3.5	2.9	2.3	3.0	13	0.40
	2	0.3	0.9	3.4	7.9	13	26	0.12
	3	<0.1	<0.1	0.2	0.7	1.6	2.5	-
2/19	1	0.4	0.2	0.6	0.9	2.6	4.7	<0.02
	2	0.2	0.9	2.0	2.1	3.8	9.0	-
	3	0.4	0.8	1.2	2.6	10	15	-
2/20	1	2.9	4.2	16	12	6.6	42	<0.01
	2	0.1	0.2	1.4	3.9	12	18	0.12
	3	0.2	0.8	0.9	1.3	2.8	6.0	-

Table 51. Analyses of Discharged Water [ng/L]

	Incinerator A	Incinerator B
Date	2/6	2/19
Samples	1	2
Cl_4DD	<0.05	<0.05
Cl_5DD	<0.1	<0.1
Cl_6DD	<0.08	0.36
Cl_7DD	<0.03	0.79
Cl_8DD	<0.2	1.9
PCDD	<0.46	3.0
$2,3,7,8-Cl_5DD$	<0.08	<0.08

Table 52. Composition of Municipal Waste
[Unit: %;Caloric Value: kcal/kg]

Composition of Refuse	Incinerator A		Incinerator B	
	Average (1984)	Average (1985)	Average (1984)	Average (1985)
Water	45.5	24.3	47.7	13.5
Ash	14.8	20.1	13.2	21.2
Combustible Material	39.7	55.6	39.1	65.3
Paper/Textile	41.2	60.4	26.3	43.9
Wood/Bamboo	4.9	1.4	1.3	15.7
Plastics	20.2	17.8	16.0	18.9
Garbage	7.8	2.3	42.7	2.3
Uncombustible Material	21.6	16.6	10.7	15.4
Others	4.3	1.5	3.0	3.8
Low Calorific Value	1,794	2,830	1,816	3,536

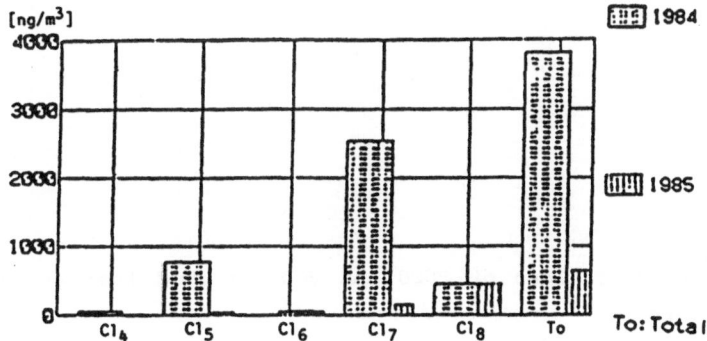

Fig. 27.   PCDD in Flue Gas; Comparison of Two Years

In another Japanese study, fly ash, bottom ash, flue gas, and atmospheric samples were collected and analyzed for four incinerators from
different locations (Asada et al., 1987).   Two of the incinerators studied were operated under continuous flow conditions while the others were
batch incinerators.   A curious choice of standards were selected to test
the flue gas and atmospheric gas sampling devices.   Di-, tri-, and tetra-
CDD isomers were used for recovery studies, yet the analyses involved the
tetra- through octa-CDD compounds only.   Good recoveries were observed,
however such information would be more encouraging if similar recoveries
were shown for the higher chlorinated species.   The average results of
the levels determined for the four Japanese municipal waste incinerators
are given in Table 53.   The isomer patterns for each type of sample were
found to be quite similar.

Table 53. Concentrations of PCDD and PCDF in Municipal Waste Incinerators

	Fly Ash [ng/g]	Bottom Ash [ng/g]	Flue Gas [ng/m^3]	Atmosphere [pg/m^3]
	Min - Max (Ave)	Min - Max (Ave)	Min - Max (Ave)	Min - Max (Ave)
2,3,7,8-Cl$_4$DD	N.D.-5.01 (1.05)	N.D. N.D.	0.28-1.20 (0.67)	N.D. N.D.
Cl$_4$DD	1.30-220 (62.3)	0.15-0.99 (0.48)	0.66-175 (53.6)	N.D.-4.9 (4.3)
Cl$_5$DD	N.D.-121 (79.1)	0.20-155 (1.00)	4.05-164 (66.0)	N.D.-2.0 (2.0)
Cl$_6$DD	N.D.-279 (121)	2.53-6.69 (4.66)	22.4-188 (101)	N.D.-16.8 (11.9)
Cl$_7$DD	14.9-427 (186)	3.16-15.9 (6.53)	30.8-249 (148)	N.D.-35.1 (18.9)
Cl$_8$DD	7.80-527 (249)	1.12-18.4 (6.44)	19.6-264 (114)	N.D.-27.2 (10.8)
Total CDDs	80.5-1250 (624)	9.25-43.1 (19.1)	101-924 (527)	N.D.-86.2 (26.8)
Cl$_4$DF	3.24-237 (66.7)	0.11-0.76 (0.44)	33.6-152 (84.5)	---
Cl$_5$DF	2.08-262 (80.4)	0.15-3.23 (1.07)	61.3-227 (145)	---
Cl$_6$DF	20.2-373 (107)	0.07-1.03 (0.52)	67.1-229 (180)	---
Cl$_7$DF	1.70-519 (117)	0.07-0.79 0.32)	27.1-109 (66.8)	---
Cl$_8$DF	5.83-441 120	N.D.-0.99 (0.40)	8.64-58.7 (34.0)	---
Total-CDF	22.4-1,510 (491)	0.62-5.89 (2.74)	231-761 511)	---

(4) Municipal Waste Incineration in Denmark

There are currently 45 municipal waste incinerators operating in Denmark, most of which are equipped with boilers for heat production (Bendixen and Nielsen, 1987). Approximately 50% of the total municipal refuse is burned in these incinerators.

Figure 28 shows the ratio of remaining concentration at time t = 2 seconds, C_t, to the initial concentration, C_o, for the thermal treatment of biphenyl, carbon monoxide (CO), and PCDD. CO has a high thermal stability (destroyed at temperatures greater than 827 °C), but PCDD has even higher stability since temperatures greater than 1,017 °C are required for destruction at a 2 second reaction time. Thus low CO emission can be considered as a prerequisite (but not an indicator) for low PCDD emission.

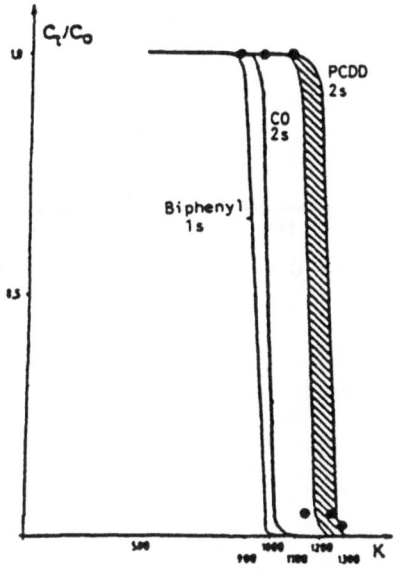

Fig. 28. Ratio of Biphenyl, Carbon Monoxide, and PCDD after Thermal Treatment

A Copenhagen incinerator equipped with refractory lined furnace, rotary kiln, and burn-out chamber was studied in an attempt to correlate PCDD emission levels with burn-out temperature (Bendixen and Nielsen, 1987). The 12 tons/hour incinerator used an energy recovery boiler and electrostatic precipitator for cooling and cleaning respectively of the flue gas.

The total concentration of PCDD and PCDF as a function of the burn-out temperature is shown in Figure 29.

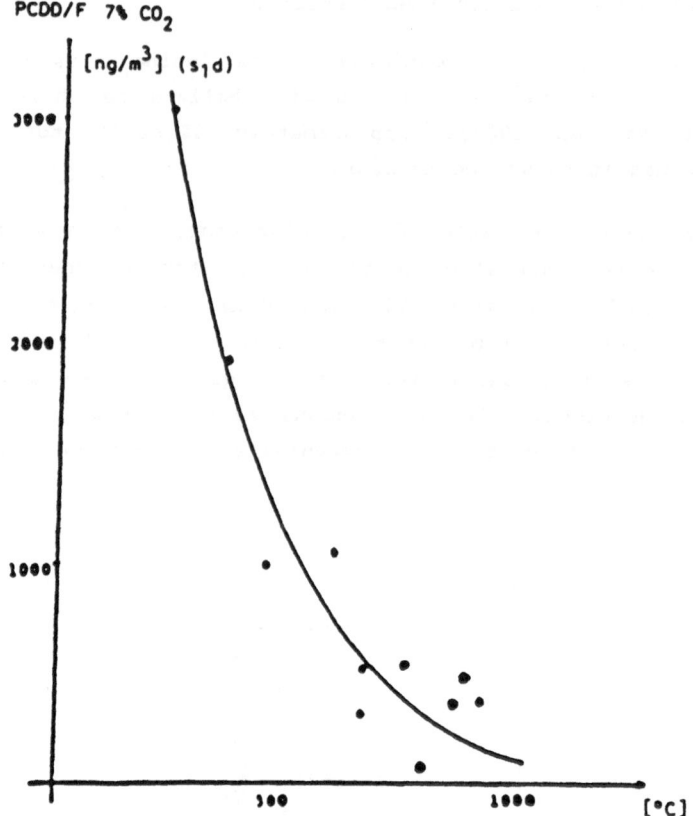

Fig. 29. PCDD and PCDF Concentrations as Function of Burn-out Temperature

The relationship between PCDD/PCDF and CO emissions was investigated for the Copenhagen incinerator and two smaller incinerators having capacities of 3.0 and 3.5 tons/hour. The smaller incinerators both had refractory lined furnaces and burn-out chambers, heat recovery boilers, and electrostatic precipitators for flue gas cleaning. The total PCDD/PCDF concentration versus the concentration of CO emitted for all three incinerators is given in Figure 30. As predicted, the lowest PCDD/PCDF concentrations were found for low CO concentration and high combustion temperature.

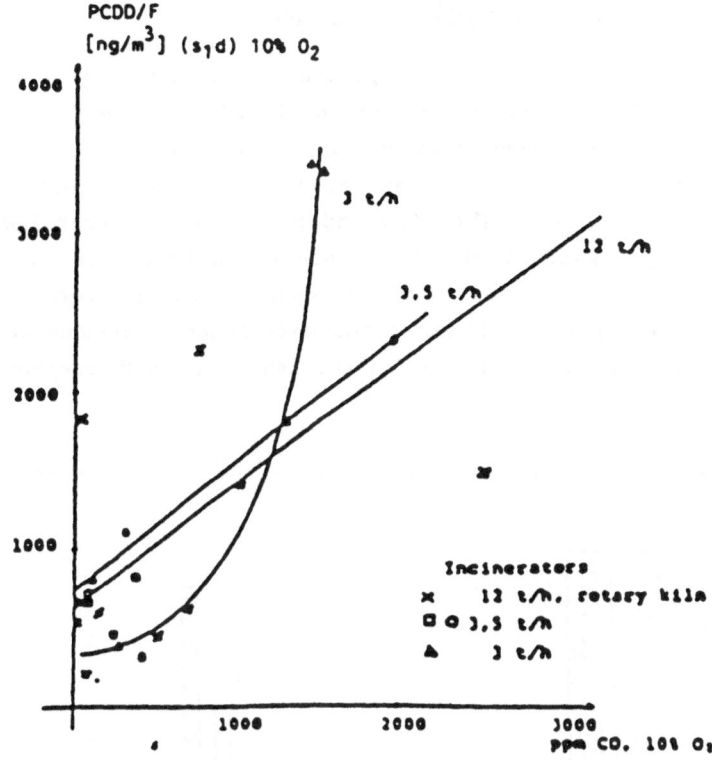

Fig. 30. Relationship between PCDD- and CO-Concentration for Three Incinerators

Guidelines for efficient incineration and reduced PCDD/PCDF emissions
were suggested by the Danish EPA:

1. Preheat combustion air;
2. Have a support burner;
3. Refractory lining of furnace and burn-out chamber;
4. Minimum 2 second dwell time in burn-out chamber;
5. No CO gradient in burn-out chamber;
6. Maintain burn-out chamber at 950 °C;
7. Start support burners if burn-out temperature drops to 875 °C;
8. Burn-out chamber temperature must be maintained at greater than
   875 °C for 95% of the time without support burners in operation;
9. Feeding of waste stopped when burn-out chamber temperature drops
   below 825 °C;
10. Continuous monitoring and control of combustion process.

353

(5) Municipal Waste Incineration in Sweden

Municipal waste incinerators operated in Sweden were studied (Tysk-lind et al., 1987) . Variations in total emissions were observed, how-ever the patterns of isomers found for individual PCDD and PCDF homolo-gues were found to be similar even for incinerators producing relatively high and low emissions. The PCDD and PCDF congener profiles for three Swedish municipal waste incinerators are shown in Figure 31. The rela-tive distributions were very similar even though the total levels were different. The highest PCDF emissions were found to be due to the Cl_4DFs and Cl_5DFs while Cl_5DDs dominated in the levels of PCDD present.

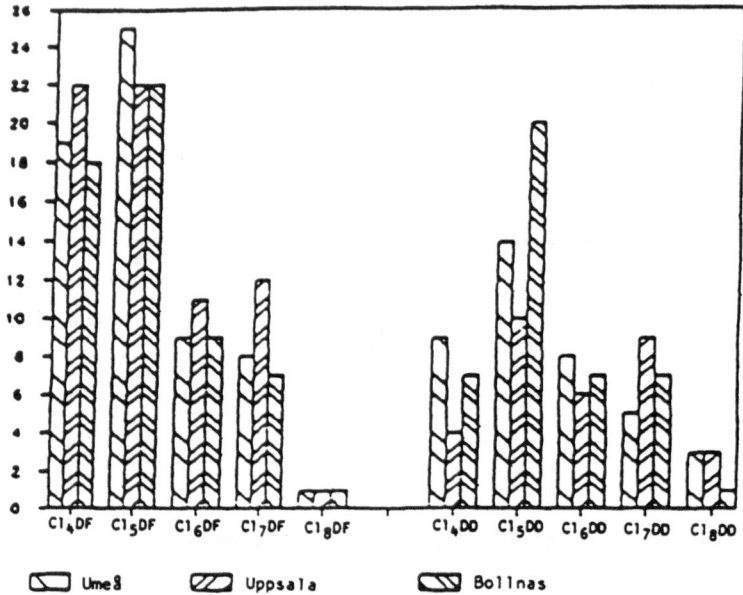

Fig. 31.   Congener Profile of PCDD and PCDF in Emissions from Three Swe-dish Municipal Waste Incinerators

All Swedish waste incinerators are combined with energy production (Bergström and Warman, 1987). The energy produced is used mostly for district heating. During waste incineration, a very efficient oxidation of the organic substances takes place. The particle size of waste varies and there is big difference in composition. Usually, 2% of the organic matter in the waste leaves the furnace with the slag and fly ash. About 75% of the chlorine in the fuel occurs as HCl in flue gas. Less than 0.1% of the chlorine produces chlorinated aromatics. Polychlorinated di-benzodioxins and furans compride only a small part of the chlorinated aromatics that are produced. Authorities in Sweden have chosen to describe the emission by a weighted mean value of 12 isomers which are considered to be especially toxic according to a model developed by Eadon (Eadon et al., 1983). The "TCDD" equivalents released with the flue gases from different incinerators amount between 1 and 50 ng/m^3 of flue gas. Taking into account the uncertainty of the method of measurement, the differences are very small. The incineration conditions have a greater influence on the emission of many other organic trace pollutants, such as polycyclic aromatic hydrocarbons, than on PCDD/PCDF.

As PCDD/PCDF are present on dust particles and emitted with the fly ash to a much higher degree than many other compounds formed during the combustion, the emission of PCDD/PCDF can effectively reduced by an efficient flue gas cleaning system. However, flue gas cleaning does not reduce the quantity of pollutants fed with the wastes or formed in the incinerator.

In Sweden, measurement were carried out on several municipal waste heating plants using the same sampling and analytic methods on each plant. Operational studies for characterization of firing conditions have been reported elsewhere (Bergström and Warman, 1987). Parameter of furnace load were calculated; furnace temperature, flow and distribution of combustion air were measured; slag and fly ash were checked during the test. Gases emitted (excess oxygen, carbon dioxide and carbon monoxide) were monitored continuously and stored in a computer, as average values for five minutes. The sampling system for organic micro-pollutants was described (Bergvall and Hult; Öberg et al., 1985). Dioxins and related compounds were determined after clean up on aluminia according to Buser (Buser, 1975) and analyzed by GC/SIM. The results are given in Table 54.

Table 54. Organic Micro-pollutants in the Flue Gas Samples; Operation with a Large Excess of Oxygen [μg/standard m^3] corrected to 10% CO_2

	Test 1	Test 2
CE	99.80	99.85
CE	99.64	99.78
Total organics (RI>900)	140	100
Total PAH	1.3	0.5
Total chlorobenzene	17	8.0
Hexachlorobenzene	5.3	2.6
Total chlorophenols	25	8.0
Pentachlorophenol	1.3	0.7
Total Cl_4DD	1.9	0.9
2,3,7,8-Cl_4DD	0.5	0.2
Total Cl_5DD	18	6.9
1,2,3,7,8-Cl_5DD	1.2	0.6
Total Cl_4DF	38	15
2,3,7,8-Cl_4DF	3.8	1.9
Total Cl_5DF	58	25
1,2,3,7,8-Cl_5DF	7.1	3.5
TCDD equivalents (Eadon)	7	4

The test results from different plants show that three factors are of great importance when limiting the production of organic micro-compounds in the furnace and the boiler. Combustion efficiency (CE) is the most important factor influencing the production of all kinds of aromatic compounds in the flue gas. Combustion efficiencies CE > 99.9%, i.e. 100 ppm CO adjusted to 10% CO_2, are considered to be a high combustion efficiency for ordinary grate fired units. During such operation conditions the frequency of CO peaks has a great influence on the production of aromatic compounds, including PCDD/PCDF. Table 55 shows the influence of CO peaks on the production of TCDD-equivalents in another waste heating plant equipped with a high efficiency flue gas cleaning. The values in Table 55 represent PCDD/PCDF concentrations after the boiler but before flue gas cleaning. After cleaning the emission of TCDD-equivalents with the flue gas was less than 0.1 ng/m^3 at 10% CO_2 in all four tests. From Table 54 it might be seen that test 2 gives lower concentrations of all compounds in the flue gas. During test 2 there was no five-minute period with CE lower than 99.78%.

Table 55. Influence of Combustion Efficiency on the Production of TCDD-Equivalents

Number of five-minute Periods with CE<99.9% during Sampling	TCDD-Equivalents [ng/m^3]
9	44
4	9
2	3
0	2

The results show a co-variation of different aromatic micro-pollutants produced in the furnace and the boiler. Seven tests in one incinerator show that combustion conditions which results in small amounts of PCDD/PCDF also produce low concentrations of PAH and the chlorobenzenes. The chloride concentrations in the flue gas after the boilers have certainly been in the range 600-1,000 mg/m^3 at 10% CO_2 calculated as HCl.

It was possible to make good prediction of the chlorinated micro-pollutants present in the flue gas using the factors CE and HCl content. A third factor that was found to be important was excess air, measured as O_2, in the flue gas. However, because CE and O_2 content are highly correlated in the range 5-12% O_2, either one could be used to predict the production of chlorinated aromatic pollutants.

(6) Waste Incineration in the United Kingdom

In the years 1985/86 four commercially available merchant high-
temperature incinerators were in operation in U.K.; these plants were at
Ellesmere Port, Pontypool, Fawley, and Hucknall (Hazardous Waste
Inspectorate, 1986).   Together they represent a notational installed
capacity of some 61,000 tons per annum, of which approximately 6,000 tons
are available for solid wastes and the balance for liquids.
Additionally, the in-house incinerator operated at West Bromwhich takes
in certain liquid combustible wastes on a commercial basis and represents
a further 4,000 tons per year of liquid waste incineration capacity.
There are currently 60 in-house chemical waste incinerators registered
under the Health & Safety (Emissions into the Atmosphere) Regulations
1983.   These plants are, in main, integral with chemical process plant
and, therefore, not subject to disposal site licensing under the Control
of Pollution Act 1974.

Perhaps the most significant current development in high-temperature
incineration is the apparent increase in the use of marine incineration
by U.K. waste producers. The trend in utilisation of this disposal route
is upward from 811 tons in 1981 to an estimated 3,500 tons in 1984.   The
actual tonnage of UK hazardous waste incinerated at sea during 1984 was
1,952 tons, only 26% of the licensed quantity, and a reduction from the
1983 total.   During 1985, however 2,584 tons of UK wastes were
incinerated at sea, and 7,850 tons are licensed for marine incineration
during 1986.

Analytical results of PCDD/PCDF analyses have not been published due
to lack of suitable analytical standards and to limitations in the
analytical method employed.   UK claims to develop standard methods of
extraction, separation and analysis of PCDD/PCDF.

In Scotland the total amount of waste is quantified to about 4.5
million tons: household waste arisings are about 1.8 million tons per
annum (ca. 40%), the other 2.73 million tons (ca. 60%) are commercial and
industrial arisings.   Most of the household waste and some of the
commercial and industrial waste are disposed of at local authorities
operated landfill sites to the amount of approximately 2.8 million tons
per annum in all.   Of the rest about 200,000 tons per annum are
incinerated to atmosphere by waste disposal authorities (1.5 million tons
are disposed of at landfill sites operated by private sector
contractors).   In 1983-4 there were 11 municipal waste incinerators
operating, which disposed of some 250,000 tons; in December 1985 only 10
plants operated about 200,000 tons of waste.   Incineration by these
municipal waste incinerators produces gaseous emissions to the

atmosphere, almost entirely of carbon dioxide. However, in Scotland, only one major incinerator has been commissioned in the past decade.

The only incineration facility for the specific disposal of wastes containing PCBs has been closed in 1984.

In future, more information has to be obtained on the impact of dioxins and furans to the environment.

(7) Resource Recovery in the United States

The solid waste disposal crisis, which arose as the result of
landfills being filled to capacity and a shortage of land for new sites
near expanding urban areas, has generated an increasing interest of
cities throughout the United States and the world to invest in various of
resource recovery. Alternative solutions to the crisis include recycling
and waste reduction and large-scale energy and materials recovery
facilities (known as resource recovery and RDF, respectively). The
discovery of dioxin and furan emissions from resource recovery plants has
led to an intensive focus on the dioxin issue by the public, the
regulators, and the builders of the plants.

Mass burning, a technology perfected in Europe for energy recovery,
is used in numerous plants now operating in the United States. Most of
the mass burn plants create steam for district heating loops and for
industrial processes; a few generate electricity. There are other plants
which employ a variety of separation and Refuse-Drived Fuel (RDF)
preparation technologies. Whilst many of the facilities in the U.S.A.
will process only relatively small quantities of refuse (less than 500
tons per day), approximately 40% of the plants will process sewage sludge
in addition to garbage. A number of usable products, including
pelletized and fluff refuse-derived fuel for different uses, ferrous
metals, aluminium, chilled water, glass, methane gas, carbon monoxide,
humus, compost, and corrugated and baled paper. Most plants have been
constructed in urban centers and dispose of municipal waste, but some
have been built on army and navy bases and on college campuses (Steisel
et al., 1987).

Due to the paucity of usable data from operating resource recovery
plants (i.e. complete combustion information and dioxin samples taken
over periods of time during which combustion conditions are held
constant) the American Society of Mechanical Engineers (ASME) has
developed a proposed protocol which is now under review. The New York
Department of Environmental Conservation researches at six resource
recovery plants of different design. Another testing program of the ASME
was established for the purpose of conducting combustion research at a
number of facilities to learn the effects of varying refuse composition
(i.e. moisture, PVC content) on emissions of dioxins and related
compounds. Dioxin testes were funded by a number of sites, i.e. New
York, Florida, California, Massachussetts and others.

(8) Influence of the Type of Refuse

A study of the influence of the type of refuse material was recently reported in which sorted and unsorted refuse were burned in a municipal incinerator (Benfenati et al., 1986). Table 56 gives the average compositions of the waste combusted in the tests. The results of the study are listed in Table 57. No conclusions were made regarding the effect of the refuse on the emissions of PCDD and PCDF. Considerable variation in the resulting levels were observed , however there does not appear to be any relationship to the nature of the refuse burned. High moisture levels in the refuse result in lower combustion temperatures and consequently in higher emissions of PCDD and PCDF.

Table 56. Composition of Combusted Waste

Parameter	Original Waste [%]		Processed Waste [%]	
	Weight	Moisture	Weight	Moisture
Paper	23.0	40	7.2	60
Soft plastics	7.6	---	11.0	---
Hard plastics	4.8	---	12.1	---
Cloth	3.6	15	8.8	76
Wood	1.6	12	5.1	15
Inert material	5.7	---	3.1	---
Organic material	30.7	57	15.0	37
Iron	4.5	---	9.3	---
Material smaller than 20 mm	18.5	58	28.4	67
Moisture	37.6	36.5		
Cinders	32.9	26.5		

Table 57. Cl_4DD and Cl_4DF [ng/Nm^3] in the Condensate before the Abatement System (upstream) and the Stack (downstream)

Sample number	Kind of refuse	Cl_4DD upstream	Cl_4DD downstream	Cl_4DF upstream	Cl_4DF downstream
1	Processed	23.6	13.3	60.8	3.1
2	Processed	<3.0	1.9	59.1	9.1
3	Unprocessed	22.7	1.2	31.7	8.0
4	Unprocessed	16.9	2.3	52.0	5.2
5	Processed	13.0	5.2	81.5	17.0
6	Processed	9.1	2.5	49.0	6.0
7	Processed	68.0	20.9	195.0	65.0
8	Processed	79.5	9.6	275.0	42.2

Table 58 summarizes the PCDD emission data taken from municipal incinerators from throughout the world (Jones et al., 1986). The total levels of PCDD emitted are found to span a very wide range of values. Contrary to what was concluded from the National Dioxin Study conducted by the U.S. EPA (des Rosiers, 1987), the plants equipped for heat recovery show low emissions as compared to other incinerators not equipped in this way. This list is not comprehensive but includes plants for which better documented studies have been performed. According to this collection of data:

1. Plants with energy recovery on average have about one-tenth the PCDD/PCDF emissions of those without energy recovery equipment;
2. The majority of the emissions for energy recovery plants are PCDF;
3. The level of 2,3,7,8-Cl_4DD is very small relative to the total PCDD/PCDF emissions;
4. Tetra-CDFs are the dominant homologue in the resource recovery emissions.

A more comprehensive list of results (although comprised of mainly earlier incinerator studies) was compiled in a study concerned with the sources and levels of PCDD and PCDF entering the environment (Ontario Ministry of the Environment, 1985). The summary of averaged values are reported in Table 59, along with the number of values averaged and the calculated standard deviations of each average. The standard deviations are always larger than the averages themselves which indicates the great variation of the reported levels of PCDD and PCDF emitted by municipal waste incinerators.

Table 58. Distribution of Worldwide PCDD Emissions Data (tetra through octa Congeners)
[ng/Nm^3, dry]

Facility	Country	All data	Plants with Heat Recovery	Best data
Montreal (Des Carrieres)	Canada	<1	<1	
Westchester RESCO	USA	24	24	
Würzburg	FRG	25	25	25
Pittsfield (Vicon)	USA	37	37	37
Chicago, NW	USA	44	44	44
Stapelfeld	FRG	49	49	49
Eksjo	Sweden	70	70	70
PEI	Canada	107	107	107
Stellinger Moor	FRG	130	130	130
Zürich	Switzerland	131	131	131
Borsigstrasse	FRG	172	172	172
Albany RDF (Sheridan Ave)	USA	310	310	310
Valmadiera	Italy	310		
Italy 1	Italy	516		
Italy 6	Italy	675		
Italy 5	Italy	746		
Niagara RDF (Occidental Chemical Co.)	USA	827	827	827
Zaanstad	Netherlands	1,294		
Hamilton/ Wentworth SWARU	Canada	4,332	4,332	
Italy 4	Italy	5,003		
Toronto	Canada	5,086		
Hampton (Langley Field)	USA	5,395	5,395	
Italy 3	Italy	8,622		
Italy 2	Italy	33,047		

Table 59. Summary of Average PCDD and PCDF Congener Concentrations and Calculated Total (Cl$_4$–Cl$_8$) Values for Municipal Waste Incinerator Emissions (Total = Calculated from sum of average isomer group values; n = number of values)

Matrix	Units		PCDD						PCDF					Total (Cl$_4$–Cl$_8$) PCDD+PCDF
		Cl$_4$DD	Cl$_5$DD	Cl$_6$DD	Cl$_7$DD	Cl$_8$DD	Total (Cl$_4$–Cl$_8$)	Cl$_4$DF	Cl$_5$DF	Cl$_6$DF	Cl$_7$DF	Cl$_8$DF	Total (Cl$_4$–Cl$_8$)	
					- Weight Basis -									
Precipitated Fly Ash	[ng/g]													
Ave.		90	151	528	322	227	1,318	151	245	379	250	47	1,072	2,390
n		44	24	35	26	45		28	14	23	14	32		
s.d.		321	394	1,265	680	344		172	271	527	348	61		
Stack-Collected Particulates	[ng/g]													
Ave.		502	723	1,003	700	546	3,474	2,321	2,345	2,076	558	114	7,414	10,888
n		5	5	5	5	5		4	4	4	5	5		
s.d.		901	925	1,170	778	787		4,123	3,787	1,786	653	154		
			- Volume Basis -					(data for Nm3 without correction (1Nm3 = 0,925 dscm)						
Stack-Collected Particulates	ng/m^3													
Ave.		49	71	1,186	104	654	2,064	150	283	218	88	240	979	3,043
n		13	11	11	11	13		11	5	5	5	13		
s.d.		112	126	3,595	189	2,005		370	476	279	121	795		
Stack-Collected Gaseous Phase	[ng/m^3] n 12													
Ave.		45	59	3,588	151	492	4,335	465	313	159	33	671	1,641	5,976
n		10	10	10	12	10	10	4	4	4	12	5		
s.d.		104	107	8,334	247	798		615	560	257	50	1,295		
Total Stack Emissions	[ng/m^3]													
Ave.		338	290	2,551	343	741	4,263	1,353	1,205	378	296	618	3,850	8,113
n		19	16	20	20	22		17	10	11	11	19		
s.d.		1,062	635	8,614	516	1,835		3,177	2,151	558	638	1,673		

2.3. SEWAGE SLUDGE INCINERATION

There are few reports of studies other than those done on municipal solid waste incinerators. In a recent study, the emissions of PCDD and PCDF were reported for a municipal sludge incinerator plant (Clement et al., 1987). Two twelve-hearth incinerators were tested. The top four hearths are used for drying the raw sludge (480-500 °C), the next four are used for incineration (700-800 °C), and the bottom hearths are used for burn-out and cooling (300-400 °C). The flue gas is cleaned by a scrubber and cooled using a sprayer system such that the scrubber exhaust gas temperature is 70 °C. Municipal sewage, which is 80-90% water, is delivered to the top hearth at a rate of 10 tons/hour.

Stack samples were collected with a modified EPA method 5 sampling train in which two florisil cartridges were placed in series after the third impinger. Samples recovered were collected as three fractions corresponding to the filter, impingers and florisil cartridges. Three separate tests (24 hour sampling) were done over a one week period. Raw sewage sludge and bottom ash were sampled periodically throughout each test in order to obtain composite samples representative of the 24 hour sampling time. All samples were fortified with [37]Cl-labelled octachloro-dibenzo-p-dioxin before extraction.

Table 60 lists the concentrations of chlorinated organics, including PCDD and PCDF, which were found in the stack gases in the three tests. The levels of PCDD and PCDF found in the stack gases for the sewage sludge incinerator were found to be low in comparison to two municipal solid waste incinerators tested at about the same time using the same sampling method. Differences in the homologue distributions in the sewage and solid waste incinerators were found, however the effects of raw waste composition could not be determined. The significance of the quantitative data is difficult to assess due to the variations in the recoveries for each fraction. The average percent recoveries, shown in Table 60, have uncertainties as large as the averages themselves. Recoveries were so low in some cases that the data were not even reported. The PCDD and PCDF concentrations were all corrected for recovery, however it is not clear how this correction was actually made.

	Raw Sludge			Bottom Ash		
	Test 1	Test2	Test3	Test 1	Test 2	Test 3
Cl_4DD	ND	ND	ND	ND	ND	ND
Cl_5DD	ND	ND	ND	ND	ND	ND
Cl_6DD	ND	ND	ND	ND	ND	ND
Cl_7DD	79	70	110	ND	ND	0.4
Cl_8DD	230	150	140	ND	ND	0.8
Cl_4DF	ND	ND	ND	ND	ND	ND
Cl_5DF	ND	ND	ND	ND	ND	ND
Cl_6DF	ND	ND	ND	ND	ND	ND
Cl_7DF	21	12	12	ND	ND	ND
Cl_8DF	25	18	9	ND	ND	ND
$^{37}Cl-Cl_8DD$	53	33	33	19	25	14
% recovery						
Total Cl_xDD	310	220	250	ND	ND	1.2
Total Cl_xDF	46	30	21	ND	ND	ND
Total CP	<1	ND	ND	ND	ND	ND
Total CB	190	77	130	5	1	2
Total PCB	4	2	1	5	21	4

ND: not detected, PCDD/PCDF single congener average detection limits range from 0.1 to 0.5 ng/g (tetra to octa congeners) for Bottom ash samples, and 1 to 3 ng/g (tetra to octa congeners) for raw sludge samples.

The concentrations of PCDD and PCDF in the raw sludge and bottom ash samples are listed in Table 61. The hepta- and octachlorinated PCDD and PCDF were the only isomers which were detected in the sewage sludge. In an earlier study, these isomers were found to be the dominant PCDD and PCDF species in the analysis of municipal refuse (Tosine et al., 1985). Similarly, the total amount of PCDD in the raw refuse was found to be higher than the total amount of PCDF. Chlorinated benzenes, found in re- latively large levels in the stack gases, were present at levels slightly less than the PCDD in the raw sludge. The bottom ash samples which were analyzed were found to contain virtually no chlorinated organic com- pounds.

Table 61. Concentrations of Chlorinated Organics in Raw Sludge and Bottom Ash. All data are [ng/g dry sample weight]

	Raw Sludge			Bottom Ash		
	Test 1	Test 2	Test 3	Test 1	Test 2	Test 3
Cl_4DD	ND	ND	ND	ND	ND	ND
Cl_5DD	ND	ND	ND	ND	ND	ND
Cl_6DD	ND	ND	ND	ND	ND	ND
Cl_7DD	79	70	110	ND	ND	0.4
Cl_8DD	230	150	140	ND	ND	0.8
Cl_4DF	ND	ND	ND	ND	ND	ND
Cl_5DF	ND	ND	ND	ND	ND	ND
Cl_6DF	ND	ND	ND	ND	ND	ND
Cl_7DF	21	12	12	ND	ND	ND
Cl_8DF	25	18	9	ND	ND	ND
$^{37}Cl-Cl_8DD$	53	33	33	19	25	14
% recovery						
Total Cl_xDD	310	220	250	ND	ND	1.2
Total Cl_xDF	46	30	21	ND	ND	ND
Total CP	<1	ND	ND	ND	ND	ND
Total CB	190	77	130	5	1	2
Total PCB	4	2	1	5	21	4

ND: not detected, CDD/CDF single congener average detection limits range from 0.1 to 0.5 ng/g (tetra to octa congeners) for Bottom ash samples, and 1 to 3 ng/g (tetra to octa congeners) for raw sludge samples.

2.4. HAZARDOUS WASTE INCINERATION

The performance of a high-temperature incinerator used for the destruction of chlorine-containing industrial wastes was evaluated with respect to its emission of dioxins (Brenner et al., 1984). The incinerator consisted of a rotating slag tap furnace, a secondary combustion chamber, a waste heat recovery boiler, and scrubbing units. Both liquid and solid industrial wastes could be delivered to the furnace. Two burners located in the front wall were used to keep the kiln temperature at about 1,000 °C or higher. Flue gases leave the kiln and enter a refractory-lined secondary combustion chamber where combustion is completed. The temperature is kept between 900 and 1,400 °C by the combustion of sufficient quantities of liquid waste using compressed-air atomizer burners. The flue gases are then cooled to 250 °C in the waste heat boiler and finally to about 50-70 °C after passage through two scrubber stages. Acid gases are scrubbed with a sodium hydroxide solution. The emissions of the plant were monitored at the stack as well as in the wash water circuit.

The operating conditions and wastes combusted in the experiments performed are summarized in Tables 62 and 63. In experiment 2b, chlorine-free waste used in the blank run was incinerated along with the chlorine-containing waste thereby reducing the chlorine feed rate in comparison to experiment 1. In all of the experiments, no PCDD were detected (detection limits ranged from 1 to 6 ng/Nm^3 for the different homologues). It must be noted that the recoveries based on ^{13}C-labelled 2,3,7,8-Cl_4DD sample spikes were quite varied and low. The workers claimed that operation of the rotary kiln at 1,000 °C and residence times of 3 to 7 seconds resulted in dioxin-free emission, however it is more likely that the PCDD were present but below the detection limit of the analytical methodology employed.

Table 62. Technical Parameters during Experiments 1 and 2

	Kiln Temperature [°C]	Residence Time of Gases	Throughout [kg/h]	% Cl	Wash Water Circuit [m^3/h]	Stack Gases HCl [mg/Nm^3]	CO	O_2
Exp.1	~1,000	3-7 sec	2,800	9	200	8	5	14.5
Exp.2a	~1,000	7 sec	2,860 [2]	0	200	0	8	14.5
		7 sec	1,700 [3]	0				
Exp.2b	~1,000	3 sec	1,100 [3]	9	200	5	5	14.5

Legend to Table 62:

[1] 2 burners in kiln (7 sec), 2 burners in secondary combustion Chamber (3 sec);

[2] blank experiment, chlorine-free waste;

[3] waste from experiment 2a was burned in the kiln during incineration of the chlorine-contain waste in the secondary combustion chamber.

Table 63. Incinerated Wastes

	C %	Cl %	Type of Waste
Exp. 1	60	9	Distillation sump from chlorinated phenols and cresols
Exp. 2a	53	0	Mixture of alcohols and aldehydes with a boiling point > 150 °C
Exp. 2b	60	9	Distillation sump from chlorinated phenols and cresols

Brenner et al. studied the emission of PCDD resulting from the high-temperature incineration of chlorine-containing industrial wastes (Brenner et al., 1986). Table 64 lists the types of chlorine-containing wastes which were incinerated in the experiments. It should be noted that additional chlorine-free waste was added to the kiln to keep the burning temperature of the kiln and secondary combustion chamber between 1,130-1,230 °C (see Table 64). In all four experiments, natural gas was also fed into the system to help regulate the temperatures. Table 65 gives the waste characteristics for the incineration experiments. PCBs and chlorinated benzenes were fed into the kiln via burners specifically designed for handling viscous wastes.

The flue gas emissions, the slags, and the wash water were all analyzed for their PCDD content. In experiments 1 and 3 the stack emissions were analyzed and in experiments 2 and 4 the unscrubbed flue gas leaving the waste heat boiler was analyzed. The gaseous emissions were sampled using impingers and the polyurethane foam plug method. The results of the experiments are summarized in Table 66 (Brenner et al., 1986).

Table 64. Technical Parameters during PCB-Burning Experiments

** burners in kiln (7 sec), secondary combustion chamber (3 sec)

+ burners 1,2,5 are compressed-air atomizer burners; burners 1a and 2a serve for dosage of pasty and viscous material (e.g. PCBs)

	kiln temp. [°C]	residence time of gases [sec]	PCB-throughput [t/h]	fire feed [t/h]	sustain no of burners in action [+]	PCB [%]	chlorine [%]	stack gases [O_2, Vol-%]
E.p. 1	1180/1230	3 - 7 **)	1.04	3.01	1,2,5 & 1a	100	about 56	7 - 12
Exp. 2	1170/1220	3 - 7 **)	0.74	2.65	2,5,1a & 2a (for pasty material)	100	about 56	9.5 - 10.5
Exp. 3	1140/1230	3 - 7 **)	"0.72"	3.26	1,2,5,1a & 2a	poly-chloro-benzene sump	about 56	10 - 12
Exp. 4	1130/1220	3 - 7 **)	0.97	3.55	1,2,5,1a & 2a	100	about 56	9 - 10.2

Table 65   Incinerated Wastes, Additional Wastes for Fire Sustain

* drums C contained polychlorobenzene sumps

	Waste	PCB [%]	Cl [%]	Type of Wastes for Fire Sustain	Remarks
Exp. 1	container A	100	about 56	distillation sump, chlorine-free, C, H, O-compounds	PCBs always fed through burner 1a for pasty material; during exp. 2 to 4 burner 2a additionally served for dosage of fire sustain material; during all four experiments natural gas was fed in order to keep the preset temperatures.
Exp. 2	container B	100	about 56	similar to exp. 1	
Exp. 3	drums C	- *)	about 56 *)	similar to exp. 1	
Exp. 4	container D	100	about 56	distillation head and sump, chlorine-free, C, H, O, N-compounds	

371

Table 66. Analytical Results for Experiment 1 to 4:
Values for Emission in [ng/g], for Slags and Wash Water in [ng/kg; ppt]

	Recovery ^{13}C-2,3,7,8-Cl$_4$DD [%]	Cl$_2$DD Σ	Cl$_3$DD Σ	2,3,7,8-Cl$_4$DD Σ	Cl$_4$DD Σ	Cl$_5$DD Σ	Cl$_6$DD Σ	Cl$_7$DD Σ	Cl$_8$DD Σ
Exp.1									
1.1. Stack emission									
1.1.1. Impinger sampling									
Water separator + probe	51	<0.1	<0.1	<0.1 **	<0.1	<0.1	<0.1	<0.1	0.4 *
Impinger 1 & 2	75	<0.1	<0.1	<0.1	<0.1	<0.1	<0.1	<0.1	<0.1
1.1.2. PFP-technique									
Water separator + probe	54	<0.1	<0.1	<0.1	<0.1	<0.1	<0.1	<0.1	<0.1
PU-foam plugs	100	1.3	0.4	<0.1	<0.1	<0.1	<0.1	<0.1	0.4
1.2. Slag	10	<50	<50	<5	<5	<50	<50	<50	<50
1.3. Wash Water	79	<5	<5	<1	<1	<5	<5	<5	<5
Exp.2									
2.1. Crude Exhaust									
2.1.1. Impinger sampling									
Water separator + probe	34	<0.1	<0.1	<0.1	<0.1	<0.1	<0.1	<0.1	<0.1
Impinger 1 & 2	90	<0.1	<0.1	<0.1	<0.1	<0.1	<0.1	<0.1	<1.0
2.1.2. PFP-technique									
Water separator + probe	46	4	3	<0.1	0.5	<0.1	<0.1	<0.1	<0.1
PU-foam plugs	114	0.3	<0.1	<0.1	<0.1	<0.1	<0.1	<0.1	<0.1
2.2. Slag	7	<10	<10	<10	<10	<10	<10	<10	<10
2.3. Wash Water	116	<5	<5	<1	<1	<5	<5	<5	<5

Table 66. Continued

Ex. 3								
3.1. Stack emission								
3.1.1. Impinger sampling								
Water separator + probe	58	0.1	<0.1	<0.1	<0.1	<0.1	<0.1	<0.1
Impinger 1 & 2	80	0.8	<0.1	<0.1	<0.1	<0.1	<0.1	<0.1
3.1.2. PFP-technique								
Water separator + probe	130	<0.1	<0.1	<0.1	<0.1	<0.1	<0.1	<0.1
PU-foam plugs	13 +	<0.1	<0.1	<0.1	<0.1	<0.1	<0.1	<0.1
3.2. Slag	21	-**	-**	<10	<10	<10	<10	<10
3.3. Wash Water	110	-**	-**	<1	<5	<5	<5	<5
Exp. 4								
4.1. Crude exhaust								
4.1.1. Impinger sampling								
Water separator + probe	88	<0.1	<0.1	<0.1	<0.1	<0.1	1.3	<0.1
Impinger 1 & 2	93	1.8	0.4	<0.1	<0.1	<0.1	<0.1	<0.1
4.1.2. PFP-technique								
Water separator + probe	95	0.2	<0.1	<0.1	<0.1	<0.1	20	10
PU-foam plugs	74	<0.1	<0.1	<0.1	<0.1	<0.1	6.0	0.3
4.2. Slag	4	<50	<50	<30	<50	<50	<50	<50
4.3. Wash Water	50	-	-	<1	<1	<1	<1	<1

† - low recovery due to experimental difficulties

* - values given below 1 ng/m³ or 1 ppt indicate trends, for reproducibility, determination and detection limits see texts

** - the sign indicates "not detected" - detection limit given.

For all experiments, only trace amounts of PCDD, usually near the detection limit, were found in the flue gases analyzed. The polyurethane foam plug method was found to be as effective as the impinger technique for trapping trace amounts of PCDD. The analyses of the slags and wash water samples revealed no detectable levels of PCDD.

The incinerator was found to effectively destroy PCBs and chlorinated benzenes without producing significant levels of PCDD (below limits usually observed for many incinerators). The good performance of this incinerator was attributed to:

1. Effective spraying and subsequent mixing of the waste with air
2. Two-step burning of the waste thereby creating an afterburner effect
3. Long residence times in the kiln and secondary combustion chamber (7 and 3 seconds respectively)
4. High temperatures in the kiln and secondary combustion chamber
5. Long residence time of the slag at the melt temperature.

The use of PCBs was banned in Japan in 1972, at which time more than 5,000 tons were confiscated and placed in storage tanks (Tsuji et al., 1987). The incineration of liquid PCB waste was studied in a series of tests in 1985. The PCB waste incinerator, capable of burning 210 kg/h, is shown in Figure 32. The residence time in the combustion chamber, which is maintained at 1,400 °C, is 2 seconds. Flue gases are passed through a treatment system consisting of a water scrubber, for gas cooling and elimination of HCl, and activated carbon beds for the removal of trace levels of hazardous by-products. The used water is passed through activated carbon and stored in a checking tank so that levels of PCDD, PCDF, and PCBs in the water could be determined. If the water in the tank contained levels higher than permissible, the water was recycled through the activated carbon beds. After neutralization, waste water is discharged into a lake.

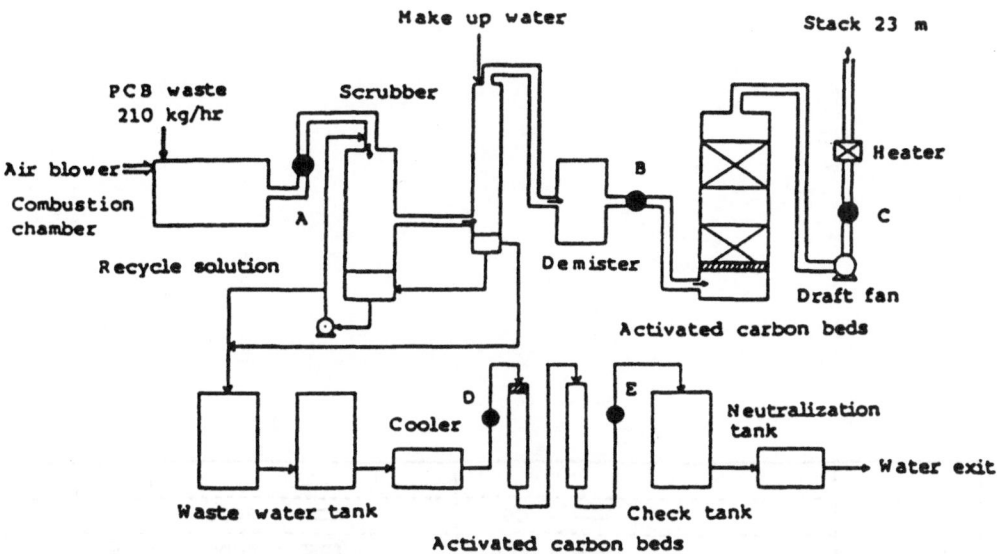

Fig. 32.   Flow Diagram of Liquid PCB Waste Incineration System
            Monitoring point for gas and water
            Samples for combustion products analysis

During the test period, concentrations of PCDD, PCDF, and PCBs in the stack emissions and waste water exiting the plant were monitored. The levels of PCBs were found to be less than the maximum permissible levels of 0.05 $\mu g/Nm^3$ in air and 0.05 $\mu g/L$ in water. The efficiency of decomposition and removal of PCBs was calculated to be greater than 99.99999%. The levels of PCDD and PCDF in the incinerator emissions were also lower than the maximum permissible levels of 0.1 ng/Nm^3 in air and 0.003 ng/L in water.

Undecomposed materials were accumulated on the activated carbon beds which were analyzed for their content of PCDD, PCDF, and PCBs. The results of these analyses are shown in Figure 33.

The levels of PCDD and PCDF in combustion effluents from hazardous waste incinerators have been reported (Ontario Ministry of the Environment, 1985; see also NATO/CCMS Report No. 172, 1988). It is interesting to note that the levels in the earlier reports are higher than those in the later studies which suggests that technological advances have been made in reducing emissions from hazardous waste incinerators.

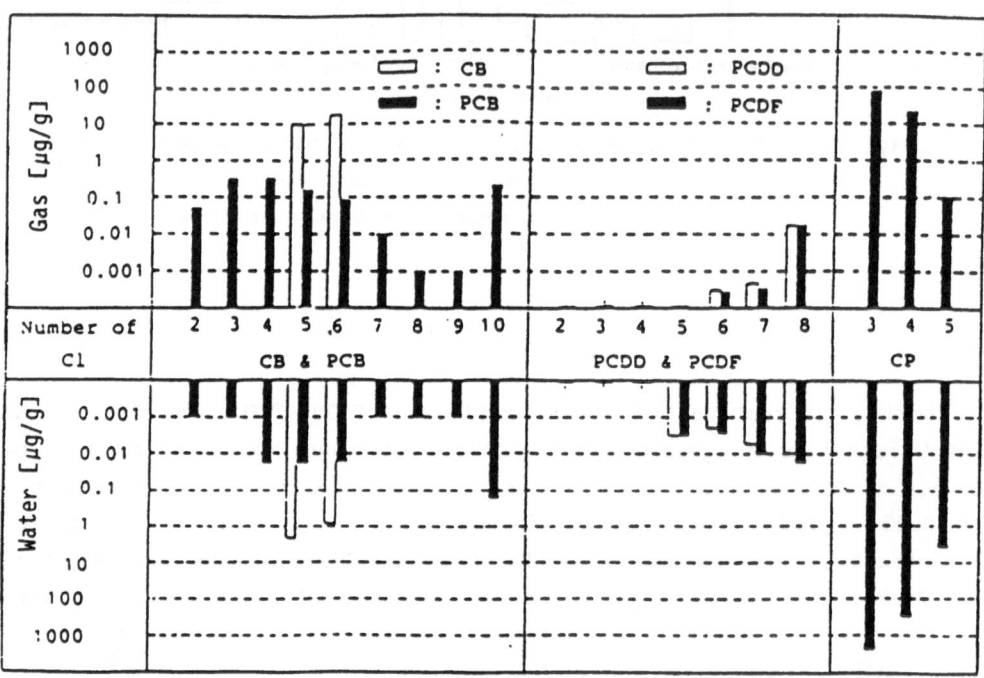

Fig. 33. Levels of Combustion Products Found in Activated Carbon.

2.5 TRANSFORMER AND CAPACITOR FIRES

PCDF have been synthesized by pyrolyzing individual PCB isomers (Hileman et al., 1985). Additional PCDF were created by chlorination or dechlorination of individual PCDF to produce at total of 110 different isomers. Chlorination was accomplished using antimony pentachloride while ultraviolet (UV) photolysis was employed for dechlorination purposes.

At elevated temperatures, such as those in transformer fires, PCBs can undergo reactions to PCDF and other by-products (Erickson et al., 1985). A study was therefore undertaken to determine the optimal conditions for PCDF formation and to study the potential for formation of PCDF and PCDD from the combustion of selected PCB-containing dielectric fluids.

Experiments were conducted using a bench-scale thermal destruction unit to determine the effects of temperature, oxygen, and residence time on the formation of PCDF. A mineral oil was spiked with three PCB isomers which form PCDF via known reaction mechanisms. Based on the results obtained from 33 test runs, it was concluded that both temperature and oxygen levels have significant effects on PCDF formation. From this study, the optimal conditions were determined to be a temperature of 675 °C with an excess oxygen concentration of 8%. The residence time did not play as important a role in the range from 0.3 to 1.5 seconds, although slightly lower yields appeared to occur at the shorter residence times.

Duplicate test runs were then performed using mineral oil and silicone oil dielectric fluids containing Aroclor 1254 at concentrations of 0, 5, 50, and 500 ppm. All runs were carried out using the pre-determined optimal conditions (a residence time of 0.8 second was used for all tests). PCDF were found in all of the combustion test runs except for the blanks (0 ppm Aroclor 1254). The results for the mineral oil runs are illustrated in Figure 34. No discussion is given by the researchers regarding why some of the points on the plot represent averages of two values while others are from single determinations. Therefore the reproducibility of the results for the levels of the various PCDF homologues formed are somewhat questionable.

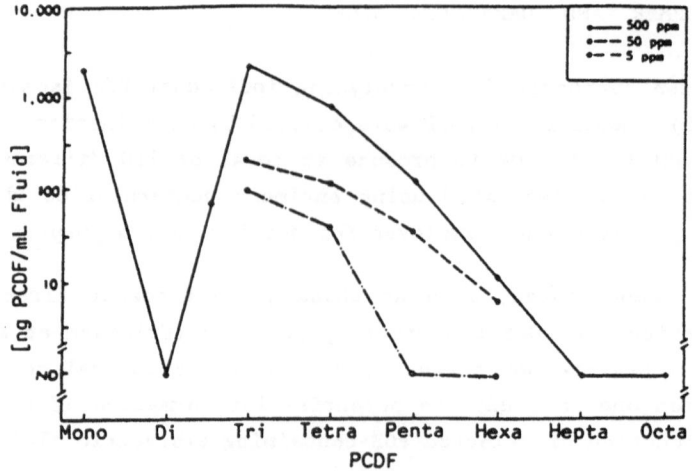

Fig. 34. PCDF Formation in Mineral Oil by Homolog
Closed symbols are averages of two values; open symbols are
single determinations

The total amounts of PCDF formed for the incineration of the mineral and
silicone oils containing various levels of PCBs are shown in Figure 35
An approximate linear relationship could be observed for the results for
both types of dielectric fluid. PCDD were reported found at low levels
in some of the samples, however, these may be due to artifacts.

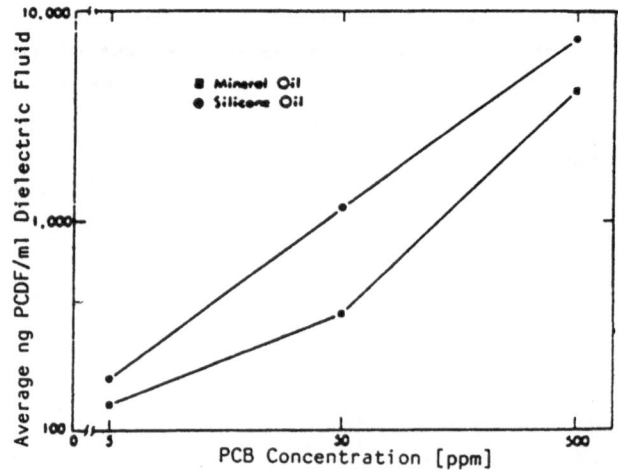

Fig. 35. Average PCDF Formation Versus PCB Concentration

A PCB-containing askarel (70% Aroclor 1260) and a non-PCB askarel (consisting of mainly trichlorobenzenes with some tetrachloro-benzenes present) were tested in duplicate under the same conditions used for the mineral and silicone oils. PCDF were produced in all of the runs, the results of which are given in Figure 36. Interestingly all eight PCDF homologues were detected in the PCB-askarel tests, with a crude bell-shaped distribution being centered about the Cl_5DFs. PCDD were found to be produced from the chlorobenzene askarel, however at levels much lower than the PCDF. It must be remembered that the conditions used were based on those optimized for PCDF formation and therefore perhaps more PCDD would be produced under different conditions.

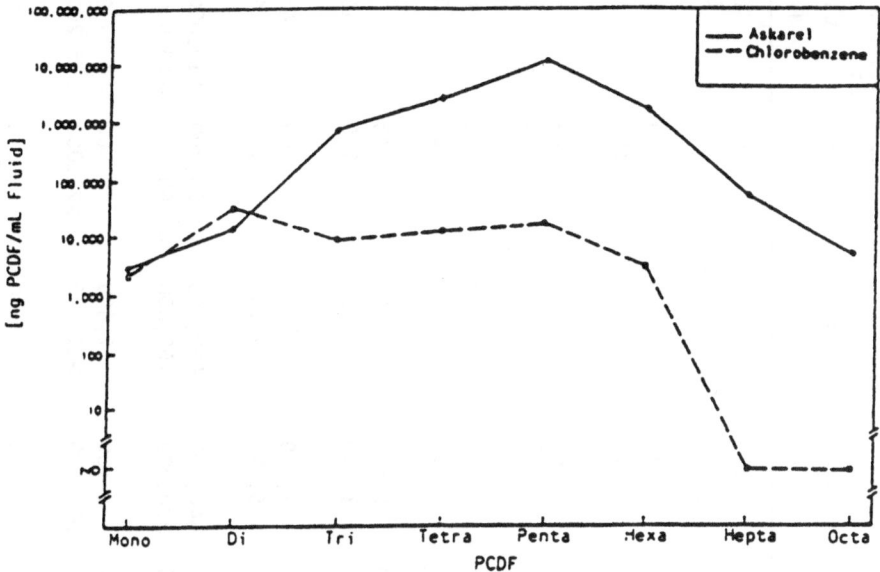

Fig. 36.   PCDF Formation from PCB Askarel and Trichlorobenzene Fluid
           Closed symbols are averages of two values; open symbols are
           single determinations

A conversion efficiency of up to 4% PCB-to-PCDF was observed for the mineral oil and silicone oil tests. The PCB-askarel fluid gave a conversion efficiency of 3%, which is quite similar to that observed for the other dielectric fluids. Conversion efficiency for the chlorobenzene askarel was much lower, 0.004% PCB-to-PCDF, than for any of the PCB-containing fluids. The conversion of tri- and tetrachlorobenzenes to PCDD was even less efficient, giving a yield of only 0.0001%.

The results of the study are summarized in Tables 67 and 68 (des Rosiers and Lee, 1986).

Table 67. PCDD Formed in Combustion Studies I

Runs	Cl_1DF [ng]	Cl_2DF [ng]	Cl_3DF [ng]	Cl_4DF [ng]	Cl_5DF [ng]	Cl_6DF [ng]	Cl_7DF [ng]	Cl_8DF [ng]	PCDF [ng]
Mineral Oil/	-[a]	-	130	49	NQ	0[b]	-	-	180
5 ppm A-1254	-	-	43	23	0	0	-	-	66
Silicone Oil/	-	-	26	0	90	0	-	-	116
5 ppm A-1254	-	-	31	9	150	0	-	-	190
Mineral Oil/	-	-	200	110	39	8.5	-	-	350
50 ppm A-1254	-	-	140	82	21	2.2	-	-	250
Silicone Oil/	-	-	290	73	62	0	0	0	420
50 ppm A-1254	-	-	530	640	83	0	0	0	1,300
Mineral Oil/	1,700	0	2,000	690	43	7	0	0	4,700
500 ppm A-1254	-	-	1,300	620	170	13	0	0	2,100
Silicone Oil/	-	-	2,000	740	340	45	-	-	3,100
500 ppm A-1254	50	1,300	5,000	2,100	170	12	0	0	8,600
70% A-1254/ 30%	810	5,100	440,000	1,400,000	6,400,000	910,000	29,000	3,400	9,200,000
Trichlorobenzene	1,900	7,000	220,000	1,100,000	4,700,000	660,000	19,000	1,300	6,700,000
Chlorobenzene Fluid	-	-	2,400	2,600	5,000	0	-	-	9,900
(tri with some tetra)	2,000	29,000	>13,000	>19,000	>22,000	5,200	0	0	>90,000
Lab Blank	0	0	0	0	0	0	-	-	0

Table 68. PCDD Formed in Combustion Studies II

Runs	Cl_1DD [ng]	Cl_2DD [ng]	Cl_3DD [ng]	Cl_4DD [ng]	Cl_5DD [ng]	Cl_6DD [ng]	Cl_7DD [ng]	Cl_8DD [ng]	PCDD [ng]
Silicone Oil/	-	-	0	0	7.7	0	-	-	7.7
500 ppm Aroclor 1254	0	0	0	0	1.7	0	0	0	1.7
70% Aroclor/	0	0	0	0	0	0	330	37	360
30% Trichlorobenzene	0	0	0	0	0	0	230	51	280
Chlorobenzene Fluids	-	-	1,100	440	0	0	-	-	1,500
(mostly tri with some tetra)	0	0	630	520	0	72	0	0	1,200

Erickson et al. studied the formation of by-products from the thermal degradation of dielectric fluids including those contaminated by PCBs
(Erickson et al., 1986).   A bench-scale thermal destruction unit, shown
in Figure 40, was used to combust the samples.   In this system, the
dielectric fluid is fed continuously with a syringe pump.   The effluent
from the combustion furnace is passed through an XAD-2 trap where the
volatilized organics are collected.   The extracts were then analyzed by
HRGC/EIMS in the selected ion monitoring mode for their PCDD and PCDF
contents.

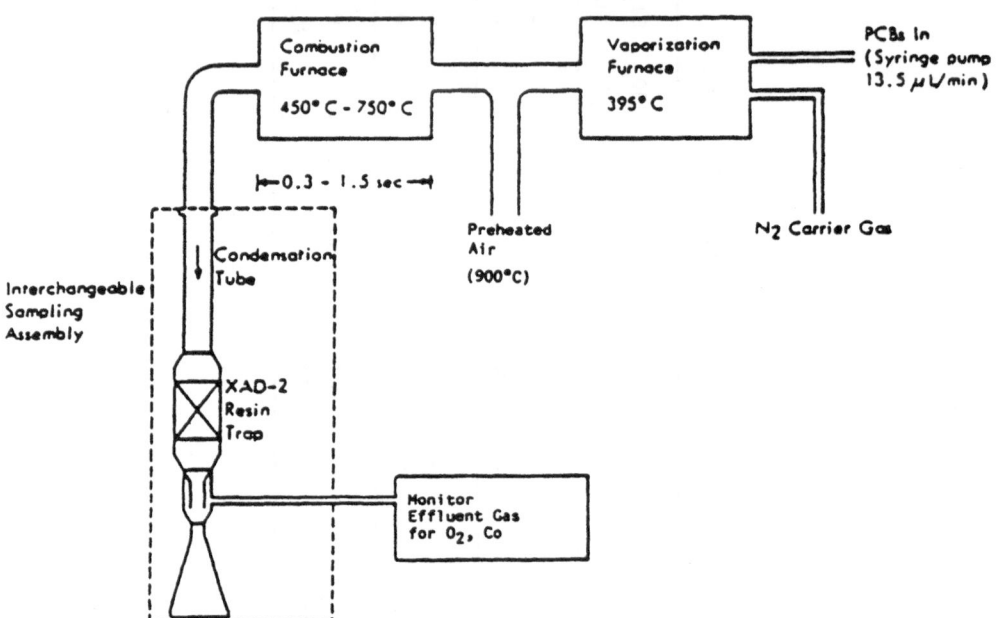

Fig. 37.   Thermal Combustion System to Model PCB Fire Conditions
           (Temperature, pressure, and flow rates are monitored through-
           out the system)

In previous work, the optimum conditions for PCDF formation from the
thermal degradation of PCBs were found to be at a temperature of 675 °C
for a residence time of 0.8 seconds (or longer), with an excess of 8%
oxygen (Erickson et al., 1985).   Using these conditions, PCDF and PCDD
formation was investigated for the combustion of PCB, trichlorobenzene,
and tetrachloroethylene transformer dielectric fluids.   The results,
including averages and variations of conversion efficiencies, are shown
in Figure 38.   The highest PCDF formation clearly arises for the thermal

destruction of PCB feed and the least efficient production is observed
for the tetrachloroethylene feed. In all runs, the amount of PCDD formed
were much lower than the amount of PCDF formed for the same feed material. The most efficient conversion of dielectric fluid to PCDD was observed for the trichlorobenzene feed while the tetrachloroethylene feed
again resulted in the lowest conversion efficiency. It should be noted,
however, that the combustion conditions used in all of the runs were
based on the optimum conditions for the formation of PCDF from PCB-containing dielectric fluids. Therefore these results do not imply that
they represent the maximum conversion yields since the optimum conditions
for the other fluids may be quite different. Similarly, the formation of
PCDD were studied under optimum conditions for the formation of PCDF from
PCBs.

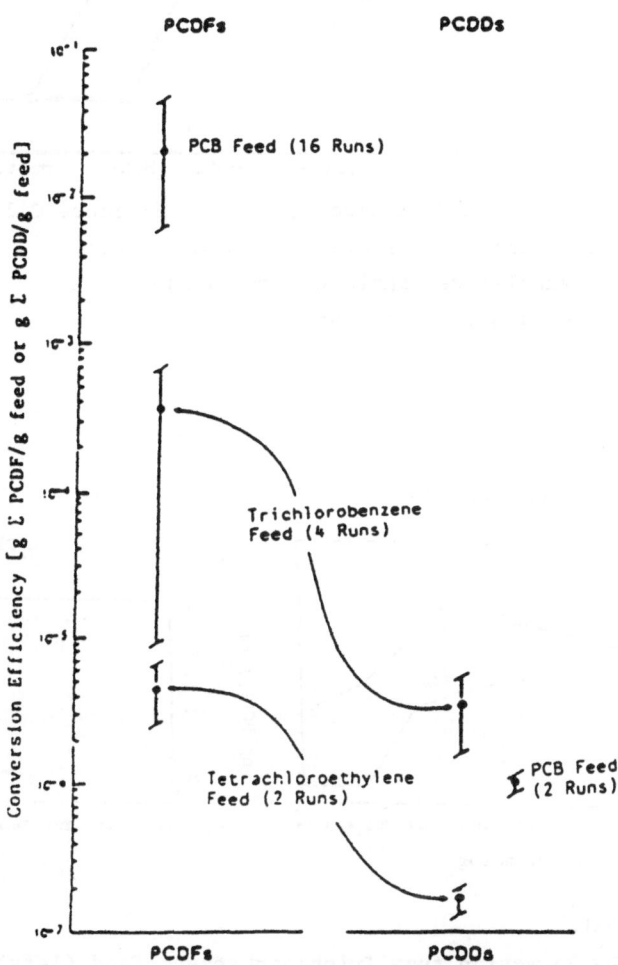

Fig. 38. Comparison of PCDD and PCDF Formation from PCB, Trichlorobenzene, and Tetrachloroethylene Feeds

Figures 39-41 show the distributions of the PCDF and PCDD homologues formed from the thermal degradation of the three dielectric fluids studied.

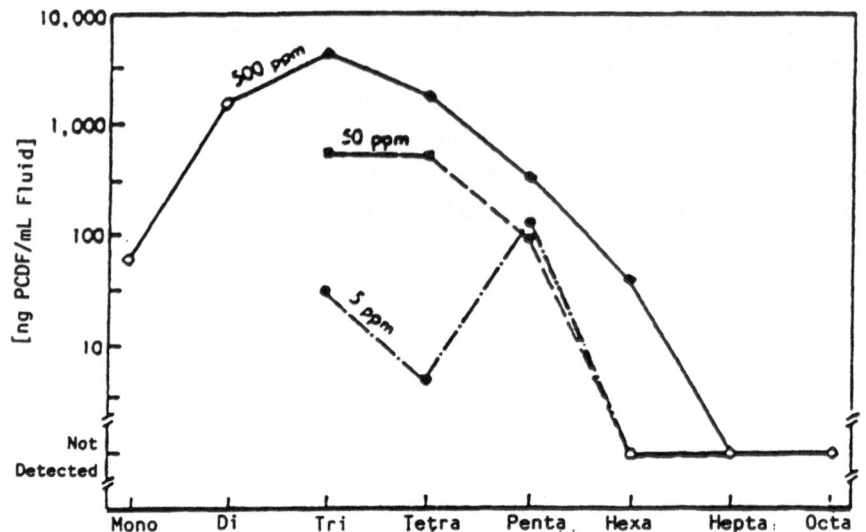

Fig. 39. PCDF Formation from Aroclor 1254 in Silicone Oil by Homolog
Closed symbols are averages of two values;
Open symbols are single determinations;
Missing points are no data.

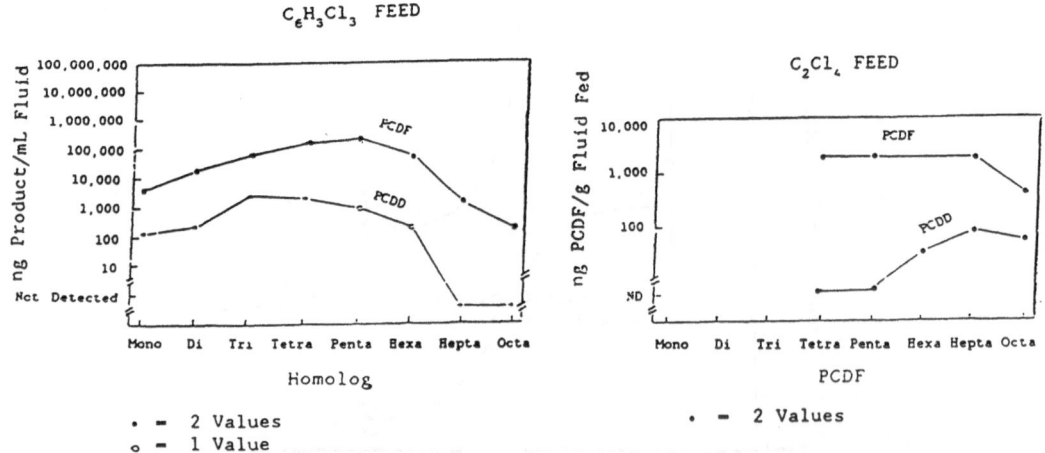

$C_6H_3Cl_3$ FEED

C_2Cl_4 FEED

• = 2 Values
○ = 1 Value

• = 2 Values

Fig. 40. PCDF Formation from Trichlorobenzene Feed (left) and Tetrachloroethylene Feed (right)
(Symbols same as above)

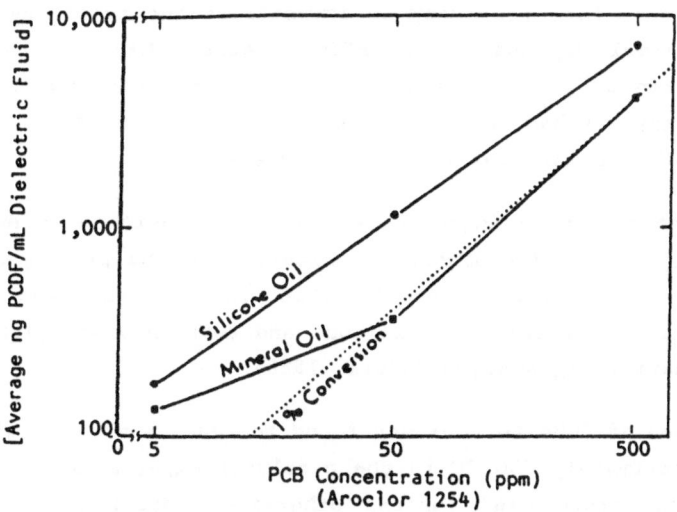

Fig. 41. Average PCDF Formation Versus PCB Concentration
(675 °C for 8 s with 8 % excess oxygen)

The major user of PCBs has been the electric utility industry (Addis,
1986). The occurrence of several PCB fires and the link between such
thermal processes and the formation of toxic by-products, specifically
PCDF, have resulted in efforts to eliminate PCBs in transformers. This
can be accomplished by either retrofilling or complete replacement of the
transformer. The former method is not easily accomplished since it is
very difficult to remove all traces of PCBs and the transformer must be
reclassified as uncontaminated (below 50 ppm PCB) before it can be used.
Even with present technology, removal of residual PCBs is not convenient
and therefore complete replacement of the transformer is often carried
out. Disposal of a transformer requires draining of the PCB and subsequent flushing with solvent. The liquids must then be destroyed in an
authorized incinerator and the transformer itself is sent to a certified
landfill.

A further problem is the roughly 2 million mineral oil transformers which
have been inadvertantly contaminated with PCBs over the years. While
most of these are contaminated at about the 50 ppm level, 10% are
contaminated at the 500 ppm level and therefore must be treated as PCB
transformers.

Studies are therefore necessary to investigate the resulting by-products from the thermal degradation of PCBs, whether they are in their concentrated state or present as contaminants in mineral and silicone oils. Tetrachloroethylene has been proposed as a replacement fluid and PCB contamination of this material is also of interest.

In a study of pyrolysis (high temperature decomposition in an oxygen-depleted atmosphere) and combustion (oxidation in the presence of an open flame and an excess of air), Aroclor 1254, in both its "neat" form and as a contaminant in mineral oil, silicone and tetrachloroethylene, was subjected to thermal degradation (Addis, 1986).

The conversion of PCBs to PCDF was found to reach a maximum at a temperature of approximately 550 °C in the pyrolysis experiments and was therefore the temperature used (unless otherwise indicated in the Tables). All runs were conducted for a 15 minute time period. A series of six runs were made for every set of conditions, all within the same day. The six runs were randomly combined prior to cleanup to yield two composite samples for analysis. Tables 69 and 70 show the resulting levels of PCDF which were formed in the the pyrolysis experiments (Table 69 shows the results in ng/g of total mixture pyrolyzed whereas Table 70 shows the results in ng/g of Aroclor 1254).

Legend to Table 69:

[1] — Each mineral oil sample represents a composite of three separate pyrolyses which were combined prior to analysis to minimize run-to-run variation. All pyrolyses at 550 °C for 15 minutes.

[2] — Includes coeluters on DB-5-column.

[3] — 650 °C.

[4] — Chlorinated fluorenes (?) at 10x higher level.

[5] — 600 °C.

[6] — Mineral oil spiked with PCDFs to test method.

[7] — 23478

[8] — Cl_8DD.

Table 69. PCDF Formed [ng/g] of "Mixture Pyrolyzed: Aroclor 1254 in Insulating Fluids"

Sample	Description	2378[2] Cl$_4$DF	Total Cl$_4$DF	12378[2] Cl$_5$DF	Base Peak Cl$_5$DF	Total Cl$_5$DF	Base Peak Cl$_6$DF	Total Cl$_6$DF	Total Cl$_7$DF	Cl$_8$DF
42582	100% Aroclor 1254	1,300	8,200	-	-	16,000	8,000	8,000	60	-
42583	100% Aroclor 1254	965	5,500	-	-	10,000	5,100	5,100	-	-
50854	5,000 ppm[1] in mineral oil	15.0	45.0	17.0	62.0	165	87.0	162	24.0	<2
50855	5,000 ppm in mineral oil	11.8	33.0	17.2	55.0	171	60.0	205	14.0	0.23
50856	500 ppm in mineral oil	1.6	-	1.8	6.7	17.2	6.3	16.8	<10	-
50857	500 ppm in mineral oil	1.1	3.1	.1.9	5.8	18.2	6.8	18.1	2.2	-
50858	50 ppm in mineral oil	-	-	0.4	1.1	3.1	1.5	3.5	-	-
50859	50 ppm in mineral oil	-	-	0.9	1.7	3.4	1.3	-	-	-
45298	5,000 ppm[3,4] in silicone	9.9	107.0	4.5	-	70.0	2.2	6.9	0.15	-
44295	5,000 ppm[5] in tetrachloroethylene	1.2	2.9	4.6	-	28.0	5.5	12.5	0.9	-
42571	Native mix[6] in mineral oil									
	Added	9.3	-	17.2[7]	-	32.4[7]				56[8]
	Found	9.1	-	17.5	-	23.0			47	

Table 70. PCDF Formed [ng] per Gram of Aroclor 1254 Pyrolyzed in Insulating Fluid

Sample	Description	2378[2] Cl$_4$DF	Total Cl$_4$DF	12378[2] Cl$_5$DF	Base Peak Cl$_5$DF	Total Cl$_5$DF	Base Peak Cl$_6$DF	Total Cl$_6$DF	Total Cl$_7$DF	Cl$_8$DF
42582	100% Aroclor 1254	1,300	8,200	-	-	16,000	8,000	8,000	60	-
42583	100% Aroclor 1254	965	5,500	-	-	10,000	5,100	5,100	-	-
50854	5,000 ppm[1] in mineral oil	3,000	9,000	3,400	12,400	33,000	17,400	32,400	4,800	64
50855	5,000 ppm in mineral oil	2,360	6,600	3,440	11,000	34,200	12,000	41,000	2,8000	46
50856	500 ppm in mineral oil	3,200	-	3,600	13,400	34,400	12,600	33,600	-	-
50857	500 ppm in mineral oil	2,200	6,200	3,800	11,600	36,400	13,600	36,200	4,400	-
50858	50 ppm in mineral oil	-	-	8,000	22,000	62,000	30,000	70,000	-	-
50859	50 ppm in mineral oil	-	-	18,000	34,000	68,000	26,000	-	-	-
45298	5,000 ppm[3] in silicone	1,980	21,400	900	-	14,000	440	1,380	30	-
44295	5,000 ppm[4] in tetrachloroethylene	240	580	960	-	5,600	1,100	2,500	180	-

1 - Each mineral oil sample represents a composite of three separate pyrolyses which were combined prior to analysis to minimize run-to-run variation. All pyrolyses at 550 °C for 15 minutes.

2 - Includes coeluters on DB-5-column.

3 - 650 °C.

4 - 600 °C.

The combustion of PCB-contaminated dielectric fluids was also studied. The results of these studies are given in Table 71. The conversion for the tri- to heptachlorodibenzofuran isomer groups was found to be near optimum for a combustion wall temperature of 550 °C and a residence time of 3 seconds. Mineral oil contaminated with PCB levels of 5000, 500, and 50 ppm produced similar levels of tri-, tetra-, and pentachlorinated furans (relative to the amount of PCBs present) and in all cases no Cl_6DFs or Cl_7DFs were detected. The PCB-contaminated tetrachloroethylene mixtures all produced higher yields of the more chlorinated PCDF isomers than observed for the mineral oil tests. Also inter-

Table 71. Combustion of Aroclor 1254 in Insulating Fluids at 550 °C; 3 Seconds

Solvent	1254 in sol. Combusted [μg/ml]	PCB's Destroyed %	% PCDF Formed[1]				
			Cl_3	Cl_4	Cl_5	Cl_6	Cl_7
Mineral Oil	5,000	82-88	0.72	0.58	0.17	-[2]	-[2]
Mineral Oil	500	87-92	0.4	0.33	0.084	-[2]	-[2]
Mineral Oil	50	88-90	0.3	0.3	0.056	-[2]	-[2]
C_2Cl_4	5,000	68	0.44	1.50	1.35	0.33	0.50
C_2Cl_4	500	92	0.24	0.7	0.66	0.23	0.70
C_2Cl_4	50	72	0.44	1.32	1.0	0.38	0.16
Silicone[3]	-	-	-	-	-	-	-

1   Based on PCB in feed
2   None detected
3   Run unsuccessful; equipment plugged with SiO_2.

esting to note was the fact that in both the pyrolysis and combustion of tetrachloroethylene contaminated with Aroclor 1254, different Cl_4DF- and Cl_5DF isomers were produced than for either the "neat" Aroclor or the contaminated mineral oil.

It must be stressed that these results cannot be used to explain results of accidental PCB fires since the conditions for such occurrences are completely random. However, these results do imply that the by-products of PCB fires will most likely include PCDF.

The arcing and subsequent burning of an electrical transformer in a Binghamton State office building on February 5, 1981, produced the first recognized case in which toxic products, PCDD and PCDF, were produced outside of the laboratory from the combustion of chlorinated aromatic compounds in the transformer fluid (Schecter, 1986). Approximately 180-200 gallons of the transformer fluid, 65 % Aroclor 1254 and 35% tri- and tetrachlorinated benzenes, leaked from the overheated transformer in the basement of the building. The ensuing fire sent a dense cloud of smoke throughout the ventilation system of the building, resulting in con-tamination of the entire building. Initial sampling revealed levels of by-products to be 2,168,000 ppb of PCDF, 20,000 ppb of PCDD, and 50,000 ppb of polychlorinated biphenylenes. As a result of this incident, the U.S. Environmental Protection Agency declared that PCB-containing transformers must be replaced in many public buildings within the following few years. The impact on the health of exposed people was not the only consideration in making this ruling. The building itself was built at a cost of about 17 million dollars; the estimated cost of cleanup was 40 million dollars.

The assessment of the impact of such accidents upon the environment as well as mankind is greatly hindered by the difficulties encountered in the analysis of the emissions. Tables 72 and 73 show the results from two different reputable laboratories which were obtained for analyses of the same samples or split samples. In Table 72, one laboratory reports a level of 2,3,7,8-Cl_4DF which is one-fifth that reported by the other laboratory and yet for the total Cl_6DF concentration, the former reports a level which is 17 times that as determined by the latter. This inconsistency of results is even more serious than that reported in Table 73 where at least one of the two laboratories gave consistently higher results than the other. The differences in the reported results does not reflect the inefficiency of the laboratories involved in determining the levels of PCDF present, but rather it illustrates the difficulties inherent with such analyses.

Table 72. Comparison of Battelle vs. New York State Department of Health (NYSDOH)

Sample	2,3,7,8-Cl_4DF	Total Cl_4DF	Total Cl_5DF	Total Cl_6DF
BSOB Wipe, Stair Tread PCDD W-13 (NYSDOH 43968) Floor 16-17	5,900	45,000	99,000	41,000
BSOB Wipe, Stair Tread PCDD W-14 (Battelle) Floor 16-17	29,500	145,000	117,000	60,700
BSOB Wipe, Vinyl Wall W-7 (NYSDOH 43969)	1,100	13,000	8,500	2,000
BSOB Wipe, Vinyl Wall PCDD W-6 (Battelle)	1,320	4,980	820	120

Table 73. Binghamton State Office Building Test Plan: Measurements of Residual Contaminants (Report September 5, 1984); Polychlorinated Dibenzofuran Analysis Laboratory Comparison (Single extract)

Substances	Stalling 1981 [$\mu g/g$]	Rappe 1981 [$\mu g/g$]	Factor
2,3,7,8-Cl_4DF	3.7	12	3.2
Cl_4DF	5.4	28	5.1
Cl_5DF	107		
Cl_6DF	512		
Cl_7DF	102	460	4.5
Cl_8DF	30	40	1.3

The analytical results of some major PCB transformer and capacitor fires which occurred within the U.S. are summarized in Table 74 (des Rosiers, 1986; des Rosiers, 1987). This list is representative of the cases in which efforts were made to determine the levels of PCDD and PCDF present. Of the fires listed, the Binghamton transformer fire is the only one in which PCDD were clearly identified. The presence of chlorinated benzenes in the transformer fluid and the intense heat of the fire provided conditions similar to those in earlier laboratory studies which produced PCDD from the pyrolysis of chlorinated benzenes.

Wipe samples from the Columbus fire showed a very small amount of PCDD, however none were quantitatively detected in the corresponding soot samples analyzed. The capacitor oil was known to contain Aroclor 1242, however it is not known if any chlorinated benzenes were present to act as precursors for the formation of the trace levels of PCDD found. The transformer involved in the San Francisco fire contained 100% Aroclor 1242 and no PCDD were detected in the soot samples analyzed. Thus the presence of chlorinated benzenes appears to be necessary for the formation of PCDD.

Table 74. Some Major Fires Involving PCB Transformers and Capacitors in the United States

Place	Analyses	PCDFs	PCDDs	Comments
Binghamton, NY - State Office Building (2-15-81) (Transformer)	Soot Samples	Total PCDF: 765-2160 µg/g 2,3,7,8-Cl$_4$DF: 12-270 µg/g	Total PCDD: 20 µg/g 2,3,7,8-Cl$_4$DD: 0.6-2.8 µg/g	Transformer filled with 1060 gals. of Pyranol (65% PCBs) (Aroclor 1254) & 35% tri- and tetrachlorinated benzenes. Very hot fire, explosion & rapture of transformer with release of 180 gals. of fluid.
Cincinnati, OH - Elementary School (12-3-80) (Capacitor)	Wipe Samples	ND	ND	Only 0.088 ml of PCBs (Aroclor 1254) was involved in the capacitor but the entire basement was contaminated at a mean level of 771 µg/100 cm^2.
Boston, MA - Office Building (1-82) (Transformer)	Soot Samples	Total PCDF: 165 µg/g 2,3,7,8-Cl$_4$DF: 3 µg/g	ND	Transformer containing Aroclor 1254
Miami, FL - Vault (4-3-82) (Transformer)	Soot Samples and Fire Residues	PCDF: 100-1000 µg/g	ND	Underground transformer vault.

Table 74. (Continued)

Place	Analyses	PCDFs	PCDDs	Comments
San Francisco, CA (Adj to High Rise) (5-15-83) (Transformer)	Wipe Samples	Cl_4DF: 15.6 µg/g		ND 3 transformers with 100% Aroclor 1242. No chlorobenzene. 50 gals. were released & smoldered for 8 hours after discovery.
Columbus, OH - Office Building (3-84) (Capacitor)	Wipe Samples	Total PCDF: up to 10,000 ng 2,3,7,8-Cl_4DF: up to 1,000 ng	Total PCDD: up to 44 ng 2,3,7,8-Cl_4DD: up to 3.0 ng	Capacitor with Aroclor 1242.
	Soot Samples	Total PCDF: 10.9 µg/g	Total PCDD: trace (Cl_4 to Cl_8). Trace amount of chlorobenzene. (Detection limit at µg/g level).	

The lack of analytical data regarding the nature of oil in the transformers or capacitors involved in fires makes evaluating data from wipe and soot samples difficult. Table 75 lists the results of analyses of the oil samples taken from fires in three parts of the U.S. No PCDD were detected in any of the oils (detection limit of 10 ng/g) and total PCDF were typical of the levels found in stock materials previously studied. Normal use of electrical equipment therefore does not appear to produce PCDF. PCBs and diluents such as chlorinated benzenes must be subjected to thermal stress in the presence of oxygen to produce PCDD and PCDF.

Table 75. Analysis of Fluids Involved in Transformer Fire Incidents

	Miami, FL[1] (May 29, 1984) [μg/g])	Binghampton, NY[2] (Feb. 5, 1981) [μg/g]	Chicago, IL[3] (Sept. 28, 1983) [μg/g]
PCDD	ND	ND	ND
Σ PCDF	6.0	16.2	3.0
Cl_4DF	0.31	0.48	<0.01
Cl_5DF	0.78	1.1	0.053
Cl_6DF	2.2	11.0	0.47
Cl_7DF	1.3	3.2	1.4
Cl_8DF	1.4	0.41	1.1
Σ Chlorobenzenes	430,000	350,000	160,000
Trichlorobenzenes	260,000 [a]	230,000 [a]	100,000 [a]
Tetrachlorobenzenes	140,000 [a]	110,000 [a]	51,000 [a]
Pentachlorobenzenes	29,000 [a]	13,000 [a]	5,500 [a]
Hexachlorobenzenes	900 [c]	45 [c]	64 [b]
Σ PCB	410,000	580,000	110,000 [b]
Monochlorobiphenyls	320 [b]	230 [b]	<100 [b]
Dichlorobiphenyls	1,100 [b]	730 [b]	<150
Trichlorobiphenyls	2,600 [b]	3,500 [c]	<200
Tetrachlorobiphenyls	2,000 [b]	88,000	<400
Pentachlorobiphenyls	31,000 [b]	300,000	11,000
Hexachlorobiphenyls	160,000 [b]	160,000 [c]	49,000 [b]
Heptachlorobiphenyls	170,000 [b]	24,000	45,000
Octachlorobiphenyls	32,000 [b]	<800	7,300
Nonachlorobiphenyls	14,000	<1,200	<1,600
Decachlorobiphenyls	<2,800 [b]	<3,500	<4,700

[a] Results of analysis at a dilution of 0.1 mg of oil/ml hexane.

[b] Results of analysis at a dilution of 2.5 mg of oil/ml hexane.

[c] Results of analysis at a dilution of 0.25 mg of oil/ml hexane.

[1] Askarel Type A: 60% PCBs with 60% chlorine (Aroclor 1260) and 40% trichlorobenzene mixture.

[2] Askarel Type D: 70% PCBs with 54% chlorine (Aroclor 1254) and 30% trichlorobenzene mixture.

[3] Mineral Oil with 25% PCBs (Aroclor 1260).

Lindane (γ-hexachlorocyclohexane, HCH) manufacturing wastes have been converted to predominantly 1,2,4-trichlorobenzene (TCB) which after purification by distillation was transformed to 1,2,4,5-tetrachlorobenzene for use in preparing 2,4,5-trichlorophenol (des Rosiers, 1987). The process involved the use of low temperature pyrolysis (200-240 °C) of the HCH wastes in the presence of a carbon catalyst in a closed reaction mantle. Residues from the mantle and the trichlorobenzene still were analyzed and were found to contain PCDD and PCDF (see Table 76). The presence of oxygen during the pyrolysis and/or distillation of chlorinated aromatics can lead to the formation of dioxins and furans.

Table 76. Dioxin Isomer Differentiation for Pyrolyzed HCH Residues and TCB Stillbottoms

Dioxin Isomer	Concentration [mg/kg] [ppm]
$2,3,7,8-Cl_4DD$	0.3
Cl_4DD	12
$1,2,3,7,8-Cl_5DD$	7
Cl_5DD	200
$1,2,3,4,7,8-Cl_6DD$	45
$1,2,3,6,7,8-Cl_6DD$	150
$1,2,3,7,8,9-Cl_6DD$	65
Cl_6DD	680
$1,2,3,4,7,8,9-Cl_7DD$	1,400
$1,2,3,4,6,7,8-Cl_7DD$	3,000
Cl_8DD	7,600

At a metal recovery site, salvaging of transformers was performed from 1972 to 1983 (des Rosiers, 1987). PCB-contaminated oil was burned in a stove and later analyses revealed the formation of PCDD and PCDF. The results of analyses performed on various samples at this site are given in Table 77.

Table 77. PCBs/PCDD/PCDF Analysis of Various Samples from the Strandley Site in Washington

Sample	PCBs [ppm]	PCDF [ppm]		PCDD [ppm]	
Soil	5 - 1,200	---		---	
Sediment	5 - 88	---		---	
Oil Sludge	5,600	---		---	
Water	4.7	---		---	
Burn area under Container	---	---		8.7	Cl_8DD
Ash on roadway bank	---	13	Cl_4DF	ND	Cl_4DD
		62	Cl_5DF	4.2	Cl_5DD
		91	Cl_6DF	1.8	Cl_6DD
		90	Cl_7DF	25	Cl_7DD
		55	Cl_8DF	43	Cl_8DD
Ash from stove in shop		545	Cl_4DF	178	Cl_4DD
		2,000	Cl_5DF	845	Cl_5DD
		2,300	Cl_6DF	1,100	Cl_6DD
		2,000	Cl_7DF	2,000	Cl_7DD
		320	Cl_8DF	890	Cl_8DD

At another scrap metal salvaging site, a small incinerator was used to burn PCBs and also provide heat to the surrounding residential community. Dust samples taken from within the building housing the incinerator were found to contain PCDF and PCDD, the latter being present in lower levels as indicated in Table 78.

Table 78. Dioxin and Furan in Dust Samples from Wire Recovery Site, Pittsburgh, PA

Homologue	Concentration [ng/g]
Cl_4DD	17 - 59
$2,3,7,8-Cl_4DD$	0.79 - 2.5
Cl_4DF	413 - 1,232
$2,3,7,8-Cl_4DF$	74 - 241

The U.S. Environmental Protection Agency (EPA) has concluded that:

1. PCDD and PCDF are not produced in PCB-containing transformers under normal operating conditions
2. Electrical arcings in transformers do not result in the formation of PCDD and PCDF
3. The amount of PCDF formed as well as the particular isomers produced are related to the concentration and types of PCBs present in the transformer fluid
4. Chlorinated benzenes must be present in transformer fluids for PCDD to be formed.

Considerable controversy has been generated by instances such as the accidental PCB transformer fire which occurred in Reims, France on January 14, 1985 (Abenhaim et al., 1987). In such cases, questions raised include:

1. What are the chemical substances involved?
2. At what concentrations are these substances present?
3. What are the exposure risks associated with the types and levels of potentially toxic materials present?

In order to help provide some of the answers to these questions, samples are invariably collected for analysis. The results of these analyses can further complicate matters as exemplified in the case of the fire in Reims. The analytical results obtained from several laboratories which analyzed the same soot sample are given in Table 79. The levels reported show tremendous deviation from one another. Analytical techniques employed today are extremely sensitive, however are not completely free of interference or errors in interpretation. There is an obvious need for better inter-laboratory analyses, most of which can be attributed to the lack of standard, validated analytical protocols.

Table 79. PCDD/PCDF Analyses (Fire in Reims); Interlaboratorial Comparison

PCDD/PCDF	Lab. #3 Sample 1 $[\mu g/m^2]$	Lab. #4 Sample 5 $[ng/g]$	Lab. #5 Sample 5 $[ng/g]$	Lab. #6 Sample 5 $[ng/g]$
Cl_4DD		35,000	220	490
Cl_5DD	330	58,000	420	820
Cl_6DD	> 500	25,000	25	190
Cl_7DD	> 900	8,000	110	4
Cl_8DD	70	3,000	160	14
Cl_4DF	Not analysed	56,000	45	ND
Cl_5DF	Not analysed	81,000	150	550
Cl_6DF	Not analysed	7,000	490	45
Cl_7DF	85	8,000	210	Not detectable
Cl_8DF	65	8,000	350	Not detectable

2.6 COMBUSTION OF WOOD

The emission of PCDD and PCDF resulting from the residential combustion of untreated wood has been studied (Clement et al. 1985). Samples of bottom and chimney ashes were collected from two wood burning stoves and an open fireplace. A sample of bottom ash was also collected from an out-door open-air burning of untreated wood. The samples collected are summarized in Table 80 and the concentrations of PCDD and PCDF found in the various ash samples are given in Table 81. PCDD were detected in all of the samples analyzed. PCDF were found in all of the ashes analyzed with the exception of one of the chimney ash samples. The relative amounts of the different congeners varied considerably from sample to sample, including the samples taken from the same wood burning stove (WB1-a and WB1-b). Perhaps the most important result of this study is the observation that the total levels of PCDD and PCDF found in the ashes are significantly lower than the levels typically observed for fly ash samples taken from municipal waste incinerators.

Table 80. Sample Description

Sample ID	Sample Type	Description	Type of Wood Burned
OA	Bottom Ash	Outside Open-Air Burning	Primarily Oak some paper used
CA-WB1	Chimney Ash	From Chimney above damper of WB1-a	same as OA
WB1-a	Bottom Ash	From Wood Burning stove	same as OA
WB1-b	Bottom Ash	Same stove as WB1-a, one year later	same as OA
WB2	Bottom Ash	From Wood Burning stove Different stove and location from WB1-a	75-80 % poplar
CA-FP	Chimney Ash	From Chimney above damper of FP	Unspecified
FP	Bottom Ash	Fireplace in home	Same as CA-FP

Table 81. PCDD/PCDF from Wood Burning

* Number in parentheses are numbers of isomers observed

ND — Not detected at average single-isomer detection limits of 10 ppt [pg/g] for Cl_4- and Cl_5 congeners, and 50 ppt for Cl_6-, Cl_7- and Cl_8DD/DF

Congener	Concentrations in [ng/g] [ppb]*						
	OA	CA-WB1	WB1-a	WB1-b	WB2	CA-FP	FP
Cl_4DD	0.8(5)	ND	ND	0.1(4)	0.1(2)	ND	ND
Cl_5DD	4.2(a)	ND	ND	3.0(10)	0.2(4)	0.5(6)	ND
Cl_6DD	7.2(6)	ND	ND	10(6)	0.7(6)	1.7(6)	0.3(6)
Cl_7DD	11(2)	0.1	0.3(2)	1.2(2)	0.5(2)	0.5(2)	2.0(2)
Cl_8DD	10	0.2	2.6	0.9	0.1	0.4	3.1
Σ PCDD	33	0.3	2.9	15	1.6	3.1	5.4
Cl_4DF	2.2(11)	ND	9.1(9)	0.4(12)	0.1(1)	0.3(8)	ND
Cl_5DF	7.6(11)	ND	2.2(7)	4.6(9)	0.2(3)	1.4(8)	ND
Cl_6DF	8.2(10)	ND	1.0(2)	9.3(8)	0.5(7)	1.7(5)	0.1(4)
Cl_7DF	11(4)	ND	0.7(1)	1.0(4)	0.3(3)	0.4(3)	0.4(4)
Cl_8DF	1.7	ND	ND	0.1	ND	0.1	0.1
Σ PCDF	31	ND	13	15	1.1	3.9	0.6

Chimney soot samples from house heating in the area of Bayreuth, Germany, have been analyzed for PCDD/PCDF (Thoma, 1988a; Thoma 1988b). High concentrations of PCDD/PCDF could be detected in chimney soots from 15 heating systems, two ovens fueled with wood/coal, 4 central heatings fueled with wood, and 9 ovens fueled with wood. The average concentrations of PCDD/PCDF of chimney soot are shown in Table 82. A difference between oven and central heating was not detected. The PCDD/PCDF contents of chimney soot from combined wood/coal burning was 5 times less than those of wood and coal burning stoves; but definite conclusions cannot be drawn since only two samples have been analyzed. The results (for fossil fuels, see Chapter 2.7) showed that house heating is a major source of PCDD/PCDF and that concentrations from wood and coal burning systems are in the range of fly ash from municipal waste incinerators.

Table 82. Average Concentrations of PCDD/PCDF in Chimney Soot [ppt]

Substance	Wood	Wood	Wood/Coal
Cl_4DD	31,279.3	10,301.4	2,410.0
Cl_5DD	21,167.3	25,979.5	1,325.0
Cl_6DD	23,264.3	18,309.0	609.1
Cl_7DD	39,689.0	13,635.0	465.5
Cl_8DD	51,480.8	10,639.2	312.3
Cl_4DF	162,696.5	76,279.7	31,300.0
Cl_5DF	60,400.0	91,048.0	8,065.0
Cl_6DF	6,953.0	56,885.9	2,335.0
Cl_7DF	13,848.0	13,281.5	600.6
Cl_8DF	2,832.3	7,268.2	146.1
$2,3,7,8-Cl_4DD$	310.1	420.7	121.2
$1,2,3,7,8-Cl_5DD$	1,429.1	860.9	107.3
$1,2,3,4,7,8-Cl_6DD$	886.9	967.7	21.8
$1,2,3,6,7,8-Cl_6DD$	1,415.2	1,399.6	63.6
$1,2,3,7,8,9-Cl_6DD$	1,262.5	1,205.0	53.5
$1,2,3,4,6,7,8-Cl_7DD$	22,065.0	7,507.4	253.3
$2,3,7,8-Cl_4DF$	11,294.3	9,374.2	1,900.0
$1,2,3,7,8-Cl_5DF$	19,658.8	10,879.5	646.6
$2,3,4,7,8-Cl_5DF$	34,818.8	10,589.0	1,287.8
$1,2,3,4,7,8-Cl_6DF$	9,422.4	4,248.6	273.1
$1,2,3,6,7,8-Cl_6DF$	8,672.9	3,262.7	251.7
$1,2,3,7,8,9-Cl_6DF$	588.9	392.2	18.3
$2,3,4,6,7,8-Cl_6DF$	4,665.1	2,200.7	100.6
$1,2,3,4,6,7,8-Cl_7DF$	8,000.6	9,230.9	377.9
$1,2,3,4,7,8,9-Cl_7DF$	1,008.3	909.2	48.1

Most of earlier studies on treated wood were conducted in pilot scale incinerators although a few studies were done on open-air burning (summary see: Ontario Ministry of the Environment, 1985; NATO/CCMS Report No. 172). In pilot scale studies, the PCDD emissions were found to decrease with increasing temperature and also with increasing residence time. These findings are similar to those reported for hazardous waste incinerators. There exists conflicting data on the combustion of untreated wood: Rudling et al. could not detect Cl_4DD and Cl_4DF whereas in later studies PCDD and PCDF were found in numerous burnings of wood carried out under a range of conditions.

2.7 FOSSIL FUEL COMBUSTION

There is conflicting and limited data regarding PCDD/PCDF formation by the burning of fossil fuels, mostly of coal. In some investigations polychlorinated dioxins were detected in ppb-amounts (Kimble and Gross, 1980; American Society of Mechanical Engineers, 1981).

PCDD/PCDF are probably formed at low levels during burning of untreated wood after being logged (Lao). Compared to municipal waste incinerators the emissions are found to be in the same range (Umweltbundesamt, FRG, 1985).

In the soot of private heating systems dioxins were found to be present but at significantly lower levels than those detected in fly-ash (Table 83) (Thoma, 1987; Hutzinger and Fiedler, 1987).

Table 83. Concentration of PCDD/PCDF in the Soot of Home Heating; Concentrations in [ppb]

Compound	MWI Fly Ash	Home Heating	
		Soot Oil	Soot Coal/Wood
Cl_4DD	19.06	3.90	1.54
Cl_5DD	37.25	0.41	5.13
Cl_6DD	115.49	3.06	5.20
Cl_7DD	275.88	1.37	2.76
Cl_8DD	598.69	1.05	2.37
Cl_4DF	79.45	28.91	50.80
Cl_5DF	120.30	16.55	30.04
Cl_6DF	116.34	6.24	11.67
Cl_7DF	108.24	1.78	3.23
Cl_8DF	42.90	0.33	0.53
$2,3,7,8-Cl_4DD$	0.60	0.10	0.21
$2,3,7,8-Cl_4DF$	2.47	1.07	1.92

Based on the results of these few studies, it appears that only coal-fired plants produce significant levels of PCDD and PCDF. It also appears that the total PCDF emission is greater than the total amount of PCDD emitted. Further work must be done to confirm these findings.

Chimney soot samples from house heating have been analyzed for its content of PCDD/PCDF (Thoma, 1988a; Thoma, 1988b). The average concentrations found in soot samples from 21 central heating systems, 7 ovens both fueled with oil and 7 ovens fueled with coal are given in Table 84. All samples from oil burning systems had detectable levels of PCDD/PCDF, but the concentrations from central heating were, on average, 10times less than in the samples of oil ovens. Generally, individual values varied within a factor of 10 to 100. In comparison to fly ash from municipal waste incinerators the concentrations in chimney soot were 10 to 100times lower.

The analyses of coal fired stoves revealed high levels of PCDD/PCDF and were in the range of those found in municipal waste incinerators and thus 10 to 100times higher than those of oil burning.

Table 84. Average Concentrations of PCDD/PCDF in Chimney Soot from Oil and Wood Fired Home Heating Systems [ppt]

Substance	Oil (Central Heating)	Oil (Oven)	Coal (Oven)
Cl_4DD	276.1	1633.8	11949.4
Cl_5DD	274.1	1681.0	7961.7
Cl_6DD	220.2	2393.4	11953.3
Cl_7DD	210.9	2712.9	11447.3
Cl_8DD	219.1	2,353.6	15,416.4
Cl_4DF	2,040.9	12,714.7	73,321.9
Cl_5DF	1,395.1	10,925.6	67,809.0
Cl_6DF	575.5	5762.1	25,493.9
Cl_7DF	243.3	2,444.4	6,329.2
Cl_8DF	98.8	598.2	839.5
2,3,7,8,-Cl_4DD	10.3	54.4	319.6
1,2,3,7,8-Cl_5DD	14.4	177.5	838.8
1,2,3,4,7,8-Cl_6DD	8.5	101.2	434.5
1,2,3,6,7,8-Cl_6DD	20.7	154.4	610.2
1,2,3,7,8,9-Cl_6DD	14.3	202.5	589.3
1,2,3,4,6,7,8-Cl_7DD	115.5	1,541.0	6,575.1
2,3,7,8-Cl_4DF	175.4	774.3	8,407.5
1,2,3,7,8-Cl_5DF	116.3	831.4	5,333.2
2,3,4,7,8-Cl_5DF	154.1	971.7	9,120.5
1,2,3,4,7,8-Cl_6CDF	79.4	732.6	3,292.6
1,2,3,6,7,8-Cl_6DF	58.8	655.3	3,176.5
1,2,3,7,8,9-Cl_6DF	6.3	68.7	211.7
2,3,4,6,7,8-Cl_6DF	34.3	411.4	1,339.0
1,2,3,4,,6,7,8-Cl_7DF	162.4	1,653.0	4,142.6
1,2,3,4,7,8,9-Cl_7DF	14.0	151.4	393.4

Emissions of fossil fuel plants revealed concentrations of the lower chlorinated dibenzofurans being 4-6times the concentrations of the respective dioxins. Within both classes lower chlorinated isomers predominated the higher chlorinated compounds.

In studies on emissions from coal- and oil-fired power plants covering the years 1978 to 1983 levels of PCDD and PCDF found in the fly ash were much lower than the levels reported in most municipal incinerator studies. No dioxins or furans were detected in the flue gas samples analyzed by Haile, however in a later study of two Canadian coal-fired plants, low levels were detected (Table 85).

Table 85. Total Particulate - Phase Stack Emission Concentrations of PCDD and PCDF from Two Coal-Fired Sources [*] [dscm]

| Source | Particulate-Borne Stack Emissions | | | | | | | | | | | |
| | PCDD | | | | | | PCDF | | | | | |
	Cl_4	Cl_5	Cl_6	Cl_7	Cl_8	Σ DD	Cl_4	Cl_5	Cl_6	Cl_7	Cl_8	Σ DF
Large coal-fired power plant	0.1	0.1	ND	0.1	ND	0.3	0	1	2	ND	ND	3
Smaller coal-fired central heating facility	ND	ND	ND	ND	ND	ND	ND	ND	ND	ND	ND	ND

[*]   measurements done in 1981
[dscm]   dry standard cubic meter
ND   Not detected

2.8. COMBUSTION OF AUTOMOBILE FUELS

In a recent report, halogenated aromatic compounds, predominantly brominated, were identified in the emissions of automobiles fueled by leaded gasoline (Müller and Buser, 1986). Dibromoethane (DBE) and dichloroethane (DCE), which are added to leaded gasoline as scavengers to prevent deposition of lead compounds in engines, are believed to become involved in the many side reactions which take place in the engine. In unleaded gasoline, where no DBE or DCE are added, no halogenated aromatics were detected.

In another recent study, the presence of PCDD and PCDF in motor oil from diesel and gasoline fueled vehicles was reported (Ballschmiter et al., 1986). Incomplete combustion and the presence of a chlorine source in the form of additives in the oil or the fuel such as dichloroethane or pentachlorophenate may lead to the formation of PCDD and PCDF. The isomeric patterns of the dioxins and furans found in the motor oils were compared to those observed from municipal incinerator emissions. Some similarities were observed, however this does not yield any understanding of the mechanisms involved. The levels of PCDD and PCDF were not reported and thus the impact of this source to the environment is not known.

A comparison of the emissions resulting from the use of leaded and unleaded gasoline in automobiles was performed in an attempt to link the formation of PCDD and PCDF to the chlorinated scavenger additive (Marklund et al., 1987). The tests were carried out using two cars which were run on unleaded gasoline which contained no dichloroethane scavenger and on four cars run on leaded gasoline. The leaded gasoline was actually the same unleaded fuel but with the addition of 0.15 g/L of tetraethyllead and 0.1 g/L of dichloroethane. The normal scavenger mixture of DBE and DCE was not used in order to avoid the formation of mixed chloro/bromo dioxins and furans which would complicate the analyses due to the large number (>5,000) of possible isomers. Carbon monoxide, hydrocarbon, and NO_x emissions were monitored and showed that all cars tested were in good condition and met the Swedish emission requirements.

The sampling was performed over two 12.4 km running cycles giving at total run distance of 24.8 km. The emissions resulting in the tests are summarized in Table 86. For the cars fueled by unleaded gasoline, no PCDD or PCDF were detected above the maximum detection limit of 0.3 ng. PCDD and PCDF were identified in all of the test cars run on leaded gasoline with the DCE scavenger. The total levels of PCDD and PCDF produced varied from car to car and therefore it is difficult to assess a representative level of emission resulting from the fuel combustion process. It should be noted that the same motor oil was used in all test cars thereby eliminating the possibility that the dioxins and furans were formed from additives or contaminants in the oil. The dichloroethane is most likely the source of the PCDD and PCDF formed, although it is possible that they were also formed in the cars run on unleaded gasoline and subsequently destroyed in the catalytic converter. The latter hypothesis seems unlikely but should be verified.

Based on their findings, the total emissions of PCDD and PCDF from all cars fueled by leaded gasoline would be equivalent to 2 to 20 municipal solid waste incinerators of average size and technology. These emissions are only considered for the exhausts produced. The levels of PCDD and PCDF in motor oil and in particulates found in mufflers still need to be estimated.

Table 86. PCDD and PCDF in Car Exhausts [ng/24.8 km]

Compounds	Cars running on non leaded gasoline (n = 2)	Cars running on leaded gasoline (n = 4)
$2,3,7,8-Cl_4DF$	ND	0.6 - 13
$\Sigma\ Cl_4DF$	ND	10 - 200
$2,3,7,8-Cl_4DD$	ND	< 0.05 - 0.3
$\Sigma\ Cl_4DD$	ND	3 - 90
$1,2,3,7,8-Cl_5DF$	ND	0.5 - 4.6
$2,3,4,7,8-Cl_5DF$	ND	0.3 - 3.9
$\Sigma\ Cl_5DF$	ND	5 - 46
$1,2,3,7,8-Cl_5DD$	ND	0.5 - 3.5
$\Sigma\ Cl_5DD$	ND	6 - 98
TCDD Equivalents	ND	0.8 - 13

Detection limits:

$2,3,7,8-Cl_4DF$	0.2	ng
$2,3,7,8-Cl_4DD$	0.05	ng
$1,2,3,7,8/2,3,4,7,8-Cl_5DF$	0.2	ng
$1,2,3,7,8-Cl_5DF$	0.3	ng

The exhaust emissions of six cars, driven through three or four cycles of the EPA urban driving cycle, have been analyzed for PCDD/PCDF (Bingham et al., 1988). Five of the cars were run on New Zealand high octane petrol containing a lead additive (0.45 g/L tetramethyllead, 0.22 g/L dichloroethane and 0.2 g/L dibromoethane). The sixth car was run on low octane lead additive-free petrol. The exhaust gases were proportionally (0.03 to 0.05 of the total inlet flow) sampled through a modified EPA Method 5 sampling train. Total Cl_4DF emissions have been detected in the range from Not determined - 1,5 ng/km for cars run on leaded fuel.

Exhaust gases from three different vehicles (two cars fueled with leaded and unleaded gasoline, respectively and a heavy diesel truck) were investigated for the presence of polybrominated dibenzo-p-dioxins (PBDD) and polybrominated dibenzofurans (PBDF) (Haglund et al., 1988). The results (Table 87) showed that leaded gasoline yields much more Br_4DD/Br_4DF than unleaded gasoline. The authors concluded that the dibromoethane added to this type of gasoline probably act as a halogen source. Generally, the levels of tetrabrominated dibenzodioxins and dibenzofurans found in this investigation were higher than the levels of the respective tetrachlorinated species reported in the above mentioned report (Marklund et al., 1987)

Table 87. Analysis of PBDD/PBDF in Vehicle Exhaust

Compounds (unit)	Car leaded gasoline	Car unleaded gasoline	Car diesel oil	Detection limits gasoline cars [ng/km]	Detection limits diesel truck [ng/L fuel]
Br_2DF [ng/km]	1,100	ND	ND	2	28
Br_2DD [ng/km]	ND	ND	ND	20	260
Br_3DF [ng/km]	180	1.4	ND	0.08	1
Br_3DD [ng/km]	19	ND	ND	0.4	5
Br_4DF [ng/km]	23	0.24	ND	0.03	0.4
Br_4DD [ng/km]	3.2	ND	ND	0.3	3
Br_5DF [ng/km]	0.98	ND	ND	0.03	0.4

2.9 PVC AS CHLORINE DONOR

PVC as chlorine donor in municipal waste incineration (Wilson, 1977) has been discussed discussed for a long time. New studies (Hawley-Fedder, 1984) and samples of fly ash from municipal waste incinerators fueled with high contents of PVC (Karasek et al., 1983) could not confirm PVC as source for the formation of high concentrations of PCDD/PCDF. Copper smelters and plants for metal recovery are to be considered separately, where cables covered with PVC are thermolyzed in the presence of e.g. copper (Hutzinger and Fiedler; 1988). The levels of PCDD/PCDF (in 100 ppb to ppm-range) found in the emissions (Eberhard and Friege, 1987; Bröker, 1987) may be due to catalytic inference of the metals present.

To get some more information on PVC as a source for the formation of PCDD/PCDF the products formed during combustion and pyrolysis of pure, commercial polyvinylchloride and PVC cable sheatings have been analyzed (Christmann et al., 1988). In simple laboratory experiments 2 g of a PVC sample were incinerated with a laboratory gas burner, the combustion products were condensed and collected on the inner walls of a cooled glas funnel placed upside down above the sample. The soot deposits were removed with toluene. Pyrolysis experiments were performed in a modified Schöninger apparatus with 50 g of PVC, heated to about 950 °C for 10 min in either air or nitrogen atmosphere. In another experimental set-up, the tip of a soldering iron was wrapped with a PVC-coated copper wire and heated for about 5 min to approximately 350 °C. In the gasflame combustion of low molecular pure PVC PCDF were found in the low $\mu g/m^2$ range (1 $\mu g/m^2$ corresponds with 1 mg/kg) in the soot (Table 88). Similar results were obtained with PVC-cable coatings. High molecular PVC gave approximately 4times higher values of PCDF, preferably of lower chlorinated PCDF, but PCDD could be detected in case of pure PVC. The analysis of the soot from the combustion of cables coated with PVC gave low values of PCDD similar to those found after fires in public buildings. The pyrolysis of pure PVC as well as PVC-cable coatings yielded dioxins and furans in the lower mg/kg range (Table 89). Here also, furans were formed predominantly. Especially high values of hepta- and octa-chlorinated dioxins and furans, ranging up to the milligram level were formed in the pyrolysis of PVC-coated cables. In the experiments carried out in the Schöninger apparatus under a nitrogen atmosphere no detectable quantities of PCDD/PCDF were formed as well as in neither combustion or pyrolysis experiments with chlorine-free polyethylene samples dioxins or furans could be detected.

Table 88. Formation of PCDD/PCDF by Combustion of PVC [$\mu g/m^2$]

PCDD/PCDF		Pure PVC High molec.		PVC-cable Low molec.	Soot from a fire
$Cl_4DD(2,3,7,8)$	*	n.d.	n.d.	n.d.	n.d.
$\Sigma\ Cl_4DD$		n.d.	n.d.	n.d.	0.2
$Cl_5DD(1,2,3,7,8)$	*	n.d.	n.d.	n.d.	n.d.
$\Sigma\ Cl_5DD$		n.d.	n.d.	n.d.	0.2
$Cl_6DD(1,2,3,4,7,8)$	*	n.d.	n.d.	0.14	n.d.
$Cl_6DD(1,2,3,6,7,8)$	*	n.d.	n.d.	0.12	n.d.
$Cl_6DD(1,2,3,7,8,9)$	*	n.d.	n.d.	0.10	n.d.
$\Sigma\ Cl_6DD$		n.d.	n.d.	0.36	0.4
$Cl_7DD(1,2,3,4,6,7,8)$		n.d.	n.d.	0.4	0.3
$\Sigma\ Cl_7DD$		n.d.	n.d.	0.14	0.4
Cl_8DD		n.d.	n.d.	0.09	0.2
$Cl_4DF(2,3,7,8)$	*	n.d.	0.04	0.38	0.2
$\Sigma\ Cl_4DF$		9.24	2.53	2.29	3.2
$Cl_5DF(1,2,3,7,8)+(4,8)$		0.65	0.17	0.35	0.3
$Cl_5DF(2,3,4,7,8)$	*	0.36	0.09	0.26	0.2
$\Sigma\ Cl_5DF$		5.36	0.98	1.76	1.5
$Cl_6DF(1,2,3,4,7,8)+(7,9)$		0.71	0.18	0.30	0.2
$Cl_6DF(1,2,3,6,7,8)$	*	0.43	0.11	0.22	0.1
$Cl_6DF(1,2,3,7,8,9)$		0.06	n.d.	0.08	0.1
$Cl_6DF(2,3,4,6,7,8)$		0.24	n.d.	0.14	0.1
$\Sigma\ Cl_6DF$		3.73	0.89	1.34	0.8
$Cl_7DF(1,2,3,4,6,7,8)$		1.55	0.40	0.43	0.2
$Cl_7DF(1,2,3,4,7,8,9)$		0.30	0.08	0.12	0.1
$\Sigma\ Cl_7DF$		2.67	0.72	0.77	0.5
Cl_8DF		1.02	0.19	0.34	0.3
TE (BGA/UBA)**		0.45	0.11	0.27	0.19
Gef Stoff V* ($\mu g/kg$)		790	240	1,220	550

*   GefStoffV: German statuatory regulation for dangerous substances, limit 5 $\mu g/kg$ for the sum of eight labeled (*) congeners. Results related to soot.
** TE: Toxic Equivalents using TEF of the German Federal Health Office.

Table 89. Formation of PCDD/PCDF by Pyrolysis of PVC [µg/kg]

| PCDD/PCDF | Schöninger Apparatus 950°C | | | | Soldering Iron 350°C |
| | Pure PVC | | PVC-cables (n = 5) | | PVC-coated copper wire |
	High molec.	Low molec.	min.	max.	
$Cl_4DD(2,3,7,8)^*$	n.d.	n.d.	n.d.	n.d.	n.d.
Σ Cl_4DD	1	24	n.d.	n.d.	n.d.
$Cl_5DD(1,2,3,7,8)^*$	n.d.	n.d.	n.d.	n.d.	7
Σ Cl_5DD	34	33		n.d.	151
$Cl_6DD(1,2,3,4,7,8)^*$	n.d.	n.d.	n.d.	n.d.	1
$Cl_6DD(1,2,3,6,7,8)^*$	n.d.	n.d.	n.d.	n.d.	13
$Cl_6DD(1,2,3,7,8,9)^*$	n.d.	n.d.	n.d.	n.d.	1
Σ Cl_6DD	34	n.d.	n.d.	n.d.	167
$Cl_7DD(1,2,3,4,6,7,8)$	27	24	10	460	105
Σ Cl_7DD	44	43	27	799	208
Cl_8DD	493	19	14	1,040	247
$Cl_4DF(2,3,7,8)^*$	n.d.	4	n.d.	n.d.	n.d.
Σ Cl_4DF	124	199	63	115	127
$Cl_5DF(1,2,3,7,8)+(4,8)$	20	27	n.d.	6	17
$Cl_5DF(2,3,4,7,8)^*$	14	15	n.d.	5	1
Σ Cl_5DF	47	190	24	370	112
$Cl_6DF(1,2,3,4,7,8)+(7,9)$	36	50	2	11	32
$Cl_6DF(1,2,3,6,7,8)^*$	32	35	n.d.	8	26
$Cl_6DF(1,2,3,7,8,9)$	7	7	n.d.	n.d.	6
$Cl_6DF(2,3,4,6,7,8)$	24	17	4	8	42
Σ Cl_6DF	156	294	32	53	216
$Cl_7DF(1,2,3,4,6,7,8)$	147	215	54	1,839	129
$Cl_7DF(1,2,3,4,7,8,9)$	30	37	n.d.	364	22
Σ Cl_7DF	236	340	72	2,857	199
Cl_8DF	143	163	64	2,016	58
TE (BGA/UBA)[**]	18.7	24.4		41	24
Gef Stoff V[*]	45	54		19	49

2.10 COMPARISON OF COMBUSTION SOURCES

Sheffield has been the only author who reviewed some years ago the emission sources of PCDD and PCDF in an attempt to estimate the levels of dioxins and furans introduced into the Canadian environment (Sheffield, 1985). From his information sources, he claimed that the largest source of PCDD emission came from forest fires. An estimated 60 kg per year are introduced into the environment from this source alone. This level was determined based on test results from residential wood combustion being extrapolated to fit the amount of wood burned during forest fires as estimated from a ten-year average. Other sources which show potentially significant releases of PCDD and PCDF include municipal incinerator fly ash and air emissions, sewage sludge incinerator air emissions, and the combustion of waste wood in wigwam burners and/or wood waste boilers. The estimated levels from each of these sources and other sources of PCDD and PCDF are given in Table 90. The levels in these tables are based on preliminary estimates from the most recent data available (at that time) and are from tests done largely by Canadian institutions.

Precipitated fly ash and stack flue gas samples have been taken from several of the municipal incinerators operating in Canada. Municipal waste incineration has been the most studied source of PCDD and PCDF in Canada, however major variations exist in the reported data and therefore a wide range of yearly total emissions have been projected in Table 90.

Precipitated fly ash samples were taken from Canadian coal-fired utilities show low concentrations of PCDD and PCDF. These levels appear to be at maximum lower than the lowest levels produced in municipal incinerators. The air emissions tested for one coal-fired plant were found to be very low and the total air emissions projected for coal-fired plants are approximately equal to the minimum levels estimated for municipal incineration. This projection must be taken with a degree of caution since the author of this work has extrapolated the results for one coal-fired plant to all such facilities in Canada. Variations from plant to plant may be observed as in the case of the municipal incinerators which have been studied.

As for coal-fired utility boilers and their air emissions, only one sewage sludge incinerator was tested and the results were extrapolated to give an estimate of yearly emissions for all such facilities in Canada.

Table 90. PCDD and PCDF Released to the Environment from Combustion
           Sources

Source	Releases [g/year]	
	PCDD	PCDF
Municipal waste incinerators		
- Fly ash	2,900 - 7,100	4,900 - 15,600
- Air emissions	250 - 13,700	550 - 21,700
Sewage sludge incineration		
- Air emissions	1,400 - 3,300	1,500 - 6,500
Coal-fired utility boilers		
- Fly ash	ND - 300	30 - 1,300
- Air emissions	300	700
Fuelwood combustion		
- Air emissions	1,800	*
Residential oil combustion		
- Air emissions	< 1	*
Residential gas combustion		
- Air emissions	900	*
Wigwam burners/Wood waste boilers		
- Air emissions	0 - 30,200	
Railway ties		
- Air emissions	6,000	
Forest fires		
- Air emissions	58,700	*
Slash burning		
- Air emissions	3,300	*
Cigarette Smoke		
- Air emissions	2 - 4	*
Motor vehicles		
- Air emissions	200	*

* - not investigated

414

The estimated emissions from fuelwood combustion were derived from the average concentrations of PCDD reported from American tests performed on residential wood combustion. The particulate emissions of such devices as determined from Canadian studies were used with the average PCDD concentration calculated from the American studies to give the estimate listed in Table 90. Residues from the wood protection and preservation industries, in the form of diptank sludges and wood shavings, can present a serious problem, especially when wastes contaminated with chlorinated phenols are incinerated in wigwam burners or wood waste boilers. Significant variations in the levels of PCDD and PCDF emitted were found for various studies. Contradictory results have been reported ranging from no emission of PCDD or PCDF to significant emissions of these compounds being released. Therefore it is apparent that much more work needs to be done in the investigation of this potentially significant source of dioxins and furans.

The estimated emission of PCDD from the burning of railway ties (Table 90; Sheffield, 1985) was determined based on data taken from studies concerning the combustion of pentachlorophenol-sprayed wood.

The results of the analyses of chimney soot samples (see Chapter 2.7, Fossil Fuel Combustion) have shown that house heating is a source for emission of PCDD/PCDF that should be taken into consideration (Thoma, 1988a, Thoma, 1988b). The German Hazardous Waste Regulations state that where the sum of $2,3,7,8-Cl_4DD$, $1,2,3,7,8-Cl_5DD$, $1,2,3,6,7,8-Cl_6DD$, $1,2,3,7,8,9-Cl_6DD$, $2,3,7,8-Cl_4DF$, $2,3,4,7,8-Cl_5DF$, and $1,2,3,6,7,8,-Cl_6DF$ exceed 5 ppb, the material has to be treated as hazardous waste (Chemikaliengesetz, 1988). Table 91 summarizes the levels of these toxic isomers found in chimney soot from house heating.

Table 91. Average Concentrations of the Toxic PCDD/PCDF-Isomers in Chimney Soot from House Heating [ppt]

Substance	Oil (Heating)	Oil (Oven)	Coal (Oven)	Wood (Oven)	Wood (Oven)	Wood Coal (Oven)
$2,3,7,8-Cl_4DD$	10.3	54.4	319.6	310.1	420.7	121.2
$1,2,3,7,8-Cl_5DD$	14.4	177.5	838.8	1,429.1	860.9	107.3
$1,2,3,6,7,8-Cl_6DD$	20.7	154.4	610.2	1,415.2	1,399.6	63.6
$1,2,3,7,8,9-Cl_6DD$	14.3	202.5	589.3	1,262.5	1,205.0	53.5
$2,3,7,8-Cl_4DF$	175.4	774.3	8,407.5	11,294.3	9,374.2	1,900.0
$2,3,4,7,8-Cl_5DF$	154.1	971.7	9,120.5	34,818.8	10,589.0	1,287.8
$1,2,3,6,7,8-Cl_6DF$	58.8	655.3	3,176.5	8,672.9	3,262.7	251.7
Total	448.0	2,990.1	2,306.4	59,202.9	23,849.4	3,785.1

The limits of 5 ng/g given by law is exceeded in chimney soots from wood and coal burning systems which implies that these soots should be treated like fly ash from municipal waste incinerators and be disposed of in special dump sites and not on municipal waste disposal sites. Until now there are no data to support the hypothesis that PCDD/PCDF emissions from household heating systems to the atmosphere potentially damage the environment, there is indirect evidence. Analysis of spruce needles from different regions revealed that the PCDD/PCDF concentrations in so-called "clean" areas were 10times less only compared to those around municipal waste incinerators (Reischl, 1988). All these data indicate that diffuse sources such as house heating influence the background level of PCDD/PCDF.

The National Dioxin Strategy set by the United States Environmental Protection Agency (EPA) in December 1983 was established in order to (des Rosiers, 1986; Southerland et al., 1987; National Dioxin Study, 1987a, 1987b):

1. Determine the overall extent of dioxin contamination in the environment;
2. Provide a systematic approach for dealing with dioxin decontamination problems.

The majority of the interest was centered around the most toxic dioxin, the $2,3,7,8-Cl_4DD$ isomer.

The National Dioxin Strategy was divided into seven different categories (referred to by the EPA as tiers). Tier 4 of the strategy dealt with studies conducted on the emission of dioxins and furans from a variety of combustion sources. The activities carried out in the Tier 4 study were focussed on those sources which were believed to have the greatest potential for emitting dioxins and furans. The sources were categorized and assigned testing priorities based on the information obtained from a literature review of relevant studies.

From the literature review it was concluded that the emission of PCDD from combustion sources could be attributed to:

a)   PCDD present in the materials being burned;
b)   The presence of PCDD precursors (for example, chlorinated phenols and chlorobenzenes) in the materials burned;
c)   The presence of chlorine, fuel, and combustion conditions conducive to PCDD formation;
d)   Relatively low combustion temperatures (500 to 800 °C);
e)   Short residence time in the combustion zone (less than 1-2 seconds);
f)   Lack of adequate oxygen resulting in incomplete combustion;
g)   Lack of adequate processing of fuels (for example, burning wet garbage);
h)   Lack of supplemental fuel to promote combustion efficiency.

The literature review resulted in the identification of thirteen generalized combustion sources which have been found to emit PCDD and PCDF. These sources are:

1. Municipal solid waste incinerators;
2. Sewage sludge incinerators;
3. Fossil fuel combustion;
4. Wood combustion;
5. Boilers co-firing wastes;
6. Hazardous waste incinerators;
7. Hospital incinerators;
8. Lime/cement kilns;
9. Wire reclamation;
10. PCB fires;
11. Automobile emissiions;
12. Activated carbon regeneration furnaces;
13. Experimental studies.

When the results of this literature review are summarized, it has to be taken into account that it is extremely difficult to directly compare data reported by various laboratories due to the differences in the sampling and analytical procedures. The lack of source-specific information and intangibles such as the quality of the reported work also hinder this comparison.

After completing the literature review combustion sources were ranked according to their potential for emitting dioxins. These rankings are shown in Table 92. Sources given a rank of "A" were expected to have the highest emission of dioxins. Ranks B and C have decreasing potentials for PCDD emission. Combustion sources classified as rank D have been previously subjected to a number of studies deemed to be sufficient by the EPA. The ranking was chosen for use in determining what sources would be studied in the Tier 4 program. The objective of the program was to test each rank A source three times, each rank B source once, and as many rank C source as possible. A detailed discussion is given in the project's engineering analysis report (EPA, 1987b).

Table 92. Final Ranked Source Category List

Rank A	Sewage Sludge Incinerators (3)
	Black Liquor Boilers (3)
Rank B	Industrial Incinerators (1)
	Wire Reclamation Incinerators (1)
	Carbon Regeneration Furnaces (1)
	Secondary Metal Blast Furnaces (1)
	Wood-Fired Boilers (1)
	Drum and Barrel Reclamation Furnaces (1)
Rank C	Mobile Sources[a] (1)
	Wood stoves[a] (1)
	Spreader-Stoker Coal Fired Boilers
	Commercial Boilers Burning Chlorinated Organic Wastes
	Lime Kilns
	Cement Kilns
	Hazardous Waste Incinerators
	Hospital Waste Incinerators
	Apartment House Incincerators
	Charcoal Manufacturing Operations
	Open Burning
Rank D	Municipal Waste Incinerators
	Commercial Waste Boilers

(1)   Indicates that the source category was tested once under Tier 4.

(3)   Indicates that the source category was tested three times under Tier 4.

[a]   Indicates that the source category was tested once under Tier 4 in conjunction with another test program.

Dioxins and furans were found to be emitted by every source tested with the exception of the woodstove for which no valid data was available because of analytical problems encountered. Considerable variations in the average total PCDD and PCDF emissions were observed for this set of test sites. It should be noted however that the variations from test-to-test for each site were generally less than the site-to-site variations.

The total PCDD and PCDF emissions were found to be roughly the same order of magnitude for each of the sites tested, however for most sites the total PCDF emissions were higher than the total PCDD emissions.

Detectable levels of 2,3,7,8-Cl$_4$DD were found in the outlet emissions from seven of the test sites. The amount of 2,3,7,8-Cl$_4$DD contributed only a small percentage to the total PCDD concentration. The amount of 2,3,7,8-Cl$_4$DD relative to the total amount of dioxins emitted was found to vary considerably from one site to another (Table 93).

Table 93. Average Outlet PCDD/PCDF Mass Emission Rates for the Tier 4 Test Sites (Sites Ranked by Total PCDD Mass Emission Rates)

Outlet Total PCDD Mass Emission Rate Rank	Source Type/ Control Device Type	Average Mass Emission Rates [μg/hr]		
		2378-Cl$_4$DD	Total PCDD	Total PCDF
1	Secondary Copper Cupola Furnace/AB BH	5,360	283,000	1,420,000
2	Salt-Laden Wood Fired Boiler/BH	5.8	4,190	1,800
3	Industrial Solid Waste Incinerator/AB	13.5	1,810	6,780
4	Sewage Sludge Incinerator/SCR	2.1	780	6,570
5	Black Liquor Boiler/ESP	ND(1.6)	667	330
6	Wire Reclamation Incinerator/AB (wire and transformer feed)	0.1	522	626
7	Black Liquor Boiler/ESP	ND (5.1)	230	214
8	Black Liquor Boiler/ESP	ND (7.2)	149	114
9	Wire Reclamation Incinerator/AB (wire-only feed)	0.1	113	205
10	Sewage Sludge Incinerator/SCR	0.1	40.5	90.4
11	Carbon Regeneration Furnace/AB, SD, BH	ND (2.0)	31.6	28.9
12	Drum and Barrel Incinerator/AB	0.3	23.8	129
13	Sewage Sludge Incinerator/SCR	ND (1.0)	11.6	194
14	Woodstove/Uncontrolled	NR	NR	NR

ND   - not detected (average detection limit in parenthesis)
AB   - afterburner
BH   - baghouse
SCR - scrubber
SD   - spray dryer
ESP - electrostatic precipitator
NR   - no valid analytical data were reported

The flue gas was also sampled at the inlet of air pollution control devices for some of the sites. Dioxins and furans were found in the inlet flue gas for all seven of the sites tested. For most sites the total PCDD/PCDF emissions were higher at the inlet of the pollution control device than at the stack outlet. This was also found to be true for the respective levels of $2,3,7,8\text{-}Cl_4DD$. Higher levels are expected at the inlet since the pollution control devices are intended to remove dioxins and furans from the flue gas (and hence prevent them from escaping directly into the atmosphere). As for the outlet concentrations, the total levels of PCDD and PCDF are roughly the same order of magnitude although there tends to be a larger amount of furans in the flue gas.

The variation in PCDD/PCDF mass flow rates is much greater for the outlet emissions than for the control device inlets of the sites. The relationships between the emission rates of $2,3,7,8\text{-}Cl_4DD$, total dioxins, and total furans for an individual test site are the same since the mass emission rate is directly proportional to the mass emission concentration.

From the homologue distribution patterns for the inlet and outlet flue gas samples for the combustion sites tested it was concluded that:

1. Each test site has a characteristic PCDD/PCDF homologue "fingerprint". Reasonably good precision between test runs was observed for an individual site. No correlation between the homologue distribution and the precursors in the feed material was apparent.

2. Considerable variation in homologue distribution patterns was observed amongst test sites.

3. Homologue distributions of PCDD did not always resemble patterns observed for PCDF for the same site.

4. Homologue distributions do not shift significantly across the control devices (for those in which both inlet and outlet emission data are available). This suggests that there are no significant differences in removal efficiencies for dioxin or furan homologues.

Based on the literature review, there are numerous variables which are suspected to influence the levels of the PCDD and PCDF emissions. The observations drawn from the data gathered from the inlet and outlet flue gas samples for the various sites are summarized below:

1. The outlet emissions of $2,3,7,8-Cl_4DD$ and total PCDD are directly related but not proportionally. Sites having high levels of total PCDD also have relatively high levels of $2,3,7,8-Cl_4DD$. The same general trend is also observed for $2,3,7,8-Cl_4DF$ and total PCDF levels.

2. Excluding the secondary copper smelter, there appears to be a correlation between the outlet flue gas total chloride concentration and the total outlet PCDD/PCDF emissions. For the secondary copper smelter the total dioxin and furan emission concentrations were extremely high yet relatively low flue gas total chloride concentrations were observed.

3. Other variables considered (such as maximum combustion temperature, CO, THC, total feed chloride, and TOX) did not show any significant correlation with the outlet emissions of dioxins and furans.

4. For the inlet flue gas, the total PCDD and PCDF concentrations appear to be inversely related to the maximum temperature.

5. Total PCDD and total PCDF emissions appear to be related since both are roughly of the same order of magnitude for each site tested. Sites emitting relatively high concentrations of PCDD also emit relatively large amounts of PCDF.

6. Other variables considered do not appear to show significant association with inlet flue gas PCDD and PCDF levels.

7. Maximum dioxin and furan flue gas concentrations for a given site appeared to be associated with higher TOX levels in the feed material.

8. Other variables did not appear to significantly influence maximum PCDD/PCDF flue gas concentrations for a given site.

Although these observations were made, they were not deemed to be conclusive. Further testing was recommended.

Analysis of data reported in the literature for municipal waste incinerators was also carried out as part of the overall study program even though such sources were not directly tested. The results of the analysis of this data are:

1. Flue gas PCDD/PCDF emissions appear to be inversely related to furnace temperature, which itself is influenced directly by the presence of excess air and the heat content of the material combusted. Maximum emissions for a given site were found to be about 10 times higher when the feed was wet (relative to normal dry conditions).

2. Higher levels of dioxins and furans were found in the flue gases emitted from incinerators with flawed design and those which are poorly maintained.

3. In general, large amounts of PCDD and PCDF in flue gas are found for incinerators having large amounts of dioxins and furans in the flyash.

4. Levels of PCDD and PCDF in municipal incinerator flyash are a function of various design and operating conditions and not just the furnace temperature.

5. PCDD/PCDF emissions for European incinerators are generally higher than for American and Canadian facilities. There are no obvious reasons for this trend, although the differences may be due to differences in incinerator design and operation, incinerator age, as well as sampling and analytical methodologies.

The levels of dioxins and furans found in flyash samples of the Tier 4 test sites were also determined. Comparison of the average values of the PCDD and PCDF contents of the ash samples and in the flue gas for the sites showed that high levels in the flyash tend to correspond with high levels in the flue gas, however a direct relationship is not clearly evident.

By combining the emission data obtained from the Tier 4 test sites with that extracted from studies reported in the literature, a summary of PCDD/PCDF stack emissions for a variety of sources was compiled (see Table 94).

The secondary copper smelter provided the highest emission of PCDD and PCDF while waste incinerators of all types were found to produce much lower levels. Variation among incinerators was widespread and although one cannot make any conclusions regarding the relative emissions of hazardous and municipal waste incinerators, it does appear that sewage sludge incinerators generally have the lowest emissions of the three. Kraft paper recovery boilers and coal-fired utility boilers all tend to emit relatively low levels of PCDD and PCDF.

Based on the results of the Tier 4 study and information extracted from the literature, the following conclusions were made:

1. PCDD and PCDF emissions appear to be higher for lower temperature combustion processes whose functions are to recover energy or other resources such as in the case of the secondary copper smelter.
2. Total organic halogen (TOX) levels showed the strongest association with the total level of PCDD and PCDF emitted whilesites with high percentages of plastics in the feed were found to yield high PCDD emissions in the Tier 4 sites studied.
3. Afterburners operating above 800 °C do not necessarily destroy all PCDD and PCDF since factors such as residence time and mixing are also significant.
4. Distribution profiles for the various PCDD and PCDF isomers were found to vary considerably amongst the sites tested in the Tier 4 program, however there were some similarities observed within sources of the same type. No universal pattern for combustion sources was found to exist.

Table 94. Summary of PCDD/PCDF Stack Emissions by Source Category

Source Category	Range of PCDD Emissions [ng/m³]	Range of PCDF Emissions [ng/m³]
1. Municipal Waste Incinerators		
European	71 - 48,997	37 - 9,831
U.S. and Canada	3.3 - 11.686	8.5 - 22,000
2. Boilers Cofiring Waste		
Commercial	1,400 - 17,000	170
Industrial	<0.002 - 76.4	<0.002 - 5.5
3. Secondary Copper Cupola Furnace	11,900	60,700
4. Wood Combustion		
PCP-Treated Wood	<17 - 1.520	<17 - 587
Salt-Laden Wood-Fired Boiler	195	83.2
5. Sewage Sludge Incinerators	ND - 812	ND - 1.374
6. Wire Reclamation Incinerator (wire and transformer feed)	704	866
7. Industrial Solid Waste Incinerator	625	2.390
8. Wire Reclamation Incinerator (wire-only feed)	173	305
9. Hospital Incinerators	15 - 69	25 - 156
10. Hazardous Waste Incinerators		
Rotary Kiln	7.7 - 8.6	11.2 - 19
11. Drum & Barrel Reclamation Incinerator	5	27
12. Carbon Regeneration Furnace		
without Afterburner	0.18	0.3
with Afterburner	1.6 - 3.7	0.05 - 3.3
13. Black Liquor Boiler	0.8 - 2.9	0.6 - 2.1
14. Cement Kilns	<1 - 1.35	<1 - 0.74
15. Lime Kilns	<0.34 - <2.0	- - - -
16. Utility Boiler		
Co-firing Waste	<0.031 - <0.10	<0.31 - <0.10
17. Fossil Fuel Combustion		
Coal-Fired Utility	<0.10 - <0.70	<0.10 - <0.70
Pulverized Coal	<4.2 - <7.9	<0.67 - <1.3
Oil-Fired Utility	<4.2 - <7.9	<0.67 - <1.3
18. Incinerator Ship	<0.0009 - <0.086	<0.3 - <3.0

2.11 CONCLUSIONS

Initially most of the interest in the emission of dioxins and related compounds involved the studies performed on municipal incinerator fly ash where such compounds were readily detected. Sampling of flue gases has always been difficult and therefore such samples have not been as thoroughly studied in comparison to fly ash. After the initial report of dioxins present at trace levels in incinerator fly ash, many other combustion sources have been tested and have been found to emit both PCDD and PCDF. Many studies have been performed in attempt to help unravel the complex mechanisms leading to the formation of these compounds in thermal processes including incineration. While a large volume of data has been generated, the quality of which has varied as much as the reported levels of PCDD and PCDF emissions. Difficulties encountered in the sampling procedures as well as tedious clean-up methods have undoubtedly led to reports of false positives or negatives. These conflicting reports have only complicated matters and have resulted in much confusion and controversy within the scientific community. Clearly there is a need for validated methods, including sampling and clean-up techniques, as well as more uniformity in the interpretation and reporting of results. This can only be accomplished by increased inter-laboratory collaboration.

Much information has been gathered on the sources and emissions of dioxins and related compounds but there is much more information which needs to be obtained. Many studies have provided evidence for potential mechanisms of dioxin and furan formation, however the mechanisms are still not fully understood.

Improvements in instrumental analysis techniques are lowering the limits of detection of these compounds and consequently advances have been made in pollution control technology. The levels of emissions appear to be decreasing in recent years. The impact of these emissions on the environment have not been clearly demonstrated and therefore it is likely that the previously mentioned trends will continue.

REFERENCES

Abenhaim, L., Suissa, S., and Bard, D., 1987, Criteria for public health decision making under uncertainty: The French experience of PCB electrial equipment fires, Chemosphere, 16:2129

Addis, G., 1986, Pyrolysis and combustion of Aroclor 1254 contaminated dielectric fluids, Chemosphere, 15:1265

Asada, S., Matsushita, H., Morita, M., and Hamada, Y., 1987, Determination of chlorodibenzodioxins and chlorodibenzofurans discharged from several municipal incinerators in Japan, Chemosphere, 16:1907

Ballschmiter, K., Buchert, H., Niemczyk, R., Munder, A., and Swerev, M., 1986, Automobile exhausts versus municipal-waste incineration as sources of the polychloro-dibenzodioxins (PCDD) and -furans (PCDF) found in the environment, Chemosphere, 15:901

Bendixen, E.L., and Nielsen, P.R., 1987, A guideline for incineration of waste based on theory and practical experiences, Chemosphere, 16:1943

Benfenati, E., Pastorelli, R., Castelli, M.G., Fanelli, R., Carminati, A., Farneti, A., and Lodi, M., 1986, Studies on the tetradichloro-dibenzo-p-dioxins (TCDD) and tetrachlorodibenzofurans (TCDF) emitted from an urban incinerator, Chemosphere, 15:557

Bergström, J.G.T., and Warman, K., 1987, Production and characterization of trace emissions in Sweden, Waste Management & Research, 5:395-435

Bergvall, G., Hult, J., Technology, Economics and Environmental Effects of Solid Waste Treatment, Final report from DRAV-project, Publication 85:11, the Swedish Association of Public Cleansing and Solid Waste Management, Malmö, Sweden

Bingham, A.G., Edmunds, C.J., Gibson, J.J., Graham, B.W., and Jones, M.T., 1988, Determination of PCDDs and PCDFs in car exhaust, Dioxin'88, 8th International Symposium on Chlorinated Dioxins and Related Compounds, 21-28 Aug. 1988, Umea, Sweden

Brenner, K.S., Dorn, I.H., and Herrmann, K., 1986, Dioxin analysis in stack emissions, slags and the wash water circuit during high- temperature incineration of chlorine-containing industrial wastes - II, Chemosphere, 15:1193

Brenner, K.S., Mader, H., Steverle, H., Heinrich, G., and Womann, H., 1984, Dioxin analysis in stack emissions and in the wash water circuit during high-temperature incineration of chlorine-containing industrial wastes, Bull. Environ. Contam. Toxicol., 33:153

427

Bröker, G., 1987, Maßnahmen zur Verminderung von Dioxinemissionen an Müllverbrennungsanlagen, in: VDI Berichte 634 - Dioxine, pp 515, VDI Verlag, Düsseldorf, 1987

Bumb, R.R., Crummett, W.B., Artie, S.S., Gledhill, J.R., Hummel, R.H., Kagel, R.O., Lamparski, L.L., Luoma, E.V., Miller, D.L., Nestrick, T.J., Shadoff, L.A., Stehl, R.H., and Woods, J.S., 1980, Trace chemistries of fire: A source of chlorinated dioxins, Science, 210:385

Buser, H.-R., 1975, Analysis of polychlorinated dibenzo-dioxins and dibenzofurans in chlorinated phenols by Mass Fragmentography, J. Chromatgr., 107:295

Choudhry, G.C. and Hutzinger, O., 1983, "Mechanistic Aspects of the thermal formation of halogenated organic compounds including PCDD", pp 71, Gordon & Breach, New York - London - Paris - Montreux - Tokyo

Christmann, W., Kasiske, D., Klöppel, K.D, Parscht, H., and Rotard, W., 1988, Combustion of polyvinylchloride - an important source for the formation of PCDD/PCDF, Dioxin'88, 8th International Symposium on Chlorinated Dioxins and Related Compounds, 21-28 Aug. 1988, Umea, Sweden

Clement, R.E., Tosine, H.M., and Ali, B., 1985, Levels of polychlorinated dibenzo-p-dioxins and dibenzofurans in wood burning stoves and fireplaces, Chemosphere, 14:815

Clement, R.E., Tosine, H.M., Osborne, J., Ozvacic, V., Wong, G., and Thorndyke, S., 1987, Emissions of chlorinated organics from a municipal sewage sludge burning incinerator, Chemosphere, 16:1895

Cook, R.J., 1987, Solid and hazardous waste incineration: An analysis for citizens and policymakers, Draft, prepared for Toxic Substance Control Commission, State of Michigan

Crummett, W.B., 1982, in: "Chlorinated dioxins and related compounds: impact on the environment", pp 253, O. Hutzinger, R.W. Frei, E. Merian, and F. Pocchiari, eds., Pergamon Press, Oxford

Crummett, W.B., and Townsend, D.I., 1984, The trace chemistries of fire hypothesis: Review and update, Chemosphere, 13:777

De Fré, R., 1986, Dioxin levels in the emissions of Belgian municipal incinerators, Chemosphere, 15:1255

des Rosiers, P.E., 1986, The National Dioxin Study, U.S. EPA, Washington, DC

des Rosiers, P.E., and Lee, A., 1986, PCBs fires: Correlation of chlorobenzene isomer and PCB fluids with PCDD and PCDF contents of soot, Chemosphere, 15:1313

des Rosiers, P.E., 1987, Chlorinated combustion products from fires involving PCB transformers and capacitors, Chemosphere, 16:1881

Dickson, L.C. and Karasek, F.W., 1987, Mechanism of formation of polychlorinated dibenzo-p-dioxins produced of municipal incinerator fly ash from reactions of chlorinated phenols, _J. Chromatogr._, 389:127

Dickson, L.C., Lenoir, D., and Hutzinger, O., 1988, Surface-catalyzed formation of chlorinated dibenzodioxins and dibenzofurans during incineration, Dioxin'88, 8th International Symposium on Chlorinated Dioxins and Related Compounds, 21-28 Aug. 1988, Umea, Sweden

Eadon, G., Aldons, K., Hilker, D., Keefe, P.O., and Smith, R., 1983, Chemical data on air samples from the Binghamton State Office Building, Center for Laboratories and Research, New York State Department of Health, Albany, 7 July 1983

Eberhard, H., Friege, H., and Schumacher, E., 1986, PVC in refuse incineration, _Müll und Abfall_, 10:377

Eiceman, G.A., and Rghei, H.O., 1984, Presence of nitro-chlorinated dioxins on fly ash from municipal incinerators and laboratory production through reactions between NO_2 and T_4CDD on fly ash, _Chemosphere_, 13:1025

Eiceman, G.A., Clement, R.E., and Karasek, F.W., 1979, Analysis of fly ash from municipal waste incinerators for trace organic compounds, _Anal. Chem._, 51:2343

Eklund, G., Pederson, J.R., and Stromberg, B., 1986, Phenol and HCl at 550 °C yield a large variety of chlorinated toxic compounds, _Nature_, 320:155

EPS Canada, 1985, The National Testing and Evaluation Program: Two-stage combustion (Prince Edward Island), EPS 3/UP/1

EPS Canada, 1986, The National Testing and Evaluation Program: Air pollution control technology, EPS 3/UP/2

EPS Canada, 1988, The National testing and Evaluation Program: Environmental characterization of mass burning incinerator at Quebec city, EPS 3/UP/5

Erickson, M.D., Cole, C.J., Flora, J.D.,jr., Gorman, P.G., Haile, C.L., Hinshaw, G.D., Hopkins, F.C., Swanson, S.E., and Heggem, D.T., 1985, PCDF formation from PCBs under fire conditions, _Chemosphere_, 14:855

Erickson, M.D., Swanson, S.E., Sack, T.M., and Heggem, D.T., 1986, Products of thermal degradation of dielectric fluids, _Chemosphere_, 15:1261

Fiedler, H. and Hutzinger, O., 1987, Leitstudie - Dioxine, _in_: VDI Berichte 634, pp. 299, VDI Verlag, Düsseldorf, 1987

Hagenmaier, H., Brunner, H., Haag, R., and Kraft, M., 1987a, _Environ. Sci. Technol._, 21:1085

Hagenmaier, H., Kraft, M., Brunner, H., and Haag, R., 1987b, _Environ. Sci. Technol._, 21:1080

Haglund, P., Egebäck, K.-E., and Jansson, B., 1988, Analysis of PBDD/F in vehicle exhaust, Dioxin'8, 8th International Symposium on Chlorinated Dioxins and Related Compounds, 21-28 Aug. 1988, Umea, Sweden

Hahn, J.L., VonDemfange, H.P., and Velzy, C.O., 1986, Effect of boiler operation and RDF feedstock on emissions of dioxins and furans from an RDF fired spreader-stoker system in Albany, NY, Chemosphere, 15:1239

Hawley-Fedder, R.A., Ph.D.Thesis 1984, Department of Chemistry, Arizona State University

Hay, D.J., Finkelstein, A., Klicius, R., Bridle, T., 1987, Chemosphere, 16:1923

Hileman, F.D., Hale, M.D., Mazer, T., and Noble, R.W., 1985, Chemosphere, 14:601

Hiraoka, M., Takizawa, Y., Masuda, Y., Takeshita, R., Yagome, K., Tanaka, M., Watanabe, Y., and Morikawa, K., 1987, Chemosphere, 16:1901

Hutzinger, O., Blümich, M.J.,.v.d. Berg, M., and Olie, K., 1985, Chemosphere, 14:581

Hutzinger, O. and Fiedler, H., 1987, Sources and Emissions of PCDD/PCDF, Dioxin'87, 7th International Symposium on Chlorinated Dioxins and Related Compounds, Las Vegas, Oct. 4-9, 1987

Hutzinger, O., and Fiedler, H., 1988, Pilot Study on International Information Exchange on Dioxins and Related Compounds, Report Number 173, August 1988, North Atlantic Treaty Organization, Committee on the Challenges of Modern Society

Jansson, B., and Sundstrom, G., 1982, in: "Chlorinated Dioxins and Related Compounds: Impact of the Environment", pp 253, Hutzinger, O., Frei, R.W., Merian, E., and Pocchiari, F., eds, Pergamon Press, Oxford

Jones, K.H., Walsh, J., and Alston, D., 1987, Chemosphere, 16:2183

Karasek, F.W. and Dickson, L.C., 1987, Science, 237:754

Krishnan S. and Hites, R.A., 1980, Chemosphere, 9:679

Lao, R.C., Lanoy, M., and Lee, S.W., 1983, Polynuclear aromatic hydrocarbons, 7th Int. Symp. 1982, Battelle Press

Liberti, A., and Brocco, D., 1982, Formation of polychlorodibenzodioxins and polychlorodibenzofurans in urban incinerators emissions, in "Chlorinated Dioxins and Related Compounds: Impact of the Environment", pp 245, Hutzinger, O., Frei, R.W., Merian, E., and Pocchiari, F., eds. Pergamon Press, Oxford

Liberti, A., Goretti, G., and Russo, M.V., 1983, PCDD and PCDF in the combustion of vegetable wastes, Chemosphere, 12:661

Lindahl, R., Rappe, C., and Buser, H.R., 1980, Formation of polychlorina-
ted dibenzofurans (PCDFs) and polychlorrinated dibenzio-p-dioxins
(PCDDs) from the pyrolysis of polychlorinated diphenyl ethers,
Chemosphere, 9:351

Lustenhouwer, J.K., Olie, K., and Hutzinger, O., 1980, Chlorinated
dibenzo-p-pdioxins and related compounds in incinerator effluents,
Chemosphere, 9:501

Mahle, N.H., and Whiting, L.F., 1980, The formation of chlorodibenzo-p-
dioxins by air oxidation and chlorination of bituminous cole,
Chemosphere, 9:693

Marklund, S., Rappe, C., Tysklind, M., and Egeback, K.-E., 1987, Identification of polychlorinated dibenzofurans and dioxins in exhausts
from cars run on leaded gasoline, Chemosphere, 16:29

Müller, M.D., and Buser, H.-R., 1986, Halogenated aromatic compounds in
automotive emissions from leaded gasoline additives, Environ. Sci.
Technol., 20:1151

National Dioxin Study Tier 4 - Combustion Sources, Project summary the
de-novo synthesis on fly ash of municipal waste incinerators, Presented at Dioxin 87, Las Vegas, Oct. 4-9, 1987 (Chemosphere, in
press)

National Dioxin Study Tier 4 - Combustion sources, Project summary
report, EPA-45/4-84-014h, United States Environmental Protection
Agency 1987

Nestrick, T.J., Lamparski, L.L., and Stehl, R.H., 1979, Synthesis and
identification of the 22 tetrachlorodibenzo-p-dioxin isomers by high
performance liquid chromattography and gas chromatography, Anal.
Chem., 51:2273

Nestrick, T.J., and Lamparski, L.L., 1982, Issomer-specific determination
of chlorinated dioxins for assessment of formation and potential
environmental emission from wood combustion, Anal.Chem., 54:2292

Nielsen, K.K., Moeller, J.T., and Rasmussen, S., 1986, Reduction of dioxins and furanes by spray dryer absorption from incinerator flue gas,
Chemosphere, 15:1247

Nilsson, C.A., Anderson, K., Rappe, C., and Westermark, S.O., 1974, J.
Chromatogr., 96:137

Norstrom, A., Andersson, K., and Rappe, C., 1977, Chemosphere, 6:241

Nottrodt, I.A., and Ballschmiter, K., 1986, Causes for, and reduction
strategies against emissions of PCDD/PCDF from waste incineration
plants - interpretation of recent measurements, Chemosphere, 15:1225

Öberg, T., Aittola, J.-P., and Bergström, J.G.T., 1985, Chlorinated aromatics from the combustion of hazardous waste, Chemosphere, 14:215

Olie, K., Vermeulen, P.L., and Hutzinger, O., 1977, Chemosphere, 6:455

Ontario Ministry of the Environment, 1985, Scientific criteria document
for standard development, No. 4-84, Polychlorinated dibenzo-p-dioxins
(PCDDs) and polychlorinated dibenzofurans (PCDFs), prepared for
Intergovernmental Relations and Hazardous Contaminants Coordination
Branch

Palmer, T.Y., 1976, Nature, 263:44

Quellmatz, E., 1988, Das neue Chemikaliengesetz - Handbuch der gefährlichen Abfallstoffe, Rechtsstand Juli 1988, WEKA Fachverlage, Kissing, FRG

Rappe, C., Marklund, S., Buser, H.R., and Bosshardt, H.P., 1978, Chemosphere, 7:269

Rappe, C., Marklund, S., Kjeller, L.-O., and Tysklind, M., 1986, PCDDs
and PCDFs in emissions from various incinerators, Chemosphere,
15:1213

Reischl, A., Reissinger, M., and Hutzinger, O., 1987, Accumulation of
organic constituents by plant surfaces: Part III, Occurrence and
distribution of atmospheric organic micropollutants in conifer needles, Chemosphere, 16:2647

Rghei, H.O. and Eiceman, G.A., 1985a, Reactions of 1,2,3,4-TCDD on fly
ash in mixed gases of H_2O, NO_2, HCl, and SO_x ιξ αιο, Chemosphere,
14ε259

Rghei, H.O. and Eiceman, G.A., 1985b, Effect of matrix on heterogeneous
phase chlorine substitution reactions for dibenzo-p-dioxin and HCl in
air, Chemosphere, 14:167

Sandermann, W., Stockmann, H., and Casten, R., 1957, Chem. Ber., 90:690

Schecter, A., 1986, The Binghamton state office building PCB, dioxin and
dibenzofuran electrical transformer incident: 1981-1986, Chemosphere, 15:1273

Sheffield, A., 1985, Sources and releases of PCDD's and PCDF's to the
Canadian environment, Chemosphere, 14:811

Southerland, J.H., Kuykendal, W.B., Lamason II, W.H., Miles, A., Ober-
acker, D.A., 1987, Assessment of combustion sources as emitters of
chlorinated dioxin compounds: A report on the results of tier 4 of
the national dioxin strategy, Chemosphere, 16:2161

Steisel, N, Morris, R., and Clarke, M.J., 1987, The impact of the dioxin
issue on resource recovery in the United States, Waste Management &
Research, 5:381

Stieglitz, L. and Vogg, H., 1987, On formation conditions of PCDD/PCDF
in fly ash from municipal waste incinerators, Chemosphere, 16:1917

Stieglitz, L., Zwick, G., Beck, J., Roth, W., and Vogg, H., 1987, On the de novo synthesis of PCDD/PCDF on fly ash of municipal waste incinerators, Dioxin'87 7th International Symposium on Chlorinated Dioxins and Related Compounds, 4-7 Oct. 1987, Las Vegas, U.S.A. (submitted to Chemosphere)

The Hazardous Waste Inspectorate, Hazardous Waste Management: "...Ramshackle & Antediluvian"?, Second Report, Department of the Environment, Welsh Office, Scottish Office, N. Ireland, July 1986

Thoma, H., 1987, PCDD/F-Konzentrationen in Kaminaschen bei Hausfeuerung, in: VDI Berichte 634, pp. 53, VDI Verlag, Düsseldorf, 1987

Thoma, H., 1988a, PCDD/F concentrations in chimney soot from house heating systems, Chemosphere, 17:1369

Thoma, H., 1988b, PCDD/F-concentrations in chimney soot from house heating systems, Dioxin '88, 8th International Symposium on Chlorinated Dioxins and Related Compounds, 21-28 Aug. 1988, Umea, Sweden

Tong, H.Y. and Karasek, F.W., 1986, Comparison of PCDD and PCDF in fly ash collected from municipal incinerators of different countries, Chemosphere, 15:1219

Tosine, H.M., Clement, R.E., Ozvacic, V., and Wong, G., 1985, Levels of PCDD/PCDF and other chlorinated organics in municipal refuse, Chemosphere, 14:821

Tsuji, M., Nakano, T., and Okuno, T., 1987, Measurement of combustion products from liquid PCB waste incinerator, Chemosphere, 16:1889

Tysklind, M., Marklund, S., and Rappe, C., 1987, Experiences from the Swedish MSW moratorium, Chemosphere, 16:1937

Vogg, H., and Stieglitz, L., 1986, Thermal behavior of PCDD/PCDF in fly ash from municipal incinerators, Chemosphere, 15:1373

Zoller W., and Ballschmiter, K., 1986, Fresenius Z. Anal. Chem., 323:19

3.   FORMATION OF DIOXINS AND RELATED COMPOUNDS IN INDUSTRIAL
      PROCESSES

3.1 INTRODUCTION

Because they are potentially dangerous compounds, polychlorinated dibenzo-p-dioxins (PCDD) and dibenzofurans (PCDF) have in recent years
aroused increasing public interest. Most technologically developed countries have taken a number of regulatory actions to control toxic substances such as polychlorinated dibenzo-p-dioxins and polychlorinated dibenzofurans (US EPA, 1986; Quellmalz, 1988). On basis of such directives
federal governments may ban the production, distribution and use of chemicals contaminated by PCDD and PCDF. In addition to the ban on dioxin-
containing substances attemps were made to include limit values for
PCDD/PCDF levels in substances, preparations and products.

Similar reactions, as they were documented by the US-EPA study (Esposito et al., 1980) which are focused primarily on chlorinated substances,
are known to occur in the manufacture of brominated chemicals (US EPA,
1985; Versar, Inc.; Lee et al., 1986). In order to make sensible risk
assessments and to undertake measures to limit the health risk these substances pose, it is advisable to first research sources of formation of
polyhalogenated dibenzodioxins (PXDD) and dibenzofurans (PXDF). With regard to this, special attention is to be directed to the industrial processes of chlorochemistry. Much less information than on the chemistry
of chlorinated dibenzo-p-dioxins and dibenzofurans is available about
their brominated counterparts.

In the search for sources of chlorinated and/or brominated dibenzodioxin and diobenzofuran, the following processes have been listed according to their significance as potential sources (Heindl and Hutzinger,
1986):
1. Processes of the chemical industry:
    - Processes to manufacture chlorophenols and their derivatives,
    - Processes to manufacture chlorobenzenes and substituted chloro-
      benzenes,
    - Synthesis of aliphatic chlorine compounds,
    - Methods involving chlorine-containing intermediates,
    - Inorganic chlorochemical processes,
    - Processes applying chlorinated catalysts and solvents.
2. Processes of the pulp and paper industry;
3. Metallurgical Processes;
4. Processes for reactivation of granular carbon;
5. Processes to manufacture brominated flame-retardants (biphenyls, di-
    phenylethers, etc.).

3.2 THE CURRENT STATE OF KNOWLEDGE

Much theoretical work has led to ranking manufacturing processes of aromatic chlorine compounds according to the probability of their producing PCDD/PCDF. The most significant result of this work is a list of priorities published by the EPA (Esposito et al., 1980; German translation: UBA, 1985).

Analytical work about chlorophenol derivatives is cited in which PCDD/PCDF were detected in the two above mentioned reports (Esposito et al. 1980; UBA, 1985). Further publications are mentioned in the following sections.

Considerably less analytic data has been published about other chloroaromatics (i.e. excluding chlorophenols and chlorophenol derivatives), such as: dichlorobenile, difluorobenzurone, tetradifone (van Daalen), chloroanile (Hess et al., 1982), hexachlorobenzene (Villanueva et al., 1975) and PCB (Vos et al., 1970; Roach and Pomerantz, 1974; Morita et al., 1977; Bauer et al., 1978).

3.2.1 Chlorophenols and Their Derivatives

Because PCDD/PCDF are often found in chlorophenols and their derivatives, production processes involving these chemicals rank first on the priority list. A short summary about PCDD/PCDF formation in these processes will be given in the following section.

(1) Chlorophenols

Chlorophenols are either directly used as biocides or are largely used as starting compounds for pesticide production. 2,4,5-Trichlorophenol and 2,4,6-trichlorophenol act as effective fungicides, herbicides and defoliants (Hawley, 1971), whereas 2,4,5,6-tetrachlorophenol is used as an insecticide and as a wood and leather preservative. Because of its toxic effect on bacteria, mold, algae, fungi and yeast, pentachlorophenol and its sodium salt have been widely applied. In Germany respectively 226 and 56 tons of these compounds have been used annually (UBA, 1985).

Production

There are two important manufacturing processes for chlorophenols:

(a) Phenol Chlorination

Phenol is chlorinated directly by chlorine gas and aluminium chloride as catalyst in counter current (US Patent, 1960).

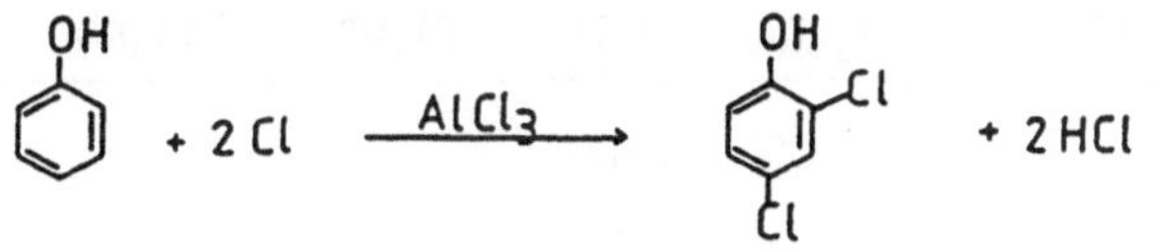

Depending on the amount of chlorine of starting materials applied, various types of higher chlorinated phenols are formed.

With this method, 2- and 4-monochlorophenol, 2,4-dichlorophenol, 2,4,6-trichlorophenol, 2,4,5,6-tetrachlorophenol and pentachlorophenol are produced. Products are purified by distillation.

(b) Basic Hydrolysis of Chlorobenzenes

In this method, the corresponding chlorobenzene and aqueous/methanolic sodium hydroxide are heated to 170 -200 °C (US Patent, 1967).

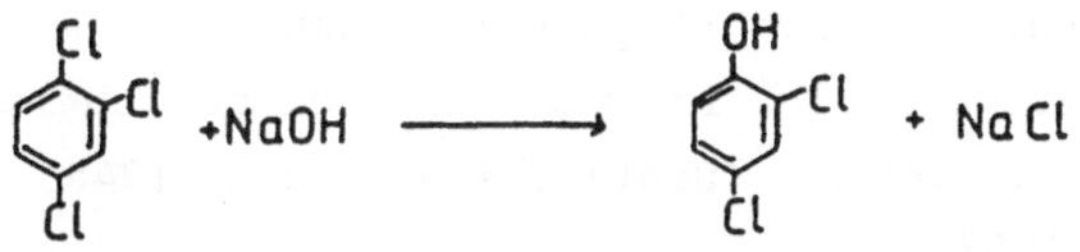

(2) PCDD/PCDF Formation during Chlorophenol Manufacture

PCDD/PCDF found in chlorophenols are compiled in Table 1.

Table 1.   Contents of PCDD/PCDF in Chlorophenols [ppm]

Chloro-phenol	Cl_2DD	Cl_3DD	Cl_4DD	Cl_5DD	Cl_6DD	Cl_7DD	Cl_8DD
				PCDD			
2-[*]			0.037				
2,4-[*]			-				
2,6-[*]			-				
2,4,5-[*]	0.72		0.30	1.5			
2,4,6-[*]		93	49				
2,3,4,6-[*]					4.1-29	5.1	0.17
penta [*+]			1.25	0.08	42	870	3,660

Chloro-phenol	Cl_2DF	Cl_3DF	Cl_4DF	Cl_5DF	Cl_6DF	Cl_7DF	Cl_8DF
				PCDF			
2,4,6-[#]		1.5	2.3-17.5	0.7-36	0.02	4.8	
2,3,4,6-[#]		0.5	10	70	70	10	
penta [#+]			0.4-0.9	4-40	0.03-90	0.8-400	1.3-260

[*]   Esposito et al, 1980
[#]   Crosby, 1981
[+]   Rappe and Buser, 1980.

By the direct chlorination of phenols, PCDD form when chlorophenols are purified by distillation.   In this case, either thermal condensation of the chlorophenols occurs (Deniville et al., 1948; Gribble, 1947; Sandermann et al., 1957)

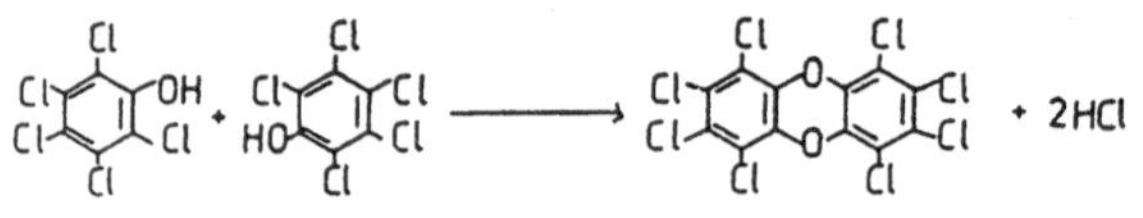

300°C. 4 %

438

or the chlorophenol is thermally condensed with hexachlorocyclohexa-
dienone which forms by excessive chlorination of the applied phenol (Kulka, 1965).

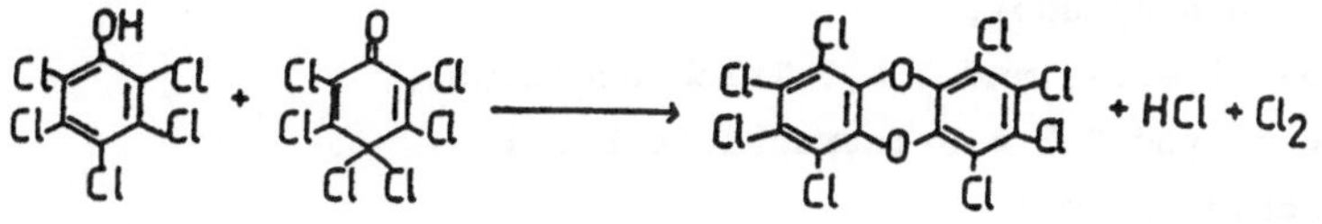

250 - 300°C, 83%

During basic hydrolysis, PCDD are formed through chlorophenolate condensation:

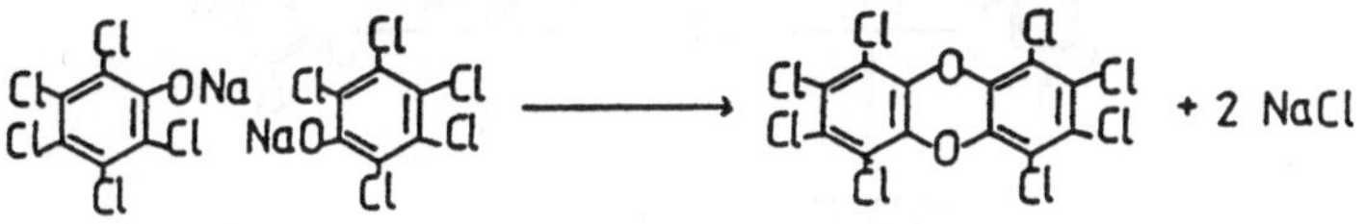

170 - 200°C

Furthermore, PCDD can form by means of radical reactions, because
chlorine can be homolytically split during phenol chlorination at higher
temperatures and can therefore act as the radical starter (Kulka, 1961).

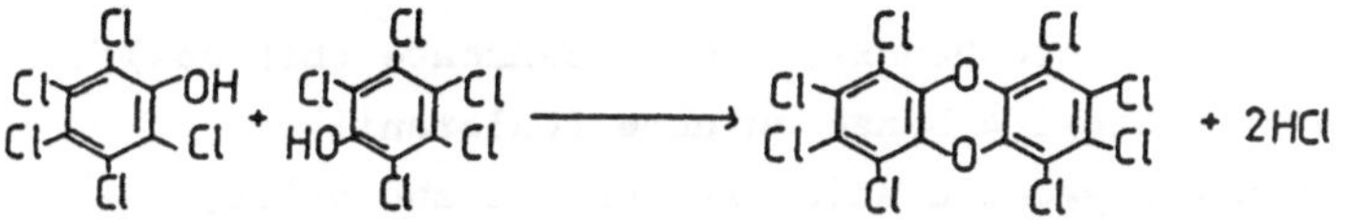

Cl$_2$, 1,2,4-Trichlorobenzene, 83%

To summarize, high temperatures and/or alkaline conditions enhance
PCDD/PCDF formation. In addition, halogens together with high temperatures as well as the presence of halogens and UV radiation or other radical starters during chlorophenol manufacture promote PCDD/PCDF formation.

Since all chlorophenol syntheses use one of the above mentioned processes, PCDD/PCDF can be formed in all of them.

Technically significant chlorophenols are listed as follows (UBA, 1985):

2-, 3-, and 4-Monochlorophenol,
2,4-Dichlorophenol,
2,4,5-, 2,4,6-, and 3,4,5-Trichlorophenol,
2,3,4,5- and 2,3,4,6-Tetrachlorophenol,
Pentachlorophenol.

PCDD/PCDF can also be formed from chlorophenols during manufacture of herbicides, insecticides and bactericides. The synthesis of the herbicide Silvex is depicted as an example of this manufacturing procedure.

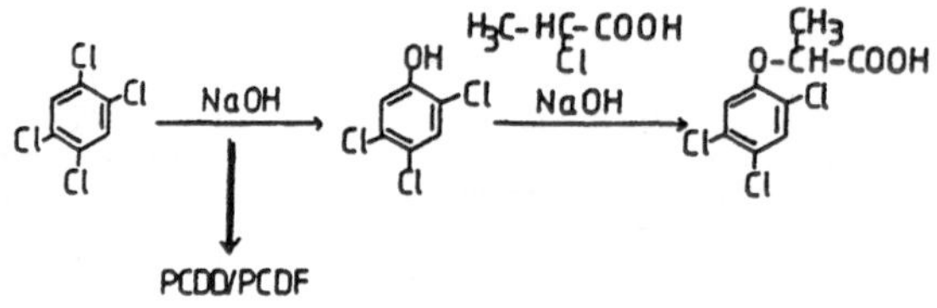

PCDD/PCDF were found in the following biocides (Esposito et al., 1980; Cochrane, 1980; Cochrane et al., 1982; Helling and Isensee, 1973; Rappe et al., 1983; Rappe, 1979):

Methyl-5-(2,4-dichlorophenoxy)-2-nitrobenzoate (Bifenox);
2,3,5,6-Tetrachloro-1,4-benzo quinone (Chloranil);
2,4-Dichlorophenoxyacetic acid, its esters and salts;
3,6-Dichloro-2-methoxybenzoic acid (Dicamba);
4-(2,4-Dichlorophenoxy)butanoic acid and its salts (2,4-DB);
O-(2-Chloro-4-nitrophenyl)O,O-dimethylphosphorothioate (Dicapthon);
O-(2,4-Dichlorophenyl)O,O-diethylphosphorothioate (Dichlofenthion);
2-(2,4-Dichlorophenoxy)ethyl hydrogen sulfate (Disul);
2-(2,4-Dichlorophenoxy)propanoic acid (Dichlorprop, 2,4-DP);
2-(2,4,5-Trichlorophenoxy)-ethyl-2,2-dichloropropionate (Erbone);
2,2'-Methylene-bis-(1-hydroxy-3,4,6-trichlorophenol) (Hexachlorophene);
2,4-Dichlorophenyl-4-nitrophenylether (Nitrophen);
O,O-Dimethyl-O(2,4,5-trichlorophenyl)phosphorothioate (Ronnel/Fenchlorphos);
2-(2,4,5-Trichlorophenoxy)-propionic acid (Silvex);

2,4,5-Trichlorophenoxyacetic acid, its esters and salts (2,4,5-T);

and chlorophenols:

Pentachlorophenol and its salts (PCP);
2,3,4,6-Tetrachlorophenol;
2,4,5-Trichlorophenol.

According to the cited literature data, all processes involving the manufacture of chlorophenol derivatives also are suspected of producing PCDD and PCDF. Most likely all technical chlorophenols and chlorophenol derivatives contain small amounts of PCDD and PCDF.

3.2.2 Chlorobenzenes and Substituted Chlorobenzenes

The probability of PCDD/PCDF formation during industrial production varies within the class of chloroaromatics. Criteria are:

- Oxygen as a nuclear substituent;
- The production and/or purification of the substance under alkaline conditions;
- Reaction temperatures above 150 °C.

(1) Chlorobenzenes with Oxygen and Other Groups on the Aromatic Nucleus

With less probability than in chlorophenol manufacture, PCDD/PCDF can form during the production of chloroaromatics which carry oxygen or other groups on the aromatic nucleus. This primarily concerns processing conditions which deviate from the normal operation (UBA, 1985).
The suspected products are listed as follows (Esposito et al., 1980):

2-Benzyl-4-chlorophenol
3,5-Dibromo-4-hydroxybenzonitrile (Bromoxynil)
S-(4-Chlorophenylthio)methyl-O,O-diethyl-phosphorodithioate
  (Carbophenothion)
Dimethyl-(2,3,5,6-tetrachloro-1,4-benzodicarbonate)
4-Chloro-2-methylphenoxy acetic acid (MCPA)
4-(4-Chloro-2-methylphenoxy)-butyric acid (MCPB)
2-(4-Chloro-2-methylphenoxy)-propanoic acid (Mecoprop)
O,O-Diethyl-O-nitro phosphorothioate (Parathion)
4-Chlorophenyl-2,4,5-trichlorophenylsulphone (Tetradifon)
1,4-Dichloro-2,5-dimethoxybenzene (Chloroneb).

(2) Chloroaromatics without Nuclear Oxygen

Processes involving the manufacture of chloroaromatics rank second on the previously cited priority list. Because PCDD/PCDF have been found in hexachlorobenzene and polychlorinated biphenyls, this classification is warranted with respect to the probability of PCDD/PCDF formation.

Application

Chlorobenzenes are primarily employed as starting compounds for the synthesis of other organic compounds. Polychlorinated biphenyls (PCB) are no longer produced in technically advanced countries but are still in use in electrical application.

Manufacture

The chlorination of aromatics such as benzene, biphenyl and naphthalene is carried out by the following methods:

a) chlorination with chlorine gas together with metal chlorides as catalysts is most commonly applied:

A

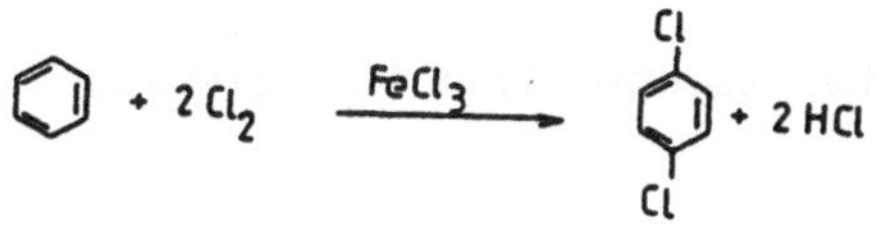

$$\text{benzene} + 2\,Cl_2 \xrightarrow{FeCl_3} \text{dichlorobenzene} + 2\,HCl$$

b) oxychlorination to produce lower substituted chlorobenzenes:

B

$$\text{benzene} + HCl + \tfrac{1}{2}O_2 \longrightarrow \text{chlorobenzene} + H_2O$$

c) the manufacture of lindane from chlorobenzene by means of dehydrochlorination of the non-insecticidial hexachlorocyclohexane isomers (De Bruin, 1978):

C

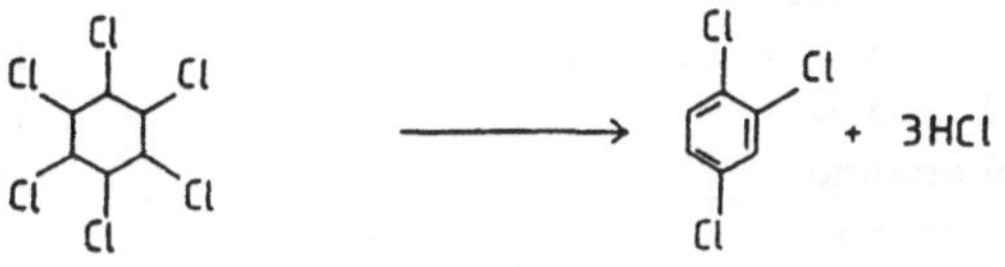

$$\text{hexachlorocyclohexane} \longrightarrow \text{trichlorobenzene} + 3\,HCl$$

PCDD/PCDF formation is not specifically derived from the three processes mentioned above but probably occurs during purification where often alkaline conditions are used (Offhaus, 1983).

In hexachlorobenzene, 0.3-58.3 ppm Cl_8DF and 0.05-212 ppm Cl_8DD were found, whereas 0.8-13.6 ppm PCDF were detected in polychlorinated biphenyls (UBA, 1985; Bowes, et al., 1975).

These values show, that all aromatic chlorinations are suspect of causing PCDD/PCDF formation. Moreover, radical side-chain chlorinations such as the chlorination of toluene are also relevant, because a small degree of nuclear chlorination can also occur. The following products are manufactured by chlorination or radical chlorination:

Chloronitrobenzenes
Chloroaniline
2,3,6-Trichlorophenylacetic acid (Chlorfenac).

(a) Trichlorobenzene

Hagenmaier et al. (Hagenmaier, in press) investigated trichlorobenzene as the starting material (reactant) for the production of PCP as well as residues from the production of PCP-Na after filtering. In all samples detectable amounts of PCDD/PCDF (including Cl_4DD, Cl_4DF) were found (see Tables 2-5).

Table 2.   Concentration of Chlorobenzenes, γ-HCH, Decachlorobiphenyl in a Commercial Sample of Trichlorobenzene (Sample taken: 29.6.1984)

Sample	Trichlorobenzene [g/kg]
Trichlorobenzenes	
1,3,5	42.3
1,2,4	262.3
1,2,3	166.7
Tetrachlorobenzenes	
1,2,3,5 + 1,2,4,5	159.4
1,2,3,4	160.5
Pentachlorobenzene	48.5
Hexachlorobenzene	4.8
γ-HCH	18.5
Decachlorobiphenyl	< 0.01

Table 3.   Concentrations of PCDD/PCDF in a Commercial Sample of Trichlorobenzene (Sample taken: 29.6.1984)

Sample	Trichlorobenzene
Σ Cl_4DD	26.6
$-2,3,7,8-Cl_4DD$	0.095
Σ Cl_5DD	139.8
$-1,2,3,7,8-Cl_5DD$	4.55
Σ Cl_6DD	258.8
$-1,2,3,4,7,8-Cl_6DD$	6.29
$-1,2,3,6,7,8-Cl_6DD$	27.46
$-1,2,3,7,8,9-Cl_6DD$	12.59
Σ Cl_7DD	253.00
$-1,2,3,4,6,7,8-Cl_7DD$	134.67
Cl_8DD	80.6
Σ PCDD	758.8
Σ Cl_4DF	735.7
$-2,3,7,8-Cl_4DF$	42.68
Σ Cl_5DF	272.5
$-1,2,3,7,8-Cl_5DF$	28.21
$(+1,2,3,4,8-Cl_5DF)$	
$-2,3,4,7,8-Cl_5DF$	14.73
Σ Cl_6DF	90.8
$-1,2,3,4,7,8-Cl_6DF$	11.19
$(+1,2,3,6,7,9-Cl_6DF)$	
$-1,2,3,6,7,8-Cl_6DF$	4.26
$-1,2,3,7,8,9-Cl_6DF$	ND
$-2,3,4,6,7,8-Cl_6DF$	0.47
Σ Cl_7DF	29.9
$-1,2,3,4,6,7,8-Cl_7DF$	23.51
$-1,2,3,4,7,8,9-Cl_7DF$	1.34
Cl_8DF	16.4
Σ PCDF	1,145.3

Table 4. Concentration of Organic Pollutants in the Sludge (after filtering) from PCP-Na Production (Sample taken: 5.4.1984); Concentrations in [g/kg dry weight]

Sample	Residues PCP-Na Production
Hexachlorobenzene	355
Cl_8DD	8
Cl_8DF	1
Decachlorobiphenyl	66
Decachlorodiphenylether	3
Pentachlorophenol	51

Table 5. Concentrations of PCDD/PCDF in the Sludge from PCP-Na Production (Sample taken: 5.4.1984); [mg/kg dry weight]

Sample	Residues PCP-Na Production
$\Sigma\ Cl_4DD$	0.333
$-2,3,7,8-Cl_4DD$	0.035
$\Sigma\ Cl_5DD$	0.484
$-1,2,3,7,8-Cl_5DD$	0.058
$\Sigma\ Cl_6DD$	0.852
$-1,2,3,4,7,8-Cl_6DD$	0.095
$-1,2,3,6,7,8-Cl_6DD$	0.077
$-1,2,3,7,8,9-Cl_6DD$	0.062
$\Sigma\ Cl_7DD$	435
Cl_8DD	7,656
$\Sigma\ PCDD$	8,092.669

Table 5. Continued

Sample	Residues PCP-Na Production
Σ Cl_4DF	0.028
-2,3,7,8-Cl_4DF	<0.005
Σ Cl_5DF	0.305
-1,2,3,7,8-Cl_5DF	0.017
(+1,2,3,4,8-Cl_5DF)	
-2,3,4,7,8-Cl_5DF	0.007
Σ Cl_6DF	0.505
-1,2,3,4,7,8-Cl_6DF	0.152
(+1,2,3,6,7,9-Cl_6DF)	0.048
-1,2,3,6,7,8-Cl_6DF	4.26
-1,2,3,7,8,9-Cl_6DF	0.003
-2,3,4,6,7,8-Cl_6DF	0.024
Σ Cl_7DF	52
Cl_8DF	870
Σ PCDF	922.838

Further investigations of 26 distillation residues from production sites considered suspect of containing PCDD/PCDF, e.g. recycling of eluents, did not reveal such industries as sources. Only in residues involving use of pentachlorophenol PCDD/PCDF have been detected.

(b) Other Chloroaromatics

Concerning PCDD/PCDF formation, the EPA priority list of 80 of the most likely to contain PCDD/PCDF aromatics was compiled under consideration of these prerequisites (Esposito et al., 1980).

According to tests performed at the University of Bayreuth, Dept. of Ecological Chemistry, on chlorobenzenes, chlorotoluenes and chloroanilines, four of the total six examined chlorobenzenes contained PCDD/PCDF (see Table 6). The table demonstrates, that with a decreased level of chlorination, the occurrence of lower chlorinated PCDD/PCDF increases. Up to now only the analytic results from Villanueva (Villanueva et al., 1975) concerning the PCDD/PCDF content of hexachlorobenzene were published which are similar to our results.

Table 6. PCDD/PCDF Concentrations in Chlorobenzenes

	Cl$_8$DD	Cl$_8$DF	Cl$_7$DD	Cl$_7$DF	Cl$_6$DD	Cl$_6$DF [ppb]	Cl$_5$DD	Cl$_5$DF	Cl$_4$DD	Cl$_4$DF	Det.Limit [ppb]
Hexachloro-benzene (97%)	6,700	2,830	470	455	-	-	-	-	-	--	20
Pentachloro-benzene (pure, 98% Cl)	0.05	0.1	0.02	0.1	0.02	-	-	-	-	0.02	0.02
1,2,4,5-Tetra-chlorobenzene (99% Cl)	0.4	2.1	0.8	1.5	0.5	0.8	0.2	0.2	-	0.03	0.02
1,2,4-Trichloro-benzene (pure)	-	-	-	-	-	-	-	-	-	-	0.1
1,2-Dichloro-benzene (for synthesis)	-	-	-	-	-	-	-	0.5	0.3	-	0.02
Chlorobenzene	-	-	-	-	-	-	-	-	-	-	0.02

Table 7. PCDD/PCDF Concentrations in Chlorotoluenes

	Cl$_8$DD	Cl$_8$DF	Cl$_7$DD	Cl$_7$DF	Cl$_6$DD	Cl$_6$DF [ppb]	Cl$_5$DD	Cl$_5$DF	Cl$_4$DD	Cl$_4$DF	Det.Limit [ppb]
2-Chlorotoluene (for synthesis)	-	-	-	-	-	-	-	-	-	-	0.05
4-Chlorotoluene (98%)	1.1	-	-	-	-	-	-	-	-	-	0.05
3,4-Dichloro-toluene (pract.)	-	-	-	-	-	-	-	-	-	-	0.05

Table 8. PCDD/PCDF Concentrations in Chloroanilines

	Cl$_8$DD	Cl$_8$DF	Cl$_7$DD	Cl$_7$DF	Cl$_6$DD	Cl$_6$DF [ppb]	Cl$_5$DD	Cl$_5$DF	Cl$_4$DD	Cl$_4$DF	Det.Limit [ppb]
2-Chloroaniline	-	-	-	-	-	-	-	-	-	-	0.02
4-Chloroaniline (pract.)	-	-	-	-	-	-	-	-	-	-	0.02
2,4,5-Trichloro-aniline (techn.)	-	-	-	-	-	-	-	-	-	-	0.02

The test results of chlorotoluenes are compiled in Table 7. The occurrence of Cl_8DD in 4-monochlorotoluene is astonishing, in that none of the above mentioned criteria necessary for PCDD/PCDF formation apply during the production of this substance. Therefore, it must be assumed, that the dioxin is introduced through the employed iron(III)chloride catalyst.

No PCDD/PCDF were detected in any of the samples of 2-chloroaniline, 4-chloroaniline and 2,4,5-trichloroaniline (detection limit: 20 ppt). Moreover, because of the production methods applied (hydrogenation of chloronitrobenzenes with noble metal catalysts or noble metal-sulfide catalysts at 50-100 °C with dehalogenation occurring as a side reaction), their formation was not anticipated (see Table 8).

The analysis of chlorophenols, their derivatives as well as chlorinated benzenes shows, that their classification in the compiled priority list is justified with respect to the PCDD/PCDF content. Furthermore, the analytic results may eventually indicate the mechanism of PCDD/PCDF formation. In the following sections the chlorophenol-derivated pesticides, bromophos and hexachlorophene are compared with other pesticides which are not chlorophenol derivatives, namely; lindane, daconil, diuron and ustinex PA.

- Bromophos

In 1964, bromophos was introduced as a widely effective, agricultural contact insecticide having low toxicity for warm blooded animals and was used to combat hygienic pests. Bromophos is manufactured by heating 0,0-dimethyl phosphorochloridothioate with sodium 4-bromo-2,5-dichloropheno-late at 100 °C (Boehringer, 1961a,b).

PCDD/PCDF formation is expected here, because of the reaction of chlorophenol and NaOH.

During processing, bromophos is hydrolyzed with KOH to the correspon-ding chlorobromophenolate and is separated by means of extraction. The separations correspond to the methods described elsewhere (Nestrick and Lamparski, 1980), and the detection limit for PCDD/PCDF lies at 5 ppb.

Although no detectable amounts of Cl_4DD, Cl_4DF, Cl_5DD and Cl_5DF were found in bromophos, it contained, however, 25 ppb Cl_6DD, 400 ppb Cl_7DD, 800 ppb Cl_8DD, 10 ppb Cl_6DF, 30 ppb Cl_7DF and 45 ppb Cl_8DF.

- Hexachlorophene

Used as an industrial and houshould disinfectant, hexachlorophene is also an effective agricultural bactericide and fungicide.

It is manufactured by reacting 2,4,5-trichlorophenol with formalde-hyde in an acidic medium (Moye, 1972) at either 5, 50-100 or 135 °C, de-pending on the applied procedure (Kimbrough, 1974).

Because of the acidic reaction medium and temperatures which are too low for PCDD/PCDF formation (UBA, 1985), it can be assumed, that PCDD/PCDF do not occur during hexachlorophene production. In the hexachlorophene sample we tested, 0.5 ppb 2,3,7,8-Cl_4DD were detected, which probably stems from contamination of the employed 2,4,5-trichlorophenol.

- Lindane

The gamma isomer of hexachlorocyclohexane is considerably significant technically as an insecticide. Worldwide, an estimate of 200,000 tons are applied per year.

Production takes place through addition of chlorine onto benzene:

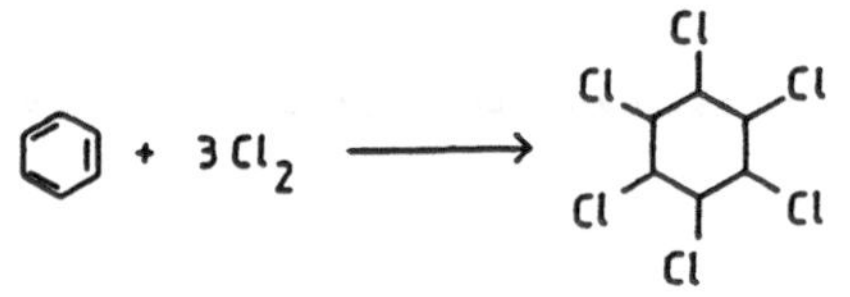

Because hexachlorobenzene can form as a by-product during this process, PCDD/PCDF may then arise during the purification of the product.

No PCDD/PCDF were found in the lindane examined at the University of Bayreuth.

- Daconil

Introduced by the firm Diamond Shamrock in 1963, daconil is used as an agricultural fungicide.

It is synthesized by reacting tetrachloro-isophthaloyl chloride with ammonia to the diamide and then dehydrating with phosphoryl-chloride:

$$\text{(tetrachloro-isophthaloyl chloride)} \xrightarrow{NH_3} \text{(tetrachloro-isophthalamide)} \xrightarrow{POCl_3} \text{(tetrachloro-isophthalonitrile)}$$

No PCDD or PCDF should form according to this reaction equation. The daconil sample we tested contained 67 ppb Cl_8DF. This analysis demonstrates, that ranking chloroaromatics without nuclear oxygen in the list of priorities is justified.

- Diuron

Diuron (3-(3,4-dichlorophenyl)-1,1-dimethyl urea), a herbicide which was first marketed in 1955, is the most important herbicide of its class today. The West German production of diuron amounted to 13600 tons in 1982. To manufacture diuron, dichloroaniline is reacted with urea to dichlorophenylisocyanate which is then added to dimethylamine:

$$\text{(3,4-dichloroaniline)} + H_2N-\underset{\underset{O}{\|}}{C}-NH_2 \longrightarrow \text{(3,4-dichlorophenyl isocyanate, } N{=}C{=}O) + 2\,NH_3$$

$$\text{(3,4-dichlorophenyl isocyanate)} + H-N\begin{smallmatrix}CH_3\\CH_3\end{smallmatrix} \longrightarrow \text{(diuron)}$$

Another method of synthesis is the reaction of dichloroaniline with phosgene to dichlorophenylisocyanate:

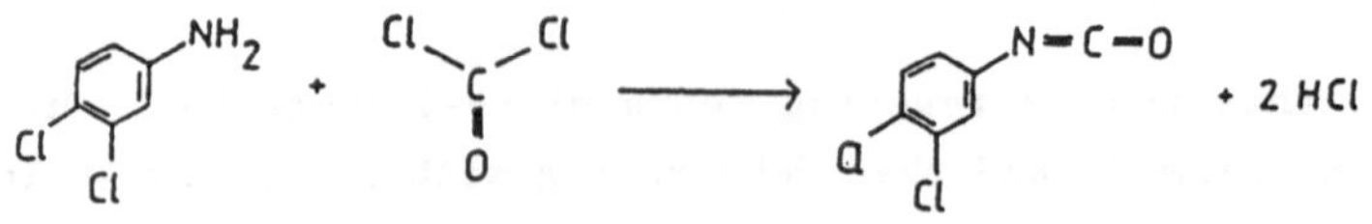

The examination of diuron showed 7 ppb Cl_8DD.

- Ustinex PA

Ustinex PA, a herbicide used against weeds and moss, contains 30% amitrol (3-amino-1,2,4-triazole) and 56% diuron 3-(3,4-dichlorophenyl)-1,1-dimethylurea as active ingredients.

In one diuron sample, 7 ppb Cl_8DF and 2 ppb Cl_8DD were found. With a detection limit of 30 ppt, no PCDD or PCDF were discovered in marketed ustinex PA dusting powder used against weeds and moss.

3.2.3 Chlorinated Aliphatics

The classification of chloroaliphatics in third place on the priority list of probability of PCDD/PCDF formation is arbitrary, because to date PCDD/PCDF have been detected in chloroaliphatics only in a few cases. Because the products of aliphatic chlorine chemistry are mostly quite volatile, PCDD/PCDF would be expected in the residues rather than in the products.

Application

Produced in large quantities, the products of aliphatic chlorine chemistry are diverse monomers, serve as starting materials for the synthesis of other organic compounds products or are used as solvents.

454

(1) Manufacture

Chloroaliphatics and derivatives are produced by the following methods:

a) Chlorination: through substitution

$$H_3CCl + Cl_2 \longrightarrow H_2CCl_2 + HCl$$

as well as by addition;

$$H_2C=CH_2 + Cl_2 \longrightarrow ClH_2C-CH_2Cl$$

b) Hydrochlorination with HCl (addition onto a double bond);

$$H_2C=CH_2 + HCl \longrightarrow H_3C-CH_2Cl$$

c) Oxychlorination: advantageous because cheaper HCl and atmospheric oxygen can be used instead of expensive chlorine;

$$H_2C=CH_2 + 2\ HCl + 1/2\ O_2 \longrightarrow ClH_2C-CH_2Cl + H_2O$$

d) Dehydrochlorination: a C=C double bond is formed through elimination of HCl;

$$ClH_2C-CHCl-CH=CH_2 + NaOH \longrightarrow H_2C=CCl-CH=CH_2 + H_2O + NaCl$$

e) Alkaline hydrolysis to transform the chloroaliphatics into aliphatic alcohols;

$$H_2C=CH-CH_2Cl + NaOH \longrightarrow H_2C=CH-CH_2OH + NaCl$$

f) Chlorohydrocarbon residues are recycled by means of chlorolysis, whereby chloroaliphatic by-products are used to produce perchloro-ethylene and carbon tetrachloride.

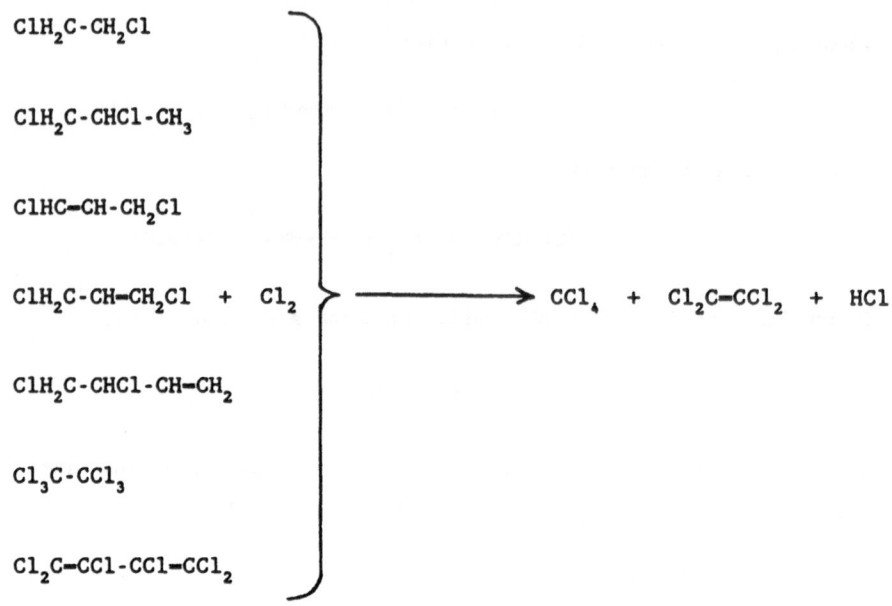

ClH_2C-CH_2Cl

$ClH_2C-CHCl-CH_3$

$ClHC=CH-CH_2Cl$

$ClH_2C-CH=CH_2Cl \quad + \quad Cl_2$

$ClH_2C-CHCl-CH=CH_2$

Cl_3C-CCl_3

$Cl_2C=CCl-CCl=CCl_2$

$\longrightarrow \quad CCl_4 \quad + \quad Cl_2C=CCl_2 \quad + \quad HCl$

(2) Possible PCDD/PCDF Formation Routes

PCDD/PCDF formation mechanisms in aliphatic chlorine chemistry have not yet been reported in the literature. It is assumed, that PCDD/PCDF may first occur by way of aromatics or chlorinated aromatics that are formed from the aliphatics. The incorporation of oxygen needed for the PCDD/PCDF structure could occur by substitution of a chlorine atom. Sodium hydroxide (during processing of the raw product) or simply atmospheric oxygen could hereby act as oxygen source.

The occurrence of (chlorinated) aromatics in a side reaction is necessary for this formation mechanism. The following equilibrium results by the pyrolysis of chlorinated hydrocarbons like CCl_4, C_2Cl_4, C_2Cl_6, C_4Cl_6 and C_6Cl_6:

400 - 450°C

Although the manufacture of chloroaliphatics does not actually represent a pyrolysis, these reactions can certainly occur as the result of high temperatures needed for production.

(3) Short Chain Chloroaliphatic Compounds

Two samples of hexachlorobutadiene were found to contain 425 and 360 ppt of Cl_8DF, respectively (Heindl and Hutzinger, 1987).

Mechanistic studies in the formation of PCDD/PCDF from short chain chloroaliphatic compounds indicate dichloroacetylene to be a key intermediate in the production of chlorinated aromatics.

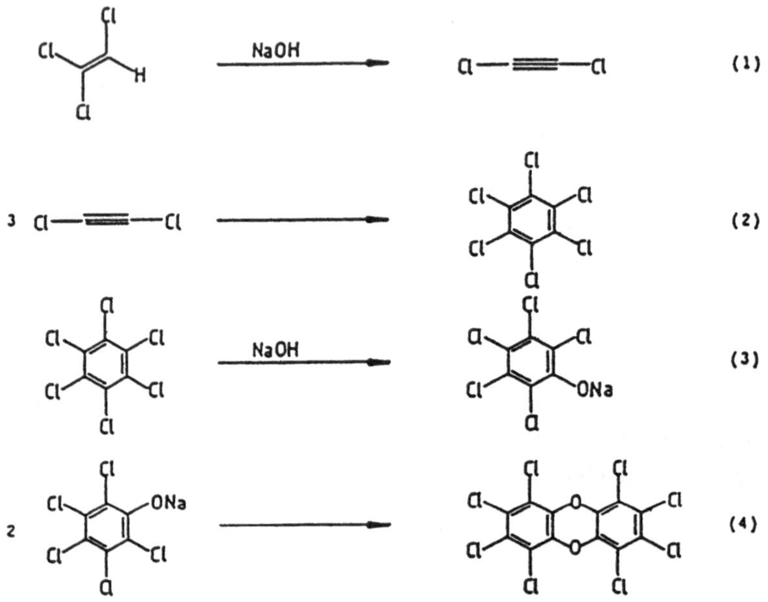

Fig. 1. Suggested Pathway for PCDD/PCDF Formation in the Reaction of Trichloroethylene with NaOH (Heindl and Hutzinger, 1987)

When trichloroethylene is heated with 2 N sodium hydroxide some higher chlorinated dioxins and furans were formed at levels 100 to 1,300 ppt in addition to the chlorinated compounds tentatively identified and presented in Figure 1.

Hydroxy radical donors such as Fenton's Reagent ($FeSO_4/H_2SO_4/H_2O_2$) hydroxylate chloroaromatic compounds such as chlorobenzenes and chlorinated biphenyls. 1,2,4-Trichlorobenzene under these conditions gives 180 ppm Cl_5DF and 1,160 ppm Cl_4DF (Heindl and Hutzinger, 1987).

458

3.2.4 Processes with Chlorinated Intermediates

All industrial chemical processes by which chlorine or hydrogen chloride is cleaved from intermediate products must be critically regarded with respect to PCDD/PCDF formation. Such chemicals and substances which contain no chlorine in the end product (in particular those resulting from eliminations, dehydrochlorinations and alkaline hydrolyses) should be especially considered.

3.2.5 Inorganic Chlorine Chemistry

The processes to manufacture metal chlorides, chlorine, NaOCl and hydrochloric acid are suspected of causing PCDD/PCDF formation.

Tests of metal chlorides which were conducted at the University of Bayreuth show, that anhydrous iron(III)chloride and aluminium- chloride contain detectable amounts of Cl_8DF and Cl_7DF (see below). Although copper(I)chloride and copper(II)chloride contain considerably lower amounts of these substances, they have the higher chlorinated dioxins Cl_8DD and Cl_7DD. With a detection limit of 30 ppt, no PCDD/PCDF could be found in the rest of the tested metal chlorides; namely, in titanium(IV)chloride, silicon(IV)chloride and zinc(II)chloride (see Table 10).

The metal chlorides were furthermore examined for their content of chloroaromatics such as tetrachlorobenzene, pentachlorobenzene, hexachlorobenzene, octachlorostyrene, tetrachlorobiphenyl, nonachlorobiphenyl and decachlorobiphenyl. The results infer, that the amount of hexachlorobenzene correlates with the amount of Cl_8DF (and Cl_8DD). It has already been suggested (Olie et al., 1977), that a group of stable substances including hexachlorobenzene and PCDD/PCDF forms during combustion processes.

(1) Chlorine and Chlorine Derivatives

The results of the PCDD/PCDF analyses of chlorine gas, sodium hypochlorite solutions and technical hydrochloric acid are shown in Table 9.

Chlorinated aromatics could only be detected in small amounts in
chlorine gas. It is obvious, however, that the lower chlorinated instead
of the fully chlorinated compounds dominate. The analyses demonstrate,
that chloroaromatics are introduced by applying chlorine gas during
industrial syntheses. Under the given reaction conditions, these compounds can further react to either the fully chlorinated species or also
to PCDD/PCDF precursors.

Table 9.   Amounts of PCDD/PCDF and Chloroaromatics in Chlorine Gas,
Sodium Hypochlorite Solutions and Hydrochloric Acid.

$Cl_{10}BP$ — Decachlorobiphenyl Cl_9BP — Nonachlorobiphenyl
Cl_8Sty — Octachlorostyrene Cl_6B — Hexachlorobenzene
Cl_4B — Tetrachlorobenzene

	PCDD/PCDF	$Cl_{10}BP$	Cl_9BP [ppt]	Cl_8Sty	Cl_6B	Cl_4B	Det.Limit [ppt]
Cl_2 (No. 165265)	-	40	80	80	-	210	4
NaOCl (techn.,13%)	-	-	-	-	3	-	1
NaOCl (10%)	-	-	-	-	1	4	1
NaOCl	-	-	-	-	1	4	1
HCl (techn.,31-33%)	-	-	-	-	-	-	1

(2) Sources of PCDD/PCDF and Other Chloroaromatics in Metal Chlorides

Because no PCDD or PCDF could be detected in chlorine gas, the use of Cl_2 in industrial processes cannot explain the occurrence of chloroaromatics. The source of metal chloride contamination can certainly be the metals used during synthesis. For example, iron wastes and scrap are applied for the production of iron(III)chloride, whereas aluminium and aluminium wastes are employed for the manufacture of aluminium chloride. These wastes originate from the metal processing industry and are contaminated with cutting, lubricating and cooling oils. These fats and oils can represent PCDD/PCDF precursors.

The addition of coal during the manufacture of titanium(IV)chloride and silicon-chloride could be responsible for the formation of chlorinated aromatics.

$FeCl_3$, $AlCl_3$, $CuCl_2$, CuCl, $SiCl_4$, and $TiCl_4$ were analysed for their content of PCDD/PCDF and other chloroaromatic compounds. $FeCl_3$ and $AlCl_3$ contain Cl_8DF and Cl_7DF in ppb- and hexachlorobenzene in ppm-range; $CuCl_2$ and CuCl contain lower concentrations of Cl_8DF, Cl_7DF and additionally Cl_8DD, Cl_7DD and hexachlorobenzene in ppb/ppt-range. $SiCl_4$ and $TiCl_4$ contain no measurable quantities of PCDD/PCDF and hexachlorobenzene. A correlation between the PCDD/PCDF- and hexachlorobenzene-content is postulated for products of industrial syntheses involving high temperatures. The origin of the organic substances in a process of inorganic chemistry could be oils on the waste metals which are used in the manufacture of metal chlorides (Heindl and Hutzinger, 1986).

Table 10. PCDD/PCDF and Chloroaromatic Compounds in Several Metal Chlorides. Amounts in [ppb]; limit of detection 20 ppt.

$Cl_{10}BP$ — Decachlorobiphenyl Cl_9BP — Nonachlorobiphenyl
Cl_8Sty — Octachlorostyrene Cl_6B — Hexachlorobenzene
Cl_4B — Tetrachlorobenzene

	Cl_8DF	Cl_7DF	Cl_8DD [ppb]	Cl_7DD	$Cl_{10}BP$	Cl_9BP	Cl_8Sty [ppb]	Cl_6B	Cl_4B
$FeCl_3$	42	12	-	-	830	-	280	4000	-
$AlCl_3$ (A)	-	-	-	-	-	-	-	-	-
$AlCl_3$ (B)	34	-	0.1	-	-	200	-77	1100	-
$CuCl_2$	0.5	0.1	0.6	0.03	0.08	0.09	0.04	4	-
CuCl	0.2	0.08	0.03	-	-	-	0.3	1	0.2
$TiCl_4$	-	-	-	-	0.02	-	-	-	-
$SiCl_4$	-	-	-	-	0.02	-	-	-	-

3.2.6 Processes Using Chlorinated Catalysts and Solvents

From the previous statements follows, that all processes that take place with chlorinated metal catalysts or which occur in chlorinated solvents must be critically considered with respect to PCDD/PCDF formation (e.g. Friedel-Crafts alkylations, aromatic substitutions).

3.3 ANALYSIS OF ALIPHATIC CHLORINE COMPOUNDS

Aliphatic chlorine compounds are of all industrial organic chlorine compounds produced in largest quantities and are employed as monomers for plastics, solvents, cleaning agents and as precursors for chemical syntheses.

Because the most industrially significant, aliphatic chlorine compounds have a low boiling point, PCDD/PCDF eventually formed during their manufacture should mostly accumulate in production residues.

3.3.1 Chloroparaffins

Chloroparaffins are produced by chlorinating hydrocarbons at 90 °C. The technical chloroparaffins which form are mixtures of products having different numbers of carbon and chlorine atoms. Because the products are not processed, no residues arise, and all by-products which could eventually contain PCDD or PCDF stay in the chloroparaffin. PCDD or PCDF could not be found in any of the three chloroparaffins tested (detection limit: 0.4 ppb) (Heindl and Hutzinger, 1987).

3.3.2 Volatile Aliphatic Chlorine Compounds

With an average detection limit of 10 ppt, no PCDD/PCDF were found in trichloroethane, trichloroethylene and allyl chloride. Parts-per-trillion amounts of highly chlorinated PCDD/PCDF were contained in every sample of 1,2-dichloroethane, tetrachloroethylene and epichlorohydrin. This is unusual, insofar that here relatively easily volatile chloroaliphatics are involved which are simply separated from PCDD/PCDF during distillation. Local overheating which regularly occurs in large industrial installations could be responsible for PCDD/PCDF formation. It is noteworthy that one sample of technical tetrachloroethylene exhibited no PCDD or PCDF, whereas a sample labeled as 98% pure did contain PCDD/PCDF.

Table 11. PCDD/PCDF Concentrations in Aliphatic Chlorine Compounds

	Cl$_8$DD	Cl$_8$DF	Cl$_7$DD	Cl$_7$DF	Cl$_6$DD	Cl$_6$DF [ppb]	Cl$_5$DD	Cl$_5$DF	Cl$_4$DD	Cl$_4$DF	Det.Lim. [ppb]
1,2-Dichloroethane											
(for synthesis)	-	55	-	-	-	-	-	-	-	-	10
(90%)	-	-	-	-	-	-	-	-	-	-	5
(protein sequencing)	-	-	-	-	-	-	-	-	-	-	5
(99%)	-	-	-	-	-	-	-	-	-	-	5
(distilled in glass)	-	-	-	-	-	-	-	-	-	-	5
1,1,1-Trichloro-ethane											
(90-95%)	-	-	-	-	-	-	-	-	-	-	10
(technical)	-	-	-	-	-	-	-	-	-	-	5
	-	-	-	-	-	-	-	-	-	-	10
Trichloroethylene											
(99+%)	-	-	-	-	-	-	-	-	-	-	20
(99%)	-	-	-	-	-	-	-	-	-	-	5
(for synthesis)	-	-	-	-	-	-	-	-	-	-	5
(technical)	-	-	-	-	-	-	-	-	-	-	10

Table 11. Continued

	Cl$_8$DD	Cl$_8$DF	Cl$_7$DD	Cl$_7$DF	Cl$_6$DD	Cl$_6$DF [ppb]	Cl$_5$DD	Cl$_5$DF	Cl$_4$DD	Cl$_4$DF	Det.Lim. [ppb]
Tetrachloroethylene											
(neat)	-	-	-	-	-	-	-	-	-	-	10
(99%)	-	-	-	-	-	-	-	-	-	-	5
(99%)	47	-	-	-	-	-	-	-	-	-	5
(technical)	-	-	-	-	-	-	-	-	-	-	10
Allyl chloride											
(for synthesis)	-	-	-	-	-	-	-	-	-	-	10
(9E%)	-	-	-	-	-	-	-	-	-	-	10
Epichlorohydrin											
(99+%)	47	15	28	18	13	-	-	-	-	-	10
(98%)	-	-	-	-	-	-	-	-	-	-	10
	-	-	-	-	-	-	-	-	-	-	10

3.4 FORMATION OF PCDD/PCDF IN OTHER INDUSTRIAL PROCESSES

3.4.1 Phthalocyanine Dyes

The PCB-content of some commercial phthalocyanine dyes suggests that the industrial synthesis of phthalocyanines could lead to formation of PCDD/PCDF. The presence of PCBs is due to use of high boiling chlorinated aromatic solvents, such as dichlorobenzene and trichlorobenzene (US EPA, 1985). For the manufacture of the blue Cu-phthalocyanine, a mixture of phthalic acid anhydride, urea, copper(II)chloride, and catalytic amounts of ammonium molybdate in trichlorobenzene is heated to 200 °C for one hour. Heating of commercial 1,2,4-trichlorobenzene to 200 °C for one hour yields no PCDD/PCDF. After addition of catalytic amounts of copper(II)chloride PCDD/PCDF could be analyzed in ppt/ppb-range. PCDD/PCDF could be detected down to a temperature of 180 °C, but not at 160 °C, 140 °C, 120 °C, and 100 °C (Heindl and Hutzinger, in press). Several reaction mechanisms have to be discussed concerning the introduction of oxygen into the molecule. It is well known that transition metals catalyze oxidation reactions (Fallab, 1967). The oxygen of the air is transformed via electron transfer reactions to hydroperoxide and hydroxyl radicals. Reaction of the latter with 1,2,4-trichlorobenzene then form PCDD/PCDF (Heindl and Hutzinger, 1987). In addition, phthalocyanine dyes themselves are oxidation catalysts (Kropf, 1960) initiating the autoxidation of organic compounds. Therefore, heating of 1,2,4-trichlorobenzene with catalytic amounts of Cu-phthalocyanine instead of copper(II)chloride leads to PCDD/PCDF, but in comparison to the metal catalyst different isomers were found.

Analyses of several commercial phthalocyanine dyes showed that in one sample of Ni-phthalocyanine PCDD/PCDF could be detected in the ppt/ppb-range (see Table 12).

Table 12. PCDD/PCDF in Phthalocyanine Dyes; Amounts Given in [ppb]; Detection Limit: 100-500 ppt

	Cu-phthalo-cyanine (2 samples)	Co-phthalo-cyanine	Ni-phthalo-cyanine
Cl_4DD	- - -	- - -	- - -
Cl_5DD	- - -	- - -	2.7
Cl_6DD	- - -	- - -	0.5
Cl_7DD	- - -	- - -	- - -
Cl_8DD	- - -	- - -	- - -
Cl_4DF	- - -	- - -	0.5
Cl_5DF	- - -	- - -	1.7
Cl_6DF	- - -	- - -	0.5
Cl_7DF	- - -	- - -	- - -
Cl_8DF	- - -	- - -	- - -

Considering the US-regulation that the PCB content of products is limited to 25 ppm on average over a year and to 50 ppm in any individual batch the purification of the product is improved. Any possibly formed PCDD/PCDF should be accumulated in the residues.

3.4.2 Investigation of Motor Oils, Waste Oils, and Recycled Oils

In autumn 1985 German media (Anonymous, 1985a,b)) reported on the presence of PCDD/PCDF in waste oils and recycled oils. Analyses revealed the following: in fresh motor oils and after 10,000 km of operation (defined as motor oils in Table 13) no PCDD/PCDF could be detected (detection limit: 0.05 ng/g). The other samples contained detectable amounts of PCDD/PCDF. In waste oils and recycled oils hexa- through octachlorinated dibenzodioxins and hepta- through octa dibenzofurans were found.

Table 13. PCDD/PCDF in Motor Oils, Waste Oils and Recycled Oils; Concentrations in [ng/g] (Hagenmaier, 1987)

Sample	Waste Oil 1	Waste Oil 2	Recycl. Oil 1	Recycl. Oil 2	Motor Oil
Σ Cl_4DD	ND	ND	ND	ND	ND
Σ Cl_5DD	ND	ND	ND	ND	ND
Σ Cl_6DD	2.5	2.2	3.4	1.5	ND
Σ Cl_7DD	7.5	9.3	14.7	12.8	ND
Cl_8DD	14.8	57.6	27.6	12.8	ND
Σ PCDD	24.8	69.1	45.7	27.1	---
Σ Cl_4DF	ND	ND	ND	ND	ND
Σ Cl_5DF	ND	ND	ND	ND	ND
Σ Cl_6DF	ND	ND	ND	ND	ND
Σ Cl_7DF	1.2	1.4	1.8	0.7	ND
Cl_8DF	2.2	3.4	1.2	1.2	ND
Σ PCDF	3.4	4.8	2.8	1.9	---

ND = Not detected

Analysis of the distribution of isomers revealed that the PCDD/PCDF detected are due to the presence of PCP and PCP-Na, respectively. According to the German Society of Chemical Producers (Verband der Chemischen Industrie, VCI) (BUA, 1985) 5 to 60 tons of PCP and PCP-Na were used by the mineral oil industry during the years 1979-1984 every year (1984: 20 t PCP-Na = 50% of the total PCP-Na production). They are used in cooling lubricants as bacteriocides and fungicides. These contaminated oils themselves are starting materials for recycled oils; therefore, they will enter the cycle of use again. The search for other sources of PCDD/PCDF in waste and recycled oils requires further research.

3.4.3 Search for PCDD/PCDF in Commercial Products of PCP and PCP-Na

All results and data (from German production sites) given in the following chapter have to be regarded with respect to the fact that in the Federal Republic of Germany production of PCP-Na was cancelled in the summer of 1985. Nevertheless, within the European Community (EEC) about 8,000 tons and in the United States of America about 23,000 tons of PCP and PCP-Na, were produced and used. Worldwide (without the Eastern Block Countries) the total production is about 35,000-40,000 tons per year (VCI).

Hagenmaier et al. described an analytic method for the determination of PCDD/PCDF in samples of PCP and PCP-Na (Hagenmaier et al., 1986a). The results from two samples of PCP and two samples of PCP-Na are given in Table 14. One could observe that both samples of PCP-Na (from different producers) contained $2,3,7,8$-Cl_4DD, whereas in the samples of PCP no $2,3,7,8$-Cl_4DD could be detected (detection limit at 0.05 ppb). In another sample of PCP-Na $2,3,7,8$-Cl_4DD was found, too (Hagenmaier et al., 1986b,c,d). These results are slightly different from those reported in the PCP report of the Society of the German Chemical Manufacturers (see Table 15).

Table 14. PCDD/PCDF in Samples of PCP and PCP-Na [mg/kg] (Hagenmaier, 1987)

Sample	Pentachlorophenol		Na-Pentachlorophenolate	
$\Sigma\ Cl_4DD$	0.0019	0.0004	0.027	0.052
-$2,3,7,8$-Cl_4DD	<0.00003	<0.00005	0.00023	0.00051
$\Sigma\ Cl_5DD$	0.0065	0.0152	0.213	0.031
-$1,2,3,7,8$-Cl_5DD	0.001	0.002	0.0182	0.0032
$\Sigma\ Cl_6DD$	1.7	3.3	3.9	0.23
-$1,2,3,4,7,8$-Cl_6DD	<0.001	<0.001	0.0283	0.0133
-$1,2,3,6,7,8$-Cl_6DD	0.831	1.480	2.034	0.0530
-$1,2,3,7,8,9$-Cl_6DD	0.028	0.053	0.282	0.0190
$\Sigma\ Cl_7DD$	154	198	18.5	5.8
-$1,2,3,4,6,7,8$-Cl_7DD	78	99	9.1	2.8
Cl_8DD	733	790	41.6	32.4
Σ PCDD	888.7084	991.3156	64.24	38.513

Table 14. Continued

Sample	Pentachlorophenol		Na-Pentachlorophenolate	
Σ Cl_4DF	0.0008	0.0004	0.082	0.012
-2,3,7,8-Cl_4DF	<0.0001	<0.0001	0.0018	0.00079
Σ Cl_5DF	0.141	0.343	0.137	0.027
-1,2,3,7,8-Cl_5DF	0.0005	0.0002	0.0082	0.0019
(+1,2,3,4,8-Cl_5DF)				
-2,3,4,7,8-Cl_5DF	0.0015	0.0009	0.0066	0.0011
Σ Cl_6DF	4.3	13.9	3.0	0.09
-1,2,3,4,7,8-Cl_6DF	0.125	0.163	0.048	0.0046
(+1,2,3,6,7,9-Cl_6DF)				
-1,2,3,6,7,8-Cl_6DF	<0.001	<0.001	0.069	0.0013
-1,2,3,7,8,9-Cl_6DF	0.032	0.146	<0.001	0.0013
-2,3,4,6,7,8-Cl_6DF	<0.001	<0.001	0.087	0.0046
Σ Cl_7DF	74	127	13.2	0.86
-1,2,3,4,6,7,8-Cl_7DF	11.28	19.94	0.699	0.197
-1,2,3,4,7,8,9-Cl_7DF	0.637	0.980	0.675	0.036
Cl_8DF	118	137	37.2	4.25
Σ PCDF	196.4418	278.2434	53.619	5.239

Table 15. PCDD/PCDF in Industrial Grade Products of PCP and PCP-Na
Concentration [mg/kg]   (BUA, 1985)

Sample	PCP	PCP-Na from HCB[*]	PCP-Na from PCP
$2,3,7,8-Cl_4DD$	0.0009	<0.001 0.00025	<0.003
$1,3,6,8-Cl_4DD$	<0.02	0.02	<0.003
other Cl_4DD	<0.02	<0.02	<0.003
$1,2,4,7,8-Cl_5DD$	<0.02	<0.02	
$1,2,3,7,8-Cl_5DD$	<0.02	<0.02	<0.1
other Cl_5DD	<0.02	<0.02	
$1,2,4,6,7,9-Cl_6DD$	0.5	<0.02	<0.1
$1,2,3,4,6,8-Cl_6DD$	<0.02	0.03	<0.1
$1,2,3,6,8,9-Cl_6DD$	1.5	0.02	up to 1
$1,2,3,6,7,8-Cl_6DD$	2.2	0.01	up to 0.9
$1,2,3,7,8,9-Cl_6DD$	0.1	0.01	<0.1
$\Sigma\ Cl_6DD$-Isomers	3.5 ± 2 4.5	0.06	up to 2
$1,2,3,4,6,7,9-Cl_7DD$	35 - 30	0.5	40
$1,2,3,4,6,7,8-Cl_7DD$	100 - 110	1	50
$\Sigma\ Cl_7DD$	130 - 70	1.5	90
Cl_8DD	600 ± 200	4.5	150 ± 100
$2,3,7,8-Cl_4DF$	<0.001	<0.00025	<0.003
other Cl_4DF	<0.02	<0.02	
$\Sigma\ Cl_4DF$	<0.02	<0.02	<0.1
$1,2,3,7,8-Cl_5DF$	0.07	<0.001	<0.1
$2,3,4,7,8-Cl_5DF$	0.002	<0.0005	
other Cl_5DF	0.08	0.04	
$\Sigma\ Cl_5DF$	0.2	0.04	
$1,2,3,4,7,8-Cl_6DF$	0.30	0.02	
$1,2,3,6,7,8-Cl_6DF$	1.2	<0.02	up to 1.2
other Cl_6DF	6 - 30	0.08	
$\Sigma\ Cl_6DF$	20	0.1	
Cl_7DF not identified Isomers			
Isomer 1	1	<0.1	<0.1
Isomer 2	15	0.3	5
Isomer 3	40 - 130	0.02	20
Isomer 4	1.5	0.02	<0.1
$\Sigma\ Cl_7DF$-Isomers	60 ± 30 62	0.4	
Cl_8DF	150 ± 60	1.8	30 ± 20

[*]   no longer produced in FRG.

Estimations of the input of PCDD/PCDF due to production of PCP and PCP-Na are given in Tables 16 and 17.

Table 16. Emissions of PCDD/PCDF from Production of PCP (Eckert and Schlöbohm, 1982) (estimated for the Federal Republic of Germany 1979-1984); (* about 2,000 tons of PCP/a)

Gaseous emissions	Cl_8DD	max. 0.2	[kg/a] *
Residues (average)	Cl_4DD	max. 1.8	[mg/a]
	Cl_5DD	max. 2.5	[mg/a]
	Cl_6DD	max. 0.013	[kg/a]
	Cl_7DD	max. 0.13	[kg/a]
	Cl_8DD	max. 0.98	[kg/a]

Table 17 PCDD/PCDF from Production of PCP-Na from Hexachlorobenzene (FRG, 1984: 2,000 tons)

Gaseous emissions	Cl_8DD	max. 200	[g/a]
Waste water	Cl_6DD	max. 0.001	[g/a]
	Cl_7DD	max. 0.1	[g/a]
	Cl_8DD	max. 0.34	[g/a]
	Cl_6DF	max. 0.002	[g/a]
	Cl_7DF	max. 0.026	[g/a]
	Cl_8DF	max. 0.1	[g/a]

Average amounts of sludge from PCP-Na production were about 40 tons per year. Estimations of the PCDD/PCDF input from this source are given in Table 18 (calculated from one sample analyzed (Hagenmaier, 1988).

Table 18. Chloroorganics and PCDD/PCDF in the Sludge from a PCP-Na Production Site

	Dioxins		Furans		Others	
PCP	-	-	-	-	900	[kg/a]
HCB	-	-	-		6,000	[kg/a]
Decachlorobiphenyl	-	-	-	-	3,400	[kg/a]
Decachlorophenoxybenzene	-	-	-	-	44	[kg/a]
Cl_8DD/F	670	[g/a]	670	[g/a]	-	-
Cl_7DD/F	170	[g/a]	45	[g/a]	-	-
Cl_6DD/F	92	[g/a]	15	[g/a]	-	-
Cl_5DD/F	16	[g/a]	5	[g/a]	-	-
Cl_4DD/F	7	[g/a]	1.4	[g/a]	-	-
$2,3,7,8,-Cl_4DD/F$	0.7	[g/a]	-	-	-	-

The US Environmental Protection Agency has taken a number of regulatory actions to control toxic substances such as polychlorinated dibenzo-p-dioxins and polychlorinated dibenzofurans. Actions by the Office of Pesticides Programs have efficiently banned the use of trichlorophenol-based pesticide substances such as 2,4,5-T, Silvex, and others in which $2,3,7,8-Cl_4DD$ has been found. The Office of Solid Waste and Emergency Response, acting under the Resource Conservation and Recovery Act (RCRA), has proposed new rules for PCDD- and PCDF-contaminated wastes specifying stringent waste treatment requirements (US EPA, 1986). Similar reactions, as they were documented by the US-EPA study (Esposito et al., 1980) which are focused primarily on chlorinated substances, are known to occur in the manufacture of brominated chemicals. Therefore, polybrominated dibenzo-p-dioxins (PBDD) and polybrominated dibenzofurans (PBDF) may be formed as undesirable contaminants in certain brominated chemical substances (US-EPA, 1985; Versar, 1968). Under contract of the US-EPA, the suggested formation mechanisms for PCDD/PCDF contaminants in the production of specific halogenated organic chemicals were documented (Lee et al., 1986). This report covers a period from October 1985 to May 1986 (see below).

3.4.4   PCDD Formation in Pentachlorophenol Manufacture

In the United States, PCP is produced by the chlorination of phenol with aluminium chloride as the catalyst, carried out under anhydrous conditions.   Table 19 lists the various substances in technical grade pentachlorophenol (Herrick et al.).

Table 19. Analysis of Technical Grade Pentachlorophenol

Pentachlorophenol	88.4	[%]
2,3,4,6-Tetrachlorophenol	4.4	[%]
2,4,6-Trichlorophenol	< 0.1	[%]
Chlorinated phenoxyphenols	< 6.2	[%]
Cl_6DD	4	[ppm]
Cl_7DD	125	[ppm]
Cl_8DD	2,500	[ppm]
Cl_6DF	30	[ppm]
Cl_7DF	80	[ppm]
Cl_8DF	80	[ppm]

3.4.5 PCDD/PCDF in 1,2,4-Trichlorobenzene Production

The process utilized in Europe converts lindane manufacturing wastes (α,β,γ-hexachlorocyclohexane, HCH) to predominantly (70-75%) 1,2,4-trichlorobenzene (1,2,4-TCB) (Jürgens, 1985). After purification, the 1,2,4-TCB is chlorinated to 1,2,4,5-tetrachlorobenzene, which is used as feedstock for the production of 2,4,5-trichlorophenol. The procedures involves the low-temperature (200-240 °C) pyrolysis of the lindane wastes with a carbon catalyst using indirect heating of a closed reaction mantle. The residues from both the mantle and the trichlorobenzene showed high concentrations of dioxin contaminations (US Patent, 1960) (see Table 20).

Table 20. PCDD Isomer Differentiation for Pyrolyzed HCH Residues and TCB Stillbottoms

PCDD isomer	Concentration [mg/kg = ppm]
2,3,7,8-Cl_4DD	0.3
Cl_4DD	12
1,2,3,7,8-Cl_5DD	7
Cl_5DD	200
1,2,3,4,7,8-Cl_6DD	45
1,2,3,6,7,8-Cl_6DD	150
1,2,3,7,8,9-Cl_6DD	65
Cl_6DD	680
1,2,3,4,6,7,9-Cl_7DD	1,400
1,2,3,4,6,7,8-Cl_7DD	3,000
Cl_8DD	7,600

3.4.6 Occurrence of PCDD/PCDF in the Effluents of Pulp and Paper Mills and Consumers' Paper Products

PCDD/PCDF may result from condensation of chlorophenols formed by chlorination of natural occurring phenolic compounds such as lignin (25% in wood) and other plant constituents. Chlorinated phenoxyphenols, potential precursors of PCDD/PCDF, may be formed during industrial production of chlorophenolic compounds, particularly chlorobleaching of wood pulp in paper industry (Christmann et al., 1984; Knuutinen et al., 1983; Leuenberger et al., 1983; Paasivirta et al., 1986). Investigations of wastewater effluents from paper mills yielded detectable levels of PCDD/PCDF; furthermore PCDD/PCDF were accumulated in sediments, fish, and crabs (Rappe, 1988; Beck, 1988).

Chlorobleaching in the wood processing industry is suspect to produce PCDD/PCDF besides a wide range of other chlorinated compounds. In samples of pulp and board products PCDD and PCDF were detected between 0 to 40 (Eadon) or 0 to 30 (US-EPA 1986) Toxic Equivalents. The major PCDD were hexa-, penta- and tetrachlorinated congeners containing at the most few pg/g $2,3,7,8-Cl_4DD$ while the PCDF were mostly tetra chlorinated isomers of which $2,3,7,8-Cl_4DF$ was the major isomer (Kitunen and Salkinoja-Salonen, 1988).

Investigations on the content of PCDD/PCDF in filter paper (for laboratory use and coffee filtering), recycled scribbling paper, cosmetic tissues, and newsprint papers revealed that all PCDD/PCDF could be detected in the ppt range. Congeners of all chlorination degrees from tetra to octa are present. Above all the relatively high amounts of $2,3,7,8-Cl_4DD$ must be emphasized, whereas the hepta and octa levels are negligible (in the range of the blanks). The toxic tetra, penta, and hexa 2,3,7,8-substituted isomers contribute to an average of 20% to the total PCDD/PCDF content (Beck et al., 1988). The high amounts of PCDD/PCDF in recycled paper is striking. While the concentrations of the toxic tetra and penta congeners are comparable to those of the "white" papers the sum values exhibit the highest concentrations for all chlorination degrees of dioxins, especially for the Cl_6DDs; in particular the 1,2,3,6,7,8- and 1,2,3,7,8,9-species determine the high TCDD equivalent values. The newsprint paper exhibits $2,3,7,8-Cl_4DD$ values near the detection limit and furthermore the lowest values for the TCDD equivalent concentrations. In combination with the fact that newsprint contains up to 75% wood pulp the low values for PCDD/PCDF support the assumption that pulp bleaching is the responsible technological step for their occurrence in paper.

Table 21. Concentrations of PCDF and PCDD in Paper [ppt][*]

Congener	Newsprint	Laboratory-filter	Coffee-filter	Cosmetic tissue	Recycled Scr.paper
$2,3,7,8\text{-}Cl_4DF$	-	4.0	5.7	13	13
$\Sigma\ Cl_4DF$	2.6	12	10	39	26
$2,3,7,8\text{-}Cl_4DD$	-	0.3	1.0	1.1	0.6
$\Sigma\ Cl_4DD$	<0.4	2.0	1.0	6.3	8.8
$1,2,3,7,8\text{-}Cl_5DF$	-	-	-	0.4	-
$2,3,4,7,8\text{-}Cl_5DF$	-	-	-	0.4	-
$\Sigma\ Cl_5DF$	<0.4	9.5	3.7	9.5	2.8
$1,2,3,7,8\text{-}Cl_5DD$	-	0.2	-	0.6	0.9
$\Sigma\ Cl_5DD$	1.6	16	2.1	33	48
$2,3,4,6,7,8\text{-}Cl_6DF$	-	0.1	0.2	-	0.2
$\Sigma\ Cl_6DF$	0.8	2.0	1.2	2.9	3.2
$1,2,3,6,7,8\text{-}Cl_6DD$	1.2	3.2	-	12	48
$1,2,3,7,8,9\text{-}Cl_6DD$	-	1.5	-	4.2	19
$\Sigma\ Cl_6DD$	7.7	23	2.1	79	335
$\Sigma\ Cl_7DF$	1.8	1.1	0.3	2.9	1.5
$\Sigma\ Cl_7DD$	4.3	4.2	1.0	15	37
Cl_8DF	1.9	0.3	-	2.4	1.5
Cl_8DD	37	6.4	1.9	56	58

[*] The sum values include the toxic congeners

Table 22. Cl_4DD Equivalent Concentrations [ppt] Calculated Using the Toxicity Equivalent Factors (TEF)*

Congener	TEF	Newsprint	Laboratory-filter	Coffee-filter	Cosmetic tissue	Recycled scr.paper
$2,3,7,8-Cl_4DF$	0.1	-	0.4	0.57	1.3	1.3
Σ other Cl_4DF	0.01	0.026	0.08	0.043	0.26	0.13
$2,3,7,8-Cl_4DD$	1.0	-	0.3	1.0	1.1	0.6
Σ other Cl_4DD	0.01	-	0.017	-	0.052	0.082
$1,2,3,7,8-Cl_5DF$	0.1	-	-	-	0.04	-
$2,3,4,7,8-Cl_5DF$	0.1	-	-	-	0.04	-
Σ other Cl_5DF	0.01	-	0.095	0.037	0.087	0.028
$1,2,3,7,8-Cl_5DD$	0.1	-	0.02	-	0.06	0.09
Σ other Cl_5DD	0.01	0.016	0.158	0.021	0.324	0.471
$2,3,4,6,7,8-Cl_6DF$	0.1	-	0.01	0.02	-	0.02
Σ other Cl_6DF	0.01	0.008	0.019	0.01	0.029	0.03
$1,2,3,6,7,8-Cl_6DD$	0.1	0.12	0.32	-	1.2	4.8
$1,2,3,7,8,9-Cl_6DD$	0.1	-	0.15	-	0.42	1.9
Σ other Cl_6DD	0.01	0.065	0.183	0.021	0.628	2.68
Σ Cl_7DF	0.001	0.0018	0.0011	0.0003	0.0029	0.0015
Σ Cl_7DD	0.001	0.0043	0.0042	0.001	0.015	0.037
Cl_8DF	0.001	0.0019	0.0003	-	0.0024	0.0015
Cl_8DD	0.001	0.037	0.0064	0.0019	0.056	0.058
Total		0.3	1.8	1.7	5.6	12.2

*The sum values include no toxic congeners

478

In June 1988, the Swedish Minister of the Environment and Energy declared that all consumers' paper products in Sweden should be "dioxin-free" within one year. Results, presented at an international meeting revealed that PCDD/PCDF were found in all paper products analyzed (Wiberg et al., 1988).

The major contribution to the Cl_4DD-Equivalent levels [ppt] were made by:

Unbleached shopping bag	$1,2,3,6,7,8-Cl_6DD$	5.2
	$1,2,3,7,8,9-Cl_6DD$	2.8
Bleached shopping bag	$1,2,3,6,7,8-Cl_6DD$	1.6
	$1,2,3,7,8,9-Cl_6DD$	9.3
	$1,2,3,4,6,7,8-Cl_7DD$	60
Cigarette paper	$1,2,3,6,7,8-Cl_6DD$	6.7
	$1,2,3,7,8,9-Cl_6DD$	3.0

Cotton and cloth diapers had no Cl_4DD and Cl_4DF, but traces of higher chlorinated congeners were found in these products. Generally, the levels were lower in unbleached products than in bleached.

An important conclusion from the National Dioxin Study (US-EPA, 1986) was the correlation between unusually high concentrations of $2,3,7,8-Cl_4DD$ contamination in fish and the frequency of pulp and paper manufacturing plants located within the watersheds from which the fish were collected. Although aqueous discharge from pulp and paper mills must now meet federal and state requirements for maximum concentrations of total suspended solids in discharge water, sludge discharge was not fully regulated until the mid of 1970s. In the sludge from seven different pulp and paper mills $2,3,7,8-Cl_4DD$ levels were in the range from not detected at 1 pg/g to over 400 pg/g (Kuehl et al., 1987).

Table 23. 2,3,7,8-Cl$_4$DD Concentrations Found in Pulp and Paper Mill Sludges

Pulp/Paper Mill	Conc. of 2,3,7,8-Cl$_4$DD [pg/g]
A	150
B	2.5
C	ND
D	53
E	37
F	ND
G	414

ND = Not detected

The results of the characterization of sludge from Mill A for tetra-through octachloro isomers are provided in Table 24. It is obvious that the number of isomers was small compared with the number usually found in municipal incinerator fly ash. The isomer of the highest concentrations was Cl$_8$DD.

Table 24. Quantification of PCDD/PCDF in Pulp Mill A Sludge

Chemical	Concentration [pg/g]	Isomers
2,3,7,8-Cl$_4$DD	150	-
total other Cl$_4$DD	ND	0
2,3,7,8-Cl$_4$DF	880	-
total other Cl$_4$DF	640	4
1,2,3,7,8-Cl$_5$DD	ND	-
total other Cl$_5$DD	ND	-
1,2,3,7,8-Cl$_5$DF	29	-
total other Cl$_5$DF	140	6
1,2,3, 6,7,8-Cl$_6$DD	17	-
total other Cl$_6$DD	62	2
1,2,3,4,7,8-Cl$_6$DF	5	-
total other Cl$_6$DF	30	3
1,2,3,4,6,7,8-Cl$_7$DD	110	-
1,2,3,4,6,7,9-Cl$_7$DD	82	-
total Cl$_7$DF	5	2
Cl$_8$DD	1,860	-
Cl$_8$DF	53	-

ND = Not detected

Further investigation will be done to determine the in-plant source of PCDD/PCDF in the sludge.

3.4.7 Metallurgical Processes as Sources of PCDD/PCDF

Metallurgical industries, e.g. scrap metal melting processes (Tysklind et al., 1988; Hagenmaier, 1988, Antonsson et al., 1988), production
of magnesium and refined nickel (Oehme and Mano, 1988), have been found
to be one source for chlorinated dibenzodioxins and -furans. Isomer specific analyses of the flue gases from scrap metal melting processes show
the presence of PCDD and PCDF as well as other chlorinated aromatics.
The levels vary between 0.1-4.0 ng/Nm^3 dry gas, expressed as TCDD-Equivalents (Eadon), after the cleaning system. The higher value is obtained,
when PVC is added to the scrap metal.

Production of magnesium and refined nickel (Lykins et al., 1988) were
identified as sources for PCDD/PCDF in Norway. Magnesium is produced
from waterfree magnesium chloride by electrolysis. The highest amounts
of PCDD/PCDF are formed during the conversion of magnesium oxide to
magnesium chloride by heating coke/MgO pellets to 700-800 °C in a chlorine gas atmosphere. The annual emissions with the waste water are in
the order of several hundred grams of 2,3,7,8-TCDD-equivalents (Eadon)
per year and about 6 g/year to air.

The nickel refining process is very complex and uses chlorine at many
stages. Although the temperatures are low (in the range of 150-200 °C)
the formation of PCDD/PCDF will be catalyzed by the presence of other
metals, e.g. Co, Cu and Pb. The currently used technique emits only
minor amounts. High emissions of PCDD/PCDF were caused by an earlier
used high temperature process which converted NiCl$_2$ to NiO. In sediments
(0-1 cm; 1 km distance from the source) mainly high levels of PCDF (15
ppb) and only little PCDD (10-50 times lower) were identified (1 ppt
2,3,7,8-Cl$_4$DD and 11 ppt 1,2,3,7,8-Cl$_5$DD). Within the isomers the positions 1,2,3,7, and/or 8 were dominant with a pattern different from incineration and other known sources.

In some processes where PCDD/PCDF are formed, the dioxins pass
through the work environment before being emitted to the outdoor environment. One of the processes from which dioxins are emitted is the melting
of scrap iron which involves some steps that release large amounts of
dust. Samples were taken at four steel mills at three different points:
a) close to the electric furnace, b) high in the furnace hall, and c) in
the cabin of the overhead crane (equipped with a filter) (Oehme and Mano,
1988). The dioxin concentrations (according to the Nordic Toxic Equivalent Factors) have been of the same order of magnitude in the companies
investigated. The measurements confirmed that the dioxin concentration

should follow the dust concentration and be highest high up in the furnace hall, at the overhead crane level. The dust filter in the cabins proved to capture 50-75% of the dioxins, the rest being let into the cabin. The results are given in Table 25.

Table 25. Concentrations of PCDD/PCDF (Nordic Equivalents) in the Work Environment in Four Steel Mills
The proposed Nordic tolerable weekly intake is 0-35 [pg/kg,w]

Measuring point	Concentrations Cl_4DD-Eq. $[pg/m^3]$	Weekly Intake Cl_4DD-Eq. $[pg/kg,w]$
Close to furnace	6.4	3.2
Overhead crane	6.8	3.4
Crane cabin	4.0	2.0
Close to furnace	4.4	2.2
Overhead crane	14	7.0
Crane cabin	5.6	2.8
Close to furnace	0.8	0.4
Overhead crane	1.8	0.9
Close to furnace	1.5	0.75
Overhead crane	4.8	2.4
Crane cabin	2.8	1.4

3.4.8 Discharge of PCDD/PCDF during Carbon Reactivation

As there has been a renewed interest in the use of granular activated carbon (GAC) by drinking water utilities and on-site reactivation of exhausted carbon. Studies were initiated to evaluate the production, discharge, and health effects of PCDD/PCDF during reactivation of exhausted granular activated carbon (Lykins et al., 1988). Granular activated carbon is an adsorbent, and once its capacity is exhausted, it needs to be replaced or reactivated thermally. On-site thermal reactivation systems studied were a fluidized-bed furnace and an infrared furnace. Associated with the reactivation is the potential of producing by-products like PCDD/PCDF from the precursor materials or organics adsorbed on the carbon. The fluidized-bed furnace provided temperatures at 1,038 °C in the combustion chamber. Either fuel oil or natural gas could be used. The reactivation section is separated from the combustion chamber by a stainless steel diaphragm. In its incineration zone temperatures averaged about 816 °C. Off-gases passed through a venturi and tray scrubber before discharge to the atmosphere. The infrared furnace experienced the classical thermal reactivation steps during drying (temperatures of 260-316 °C), pyrolysis (760 °C), and reactivation (927 °C). The furnace off-gases entered an afterburner where they were mixed with air and burned at 1,010 °C.

Studies with carbon, being heavily loaded with total organic chlorine, have shown that low levels of PCDD/PCDF were emitted into atmosphere during reaction of granular activated carbon. The $2,3,7,8-Cl_4DD$ isomer was seen in the particulate stack emission from the fluidized-bed furnace discharges in the range of 0.001-0.02 ppt by volume. Relatively high concentration of all Cl_4DD isomers were detected in the cyclone (0.2-3.3 ppt). In most tests $2,3,7,8-Cl_4DF$ was detected in the stack emission particulates (0.004-0.02 ppt by volume) and in the cyclone catch (0.06-2.7 ppb by weight).

When no chlorine is added prior to GAC adsorption, no $2,3,7,8-Cl_4DD$ was found. Concentrations of hepta- and octa-CDDs ranged from 0.001-0.05 ppt and 0.006-0.28 ppt, respectively in the spent GAC, reactivated GAC, scrubber water , afterburner or stack emissions. Chlorinated PCDF were detected in the afterburner for all of the homologues evaluated (tetra through octa); only occasional small concentrations of PCDF were detected in the spent GAC and stack emissions.

During infrared reactivation of virgin GAC no $2,3,7,8-Cl_4DD$, $2,3,7,8-Cl_4DF$, total Cl_4DD or total Cl_4DF concentrations were detected in any stack effluent or process stream samples. 0.18 ppt Cl_4DD were found in the scrubber water. Concentrations for hexa-, hepta- and octa-CDD for the particulate and gaseous portions of the stack emissions were low (not detected-0.025 ppt). All of the PCDF homologues were detected in the gaseous stack emissions. Tetra-, hepta- and octa-CDF concentrations ranged from 0.03-0.40 ppt in the scrubber water. Spent GAC contained an average of 0.17 ppb of Cl_4DF but none was found in reactivated GAC, quench water, or drinking water samples.

3.5 POLYCHLORINATED DIBENZOFURANS FROM PCBs

PCDF has been found as contaminants of polychlorinated biphenyls, chlorinated phenols, and as combustion products in fires involving PCB-containing transformers and capacitors. Besides the formation mechanisms discussed for PCDD, PCDF can also be formed from the intramolecular conversion of PCBs.

(1) **Loss of Ortho-Cl$_2$**

Example:

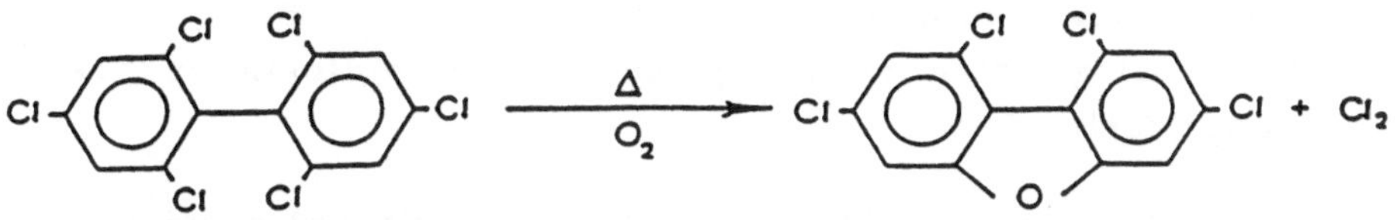

2,4,6,2',4',6'-Hexachlorobiphenyl   1,3,7,9-Tetrachlorodibenzofuran

(2) **Loss of HCl Involving 2,3-Chlorine Shift at the Benzene Nucleus**

Example:

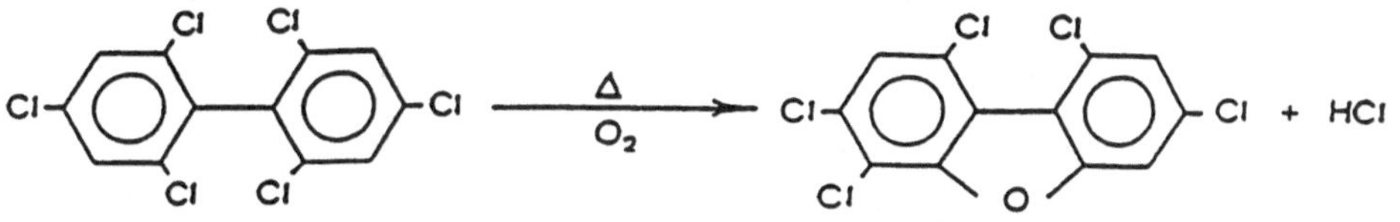

2,4,6,2',4',6'-Hexachlorobiphenyl   1,3,4,7,9-Pentachlorodibenzofuran

(3) **Loss of Ortho-HCl**

Example:

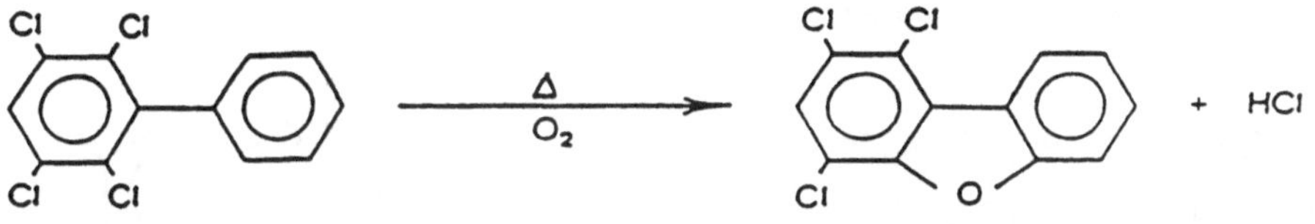

2,3,5,6-Tetrachlorobiphenyl   1,2,4-Trichlorodibenzofuran

(4) **Loss of Ortho-H$_2$**

Example:

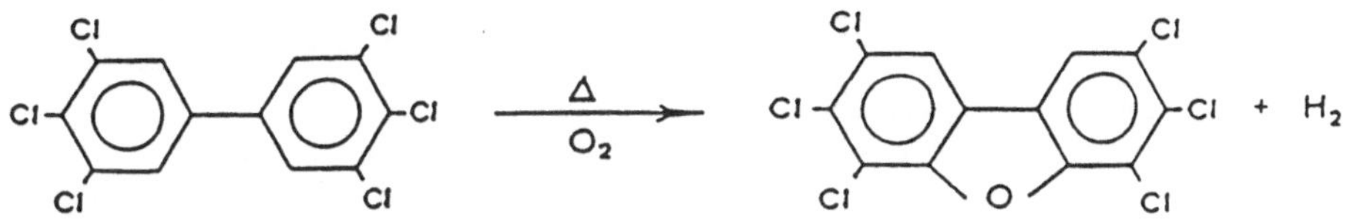

3,4,5,3',4',5'-Hexachlorobiphenyl   2,3,4,6,7,8-Hexachlorodibenzofuran

Fig. 2. Reaction Mechanisms Involved in the Formation of PCDF Isomers from Various PCB Congeners (Buser and Rappe, 1979).

3.6 FORMATION OF POLYBROMINATED DIBENZODIOXINS AND DIBENZOFURANS

Much less information than on the chemistry of chlorinated dibenzo-p-dioxins and dibenzofurans is available about their brominated counterparts. Brominated dioxins have been made in laboratory experiments, and thus may be formed in the commercial manufacture of brominated compounds. Unlike the chlorination process, bromination of organic compounds is a low-temperature process. The low-temperature conditions do not favor formation of dibenzodioxins.

Classical laboratory preparations of polyhalodibenzo-p-dioxins involve:
1) intermolecular condensation of polyhalophenols, and
2) direct halogenation of the parent dibenzo-p-dioxin or lower halogenated derivatives.

The first method is the one in which chlorinated dioxins form as a contaminant in commercial products (Poland and Young 1972; Aniline, 1973). Direct bromination of dibenzo-p-dioxin according to the second method gives both 2,8- and 2,7-dibromodibenzo-p-dioxin; under forcing conditions, the 2,3,7,8-tetrabromo derivative is formed (Gilman and Dietrich, 1957).

Formation of brominated dibenzo-p-dioxins

a)   Condensation of catechols with a polyhalobenzene in boiling dimethyl sulfoxide; e.g. 1,2,4,5-tetrabromobenzene as acceptor yielded 2,3,dibromodibenzo-p-dioxin (O'Keefe, 1979):

b) Chlorination of 2,3-Br$_2$DD gave 2,3-Br$_2$-7,8-Cl$_2$DD, identical with the product prepared by bromination of 2,3-Cl$_2$DD (Kende et al., 1974):

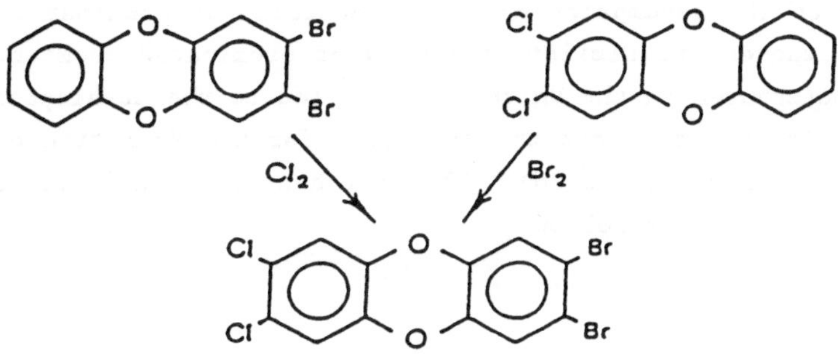

Brominated phenolic derivatives and diphenyl ethers used as flame retardants are precursors of dibenzodioxins and -furans. Pyrolysis and combustion experiments revealed high concentrations of PBDD/PBDF (Buser, 1986; Thoma, 1986; Thoma et al., 1987). In the presence of PVC bromine-chlorine exchange is observed. Experiments revealed stepwise exchange of Br ⟷Cl resulting in the completely chlorinated PCDD/PCDF. Formation of mixed halogenated dibenzodioxins and dibenzofurans is postulated, when both, chlorine and bromine donors, are present at the same time. Type and amount of the mixed species will be influenced not only by the Br/Cl ratio but also by the number of halogen atoms entering the molecule and other factors, e.g. reactivity of the precursors, thermal stability of products, etc. (Buser, 1987). Recent results (Thoma et al., 1987) showed that during pyrolysis bromine atoms on an aromatic system, such as dibenzodioxin, can be exchanged with chlorine which originated from inorganic chlorine precursors, such as HCl and NaCl, or originally bound forms, such as PVC. Moreover, the results demonstrated that this exchange is facilitated at higher temperature (800-900 °C).

3.6.1 Possible Formation of PBDD/PBDF during the Manufacture of Brominated Chemicals

Based on the assumption that the chemistry of brominated and chlorinated substances are similar, a review of brominated chemicals, reported to be manufactured in the United States, led to the development of a list of chemicals that may have the potential for the formation of brominated dioxins and furans (Table 26). The list has been classified according to structural classes as follows:

Table 26. List of Brominated Compounds with Potential Formation of PBDD/PBDF

(1) Brominated Phenols

4-Bromo-2,5-dichlorophenol (2,5-Dichloro-4-bromophenol; Leptophos phenol; Phosvel phenol)

o-Bromophenol (2-Bromophenol)
  Synthesis of resorcinol, other organics

m-Bromophenol (3-Bromophenol)

p-Bromophenol (4-Bromophenol)
  Antiseptic, synthesis of other compounds

2,6-Dibromo-4-nitrophenol (4-Nitro-2,6-dibromophenol)
  Production: >5,000 lbs

2,4-Dibromophenol
  Flame retardant intermediate

2,6-Dibromophenol
  <1,000 lbs/yr (EPA estimate)

Pentabromophenol (Flammex 5BP)
  Flame retardant intermediate; Molluscicide (experimental)

Tetrabromocatechol (Tetrabromopyrocatechol)

2,4,6-Tribromo-m-cresol (2,4,6-Tribromo-3-methylphenol; Triphysan; Triphysol)

2,4,6-Tribromophenol (Bromkal Pur 3; Bromol; Flammex 3BP; Great Lakes PH-73)
  Flame retardant intermediate; Anti-fungal agent; Chemical intermediate; Antiseptic germicide;
  Production: >2.27 t (1979), >4.54 t (1981).

Table 26. (Continued)

(2) Brominated Compounds

Bromobenzene (Monobromobenzene; Phenylbromide)
    Solvent; Top-cylinder compound; Chem. intermediate; Prod.: >5,000 lbs
    (USITC 1984)

o-Bromofluorobenzene (1-Bromo-2-fluorobenzene; o-Fluorobromobenzene; 2-
    Fluorobromobenzene; 1-Fluoro-2-bromobenzene)
    Production: <1,000 lbs/y (EPA CBI Aggregate)

Bromophenetole (relates to p-compounds: 1-Bromo-4-ethoxybenzene;
    p-Ethoxybromobenzene; p-Ethoxyphenylbromide; 4-Ethoxyphenylbromide;
    and ω-compound: Phenoxyethylbromide)

Decabromodiphenyloxide (Berkflam B10E; Bis(pentabromophenylether); Brom-
    kal 81; Bromkal 82-ODE; Bromkal 83-10DE; BR55N; DE 83; DE 83R; Deca-
    brom; Decabromobiphenylether; Decabromobiphenyloxide; Decabromodi-
    phenyl ether; Decabromophenoxybenzene; Decabromophenyl ether; EB10FP;
    EBR 700; FR 300; FR 300BA; FRP 53; FR-PE; Planelon DB100; Saytex 102;
    Saytex 102E
    Flame retardant; Production: 2-8 million lbs/yr (1976)

1,3-Dibromobenzene (m-Dibromobenzene)
    Ingredient of fire extinguishers; Flame retardant;
    Ingredient of heat transfer fluids

2,4-Dibromofluorobenzene (1,3-Dibromo-4-fluorobenzene)
    Intermediate for agricultural and pharmaceutical chemicals

2,6-Dibromofluorobenzene

Hexabromobenzene (Benzene hexabromide; HBB; Perbromobenzene)
    Flame retardant; Production: >454 kg (1975)

Pentabromochlorocyclohexane (Chloropentabromocyclohexane; FR 651A;
    1,2,3,4,5-Pentabromo-6-chlorocyclohexane)
    Flame retardant; Production: 10^6-10^7 lbs/yr (EPA estimate)

Pentabromoethylbenzene (EB 80; Saytex 105)
    Flame retardant

Pentabromotoluene (Flammex 5BT; Pentabromomethylbenzene)
    Flame retardant

Tetrabromophthalic Anhydride (Bromphthal; EG 4000; FireMaster PHT4; Great
    Lakes PHT4; Saytex RB-49)
    Flame retardant; Production: >6.81 t (1981)

Tribromobenzene (1,2,4, or 1,3,5)
    Oil Additives.

Table 26.   (Continued)

(3) **Brominated Bisphenols**

1,2-Bis(tribromophenoxy)ethane (BTBPE; FireMaster 680)
  Flame retardant

2,6-Dibromobisphenol A
  Flame retardant

Octabromodiphenyloxide (DE-79; FR 143; Octabromodiphenyl ether; 1,1'-Oxy-
  bis-octabromobenzene; Tardex 80)
  Flame retardant

Pentabromodiphenyloxide (DE-71; 1,1'-Oxybis-pentabromobenzene;
  Pentabromodiphenyl ether)
  Flame retardant

Tetrabromobisphenol A (BA-59; BA-59P; 2,2-Bis(3,5-dibromo-4-hydroxy-
  phenyl)propane; Bromdian; FireGuard 2000; FireMaster BP4A; 4,4'-Iso-
  propylidene bis-(2,6-dibromophenol); Saytex RB-100; Tetrabromdian;
  Tetrabromodihydroxy diphenyl propane)
  Flame retardant; Production: >10^6 lbs/yr

Tetrabromobisphenol A, allylether (1,1-(1-Methylethylidene)-bis(3,5-di-
  bromo)-4-(2-propenyloxy)-benzene)
  Flame retardant

Tetrabromobisphenol A, bis-2,3-dibromopropyl ether (2,3-Bis[3,5-dibromo-
  4-(2,3-dibromopropoxy)phenyl]-propane; FireGuard 3100; Great Lakes
  PE-68)
  Flame retardant

  Tetrabromobisphenol A, bisethoxylate
  Flame retardant

Tetrabromobisphenol A, bismethylether
  Flame retardant

Tetrabromobisphenol A, diacrylate (4,4'-Isopropylidene-bis(2,6-dibromo-
  phenyl)-acrylate)
  Flame retardant

Tetrabromobisphenol B
  Flame retardant.

(4) **Pesticides**

  Bromophos (Brofene; o-(4-Bromo-2,5-dichlorophenyl)-o,o-dimethyl
  phosphorothioate; Bromofos; Brophene; Nexion; Nexion 40; S1942)
  Insecticide

Bromoxynil butyrate (2,6-Dibromo-4-cyanophenyl butanoic ester)
  Insecticide

Table 26. (Continued)

Bromoxynil octanoate (Bronate; Buctril; 3,5-Dibromo-4-octanoyl-oxybenzo-
nitrile
Herbicide
Profenofos (o-(4-bromo-2-chlorophenyl)-o-ethyl-S-propylphosphorothioate;
CGA 15324;
Insecticide
Curacron, Polycron; Selecron)
Insecticide.

It is assumed that the mechanisms of dioxin and furan formation for brominated substances are similar to those for chlorinated substances.
Thus, bromine analogs of chlorinated compounds which have been associated
with dioxin contamination, could reasonably be expected to be associated
with PBDD and PBDF contamination. For 18 industrial brominated chemicals
(see list below) information on the manufacturing process, possible contaminants, possible PBDD/PBDF formation pathways and a product analysis
are given.

(a) 2,4-Dibromophenol

Manufacturing process: Bromination of phenol in water or another polar
hydroxylic solvent; $FeBr_3$ as catalyst.

Possible contaminants: 2,4,6-Tribromophenol; 2,4,5-tribromophenol; possibly tetrabromophenols and pentabromophenol.

Possible PBDD/PBDF formation pathways (most likely to 2,7-dibromodibenzo-
p-dioxin):

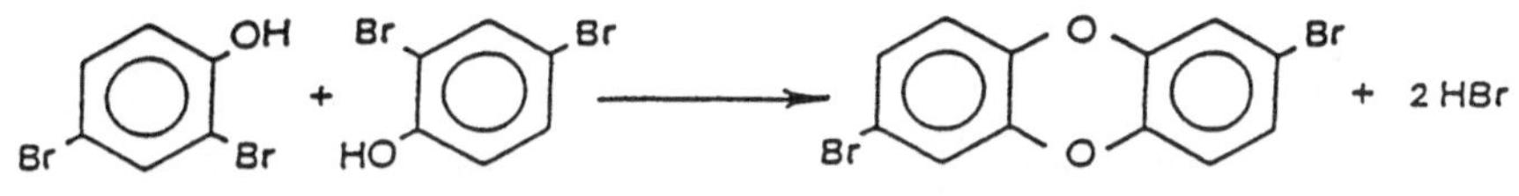

2,4-Dibromophenol   2,4-Dibromophenol   2,7-Dibromodibenzo-p-dioxin

Product analysis: Not available.

(b) 2,4,6-Tribromophenol

Manufacturing process: Bromination of phenol in the presence of water or other polar hydroxylic solvent; reaction is uncatalyzed.

Possible contaminants: 2,4,5-Tribromophenol; tetrabromophenols.

Possible PBDD/PBDF formation pathways (most likely to 1,3,6,8-tetrabromo-dibenzo-p-dioxin)

(a)

2,4,6-Tribromo-   2,4,6-Tribromo-   2,4,7,9-Tetrabromodibenzo-p-
    phenol   phenol   dioxin

(b)

2,4,5-Tribromo-   2,4,5-Tribromo-   2,3,7,8-Tetrabromodi-
    phenol   phenol   benzo-p-dioxin

Product analysis: HRGC/MS analysis (Thoma et al., 1986) of technical grade extracts shows the following:

Dibromodibenzo-p-dioxin	Tribromodibenzofuran
Tribromodibenzo-p-dioxin	Tetrabromodibenzofuran
Tetrabromodibenzo-p-dioxin	Pentabromodibenzofuran
(main component, ~90 ppb)	Hexabromodibenzofuran
Pentabromodibenzo-p-dioxin	Heptabromodibenzofuran
Dibromodibenzofuran	Octabromodibenzofuran.

(c) Pentabromophenol

<u>Manufacturing process</u>: a) Bromination of phenol using excess Br_2 in the presence of $FeBr_3$ at 46 °C, followed by an increase of incubation temperature to 60 °; and b) Bromination of phenol with Br_2 to yield intermediate tribromophenol, followed by perbromination with Br_2 in the presence of $FeBr_3$.

<u>Possible contaminants</u>:   2,3,4,6,-; 2,3,4,5; 2,3,5,6-Tetrabromophenol, 2,4,6- or 2,4,5-tribromophenol.

(d) Bromophenols (o, m and p isomers)

<u>Manufacturing process</u>:   Bromination of phenol in water or another polar hydroxylic solvent.

<u>Possible contaminants</u>:   Phenol; dibromophenol.

<u>Possible PBDD/PBDF formation pathways</u> (most likely to 2,4,7,9-tetrabromo-dibenzo-p-dioxin):

o-Bromophenol

b) Condensation of dibromophenols as shown for 2,4-dibromophenol.

<u>Product analysis</u>: Not available.

(e) 2,6-Dibromo-4-nitrophenol

Manufacturing process: Bromination of p-nitrophenol with excess Br_2; catalyst is probably not necessary.

Possible contaminants: 2,5-Dibromo-4-nitrophenol.

Possible PBDD/PBDF formation pathways:

2,6-Dibromo-4-nitrophenol   1,6-Dibromo-3,8-dinitrodi-
benzo-p-dioxin

PBDD/PBDF isomers are not expected due to steric hindrance and other effects from NO_2 groups on aromatic ring.

Product analysis: Not available.

(f) Decabromodiphenyloxide

Manufacturing process: Perbromination of diphenyloxide with at least 150% excess Br_2 in the presence of $AlBr_3$ at 35°C, followed by an increase in temperature to around 60 °C.

Phenol   Benzene   Diphenyloxide   Decabromodiphenyloxide

Possible contaminants: Diphenyloxide with varying degrees of bromination.

494

<u>Possible PBDD/PBDF formation pathways</u> (most likely to octabromodibenzo-furan):

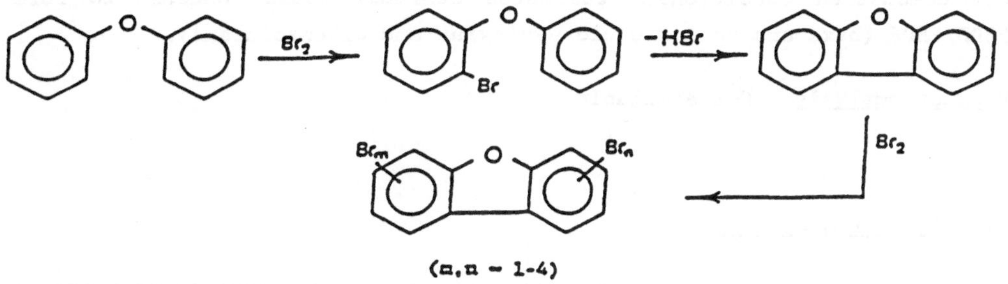

(m,n ~ 1-4)

<u>Product analysis</u>: Not available.

<u>(g) Octabromodiphenyloxide</u> (see Decabromodiphenyloxide)

<u>(h) Bromobenzene (Mono and Di)</u>

<u>Manufacturing process</u>:   Bromination of benzene with Br_2 and $FeCl_3$ as catalyst (80-100 °C).

<u>Possible contaminants</u>: Varying degrees of brominated benzene.

<u>Possible PBDD/PBDF formation pathways</u>: None under normal manufacturing process.   Under combustion conditions brominated benzenes could condense to PBDD/PBDF similar to chlorinated benzenes.

<u>Product analysis</u>: Not available.

<u>(i) Hexabromobenzene</u>

<u>Manufacturing process</u>:   Direct bromination of benzene with excess Br_2 using Friedel-Crafts catalyst (e.g. $FeBr_3$, $AlBr_3$); another patented method employs perbromination with bromine chloride under high pressure (US Patent, 1974).

<u>Possible Contaminants</u>: Tribromobenzenes; tetrabromobenzenes; pentabromo-benzene.

Possible PBDD/PBDF formation pathways: None under normal manufacturing conditions. In case of plant upset where excess oxygen is introduced under combustion conditions, brominated benzenes could condense to form PBDD/PBDF (hexa and octa dibenzo-p-dioxins and dibenzofurans).

Product analysis: Not available.

(j) Pentabromotoluene

Manufacturing process: Perbromination at room temperature in presence of catalyst (US Patent, 1981).

Possible contaminants: Varying brominated toluenes.

Possible PBDD/PBDF formation pathways: None under normal conditions. However, PBDD/PBDF could be formed under combustion conditions similar to that of the chlorinated benzenes in PCB transformers (depending on the condensation of the particular brominated toluenes, several dimethyl polybrominated dioxins and furans are possible).

Product analysis: Not available.

(k) Tetrabromophthalic Anhydride

Manufacturing process: Phthalic anhydride, in 50-80% oleum, is brominated using halogenation catalyst at 80-90°C, then the temperature is raised to 100-110 °C at end of Br_2 addition; then heat to 120-150°C to get rid of excess SO_3 and Br reactants (US Patent, 1968a).

Possible contaminants: Phthalic anhydride with varying degrees of bromination.

Possible PBDD/PBDF formation pathways: Highly improbable.

Product analysis: Not available.

(l) 1,2,4-Tribromobenzene

Manufacturing process: By-product from the preparation of 1,3,5-tri bromobenzene.

Other items: see Bromobenzene and Pentabromotoluene.

(m) 1,3,5-Tribromobenzene

__Manufacturing process__: Bromination of benzene with Br_2 using $AlBr_3$ as catalyst; since 1,3,5-isomer is most stable, will tend toward this configuration (yield 1,3,5-isomer >50%).

Other items: see Tribromobenzene.

(n) Pentabromodiphenyloxide

__Manufacturing process__: Bromination of diphenyloxide in ethylene bromide with bromine and chlorine in presence of iron powder and at high temperatures (US Patent, 1968b).

Other items: see Decabromodiphenyloxide.

(o) Tetrabromobisphenol A

__Manufacturing process__: Bromination of bisphenol A in organic solvent with Br_2 at 20-40 °C and then increasing the incubation temperature to 65-70 °C.

__Possible contaminants__: Brominated phenols.

Possible PBDD/PBDF formation pathways (most likely to octabromodibenzo-p-dioxin):

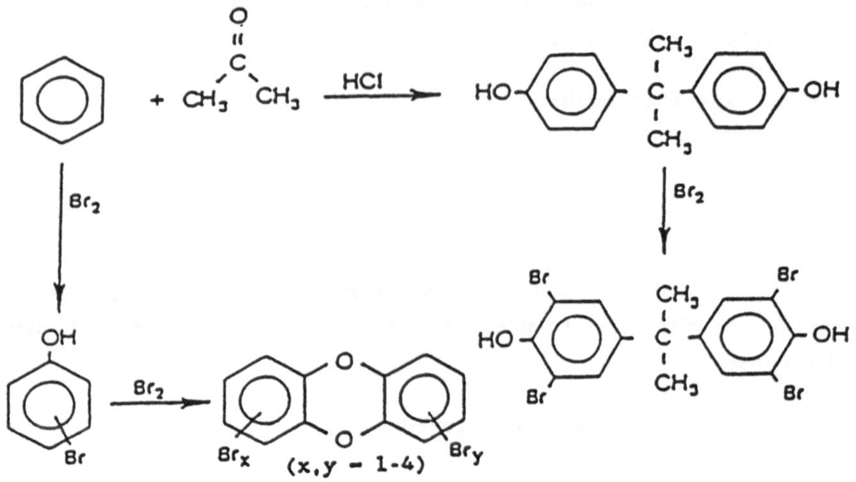

Product analysis: HRGC/MS analysis (Kulka, 1961) of technical grade extracts shows the following:

    Pentabromodibenzofuran
    Hexabromodibenzofuran
    Heptabromodibenzofuran
    Octabromodibenzofuran (main component, ~30 ppb).

(p) Bromophos

Manufacturing process: Reaction of 4-bromo-2,5-dichlorophenyl sodium with o,o-dimethylthiophosphoric acid chloride at 100 °C for 8 hours. Potassium bromide is used as a catalyst and chlorobenzene is the solvent.

Possible contaminants: Brominated phenols.

Possible PBDD/PBDF formation pathways (most likely to 2,7-Dibromo-3,8-dichlorodibenzo-p-dioxin):

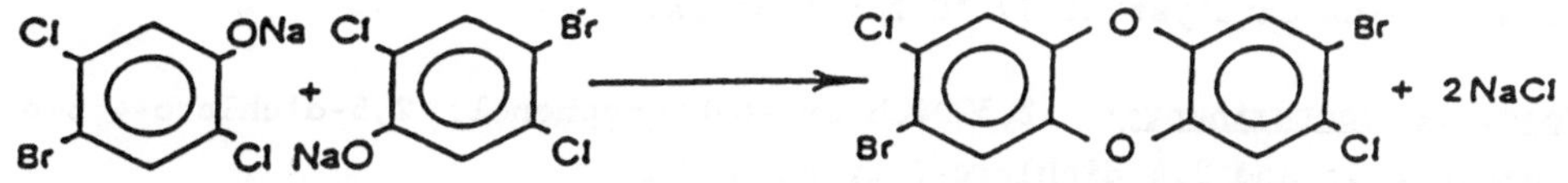

4-Bromo-2,5-dichlorophenyl
sodium

2,7-Dibromo-3,8-dichloro-
dibenzo-p-dioxin

Product analysis: Not available.

(q) 1,2-Bis(tribromophenoxy)ethane

Manufacturing process:

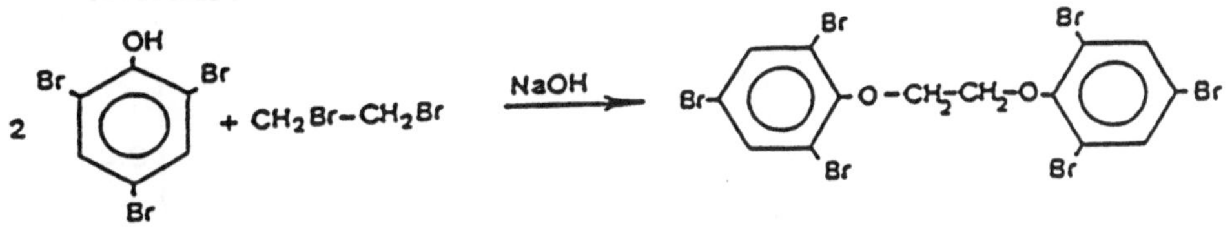

Possible contaminants: 2,4,5-Tribromophenol; tetrabromophenol.

Possible PBDD/PBDF formation pathways (most likely to 1,3,6,8-Tetrabromo-dibenzo-p-dioxin):

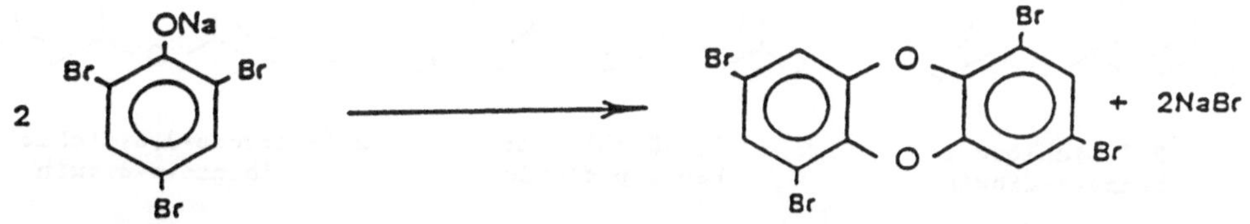

2,4,6-Tribromophenol
sodium salt

1,3,6,8-Tetrabromodibenzo-p-dioxin

Product analysis: Not available.

(r) 2.5-Dichloro-4-bromophenol

Manufacturing process: Bromination of 2,5-dichlorophenol in the presence
of an iodine catalyst at 77 °C for 4 hours.

Possible contaminants: 2,5-Dichloro-6-bromophenol; 2,5-dichloro-4,6-di-
bromophenol; and 2,4-dichloro-6-bromophenol.

Possible PBDD/PBDF formation pathways (most likely to 2,7-Dichloro-
dibenzo-p-dioxin; 3,7-dichlorodibenzo-p-dioxin; and 2,7-dibromo-3,8-di-
chlorodibenzo-p-dioxin.

Product analysis: Not available

500

3.6.2 Polybrominated Biphenyls

Formation of polybrominated Dibenzofurans (PBDF) may occur in situations involving combustion of products containing polybrominated biphenyls: tetra and penta dibenzofurans were identified in the residues of combustion studies with polybrominated biphenyls (O'Keefe, 1979).

Polybrominated biphenyls are produced by the direct bromination of biphenyl, and are the primary constituent of various fire retardants such as FireMaster FF-1, FireMaster BP-6, and Bromkal 80. FireMaster BP-6 was marketed as technical grade hexabromobiphenyl although up to 24 different biphenyl isomers have been identified in this product. The content of hexabromobiphenyl in commercial FireMaster BP-6 ranged from less than 60% to 90% with 2,4,5,2',4',5'-hexabromobiphenyl being the main isomer (Sundstorm et al., 1976, DiCarlo et al., 1978). FireMaster FF-1 is identical to FireMaster BP-6, with the addition of 2% calcium silicate as an anti-caking agent (DiCarlo et al., 1978).

A sample of technical grade octabromobiphenyl from a U.S. producer contained primarily nonabromobiphenyl (60%), with octabromobiphenyl (33%), decabromobiphenyl (6%), and heptabromobiphenyl (1%) isomers, respectively (Waritz et al., 1977). The German octabromobiphenyl product, Bromkal 80, consisted of octabromobiphenyl (72%), heptabromobiphenyl (27%), hexabromobiphenyl (1%), and a trace of nonabromobiphenyl (Norstrom et al., 1976). In other samples of commercial octabromobiphenyls 45.2% octabromobiphenyl, 47.4% nonabromobiphenyl, 5.7% decabromobiphenyl, and 1.8% heptabromobiphenyl were detected (diCarlo, 1978).

Decabromobiphenyl compositions have assayed at 96.8% to more than 98% decabromobiphenyl, the rest being nonabromobiphenyl and octabromobiphenyl (DiCarlo et al., 1978; Neufeld et al., 1977).

A number of brominated naphthalenes have been found as impurities in PBB mixtures: approximately 25 ppm hexabromonaphthalene, 1 ppm tetrabromonaphthalene and traces of pentabromonaphthalene have been analyzed in FireMaster FF-1 (O'Keefe, 1979). FireMaster BP-6 has been found to contain 150 ppm pentabromonaphthalene and 70 ppm hexabromonaphthalene (Hass et al., 1978).

PBDD or PBDF have not been detected in neither FireMaster BP-6 nor FireMaster FF-1 (O'Keefe, 1979; Hass et al., 1978).

The synthesis of polybrominated biphenyls generally involves the bromination of biphenyl. Patents were obtained by a German firm in 1964 and 1966 for the bromination of aromatic compounds, specifically biphenyl and biphenyl ether (Neufeld et al., 1977). A U.S. patent from 1973 involves the reaction of an aromatic compound with bromine catalyzed by $AlCl_3$ or $AlBr_3$; this process was used by manufacturers of PBBs in the 1970s in the State of New Jersey. Another process consists of the bromination of biphenyl with bromine chloride in the presence of iron or a Friedel-Crafts catalyst in a closed vessel (Neufeld et al., 1977; Mumma et al., 1975). In laboratory scale polybrominated biphenyls are synthesized through diazo coupling of brominated aniline with an excess of a brominated benzene (Robertson et al., 1984).

In 1970, a U.S firm began commercial production of a flame retardant additive that contained various isomers of hexabromobiphenyl as the principal constituent. The U.S. production of polybrominated biphenyls is given in Table 27 (Neufeld et al., 1977).

Table 27. U.S. Production of PBBs [in tons]

PBB	1970	1971	1972	1973	1974	1975	1976
Hexabromo-biphenyl	9.5	84.0	100.83	1,765.6	2,216.4	0	0
Octa- and Deca-bromobiphenyl	14.1	14.1	14.5	163.0	48.1	77.2	365.5

On November 20, 1974 the sole U.S. manufacturer of hexabromobiphenyl ceased production, after producing a total of 5.08 million kg of this compound since 1970 (Archer et al., 1979). This action was precipitated by an incident in 1973, when 230 to 450 kg of hexabromobiphenyl (as FireMaster BP-6) were added to animal feed in Michigan and as a result, thousands of cattle, hogs, sheep and million of chickens were destroyed (DiCarlo et al., 1979). All remaining inventory of FireMaster BP-6 held by the manufacturer was depleted in April 1975 (Neufeld et al., 1977).

Imports of PBBs to the United States have been negligible. No exposure to PBBs is expected by use of impregnated plastic granules, which have been imported to the States (US-EPA).

(1) Octabromobiphenyl and Decabromobiphenyl

Commercial quantities of octabromobiphenyl and decabromobiphenyl were produced in the U.S. until 1979. Both were produced by one firm since 1970 (Archer et al., 1979); another manufacturer produced decabromobiphenyl from 1973 to 1977 (Neufeld et al., 1977). In the period 1970-1976 a total of 696 tons of decabromobiphenyl and octabro-mobiphenyl were produced in the United States; 365 tons of them being exported to Europe (Neufeld et al., 1977).

In the United States, other PBB isomers, from monobromobiphenyl to decabromobiphenyl are only produced in laboratory quantities for research purposes (Neufeld et al., 1977; Stratton and Whitlock; Anonymous, 1985). Octabromobiphenyl is produced commercially by one manufacturer in West Germany (Norstorm et al., 1975; Neufeld et al., 1977). In the United Kingdom, one firm made available octabromobiphenyl, another produced "polybromobiphenyl" as of 1980 (Anonymous, 1980). In France commercial quantities of decabromobiphenyl were produced by one company.

Most of the produced hexabromobiphenyl was used to produce flame retardant resins (Mumma and Wallace, 1975). These plastics (with a content of about 10%) were mostly consumed by business machines, industrial equipment (as typewriters, calculators, microfilm reader and business machine housings) and the electrical industry (radio, television, thermostats, shavers, hand tool housings) (Neufeld et al., 1977). As a result of the Michigan feed contamination incident, all use of hexabromobiphenyl in the United States was suspended in 1974 (Neufeld et al., 1977).

Octabromobiphenyl is the principal component of a flame retardant mixture manufactured by a German firm (Norstrom et al., 1975). Decabromobiphenyl is presumably used in similar flame retardant products.

(2) Predicting PBDD/PBDF Contamination of Brominated Products

Formation of a variety of PBDD/PBDF are theoretically possible on the basis of the structural configuration of the reactants, by-products and products.   Factors that may influence manufacturing processes to form PBDD/PBDF include: temperature, pH value, catalyst, and reaction kinetics. The effect of temperature on the formation of halogenated dioxins and furans has been calculated as (Wu, 1985):

$$y = 0.025 \ t^2 \ e^{-3 \ (t-200)/200}$$

where   y = dioxin concentration
        t = temperature

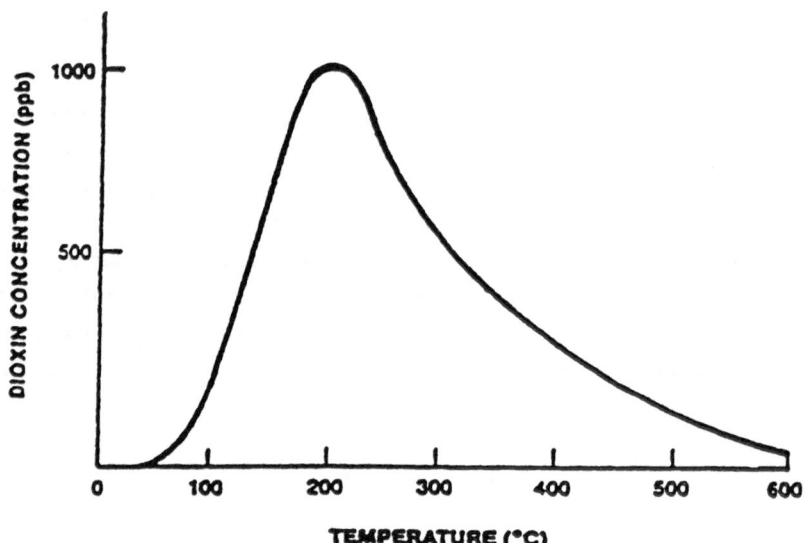

Fig. 3.   Dioxin Formation as a Function of Temperature

The graphical representation of this function (Figure 2) shows, that dioxin formation peaks at 200 °C and decreases unsymmetrically with increasing temperature.   The use of this predictive model also assumes that:

- Impurities in the feedstock (including any dioxins) are expected to be carried through to the final product on the chemical stability and low concentration of dioxins formed.

- The process temperature is 200 °C, unless specific information is available.

- There is a special catalyst present and no purification of the product.

504

3.7 GUIDELINES FOR THE DETERMINATION OF POLYHALOGENATED DIBENZO-p-DIOXINS AND DIBENZOFURANS IN COMMERCIAL PRODUCTS

In 1984, the US EPA was petitioned to issue rules and orders to prevent and reduce environmental contamination by polyhalogenated dibenzodioxins and dibenzofurans (PXDD/PXDF; specifically chlorinated and/or brominated) (US EPA, 1985b). The development of analytical methodologies for the determination of PCDD/PCDF in commercial products is complicated by the diverse range of sample matrices, the potentially large number of brominated, chlorinated, and mixed brominated/chlorinated congeners of dibenzodioxins and dibenzofurans, and by possible interferences of other substances.

3.7.1 Possible Number of PXDD/PXDF

Both the dibenzodioxin and the dibenzofuran molecule have eight positions where bromine or chlorine can substitute. There are 1,700 halogenated dibenzodioxins (PXDD) and 3,320 halogenated dibenzofurans (PXDF) for bromo, chloro and bromo/chloro-substitution (Buser, 1987). Within these 5,020 halogenated compounds there are 351 PXDD and 667 PXDFs with substitution in the 2,3,7,8-positions. Although the theoretical number of isomers is high, the actual number of PXDD/PXDF in commercial products may be limited to relatively few.

3.7.2 Analytical Methods for Determination of PXDD/PXDF in Commercial Products

In contrast to biological and environmental samples where the matrices are (mostly) different enough from the analyte, the commercial products in most cases may be structurally similar to the analyte. Therefore, the separation is complicated and complete removal of the matrix is necessary to avoid interferences in the final determination. Most of the analyses of commercial products for polychlorinated dibenzodioxins and dibenzofurans, polychlorinated and polybrominated biphenyls have employed (high resolution) mass spectrometry detection. (HR)MS will provide both qualitative and quantitative data for PXDD/PXDF.

Figure 4 shows a flowchart for the general guideline for the analysis of PXDD/PCDF in commercial products (US EPA, 1985b). This scheme, adapted from general separation principles and techniques reported in the literature, is based on the physical/chemical properties of the commercial product; therefore, the exact extraction and clean-up procedure is dependent upon the nature of the product or formulation.

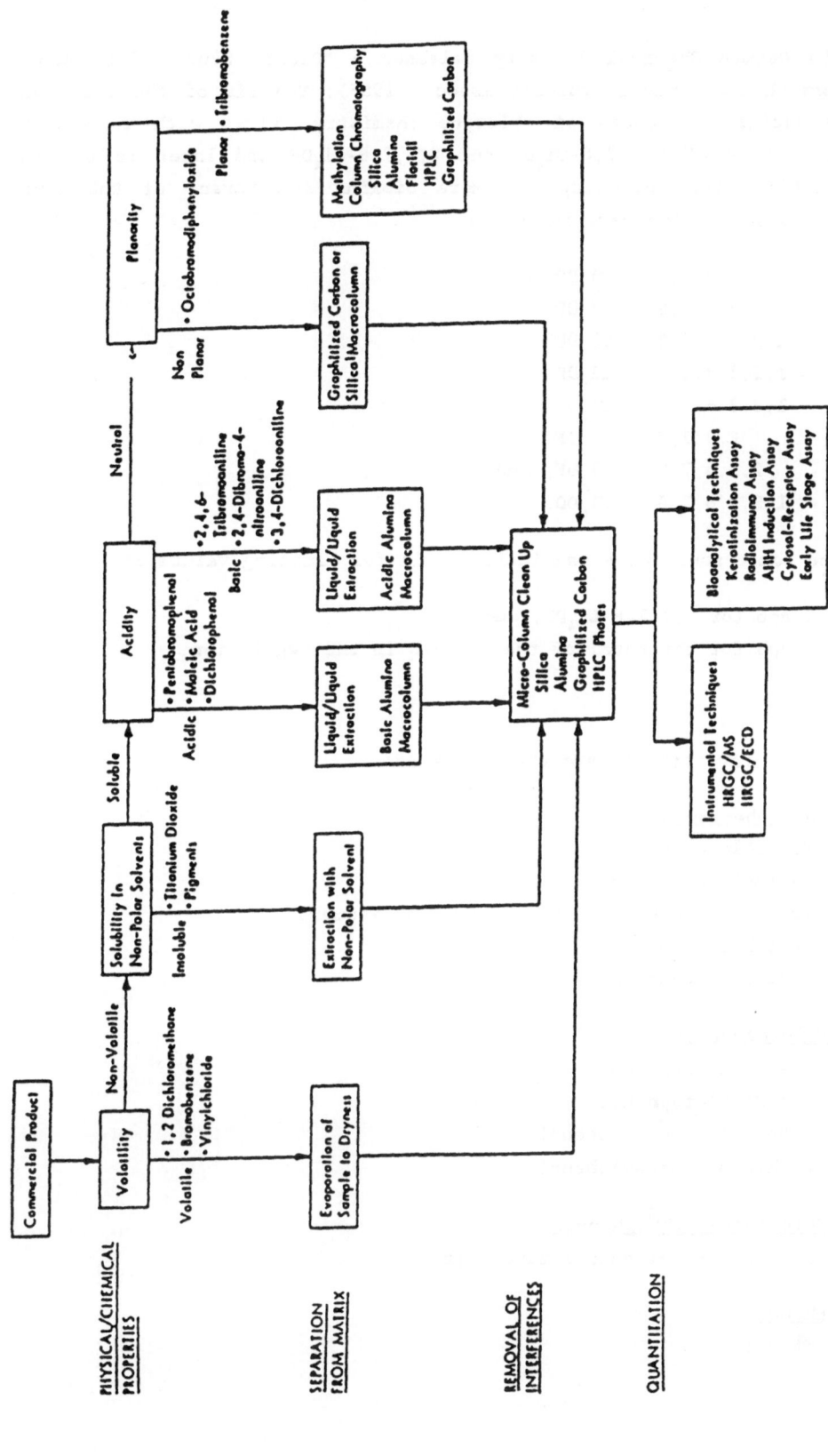

Fig. 4.   Tentative Scheme for the Analysis of PXDD/PXDF in Commercial Products

507

3.8 INVESTIGATIONS OF INDUSTRIAL PRODUCTS

The German Chemical Industry released a "Dioxin Study of Products"
(german Chemical Manufacturer's Assoc., 1986). The aim of the study was
to investigate products of chlorine chemistry, first with respect to
their content of $2,3,7,8-Cl_4DD$ and $2,3,7,8-Cl_4DF$ and later in the investigations the following six more dioxins and furans of the toxic
materials regulation were included:

2,3,7,8	Cl_4DD
1,2,3,7,8	Cl_5DD
1,2,3,6,7,8	Cl_6DD
1,2,3,7,8,9	Cl_6DD
1,2,3,4,7,8	Cl_6DD
2,3,7,8	Cl_4DF
2,3,4,7,8	Cl_5DF, and
1,2,3,6,7,8	Cl_6DD.

The toxic substances regulations call for limiting values of:

2 ppb for $2,3,7,8-Cl_4DD$, and
5 ppb for the total of the isomers in the regulation.

Substances investigated are the following:

1.   Chlorobenzenes
     Monochlorobenzene
     p-Dichlorobenzene
     Tri-/Tetrachlorobenzene (Mixture of isomers)
     Hexachlorobenzene
     Pentachlorobenzene

2.   Chlorophenols
     2,4-Dichlorophenol
     2,5-Dichlorophenol
     2-Benzyl-4-chlorophenol
     4-Chloro-3-methylphenol

3.   Chlorinated Thiophenols
     Pentachlorothiophenol zinc salt

4.   Phenols
     Phenol

5. Alkoxyanilines
2-Methoxyaniline

6. Polychlorinated Biphenyls
Chlorinated biphenyl with 60% chlorine

7. Chlorinated Alkenes
2-Chloro 1,3-butadiene

8. Chlorinated Rubber Products
Chlorinated rubber types

9. Nitrophenols and Derivatives
o-Nitrophenol
Thiophosphoric acid tris-p-nitrophenyl ester

10. Chlorophenylisocyanates
3,4-Dichlorophenylisocyanate

11. Phenoxycarboxylic Acids
2,4-Dichlorophenoxyacetic acid
4-Chloro-2-methylphenoxyacetic acid
2-(4-chloro-2-methyl)-phenoxyacetic acid

12. Carbamates
3,5-Dimethyl-4-methylmercaptophenyl-N-methyl carbamate
Methoxycarbonylaminobenzimidazole

13. Carboxylic Acids Esters
1-(3,4-Dichlorophenyl)-2,2,2-trichloromethyl acetate

14. Ureas
3-(3,4-Dichlorophenyl)-1,1-dimethylurea
1-(3-chloro-2,2,3-trifluoro-1,4-benzodioxan-6-yl)-3-(2,6-difluorobenzoyl)-urea

15. Phosphoric Acid Esters
O-(2,4-Dichlorophenyl)-O-ethyl-S-propyldithiophosphate

16. Heterocycles
N-(1,1,2,2-Tetrachloroethylthio)-3,6,7,8-tetrahydrophthalimide

The results of this study indicate that in no case did the concentrations exceed the levels stated the toxic materials regulations.

REFERENCES

Aniline, O., 1973, Preparation of chlorodibenzo(p)dioxins for toxicological evaluation in chlorodioxins, origin and fate; a symposium, Adv. Chem. Ser., 120:126.

Anliker, R., 1981, Problems of polychlorinated bphenyls (PCBs) in synthetic organic pigments, Swiss. Chem., 3:17.

Anonymous, 1980, Chem. Sources-Europe, Mountain Lakes, NJ, Chemical Sources Europe.

Anonymous, 1985, Chem. Sources-USA, Ormond Beach, FL, Directories Publishing Co., Inc.

Anonymous, Stern, 1985, vol. 44.

Anonymous, Monitor (German TV, WDR), 29.10.1985.

Antonsson, A.-B., Runmark, S., and Kjeller, L.-O., 1988, Dioxins in the working environment in steel mills, SOU19, presented at DIOXIN'88, Umea, Sweden, August 21-26, 1988.

Archer, S.R., Blackwood, T.R., and Collins, C.S., 1979, Status Assessment of Toxic Chemicals: Polybrominated biphenyls, EPA-600/2-79-210k, Cincinnati, OH, U.S. Protection Agency.

Bauer, H.R., Rappe, C., and Gara, A., 1978, Polychlorinated dibenzofurans (PCDFs) found in Yusho oil and in used Japanese PCB, Chemosphere, 5:439.

Beck, H., 1988, personal communication.

Beck, H., Eckart, K., Mathar, W., and Wittkowski, R., 1988, Occurrence of PCDD and PCDF in different kinds of paper, Chemosphere, 17:51.

Boehringer, 1961, Belgisches Patent 625298.

Boehringer, 1961, DAS 1174104.

Bowes, G.W., Mulvihill, M.J., Simoneit, B.R.T., Burlingame, A.L., and Risebrough, R.W., 1975, Identification of chlorinated dibenzofurans in American polychlorinated biphenyls, Nature, 256:305.

BUA, GDCh-Gremium für umweltrelevante Altstoffe, 1985, Stoffbericht: Pentachlorphenol, December 1985.

Buser, H.R., and Rappe, C., 1979, Formation of polychlorinated dibenzofurans (PCDFs) from the pyrolysis of induced PCB Isomer, Chemosphere, 8:157.

Buser, H.R., 1986, Polybrominated dobenzofurans and dibenzo-p-dioxins: Thermal reaction products of polybrominated diphenyl ether flame retardants, Environ. Sci. Technol., 20:404.

Buser, H.-R., 1987, Bromierte und gemischt bromiert/chlorierte Dibenzodioxine und Dibenzofurane, VDI-Bericht 634, pp 243.

Buser, H.R., 1987, Brominated and Bbrominated/chlorinated dibenzodioxins and dibenzofurans: Potential environmental contaminants, Chemosphere, 16:713.

Christmann, W., Erzmann, M., and Irmer, H., Dez. 1984, Abwassersituation der Zellstoffindustrie, WaBoLu-Hefte, 2/1985.

Cochrane, W.P., Chlorinated Dioxins and Related Compounds, O. Hutzinger, R. W. Frei, E. Merian, F. Pocciari, eds., Proc. Workshop 22-23 Oct. 1980, Rome; Pergamon Press, Oxford.

Cochrane, W.P., and Miles, W., 1982, Pest. Chem. Hum. Welf. Env. J. Miyamoto, P. Kearny, Proc. 5th Int. Congr. 1982, 341.

Crosby, D.G., 1981, Environmental chemistry of pentachlorophenol, Pure Appl. Chemistry, 53:1052.

Daalen, J.J. van, De produktie van gewasbeschermingsmiddelen door Duphar B. V. in Amsterdam, Duphar-Schrift

De Bruin, J., 1978, Commission of the European Communities, Environment and Consumer Protection Service ENV/223/74, ENV/313/79.

Denivelle, L., Ford, R., and Pham, V.H., 1958, Compt. Rend., 248:2766.

DiCarlo, F.J., Seifter, J., and De Carlo, V.J., 1978, Assessment of the hazards of polybrominated biphenyls, EPA-560/6-77-037, Washington, DC, U.S. Protection Agency.

Eckert, W.R., and Schlöbohm, A.M., 1982, unpublished results of NATEC, West Germany.

Esposito, M.P., Tiernan, T.O., and Dryden, F.E., 1980, Dioxins, EPA/600/2-80-197.

Fallab, S., 1967, Reactions with molecular oxygen, Angew. Chem.. Internat. Ed., 6:496.

German Chemical Manufacturers Association Test Program for Dioxin, September 1986, published by NATO/CCMS (TR-87-0021).

Gilman, H. and Dietrich, J.J., 1957, Halogen derivatives of dibenzo-p-dioxin, J. Amer. Chem. Soc., 79:1439.

Gribble, W.G., 1947, TCDD a deadly molecule, Chemistry, 47:15.

Guidelines for the Determination of Polyhalogenated Dibenzo-p-dioxins and Dibenzofurans in Commercial Products, June 1985, Draft Final Report, EPA Prime Contract No. 68-02-3938, Prepared for U.S. Environmental Protection Agency, Office of Pesticides and Toxic Substances, Washington, DC.

Hagenmaier, H., Brunner, H., Haag, R., and Kraft, M., 1986a, Selective determination of 2,3,7,8-tetrachlorodibenzo-p-dioxin in the presence of a arge excess of other polychlorinated dibenzodioxins and polychlorinated dibenzofurans, Fresenius Z. Anal. Chem., 323:24.

Hagenmaier, H., 1986b, Determination of 2,3,7,8-tetrachlorodibenzo-p-dioxin in commercial chlorophenols and related products, Fresenius Z. Anal. Chem., 325:603.

Hagenmaier, H., and Berchthold, A., 1986c, Analysis of waste from pro-
    duction of sodium-pentachlorophenolate for polychlorinated dibenzo-
    dioxins (PCDD) and dibenzofurans (PCDF), Chemosphere, 15:1991.

Hagenmaier, H, and Brunner, H., 1988d, submitted to Chemosphere.

Hagenmaier, H., 1988, Report on a PCDD/PCDF contaminated area in south-
    west Germany, SOU22, presented at DIOXIN'88, Umea, Sweden, August
    21-26, 1988.

Hagenmaier, H., 1988, Belastung der Umwelt mit Dioxinen, Chemosphere
    (in press).

Hagenmaier, H., 1988, University of Tübingen, personal communication.

Hass, J.R., McConnell, E.E., and Harvan, D.J., 1978, Chemical and toxi-
    cological evaluation of FireMaster BP-6, Agric. Food Chem., 26:94.

Hawley, G., 1971, Condensed Chemical Dictionary, 8th ed; Van Nostrand-
    Reinhold Co., New York.

Heindl, A., and Hutzinger, O., 1986, Search for industrial sources of
    PCDD/PCDF, Chemosphere, 15:2001.

Heindl, A. and Hutzinger, O., 1986, Search for industrial sources of
    PCDD/PCDFs: II. Metal chlorides, Chemosphere, 15:653.

Heindl, A. and Hutzinger, O., 1987, Search for industrial sources of
    PCDD/PCDF. III. Short chain chlorinated hydrocarbons, Chemosphere,
    16:1949.

Heindl, A. and Hutzinger, O., 1989, Search for industrial sources of
    PCDD/PCDFs: IV. Phthalocyanine dyes, Chemosphere, (in press).

Helling, C.S. and Isensee, A.R., 1973, Chlorodioxins in pesticides,
    soils, and plants, J. Environ. Qual., 2:171.

Herrick, E.C., Goldfarb, A.S., Fong, C.V., Konz, J., and Walker, P.,
    Chlorophenols by chlorination of phenol, in: Hazards Associated with
    Organic Chemical Manufacturing, MITRE Technical Report, MTR-78
    W00364-05.

Hess, P., Assauer, J., and Holländer, H., 1982, The possible formation
    of tetrachlorodibenzo-p-dioxins in the production of chloranil,
    Ecotoxicol. Environ. Safety, 6:336.

Institut für Organische Chemie der Universität Tübingen, Juni 1987, Be-
    lastung der Umwelt mit Dioxinen, Abschlußbericht für das Ministerium
    für Ernährung, Landwirtschaft, Umwelt und Forsten Baden-Württemberg.

Jürgens, H., 1985, Personal communication (DEKONTA/Mainz, FRG) to Paul E.
    des Rosiers, (US-EPA).

Kende, A.S. et al., 1974, Synthesis and fourier transform carbon-13
    nuclear magnetic resonance spectroscopy of new toxic polyhalodibenzo-
    dioxins, J. Org. Chem., 39:932.

Kimbrough, R.D., 1974, The toxicology of polychlorinated polycyclic com-
    pounds and related chemicals, CRC Critical Rev. Toxicol , 2:45.

Kitunen, V.-H., and Salkinoja-Salonen, M.S., 1988, Occurrence of PCDDs and PCDFs in pulp and board products, SOUP05 presented at DIOXIN'88, Umea, Sweden, August 21-26, 1988.

Knuutinen, J., Salovaara, J., Tarhanen, J., Passivirta, J., Virkki, L., Lahtiperä, M., Humppi, T., Laitinen, R., and Kantolahti, E., 1983, Chemosphere, 12:511.

Kropf, H., 1960, Ann. Chem., 637:111.

Kuehl, D.W., Butterworth, B.C., DeVita, M.W., and Sauer, C.P., 1987, Environmental contamination by polychlorinated dibenzo-p-dioxins and dibenzofurans associated with pulp and paper mill discharge, Biomed. Environ. Mass Spectrom., 14:443.

Kulka, M., 1961, Octahalogenodibenzo-p-dioxins, Can. J. Chem., 39:1973.

Kulka, M., 1965, Canadian Patent 702.144.

Lee, A., Campbell, B., and Kelly, W., 1986, Dioxin and furan contamination in the manufacture of halogenated organic chemicals, EPA/600/2-86/101 (prepared for Hazardous Waste Engineering Research Laboratory, Office of Research and Development, US Environmental Protection Agency, Cincinnati, OH 45268, Contract No. 68-03-3274).

Leuenberger, C., Coney, R., Graydon, J.W., Molnar-Kubica, E., and Giger, W., 1983, Persistent organic chemicals in pulp-mill effluents: Ocurrence and behavior in a biological treatment plant, Chimia, 37:345.

Lykins Jr., B.W., Clark, R.M., and Cleverly, D.H., 1988, Polychlorinated dioxin and furan discharge during carbon reactivation, J. Environ. Engineering, 114:300.

Morita, M., Nakagawa, J., Agiyama, K., Mimura, S., and Isono, N., 1977, Detailed examination of polychlorinated dibenzofurans in PCB preparations and Kanemi Yusho oil, Bull. Environ. Contam. Toxicol., 18:67.

Moye, A.L., 1972, Experiencing relevancy in organic chemistry. Hexachlorophene. Manufacturing the great clean-all, J. Chem. Education, 49:770.

Mumma, C.E. and Wallace, D.D., 1975, Survey of industrial processing data, Task II: Pollution potential of polybrominated biphenyls, EPA-560/3-73-004, Washington, DC, U.S. Government Printing Office.

Nestrick, T.J. and Lamparsky, L.L., 1980, Determination of tetra-, hexa-, hepta-, and octachlorodibenzo-p-dioxin isomers in particulate samples at parts per trillion levels, Anal. Chem., 52:2045.

Neufeld, M.L., Sittenfield, M., and Wolk, K.F., 1977, Market input/output studies, Task IV: Polybrominated biphenyls, EPA-560/6-77-017, Washington, DC, U.S. Government Printing Office.

Norstrom, A., Andersson, K., and Rappe, C., 1975, Major components of some brominated aromatics used as flame Rretardants, Chemosphere, 4:255.

O'Keefe, P.W., 1979, Trace contaminants in a polybrominated biphenyl fire retardant and a search for these compounds in environmental samples, Bull. Environ. Contam. Toxicol., 22:420.

Oehme, M., and Mano, S., 1988, Metallurgical processes as new sources for PCDF and PCDD and their implement on the environment, SOU20, presented at DIOXIN'88, Umea, Sweden, August 21-26, 1988.

Offhaus, E., 1983, Umweltbundesamt, Az. I 4-97 061/61.

Olie, K., Vermeulen, P., and Hutzinger, O., 1977, Chlorodibenzo-p-dioxins and chlorodibenzofurans are trace components of fly ash and flue gas of some municipal incinerators in the Netherlands, Chemosphere, 6:455.

Paasivirta, J., Tarhanen, J., and Soikkeli, J., 1986, Occurrence and fate of polychlroinated aromatic ethers (PCDE, PCA, PCV, PCPA and PCBA) in environment, Chemosphere, 15:1429.

Poland, A.E., and Yang, G.C., 1972, Preparation and characterization of chlorinated dibenzo-p-dioxins, J. Agric. Fd. Chem., 20:1093.

Quellmalz, E., 1988, Das neue Chemikaliengesetz - Handbuch der gefährlichen Arbeitsstoffe, Bd. 1-3, Weka-Verlag, Kissing (for the Federal Republik of Germany).

Rappe, C., 1979, Ann. N. Y. Acad. Sci., 1.

Rappe, C., Buser, H.R., and Boschardt, H.P., 1978, Identification and quantification of polychlorinated dibenzo-p-dioxins (PCDDs) and dibenzofurans (PCDFs) in 2,4,5-T-ester formulations and herbicide Orange, Chemosphere, 5:431.

Rappe, C., 1988, personal communication.

Rappe, C., and Buser, H.R., 1980, Halogenated biphenyls, terphenyls, naphthalenes, dibenzodioxins and related compounds; pp 47 R. Kimbrough, ed.

Roach, J.A.G., and Pomerantz, H., 1974, The finding of chlorinated dibenzofurans in a Japanese polychlorinated biphenyl samples, Bull. Environ. Contam. Toxicol., 12:338.

Robertson, L.W. et al., 1984, Synthesis and Iidentification of highly toxic polybrominated biphenyls in the fire retardant FireMaster BP-6, J. Agric. Food Chem., 32:1107.

Sandermann, W., Stockman, H., and Casten, R., 1957, Chem. Ber., 90:690.

Stratton, C.L., and Whitlock, S.A., A Survey of polybrominated biphenyls (PBBs) near sites of manufacture and use in northeastern New Jersey, Washington, DC, U.S. Environmental Protection Agency.

Sundstroem, G., Hutzinger, O., and Safe, S., 1976, Identification of 2,2',4,4',5,5'-hexabromobiphenyl as the major component of flame retardant FireMaster BP-6, Chemosphere, 5:11.

Thoma, H., Rist, S., Hauschulz, G., and Hutzinger, O., 1986, Polybrominated dibenzodioxins and dibenzofurans in some phenolic flame retardants, Chemosphere, 15:1649.

Thoma, H., Hauschulz, G., and Hutzinger, O., 1987, PVC-induced chlorine-bromine exchange in the pyrolysis of polybrominated diphenyl ethers, -biphenyls, -dibenzodioxins and dibenzofurans, Chemosphere, 16:297.

Thoma, H., Hauschulz, G., and Hutzinger, O., 1987, Chlorine-bromine exchange during pyrolysis of 1,2,3-tetrabromodibenzodioxin with various chlorine donors, Chemosphere, 16:1579.

Tysklind, M., Söderström, G., Rappe, C., Hägerstedt, L.-E., and Burström, E., 1988, PCDD and PCDF emissions from scrap metal melting processes at a steel mill, SOU21, presented at DIOXIN'88, Umea, Sweden, August 21-26, 1988.

U.S. Patent No. 3382254, 1968a, Chemische Fabrik Kalk GmbH DE.

U.S. Patent No. 3285965, 1968b, Chemische Fabrik Kalk GmbH DE.

U.S. Patent No. 3,845,146, 1974, The Dow Chemical Company.

U.S. Patent No. 4287373, 1981, Great Lakes Chemical Corporation.

U.S. Environmental Protection Agency, Submission of Notice of Manufacture or Importation of PBBs and Tris, Federal Register 45:70728.

U.S. Environmental Protection Agency, January 14, 1986, Proposed rules for PCDDs- and PCDFs-contaminated wastes, Federal Register 51 (9), 1602-1766.

U.S. Environmental Protection Agency, January 14, 1986, Proposed rules for PCDDs- and PCDFs-contaminated wastes, Federal Register 51(9):1602-1766.

U.S. Environmental Protection Agency, April 1986, The National Dioxin Study: Tier 3,5,6,7, Office of Water Regulations and Standards (WH-553), Washington, DC.

U.S. Environmental Protection Agency, 1985, Polyhalogenated dibenzo-p-dioxins/dibenzofurans; Testing and reporting requirements; Proposed rule, Federal Register, 50 (244):51794-51823.

U.S. Patent Office, 1960, 2.922.811 und 2.947.790.

U.S. Patent Office, 1967, 3.347.937.

UBA-Bericht, 1985, Sachstand Dioxine.

USITC (1984); cited from Lee et al.

VCI (Society of the Chemical Industry) (FRG) from data given by the producers and manufacturers of PCP.

Versar, Inc., List of chemicals contaminated or precursors to contamination with identically generated polychlorinated and polybrominated dibenzodioxins and dibenzofurans, EPA Contract No. 68-02-3968, Task No. 48.

Villanueva, E.C., Jennings, R.W., Burse, V.W., and Kimbrough, R.D., 1975, A comparison of analytical methods for chlorodibenzo-p-dioxins in pentachlorophenol, Agric. Fd. Chem., 23:1089.

Vos, J.G., Koeman, J.H., Maas, H.L. van der, Noever de Brouw, M.C. ten, and Vos, R.H. de, 1970, Identification and toxicological evaluation of chlorinated dibenzofuran and chlorinated naphthalene in two commercial polychlorinated biphenyls, Food Cosmet. Toxicol., 8:625.

Waritz, R.S., Aftosmis, J.G., Culik, R., Dashiell, O.L., Faunce, M.M., Griffith, F.D., Hornberger, C.S., Lee, K.P., Sherman, H., and Tayfun, F.O., 1977, Toxicological evaluations of some brominated biphenyls, Am. Ind. Hyg. Assoc. J., 38:307.

Wiberg, K., Lundström, K., Glas, B., and Rappe, C., 1988, PCDDs and PCDFs in consumers' paper products, Poster SOUP10, presented at DIOXIN '88, Umea, Sweden, August 21-26, 1988.

Wu, J., 1985, A quantitative method for evaluating the potential of chemicals for dioxin contamination, Rickover Science Institute.

4. METHODS FOR DEGRADATION, DESTRUCTION, AND DETOXIFICATION OF
   DIOXINS AND RELATED COMPOUNDS

4.1 INTRODUCTION

An intrinsic element of managing risk exposure to dioxins and related chemicals is to reduce release into the environment. Such efforts include modifying manufacturing processes and/or adding waste reduction and recovery technologies such as producing hexachlorophene from dichlorophenols, modifying pulp and paper mill bleaching processes, and pyrolyzing chlorinated still bottoms for HCl recovery.

In the past, dioxin and related compounds have entered the biosphere through accidents, spills, in products and wastes. There are a number of cases known where environmental concentrations of dioxins reached a level that calls for action to remove these persistent and, depending upon the congener-species, partly highly toxic compounds from water and soil. The goal of these actions is to decrease any related risk to human health by means of degradation, destruction, detoxification and disposal of dioxins containing materials. Realistic health-based risk assessments are a vital tool in selecting proper methodologies from a wide-variety of treatment technologies that have been tested in recent years for their applicability. These incorporate thermal destruction, physical methods, biological degradation, chemical dechlorination, fixation/vitrification and safe ultimate disposal. This part of the NATO CCMS Pilot Study on Dioxins covers proven methods as well as promising new innovative and research options.

This document presents information describing how successful (or unsuccessful) different technologies have performed relative to degradation, destruction, detoxification or disposal of dioxin-contaminated media (soil, sludge, water, liquids, etc.).

Full-scale technologies are considered, such as rotary kiln and infrared methods, are discussed in detail. New and emerging technologies--pilot-scale units--are covered and include UV photolysis, UV oxidation, supercritical oxidation, fluidized bed incineration, electric pyrolysis, pyroplasma, and catalytic decomposition; their effectiveness is explained using the latest experimental and engineering data. More recently, KPEG chemical detoxification technology has been field-tested on liquid PCBs, PCP-oil, and spent solvent waste with excellent results; currently, a pilot-scale unit is being evaluated on PCBs-contaminated soil.

Critical data in this document can be employed to determine health risk assessments and, if applied appropriately, significant risk reduc-

tion at contaminated sites can be achieved with the full-scale methods
and some of the soon-to-be-demonstrated new and emerging techniques by
eliminating and/or reducing levels of dioxins and furans in the biosphere
to or below regulatory limits.

4.2 BIOLOGICAL METHODS

Microorganisms can degrade an amazingly wide spectrum of organic
materials with various molecular structures by means of aerobic or anaerobic metabolic processes. The microorganisms utilize the organic compounds as a source of carbon as well as an energy source for cellular
metabolism. All decaying processes, including the natural self-purifica-
tion water bodies, are based on such degradation processes. Biological
wastewater treatment represents a familiar example of the technical use
of microorganisms.

In recent years, examples of detoxification have become known utilizing microbic treatment of soil and groundwater contaminated with organic
chemicals. Substrate-adapted microorganisms, capable of selectively decomposing contaminants such as mineral oil and solvents, are mixed with
the contaminated material and degrade the polluting molecules.

Selective biological purification methods seem to offer an advantage
over the other treatment options, whenever the environmental contamination encompasses large areas or volumes. In such cases other technical
procedures, by which the total contaminated material must be treated
(e.g., thermal treatment), are expected to be relatively more expensive.

Many such cases of wide-spread environmental contamination involve
polychlorinated dibenzo-p-dioxins: i.e., the extensive spraying of waste
oil for dust control on unpaved roadbeds in the U.S.A. or voluminous rehabilitation projects such as Georgswerder, Malsch and Karsau landfills
in the Federal Republic of Germany.

At the present time, there are no published cases about the successful detoxification of dioxin-contaminated soil, sludge or sewage through
biological degradation. However, this area seems to promise new and
potentially innovative cleanup options for the future. Different
approaches are therefore discussed.

4.2.1 Extraction of Natural Microorganisms Adapted for Dioxin Degradation

Because of the rapid reproduction of microorganism populations which are adapted to particular substrates, the random possibility exists, that daughter cells having altered enzyme systems arise through spontaneous mutations. These cells are capable of gaining a competitive advantage in the cell-specific use of energy and nutrients by degrading other materials found in their surroundings. Therefore, one can extract, for example, microorganisms from oil-contaminated soil which can degrade aliphatic as well as aromatic hydrocarbons and which can be used to decontaminate other oil-polluted sites.

As yet, no microorganisms which can clearly degrade dioxins have been extracted from dioxin-contaminated soil and sewage. One reason for this can be the low solubility of dioxins in water, which allows only minute available substrate concentrations. The dioxin concentrations in contaminated soil and sewage generally lie in the ppb range. Because the nutrient availability is hereby limited for microorganisms capable of degrading dioxins, these organisms are quickly overrun by other organisms having more available substrates and consequently disappear.

Positive results about the anaerobic degradation of dioxins in a fixed bed reactor (activated charcoal) innoculated with leachate from a dioxin-contaminated landfills need further confirmation due to analytical difficulties and because of activated charcoal adsorption effects in the low substrate range (Frahne, 1984). True destruction is difficult to quantify because the dioxin may be adsorbed to the carbon, which is suggested by poor analytical recovery efficiencies.

4.2.2 Enzyme Systems for Degradation of Structurally Related Compounds

Co-metabolism is an effect by which a complete or partial elimination of simultaneously present, structurally related compounds also can be detected during the enzymatic degradation of a substrate.

EPA-sponsered research is nearing fruition in applying *Phanerochaete chrysosporium*, known also as white rot fungus, to degrade toxic halo-organic waste constituents. The fungus' lignin-degrading enzyme is one of the strongest oxidizing enzyme systems known. It is nonspecific and thus capable of oxidizing widely divergent organic compounds with different attached functional groups. Lignin represent a high-molecular molecule which contains the nonchlorinated dioxin configuration as a structural element. It was proven that hydrogen peroxide-active oxidases of *Phanerochaete chrysosporium* not only cleave the lignin structure, but can also degrade several chlorinated organic compounds (Bumpus et al., 1985a; Eaton, 1985). For optimal effectiveness, however, nutrients must be added. It is currently being developed by EPA as an in situ degrader for contaminated soils and for treating aqueous wastes in reactors. Laboratory-scale tests reduced aqueous concentrations of pentachlorophenol from 250 mg/l to 5 mg/l in 24-hours (desRosiers, 1987c). The Agency plans to conduct small-scale field tests in the near future.

In the literature it is reported, that in 30 days 2% and in 60 days ca. 4% of the applied, C^{14}-labeled 2,3,7,8-TCDD substrate was metabolized to $^{14}CO_2$ (Bumpus et al., 1985a, 1985b).

The ability of the enzyme system of *Phanerochaete chrysosporium* to dehalogenate and break aromatic ring systems conceivable raises optimistic prospects of decontaminating dioxin-polluted soils. However, because of the low, laboratory elimination rates measured, one must wait for the results of the planned, semi-technical field tests in the U.S.A., in order to make a final judgment possible.

4.2.3 Achieving Desired Biodegradation Through Genetic Engineering

As explained in the previous section, naturally occurring microorganisms are known which are able to split C-Cl bonds or open aromatic ring systems. Such "specialists," however, are often not able to survive under normal environmental conditions and become overrun by more robust species which find better substrate conditions available.

Gene-technology meanwhile offers possibilities to combine both traits: i.e., the destruction of chlorinated aromatics and growth under conditions which occur in contaminated soil or in landfill leachate.

The corresponding procedure to produce microorganisms which degrade 2,4,5-trichlorophenoxyacetic acid is described in the literature. Natural microorganisms possibly degrade 2,4,5-T to a small degree by co-metabolism (Alexander, 1981). Chakrabarty and co-workers, however, succeeded in transplanting plasmid genes, which determine the degradation of aromatic chlorine compounds, from landfill microorganisms to robust *Pseudomonas* bacteria (Kellog et al., 1981; Chakrabarty et al., 1982).

Through slow adaptation, it was found that bacteria from the *Pseudomonas* strain finally employed 2,4,5-T and 2,4,5-trichlorophenol exclusively as carbon and energy sources. Furthermore, these bacteria were able to degrade brominated and fluorinated phenols (Chakrabarty et al., 1983). Successful tests regarding the purification of soil contaminated with up to 2% 2,4,5-T were reported.

Although no successful genotechnological advances have yet been reported for dioxins concerning the culturing of 2,3,7,8-TCDD-degrading microorganisms, attempts to solve this problem should be further pursued in view of the positive results obtained with other chlorinated aromatic compounds.

4.3 CHEMICAL METHODS

Chemical methods of treating dioxin-containing materials are focussed on dechlorination by means of alkaline reagents. The resulting products from dechlorination of organic molecules are inorganic chlorides and a mixture of chlorine-free organics to be further treated. Incineration is one possible option; recycling into production processes appears also to be feasible.

4.3.1 Potassium Polyethylene Glycolate (KPEG)

A chemical destruction technique known as KPEG (potassium polyethylene glycolate) has been used successfully in the states of Montana and Washington to detoxify pentachlorophenol (PCP)-oil and spent solvent waste contaminated with dioxin (desRosiers, 1987a).

A mobile treatment unit, mounted on a 45-foot trailer, was employed to process 8,650 gallons of PCP wood-treating chemical waste at the Montana Pole site in Butte, Montana, in July 1986.

Subsequently, the equipment was transported to the EPA Superfund Western Processing site in Kent, Washington, where it successfully processed spent solvent.

The chemistry used employs two basic ingredients: Potassium hydroxide and polyethylene glycol, which combine to form the reactive agent, KPEG. KPEG chemically removes chlorine atoms from the CDD molecule to form potassium chloride, thus rendering the dioxin molecule non-toxic. A battery of bioassay test, conducted at EPA laboratories in Research Triangle Park, North Carolina, and Duluth, Minnesota, was used to ascertain whether KPEG by-products:

(a)   bioaccumulated in tissues of organisms;

(b)   caused cell mutations; or

(c)   caused immediate harm to fish or mammals.

There was no evidence that the by-products were toxic in any of the tests performed.

Pentachlorophenol-Oil

The site at Butte, Montana, is an inactive wood-treating facility located on a 20-acre, sloping, abandoned mining site where contamination by dioxins (CDDs) and furans (CDFs) reached an adjacent creek, including groundwater and surface soil. The waste is generated as the oily phase of groundwater pumped from 21-foot deep wells; after separation by decantation, approximately 3% pentachlorophenol (PCP) in a diesel-like oil is obtained at the rate of 30 to 50 gallons per day (gpd). The PCP-oil waste contained CDD/CDF homologs ranging from 147 ppb of tetra- to 83,923 ppb of octa-congeners.

Laboratory conducted parametric studies on soil and oil samples to determine treatment conditions were carried out (under EPA contract by the Brehm Laboratory at Wright State University in Dayton, Ohio) to effect optimum decontamination. Table 1 shows that the waste oil was effectively decontaminated by the KPEG reagent in the laboratory at conditions as mild as 70 °C.

The PCP-oil was processed in five batches, each batch consisting of approximately 1,400 to 1,850 gallons of waste oil together with the KPEG reagent. The mixture was heated to 150 °C and allowed to react for 90 minutes before cooling. Table 2 summarizes both the batch and reagent sludge analytical findings and the destruction efficiencies for CDD/CDF homologs. The data indicate that all CDDs and CDFs were destroyed to concentrations below detection limits, which were, on the average, less than 1 ppb (Ehreth, 1986 and Peterson, 1986).

Dioxin Contaminated Spent Solvent

Western Processing, Inc., located in Kent, Washington, was another site where the KPEG technology was employed. The facility was listed as a Superfund site and underwent extensive remedial efforts in 1984. One remaining necessary remedial action was the treatment of a tank of waste solvents containing low levels of 2,3,7,8-TCDD. The KPEG technology equipment utilized at Montana Pole was transported to the Western Processing site in Kent, Washington, where it successfully processed 7,550 gallons of spent solvent containing an oily waste with a high moisture content (28%) and total chlorides of 20,700 ppm; the 2,3,7,8-TCDD content was 120 ppb (Peterson, 1986) (Table 3). Bench-scale studies were performed (Table 4) to determine optimum conditions under which to conduct the field operations. Field reaction conditions, however, were less than optimal: 115 °C for 12 hours. No 2,3,7,8-TCDD was found at a detection limit of 0.3 ppb (using EPA method 8280 as the analytical protocol) in any of the processed batches. Spent reagent was transported to the Chemical Waste Management PCB-permitted incinerator in Chicago, Illinois, for final disposition.

Table 1.   Destruction Efficiency (DE) (%)[a,b] of KPEG-Treated PCP-Oil;
(Ehreth, 1986, desRosiers, 1986b, and Roulier, 1987)

	Raw Oil	KPEG Treated Oil 70°C		100°C	
		Elapsed Time = 30 minutes			
CDD/CDF	ppb	ppb	DE	ppb	DE
2,3,7,8-TCDD[c]	28.2	ND (0.738)	>97.7828	ND (0.544)	>96.2414
TCDDs	422	ND (0.373)	>99.9180	ND (0.274)	>99.8881
PCDDs	822	ND (0.922)	>99.9319	ND (0.411)	>99.9814
HxCDDs	2982	ND (1.83)	>99.9524	ND (2.43)	>99.9031
HpCDDs	20671	2.14	99.9846	1.43	99.9832
OCDD	83923	4.01	99.9942	2.56	99.9936
2,3,7,8-TCDF[c]	23.1	1.28	93.9583	ND (0.282)	>98.4042
TCDFs	147	16.3	87.3189	ND (0.351)	>99.5015
PCDFs	504	ND (0.468)	>99.9355	ND (0.288)	>99.8888
HxCDFs	3918	2.98	99.9284	ND (0.757)	>99.9671
HpCDFs	5404	2.63	99.9144	ND (1.06)	>99.9624
OCDF	6230	ND (3.63)	>99.9370	ND (1.63)	>99.9285

[a] These values are not corrected for small concentrations of HpCDDs, OCDD, and HpCDFs detected in lab blanks, so actual DEs for these isomers are actually higher than indicated.

[b] The notation > indicates that the isomer(s) were not detected and the lower limit cited for the DE is based on the analytical detection limit.

[c] Source of the 2,3,7,8-TCDD/TCDF is PCBs, which contain chlorobenzenes, and have experienced thermal stress, namely, pit fires.

( ) = Detection limit

Table 2. Results of KPEG Chemical Detoxification of PCP-Oil, Montana Pole Site, Butte, Montana[b] (Ehreth, 1986 and Peterson, 1987)

CDD/CDF Homolog	Raw PCP-Oil (ppb)	KPEG-Treated Oil @ 150°C, 90 min (ppb) Batch No.: 1	2	3	4	5	Reagent Sludge	DLs avg[a]	DEs χ[a]
TCDDs	422	(0.64)	(0.68)	(0.82)	(0.75)	(0.80)	(0.12)	(0.73)	>99.98
PCDDs	822	(0.97)	(0.49)	(0.78)	(0.52)	(0.74)	(1.0)	(0.70)	>99.91
HxCDDs	2,982	(0.92)	(0.63)	(0.43)	(0.70)	(0.90)	(0.44)	(0.76)	>99.97
HpCDDs	20,671	(0.41)	(0.52)	(0.32)	(0.39)	(0.47)	(0.18)	(0.48)	>99.99
OCDD	83,923	(1.3)	(0.87)	(1.1)	(0.54)	(1.1)	(1.0)	(0.98)	>99.99
TCDFs	147	(0.77)	(0.59)	(0.83)	(0.62)	(1.5)	(0.14)	(0.86)	>99.41
PCDFs	504	(1.3)	(0.58)	(0.66)	(0.67)	(1.1)	(0.51)	(0.81)	>99.84
HxCDFs	3,918	(0.84)	(0.53)	(2.2)	(0.67)	(2.1)	(0.49)	(1.27)	>99.97
HpCDFs	5,404	(0.49)	(1.0)	(1.6)	(0.56)	(0.83)	(0.28)	(1.09)	>99.98
OCDF	6,230	(0.53)	(1.3)	(1.1)	(0.61)	(0.83)	(0.22)	(0.87)	>99.99

[a] Detection limit (DL) averaged for five batches; destruction efficiency (DE) based on average DL of homolog. (The notation > indicates that the homologs were not detected and the lower limit cited for the DE is based on the average of the analytical detection limit.)

[b] Oil processed July 25-31, 1986; in addition, wipe samples of equipment after decontamination procedures completed showed: no TCCDs/TCDFs, PCDFs, HxCDDs/HxCDFs, or OCDF detected at or above DL - 3.8 ng/m^2 and OCDD - 217 ng/m^2 were found above the DL. These positives values are well below any concentration considered to be of concern according to EPA ORD ECAO-Cincinnati and EMSL-Las Vegas laboratories.

( ) - Detection limit. Analyses performed by IT Corporation, Knoxville, TN; QA/QC performed by EPA EMSL, Las Vegas, NV.

Table 3.   Spent Solvent-Oil Waste Analysis, Western Processing Site, Kent, Washington (Peterson, 1986)

Parameter	Concentration	Comments
oils	42 %w	heterogeneous mixture
solids	10-50 %w	
water	28 %w[a]	
pesticides	6 %w	includes phosphate esters
total chlorides	20,700 ppm	organic + inorganic
BTU value	9,660 BTU/lb	
2,3,7,8-TCDD	120 ppb	analyzed by Chemical
Waste		
		Management

[a] Water content, as determined by the Karl Fisher method, may be misleadingly high due to presence of glycols and alcohols.

Table 4.   Bench-Scale Parametric Studies of KPEG Treatment of Solvent-Oil, Western Processing, Kent, Washington (Peterson, 1986)

Operating Conditions		KPEG Concentration[a]	
		25g	50g
Temperature °C	Reaction Time hrs	2,3,7,8-TCDD in Treated Oil (ppb)	
115	0	120	120
115	5.5	5.5	-
115	6.5	-	2.5
115	12	-	< 0.3

[a] KPEG dosage either 25 g (standard reagent) or 50 g (double KOH), combined with 25 g PEG-400 + 25 g DMSO per 250 ml solvent-oil.

Polychlorinated Biphenyls

As mentioned previously, the KPEG technology was first developed to treat polychlorinated biphenyls (PCBs) (Kornel and Rogers, 1985). Reagent recycle tests from pilot plant operations were carried out by the Galson Research Corporation during 1983-1984. Typical data from two such recycle runs are given in Table 5. These data indicate that eight batches of PCB-contaminated oil (112-200 ppm PCBs) were successfully treated with a single dose of KPEG reagent during Recycle Test Run A and that five batches of contaminated oil (398-1080 ppm PCBs) were successfully treated with a separate, single dose of fresh reagent during Recycle Test Run B (Adams and Peterson, 1986).

More recently, treatment of PCB-contaminated soil has been achieved by mixing the soil with an equal volume of hot (150 °C) KPEG reagent in a rotating industrial mixer (similar to a large cement mixer) (Figure 1). Soil moisture is volatilized and recovered for later use in the process. The reagent employed comprises a mixture of polyethylene glycol (PEG) and polyethylene glycol monomethylether (PEGM), potassium hydroxide (KOH), and dimethylsulfoxide (DMSO) (U.S. Patent 4,574,013). At the termination of the reaction, usually 30-120 minutes, the bulk of the reagent (>80 percent in the field tests) is decanted from the treated soil. The residual reagent and dechlorinated by-products are removed from the soil by mixing the soil with an equal volume of water and decanting the water. This washing is done two or three times and provides >99% overall recovery of reagent. The washwater from the last wash is passed through a bed of activated carbon, which preferentially removes the dehalogenated products. The contaminated carbon is burned in a PCB-approved incinerator.

This equipment can be assembled in a 6' X 8' trailer. Process kinetics have been found to be affected by: (a) soil organic carbon content; (b) soil particle size distribution; (c) PCB isomer distribution; (d) soil moisture content; (e) reaction temperature; (f) reagent formulation; and to a lesser extent, (g) reaction time at the reaction temperature (Carpenter, 1986). Furthermore, at reaction temperatures below 100 °C, the water content of the soil affects the rate of reaction. It is necessary to determine experimentally the optimum reagent formulation for each soil. Under optimum reaction conditions, PCBs are reduced from 500 ppm to <0.1 ppm in 0.5-2.0 hours.

Most laboratory-scale trials of this process have been applied to soils contaminated with 1,2,3,4- or 2,3,7,8-TCDD. The small amount of PCB work in this area has been done as a preliminary to the dioxin studies. 2,3,7,8-TCDD levels in treated soil have been reduced to <0.06

ppb, although the usual detection limit has been in the area of 1.0 ppb. Pilot testing with PCBs was viewed primarily as a support activity under the EPA National Dioxin Study (desRosiers, 1986b). The KPEG reagent formulation used was optimized for CDD content rather than for PCBs.

In summary, conditions for successful processing of PCB-contaminated soils to <1 ppm require reaction temperatures on the order of 100°-150 °C and reaction times of 0.5-2 hours. Total cycle times of 4-8 hours are probable. Reagent is applied to the soil in a 1:1 ratio (dry basis) followed by reaction and reagent recovery. Reagent recoveries in excess of 99% have been obtained in PCB field testing. Initial soil water concentration is not relevant to the process, except as it affects process economics because moisture is volatilized during the heating phase.

Table 5.   Recycle Test Run Data for KPEG Treatment of PCB-Contaminated Transformer Oil[a] (Adams and Peterson, 1986)

Recycle Test Run[b]	% PEG	% TMH[c]	% DMSO	%KOH	Reaction Time (min)	KPEG-Treated Oil @120°C PCB Conc. (ppm) Initial	Final
A-1	5	5	10	9.7	30	140	<1
A-2	recycle	recycle	recycle	recycle	60	200	<1
A-3	recycle	recycle	recycle	recycle	290	200	1.2
A-4	5.8	5.8	11.6	10.8	230	150	1.4
A-5	recycle	recycle	recycle	recycle	120	151	2.4
A-6	recycle	recycle	recycle	recycle	210	144	3.7
A-8	recycle	recycle	recycle	recycle	570	112	1.8
A-9	recycle	recycle	recycle	recycle	180	112	30
B-1	5	5	10	9.7	150	398	0.6
B-2	recycle	recycle	recycle	recycle	120	790	<1
B-3	recycle	recycle	recycle	recycle	120	1,049	<1
B-4	recycle	recycle	recycle	recycle	360	1,080	<1
B-5	recycle	recycle	recycle	recycle	240	1,080	<1
B-6	recycle	recycle	recycle	recycle	300	838	7.3

[a] Transformer oil comprised Aroclor 1260.

[b] Each test run consisted of 40-gallon batches treated with one batch of reagent mixture as shown.

[c] THM — Triethylene glycol monomethylether homologs.

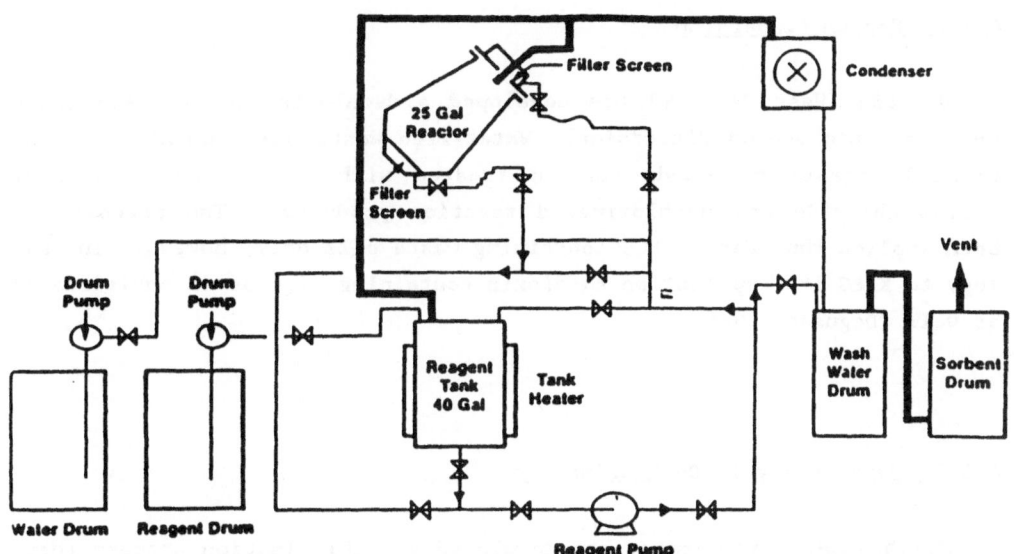

Fig. 1   General Pilot Plant Design of KPEG Dechlorination Process

More recently, EPA has constructed a 2-m^3 transportable, horizontal industrial mixer reactor for use in field tests of soils contaminated with PCBs. The reactor is currently being evaluated on the islands of Guam and later Saipan in the Pacific, on PCBs-contaminated soil (2,500-4,500 ppm) in 30-day field tests.

4.3.2 Sodium Dispersion

DEGUSSA (West Germany) has developed a dechlorination process based on an organic sodium dispersion. Water-free waste oils containing chlorinated organic compounds can be treated with this reagent yielding sodium chloride and dechlorinated reaction products. The process has been applied thus far to PCB containing waste oils only; however, in analogy to KPEG its application to dioxin containing liquids is envisionable as well (Degussa, 1986).

4.3.3 Pyrohydrolytic Dechlorination

NUKEM GmbH, West Germany has developed a dechlorination process (Maurer and Schöner, 1985) under support by the German Ministry of Research and Technology (BMFT) that utilizes the direct decomposition of chlorinated hydrocarbons under inert gas conditions (N_2) in the temperature range of 600-800 °C. Generation of HCl as well as polychlorinated dibenzo-p-dioxins or dibenzofurans is precluded. The strong dechlorination action is applicable for thermo-chemical destruction of the latter components themselves.

Having tested a variety of calcium-oxide based reactant mixtures, Nukem proposes the use of a solid porous CaO/SiO_2 combination in which the active reaction partner CaO is fixed in a silicate matrix that remains stable after the dechlorination reaction took place.

Generated $CaCl_2$ is embedded in the porous SiO_2-structure and the reaction mixture will not fuseenabling the use of the granulated material in a continuous forward-flow reactor.

Figure 2 depicts a process flow diagram. Nitrogen serves as inert gas. The dechlorinating CaO/SiO_2 granules are fed from the top into the electrically heated (600-800 °C) flowthrough reactor which is mechanically stirred. Chlorinated hydrocarbons are added, the thermo-chemical decomposition of which is either exo- or endothermal, depending upon their chlorine content.

Average residence times for aliphatic chlorinated compounds are up to 5 seconds; for aromatic chlorinated compounds 10-20 seconds. The solid phase, removed with a screw-extractor at the bottom of the reactor, contains carbon, $CaCl_2$, and excess CaO/SiO_2 reactant.

The emerging gas contains the following components, depending upon the chemical structure of reactor feed:

Reactor Waste Feed	Emerging Gas
Polychlorinated aliphatic compounds	CO, CO_2
Partially chlorinated aliphatics	H_2, CH_4, C_2H_6, CO, CO_2
Chlorinated aromatic compounds	H_2, CH_4, CO, CO_2, C_2H_6, dechlorinated + condensed aromatics

Decomposition of PCBs under inert-gas conditions as well as in the presence of oxygen did not yield polychlorinated dioxins or furans, neither in the solid residue nor in the reactor off-gas. The direct decomposition of OCDD and OCDF was carried out in a lab-scale test unit in the mg-range. Destruction occurred down below detection limits with destruction and removal efficiencies (DREs) > 99.99%.

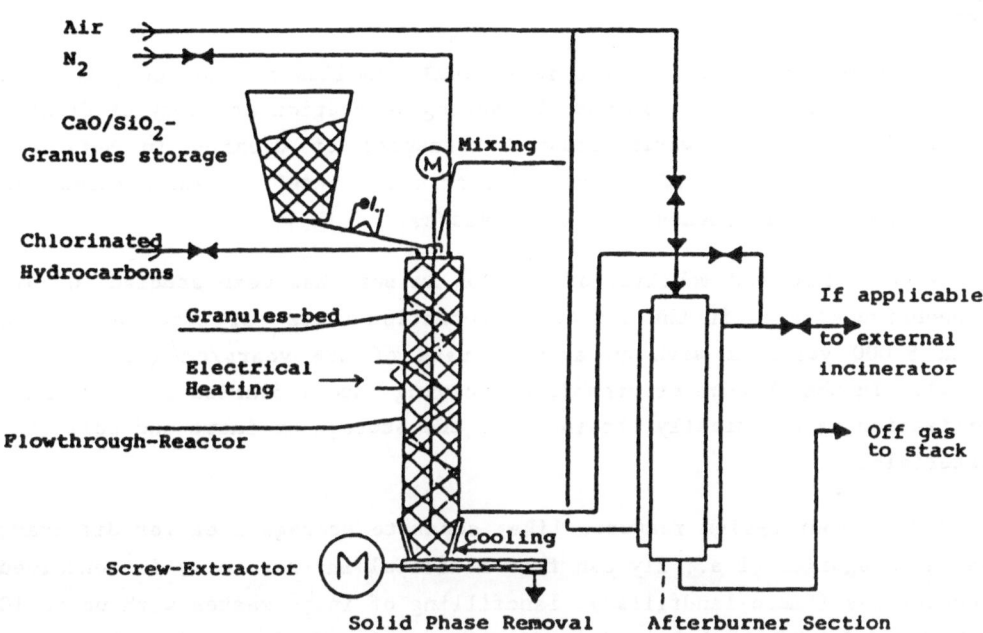

Fig. 2.   Thermo-Chemical Decomposition of Chlorinated Hydrocarbons in a Stirred Flowthrough-Reactor

4.4 DISPOSAL/FIXATION

Proper waste management of dioxin-contaminated materials aims for permanent reduction of the risk originating from the source of pollution. Appropriately applied disposal, fixation and vitrification technologies can play a significant role in removing dioxins from the accessible environment. Although these technologies usually do not materially destroy dioxins, they do, in many cases, represent the best solution, under technological and economical constraints, for low-contamination level solid wastes.

4.4.1 Landfilling

Primary concern for deposition of dioxin containing wastes in landfills has to be focussed on exclusion of possible remobilization. This has to encompass solubilization by organic solvents as well as dust blowing from wind-exposed surfaces and overland transport through rainfall.

Water solubility of dioxins is minimal and lies in the low ppt range (8-19 ppt). Taking into account the strong adsorption affinity of dioxins to surfaces (soil), water induced transport of dioxins in soil can practically be disregarded; the "penetration rate" in subsurfaces was predicted to be in excess of 1,000 years/cm!

Solvent enhanced mobility of dioxins in soil has been studied in comprehensive studies in the U.S.A. Penetration rates vary from 4 to more than 5,000 years/cm with an average rate of 388 years/cm (desRosiers, 1984). In the absence of organic dissolving intermediaries desired immobilization is practically attained for the storage of inert and soil-like materials.

With proper design features like: separate storage area for different waste categories (i.e., fly ash from municipal incinerators) in contained receptacles ("mono-landfills"), landfilling of inert wastes with up to 10 ppb 2,3,7,8-TCDD is believed to be acceptable (Paustenbach and Murray, 1985). Depositing such wastes in large, sturdy (1 m^3) "super sacks" made from nylon or polypropylene fabrics is also deemed applicable for wastes of the above indicated contamination level.

For an average level of dioxin contamination of inert materials and combustion residues up to 100 ppb 2,3,7,8-TCDD, hazardous waste landfills with double-lining and leachate control systems should be used. This also applies to sludges with mostly inorganic components, which can be immobilized with suitable fixing materials such as alkaline fly ash, cement, lime made water repellant by organic coating, etc., which must be added according to waste-specific requirements. At the present time, regulatory agencies in different countries have chosen differing approaches as to permissible levels of dioxin in landfills.

In 1986, fire debris with dioxin concentrations in the ppb range from a PVC-processing factory in West Germany (Refrath near Cologne) was stored in a hazardous waste landfill. Serving as a model project for the disposal of dioxin-contaminated waste, this landfilling was conducted as follows: An 80-cm thick clay trough was laid out on a landfill area having a double-lined base. Solid fire debris, contaminated plastic parts and drums filled with ashes/dust were carefully piled up in the trough. A construction filler material (Ca-anhydrite) was finally poured to fill in all empty spaces. A solid block of ca. 1,500 m^3 resulted after the construction filler hardened. The surface of the clay trough was covered with a 50 cm thick (KF = 10^{-10} cm/second), slanted clay roof. Consequently, the waste lays as a closed inert block in the landfill. Such a procedure, however, can only be considered reasonable for highly contaminated wastes greater than 100 ppb 2,3,7,8-TCDD (Fuhr, 1987).

4.4.2 In Situ Stabilization and Solidification

Stabilization and solidification have become commonplace treatment technologies in recent years. The process, also known as chemical fixation, usually involves the use of portland cement or lime and other materials such as ash, cement kiln dust, and blast furnace slag. The wastes are mixed with these materials in a liquid state, and when allowed to harden, the hazardous constituents are physically incorporated within the solid matrix. The resulting mass is less permeable to leaching by water, but is susceptible to breakdown by acids. Because of the high pH of the mixture (normally 9-11), metals precipitate as relatively insoluble hydroxides, carbonates, or silicates. Soluble silicates are sometimes added to enhance the chemical fixation of heavy metals.

Three Missouri sites were selected for evaluation of the feasibility of cementitious and asphaltic stabilization techniques: the Minker site, the site at Piazza Road, and the Sontag Road site. After mix-design work was completed on the Minker and Piazza Road sites, it became apparent that emulsified asphalt per se would not provide an effective method for stabilization of the test soils. As a result, calcitic lime--calcium hydroxide--was employed to modify the soil prior to asphalt addition. Fixation of soil contaminated up to 700 ppb 2,3,7,8-TCDD resulted in leachate concentrations of only 2-3 ppt 2,3,7,8- TCDD (Ellis et al., 1985; Vick et al., 1985) (using EPA Solid Waste Leaching Procedure SW-924, 1982).

Preliminary observations suggest that the asphalt/lime combination acted effectively to flocculate fine particles and reduce the percentage of "erodible" particles (<50 lm) relative to the native soil and will be effective as an interim remedial measure for solidification of dioxin-contaminated soils and concomitant mitigation of further environmental pollution (desRosiers, 1987c). Leaching experiments with these specimens are complete and the extracts were subjected to GC/MS analysis. Examination of leachate concentration data from these stabilized soils revealed that 2,3,7,8-TCDD levels were not statistically different from those obtained for the corresponding unstabilized soils. This was true for leachates from both the parent soil-cement specimens, as well as for the dissociated particulate material. However, the absence of 2,3,7,8-TCDD at detectable levels in virtually all soil-asphalt leachates implied that soil stabilization effectively reduced solubilization of the 2,3,7,8-TCDD contaminant.

4.4.3 In Situ Vitrification

Battelle scientists at the Pacific Northwest Laboratory (PNL), Richland, Washington, have developed an innovative soil-melting technology, with potentially broad application to hazardous waste treatment, soil stabilization, and construction needs. This new technology is termed in-situ vitrification (Technical Resources, 1987 and Timmerman, 1984).

The process utilizes an electric current passed between electrodes placed in the ground to convert soil and contaminated waste to a stable, glasslike material. Heat from the electric current melts the soil and decomposes the organic materials. During the process, as temperatures reach 20,000 °C, metallic and other inorganic materials are dissolved into or are encapsulated within the vitrified mass. Gases evolved from the melt are collected and treated on the surface (see Figure 3). Convective currents within the melt uniformly mix materials that are present in the soil. When the electric current ceases, the molten volume cools and solidifies.

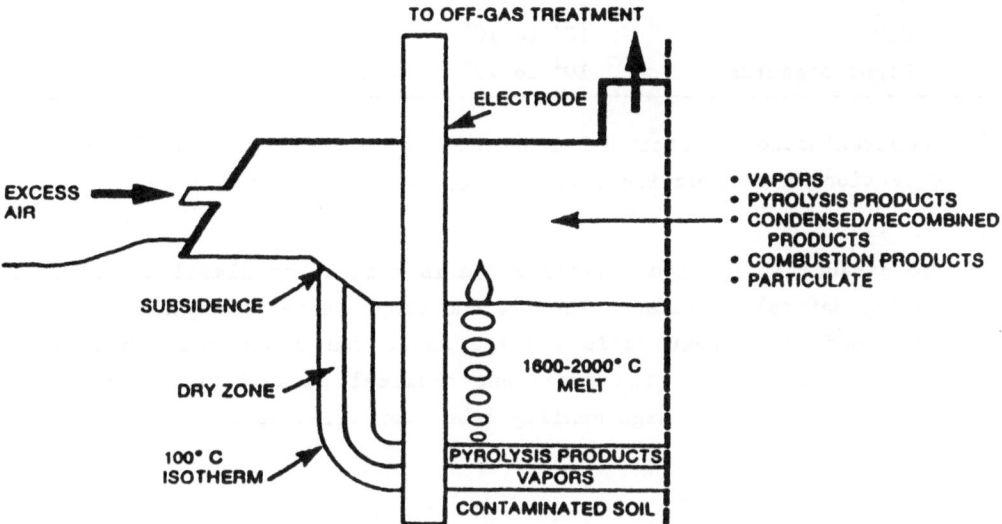

Fig. 3.   Disposition of Materials during Processing, Using In-situ Vitrification Technology

Examples of decontamination factors[*] achieved on wastes, such as zirconia/lime sludge, dioxin- and PCBs-contaminated soil, liquid and solid organics, etc., treated in bench, pilot, and field tests, include:

Contaminant	Soil	Off-Gas	Overall
Mo, Sr, Pu, Am, U	10^3 to 10^5	10^5	10^8 to 10^{10}
Sb, Te, Ru, Cs	10^2 to 10^3	10^4	10^6 to 10^7
Cd, Pb	10	10^4	10^5
F	10^2	10^5	10^7
NO_x	10^2	10^3	10^5
SO_2	1	10^3 to 10^4	10^3 to 10^4
PCBs	10^3 to 10^4	$>10^3$	$>10^6$
Light organics	10^2 to 10^4	$>10^3$	$>10^5$

[*] Decontamination factor is measured as $1/(1-Rf)$, where Rf is the retention or destruction factor (e.g., 99.999% removal = 10^5).

The product of in-situ vitrification is a block of glasslike material resembling natural obsidian. Tensile and compressive strengths of blocks that are made from most soils average about ten times the strength of unreinforced concrete. The excellent chemical durability of vitrified blocks is comparable to high-quality laboratory glassware.

4.5 PHYSICAL METHODS

4.5.1 Clay Adsorption

Research is underway on the use of clay adsorbents for the removal of dioxin from industrial wastewaters. Three types of chemically modified clays that have been tested are:

(1) Hydroxy aluminum montmorillonite (HYDAL-clay), a zeolite-type microsporous adsorbent
(2) Cetylpyridinium montmorillonite (CPC-clay), an organoclay
(3) Cetylpyridinium hydroxy aluminum montmorillonite (CPC-HYDAL-clay).

The Henry's Law constants from the equilibrium, batch-type binding of dioxins to the three modified clay sorbents are listed in Table 6.

Table 6. Binding of Dioxins to Clay-Based Sorbents (Srinivasan and Fogle, 1987)

SORBENT	Partition Coefficient, PC x 10^{-3} (dm^3/kg)	
	OCDD	2,3,7,8-TCDD
HYDAL-clay	100.00	67.5
CPC-clay	105.0	----
CPC-HYDAL-clay	97.8	----

These studies showed that OCDD and 2,3,7,8-TCDD strongly bind to HYDAL-clay. OCDD also binds strongly to CPC-clay. Furthermore, it was concluded that:
- The adsorption rate of OCDD on HYDAL-clay is very rapid and follows a Freudlich-type sorption isotherm;
- Packed-bed sorption tests indicate no breakthrough of OCDD, suggesting multilayer-adsorption; and
- OCDD-THF (tetrahydrofuran) elution experiments indicate that effective regeneration can be accomplished.

At this point in time, these modified clays still lack the adsorptive properties of most activated carbons. In indirectly related estudies by Marple et al. (1987), ppt levels of 2,3,7,8-TCDD in industrial lagoon wastewater were successfully removed by alum flocculation to an average of 0.36 ppt in ten experiments. Hence, it may be concluded that the standard series of coagulation, flocculation, sedimentation and filtration should be capable of removal of 2,3,7,8-TCDD at its solubility limit of approximately 8 ppt to parts per quadrillion (ppq) levels.

4.5.2 Underground Storage In Mines

The underground depot of Herfa Neurode in the Federal Republic of Germany (Hessen) represents an example for the storage of hazardous wastes in salt mines (Herfa-Neurode, 1986). Predetermined for their suitability (not gas-emitting, not liquid etc.), wastes are orderly stocked in steel barrels in excavated salt strata. Because their position is exactly determined in the mines, wastes can even be reclaimed and recycled at a later point in time, as already practiced in this mine.

Due to the constantly controllable storage conditions, such underground depots are suited for storage of solid wastes without concentration restrictions. Dioxin-containing wastes need no special pretreatment besides proper packaging and labelling. Because of the restricted available space and also the limited hauling possibilities in the shafts, this disposal method is limited to hazardous wastes with high dioxin-concentrations, however.

Considerably larger volumes of wastes (soil, sludge) can be stored in massive salt caverns. Cavities of 150,000 m^3 to 250,000 m^3 which are toay used in West Germany to stockpile the national oil reserves, are leached out of geological-formed salt deposits. The Cavern Construction and Management Corporation, FRG (Kavernen Bau-und Betriebsgessellschaft), together with the Society for Radiation and Environmental Research, Department of Deep Storage of Wastes (Gesellschaft fuer Strahlen- und Umweltforschung, Abt. Tieflagerung von Abfaellen) are currently working out waste storage concepts which could offer disposal alternatives for dioxin-contaminated fly ash, dust, etc. (Schneider, 1987). Similar concepts are presently developed in the state of Texas, U.S.A.

4.6 Treatment Methods

Treatment methods for dioxins and related compounds are engineered for maximum destruction rates of the respective molecular structures leading to harmless reaction products. At present, there are a wide variety of treatment technologies under consideration such as:

- High temperature incineration
- Infrared destruction technology
- UV-Photolysis
- Supercritical oxidation
- Electric pyrolysis
- Pyroplasma technology
- Catalytic decomposition

Only a few of these methods are fully tested and routinely applied in practice (2.5.1). The larger number has to be termed "experimental innovative," with promising outlook (Freeman and Olexsey, 1986) for specific application areas (5.2). Not all technologies in development can be covered in this chapter in detail. With future availability of field test data, processes not mentioned at this point in time may also be considered for use in treating dioxin-containing wastes.

4.6.1 Full-Scale Technologies

Rotary Kiln Incineration

Incineration of wastes in rotary kilns is commonplace practice in many industrialized nations. Rotary kilns are best suited to accept a wide variety of wastes ranging from solids through pasty wastes to liquids (Fabian et al., 1979, Fuhr, 1985a,b). In Europe, stationary rotary kiln incinerators have been used to burn dioxin containing wastes. One of the first recorded cases has been the thermal destruction of the wastes originating from the Icmesa reactor in Seveso, Italy. The preparation and execution of this burn in Basel, Switzerland, has been reported, in detail, by Swiss authorities (Bundesamt für Umweltschutz, Bern, 1985). The rotary kiln used had first been tested for destruction and removal efficiency with trichlorodibenzo-furan, a toxicologically harmless substance (Fuhr, 1987). Sampling the untreated flue gas verified DEs of 99.9999%, and combined with flue gas scrubbing, DREs of 99.99999%. The actual burn of 2,500 kg of contaminated wastes containing 600 g 2,3,7,8-TCDD resulted in dioxin levels below detection limits (0.05-0.2 $\mu g/m^3$).

In the United States, 29 mines in the state of Missouri have been examined for their suitability to store dioxin-contaminated soil (EPA, 1985). The project is also supposed to eventually examine the feasibility of digging new underground depots to store dioxin-contaminated soil as opposed to using old mines.

Shallow, underground limestone/dolostone (dolomite) mines were found to offer distinct technical advantages over other types of deeper underground mines, i.e., sandstone, lead-zinc, iron, or coal. These advantages are related principally to dryness, structural stability, potential size, location, and accessibility factors.

Packaging options evaluated included rectangular steel vaults, steel drums, and woven polypropylene sacks (supersacks). Supersacks are considerably less expensive than metal drums (plastic drums were not used in cost estimates), and metal drums are less expensive than metal vaults. Supersacks allow more soil to be stored in the same available space than the other packaging systems.

A second case is reported from the Federal Republic of Germany. Chlorinated dibenzo-p-dioxins and dibenzofurans are contaminants of oil that leaks out of a large landfill site at the city of Hamburg, West Germany. The contamination probably originates from chlorophenol wastes deposited in the landfill which are gradually leached out by mineral oils contained in the waste mixture deposited. Due to lack of treatment options, this waste oil did accumulate over several years and was stored in 55 gallon drums at the site.

In late 1987, a chemical company operating an incinerator licensed to burn dioxins agreed to thermally destroy the dioxin containing wastes. The oil contained a wide variety of chlorinated organic compounds such as chlorobenzenes, chlorophenols, and PCBs.

The average concentration for dibenzo-p-dioxins/dibenzofurans were:

2,3,7,8 - TCDD		32 μg/kg
Total	TCDD	60 μg/kg
Total	CDDs/CDFs	42530 μg/kg

Dioxin destruction took place in a rotary kiln incinerator designed for hazardous waste incineration, i.e., PCBs. Two analytical control groups were included to independently monitor concentrations of dioxins/furans in the off-gas.

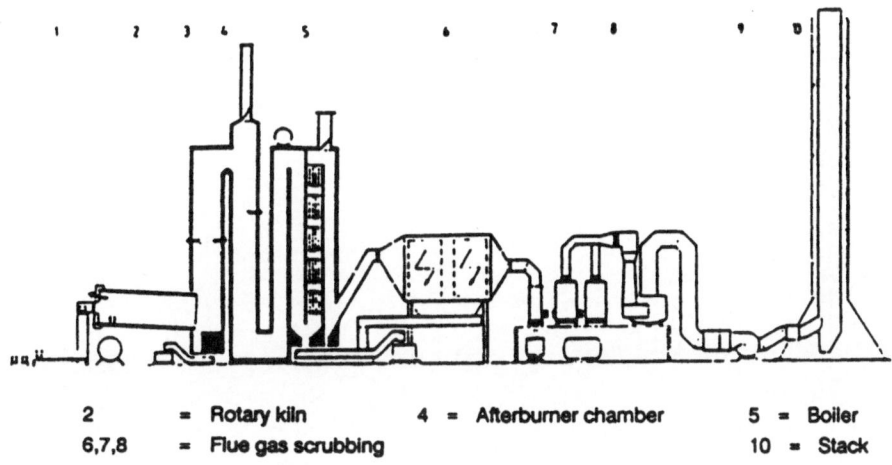

2	= Rotary kiln	4 = Afterburner chamber	5 = Boiler
6,7,8	= Flue gas scrubbing		10 = Stack

Fig. 4.   A schematic of the incineration plant.

The oil was atomized through a burner nozzle into the rotary kiln which was operated above 1,100 °C. Flue gases enter at the afterburner chamber by induced draft. Temperatures in the afterburner chamber were set to (greater than or equal to) 1,200 °C. Fuel oil was used to maintain the desired temperature level in the plant. Prior to incineration of the dioxin-containing oils, a baseline test was run with burning only fuel oil for a period of several hours.

Sampling was carried out at four different sampling points (two per analytical team).

Incineration took place at a rate of 550 kg/h of waste oil. C^{13} - labeled dibenzofurans and dibenzo-p-dioxins were used as internal standards in order to calibrate for retrieval losses in the analytical/sampling procedure.

The results of the actual burn are given in Table 7 (Hagenmaier, 1988) (All results are given in ng/m^3).

Table 7.   Incineration of Georgswerder Landfill Oils

A1 = Baseline level (fuel oil burnt) - Analytical Group A
A2 = Actual levels during waste oil incineration - Analytical Group A
B1 = Baseline level (fuel oil burnt) - Analytical Group B
B2 = Actual levels during waste oil incineration - Analytical Group B

	A1	A2	B1	B2
Total - Cl_4DD	0.35	0.30	0.40	0.50
-2,3,7,8-Cl_4DD	0.016	0.011	0.017	0.023
Total - Cl_5DD	0.48	0.47	0.68	0.78
-1,2,3,7,8-Cl_5DD	0.074	0.076	0.095	0.103
Total - Cl_6DD	0.51	0.56	0.70	0.78
-1,2,3,4,7,8-Cl_6DD	0.057	0.058	0.076	0.067
-1,2,3,6,7,8-Cl_6DD	0.047	0.052	0.072	0.083
-1,2,3,7,8,9-Cl_6DD	0.038	0.047	0.051	0.057
Total - Cl_7DD	0.63	0.68	0.92	1.11
-1,2,3,4,6,7,8-Cl_7DD	0.29	0.31	0.45	0.52
Cl_8DD	0.89	2.98	0.87	1.83
Total - Cl_4DF	2.63	2.69	2.64	2.20
-2,3,7,8-Cl_4DF	0.043	0.034	0.025	0.035
Total - Cl_5DF	4.78	3.52	2.98	2.46
-1,2,3,7,8-Cl_5DF	0.40	0.31	0.29	0.21
(+1,2,3,4,8-Cl_5)				
-2,3,4,7,8-Cl_5DF	0.13	0.08	0.10	0.09
Total - Cl_6DF	2.75	2.80	3.19	2.49
-1,2,3,4,7,8-Cl_6DF	0.34	0.33	0.40	0.31
(+1,2,3,4,7,9-Cl_6DF)				
-1,2,3,6,7,8-Cl_6DF	0.29	0.28	0.34	0.26
-1,2,3,7,8,9-Cl_6DF	n.n	n.n	n.n	n.n
-2,3,4,6,7,8-Cl_6DF	0.10	0.12	0.11	0.11
Total - Cl_7DF	1.91	1.49	2.54	2.59
-1,2,3,4,6,7,8-Cl_7DF	1.61	1.33	2.36	2.32
-1,2,3,4,7,8,9-Cl_7DF	n.n	n.n	n.n	n.n
Cl_8DF	0.58	0.65	5.10	2.65

Retrieval rate for the sampling process in [%]:

^{13}C-1,2,3,4-TCDD	34	98	65	99
^{13}C-1,2,3,4,6,7,8-HexaCDF	51	78	59	96
^{13}C-OctaCDD	48	103	80	98

In the U.S., transportable (modular) rotary kilns are viewed as a viable option to do remediation clean-ups on-site. After dioxin contaminated wastes have been treated, the unit can be disassembled and moved to another site.

Most well-known is the EPA Mobile Incineration System (MIS), which to date has treated over 1,340,000 kg of dioxin-contaminated soils and other solids and over 80,000 kg of liquid waste. This system has achieved a destruction and removal efficiency (DRE) exceeding 99.9999% (Skinner and Lindsey, 1987). The EPA MIS was built by mounting a rotary kiln unit and its accessories on four flatbed truck trailers. The first trailer holds a 4.9-meter-long, 1.2 meter-diameter kiln that operates up to 1,000 °C, with a nominal-solids retention time to 60 minutes. On the second trailer is a secondary combustion chamber that operates to 1,300 °C, with a gas retention time of 2-3 seconds. The third trailer holds a water quench, a wet electrostatic precipitator (WEP), and an alkaline scrubber for off-gas treatment. The fourth trailer contains monitoring instruments and controls.

Although EPA has invested over $10 million in the development of this system, a duplicate could be built for roughly half the cost because of experience gained. A number of engineering changes have recently been made that should significantly improve the operation of the system. The feed system has been enlarged, increasing the maximum feed rate to the kiln to 5,000 lbs per hour. The overall gas velocity leaving the kiln has been reduced by enriching the combustion air with oxygen. A cyclone has been installed between the kiln and afterburner to collect the heavy load of fine dust carry-over that otherwise accumulates in the afterburner. Such carry-over from burning dusty soils had caused considerable downtime in the incinerator. A venturi scrubber has been installed in conjunction with the WEP, thus reducing particulate emissions. Operating costs are expected to decline dramatically as capacity and reliability are increased and additional operating experience is obtained. Over one-third of the original costs were for sampling and analysis necessary to prove that the process waters and kiln ash should no longer be considered a hazardous waste and thus were RCRA-delistable.

A compilation of test data from trial burns in dioxins are given in the following table (desRosiers, 1986a) (Note that the following data is for the EPA MIS only).

Table 8. Mobile, Rotary Kiln Incinerator for Dioxin Combustion. Results of EPA MIS Dioxin Trial Burns in Missouri

Parameter	Permit Limit	Test Number 2-2[a]	2-3	2-4	2-5	3-1,-2,-3[b]
Rotary kiln						
Temp. (°C)	760-1040	845	826	902	918	852
Secondary						
Temp. (°C)	1120-1315	1153	1199	1194	1199	1187
Ret. time (sec.)		2.5	3.2	2.5	2.6	-
O_2 (vol. %)	>4	7.96	6.61	6.44	6.42	5.4
CO (ppm_v)	<100(6 min.)	7.7	1.3	2.3	2.5	6.8
CO_2		10.9	11.3	11.4	11.1	-
C.E. (%)		99.993	99.999	99.998	99.998	-
TH (ppm_v)		0.1	0.5	0.5	0.9	-
NO_x (ppm_v)		139	132	126	166	-
Liquid waste feed						
Total flow (#/hr)		233	236	234	247	-
TCDD (ppm)	<400	249	357	264	225	-
TCDD (g/hr)	<27.2	26.3	38.3	28.0	25.2	-
Solids feed						
Total flow (#/hr)	<2000	1158	1163	2068	1322	973
TCDD (ppb)		101	382	1010	770	-
TCDD (g/hr)		0.05	0.24	0.95	0.46	-
Stack						
Emission rate:						
TCDD[c] (mg/day)		<0.169	<0.127	<0.031	<0.064	-
Particulates (mg/Nm³@7%O_2)	<180	134.3	147.3	145.6	201.5	-
DRE (%)	>99.9999	>99.99997	>99.9998	>99.9999	>99.99998	-

TCDD = 2,3,7,8-TCDD; C.E. = combustion efficiency;

R.E. = removal efficiency;

DRE = destruction and removal efficiency; TH = total hydrocarbons.

[a] Test 2: liquid feed: dioxin-contaminated TCP still bottoms solid feed: TCP still bottoms + contaminated soil.

[b] Test 3: liquid feed: fuel oil; solids feed: dioxin-contaminated/brominated naphthalene-contaminated lagoon sludge; combined results for three one-half hour runs.

[c] No TCDD detected at detection limits in effect; hence, emission rates denoted are not measured quantities.

Other rotary kiln incinerators for dioxin treatment are now on the market. Recently, the U.S. Air Force conducted test burns of a full-scale rotary kiln incinerator developed by the ENSCO Corporation, Little Rock, Arkansas (Stoddart, 1987). The ENSCO MWP-2000 rotary kiln incinerator is designed to treat approximately 100 tons per day of dioxin-contaminated soil.

Basically, contaminated soil is fed into the rotary kiln and heated to 1,000 to 1,800 °F. The heated soil then exits the rotary kiln into a water-sealed, heated, soil quench. A chain-drag conveyor discharges the soil into a large, roll-off bin. The off-gas from the rotary kiln is drawn into the secondary combustion chamber (SCC), where it is subjected to temperatures of 2,000-2,400 °F for a minimum residence time of 1.65 seconds in an excess-oxygen atmosphere. Acidic brine produced in the scrubber is passed through tandem activated-carbon filters and is neutralized. A countercurrent flow of water in the packed tower removes 99% of the acid gases from the released combustion air by scrubbing.

After passing thorugh the SCC, the hot gases flow through a boiler to produce steam that is used either within the system or vented. Then, the gases are quenched and passed through a packed-bed scrubber. A countercurrent flow of water in the packed tower removes 99% of the acid gases from the released combustion air by scubbing. Acidic brine generated in the scrubber is passed through tandem activated-carbon filters and is neutralized. The eductor not only removes particulates in the sub-micron range, but also provides sufficient draft for the incinerator (the kiln and SCC operate at negative pressure). After passing through a demister, the flue gases exit via the stack.

The data collected during a test run by the U.S. Air Force at the Naval Construction Batallion Center, Gulfport, Mississippi (NCBC) (Stoddart, 1988) indicate that the rotary kiln incinerator is capable of removing 2,3,7,8-TCDD and other organic substances from complex soil matrices at feed rates up to 6.3 tons per hour. Data from stack-gas samples indicate that when sufficient contaminants are present in the feedstock, a DRE of greater than 99.9999% can be obtained. This unit recently underwent a second trial burn with surrogates used for 2,3,7,8-TCDD (hexachloroethane and trichlorobenzene) to firmly establish DREs in excess of 99.9999% prior to commencement of field operations.

The data indicate 2,3,7,8-TCDD is present in surface soils at concentrations averaging 14.3 ppb, based on 1,639 samples, with a range of less than 0.1 ppb to 500 ppb. Profiles of the NCBC site indicate that most of the contamination is contained in the upper 30 cm of soil. Trace levels of TCDD were detected at a maximum depth of one meter. Based on the surface and profile data there are an estimated 17,000 tons of soil containing 2,3,7,8-TCDD contamination in excess of one part per billion.

EPA Region IV provided final project approval on November 23, 1987. The field trial commenced on November 2, 1987, at 23.00 hours. As of May 12, 1988, 11,100 tons of contaminated soil have been treated. All treated soil has met the 1 ppb treatment standard. Figure 5 is a summary, in graphical form, of operations to date.

One principal difficulty has been the carry-over of fine particulates from the rotary-kiln into the secondary combustor, boiler and packed tower. This situation has created excessive downtime. Engineering investigations have been conducted to determine the cause of the problem. Actions taken include slowing of the kiln rotation speed and reduction of gas velocity in the kiln. As a result of these operational improvements, the incinerator was able to operate for 51 days without the need for shutdown to clean out the secondary combustor.

Additionally assessment of the total quantity of soil has been made difficult by the following:

- variable bulk density;
- variable moisture content; and
- variable site characteristics.

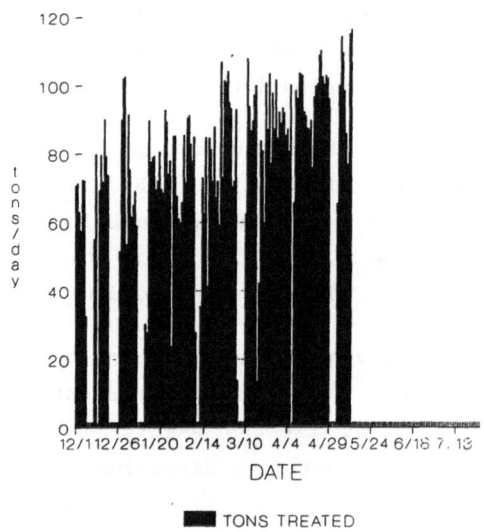

TONS SOIL TREATED/DAY
(FROM 01 DEC 87 TO PRESENT)

NCBC INCINERATION PROJECT

Fig. 5.   Rate of Treatment of Soil, NCBC, Gulfport, Mississippi.

Infrared Technology

Shirco Infrared Systems, Inc., (since late 1987: Ecova, Redmond, Washington), offers a thermal treatment process consisting of a primary chamber heated by infrared energy provided by electrically powered silicon carbide rods. Temperatures in the primary chamber can reach 1,000 °C, and residence times can range from 10-90 minutes. Organic materials, volatilized in the primary chamber, enter a secondary chamber, where combustion is completed at temperatures up to 1,260 °C during a 2-second gas residence time. Exhaust gases then pass through a venturi scrubber spray tower.

Shirco has built three different portable pilot units, any of which can be housed in a 14-meter-long truck trailer and are capable of processing between 10 and 45 kg per hour. The technology has been tested at several hazardous waste sites including Times Beach, Missouri. Several larger units, capable of processing 100 tons per day, have recently been tested with varying success on soils contaminated with 10-670 ppm PCBs and high levels of Pb (up to 6,000 ppm) (i.e., Peak Oil site, Brandon, Florida; Florida Steel site, Indiantown, Florida; and Demode Road site, Rose Township, Michigan), with varying degrees of success.

The infrared furnace is designed to desorb soils or sludge contaminated with hazardous organic constituents, such as PCBs, pesticides, CDDs, and CDFs. At a problem site cleanup, the portable system achieved a DRE of 99.999996% on soil containing 227 ppb 2,3,7,8-TCDD. In another application, the portable infrared furnace was used to decontaminate creosote wood-treatment sludges; a DRE of 99.99999% was achieved for pentachlorophenol and 99.99% for naphthalene.

In 1984, a chemical production site was closed at Hamburg, Federal Republic of Germany. At this site, dioxin contamination was found in the ground due to past chlorobenzene/chlorophenol handling and production. This site is presently being prepared for cleanup by DEKONTA, a C. H. Boehringer-Ingelheim subsidiary. After a number of process evaluations, DEKONTA decided to make use of the Shirco Infrared System to decontaminate the soil. The results of pilot scale test studies (1 t/day unit) have been presented in a report by (DEKONTA, 1988a). The other site at Ingelheim has also been studied (DEKONTA, 1988b).

Table 9.   DEKONTA Test Burns (Shirco Process)

Initial Concentrations	Compound	Final Concentration After Treatment
Up to 40 ppm	Chlorophenols	0.001 - 0.008 ppm
Up to 16,600 ppm	Chlorobenzenes	0.005 - 0.080 ppm
Up to 4,450 ppm	Hexachlorocyclohexane	0.005 - 0.050 ppm
2 - 96 ppb	2,3,7,8-TCDD	0.002 - 0.012 ppb
200 - 7,200 ppb	OCDD	0.010 - 0.150 ppb
35 - 9,500 ppb	OCDF	0.010 - 0.030 ppb
Up to 106 ppb	TCDD	0.002 - 0.026 ppb
Up to 95 ppb	TCDF	0.002 - 0.160 ppb

4.6.2 New and Emerging Technologies (desRosiers and Skinner, 1987d)

UV Photolysis

The International Technologies Corporation (IT), Knoxville, Tennessee, offers a thermal desorption/UV photolysis system for treatment of dioxin-contaminated soils. The thermal desorption process involves passing contaminated soils through an indirectly fired rotary kiln operating at a temperature of range of 450-600 °C. In the kiln, 2,3,7,8-TCDD and other organics are volatilized and entrained in an inert purge gas (N_2), collected in an organic solvent, and destroyed by ultraviolet light photolysis.

A pilot-scale IT UV photolysis unit was tested at the Naval Construction Battalion Center (NCBC), Gulfport, Mississippi, and Johnston Atoll in the Pacific Ocean. Data from four tests indicate the thermal desorption/UV destruction system is capable of reducing 2,3,7,8-TCDD in contaminated soils from 34-57 ppb to a nondetectable level (detection limit, 0.1 ppb). Optimal solid temperature was 560 °C and residence time ranged from 10.5-40 minutes. Although the residual scrubber solvent must be handled as a hazardous waste, 2,3,7,8-TCDD levels were greatly reduced, thus lowering the toxicity of the residual scrubber solvent. The application of this technology as a volume reduction process is feasible (Stoddart, 1987a; Helsel et al., 1987).

UV Oxidation

Virontec, Inc., Irvine, California, has developed the RADINOX®, process that combines ultraviolet energy with oxidizing agents, ozone or hydrogen peroxide, to convert organic contaminants to nontoxic forms, such as carbon dioxide and water. The RADINOX, process is a proprietary technology designed to oxidize naturally recalcitrant organics in contaminated water. The ultraviolet energy energizes chemical bonds, causing them to break. The oxidant produces free charged hydroxyl ions that readily associate with the broken bonds, causing carbon-carbon bonds to be split and oxygenated, forming shorter chain molecules. Driven to completion, hydrocarbons are converted to carbon dioxide and water, and halogenated hydrocarbons are converted to carbon dioxide and halide, i.e., chloride ion. The RADINOX®, process is targeted for treating wastewater containing refractory organic contaminants such as halogenated solvents, pesticides, and polynuclear aromatic hydrocarbons, as well as other organics (Lee, 1985). Data from bench-scale treatability tests on pentachlorophenol (PCP) indicate the Virontec process is capable of reducing PCP concentrations from 10 mg/l to 0.5 mg/l in a 30-gpm continuous flow system. As a result, a full-scale treatment system was scheduled for startup in late August 1987, at a PCP-contaminated wood-preserving site in Oregon. Waste flow treatment in the UV oxidation reactor will be followed by carbon adsorption to remove the residual 0.5 mg/l PCP (Schmitt and Cole, 1987).

Supercritical Oxidation

MODAR, Inc., Houston, Texas, has developed a supercritical water (SCW) oxidation system to destroy liquid wastes. It utilizes water at a temperature and pressure above its critical point (so-called "supercritical" water with temperature above 374 °C and pressure over 215 bars) to break down organic substances to carbon dioxide and water. Some inorganic compounds, such as those containing halogens, phosphorus, and sulphur, may be decomposed by a similar process and converted from highly toxic, complex forms to weak acids. Supercritical water has a low density (0.05 to 0.30 g/ml), essentially possesses no hydrogen bonding, and has a very low dielectric constant (less than 2). These properties make this form of water an excellent solvent for organic substances and oxygen gas.

Over the past five years, MODAR's laboratory unit has been tested on more than 50 types of actual and synthesized wastes. In 1985, a series of bench-scale tests were performed on soils and liquid wastes contaminated with chlorobenzenes and CDDs for a client. 2,3,7,8-TCDD was reduced in soil from 110 ppb to <0.23 ppb for a DE >99%, whereas the DE, for chlorobenzenes was >99.98%. Supercritical oxidation of liquid wastes contaminated with 2,3,7,8-TCDD, TCDDs and OCDD showed DEs of 99.99999% (Thomason, 1987). At a demonstration on July 21, 1986, at CECOS International Inc.'s Niagara Falls, New York, site, a 50-fold upscale MODAR unit was used to detoxify hazardous wastes that included PCBs. Results of the demonstration indicate a 99.99% DRE. A full-scale MODAR unit with a capacity up to 25,000 gallons per day is now being designed (Lee, 1985).

Fluidized Bed Incineration

Ogden Environmental Services, San Diego, California, offers an advanced fluidized bed incinerator that uses high-velocity air flow to suspend a bed of solid particles in the combustion chamber. This creates a highly turbulent combustion zone, with temperatures ranging from 800-1,100 °C, into which solid or liquid waste can be introduced. Solid materials have a 30-minute residence time and gases have a 2-second residence time; there is no afterburner. The suspended bed materials (and waste solids) that leave the combustion chamber are recovered in a cyclone and recirculated through the furnace. The exhaust gases pass through a convective gas cooler and flue gas filter.

There are approximately 30 circulating-bed combustors operating worldwide and burning materials such as high-sulfur coal, peat, wood, waste oils, and municipal waste. A transportable pilot unit, with a soil feed rate of 180 kg per hour, achieved a DRE of 99.9999% for soils contaminated with 12,000 ppm PCBs. Test burns were also conducted on a mixture of organic liquids and several halogenated compounds. DREs of at least 99.99% were achieved for Freon 113, carbon tetrachloride, ethylbenzene, and xylene. Ogden has recently received a National TSCA PCB permit from the EPA Office of Toxic Substances.

Waste-Tech Services, Inc. (WTS), Lakewood, Colorado, offers fluidized bed incineration systems which are in operation at 50 sites burning waste and recovering energy. The primary combustion chamber consists of two concentric, inverted cones. Aluminum silicate firebrick particles fill the inner cone, which is heated to designated operating temperatures ranging from 750-1,200 °C, depending on the waste type. Pressurized air is introduced to maintain maximum turbulence of the bed mass. Solid wastes (preground to less than 3 inches in diameter) and liquid wastes are injected directly into the inner cone. Waste-to-bed mass is maintained below 3% (3% waste to 97% bed mass) at all times to ensure minimum influence of the heat released from the waste to the bed, and the bed temperature is allowed to vary only within a 2.5 °C range. The secondary reaction chamber, a U-tube device, maintained at 870-1,200 °C, provides gases and suspended particulates a 2-second residence time to ensure a high DRE. Off-gases are treated by a wet scrubber to remove particulate material and to neutralize the acidic compounds.

Electric Pyrolysis

Pyrolysis is another form of thermal treatment for waste. Westinghouse Waste Technology Services Division, Madison, Pennsylvania, has developed an electric pyrolysis unit that is designed to destroy organic hazardous wastes. The pyrolyzer is designed to operate at temperatures up to 1,800 °C. Gas residence time can be controlled, as necessary, to destroy organic materials. Exhaust gases are treated in a water-cooled cyclone, a baghouse filter, and a venturi scrubber.

In addition to destroying organics, the pyrolysis system reduces metals to their elemental states; all inorganic oxides, sulfides, and halides are fixed as liquid silicates. Solids fall into a molten bath which, when cooled, forms a vitrified material. A prototype system has been constructed that can process 5-18 tons per day of solid waste material consisting of up to 10% organics and 25% water. However, the system has not been tested on dioxin-contaminated wastes to date.

J. M. Huber Company, Borger, Texas, also offers a pyrolysis technology known as the Advanced Electric Reactor (AER). The AER process involves passing finely ground soil or other waste through a heated, graphite core maintained at 2,200 °C. The waste is isolated from the reactor core by a gaseous fluid wall of nitrogen. Waste containing 2,3,7,8-TCDD and other organics is heated rapidly via thermal radiative coupling during the brief contact period. Since energy transfer is through radiation, retention times are in the order of milliseconds. The waste exits the reactor and falls into a sealed steel receptacle. The nitrogen off-gas is vented through a gas-tight baghouse and subsequently through tandem activated-carbon cylinders. Metals, volatilized by the treatment process, are apparently concentrated in the baghouse particulate. It may be possible to employ the AER to treat mixed organic/metals wastes, subsequently recovering the metals in the baghouse for possible recycling or ultimate disposal. Carryover of metals in the gas stream is mitigated due to the extremely low flue gas velocities and low temperatures (near ambient).

This system was recently tested at the Naval Construction Battalion Center, Gulfport, Mississippi, as part of the U.S. Air Force environmental restoration technology evaluation and demonstration program. Data from a single test run indicated that the total TCDD content was reduced from 113 ppb to <0.036 ppb (Stoddart, 1987b; Helsel and Thomas, 1988).

Pyroplasma System

Westinghouse Waste Technology Services also offers a technology that pyrolyzes wastes using a thermal plasma field. The heart of the system is a plasma arc, which produces a thermal plasma with temperatures greater than 5,000 °C. Waste liquids are injected directly into the plasma, where the molecules are broken into their atomic states. The atoms then recombine to produce hydrogen, carbon monoxide, nitrogen, hydrogen chloride, particulate carbon, carbon dioxide, ethylene, and acetylene. The product gas is scrubbed with caustic soda to neutralize and remove acid gases and particulates.

Although the pyroplasma system has not been tested on dioxins, it has been tested on PCBs contaminated with CDDs and CDFs, and other chlorinated wastes. The unit is not designed to destroy solids, although it can handle up to 40% solids if they are pumpable and can pass through a 200-mesh screen. Heavy metals pass through the system into the scrubber water. A mobile unit mounted on a 15-meter-long truck trailer is available and can process 8-11 liters per minute.

Catalytic Decomposition

In 1972, experiments sponsored by the U.S. Air Force were conducted by Transvaal, Inc., Jacksonville, Arkansas, to investigate the potential for separation of n-butyl esters of 2,4-D from 2,4,5-T in Agent Orange by fractionation. Overheads were found to contain 93% 2,4-D ester, whereas bottoms comprised 93% 2,4,5-T ester. It was not possible to determine 2,3,7,8-TCDD closure by materials balance, as the stainless steel packing and steel column, it was hypothesized, initiated some catalytic destruction of the 2,3,7,8-TCDD. Greater than 93% of the total 2,3,7,8-TCDD was found in the bottoms fraction (2,4,5-T) compared with 0.15 ppm 2,3,7,8-TCDD in the overhead fraction (2,4-D), demonstrating that an effective and efficient separation of the esters was feasible (desRosiers, 1983).

Hagenmaier et al. (1987) investigated the effects published by Vogg and Stieglitz (1986) about the heat-treatment of CDD/CDF-containing fly ashes from municipal waste incineration who detected a buildup of dioxins on fly ash. In contrast to Vogg and Stieglitz, they could not find, in every case, the build up of higher concentrations of CDD/CDF in comparison to the original fly ash composition during a heating cycle, followed by a final decomposition at higher temperatures/longer treatment times. They report instead that under oxygen depletion conditions, thermal treatment of fly ashes can yield a dramatic decrease in CDD/CDF concentrations. The effect was explained using the catalytic HCl/Cl_2 conversion of the Deacon process as a model according to:

$$CuCl_2 + 1/2\ O_2 \longrightarrow CuO + Cl_2$$
$$CuO + 2\ HCl \longrightarrow CuCl_2 + H_2O$$

$$2\ HCl + 1/2\ O_2 \longrightarrow H_2O + Cl_2$$

The net reaction is the HCl-conversion into active Cl_2 which can form de novo CDD/CDF from organic precursors. With Cu-salts and/or other transition metals present in ashes of this kind, the increase in CDD/CDF concentration can be explained and in due course the absence of oxygen should, under heat treatment conditions, explain the thermal degradation of initially present CDD/CDF by intramolecular decay-mechanisms.

Technologies that will probably be employed at some of the dioxin-contaminated Superfund sites in the U.S.A. are listed below:

Table 10. Alternative Technologies Proposed or Used for Certain Dioxin-Contaminated Superfund Sites

Dioxin-Contaminated Superfund Sites	Technologies to be Applied
Tibbetts Road, Barrington, NH	Infrared
Brady Metals, Newark, NJ	Removal and Storage
Diamond Shamrock, Newark, NJ	Removal and Storage
Denney Farm, McDowell, MO	EPA Mobile Incineration System
Times Beach, MO	(Technology options under study)
Vertac Chemical, Jacksonville, AR	(Technology options under study)
Montana Pole, Butte, MT	KPEG
Western Processing, Kent, WA	KPEG
NCBC, Gulfport, MS	ENSCO Mobile Incineration
Syntex, Verona, MO	UV Photolysis/EPA MIS
Hyde Park, Niagara Falls, NY	(Technology options under study)
Love Canal, Niagara Falls, NY	(Various thermal treatment technologies under consideration)
A. P. Hill, Fredericksburg, VA	Removal and Storage

REFERENCES

Behörde für Bezirksangelegenheiten, Naturschutz und Umweltgestaltung, July 12, 1984, Freie und Hansestadt Hamburg.

Bürgerschaft der Freien und Hansestadt Hamburg, 11. Wahlperiode, Drucksache 11/3374, February 26, 1985.

Deegan, J., Jr., 1987, Looking Back at Love Canal, Environ. Sci. Technol., 21:328.

Deegan, J., Jr., 1987, Looking Back at Love Canal, Environ. Sci. Technol., 21:421.

Der Hessische Minister für Arbeit, Umwelt und Soziales, im März 1985, Wiesbaden.

des Rosiers, P.E., 1987, National Dioxin Study, Chapter 3, U.S. Environmental Protection Agency (RD-681), Washington, D.C. 20460.

Deutsche Einheitsverfahren zur Wasser-, Abwasser- und Schlammuntersuchung.

Dr. Schreiner & Partner, Zetel 1.

Dynamit/Nobel, Troisdorf.

Eckrich, W., April 1985, Analytik von schwerflüchtigen chlororganischen Verbindungen, pp. 100, Seminar über Altlasten und kontaminierte Standorte - Erkundung und Sanierung, Ruhr-Universität Bochum.

EPA, 1985, Uptake and Depuration Studies of PCDDs and PCDFs in Freshwater Fish, EPA Environmental Research Laboratory, Duluth, Minnesota.

ERGO-Forschungsgesellschaft mbH, September 7, 1984.

Esposito, M.P., Tiernan, T.O., Dryden, F.E., 1980, Dioxins, EPA-600/2-80-197.

Freeman, R.A. and Schroy, J.M., 1988, Comparison of the rate of TCDD transport at Times Beach and Eglin AFB, Chemosphere, (in press).

Freie und Hansestadt Hamburg, Juni 1984. Internationales Symposium Deponie Georgswerder, Wortprotokoll, Hamburg.

Hagenmaier, H., Gutachtliche Stellungnahme, Untersuchung und Bewertung des Gefahrenpotentials dioxinhaltiger Abfälle auf der Sondermülldeponie Münchehagen.

Hagenmaier, H., Brunner, H., Knapp, W., and Weberruß, U., 1988, Untersuchung der Gehalte an polychlorierten Dibenzodioxinen, polychlorierten Dibenzofuranen und ausgewählten Chlorkohlenwasserstoffen in Klärschlämmen, Umweltforschungsplan des Bundesministers des Innern, Abfallwirtschaft, Forschungsbericht 103 03 305.

Hutzinger, O., and Fiedler, H., August 1988a, Emissions of Dioxins and Related Compounds from Combustion and Incineration Sources, NATO/CCMS Pilot Study on International Information Exchange on Dioxins and Related Compounds, Report Number 172, Bayreuth.

Hutzinger, O., and Fiedler, H., August 1988b, Formation of Dioxins and Related Compounds in Industrial Processes. NATO/CCMS Pilot Study on International Information Exchange on Dioxins and Related Compounds, Report Number 173, Bayreuth.

Jager, J., and Pflug, G., 1985, Freisetzung von Dioxinen bei der Ablagerung von Abfällen, in: "Abfallwirtschaft in Forschung und Praxis 14; Dioxine - Entstehung - Wirkungen - Beseitigung", pp. 59, FGU Berlin, ed., E. Schmidt, Berlin.

Jager, J., Pflug, G., Sünderhauf, W., 1986, Auslaugung und Deponierbarkeit von Flugaschen, Internationaler Recycling Congress, Berlin.

Jager, J., 1987, unpublished results. Source: Hagenmaier, H., Gutachtliche Stellungnahme, Untersuchung und Bewertung des Gefahrenpotentials dioxinhaltiger Abfälle auf der Sondermülldeponie Münchehagen.

Jürgens, H.-J., Roth, R., Schlesing, H., Fallstudie und Vorschläge zur Dekontamination des Geländes einer stillgelegten Herbizidfabrik, 1988, in: "Altlastensanierg '88", Bd. 2, pp. 1067-1073, K. Wolf, W.J. van den Brink, F.J. Colon, eds., Kluwer Academic Publishers, Dordrecht-Boston-London.

Lamparski, L.L., Nestrick, T.J., and Stenger, V.A., 1984, Presence of chlorodibenzo dioxins in a sealed 1933 sample of dried municipal sewage sludge, Chemosphere, 13:361.

Landratsamt Rhein-Neckar-Kreis, August 9, 1984, Heidelberg.

Landkreis Nienburg/Weser, 1987, Vorschlag für die Sanierung der SAD Münchehage, Geb. 20.

New York State Department of Health, September 1978, Love Canal, Public health time bomb., Albany.

Schumacher, E., 1985, Dioxine in der Deponie Georgswerder, Bekanntwerden, Untersuchungen, Gefährdungsabschätzung, in: "Abfallwirtschaft in Forschung Praxis 14; Dioxine - Entstehung - Wirkungen - Beseitigung", pp. 81, FGU Berlin, ed., E. Schmidt, Berlin.

Schumacher, E., 1987, Dioxin - Eine Technische, Analytische und Toxikologische Herausforderung, VDI Berichte 634, pp. 291, VDI Verlag, Düsseldorf.

Selenka, F., 1985, Gutachten, Department of Hygiene, University of Bochum, FRG.

Sievers, S. and Friesel, P., 1988, Soil contamination patterns of chlorinated compounds: Looking for the source, Chemosphere (in press).

Tabasaran, O., Ermittlung des Kontaminationspotentials eines Industriestandortes in der BR Deutschland, 1988, in: "Altlastensanierung '88", Bd. 2, pp. 999-1008, K. Wolf, W.J. van den Brink, F.J. Colon, eds., Kluwer Academic Publishers, Dordrecht-Boston-London.

Tabasaran, O., and Thomanetz, E., 1985, Wesentliche Untersuchungsergebnisse in Kurzfassung und Sanierungsvorschläge für die Sonderabfalldeponie Gerolsheim, in: "Abfallwirtschaft in Forschung und Praxis 14; Dioxine - Entstehung - Wirkungen - Beseitigung", pp. 71, FGU Berlin, ed., E. Schmidt, Berlin.

Technischer Überwachungsverein Berlin, 1985, Technischer Bericht Nr. D-85/534.

The Hazardous Waste Inspectorate, July 1986, Second Report: Hazardous Waste Management: "Ramshakle & Antediluvian"?, Department of the Environment, Welsh Office, Scottish Office, N. Ireland.

Thoma, H., Carsch, S., Hutzinger, O., 1986, Leaching of polychlorinated dibenzo-p-dioxins and dibenzofurans from municipal waste incinerator fly ash by water and organic solvents, Chemosphere, 15:1927.

Thomanetz, E., 1987, Vorgehensweise bei der Einschätzung von Altablagerungen und Erarbeitung von Sanierungskonzepten am Beispiel der Industriedeponie Karsau, Institut für Siedlungsbau, Wassergüte- und Abfallwirtschaft, Universität Stuttgart.

U.S. Environmental Protection Agency, November 28, 1983, Dioxin Strategy, Washington, D.C.

U.S. Environmental Protection Agency, April 1985, The National Dioxin Study - Guidelines for Review of 2,3,7,8-TCDD Data. Prepared by National Dioxin Study QTAG.

U.S. Environmental Protection Agency, April 1986, The National Dioxin Study, Tiers 3, 5, 6 and 7, Washington, D.C.

U.S. Environmental Protection Agency, January 1986, National Dioxin Strategy, Tier 1 and 2 Accomplishments. Draft Technical Support Document, Office of Solid Waste and Emergency Response, Washington, D.C.

Umweltbehörde Hamburg, Amt für Altlastensanierung, April 1988, Sanierung der Deponie Georgswerder.

Yanders, A.F., Kapila, S., and Schreiber, R.J., 1986, Dioxin: Field research opportunities at Times Beach, Missouri, in: "Chlorinated Dioxins and Dibenzofurans in Perspective", C. Rappe, G. Choudhary and L.H. Keith, eds., Chap. 17, pp 237-239, Lewis Publishers, Chelsea, Michigan.

Yanders, A.F., Orazio, C.E., Puri, R.K., and Kapila, S., 1988, On translocation of 2,3,7,8-Tetrachlorodibenzo-p-dioxin: Time dependent analysis at the Times Beach Experimental site, Chemosphere, (in press).

van Zorge, J.A. (Ministerie can Volkshuisvesting, Leidschendam), July 11, 1988, Personal Communication to H. Fiedler.

Vanna, N.J., 1983, The Love Canal: Issues and Problems, Chemosphere, 12:705.

Weerasinghe, N.C.A., Gress, U.L., and Lisk, P.J., 1985, Polychlorinated Dibenzodioxins and Dibenzofurans in Sewage Sludges, Chemosphere, 14:557.

Wolf, K., 1985, Dioxine in der Deponie Georgswerder - Geschichte der Deponie, Sanierungsalternativen, in: "Abfallwirtschaft in Forschung Praxis 14; Dioxine - Entstehung - Wirkungen - Beseitigung", pp. 85, FGU Berlin, ed., E. Schmidt, Berlin.

5. WASTE DISPOSAL SITES CONTAMINATED WITH DIOXINS AND RELATED COMPOUNDS

5.1 INTRODUCTION

The following report discusses environmental contamination problems arising from the disposal of waste containing polychlorinated dibenzo-p-dioxins (PCDD) and polychlorinated dibenzofurans (PCDF).

Landfill and surface dumping have historically been inexpensive means of waste disposal, but within the last 10-15 years it has become obvious that there are many hidden costs associated with this practice. Largely unregulated in the past, landfilling and dumping have created many locations where groundwater and surface water, soils, wildlife, and air have been contaminated or are threatened with contamination by a variety of undesirable and often toxic substances.

The industrial accident at Seveso, the Love Canal controversy, the Yusho rice oil poisoning in Japan and the Agent Orange experience have heightened awareness in both the scientific community and the public concerning the dangers to man and environment arising from polychlorinated dioxins and furans. These compounds result from certain chemical processes (Hutzinger and Fiedler, 1988), specifically in the production of chlorophenols, as well as in combustion processes (Hutzinger and Fiedler, 1988). Historically, the residues from these processes have been landfilled. Today, some of the production and waste sites are already having an adverse effect on the health and well-being of the surrounding environment. But more commonly the PCDD/PCDF containing wastes pose a long-term threat that is difficult to evaluate.

This report presents brief case studies of sites that have been contaminated with PCDD/PCDF. It must be emphasized that the information presented here is limited to that which was made available to the authors in the course of this project (1985-1988). For this reason, the American sites are under-represented. Apart from a summary of the National Dioxin Study, only a brief case descriptions are presented although there are many significant and extensively investigated sites. In other countries little investigation has been carried out to date for political or technical reasons, even though contaminated sites are likely. Therefore, no conclusions should be drawn from this report regarding the international distribution of dioxin wastes. Rather, it is intended as a collection of experiences which describes different problems that have arisen and how they have been dealt with.

5.2 WASTE SITES IN WEST GERMANY

5.2.1 Hamburg-Moorfleet Chemical Plant

In 1945 a German chemical company opened a plant in Hamburg beside the Moorfleet Canal and began producing lindane. Starting in 1952 the residues from lindane production were used for the production of 2,4,5-T. The plant was closed in 1984 following controversy about dioxin emissions. Not only was the production site found to be polluted; the wastes from this one chemical plant are implicated in the majority of the German landfills known to be severely contaminated with $2,3,7,8-Cl_4DD$ (Georgswerder, Müggenburger Straße, Malsch, Gerolsheim sites). A brief summary of the production processes and the source of the PCDD/PCDF is therefore warranted (Bürgerschaft Hansestadt Hamburg, 1985; Tabarasan, 1988; Jürgens et al., 1988; Landratsamt Rhein-Neckar-Kreis, 1984; Behörde für Bezirksangelegenheiten, Hamburg, 1984).

Origins of PCDD/PCDF in the Production Process

The PCDD/PCDF waste originated primarily in the 2,4,5-T production line. The process sheet is shown in Figure 1.

Lindane was produced from benzene by addition of chlorine. The resulting impurities, the alpha and delta isomers of hexachlorocyclohexane (HCH), were known as alpha crystals and delta paste. They were feedstock for the 2,4,5-T production. In the first step they were thermally degraded to form technical grade trichlorobenzene. During this process some $2,3,7,8-Cl_4DD$ and large quantities of higher chlorinated dioxins were formed. Some were removed at this point in the so-called Zersetzerückstände or degradation residues (primarily HCH-isomers). The typical PCDD content of this waste is given in Table 1.

Table 1.   PCDD Concentrations in Degradation Wastes

$2,3,7,8-Cl_4DD$	0.5	mg/kg
Sum of Cl_4DD	5	mg/kg
Sum of Cl_5DD	560	mg/kg
Sum of Cl_6DD	4,200	mg/kg
$1,2,3,4,6,7,9-Cl_7DD$	2,800	mg/kg
$1,2,3,4,6,7,8-Cl_7DD$	6,900	mg/kg
Cl_8DD	32,000	mg/kg

Some of the PCDD continued on to the next step, the addition of chlorine to form tetrachlorobenzene. As only the 1,2,4,5 isomer was required, the other isomers were removed at this point and either sold or disposed of. The PCDD content of this waste is not known.

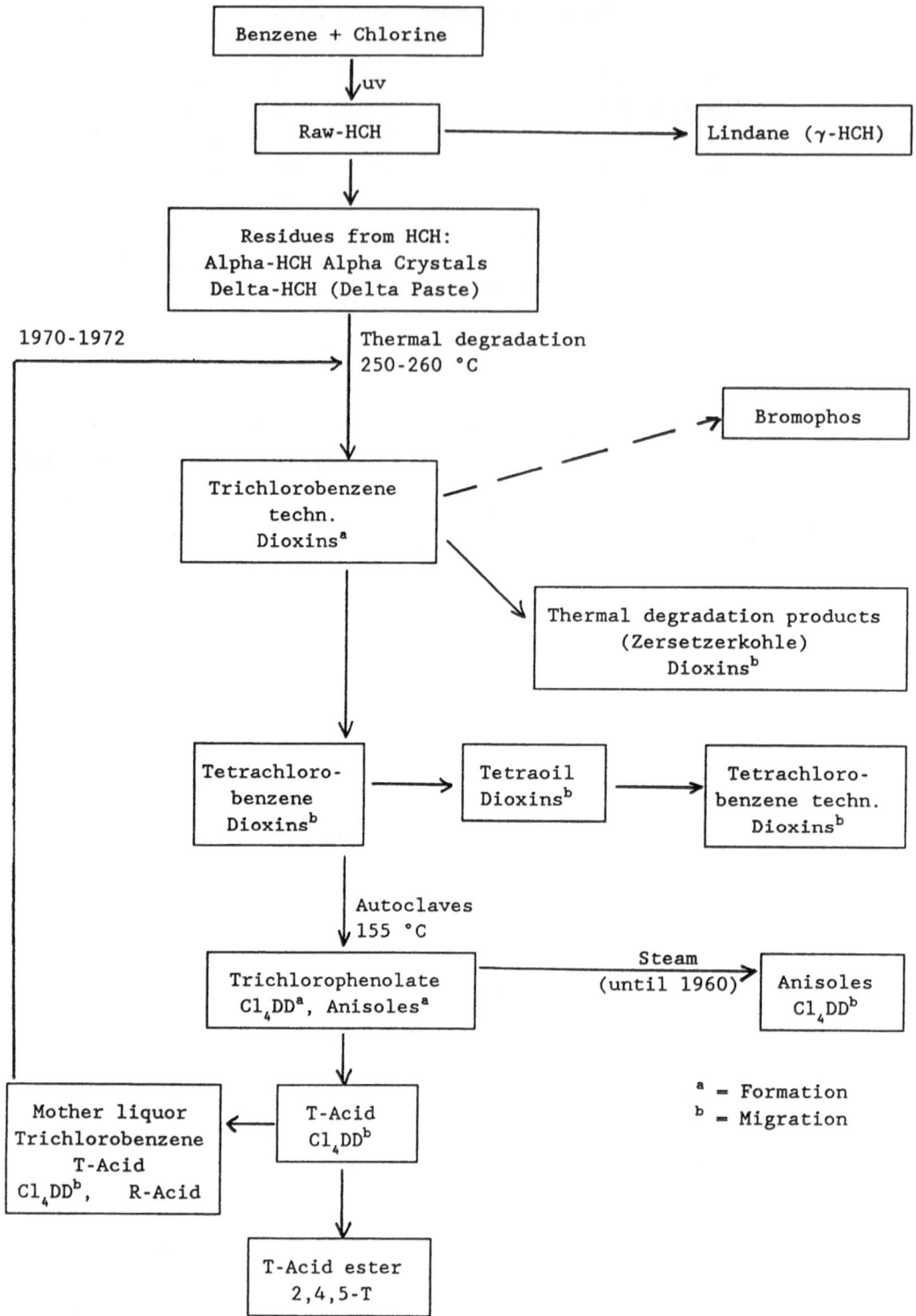

Fig. 1.   PCDD/PCDF in the Moorfleet Plant 2,4,5-T Production Process

The 1,2,4,5-trichlorobenzene then underwent an autoclave reaction in the presence of NaOH and methanol to form 2,4,5-trichlorophenol. Most of the 2,3,7,8-Cl_4DD was formed in this step, which was characterized by the fact that the 2,3,7,8 isomer comprised more than half of the total Cl_4DD synthesized. The amount of Cl_4DD produced in this reaction increases rapidly with temperature, yet this reaction was commonly carried out at 155 °C, above the maximum allowed by the authorities. Some of the resulting contaminant was present in the residues from this step, the anisoles.

In the next step T-acid was formed by reacting the phenol with chloroacetic acid. The residues from this process, known as "mother liquor", were a mixture of residual acids (R-acid) and trichlorobenzene and contain the majority of the 2,3,7,8-Cl_4DD formed in the autoclave reaction.

The degradation residues, anisoles, and R-acid were generally disposed of in landfills. The typical 2,3,7,8-Cl_4DD concentrations in the various wastes are summarized in Table 2. This distribution was somewhat different during the years 1970-72 when the mother liquor was recycled into the thermal degradation step, resulting in much higher contamination of the degradation residues.

Table 2.   2,3,7,8-Cl_4DD in the 2,4,5-T Wastes of the Moorfleet Plant

Waste Type	Concentration [mg/kg]
Anisoles	0.37
R-acid	80
Degradation Residues	0.5
Tetrachlorobenzene	?
Delta Paste	?

Site Investigation

A geological investigation found that the 8.5 ha plant site is located on a six meter thick artificial deposit of sand, gravel, silt, and rubble. This lies on a 0.4-4 m thick clay and silt formation. A 15-20 m thick sand and gravel aquifer follows. It serves as an important drinking water source, and the Kaltehofe waterworks is located close by. The aquifer rests on a 8-37 m thick clay formation. This barrier is broken by sand and gravel filled trenches carved by the glaciers. They have been found to extend up to 157 m below the surface. The aquifer ground water flow depends strongly on the draw down at the waterworks. Under normal conditions it moves at 22-32 m/year from the site towards the wells.

An extensive on-site sampling program was conducted. A total of 196 cores amounting to 4,081 bore-meters were collected, from which 2,652 samples were taken. A further 91 bore hole water samples were collected. Of the 913 bore cores analysed for $2,3,7,8-Cl_4DD$, 31 had concentrations in excess 5 μg/kg. The Cl_4DD contamination was limited to the near surface cores and concentrated around the former production facility and storage tank. A few contaminant hot spots were found outside of this area and are thought to have come from leaking pipes or trucks.

A groundwater contaminant migration model was assembled using the wealth of bore core and water data collected. It predicted that 38-55 kg of chlorinated pollutants are leaving the site each year and that the contaminant plume would reach the waterworks within ten years. However, there is no evidence yet of these compounds at the waterworks, and core and water samples collected off of the site indicate that the migration is much slower than predicted in the model. It is hypothesized that this is because the contaminants are often in hydrophobic soil environments which the groundwater flows around and not through. This theory is supported by the very large concentration differences observed between adjacent bore holes and the oily appearance of some of the core samples collected.

Contaminants have also left the site by way of the Moorfleet Canal, which has twice been the subject of clean-up projects. It is thought that surface water contamination could reach the waterworks by the Moorfleet Canal/Elbe River/groundwater route, but it is difficult to confirm this hypothesis since the Elbe has a significant organochlorine load of its own.

Remedial Measures

The near surface hot spots were immediately excavated, in some cases to a depth of four meters, and the material (2,699 m^3) was stored.

Since 1986 organic phase, mostly chlorinated benzenes, has been pumped out of two wells on the site. It is estimated that more than seven tons of this material will have been pumped out of the site within three years.

It is also planned to excavate the contaminated soil and submit it to desorption/destruction decontamination. The soil would then be redeposited on the site. Initial tests were successfully completed in 1986 using an infrared oven for desorption followed by an after-burner (for details see: Fuhr and des Rosiers, 1988). A pilot plant is now under construction and is targeted to start operation at the beginning of 1989.

5.2.2   Georgswerder Landfill (Hansestadt Hamburg)

Background

In 1948 the city of Hamburg began depositing waste at the Georgswerder landfill. The 44 ha site, which also served as a clay pit for brick making until 1962, is located on an island in the Elbe River. Before being closed in 1979 it received ca 14 million m^3 rubble and household waste resulting in a hill over 40 m high. From 1967-74 it also served as the primary industrial waste disposal site for Hamburg. Over 150,000 m^3 of liquid hazardous waste and 100,000 drums were disposed of in 10 pits and 4 drum disposal sites excavated out of the existing waste (Bürgerschaft Hansestadt Hamburg, 1985; Umweltbehörde Hamburg, 1988; Int. Symposium Deponie Georgswerder, 1984; Schumacher, 1985, 1987; Wolf, 1985). The Moorfleet chemical plant is known to have delivered waste containing between 13 and 26 tons of PCDD/PCDF to the site, of which 4.5 kg is estimated to be $2,3,7,8-Cl_4DD$ (see Table 3).

Table 3.   Moorfleet Plant Waste Disposed of at Georgswerder

Waste Type	Quantity [tons]	$2,3,7,8-Cl_4DD$ [grams]
Degradation Residues	1,750	4,400
Anisoles	325	60
Tetrachlorobenzene	3,375	?

The oily waste in the pits mobilized the hydrophobic dioxins, and the oil mobility was enhanced by the presence of large quantities of emulsifiers in the site. Efforts to turn the closed site into a recreational area were terminated in 1983 following the discovery of $2,3,7,8-Cl_4DD$ in the oily phase of site leachate.

Site Investigation

A cross-section of the Georgswerder site is shown in Figure 2. The waste lies directly on a 2-5 m thick clay bed which overlies a sand/gravel aquifer. The clay be is considerably thinner in places due to the earlier mining activity. It is further weakened by some 100 World War II bomb craters which cover about 0.4% of the site area and are estimated to increase the average clay permeability of 10^{-9} m/s by a factor of 1.3. Formation weaknesses are thought to increase the permeability in some locations by a factor of 100.

Upon closure the site was covered with two m of earth and seeded with grass. This porous cover resulted in extensive penetration of rainwater into the site. The low permeability of the underlaying clay and the presence of a containment wall in the middle of the site resulted in accumulation of the rainwater. An immense reservoir of saturated waste was formed, extending up to 14 m above the site bed and causing an average hydrostatic pressure difference of 6 m across the clay layer. Due to the higher lateral permeability, large quantities of leachate (7 L/s) flowed out of the sides of the hill and were collected in the innermost of two trenches surrounding the site. This oil and water mixture was originally dumped directly into the Georgswerder Wettern, a water body leading to the Elbe River. Beginning in 1981 it was passed through an oil separator before being discharged into the city sewer system. The principle liquid fluxes are indicated on Figure 2, and Table 4 contains an estimate of the annual water balance for the waste site.

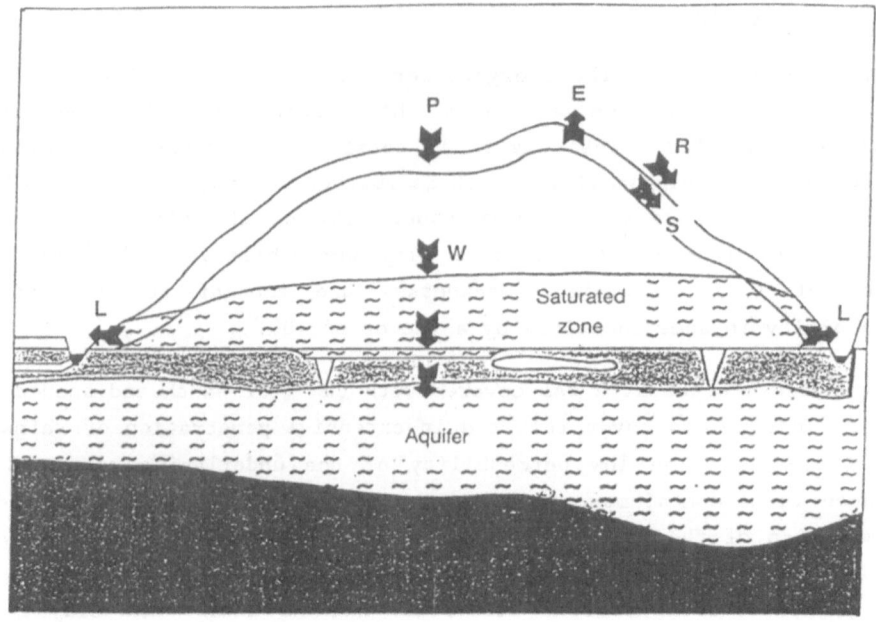

Fig. 2.   Section View of the Georgswerder Landfill with Water Flows

Table 4.   Estimated Water Budget for the Georgswerder Landfill

Component	Flux [m³/y]
P: Precipitation	305,00
E: Evaporation	220,00
R: Surface Runoff	5,000
S: Sub-Surface Runoff	0
W: Waste Infiltration	80,000
L: Leachate Collected	60,000
A: Aquifer Infiltration	20,000

The site investigation included sampling of surface material, leachate, groundwater, surface water, and biota.

Soil samples take from the leachate seeps in the southeastern part of the site were found to contain 2,3,7,8-Cl_4DD at levels as high as 850 ng/kg. However, neither it nor other chlorinated hydrocarbons were identified at seeps elsewhere on the site. A leachate sample from a southeastern seep had a 2,3,7,8-Cl_4DD concentration of 0.2 ng/L and PCDD/PCDF levels of 193 ng/L and 151 ng/L respectively. In addition, sediments taken from the trenches draining this area had PCDD/PCDF levels (2,3,7,8-Cl_4DD = 58 ng/kg, PCDD = 59 ng/kg, PCDF = 29 μg/kg) about 50 times higher than regional background levels. It is clear that the southeastern part of the landfill is the main source of PCDD/PCDF on-site surface contamination. This is thought to be related to two of the liquid waste pits located in this area that received about 20% of the hazardous organic compounds disposed of at Georgswerder.

Further investigations demonstrated that some of the contaminants have left the site. Analysis of sediments in the sump of the pump delivering leachate to the sewer system showed very high levels (2,3,7,8-Cl_4DD = 790 ng/kg, PCDD = 579 μg/kg, PCDF = 267 μg/kg) while analysis of the site effluent showed levels of PCDD at 7.7 ng/L. Thus it is not surprising that sediments in the Georgswerder Wettern and the Elbe River are heavily contaminated with oily leachate residues containing PCDD/PCDF (see Table 5).

Table 5. PCDD/PCDF in Sediments from the Elbe River and Associated Water Bodies; Concentrations in [ng/kg]

Gewässer PCDD/PCDF	Alster unterh.Br.	Bille- kanal	Kohl- brand	Reiher- stieg	Äußerer Veringk.	Moorfleet. Kanal
Cl_8DD	31	350	30	<30	250	77
Σ Cl_7DD	550	6,200	<24	<24	580	850
Σ Cl_6DD	1,250	4,300	640	580	7,500	4,300
Σ Cl_5DD	200	290	470	260	2,700	930
Σ Cl_4DD	23	130	150	80	1,700	700
2,3,7,8-Cl_4DD	<2.7	88	4	3.4	1,500	320
Σ Cl_4DF	85	35	85	50	21	95
Σ Cl_5DF	480	720	4,100	1,300	2,100	2,200
Σ Cl_6DF	640	2,800	1,000	930	8,600	2,800
Σ Cl_7DF	440	3,100	140	<32	2,800	820
Cl_8DF	1.5	2.0	<13	<13	<18	<2
Σ PCDD/PCDF	3,700	17,927	7,170	3,500	20,400	13,077

Migration to the surrounding surface environment does not appear to have occured. Measurements of sediments from the outer containment ditch showed normal background levels of all contaminants including PCDD/PCDF. The PCDD/PCDF concentrations in soil samples from garden plots adjacent to the site were close to background levels. And although mice living on the site had high containment concentrations, it is estimated that the biological export of 2,3,7,8-Cl_4DD from the landfill amounted to only 140 ng/year.

The groundwater surrounding the site has been monitored since 1967, and today there is a network of some 100 wells. Cl_4DD has not been detected in any sample to date, but there is evidence of other organic contaminants including lindane at the southeast edge of the site. The mean groundwater flow is about 20 m/year towards the southwest, so it may be some time before site leakage from the southeast is detected leaving the site.

Remedial Measures

The extensive remedial work being undertaken at Georgswerder focusses on protecting the groundwater supply, while upholding the principle of on-site destruction or immobilization of wastes.

One priority is to immobilize the waste by drying out the site. A multi-layered cap including two drainage beds separated by a clay barrier and a 1.5 mm HDPE sheet has been installed on the upper part of the site above the leachate seeps (see Figure 3). The inner drainage layer is intended for excess gas collection, while the outer is to collect adsorbed rainwater. The surface runoff is discharged directly into the Georgswerder Wettern, while the subsurface drainage water, where gas contamination is considered a possibility, is discharged into the sewer. A treatment plant will be built should this prove necessary.

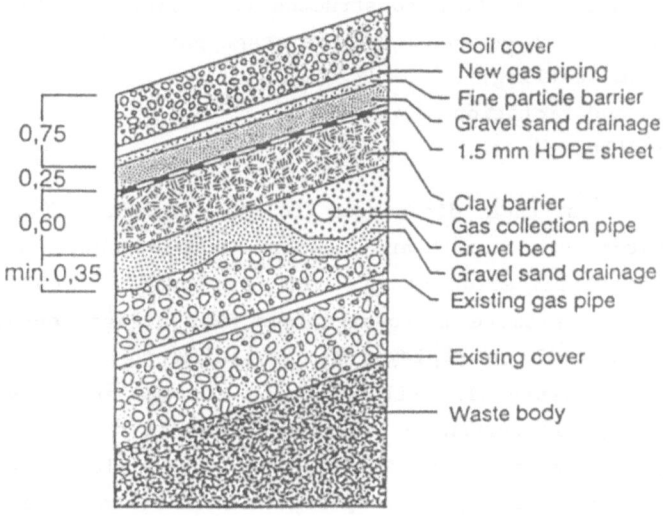

Fig. 3.   Upper Cap Design for the Georgswerder Landfill

A similar cover will be applied to the lower slopes of the site, but this will also involve the construction of a leachate collection system. A research contract has been let to develop a system to remove the heterogeneous mixture of aqueous phase liquid from the 14 m deep saturated zone. In the meantime a temporary external leachate collection system has been installed.

In addition to these efforts to dry out the site, it is also planned to remove and destroy some waste. The two buried liquid waste pits that are suspected of being the major cause of the surface pollution are also thought to pose a special danger to the groundwater due to a local weakness in the underlaying clay bed. The intent is to remove this material, but since an excavation of this kind has never been undertaken, extensive planning and development are necessary. This has begun, but it will be several years before excavating starts.

A treatment plant has been constructed to handle site leachate. The liquid is currently subjected to a three stage physical process:

1. flotation and settling;
2. floculation and flotation;
3. quick sand filter.

The effluent is then discharged to the sewer. Two additional biological batch reactors are under development, including a nitrification/denitrification stage. It is planned to control reactor conditions so as to promote micro-organisms that digest contaminants. A membrane diffuser will be employed for aeration to avoid stripping of volatiles, and all waste air will be passed through an activated carbon filter. The sludge is currently centrifuged and in the future will also undergo aerobic digestion. It will then be burned along with excavated liquid and solid waste in a high temperature incinerator, which is in the planning stage.

The gas collected at the site (3 million m^3/year) is pumped to a nearby refinery where it is burned in production processes. No PCDD/PCDF has been detected in the landfill gas.

Direct measures to protect the groundwater such as barrier walls through the aquifer, a barrier under the site, or hydrological measures to modify groundwater flow were considered. However, due to very high costs and the absence of significant off-site contamination they were discarded.

A plan to control groundwater contamination in the southern part of the site using wells is being developed so that emergency action can be taken if the situation there suddenly worsens. However, it is possible that lowering the hydraulic head below the site could increase the flow out of it.

Georgswerder is the most extensively analysed and remediated dioxin site in West Germany. Some 143 million DM have been committed to date for site investigation and remediation, but it is doubtful that this will be sufficient.

5.2.3 Müggenburger Straße Landfill (Hansestadt Hamburg)

The Müggenburger Straße landfill is located in Hamburg, just several hundred meters from the Georgswerder landfill (Bürgerschaft Hansestadt Hamburg, 1985; Behörde für Bezirksangelegenheiten, Hamburg, 1984). The following analyses showed massive contamination with organochlorine compounds bearing suspicious similarities to products, intermediates, and residues of the Morfleet plant production process (see Figure 1) (Sievers and Friesel, 1988).

Table 6. Maximum Content of Organochlorine Compounds in the Waste Disposal Site "Müggenburger Straße"

Σ HCH	182,000	mg/kg
Chlorobenzenes	13,900	mg/kg
Chlorophenols	330	mg/kg
$2,3,7,8-Cl_4DD$	200	mg/kg
Σ PCDD	2,120	mg/kg
Σ PCDF	1,550	mg/kg

Indeed, from 1964-67 over 2,000 tons of waste from the Moorfleet plant were disposed of at the site (Table 7).

Table 7. Moorfleet Plant Waste in the Müggenburger Straße Site

Waste Type	Quantity [tons]	$2,3,7,8-Cl_4DD$ [grams]
R-acid	127	10,160
Degradation Products	> 600	300
Anisoles	> 260	95
Tetrachlorobenzene	300	?
Delta Paste	> 700	?

The landfill is estimated to contain at least 10.5 kg of $2,3,7,8-Cl_4DD$ and 13-16 tons of PCDD/PCDF. It was closed in 1984 and these residues have been declared harmless.

5.2.4 Moorfleeter Brack Contaminated Site (Hansestadt Hamburg)

A contaminated site in Hamburg, the "Moorfleeter Brack", was used as a waste dump likely up to the early sixties. Maximum contaminant concentrations found at this site are given in Table 8 (Sievers and Friesel, 1988).

Table 8. Maximum Concentrations of Organochlorine Compounds at the Moorfleeter Brack Site

Σ HCH	211,200	mg/kg
Chlorobenzenes	5,890	mg/kg
Chlorophenols	330	mg/kg
2,4,5-T	2,700	mg/kg
Trichloroanisole	860	mg/kg
2,3,7,8-Cl_4DD	874	mg/kg
Σ PCDD	1,014	mg/kg
Σ PCDF	215	mg/kg

The PCDD/PCDF pattern was dominated by 2,3,7,8-Cl_4DD with hardly any other Cl_4DDs present. This is a pattern that has been correlated to wastes stemming from the 2,4,5,-T production. Results from the few samples analysed to date for chloroanisoles indicate that there may be a correlation between 2,3,7,8-Cl_4DD and trichloroanisole in this type of waste.

5.2.5 Malsch Hazardous Waste Disposal Site (Baden-Württemberg)

Background

The Malsch Hazardous Waste Site is located in a clay depression close to Heidelberg. The Moorfleet plant deposited 800 tons of waste here between 1972 and 1976 (Table 9) (Bürgerschaft Hansestadt Hamburg, 1985; Landratsamt Rhein-Neckar-Kreis, 1984; Dynamit/Nobel; ERGO, 1984).

Table 9.   Moorfleet Plant Waste Disposed of at Malsch

Waste Type	Quantity [tons]	$2,3,7,8\text{-}Cl_4DD$ [grams]
R-acid	37	2,960
Degradation Products	107	53
Anisoles	350	?
Tetrachlorobenzene	140	?
Delta Paste	> 160	?

Thus an estimated 3 kg of $2,3,7,8\text{-}Cl_4DD$ and a further 2.5-5 tons of PCDD/PCDF are in the site. A leak was discovered in 1984 and traces of PCDD were detected in the leachate.

Site Investigation

Leachate from the Malsch site was separated into aqueous and oily fractions and analysed for PCDD/PCDF along with surface water samples collected around the site. At a detection limit of 0.2 ng/L $2,3,7,8\text{-}Cl_4DD$ was detected in all of the samples. The concentration of the other PCDD/PCDF isomers was generally below the detection limit. The results of the analyses are given in Table 10.

Table 10. PCDD/PCDF in Leachate and Surface Water at the Malsch Landfill

Sample	water-leachate oil layer	water leachate without oil	monitor. site	monitor. site	monitor. site	Site No. 14
	[µg/kg]	[ng/l]	[ng/l]	[ng/l]	[ng/l]	[ng/l]
$2,3,7,8-Cl_4DD$	0.6	6	< 2	< 2	< 2	< 2
Sum of Cl_4DD	0.6	14	n.d.	n.d.	n.d.	n.d.
Sum of Cl_5DD	n.d.	n.d.	n.d.	n.d.	n.d.	n.d.
Sum of Cl_6DD	n.d.	n.d.	n.d.	n.d.	n.d.	n.d.
Sum of Cl_7DD	3.5	30	n.d.	n.d.	n.d.	n.d.
Cl_8DD	5.0	140	n.d.	n.d.	n.d.	n.d.
$2,3,7,8-Cl_4DF$	n.d.	n.d.	n.d.	n.d.	n.d.	n.d.
Sum of Cl_4DF	n.d.	n.d.	n.d.	n.d.	n.d.	n.d.
Sum of Cl_5DF	n.d.	n.d.	n.d.	n.d.	n.d.	n.d.
Sum of Cl_6DF	n.d.	n.d.	n.d.	n.d.	n.d.	n.d.
Sum of Cl_7DF	1.2	n.d.	n.d.	n.d.	n.d.	n.d.
Cl_8DF	2.6	n.d.	n.d.	n.d.	n.d.	n.d.

Remedial Measures

Remedial efforts commenced in 1984. It was planned to install containment walls including a 2.5 mm HDPE barrier. Work began on testing different wall materials. The form of the final solution was to depend on the results of these tests.

5.2.6 Gerolsheim Hazardous Waste Disposal Site (Rheinland-Pfalz)

Background

During the years 1969-1972 production waste from the Moorfleet chemical plant was disposed of at the Gerolsheim hazardous waste site (Bürgerschaft Hamburg, 1985; Tabasaran and Thomanetz, 1985). About 4,000 drums containing hazardous pasty residues were brought to the dump, and half of them are believed to be contaminated with PCDD/PCDF. According to a company declaration 403 tons of dioxin-containing sludge from the production of R-Acid were disposed of. This would result in a total of 4-22 kg of $2,3,7,8-Cl_4DD$ in the landfill.

Site Investigation

Although it is known that $2,3,7,8-Cl_4DD$ containing wastes are disposed at the site, the presence of $2,3,7,8-Cl_4DD$ could not be confirmed in 47 monitoring wells and 11 bore core samples.

Because of offensive odor, four torches to destroy gaseous emissions from the waste dump (mostly H_2S, CH_4) had been installed in 1980 (with a suction capacity of 250 m^3/h). Measurements of the waste site gas and condensate from the flares showed no evidence of PCDD/PCDF. In addition, no PCDD/PCDF was detected in samples of soil and grass from around the flares or in liver and fat samples from local rabbits.

Another study of landfill flaring came to other conclusions. Analyses of the gases from a waste disposal site at Braunschweig confirmed the presence of PCBs, HCH, and HCB at concentrations of 1-5 $\mu g/m^3$, but PCDD/PCDF could not be detected. Yet PCDD/PCDF was detected in the flare flue gas at concentrations similar to those in the stack emissions of waste incinerators (0.5-5 ng/m^3 of $2,3,7,8-Cl_4DD$ before the scrubber or filter; 0.05-1.5 ng/m^3 of $2,3,7,8-Cl_4DD$ after the cleaning systems) (see Table 11). The similarities between the isomer patterns from the flare gas and municipal waste incinerators suggests that they may have the same precursors or similar formation mechanism. It was concluded that the use of flares to minimize volatile emissions from waste sites should be limited to special conditions in order to limit PCDD/PCDF emissions.

Remedial Measures

Because of the large size of the site (3 million m^3), destruction or removal of the PCDD/PCDF contamination is not considered feasible for economic and safety reasons (Technischer Überwachungsverein Berlin, 1985).

Table 11. Concentrations of Low Volatile Pollutants in Landfill Gas and Flare Gas (Dump at Braunschweig)
(Flare temperature: 890-920 °C)

Compound	Concentration [ng/m^3]		
	Decomposition Gas	Flue Gas of Torch	
		absolute	rel. to 7% O_2
Sum of HCH	175.4	37.2	26.8
HCB	56.6	6.7	4.8
Sum of PCB	1,600	260	187
Sum of Cl_3DD	< 0.01	0.71	0.51
Sum of Cl_3DF	< 0.01	5.75	4.14
Sum of Cl_4DD	< 0.01	0.99	0.71
Sum of Cl_4DF	< 0.01	4.81	3.46
2,3,7,8-Cl_4DD	< 0.005	0.35	0.25
2,3,7,8-Cl_4DF	< 0.005	0.64	0.46
	< 0.01	2.57	1.85
	< 0.01	0.96	0.69

5.2.7   Münchehagen Hazardous Waste Disposal Site (Rehburg-Loccum)

Background

The hazardous waste disposal site at Münchehagen close to Hanover opened in 1977. Between 1977 and 1983 350,000 m^3 of hazardous waste were disposed of in three pits excavated in a clay formation (see Figure 4). Capacity was later expanded with the excavation of two further pits (IV with 85,000 m^3 and IVb with 50,000 m^3) in 1982, but they were never used as the site was closed in 1983 following the Seveso waste scandal.

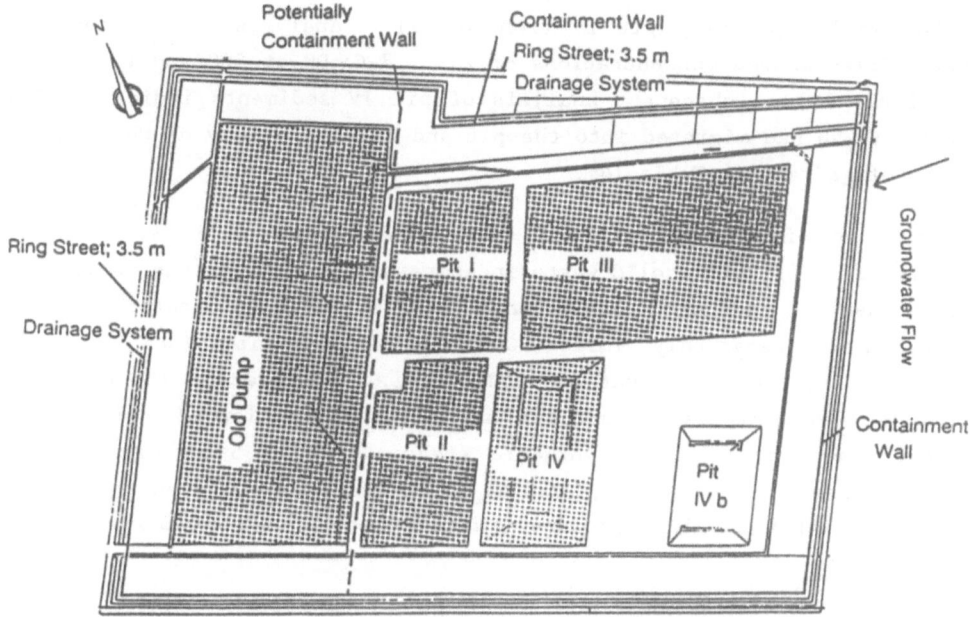

Fig. 4.   Layout of the Münchehagen Waste Disposal Site

After closure, water from surface run-off and from seepage throughout the pit embarkments began to accumulate in pit IV. Since some of the seepage was observed to originate from the wall separating pits II and IV, it was thought that the water could be contaminated with hazardous waste leachate. Thus water and sediment samples from pit IV were analysed for PCDD/PCDF in Autumn, 1984. No PCDD/PCDF could be found in water at a detection limit of 5 ng/L, and although some isomers were found in the sediment, Cl_4DD was below the detection limit of 0.2 ng/g (see Table 12).

Reassured by these results, the authorities began trucking the water from pit IV to the local sewage treatment plant in June, 1985, and the water level was reduced. On August 22, 1985 an oil seep was discovered in the wall separating pit IV from pit II. Analysis by two separate laboratories showed concentrations of $2,3,7,8-Cl_4DD$ at 560 and 1,125 $\mu g/g$ (see Table 13). Subsequent analysis of pit IV sediments indicated that $2,3,7,8-Cl_4DD$ had migrated into the pit and likely, by way of the trucked water, off site (see Table 14).

Analysis of disposal records showed that there were two possible sources of PCDD/PCDF: solid waste incinerator fly ash, or the waste of a French chemical producer. It is known that fly ash from locations around the world show remarkably similar PCDD/PCDF concentrations (see Tables 15 and 16). It is evident that the $2,3,7,8-Cl_4DD$ measured in the oil seep could not have originated from fly ash. The production of 2,4,5-T and related compounds (earlier synthetic steps; e.g. 2,4,5-phenol) are the only processes known where $2,3,7,8-Cl_4DD$ is the main component of Cl_4DD. Thus, the French chemical producer was implicated as the source of the contamination.

Analysis of disposal records showed that there were two possible sources of the PCDD/PCDF: solid waste incinerator fly ash, or the waste of a French chemical producer. It is known that fly ash from locations around the world show remarkably similar PCDD/PCDF concentrations (see Tables 15 and 16). It is evident that the $2,3,7,8-Cl_4DD$ measured in the oil seep could not have originated from fly ash. The production of 2,4,5-T and related (earlier synthetic steps; e.g. 2,4,5-chlorophenol) are the only processes known where $2,3,7,8-Cl_4DD$ is the main component of Cl_4DD. Thus the French chemical producer was implicated as the source of the contamination.

Site Investigation

The site investigation focussed on determining the degree of PCDD/PCDF contamination in the landfill, and establishing the potential for groundwater contamination.

A hydrogeological investigation revealed that the clay formation is characterized by zones of higher permeability near the surface, while the overall permeability decreases with depth.

A special probe was developed to collect samples at Münchehagen. Through the use of liquid nitrogen the waste, soil, sediment or water sample was frozen in situ. It could then be removed without risk to the technical personnel or the environment while at the same time maintaining the integrity of the sample.

It was known from the disposal records that the waste from the French chemical manufacturer was buried in drums in pit II close to the point where the contaminated oil seep was discovered. Thus the core sampling was concentrated in this area and along the estimated path of migration to the seep. A total of 23 bore cores were collected, from which 23 samples were analysed. The results are summarized in Table 17. From the relative concentrations of $2,3,7,8\text{-Cl}_4\text{DD}$ it is clear that the PCDD/PCDF in cores E70, F60/3, F60/4, and F60/6 from pit II originate from chemical waste. Although no 2,4,5-T was detected in the contaminated core samples, all other evidence points to the chemical manufacturer as the source of the $2,3,7,8\text{-Cl}_4\text{DD}$. It is possible that the more water soluble 2,4,5-T was leached away from the point of disposal to a greater extent than the PCDD/PCDF.

The $2,3,7,8\text{-Cl}_4\text{DD}$ level measured in core 60/4 is similar to that measured in the oil seep, which links this waste to the original seep. On the other hand, no significant quantities of PCDD/PCDF were found in the embankment leading towards the original oil seepage (the "Z" samples in Table 16) and no oil was observed in any other cores. It was therefore concluded that the original oil seep was a one time event. The possibility that it was the result of a leaking drum, a repeatable event, does not appear to have been considered.

Subsequent sampling of the on site bore hole liquids revealed PCDD/PCDF in all samples and $2,3,7,8\text{-Cl}_4\text{DD}$ in three, including 64 ng/L in borehole F60/4. These substances have not been detected in the off-site monitoring wells, although other contaminants have been detected at the edge of the site. This suggests that adsorption processes in the site have limited PCDD/PCDF mobility.

Sediment samples taken from a nearby stream that receives some site runoff indicate somewhat elevated levels of PCDD/PCDF. However, no $2,3,7,8-Cl_4DD$ was found. This suggests that there might be some off-site contamination related to fly ash, but that the chemical waste dioxins, for which $2,3,7,8-Cl_4DD$ is a marker, have not migrated this way.

The gas generation in the pits is very low, and since PCDD/PCDF migration by this route is in any case generally insignificant, this potential pathway can be safely neglected.

The investigators concluded that no significant off-site contamination was occurring or was expected to occur in the near future, but that measures to ensure the long term security of the site would be prudent.

Remedial Measures

Following the discovery of the oil seep and a report questioning the stability of the wall between pit II and IV, the contents of pit IV were pumped into pit IVb and pit IV was filled with clay. The liquid in pit IVb is now being treated in an on-site physical-chemical plant. The effluent is stored in ponds on the surface of pits I and III it has been tested and approved for transport to the local sewage treatment plant. The process sludges are solidified and stored, awaiting transport to another hazardous waste site.

A 4 m thick additional clay cap was placed on pit II following the oil seepage. Rainwater continues to infiltrate pits I and III. A cap for the whole site is planned to minimize leachate generation.

Excavation was rejected as an option due to low concentrations found in the site, the lack of proven technology, and the risk of spreading the contaminants.

A plan has been developed to prevent off-site migration of contaminated groundwater by reversing the flow into the site. An inner and an outer enclosing trench will be built, with the inner having a lower hydraulic potential than the outer. The uncontaminated water from the outer will be discharged into a nearby stream, while the contaminated leachate from the inner drain will be purified in a low maintenance combined physical-chemical/biological treatment plant.

584

The effectiveness of these measures will be monitored with an expanded well system. If performance is not satisfactory, containment walls will be built around the site to a depth of 10 m, below which injection into the formation will reduce permeability. The construction of a containment floor is not considered possible for this geological formation.

The preliminary construction cost estimate was 45 million DM.

Table 12. Concentration of PCDD/PCDF in Water and Sludge from Pit IV (Autumn 1994)

	Water [ng/l]	Sludge [ng/g]	Water [ng/l]
Sum of Cl_4DD	ND	ND	ND
Sum of Cl_5DD	ND	0.5	ND
Sum of Cl_6DD	ND	4	ND
Sum of Cl_7DD	ND	12	ND
Cl_8DD	ND	10	ND
Sum of Cl_4DF	ND	2	ND
Sum of Cl_5DF	ND	5	ND
Sum of Cl_6DF	ND	4	ND
Sum of Cl_7DF	ND	4	ND
Cl_8DF	ND	2	ND

Table 13. Concentration of PCDD/PCDF in the Oil Seep (August 23, 1985)

Laboratory	Kuhlmann $[\mu g/kg]$	Natec $[\mu g/kg]$
$2,3,7,8-Cl_4DD$	1,125	560
Sum of Cl_4DD	1,125	580
Sum of Cl_5DD	21	23
Sum of Cl_6DD	55	115
Sum of Cl_7DD	29	60
Cl_8DD	60	41
Sum of Cl_4DF	ND	48
Sum of Cl_5DF	ND	15
Sum of Cl_6DF	ND	14
Sum of Cl_7DF	ND	24
Cl_8DF	ND	20

Table 14. Concentration of PCDD/PCDF in Sludge from Pit IV (December 4, 1985)

	Sample 1 + 2 $[ng/g]$	Sample 3 + 4 $[ng/g]$
$2,3,7,8-Cl_4DD$	0.021	0.86
Sum of Cl_4DD	0.021	0.94
Sum of Cl_5DD	0.03	0.07
Sum of Cl_6DD	0.06	0.28
Sum of Cl_7DD	0.02	0.10
Cl_8DD	0.02	0.07
Sum of PCDD	0.151	1.46
Sum of Cl_4DF	0.026	0.214
Sum of Cl_5DF	ND	0.13
Sum of Cl_6DF	ND	0.12
Sum of Cl_7DF	ND	0.04
Cl_8DF	ND	0.01
Sum of PCDF	0.026	0.514

Table 15. Concentration of PCDD/PCDF in Fly Ash from Solid Waste Incinerators (50 mixed samples from 10 incinerators)

	Minimum [ng/g]	Average [ng/g]	Maximum [ng/g]
Cl_4DD	0.1	11	67
Cl_5DD	0.3	34	201
Cl_6DD	0.4	50	253
Cl_7DD	0.3	57	260
Cl_8DD	0.2	65	365
Sum of PCDD	1.3	210	861
Cl_4DF	0.7	72	477
Cl_5DF	0.8	95	494
Cl_6DF	0.3	82	404
Cl_7DF	0.1	56	386
Cl_8DF	0.02	13	174
Sum of PCDF	1.9	275	1,660

Table 16. Concentration of 2,3,7,8-Substituted PCDD/PCDF in Fly Ash from Solid Waste Incinerators (50 mixed samples from 10 incinerators)

	Minimum [ng/g]	Average [ng/g]	Maximum [ng/g]
2,3,7,8-Cl$_4$DD	0.11	0.6	2.3
1,2,3,7,8-Cl$_5$DD	0.11	3.3	11
1,2,3,4,7,8-Cl$_6$DD	0.04	3.9	15
1,2,3,6,7,8-Cl$_6$DD	0.07	7.2	30
1,2,3,7,8,9-Cl$_6$DD	0.02	5.5	24
1,2,3,4,6,7,8-Cl$_7$DD	0.02	49.9	129
2,3,7,8-Cl$_4$DF	0.03	4.6	28
1,2,3,7,8-Cl$_5$DF	0.14	9.4	29
2,3,4,7,8-Cl$_5$DF	0.12	11.0	34
1,2,3,4,7,8-Cl$_6$DF	0.09	13.5	46
1,2,3,6,7,8-Cl$_6$DF	0.09	14.5	44
1,2,3,7,8,9-Cl$_6$DF	0.01	0.8	3
2,3,4,6,7,8-Cl$_6$DF	0.10	10.5	39
1,2,3,4,6,7,8-Cl$_7$DF	0.08	76.5	294
1,2,3,4,7,8,9-Cl$_7$DF	0.02	3.0	13

Table 17.Concentration of 2,3,7,8-Cl$_4$DD and PCDD/PCDF in Bore Cores from Münchehagen

Sample #	Core Depth [m]	2,3,7,8-Cl$_4$DD [ng/g]	Σ Cl$_4$DD [ng/g]	Σ PCDD/PCDF [ng/g]
E 70	7 - 8.5	2.9	2.9	36
	8.5 - 14	28.9	28.9	186
F 30/1	13 - 14	ND	ND	70
	14 - 15	4.9	18	912
F 20/2	7 - 8	1.1	10	378
	8 - 9 upper	ND	1	113
	8 - 9 lower	ND	2	118
	9 - 12.5	1.9	2	247
F 60/3	5 - 6	ND	ND	ND
	8 - 9	440*	440*	440*
F 60/4	5.3 - 6.5	315	315	526
	6.5 - 7.5	ND	ND	112
F 60/6	6 - 7	ND	ND	49
	7 - 8	8.6	8.6	98
F 70/1	6 - 6.5	ND	ND	ND
	6.5 - 7.5	ND	ND	ND
Z 1	6 - 7	0.7	0.7	88
Z 2	6 - 8	ND	ND	13
Z 3	6 - 8	ND	ND	71
Z 5	5 - 7	ND	ND	ND
	7 - 8	ND	ND	ND
Z 6	9 - 10	ND	ND	ND
	10 - 12	ND	ND	ND

* Water sample, therefore, concentrations in [ng/L].

5.2.8 Chemical Production Site at Gernsheim (Hessen)

Background

Lindane was also produced by a German firm at Gernsheim on the Rhine River close to Darmstadt (Hessischer Minister für Arbeit, Umwelt und Soziales, 1985; Gutachten, Prof. Selenka, 1985). In 1979/80 large areas of soil, both on and off the site, were found to be contaminated with hexachlorocyclohexane (HCH). As a result agriculture was banned on 60 ha to the east of the site and restricted on a further 140 ha. In waste disposal areas close to Darmstadt highly contaminated land was designated for soil removal while monitoring was instituted on agricultural areas near a former waste water percolation field. Contaminated milk and feed were treated as hazardous waste.

It was known that trichlorobenzene had been produced from the HCH waste using a process similar to that used at the Moorfleet plant. Following the discovery of PCDD/PCDF in the Moorfleet waste the Gernsheim facility was also investigated. It was quickly established that all of the trichlorobenzene process waste had been disposed of in an underground storage site. However, there was concern that PCDD/PCDF might also have been synthesized during the production of lindane, and that the HCH contaminated areas might also contain elevated levels of PCDD/PCDF.

Site Investigation

The first goal of the site investigation was to determine if the HCH wastes contained PCDD/PCDF. To this end three residue samples and one leachate sample were taken from the methanol residue pit on the Gernsheim plant site, an area contaminated with HCH. The results are summarized in Table 18. Although no 2,3,7,8-Cl$_4$DD/F was found, significant concentrations of other PCDD/PCDF were measured.

In addition four frozen bore cores taken from the waste disposal site at Gernsheim for the earlier HCH study were re-analysed for PCDD/PCDF (see Table 19). The concentrations varied widely from sample to sample. Once again, no 2,3,7,8-Cl$_4$DD/F was found at a detection limit of 0.4 µg/kg, yet it is clear that the HCH wastes are contaminated with PCDD/PCDF.

590

Soil samples from the contaminated fields downwind of the Gernsheim plants and from the former percolation field close to Weiterstadt were analyzed for PCDD/PCDF. In Table 20 the results are compared with levels measured in the Korbacher Wald, a nearby uncontaminated area. Two samples display elevated levels of Cl_4DF and higher chlorinated PCDD/PCDF.

Remedial Measures

The initial studies indicated that a PCDD/PCDF contamination problem exists, but did not provide enough information to allow selection of appropriate remedial measures. Thus further site investigation is planned. In particular the extent of the contamination in the methanolic waste basin will be examined and the suitability of removal of the waste considered. Rhine River sediments and fish will be analysed to determine the extent of surface water contamination. The soil levels in contaminated areas where agriculture is still permitted will be sampled more intensively and an examination of food chain accumulation is planned.

Table 18. PCDD/PCDF in Samples from the Methanol Waste Basin at Gernsheim; Concentrations given in [μg/kg or μg/L]

Sample	Leachate	Waste 1-5	Waste 6-8	Waste 9-10
Cl_3DD	0.007	88	530	200
Cl_3DF	0.07	70	680	43
Cl_4DD	<0.01	3	<4	<2
$2,3,7,8-Cl_4DD$	<0.01	<2	<4	<2
Cl_4DF	0.2	160	1,200	950
$2,3,7,8-Cl_4DF$	0.05	<2	<4	<2
Cl_5DD	<0.01	<3	130	50
Cl_5DF	0.08	65	1,700	110
Cl_6DD	<0.01	6	200	18
Cl_6DF	0.05	45	850	70
Cl_7DD	<0.01	29	440	22
Cl_7DF	<0.01	17	700	19
Cl_8DD	<0.04	340	2,100	260
Cl_8DF	<0.04	37	1,400	66

Table 19. PCDD/PCDF in Bore Cores from the Waste Disposal Site at Gernsheim; Concentrations given in [μg/kg]

Sample	1	2	3	4	1	2	3	4
Cl_3DD	<0.4	<0.2	<0.4	<0.2	<0.2	0.5	<2	<8
Cl_3DF	2.0	0.9	0.6	1.8	0.6	<0.5	3	<8
Cl_4DD	<0.8	<0.2	3.4	<0.4	<0.5	<0.5	<10	<4
$2,3,7,8-Cl_4DD$	<0.8	<0.2	<0.4	<0.4	<0.5	<0.5	<10	<4
Cl_4DF	3.2	1.0	14	16	5	2	500	69
$2,3,7,8-Cl_4Df$	0.9	0.3	7.5	3.1	1	0.3	28	12
Cl_5DD	<0.8	<0.8	80	14	<0.6	<0.6	74	<6
Cl_5DF	6.4	2.5	800	110	9	1	1,100	230
Cl_6DD	7.7	5	1,100	80	7	5	1,600	130
Cl_6DF	24	8	6,900	200	10	8	2,100	280
Cl_7DD	62	30	13,200	200	16	16	3,900	270
Cl_7DF	39	10	12,500	160	7	5	3,100	220
Cl_8DD	not analyzed			100	160	26,000	5,700	
Cl_8DF	not analyzed			26	25	25,000	1	

Table 20. PCDD/PCDF in Soil Samples from HCH Contaminated Areas around Gernsheim; Concentrations given in [ng/kg]

Sample	GE-P	Korb-F	GE	Allmf.	Allmf.	Wstd.	Wstd.
Cl_3DD	0.3	<0.1	nodata	nodata	nodata	nodata	nodata
Cl_3DF	2.5	<0.1	nodata	nodata	nodata	nodata	nodata
Cl_4DD	2.8	0.7	9	<5	<5	6	47
$2,3,7,8-Cl_4DD$	<0.1	<0.1	<5	<5	<5	<5	<5
Cl_4DF	12.6	5.1	<5	<5	<5	20	150
$2,3,7,8-Cl_4DF$	1.5	0.8	<5	<5	<5	<5	<5
Cl_5DD	5.1	1.5	<5	<5	<5	<5	<5
Cl_5DF	32	7.5	<5	<5	<5	<5	<5
Cl_6DD	29	2.8	100	<5	<5	<5	<5
Cl_6DF	64	7.4	188	<5	<5	<5	<5
Cl_7DD	60	2.8	100	<5	<5	<5	<5
Cl_7DF	32	1.8	130	<5	<5	<5	<5
Cl_8DD	not analyzed		320	160	<50	<50	<50
Cl_8DF	not analyzed		<50	<50	<50	<50	86

Abbreviations of locations:

GE: Gernsheim
Korb-F: Korbach - Forest
Allmf.:   Allmendfeldt (surroundings)
Wstd.:   Weiterstadt.

5.2.9 Karsau Hazardous Waste Site (Baden-Württemberg)

Background

The hazardous waste site at Karsau was opened in 1971. It was operated by three industries from nearby Rheinfelden, one of which left the partnership in 1984. The landfill is still in operation (1987).

The site is located at the head of a small valley draining into the Rhine River. About 1.4 ha of the available 3 ha had been filled with some 142,000 m^3 of waste by 1984. About 76% consisted of inorganic mineral waste while the remainder was a mixture of refuse types. Included in this material were several tons of Cl_8DD/F and lesser quantities of other PCDD/PCDF isomers, including an estimated 6-8 g of $2,3,7,8-Cl_4DD$ (see Table 21). These contaminants originated from the production of sodium pentachlorophenol, a process which is described in detail in the report "Formation of PCDD/PCDF in Industrial Processes" (Hutzinger and Fiedler; 1988). The contaminated waste is distributed throughout the site (Thomanetz, 1987).

Table 21. Estimated Quantities of Hazardous Compounds in the Karsau Waste Site

Cyanides (highly mobile)	40	t
Mercury (metallic & ionic)	0.12	t
Hexachlorobenzene	100-200	t
Pentachlorophenol-Na	3-7	t
Cl_8DD and Cl_8DF	0.46-7.4	t
$2,3,7,8-Cl_4DD$	6-8	g
Other PCDD/PCDF-isomers (mostly Cl_7DD)	28-66	kg

Site Investigation

The waste site is located on a Keuper formation which overlies an aquifer. The Keuper formation consists of a marl layer and a gypsum layer. The thickness decreases from 27 m in the upper valley to 15 m in the lower valley. It has been estimated that the leachate from the site would break through to the aquifer in between 9 and 90 years.

The landfill generates very large amounts of leachate (1984: ca.42 m^3/day). It is thought that a significant fraction of this originates from near-surface flow draining down the slopes of the valley and entering the waste body from the side. The leachate was originally fed directly into a stream flowing into the Rhine.

The contaminant levels in the leachate have been analysed and found to vary a great deal. Although Cl_8DD/F have only occasionally been measured above the detection limit of 0.9 µg/L, pentachlorophenol (PCP) levels average 570 µg/L and hexachlorobenzene (HCB) 230 µg/L.

Remedial Measures

A temporary covering has been placed on parts of the site resulting in some reduction in leachate generation.

An active carbon treatment of the leachate was also installed but has not performed as expected. Although the HCB concentration has been reduced to well under 25 µg/L, the PCP levels in the effluent remain over 100 µg/L, at times over 500 µg/L.

The following measures have been proposed to stabilize the site:

1.   Beginning at the foot of the waste body the natural ground should be removed down to the Keuper formation and replaced with a 1 m thick clay barrier ($k_f < 10^{-9}$) extending up the sides of the valley.

2.   The barrier should be covered with a drainage layer containing a leachate collection system extended as far up the side of the valley as possible. This would then become the new waste site, restricted to non-hazardous waste.

3.   The new and old waste sites should be separated by a wall containing a drainage layer and a leachate barrier. The leachate from each site should then be collected separately and piped to one of the site operators in Rheinfelden for suitable treatment.

4.   A cap containing surface water and gas drainage layers separated by a plastic barrier should be constructed on the old site. It should be constructed to ensure good surface runoff, which should then be fed directly in the natural valley drainage system.

5.   If the cap fails to significantly reduce the leachate contaminant
      load, then consideration should be given to the construction of
      containment walls between the waste site and the valley walls.
      Otherwise perpetual leachate treatment will be required.

5.2.10. Other Waste Disposal Sites

Investigations of water leachate, groundwater, soil, and gas from
eight waste sites in Germany (Außernzell, Braunschweig, Geldern,
Kahlenberg, Poppenweiler, Schwäbisch Hall, Senne, Weilbach) revealed the
presence of the following compounds in all matrices: PCBs, HCH, HCB,
PAHs, phthalic acid esters, endosulfane, and DDT (Jager and Pflug, 1985).
Trace amounts of PCDD/PCDF could only be observed in soil at the border
of one site (Cl_8DD/Cl_8DF in the range of 0.1-0.5 ppb). Generally, the
concentration of PCBs equals the total sum of all other pollutants. The
emissions from waste sites can be summarized as follows:

1,000,000 m^3 volume of waste generates 0.2 kg of PCBs via water
leachate and 10 g of PCBs via dump gas.

In the case of incomplete incineration of the PCBs the latter can be
considered as secondary source for PCDD/PCDF (Jager and Pflug, 1985).

5.3 WASTE SITES IN THE UNITED STATES OF AMERICA

5.3.1 The National Dioxin Study

As part of its Dioxin Strategy (U.S. EPA, 1983), the U.S. Environmental Protection Agency (EPA) conducted the National Dioxin Study, an investigation to determine the extent of 2,3,7,8-tetrachlorodibenzo-p-dioxin (2,3,7,8-Cl_4DD) contamination in the environment. The EPA defined seven categories (or tiers) of sites for investigation, ranging from the most probable tier of contamination (Tier 1) to the least likely (Tier 7) as follows:

Tier 1:   2,4,5-trichlorophenol (2,4,5-TCP) production sites and associated waste disposal sites.

Tier 2:   Sites and associated waste disposal sites where 2,4,5-TCP was used as a precursor to make pesticidal products.

Tier 3:   Sites and associated waste disposal sites where 2,4,5-TCP and its derivatives were formulated into pesticidal products.

Tier 4:   Combustion sources.

Tier 5:   Sites where 2,4,5-TCP and pesticides derived from 2,4,5-TCP have been or are being used on a commercial basis.

Tier 6:   Sites where improper quality control on manufacturing of certain organic chemicals and pesticides could have resulted in the inadvertent formation of 2,3,7,8-Cl_4DD.

Tier 7:   Control sites where contamination from 2,3,7,8-Cl_4DD is not suspected.

Investigations conducted at sites in Tiers 1 and 2 are managed by the Office of Solid Waste and Emergency Response (OSWER) and are funded under the Comprehensive Environmental Response, Compensation, and Liability Act (CERCLA).

Due to lack of documentation on identity or location of storage sites, such as the contaminated site at Fort A.P. Hill in Virginia where $2,3,7,8-Cl_4DD$ was found in soil at concentrations up to 1,200 ppt, these sites were not incorporated into the overall investigation. Sites associated with the transportation and distribution of the chemicals of concern also were not investigated for the same reasons.

In total there are 99 dioxin sites in Tiers 1 and 2 (Table 22) scattered in six of the ten EPA regions (Figure 5) (U.S. EPA, 1986). All sites found to contain $2,3,7,8-Cl_4DD$ were referred to appropriate EPA Offices or other federal or state agencies for any appropriate response or follow-up action.

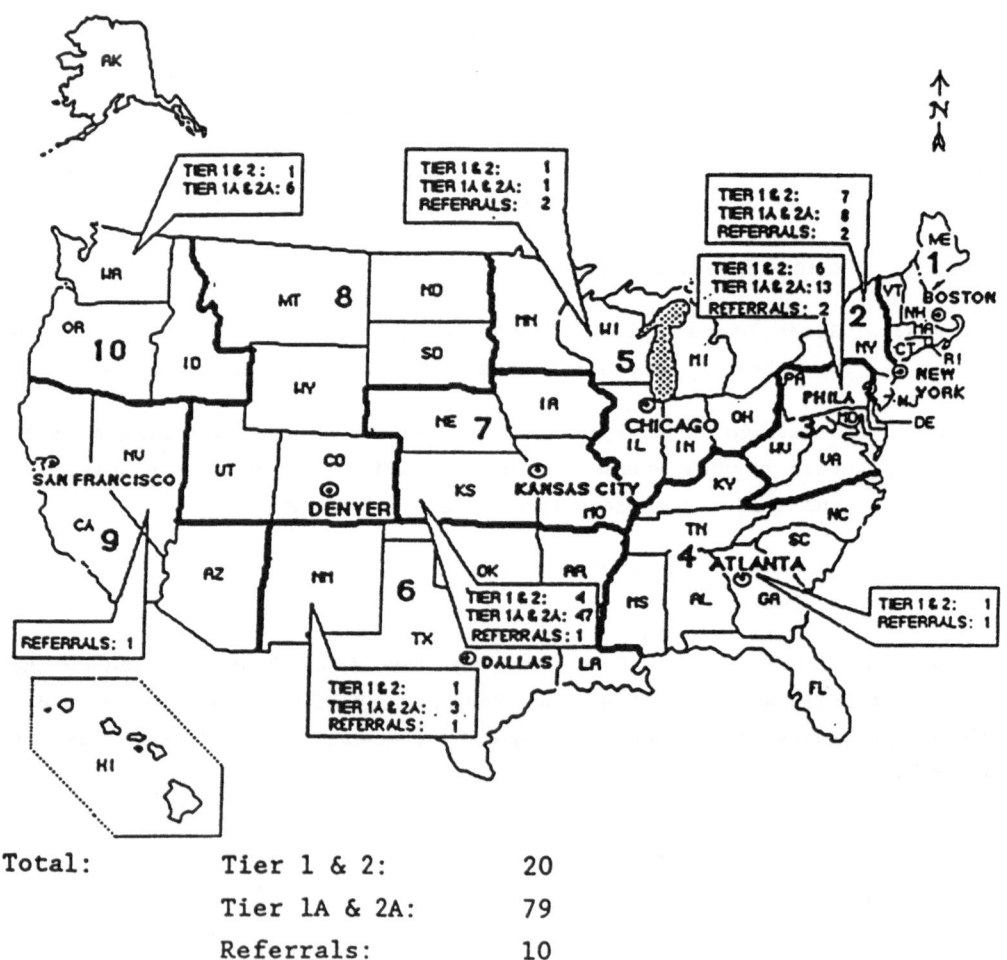

Total:	Tier 1 & 2:	20
	Tier 1A & 2A:	79
	Referrals:	10

Fig. 5. The Regional Distribution of Dioxin Tier 1/Tier 2 Sites

Table 22. Dioxin Sites by Tier

Tier	Number of Sites
1	11
1A	53
2	9
2A	26

Seventeen dioxin sites are on the current Superfund National Priorities List (NPL); four more have been proposed in Updates 2 and 3 to the NPL. They are listed in Table 23 (des Rosiers, 1987). Sites found to be contaminated in Tiers 3 through 7 are referred to the Office of Solid Waste and Emergency Response (OSWER) for possible Superfund (CERCLA) action. These sites are then considered along with the sites in Tiers 1 and 2 and all other hazardous waste sites managed under the Superfund program. Of the 99 Tier 1 production and waste disposal sites, ten are classified as requiring "no further action" based on sampling results indicating very low or undetectable levels of $2,3,7,8-Cl_4DD$.

Table 23. Dioxin Sites on the Superfund National Priorities List

Cashel Residence	Fenton, MO
Conservation Chemical	Kansas City, MO (Update 3)
Diamond Shamrock	Newark, NJ
Drake Chemical	Lock Haven, PA
Fike Chemicals	Nitro, WV
Hooker Chemical (Hyde Park)	Niagara Falls, NY
Hooker Chemical (Love Canal)	Niagara Falls, NY
Hooker Chemical (S-Area)	Niagara Falls, NY
Hooker Chemical (102nd St.)	Niagara Falls, NY
Minker Residence	Imperial, MO
Moyers Landfill	Collegeville, PA
NIES	Furley, KS (Update 2)
Quail Run	Gray Summit, MO (Update 2)
Rohm & Haas	Bristol, PA (Update 3)
Romaine Creek	Imperial, MO
Shenandoah Stables	Moscow Mills, MO
Stout Residence	Imperial MO
Sullins Residence	Fenton, MO
Times Beach	Times Beach, MO
Vertac	Jacksonville, AR
Western Processing	Kent, WA

Of the sites where dioxin was detected at levels of concern, the most common materials contaminated were soil and dust. These sites were generally the production facilities of Tiers 1 and 2 and those waste disposal sites where intact waste or scrapped equipment was stored or disposed. The majority of dioxin contamination at Tiers 1 and 2 remained on-site.

Site specific sampling plans were prepared by the regional offices for each site to be sampled in Tiers 3, 5, and 6. For Tier 7, sampling sites were statistically selected. Field sampling collection activities were managed by the regional offices and involved the participation of state and local agencies and EPA contractors. Analytical methods used by the Contract Laboratory Program (CLP; commercial laboratories under EPA contract) had a nominal detection limit of 1 part per billion (ppb), while methods used by the Troika (3 EPA laboratories at Duluth, Bay St. Louis, Research Triangle Park) and Wright State University had a detection limit of 1 part per trillion (ppt) for all media other than water (U.S. EPA, 1986).

(1) Results from Site Tier 3

Sixty-one selected sites of facilities and associated waste disposal sites where 2,4,5-T and its derivatives were formulated into pesticidal products should have been analyzed. These products were: 2,4,5-trichlorophenol (2,4,5-TCP); 2,4,5-trichlorophenoxyacetic acid (2,4,5-T); silvex; erbon; ronnel; hexachlorophene, and isobac 20. Of the 61 facilities 20 were considered ineligible, because none of the above mentioned compounds were handled. Seven other sites were covered with pavement or extensively reworked, so that no original soil could be identified for sampling. Of the 41 eligible sampled facilities, ronnel was the most widely handled pesticide (18 facilities), while 2,4,5-T and silvex were handled at 11 and 10 facilities, respectively.

Additionally, twenty-three regionally selected Tier 3 sites were sampled. Silvex, 2,4,5-T, and ronnel were the most widely handled compounds at these facilities.

None of the statistically selected sites and only two of the regionally selected sites generated wastewater that was directly discharged into surface waters. Fifty percent of the Tier 3 sites that generated wastewater discharged to publicly owned treatment works, while the rest incinerated, evaporated or deep well injected their wastewater. Most of the solid wastes were disposed of in off-site landfills.

Analytical Results

At five of the 41 statistically selected sites soil contamination greater than or equal to 1 ppb Cl_4DD or detectable levels in other media were measured (Table 24):

Table 24. Cl_4DD at the Statistically Selected Sites at Tier 3

Farmingdale, NY	17.6 ppb	(soil)
Ona, WV	0.5-2.9 ppt	(fish)
Fort Valley	23-40 ppb	(soil)
	36.7 ppt	(soil)
Chicago Heights, IL	1.1-364 ppb	(soil)
Grand Ledge, MI	0.56-1.13 ppb	(soil)

Table 25 depicts the results of the six regionally selected contaminated sites.

Table 25. Cl_4DD at the Tier 3 Regional Sites

Name of Site	Matrix	Concentration	
Alexandria, VA	Fish-whole	1.6-6.3	ppt
	Fish-filet	1.9-5.0	ppt
	Sediment	5.5-23	ppt
Norfolk, VA	Soil	10.1	ppb
Belle Glade, FL	Soil	0.2-3.0	ppb
	Sediment	20.9-515	ppt
Bedford Park, IL	Soil	1.9-2.2	ppb
Saint Joseph, MO	Soil	0.13-39.1	ppb
Santa Fe Springs, CA	Soil	2.0	ppb

Contamination, when found, was usually limited to the site area. The widely contaminated site handled 2,4,5-T, silvex, and 2,4,5-TCP with a total amount greater than 100,000 pounds.

Conclusions

At contaminated sites the extent of contamination was usually limited to one or two soil samples with concentrations of $2,3,7,8\text{-}Cl_4DD$ above 1 ppb. Only two Tier 3 sites were extensively contaminated. One statistically selected site was already under investigation by the Superfund program when it was chosen for the study. All 11 contaminated sites were at or near facilities handling 2,4,5-TCP, 2,4,5-T, and/or silvex. None of the facilities exclusively handling hexachlorophene or ronnel were contaminated with $2,3,7,8\text{-}Cl_4DD$.

Based on the limited number of sites found to be contaminated, the small number of positive samples at most of these sites, and the generally low levels of $2,3,7,8\text{-}Cl_4DD$ detected, immediate national investigation of all of the remaining Tier 3 facilities does not appear to be warranted. Since the two most contaminated facilities were both large handlers of 2,4,5-TCP, 2,4,5-T, and/or silvex, further evaluation of other large handlers of these compounds is warranted.

(2) Results from Site Tier 5

Tier 5 consisted of areas where 2,4,5-trichlorophenol (TCP) and pesticides derived from it (including 2,4,5-T and silvex) were used on a commercial basis. The objective of the Tier 5 sampling program was to determine whether $2,3,7,8\text{-}Cl_4DD$ is present at detectable levels of approximately 1 ppt in areas where major uses of these pesticides occurred. Twenty-six sites were selected for sampling in Tier 5. At these sites present or past use of 2,4,5-TCP-based herbicides or slimicides has been documented; e.g. rice fields, canals adjacent to sugarcane fields, rangelands, forests, right-of-way, and recreational areas. In most cases, suspected "hot spots" could not be identified within a pesticide use site. Therefore, the rambling sampling approach (cf. Tier 3) was used to select sampling locations within a Tier 5 site. The environmental media included soils, stream sediments, fish tissue, vegetation, and animal tissue.

Analytical Results

The herbicide 2,4,5-T was applied at 18 of the 26 Tier 5 sites, including three where silvex was also applied. $2,3,7,8-Cl_4DD$ has been detected at 15 of the 26 sites, including two right-of-way, one aquatic use site, two sugarcane fields, canals adjacent to one sugarcane field, four rice fields, three forest areas, one rangeland area, and one multiple use area. Over 40 percent of the soil and sediment samples taken at contaminated sites had $2,3,7,8-Cl_4DD$ present above the detection limit. Two sites had fish samples contaminated; at one all samples collected were positive. Detailed results are given in Tables 26 and 27.

Table 26. Levels of Cl_4DD in Tier 5 Samples

Matrix	Range [ppt]	Comments
Soil	0.6-564	67 % below 5 ppt
Sediment	0.7-200	61 % below 5 ppt
Fish filet	8-23	
Animal tissue	n.d.	
Vegetation samples	n.d.	

Table 27. Tier 5 Site Characterization

Name of Site	Site Use	Date(s) Treated	Matrix	Concentration [ppt]
Grindstone, ME	Railroad right-of-way	?-1977	Soil	8-35
Long Island Railroad, NY	Right-of-way	1970's	Soil	9
Cleveland, MS	Ricefield	1978-84	Soil	0.8-1.17
Scot, MS	Ricefield	1984	Soil	0.6-0.7
West Palm Beach, FL	Sugarcane field	unknown	Soil	0.7-26.5
Petenwell Flowage, WI	Carp fishery; closed 1983	? -1980	Fish whole Fish filet sediment	9-47 2-23 2-200
Assumption Parish, LA	Sugarcane	1982	Soil	0.3-1.1
Pointe Coupee Parish, LA	Sugarcane field, (Soybeans in 1985)	1985	Soil	1-2.5
Exper. Station Desha County, AR	Ricefield (soybean in rotation	1972,1974, 1975	Soil	3
Exper. Ranch; Spoford. TX	Rangeland	1981	Soil	0.2-0.3
Richland Parish, LA	Ricefield	1982,1983	Soil	0.3-0.4
Mark Twain National Forest, MO	Forest	1977	Soil	0.3-120
Tonto National Forest, AZ	Forest	1965-1966 1968-1969	Soil Soil	2-564 0.4-6623
Santa Ana River, CA	Multiple uses	unknown	Sediment Fish whole	0.6 4.6
Santiam State Forest, Gates, OR	Forest	1976-1977	Sediment	0.2-0.4

604

Contamination was found at a variety of the sites sampled and in various media; however, the levels were generally very low. The highest levels for each media were generally found at areas used for equipment loading and areas where contaminants would tend to accumulate. Levels were much lower in areas where the pesticides were uniformly applied (e.g. sprayed). Sludges from paper mills on the Petenwell Flowage were found to have levels of 2,3,7,8-Cl_4DD over 100 ppt, even though chlorophenol-based slimicides are no longer used.

Conclusions

Further investigation of Tier 5 spray areas, where levels were generally of no concern, does not appear to be warranted. The current source of 2,3,7,8-Cl_4DD at the one significantly contaminated Tier 5 site (Petenwell Flowage) does not appear to be related to pesticide use (use of chlorophenol-based slimicides by several pulp and paper mills along the river). Therefore, further investigations of pulp and paper mills are being conducted (see also Tier 7).

(3) Results from Site Tier 6

Tier 6 consisted of organic chemical and pesticide manufacturing facilities where improper quality control of production processes could have caused products or waste streams to be contaminated with 2,3,7,8-Cl_4DD. The production of 60 of the 125 organic and pesticide compounds mentioned in the U.S. EPA report (Esposito et al., 1980) is most likely to lead to dioxin formation. Facilities producing any of these 60 compounds were identified as Tier 6 candidates. A total of twenty-five sites were selected statistically to represent the 67 facilities (the facilities producing only brominated Tier 6 compounds were later considered ineligible).

Waste generation and disposal information is available for 11 statistically selected sites and for all three regionally selected sites. Of the 11 statistically selected sites, ten generated processed wastewater and seven generated solid waste (six generated both). Six of the ten facilities generating wastewaters discharged off-site, including three which discharged to surface waters and four which deep-well injected (one did both). All seven of the facilities generating solid wastes disposed primarily in landfills. Two of the three regionally selected sites disposed of solid waste both on-site and off-site and liquid waste on-site. The third generated no waste.

Analytical Results

Soil contamination was found at two of the 15 statistically selected eligible sites. Two additional sites had $2,3,7,8-Cl_4DD$ concentrations in soil below 1 ppb. One of the three regionally selected sites was contaminated and one additional site had detectable levels below 1 ppb (Table 28). At all three contaminated sites, soil contamination was limited to one or two samples. At the regionally selected contaminated site, groundwater contamination was also found at the 0.07 to 0.10 ppt level in three samples. The groundwater in this case is not used as a drinking water source.

Table 28. Tier 6 Site Characterization

Name of Site	Matrix	Concentration	
Somerset, NJ	Soil	34.7	ppb
Port Neches, TX	Soil	0.1-1.4	ppb
Henderson, NV	Water	0.07-0.11	ppt

Conclusions

The facility at Somerset, NJ (manufacture of mecoprop and 2,4-D salts from 1977 to 1983) has been referred to the Superfund program for further sampling. The facility at Port Neches, TX (manufacture of pentachlorophenol until 1978; additionally organic chemicals, e.g. 2,4,5-T, 2,4-D, and parathion) is under an enforcement action of the Texas Water Commission (TWC) to undertake remedial action relating to contamination found on site and in adjacent ditches. In addition soil, water, and waste samples have been collected from the contaminated regionally selected facility at Henderson, NV (manufacture of lindane from 1948 to 1956; ethyl and methyl parathion since 1958, and carbophenothion; chlorobenzenes from 1947 to 1976). Difficulties due to the complexity of the chlorinated products have delayed analysis of these samples.

606

EPA estimates that 9 percent of the 67 facilities originally identified at Tier 6 sites are contaminated. None of the three contaminated sites were extensively contaminated with $2,3,7,8-Cl_4DD$. Therefore, further investigation of Tier 6 sites for $2,3,7,8-Cl_4DD$ does not appear to be warranted.

(4) Results from Site Tier 7

Tier 7 consisted of sites that did not have any previously known sources of $2,3,7,8-Cl_4DD$ contamination. It was intended to establish prevalence of $2,3,7,8-Cl_4DD$ in the environment and to provide a basis for comparison with results from other Tier sites. Soil sampling from rural (141) and urban (221) sites as well as fish samples (90) were analyzed.

Analytical Results

Seventeen of the 221 urban soil samples were found to have detectable levels of $2,3,7,8-Cl_4DD$. Levels in the 17 sites were very low, between 0.2 and 11.2 ppt; seven of the samples were less than 1 ppt. By comparison, $2,3,7,8-Cl_4DD$ levels in the soil of Times Beach, MO, a Tier 1 disposal site which was permanently evacuated, ranged up to greater than 1,000 ppb (1,000,000 ppt). These soils were sprayed for dust control with a mixture of waste oil and concentrated wastes containing $2,3,7,8-Cl_4DD$. All 32 soil samples collected in Midland near the Dow Chemical facility (a Tier 1 site) showed $2,3,7,8-Cl_4DD$ contamination in a range between 3 and 76 ppt in open public and residential areas (U.S. Environmental Protection Agency, 1986). Only one of the 138 rural samples contained $2,3,7,8-Cl_4DD$ at a concentration of 0.5 ppt. Fish from 17 of the sampling sites had detectable levels up to 19 ppt in whole fish composite samples. Levels found in fish filets (between 0.4-41 ppt) were generally lower than levels in the whole fish samples. The two sites with the highest $2,3,7,8-Cl_4DD$ levels in whole fish (the Androscoggin-River - maximum 29 ppt, and the Rainy River - maximum 85 ppt) both have upstream pulp and paper mill discharges. Complete results are listed in Table 29.

Table 29. Results of 2,3,7,8-Cl$_4$DD Analyses of Tier 7 Soils

Location description	Type	Levels [ppt]
D.C., Washington	Urban	3.0
		2.0
		4.0
PA, Pittsburgh	Urban	5.0
		2.0
IN, Evansville	Urban	1.3
IN, Gary	Urban	0.5
		4.1
LA, Lake Charles	Urban	0.2
CA, San Francisco	Urban	2.0
WA, Tacoma	Urban	0.4
		0.5
		0.6
		0.8
		1.9
		8.7
		11.2
OR, Linn County	Rural	0.5

The EPA prepared stream profiles to identify types of industrial dischargers in the vicinity of Tier 7 fish sampling sites. This should help to determine possible associations between 2,3,7,8-Cl$_4$DD presence in fish and various sources. The major types of industrial dischargers 30 miles upstream and three miles downstream of the statistically selected and regional selected fish sites were summarized. From Table 30 it can be seen that chemical plants and pulp and paper mills were the facilities that appear to be the most frequently associated with the detection of 2,3,7,8-Cl$_4$DD in fish. Chemical plants were already under investigation under Tiers 1, 2, and 6, while pulp and paper mills had not previously been suspected as a source of dioxins.

Table 30. Summary of Industrial Dischargers at Tier 7 Fish Sites

| | Number of Sites[a] | | | | | |
| | Statistical | | Regional | | All | |
Industry[*]	Det	Not Det	Det	Not Det	Det	Not Det
No Industry	2	22	28	28	30	50
Any Industry	15	51	69	126	84	177
Pulp&Paper	7	6	27	24	34	30
Steam Electric	8	14	36	40	44	54
Organic Chemicals	3	5	27	15	30	20
Inorg. Chemicals	3	7	19	10	22	17
Superfund Sites	2	3	17	18	19	21
POTWs	13	44	65	108	78	152
Metal Discharging	6	21	46	60	52	81
Other Industrial	8	14	36	36	44	50
Misc. Industries	11	32	53	71	64	103

[a] Total number of statistically selected sites = 90; detected = 17; not detected = 73

total number of regional sites = 223; detected = 69; not detected = 154

[*] Industrial categories include according to the SIC code:
  - Pulp and paper
  - Steam electric
  - Organic chemicals, plastics, synthetic fibers and pesticides manufacturers
  - Inorganic chemicals and Pesticide manufacturers
  - Superfund sites
  - POTWs
  - Metal discharging industry = all categories including: aluminium forming, battery manufacturing, coal mining, coil coating, foundries, iron & steel, metal finishing, nonferrous metals, ore mining, paint, porcelain enameling, copper forming, ink, auto and other laundries, photographics, and electric components
  - Other industrial categories includes: leather tanning, textiles timber, pharmaceuticals, petroleum refining, misc. chemicals, agricultural chemicals, misc. petroleum products
  - Miscellaneous industries includes: food products, lumber, real estate.

Conclusions

2,3,7,8-Cl_4DD was detected infrequently and at very low levels in background soil samples. Seventeen of the 221 urban sites and 1 of 138 rural sites had detectable levels, with the highest level found being 11.2 ppt in an urban soil sample. 2,3,7,8-Cl_4DD was detected more frequently in background fish samples. The EPA estimates that 21 percent of the sites in the U.S. Geological Survey national monitoring networks have detectable levels; the frequency of detection is 31 percent at sites selected by EPA's regional offices. An even higher proportion (23 of 29) of Great Lakes fish sampling sites had detectable levels.

2,3,7,8-Cl_4DD levels found in filet samples can be a cause for concern at specific locations under certain consumption patterns. Fish and shellfish from estaurine and coastal waters were rarely contaminated. A previously unsuspected possible source for contamination in some areas appears to be pulp and paper mill dischargers. Fish contamination is a current continuing phenomenon since recent EPA studies (EPA, 1985) indicate that 2,3,7,8-Cl_4DD has a half-life of slightly less than one year in fish.

5.3.2 Love Canal

The Love Canal has come to symbolize the many problems and unresolved scientific issues relating to the disposal of industrial waste products in landfills. The Love Canal is a rectangular, sixteen-acre, below-ground level landfill site located in a residential section in the south-coast corner of Niagara Falls (Figure 6) (New York State Department of Health, 1978; Deegan, Jr., 1987). The site was excavated in the late nineteenth century as part of a proposed canal linking the Niagara River and Lake Ontario. After the Love Canal project was abandoned the area was used from 1942 until about 1953 for the disposal of chemical wastes, e.g. chlorinated hydrocarbon residues, processed sludges, flyash. Some 21,88 tons of waste were disposed (Deegan, Jr., 1987). Presently, the site is contained within the area bordered by streets and the backyards of single-family homes. The natural drainage ways offer routes for the movement of liquids underground. Depressions known as swales (about six feet deep) traverse the area and are often filled with water during wet periods. The major swale, which intersects the Canal and provides a continuous conduit for surface water flow from the Canal to peripheral areas, was eliminated in 1956. In 1958, the Hooker Chemical Company sold the Love Canal to the Niagara Falls Board of Education, and in 1955, an elementary school was opened on this property (Vanna, 1983).

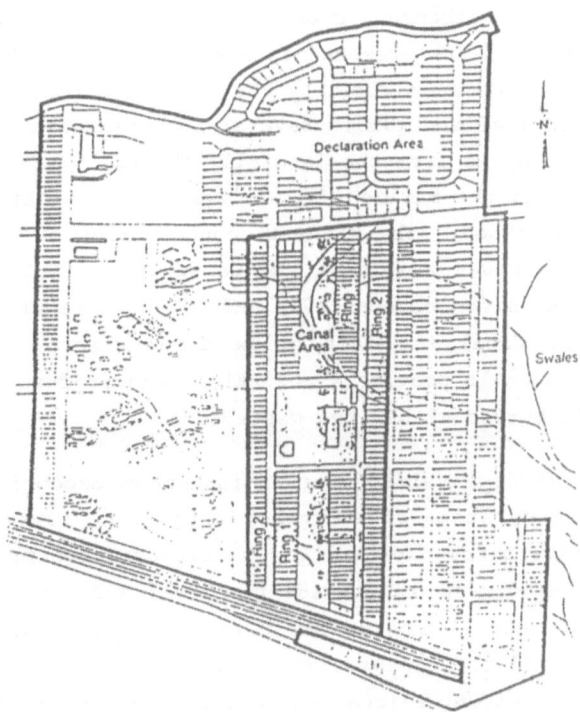

Fig. 6   Love Canal Study Area

More than 200 chemicals were identified at Love Canal, the most important
of which are listed below:

Chloroform	Pentachlorobenzene
Carbon tetrachloride	Allyl phenols
1.2-Dichloroethane	Hexachlorobenzene
1.2-Dichloroethane	2,3,7,8- and other Cl_4DDs
1.1.-Trichloroethane	Hexachlorodibenzodioxin
Trichloroethylene	Dichlorodibenzodioxin
Chlorobenzene	Chlorinated naphthalenes
Dichlorobenzene (2 isomers)	Trichlorophenols (isomers)
1,2-Dichloropropane	PCBs (isomers)
Chlorotoluene (2 isomers)	Octachlorocyclopentane
Dichlorotoluene (3 isomers)	Dichlorobenzaldehyde (isomers)
Trichlorobenzene (isomers)	1,2-Dibromoethane
Trichlorotoluene (5 isomers)	Tetrachlorobenzene (3 isomers)
Benzene	Toluene
Xylenes	Benzaldehyde
α-Chlorotoluenes	

Tetra, penta, hexa chlorinated anthracenes or phenanthrenes.

2,3,7,8-Tetrachlorodibenzo-p-dioxin was one of the chemicals identified in the landfill. Its presence is probably due to the fact that the
Chemical Company buried approximately 200 tons of trichlorophenols in the
Love Canal between 1942 and 1953. Several analyses have demonstrated its
presence in soil samples from the landfill, the backyards of nearby homes
and the sediment and the marine life of two of the Great Lakes near the
Love Canal neighbourhood. The highest concentration of dioxin found was
300 ppb in a storm sewer adjoining the Canal area. This observation is
of great importance since between 1957 and 1960 two sewer pipes were laid
right through the Canal ten feet below the surface. The gravel bed
around these pipes could provide a potential conduit for the migration of
certain chemicals buried within the waste site.

On August 7, 1978, President Carter issued an executive order decla-
ring that a man-made state of emergency existed at Love Canal. It was
the first executive order addressing a hazardous-waste problem and enabled the federal government to provide technical and financial assistance
to the City of Niagara Falls to begin clean-up and containment of the
site. To date (April 1987) approximately $ 100 million has been spent on
site remediation, resident relocation, and environmental and human health

investigations at Love Canal. In 1980 some 800 families who lived close to the landfill were relocated. In May 1980, the U.S. EPA assembled a study team which designed a monitoring program to collect and analyze environmental samples. The study should allow identification of the actual movement of Love Canal wastes into the Declaration Area.

Air transport of contaminants was probably stopped by the placement of an extensive clay cap over the landfill (Figure 7). Similarly, surface water transport was also likely eliminated as a source of contamination, although residual near-surface soil contamination outside the area subject to remediation might continue to serve as an isolated source. Since the initial site investigation work indicated that significant groundwater transport of pollutants was occuring, EPA concentrated its efforts and resources on conducting a comprehensive hydrogeological investigation of the Love Canal area. 174 wells were installed to sample groundwater (6,835 samples were collected between August and October 1980).

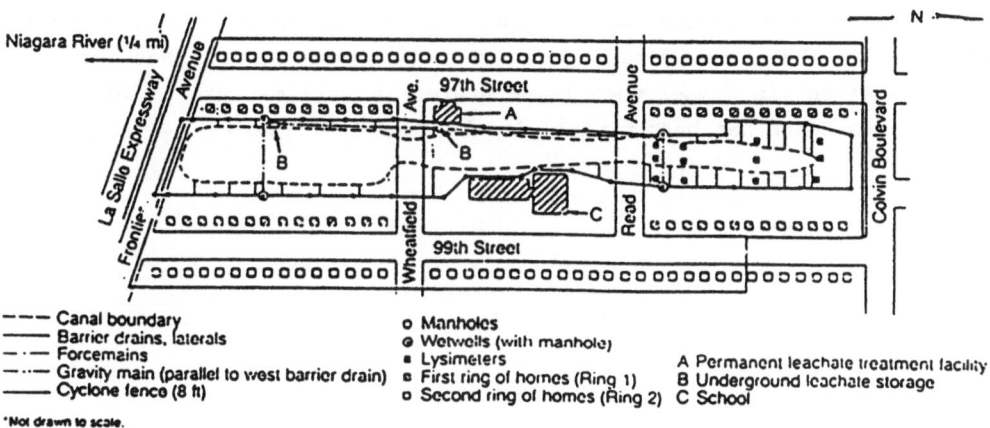

Fig. 7. Love Canal Remedial Action Project

5.3.3 Dioxin Contamination at Times Beach

Times Beach was contaminated with dioxin in 1972 when waste oil contaminated with dioxins was sprayed on unpaved roads to control dust. Following the initial contamination it is believed that spraying with uncontaminated waste oil was carried out before the streets were paved. In 1984, two blocks of the most heavily contaminated streets were set aside by the Missouri Department of Natural Resources as a Dioxin Research Facility (Yanders et al., 1986). In this section, the dioxin contamination of the soil under the asphalt averaged about 300 ppb. It is difficult to characterize the contamination at Times Beach since neither the original contamination of the waste oil with Cl_4DD nor the number of applications is known. Research plots were prepared by removing the asphalt and excavating the underlaying soil and gravel to a depth of about 8 inches (dioxin was not found at greater depths in this street). The bins are located in an area which is subject to the natural conditions of sunlight and precipitation. The concentrations of Cl_4DD at various depths. The Cl_4DD concentrations found in the soil sampled in July, 1984, are given in Table 28 (left side) (Yanders et al., 1988). The concentrations found in samples taken in 1988 and normalized are presented in Table 28 (right side). There is no indication of any vertical movement of Cl_4DD; the data suggest that the loss due to photolysis or volatilization (in summer when temperatures in soil may exceed 40 °C) at Times Times Beach has been minimal in the last four years.

Table 31. Comparison of the Concentration of 2,3,7,8-Cl$_4$DD in Soil from Three Experimental Plots at the Times Beach Dioxin Research Facility Sampled in July, 1984, and July, 1988 (Yanders et al., 1988)

Section Size and Depth			1984 Sample Southwest Corner Conc. [ppb]	1988 Sample Center (covered) Conc. [ppb]	North (open) Conc. [ppb]
5	mm	(0-5)	73	73	77
10	mm	(5-15)	70	70	68
10	mm	(15-25)	77	71	77
25	mm	(25-50)	79	90	82
25	mm	(50-75)	84	81	87
50	mm	(75-125)	73	89	78

			1984 Sample Center Conc. [ppb]	1988 Sample North Conc. [ppb]
5	mm	(0-5)	118	121
10	mm	(5-15)	126	127
10	mm	(15-25)	128	142
25	mm	(25-50)	148	128
25	mm	(50-75)	137	129
50	mm	(75-125)	124	

			1984 Sample Southwest Corner Conc. [ppb]	1988 Sample South Central (open) Conc. [ppb]
5	mm	(0-5)	149	159
10	mm	(5-15)	167	167
10	mm	(15-25)	163	158
25	mm	(25-50)	166	154
25	mm	(50-75)	176	169
50	mm	(75-125)	152	155

The present study (Yanders et al., 1988) is in agreement with calculations (Freeman and Schroy, 1988) which estimate that the levels of Cl_4DD which existed at Times Beach sites contamination sixteen years ago have been preserved more or less unchanged. This is inpart due to the extremely low water solubility of Cl_4DD and the paving of the streets, which provided an effective barrier to losses through volatilization and photolysis. Furthermore, it can be infered that all of the $2,3,7,8-Cl_4DD$ which penetrated into the soil (more than a few millimeters in depth) during the accidental application in 1972 is still there, and, under the existing natural conditions, could remain there indefinitely.

5.4 WASTE SITES IN THE UNITED KINGDOM

The Hazardous Waste Inspectorate (HWI) was formally announced in the United Kingdom on August 8, 1983. It is an advisory body with no enforcement powers. Its function is to advise waste disposal authorities (WDAs), to encourage adequate and consistent standards of control throughout the country, and to make an annual report to the Parliament. It examines the hazardous waste management at all levels from the point of generation to final disposal by visiting facilities used to handle, process, and dispose of such wastes. In 1984 there was considerable public concern associated with the incineration of PCBs at the facilities owned by Re-Chem International at Bonneybridge, in Scotland, and Pontpool, in Wales. A very limited soil sampling program was undertaken with the soils being analyzed for PCDD/PCDF in an attempt to identify the background levels of contamination throughout the country. The results obtained were of restricted value due to lack of analytical standards and to limitations in the analytical method employed. For these reasons, the results were not published. In order to develop a UK analytical method and to establish environmental background norms for PCDD/PCDF in soils a research contract was let to the University of East Anglia in October 1985. After this work has been completed it is intended to carry out environmental soil surveys in the vicinity of various types of incinerators and other possible sources of PCDD/PCDF emission (The Hazardous Waste Inspectorate, 1986).

There are no known sites of extensive PCDD/PCDF contamination in Great Britain, but it is clear that investigations are just beginning.

5.5 WASTE SITES IN THE NETHERLANDS

5.5.1 Volgermeerpolder

The Volgermeerpolder is a 100 ha landfill near Amsterdam. It is seriously contaminated with a variety of organochlorine wastes. Elevated levels of $2,3,7,8-Cl_4DD$ and PCDD/PCDF have been found in soil, sediment, fish and wildlife from the area. Some values are presented in Table 30. The planned remedial measures include construction of a leachate collection and treatment system, a containment wall and a cap for the site (van Zorge, 1988).

Table 32. $2,3,7,8-Cl4DD$ in Biota from the Volgermeerpolder Waste Site

Subject	Concentration [mg/kg]	
Eel	0.144	(fresh weight)
Worm	6.8	(dry weight)
Mouse	16.7	(dry weight)

5.5.2 Diemerzeedijk

Diermerzeedijk is a former hazardous waste incineration site. An area of about 3 ha around the site is contaminated. High levels of PCDD/PCDF have been identified in wildlife in the area, including 50-1,600 units EPA toxic equivalents on a dry weight basis in rabbit liver. Remedial measures are under study (van Zorge, 1988).

5.6 THE SEWAGE SLUDGE CONTROVERSY

Sludges have been found to contain a variety of organic compounds including chlorinated sovents, polycyclic aromatic hydrocarbons, pesticides, flame retardants, nitrosamines, PCBs, heavy metals (lead, cadmium, chromium) and other elements. As sewage sludge represents an extremely difficult matrix for trace analysis, only a few articles exist dealing with organic pollutants identified therein (Lamparski et al., 1984; Weerasinghe et al., 1985; Hagenmaier et al., 1988). Since PCDD/PCDF have been found quite ubiquitously in the industrial developed countries (Hutzinger and Fiedler, 1988a, 1988b) dioxins and related compounds as trace contaminants from various sources could be present in wastewater effluents and concentrated in treated sludge. Presently, proper disposal of sewage sludge is a very controversial issue. In Germany, application of sewage sludge to farm land and greenland is allowed up to 5 tons/ha/3 years.

Thirty samples of sewage sludge from 18 different wastewater treatment plants from thorughout Germany and an additional 13 samples from 9 plants in Baden-Württemberg (county Lörrach, county Waldshut) have been analysed for PCDD/PCDF, PCBs, and organochlorine pesticdes (Hagenmaier et al., 1988). The results are given in Table 32. It can be seen that there are large differences between the maximum and minimum values. Lindane (γ-HCH), hexachlorobenzene (HCB), p,p'-DDE (main metabolit of DDT), and PCBs could be identified in all samples in the lower ppb-range (except PCBs). The data for PCBs revealed high amounts of the higher chlorinated mixtures (Chlophen A50 and A60). Most of the samples analysed showed concentrations between 15 and 40 ppb (average: 38 ppb) for PCDD and about 1 ppb for PCDF (average: 3,1 ppb). The samples from Baden-Württemberg gave similar results with slightly higher values.

Table 33. Concentrations of Organochlorine Pesticides, PCBs, and PCDD/PCDF in Sewage Sludge

	Average [ng/g]	Minimum [ng/g]	Maximum [ng/g]
Lindane	38.1	5	120
HCB	20.7	5	42
p,p'-DDE	25	13	77
PCBs	1,304	354	4,412
PCDD (FRG)	38	8	281
PCDF (FRG)	2.8	0.1	19
PCDD (Bad.Württberg)	70.5	23	195
PCDF (Bad.Württberg)	3.1	0.6	14

Looking for the Source

The distribution of PCDD/PCDF homologues was similar to the pattern of isomers found in commercial samples of PCP and PCP-Na (hexa- to octa dibenzodioxins and -furans). Only a few samples had patterns that would indicate sources generated by combustion processes. In 1986 the total amount of sewage sludge was 2.3-2.4 million tons dry weight with a total amount of 600 kg of PCDD/PCDF. Calculations of the total input of PCDD/PCDF from PCP and PCP-Na emissions - supposing that all PCDD/PCDF is concentrated in sewage sludge - would result in an average amount of 250 μg/kg (dry weight). Analyses shwoed average concentrations of 50 μg/kg in sewage sludge, being 20% of the total immission. Therefore, mathematically it is reasonable that pentachlorophenol and pentachlorophenol-Na are the source of the PCDD/PCDF levels. Especially high contaminations with PCDD/PCDF are due to industrial effluents.

Risk Assessment

The potential hazard from PCDD/PCDF in sewage sludge can be calculated as follows: Since the accident of Seveso 5 ng/kg TCDD equivalents
(TE) have been accepted as the upper limit in soils for agricultural use.
The German sludges averaged 202 ng/kg; agricultural application would
distribute 5 tons of sewage sludge (dry weight) per ha with an assmed
depth of 30 cm (density - 1.4). Thus, the PCDD/PCDF concentration would
be "diluted" by the factor of 840; resulting in a final concentration of
0,24 ng TE/kg in soil. As application of sewage sludge is allowed every
three years the limit of 5 ng/kg TE will be reached after 21 applications
(- 61 years; potential biodegradation, photolysis, etc. are not taken
into account). Uptake of PCDD/PCDF by plants via roots is not known to
occur. Although salad has been found to be contaminated with dioxins,
this was due to particulates. Although there appears to be little danger
that PCDD/PCDF will enter the food chain as a result of application of
sewage sludge to agricultural soils, it does not to seem reasonable to
spread the pollutants again after they have been concentrated in sewer
plants.

5.7.1 Air Sampling Methods

For analysis of soil, air, water, oil and sludge samples as well as for analysis of chemical residues from waste sites a new monitoring well, the "Münchehagen Sonde", was developed (Schreiner et al.). The construction of the well is given in Figure 8. With this device samples can be taken in hazardous waste without danger of spreading contaminated material. The hazardous waste is frozen immediately at the point of sampling and removed as frozen solid material. Therefore even highly contaminated waters, sludges, and oils may be handled without emission of pollutants. Liquid nitrogen is recommended as freezing agent since its inertness will hinder further reactions.

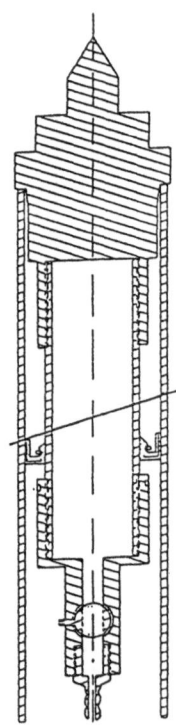

Fig. 8.   Monitoring and Sampling Well
            (μ 60/50 mm, length of filter: 1.0 m: width of slots: μ 3.0-
            0.2 mm)

5.7.2 Laboratory Leaching Experiments

Generally mobilization of organic pollutants occurs via water leachate (solubility, elution) and by volatilization of decomposition gases. Within the last few years, experiments on the elution of PCDD/PCDF adsorbed on fly ash have been performed. In batch-experiments it has been shown that detectable amounts of PCDD and PCDF (but no 2,3,7,8-Cl_4DD) can be eluted with distilled water within one day (Jager et al., 1986) using a method described in the German guideline DEV-S4 (Deutsche Einheitsverfahren zur Wasser-, Abwasser- und Schlammunter-suchung). The isomer distribution pattern is identical with water leachates from waste dumps. The findings suggest that the concentrations are not really due to dissolved PCDD/PCDF in water but to PCDD/PCDF adsorbed in microparticles (that have passed through the filter). In mixed solutions of water-waste oil and pure waste oil the same concentrations of tetra-CDD/CDF were found in the effluent, the values of Cl_8DD/Cl_8DF were elevated (due to contamination of waste oil with Cl_8DD/Cl_8DF). The extraction of PCDD/PCDF from fly ash depends on the organic solvents used (Thoma et al., 1986): < 1 % with hexane; 1-3 % with 80 % methanol/water; 40-60 % with toluene. Column experiments revealed leaching of PCDD/PCDF by toluene, hexane, and methanol/water mixtures. Small amounts of PCDD/PCDF were detected in the effluents of fly ash when treated with 1 N HCl, H_2SO_4/H_2O, and H_2O according to DEV S4 (Jager, 1987).

5.7.3 Analyses of Gaseous Emissions

Collection of low volatile substances from air is still a challenge to the analyst. The impinger is the oldest method, where the air passes through an appropriate solvent. Due to short contact times between air bubbles and liquid, recoveries are generally small. Better results are obtained by sorption on solid sorbents: air passes through a column filled with inorganic material, e.g. silica, or organic adsorbents (resins like XAD or polyurethane). Recoveries obtained are quite high but limitations concerning the air volume result in long sampling periods (for PCBs about 50 hours). In high-volume samplers (speed: 100 cm^3/hour) air is passed through a spiked glass fiber filter. The pollutants are adsorbed to the filter and afterwards extracted and analyzed (Eckrich, 1985).

For analysis of PCDD/PCDF in gaseous emissions of waste dumps special sampling methods have to be used. For concentrations in the range of 50-500 pg/m^3, 5-50 m^3 air have to be sampled. Figure 9 shows the sampling apparatus used to analyze the decomposition gas as well as the flue gas of a torch at a temperature of 900 °C. Clean-up of the samples was done according to EPA Method 613; analyses by GC/MS.

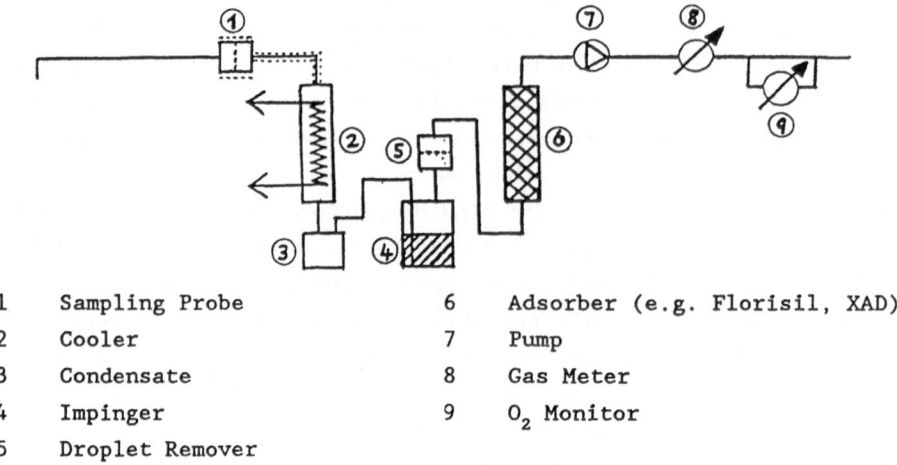

1	Sampling Probe	6	Adsorber (e.g. Florisil, XAD)
2	Cooler	7	Pump
3	Condensate	8	Gas Meter
4	Impinger	9	O_2 Monitor
5	Droplet Remover		

Fig.9.   Scheme of a Sampling Apparatus for Organic Pollutants

Analyses of the volatile compounds from the waste disposal site at Braunschweig confirmed the presence of PCBs in a concentration of 1-5 $\mu g/m^3$; HCH and HCB were found in the same range. PCDD/PCDF could not be detected within the decomposition gases (see Table 11).

REFERENCES

Behörde für Bezirksangelegenheiten, Naturschutz und Umweltgestaltung, July 12, 1984, Freie und Hansestadt Hamburg.

Bürgerschaft der Freien und Hansestadt Hamburg, 11. Wahlperiode, Drucksache 11/3374, February 26, 1985.

Deegan, J., Jr., 1987, Looking Back at Love Canal, Environ. Sci. Technol., 21:328.

Deegan, J., Jr., 1987, Looking Back at Love Canal, Environ. Sci. Technol., 21:421.

Der Hessische Minister für Arbeit, Umwelt und Soziales, im März 1985, Wiesbaden.

des Rosiers, P.E., 1987, National Dioxin Study, Chapter 3, U.S. Environmental Protection Agency (RD-681), Washington, D.C. 20460.

Deutsche Einheitsverfahren zur Wasser-, Abwasser- und Schlammuntersuchung.

Dr. Schreiner & Partner, Zetel 1.

Dynamit/Nobel, Troisdorf.

Eckrich, W., April 1985, Analytik von schwerflüchtigen chlororganischen Verbindungen, pp. 100, Seminar über Altlasten und kontaminierte Standorte - Erkundung und Sanierung, Ruhr-Universität Bochum.

EPA, 1985, Uptake and Depuration Studies of PCDDs and PCDFs in Freshwater Fish, EPA Environmental Research Laboratory, Duluth, Minnesota.

ERGO-Forschungsgesellschaft mbH, September 7, 1984.

Esposito, M.P., Tiernan, T.O., Dryden, F.E., 1980, Dioxins, EPA-600/2-80-197.

Freeman, R.A. and Schroy, J.M., 1988, Comparison of the rate of TCDD transport at Times Beach and Eglin AFB, Chemosphere, (in press).

Freie und Hansestadt Hamburg, Juni 1984. Internationales Symposium Deponie Georgswerder, Wortprotokoll, Hamburg.

Hagenmaier, H., Gutachtliche Stellungnahme, Untersuchung und Bewertung des Gefahrenpotentials dioxinhaltiger Abfälle auf der Sondermülldeponie Münchehagen.

Hagenmaier, H., Brunner, H., Knapp, W., and Weberruß, U., 1988, Untersuchung der Gehalte an polychlorierten Dibenzodioxinen, polychlorierten Dibenzofuranen und ausgewählten Chlorkohlenwasserstoffen in Klärschlämmen, Umweltforschungsplan des Bundesministers des Innern, Abfallwirtschaft, Forschungsbericht 103 03 305.

Hutzinger, O., and Fiedler, H., August 1988a, Emissions of Dioxins and Related Compounds from Combustion and Incineration Sources, NATO/CCMS Pilot Study on International Information Exchange on Dioxins and Related Compounds, Report Number 172, Bayreuth.

Hutzinger, O., and Fiedler, H., August 1988b, Formation of Dioxins and Related Compounds in Industrial Processes. NATO/CCMS Pilot Study on International Information Exchange on Dioxins and Related Compounds, Report Number 173, Bayreuth.

Jager, J., and Pflug, G., 1985, Freisetzung von Dioxinen bei der Ablagerung von Abfällen, in: "Abfallwirtschaft in Forschung und Praxis 14; Dioxine - Entstehung - Wirkungen - Beseitigung", pp. 59, FGU Berlin, ed., E. Schmidt, Berlin.

Jager, J., Pflug, G., Sünderhauf, W., 1986, Auslaugung und Deponierbarkeit von Flugaschen, Internationaler Recycling Congress, Berlin.

Jager, J., 1987, unpublished results. Source: Hagenmaier, H., Gutachtliche Stellungnahme, Untersuchung und Bewertung des Gefahrenpotentials dioxinhaltiger Abfälle auf der Sondermülldeponie Münchehagen.

Jürgens, H.-J., Roth, R., Schlesing, H., Fallstudie und Vorschläge zur Dekontamination des Geländes einer stillgelegten Herbizidfabrik, 1988, in: "Altlastensanierng '88", Bd. 2, pp. 1067-1073, K. Wolf, W.J. van den Brink, F.J. Colon, eds., Kluwer Academic Publishers, Dordrecht-Boston-London.

Lamparski, L.L., Nestrick, T.J., and Stenger, V.A., 1984, Presence of chlorodibenzo dioxins in a sealed 1933 sample of dried municipal sewage sludge, Chemosphere, 13:361.

Landratsamt Rhein-Neckar-Kreis, August 9, 1984, Heidelberg.

Landkreis Nienburg/Weser, 1987, Vorschlag für die Sanierung der SAD Münchehage, Geb. 20.

New York State Department of Health, September 1978, Love Canal, Public health time bomb., Albany.

Schumacher, E., 1985, Dioxine in der Deponie Georgswerder, Bekanntwerden, ·Untersuchungen, Gefährdungsabschätzung, in: "Abfallwirtschaft in Forschung Praxis 14; Dioxine - Entstehung - Wirkungen - Beseitigung", pp. 81, FGU Berlin, ed., E. Schmidt, Berlin.

Schumacher, E., 1987, Dioxin - Eine Technische, Analytische und Toxikologische Herausforderung, VDI Berichte 634, pp. 291, VDI Verlag, Düsseldorf.

Selenka, F., 1985, Gutachten, Department of Hygiene, University of Bochum, FRG.

Sievers, S. and Friesel, P., 1988, Soil contamination patterns of chlorinated compounds: Looking for the source, Chemosphere (in press).

Tabasaran, O., Ermittlung des Kontaminationspotentials eines Industriestandortes in der BR Deutschland, 1988, in: "Altlastensanierung '88", Bd. 2, pp. 999-1008, K. Wolf, W.J. van den Brink, F.J. Colon, eds., Kluwer Academic Publishers, Dordrecht-Boston-London.

Tabasaran, O., and Thomanetz, E., 1985, Wesentliche Untersuchungsergebnisse in Kurzfassung und Sanierungsvorschläge für die Sonderabfalldeponie Gerolsheim, in: "Abfallwirtschaft in Forschung und Praxis 14; Dioxine - Entstehung - Wirkungen - Beseitigung", pp. 71, FGU Berlin, ed., E. Schmidt, Berlin.

Technischer Überwachungsverein Berlin, 1985, Technischer Bericht Nr. D-85/534.

The Hazardous Waste Inspectorate, July 1986, Second Report: Hazardous Waste Management: "Ramshakle & Antediluvian"?, Department of the Environment, Welsh Office, Scottish Office, N. Ireland.

Thoma, H., Carsch, S., Hutzinger, O., 1986,Leaching of polychlorinated dibenzo-p-dioxins and dibenzofurans from municipal waste incinerator fly ash by water and organic solvents, Chemosphere, 15:1927.

Thomanetz, E., 1987, Vorgehensweise bei der Einschätzung von Altablagerungen und Erarbeitung von Sanierungskonzepten am Beispiel der Industriedeponie Karsau, Institut für Siedlungsbau, Wassergüte- und Abfallwirtschaft, Universität Stuttgart.

U.S. Environmental Protection Agency, November 28, 1983, Dioxin Strategy, Washington, D.C.

U.S. Environmental Protection Agency, April 1985, The National Dioxin Study - Guidelines for Review of 2,3,7,8-TCDD Data. Prepared by National Dioxin Study QTAG.

U.S. Environmental Protection Agency, April 1986, The National Dioxin Study, Tiers 3, 5, 6 and 7, Washington, D.C.

U.S. Environmental Protection Agency, January 1986, National Dioxin Strategy, Tier 1 and 2 Accomplishments. Draft Technical Support Document, Office of Solid Waste and Emergency Response, Washington, D.C.

Umweltbehörde Hamburg, Amt für Altlastensanierung, April 1988, Sanierung der Deponie Georgswerder.

Yanders, A.F., Kapila, S., and Schreiber, R.J., 1986, Dioxin: Field research opportunities at Times Beach, Missouri, in: "Chlorinated Dioxins and Dibenzofurans in Perspective", C. Rappe, G. Choudhary and L.H. Keith, eds., Chap. 17, pp 237-239, Lewis Publishers, Chelsea, Michigan.

Yanders, A.F., Orazio, C.E., Puri, R.K., and Kapila, S., 1988, On translocation of 2,3,7,8-Tetrachlorodibenzo-p-dioxin: Time dependent analysis at the Times Beach Experimental site, Chemosphere, (in press).

van Zorge, J.A. (Ministerie can Volkshuisvesting, Leidschendam), July 11, 1988, Personal Communication to H. Fiedler.

Vanna, N.J., 1983, The Love Canal: Issues and Problems, Chemosphere, 12:705.

Weerasinghe, N.C.A., Gress, U.L., and Lisk, P.J., 1985, Polychlorinated Dibenzodioxins and Dibenzofurans in Sewage Sludges, Chemosphere, 14:557.

Wolf, K., 1985, Dioxine in der Deponie Georgswerder - Geschichte der Deponie, Sanierungsalternativen, in: "Abfallwirtschaft in Forschung Praxis 14; Dioxine - Entstehung - Wirkungen - Beseitigung", pp. 85, FGU Berlin, ed., E. Schmidt, Berlin.

CHAPTER 4 - MANAGEMENT OF ACCIDENTS WORKING GROUP

Alessandro di Domenico* and A. Essam Radwan**

*Laboratorio di Tossicologia Comparata ed Ecotossicologia
Istituto Superiore di Sanita
00161 Rome, Italy

**Department of Civil Engineering, Arizona State University
Tempe, Arizona 85287

This chapter contains some of the results of Pilot Study "Dioxin Problems, a project concerning the international exchange of information on polychlorinated dibenzo-p-dioxins (PCDDs), dibenzofurans (PCDFs), and related chemicals. The project was undertaken by the Committee on the Challenges of Modern Society (CCMS) of the North Atlantic Treaty Organization (NATO). In addition to the major objective, secondary goals included assistance in identification of knowledge voids and reduction of research program duplication. Several NATO-member nations participated in the project. They were: Canada, Denmark, the Federal Republic of Germany, Italy, the Netherlands, Norway, the United Kingdom, and the United States. Other participants were international organizations as well as representatives of industrial trade associations and nongovernmental environmental groups.

Presented in this chapter are the papers delivered by the experts of Working Group on "Management of Accidents" at the NATO/CCMS Plenary Meeting held in Como, Italy, July 13-14, 1987. The Working Group is one of the three Working Groups acting within the framework of NATO/CCMS Pilot Study "Dioxin Problems" whose aim is the international exchange of information on PCDDs, PCDFs, and related compounds. Most papers stem from a common background, i.e., the renowned industrial accident occurred at the ICMESA chemical plant near Seveso (Milan, Italy) on July 10, 1976. The accident resulted in the release of 2,3,7,8-tetrachlorodibenzo-p-dioxin (2,3,7,8-TCDD) and other dangerous chemicals into the surroundings, thereafter marked by severe contamination. With reference to the actions taken at Seveso to limit man's exposure to TCDD and detoxicate the environment, the papers describe various aspects of management of accidents, that is: contingency planning, immediate emergency response, long-term rehabilitation, and exchange of information.

Much of the material presented in this report was delivered in an experimental form during the July 13-14, 1987, Plenary Meeting of NATO/CCMS Pilot Study "Dioxin Problems". The Meeting was hosted by the Centro di Cultura Scientifica "Alessandro Volta" (Como, Italy) which is supported and sponsored by:

- Provincial Administration of Como;
- Chamber of Commerce of Como;
- Municipality of Como;
- Polytechnic of Milan;
- University of Milan;
- University of Pavia.

The editors wish to indicate here their sincere gratitude and deep appreciation for the above administrative bodies for having provided a significant aid in realizing the Meeting. Gratitude is also expressed for the Italian, and Como, authorities and, specifically, for Dr. Antonio Spallino, President of "Alessandro Volta" Center. The editors are also indebted to the Istituto Nazionale di Fisica Nucleare, Section of Pavia (Pavia), and Istituto Superiore di Sanita (Rome) for their support of the Meeting.

On the occasion, the Istituto Superiore di Sanita provided also essential secretarial assistance: for that, sincere thanks are expressed to Miss Raffaella Cornacchini and Mrs. Paola Tacchi-Venturi of the Secretariat for Cultural Affairs. The kind collaboration in manuscript proofing and revising of Miss M. Celeste Schina and Miss Laura Moreschi is also acknowledged; both are from the Laboratory of Comparative Toxicology and Ecotoxicology. The editors are gratefully indebted to the Editorial Service of the same Institute for having kindly supplied help and facilities to publish the manuscript.

Last, Prof. Francesco Pocchiari, Director General of the Istituto Superiore di Sanita when the report was written, is thanked for approving of and sustaining the NATO/CCMS project and this publication. Most sorrowfully, Prof. Pocchiari died unexpectedly on January 2, 1989.

1. INTRODUCTORY COMMENT ON THE SEVESO (ITALY) INDUSTRIAL ACCIDENT

1.1 PCDDs and PCDFs

PCDDs and PCDFs are two series of tricyclic, almost planar, aromatic compounds that exhibit similar physical, chemical, and biological properties. The chemical structures and numbering of these hazardous compounds are given in Figure 1. The number of chlorine atoms in these compounds can vary between one and eight to produce up to 75 PCDD and 135 PCDF positional isomers (Table 1). The more chlorinated PCDDs and PCDFs exhibit good chemical stability, strong lipophilic character, long environmental persistence, and tendency to bioaccumulate. They can be photodegraded by sunlight.

Table 1. PCDD and PCDF Positional Isomers

Chloro-substitution degree	Number of isomers			
	PCDDs		PCDFs	
	No.	Acronym	No.	Acronym
Monochloro-	2	M_1CDD	4	M_1CDF
Dichloro-	10	D_2CDD	16	D_2CDF
Trichloro-	14	T_3CDD	28	T_3CDF
Tetrachloro-	22	T_4CDD	38	T_4CDF
Pentachloro-	14	P_5CDD	28	P_5CDF
Hexachloro-	10	H_6CDD	16	H_6CDF
Heptachloro-	2	H_7CDD	4	H_7CDF
Octachloro	1	O_8CDD	1	O_8CDF
TOTAL	75		135	

Animal studies and *in vitro* experiments have indicated that there are pronounced differences in toxic and biologic effects among the different PCDD and PCDF isomers. The isomers with the highest activity, acute toxicity and greatest carcinogenic potency are those having 4-6 chlorine atoms and all lateral (C2, C3, C7, and C8) positions substituted with chlorine (the "dirty dozen"; Table 2). 2,3,7,8-Tetrachlorodibenzo-*p*-

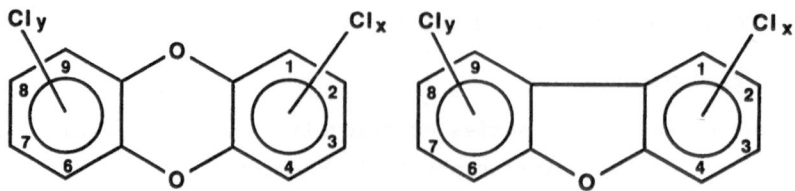

Figure 1. PCDD and PCDF Structures

dioxin (often called "dioxin" or TCDD) is the most toxic compound of both series and one of the most toxic man-made chemicals.

Due to the large number of congeners and the large variation in toxicity and biological potency among closely related congeners, it is generally accepted that risk evaluation for PCDD and PCDF mixtures be based especially on the levels of those isomers which - within their own isomer group - exhibit the highest toxicity. Chapter 2.5 of this book describes the International Toxicity Equivalency Factor (I-TEF) method for estimating risks associated with the complex mixtures of the 210 PCDD and PCDF compounds. The procedure assigns factors to the most toxic congeners, the 2,3,7,8-substituted variety, that produces an equivalent amount of 2,3,7,8-TCDD.

PCDDs and PCDFs have been found in a variety of matrices, such as: technical products, the emissions from incinerators of various types and other combustion sources, and many different environmental samples. They are sustantially ubiquitous (Bumb et al., 1980; Junk and Ford, 1980; Loustenhouwer et al., 1980; Eitzer and Hites, 1987). Low levels of PCDDs and PCDFs have been detected in pyrolysis experiments of organochlorine polymers: a similarity in congener pattern between extracts from these experiments and extracts from various incinerators is generally observed. Trace levels of the most toxic 2,3,7,8-substituted PCDDs and PCDFs have been identified in environmental samples, among which aquatic organisms and river and lake sediments (Czuczwa et al., 1984). Low background levels have also been found in human samples including adipose tissue and

Table 2. 2,3,7,8-Chlorosubstituted PCDDs and PCDFs.

PCDDs	PCDFs
The "dirty dozen":	
$2,3,7,8-T_4CDD$	$2,3,7,8-T_4CDF$
$1,2,3.7.8-P_5CDD$	$1,2,3,7,8-P_5CDF$
	$2,3,4,7,8-P_5CDF$
$1,2,3,4,7,8-H_6CDD$	$1,2,3,4,7,8-H_6CDF$
$1,2,3,6,7,8-H_6CDD$	$1,2,3,6,7,8-H_6CDF$
$1,2,3,7,8,9-H_6CDD$	$1,2,3,7,8,9-H_6CDF$
	$2,3,4,6,7,8-H_6CDF$
Additional important isomers:	
$1,2,3,4,6,7,8-H_7CDD$	$1,2,3,4,6,7;8-H_7CDF$
	$1,2,3,4,7,8,9-H_7CDF$
$1,2,3,4,6,7,8,9-O_8CDD$	$1,2,3,4,6,7,8,9-O_8CDF$

breast milk from various countries. The same congeners are generally found in all human samples (Rappe, 1986).

The $1,2,3,7,8-P_5CDD$ is detected in human samples as well as in environmental samples. As it is always found in samples from various incinerators, its use as a marker may be suggested: in fact, a correlation between various incineration sources and the observed background levels cannot be excluded.

1.2   The ICMESA Accident

On July 10, 1976, a runaway exothermic process started in the reaction batch of a 2,4,5-trichlorophenol (TCP) synthesis process at the Givaudan-Hoffmann-LaRoche ICMESA plant at Meda, 30 km north of Milan, Italy. The temperature within the chemical reactor rose far above 200°C, thereby producing a large amount of TCDD, normally present in TCP and derivatives at low ppb (10^{-9} g/g) levels. When the reactor safety valve gave way due to an increase in the internal pressure, a fluid mixture containing reactive chemicals and a yet-unquantitated amount of TCDD burst through the valve port high into the open air. Within a short time of the accident, chemicals settled to the ground or were dispersed by wind streams. A detailed account of this phase of the accident is given in Section 5.1.

Vegetation near the ICMESA plant, courtyard animals, and birds were seriously affected. Many animals died within a few days. At the same time, dermal lesions began to appear on human beings who had been exposed to the toxic alkaline cloud. Approximately 10 days after the accident, it became certain that TCDD was present in various types of samples collected near the ICMESA plant. Several difficulties had to be overcome in order to define emergency and post-emergency strategies for protection of the population and the environment. Some of these difficulties were that (1) initial TCDD distribution levels in the soil surface were seen to be highly uneven even within very short radii; (2) a number of constraints depending on sampling, extraction, and determination of the toxicant prevented its immediate extensive, rapid, and accurate monitoring; (3) no fully comprehensive analytical methods were available at the time to assay TCDD in some substrata, especially on such a tight routine basis as demanded; and (4) there were problems of coordinating national, regional, and municipal authorities.

For a full account of the event and its aftermath, we refer the reader to the comprehensive articles by Hay (1977), Homberger et al. (1979), Silano (1981), and Pocchiari et al. (1983). For quick reference, we shall recall that, shortly after the accident, the affected area was sub-divided into Zones A, B, and R, in descending TCDD and toxicologic risk levels (Figure 2; di Domenico et al., 1980). It may be useful to stress that Zone A (approximately 110 ha, over 730 inhabitants; Figure 3) was evacuated soon after the accident, whereas Zones B (270 ha) and R (1430 ha) were subjugated to area-specific hygiene regulations. These and other facts are schematically reported in Table 3 built up around a somewhat arbitrary subdivision of the time span elapsed since the accident (Cattabeni et al., 1986).

Other TCDD isomers and homologues were found in the Seveso area, but TCDD was by far the one raising the most concern and in general present in higher levels.

1.3   The Present Status

Due to the specific arguments dealt with in this chapter, additional pertinent information will be provided by the many contributions comple-menting this one and following hereafter. However, it should be pointed out that the technological aspects (land reclamation, dismantling of ICMESA plant production department, etc.) were probably the most, and perhaps main successful outcomes of the entire Seveso experience: for details, the reader is addressed to the Proceedings of the International Symposium on Technological Response to Chemical Pollutions held in Milan (Italy), September 20-22, 1984 (Lombardy Region, 1985).

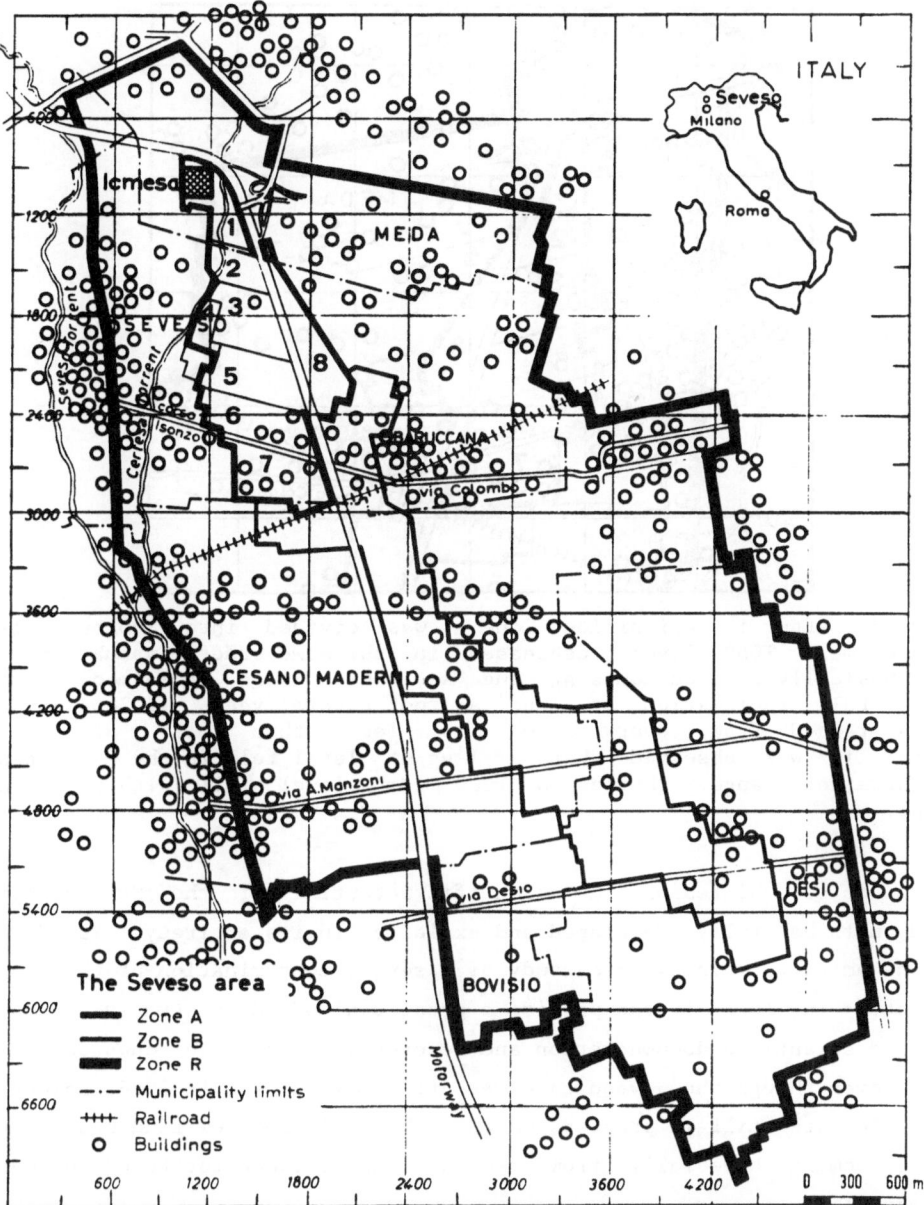

Figure 2. Zones A, B, and R at their maximum extension, showing major built-up areas (O) and surrounding farm lands. Different risk areas were defined on the basis of extensive and repeated soil assay. For that, thousands of 7-cm-thick topsoil samples were collected with 7-cm-diameter cylindrical steel samplers and analyzed. As TCDD was mostly distributed in the topsoil layer, contamination levels were expressed in $\mu g/m^2$ or ng/m^2 (detection threshold, better than 0.75 $\mu g/m^2$). The ICMESA plant appears within Meda municipality boundaries near the Meda-Seveso borderline. In spite of its actual location, the accident has generally been named after Seveso as most of Zone A (where evacuation took place) lay within the Seveso municipal boundaries. In the picture, Zone B comes within the Cesano Maderno and Desio municipality boundaries. The reference grid is oriented north-south.

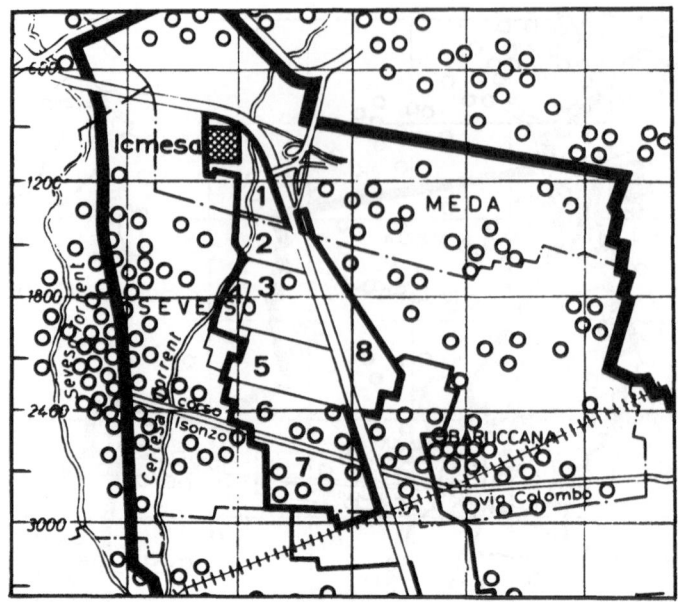

Figure 3. After its definition, Zone A was divided into Subzones 1-8. Subzone mean TCDD levels decreased in the same order. Evacuation took place basically from Subzones A_6 and A_7, which were reclaimed shortly after. Full reclamation of Zone A took several years. In Zone A, TCDD values ranged from over $2 \cdot 10^4$ to less than 0.75 $\mu g/m^2$. In general it was observed that a 1-$\mu g/m^2$ level fell in the 2- to 8-ppt concentration range, whereas a 1000-$\mu g/m^2$ level fell within 5 and 15 ppb.

Due to several reasons the scientific potential of the Seveso event could not be fully developed and exploited in its entirety. Yet Seveso remains an extraordinary case-study of a severe contamination of the environment: in fact it can provide a great deal of administrative, technical, and scientific documentation and records from the onset onwards. Indeed, even though the Seveso case was officially closed in 1986, scientific studies are still in progress based on both old and new findings - the latter coming especially from medical surveillance for tumor incidence and mortality.

Looking back at the ICMESA accident, one has the impression that the event triggered a number of long-overdue environmental actions at both the national and European level. This will be dealt with in more detail in the next papers. At any rate, the specific knowledge determined in the technological, scientific, and medical sectors by the Seveso event - as well as by other severe cases of environmental contamination due to TCDD in different parts of the world - has produced much awareness of the risks associated with exposures to PCDDs and PCDFs.

Table 3. Timeline of Events Following Seveso Accident

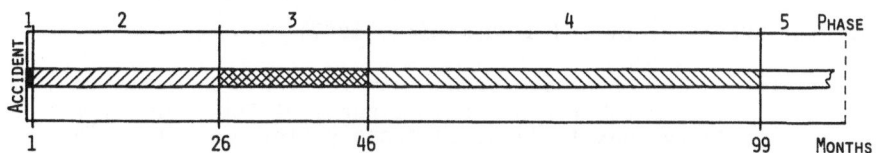

PHASE 1
JUL 10 - AUG 10, '76

A) FIRST TERRITORIAL MAPPING OF THE AREA HIT BY THE CHEMICAL CLOUD
   (AUGUST 1976 MAP);
B) DEFINITION OF SUBAREAS (ZONES A, B AND R) AT DIFFERENT (DESCEND-
   ING) TOXICOLOGIC RISK LEVELS;
C) EVACUATION OF ZONE A;
D) SETTING OF HYGIENE REGULATIONS FOR ZONE B;
E) SETTING OF HYGIENE REGULATIONS FOR ZONE R.

PHASE 2
AUG 11, '76 - SEP '78

A) START OF MICROANALYTICAL ENVIRONMENTAL MONITORING OF AREA;
B) START OF BLANKET IN-DEPTH MEDICAL SURVEILLANCE OF ALL POPULA-
   TION GROUPS AT RISK;
C) SECOND AND THIRD MAPPING OF ZONE A (SEPTEMBER 1976 AND JANUARY
   1977 MAPS), FOURTH MAPPING OF SUBZONES A1, A2, A3, AND A4 (MARCH
   1978 MAP);
D) SECOND MAPPING OF ZONES B AND R (SEPTEMBER 1977 MAP);
E) DETOXICATION OF SUBZONES A6, A7, AND A8, AND SUBJUGATION TO HY-
   GIENE REGULATIONS ALREADY ADOPTED FOR ZONE B;
F) RETURN OF EVACUATED POPULATION TO THEIR DETOXICATED HOMES IN
   SUBZONES A6 AND A7.

PHASE 3
OCT '78 - APR '80

A) CONTINUATION OF MICROANALYTICAL ENVIRONMENTAL MONITORING;
B) CONTINUATION OF BLANKET IN-DEPTH MEDICAL SURVEILLANCE;
C) DETOXICATION OPERATIONS IN SOME AREAS OF ZONES B AND R.

PHASE 4
MAY '80 - SEP '84

A) CONTINUATION OF MICROANALYTICAL ENVIRONMENTAL MONITORING;
B) MEDICAL SURVEILLANCE RESTRICTED TO SPECIFIC GROUPS (DETOXICA-
   TION WORKERS AND CHLORACNE CASES) AT RISK SINCE JULY 1982, END
   OF MALFORMATION REGISTRY, CONTINUATION OF TUMOUR INCIDENCE AND
   MORTALITY REGISTRY FOR INITIALLY MONITORED POPULATION;
C) COMPLETE REHABILITATION OF ZONE R, RESCINDING OF PREVIOUS RE-
   STRICTIONS;
D) DETOXICATION OF SUBZONES A1, A2, A3, A4 AND A5 TO BE SET UNDER
   APPROPRIATE HYGIENE REGULATIONS.

PHASE 5
OCT '84 - ?

A) CONTINUATION OF MICROANALYTICAL ENVIRONMENTAL MONITORING;
B) MEDICAL SURVEILLANCE TO BE CONTINUED UNTIL END OF 1997 FOR TUMOR
   INCIDENCE AND MORTALITY;
C) THIRD MAPPING OF ZONE B;
D) COMPLETE REHABILITATION OF ZONE B (PROJECTION);
E) COMPLETE REHABILITATION OF SUBZONES A6, A7, AND A8 (PROJECTION);
F) REHABILITATION OF SUBZONES A1, A2, A3, A4, AND A5, TO BE UTILIZED
   AS PUBLIC PARK ONLY (PROJECTION).

637

At least in Italy such awareness is at the basis of the present criteria for management of risks associated with exposure to PCDD and PCDF mixtures. This is visible for instance in Table 4 where the maximum tolerable environmental limits recently indicated by the National Toxicologic Committee are reported. Limits were set as a result of detecting PCDDs and PCDFs in soils exposed to fallout from a municipal solid waste incinerator (Berlincioni and di Domenico, 1987).

Table 4. Action Levels or Maximum Tolerable Environmental Limits[a] for Complex Mixtures of PCDDs and PCDFs (Values in "$2,3,7,8$-T_4CDD Equivalents" Units[b])

Matrix or environment	Limit[c]
1. <u>Air</u>	
1.1. General environment	40×10^{-15} g/m^3
1.2. Occupational environment	120×10^{-15} g/m^3
2. <u>Water</u>	
2.1. Surface, ground-, marine, and potable water	50×10^{-15} g/l
2.2. Municipal wastewaters after treatment	500×10^{-15} g/l
2:3. Municipal wastewaters before treatment	50×10^{-12} g/l
2.4. Industrial wastewaters after treatment	500×10^{-15} g/l
2.5. Industrial leachates, wastes	$1,000 \times 10^{-9}$ g/kg
3. <u>Soil</u>	
3.1. Farming soil	10×10^{-12} g/g
3.2. Nonfarming soil	50×10^{-12} g/g
3.3. Industrial soil	250×10^{-12} g/g
4. <u>Surfaces</u>	
4.1. Exterior walls	75×10^{-9} g/m^2
4.2. Interior walls, surfaces	25×10^{-9} g/m^2

[a] Defined on the basis of an ADI ≤ 10 pg/day·kg bw, and under the assumption that PCDDs and PCDFs be associated with nonspecific fallout from combustion sources.

[b] $2,3,7,8$-T_4CDD Toxicity Equivalency Factors as per USEPA, 1987.

[c] For proper comparison with a given limit, an assessed environmental level should be the mean of several independent determinations carried out within a short length of time. An assessed level would differ from pertaining limit only if difference is greater than analytical uncertainty.

REFERENCES

Eitzer, B.D., and Hites, R.A. 1987: Dioxins and Furans in the Ambient Atmosphere: A Baseline Study. Presented at the Seventh International Symposium on Chlorinated Dioxins and Related Compounds "Dioxin '87", October 4-9, Las Vegas, Nevada (US).

Hay, A.W.M. 1977: Tetrachlorodibenzo-p-dioxin release at Seveso. Disasters 1, 289-308.

Homberger, E., Reggiani, G., Sambeth, J., and Wipf, H.K. 1979: The Seveso Accident: Its Nature, Extent and Consequences. Annals of Occupational Hygiene 22, 327-367.

Junk, G.A., and Ford, C.S. 1980: A Review of Organic Emissions from Selected Combustion Processes. Chemosphere 9, 187-230.

Lombardy Region 1985: Technological Response to Chemical Pollutions, Ufficio Speciale di Seveso della Regione Lombardia, Ed., Centro Stampa Litho Gamas, Seregno (Milan, Italy).

Loustenhouwer, J.W.A., Olie, K., and Hutzinger, O. 1980: Chlorinated Dibenzo-p-dioxins and Related Compounds in Incinerator Effluents: A Review of Measurements and Mechanisms of Formation. Chemosphere 9, 501-522.

Pocchiari, F., di Domenico, A., Silano, V., and Zapponi, G. 1983: Environmental Impact of the Accidental Release of Tetrachlorodibenzo-p-dioxin (TCDD) at Seveso (Italy). In: Accidental Exposure to Dioxin - Human Health Aspects, pp. 5-37, Coulston, F., and Pocchiari, F., Eds., Academic Press, New York (NY).

Rappe, C. 1986: Polychlorinated Dioxins (PCDDs) and Dibenzofurans (PCDFs) - Occurrence, Environmental Levels, and Formation in Thermal Processes. Report prepared for APME Meeting, June 2-3, Brussels (Belgium).

Silano, V. 1981: Case Study: Accidental Release of 2,3,7,8-Tetrachlorodibenzo-p-dioxin (TCDD) at Seveso, Italy. In: Emergency Response to Chemical Accidents, Interim Document 1, pp. 167-203, World Health Organization, International Programme on Chemical Safety, Copenhagen (Denmark).

U.S. EPA 1987: Interim Procedures for Estimating Risks Associated with Exposures to Mixtures of Chlorinated Dibenzo-p-dioxins and Dibenzofurans (CDDs and CDFs). EPA/625/3-87/012, Risk Assessment Forum, US Environmental Protection Agency, Washington, DC (USA).

2.   CONTINGENCY PLANNING

2.1   <u>Potential PCDD and PCDF Sources in Italy: Survey, Chemical
        Analysis, Legislative, and Future Needs</u>

Domenico Brocco **and** Paolo Ciccioli

Istituto Inquinamento Atmosferico, CNR
00016 Monterotondo Stazione (Rome), Italy

Franco Merli **and** Giuseppe Viviano

Laboratorio di Igiene Ambientale
Istituto Superiore di Sanita
00161 Rome, Italy

2.1.1   <u>Abstract</u>

In recent years polychlorinated dibenzo-*p*-dioxins (PCDDs), dibenzofurans (PCDFs), and related compounds resulting from combustion processes and several type of accidents (e.g., transformer fires, chemical
plant fires, etc.) have received much scientific, political, and social
attention. In this report, the different and potential sources of the
above-mentioned compounds, and the Italian initiatives and regulations
regarding this matter are described.

2.1.2   <u>Introduction</u>

To assess the risk associated with contamination due to a given
pollutant, three fundamental steps are required: identification of the
anthropogenic activities that can act as potential sources, determination
of their emission power, and evaluation of the impact on human health and
the environment. Successful achievement of these tasks implies that:
(1) all physical-chemical processes leading to the production, transformation, and dispersion of the pollutant into the environment be understood;
(2) the various pathways through which the compound penetrates into the
human body identified; (3) toxicity levels assessed; (4) dose-response
effects understood, and (5) analytical techniques available for a precise
and accurate evaluation.

In spite of the outstanding progresses made since the Seveso accident, some of the basic information on PCDDs and PCDFs is still missing. Others are a matter of controversy. Because of these uncertainties, any actions taken to prevent contamination from these pollutants vary from country to country. Differences arise not only from the interpretation of the existing data, but also from the fact that environmental management is handled by different institutions, depending on the political and social organization of the country considered. This aspect is extremely important when laws regulate pollution problems.

The aim of this work is to briefly review the initiatives that have been taken in Italy to prevent PCDD and PCDF contamination of the environment. In particular, the efforts made to survey gaseous emissions on a national basis and elaborate suitable methodologies for the sampling and analysis of these pollutants in combustion and industrial sources are presented. The legislation regulating this matter is also discussed. Present gaps are identified and future needs highlighted.

2.1.3   Emission Survey and Present Legislation

The identification of potential areas vulnerable to PCDD and PCDF contamination requires a detailed knowledge of the geographic distribution over the whole territory of mobile and stationary sources and their potential emission power. The best present knowledge of the anthropogenic activities that are potential sources for PCDDs and PCDFs is presented in Table 1.

Sources are classified in two main classes: combustion sources, and industrial processes where PCDDs and PCDFs can be formed as impurities during the synthesis of a main product. The former have been further classified into continuous and accidental sources. The main difference between discontinuous and continuous sources is that, in the former case, the impact is unpredictable in terms of amount, duration, and transport, so that it is very difficult to quantify the risk for the population.

Vehicular emissions and various combustion and incineration processes are among the major continuous sources. Coal is not widely used either for power generation, or industrial and domestic heating, and wood burning is not very common in our country. Other activities listed in Table 1 have little impact either because of the low emission rates or for the restricted number of sources present in the whole territory.

Table 1. Potential Sources of PCDDs/PCDFs and Related Compounds.

COMBUSTION SOURCES

Continuous	Accidental
Incinerators of waste	Forest fires
Municipal	Plants fires
Industrial	Landfills fires
Hospital	
Sludge	
Motor vehicles (gasoline, diesel)	
Cigarettes	
Coal preparation plants	
Power plants (coal and peat fuels)	
Hog fuel burners	
Wire reclamation	
Railroad tie disposal	
Open-pit wood-waste burning	
Woodburning stoves	
Steel mills	
Waste burning	

CHEMICALS
(Technical products and wastes)

1)	PCBs	2)	Tetrachlorophenol
3)	2,4,5-T	4)	Trichlorophenol
5)	2,4-D	6)	Hexachlorophene
7)	Pentachlorophenol	8)	Dicamba
9)	Diphenyl ether herbicides	10)	Chlorobenzenes
11)	Related halogenated compounds		

Although PCDDs and PCDFs have been detected in the emissions of motor vehicles (Ballschmiter et al., 1986), measurements carried out in the urban area of Rome, where heavy traffic develops during rush hours, did not provide evidence that these compounds were present in urban dust at significant levels. Negative results were also obtained when diesel soot, emitted by a car run according to the European Cycle, was analyzed (Ciccioli et al., 1986). At present, there is not enough evidence that motor vehicles are a major source of PCDDs and PCDFs in Italy, and attention is mostly devoted to waste incineration processes.

The other category of potentially hazardous sources (accidental) includes discontinuous processes. They become active only when accidental fires or uncontrolled reactions occur in open air or in industrial plants, respectively. Although electric transformer fires and waste burning in uncontrolled landfills are potentially hazardous, it is unlikely that the amount of pollutants emitted is as high as that eventually formed by chemical plants where syntheses of chlorinated pesticides, herbicides, and in general organochlorine compounds are carried out. Therefore, industrial plants which manufacture such chemicals are the sources to be monitored.

To reduce their number somewhat, identification of plants bearing a greater potential for generation of PCDDs and PCDFs is necessary. This may be done through an extensive survey of the industrial activities carried out over the territory in order to reinforce controls and plant security measures for population and environmental protection in case of accidents.

Enactement of directives issued by the European Economic Commission (EEC) in 1982, No. 501, and 1984, No. 631, both aimed at preventing accidents such as the one in Seveso, should help in identifying areas subject to risk from discontinuous sources. In particular, directive No. 501 was discussed by the Italian Parliament on February 18, 1988, and likely it will be converted into a national law. According to the provisional text, it will be possible to have a complete list of industrial plants producing potentially hazardous compounds and, among them, those which can be a source of PCDDs and PCDFs. The survey of discontinuous sources will be carried out by the Ministries of Industry, Health, and the Environment. Data will be sent to the Department for Civil Protection which is in charge of handling specific emergency situations and checking whether safety measures taken by the industry meet the requirements of the law. Based on the results of this investigation, it should be possible to make specific plans aimed at preventing accidents from discontinuous sources. In any case, it may be assumed that beneficial effects of this future Italian law will be obtained within no less than three or four years.

National regulations (see Table 2 and 3; di Domenico and Pocchiari, 1986) concerning environmental pollution, and also the one dealing with waste disposal (No. 915 of 1982), have made the collection of data on incineration plants feasible. Information on these continuous sources was and is considered essential to evaluate the population risk of exposure to PCDDs and PCDFs.

Surveys of municipal waste incinerators have been carried out by the Ministry of Health (Giannico and Seller, 1982), private companies (Eco Consulting, 1986), nonprofit environmental institutions (Federambiente, 1984), and research groups (Viviano and Ziemacki, 1987). The results of these studies are fairly consistent. An example is reported in Figure 1,

Table 2. Laws and Regulations Concerning Industrial
Safety and Environmental and Occupational
Health Before and After Seveso

Object	Period
Use of toxic gases	1927
High-pressure Machinery	1927
Production/Storage of Hazardous Materials	1931
Hazardous Industries	1934
Prevention of Industrial Fires	1942, 1961°, 1965°
Dangerous Manufacturings	1955, 1959°
Occupational Environment	1956
Prevention of Atmospheric Pollution	1966, 1971°
Prevention of Water Pollution	1976
Classification, Packaging, and Labeling of Hazardous Industrial Materials	1974 1981°
Management and Disposal of Toxic and Dangerous Wastes	1982 1984°
"Seveso Directive" Concerning Prevention of Major Accidents Resulting from Certain Industrial Activities (ECC)*	1982
OECD Council Directive on Transfrontier Movements of Hazardous Waste (EEC)	1984

* Not yet enacted.
° Revisions or technical regulations.

Table 3. Some "Reference Levels" for PCDDs and PCDFs

(A) PCDDs and PCDFs in Wastes:

$1,2,3,6,7,8-H_6CDD$ or
$1,2,3,7,8,9-H_6CDD$ or
$1,2,3,7,8-P_5CDD$ or
$2,3,7,8-T_4CDD$ or
$2,3,7,8-T_4CDF$ > 1 ppb "Toxic"

PCDD or PCDF   > 500 ppb "Toxic"

(B) $2,3,7,8-T_4CDD$ in Pesticides (Active Principles) $\leq$ 1 ppb

(C) PCDDs and PCDFs in the Environment:

	$2,3,7,8-T_4CDD$	PCDDs + PCDFs
(1) Soil, Crop Growth Permitted	$\leq$ 6 ppt*	≤ 0.75 $\mu g/m^2$
(2) Soil, Crop Growth Forbidden	$\leq$ 40 ppt**	≤ 5 $\mu g/m^2$
(3) Outdoor Building Surfaces	$\leq$ 0.75 $\mu g/m^2$	≤ 0.75 $\mu g/m^2$
(4) Indoor Building Surfaces	$\leq$ 0.01 $\mu g/m^2$	≤ 1 $\mu g/m^2$
	As established by Lombardy Region at Seveso	As indicated by National Toxicology Commission

* Equivalent to 0.75 $\mu g/m^2$.
** Equivalent to 5 $\mu g/m^2$.

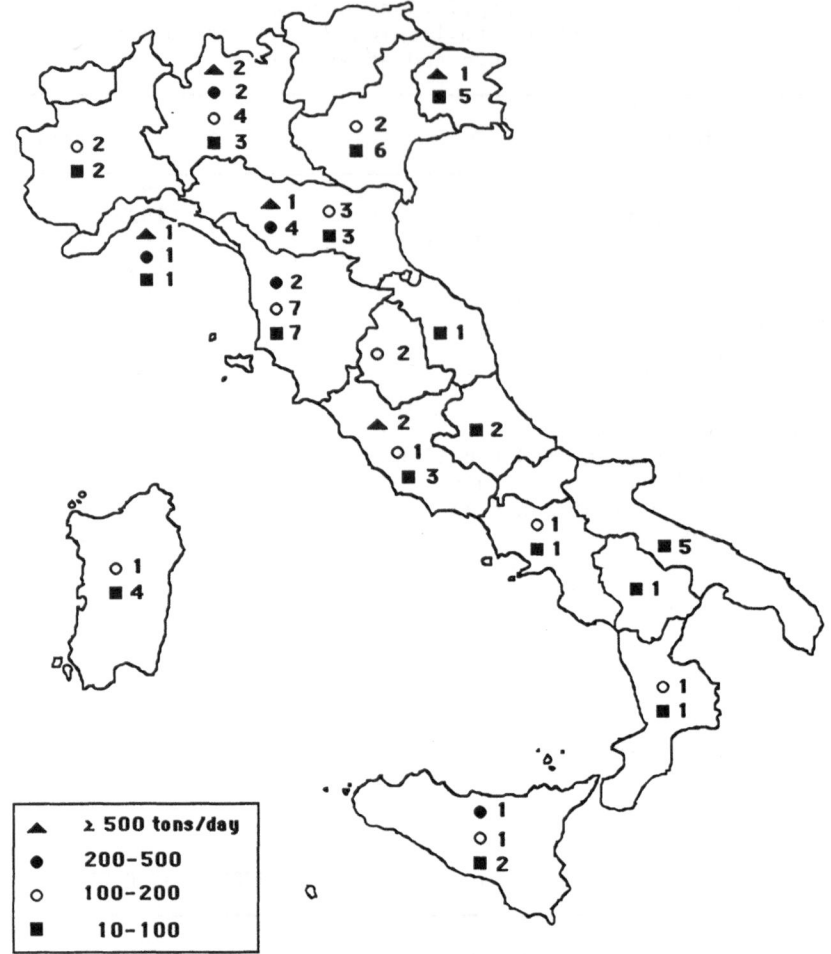

Figure 1. Waste Incineration Plants in Italy (1985)

where municipal waste incinerators were classified according to their capacity and grouped by regions. The reason for this choice is that responsibility for environmental management is attributed to regional and local authorities. Furthermore, regional authorities have the task of defining limits for PCDDs and PCDFs released into the environment.

Various studies indicate that incinerated waste material is approximately 20 percent of the total amount produced in Italy (15 million metric tons per year). This amount corresponds to an average production of 0.8 kg of waste material per person. However, the results shown in Figure 1 need to be periodically revised, as many municipal waste incineration plants are or have been shut down in the last two years because of difficulties in meeting law requirements (i.e., the use of afterburners to reach 1,200 °C when the content of organic chlorine in the waste exceeds 2 percent), or because of strong opposition of the population to incineration plants. In addition, the survey needs to be completed by including industrial incinerators, and incinerators used in small communities (hospitals, airports, etc.). For this reason, a working group was established at the Istituto Superiore di Sanita (Rome) in December 1986. The group is made up of scientists working for various research institutions, industries, and universities; experts coming from different regions are also present.

In order to begin work it was first necessary to prepare a special questionnaire to obtain the technical information on the incineration plant (Merli et al., 1988). As can be seen from Figures 2, 3, and 4, questions were asked about type, location, and potentiality of the plant, as well as working conditions of the furnace, abatement devices used, type of waste material treated, and the final destination of slags, ashes, wastewaters, and sludges. Formats were sent to managers responsible for waste incineration plants. Data received until December 1987 are shown in Figure 5. After completion of the survey, it will be possible to identify the areas where contamination from PCDDs and PCDFs is expected.

2.1.4   Sampling and Chemical Analysis

In order to obtain information on the total amount of PCDDs and PCDFs released in the areas at risk, it is extremely important to know how formation of such pollutants changes with working conditions of the plant, and the type and amount of waste material burned. This entails that the methods used for sampling, fractionation, identification, and quantification of PCDDs and PCDFs present in the emission be accurate, sensitive, and reproducible. Due to the very small amount of PCDDs and PCDFs and the different toxicity of congeners, it is also important that the most toxic compounds of the two families be separated and quantified using isomer-specific techniques. Although the Italian law on waste disposal states that chlorinated pollutants, and particularly PCDDs and PCDFs, must be measured regularly during the year, no mention is made of the procedures to be followed for sampling and analysis. For this reason, operational units in charge of pollution control have great difficulties in accomplishing their task.

To overcome this problem a Committee was formed within the framework of the Associazione per l'Unificazione nel Settore dell'Industria Chimica (UNICHIM, Milan), the Italian affiliate of the International Organization for Standardization (ISO). The Committee is made of chemists with specific experience in sampling and analysis of PCDDs and PCDFs and other

INCINERATOR PLANT SURVEY FORM

"ISTITUTO SUPERIORE DI SANITA' - TASK GROUP"

(December 1987)   # ☐☐☐☐☐ ☐ *

LOCATION OF THE PLANT

Region .... ☐☐

Province .... ☐☐

Municipality .... ☐☐☐

Location ....

Area: |1| Urban |2| Suburban

Plant located in: |01| Airport |02| Hospital/Clinic

|03| Slaughter-house |04| Cemetery |05| Holiday village

|06| Industry |07| Port |08| Power plant

|09| Water treatment |10| Disposal plant |11| Research centre

Other .... ☐☐

Plant owner .... ☐☐

Plant manager .... ☐☐☐

GENERAL CHARACTERISTIC OF THE PLANT

n° Kilns ☐   n° Stacks ☐

Plant in operation |1| Yes |2| No   Year activated |1|9| |

Authorization number |_|_|_|_|_|_|_| Date |_|_|_|_|_|_|_|

Predominat waste incinerated:

|01| Sludge |02| Special |03| Hospital |04| Urban

|05| Toxic   Other .... ☐☐

Material recovery:

Before incineration

|1| Compost |2| Glass |3| RDF |4| Paper |5| Metals

|6| Plastic   Other... ☐

After incineration

|1| Metals   Other... ☐

If plant is not activated, indicate reason:

|1| In modification |2| Testing |3| Court inquiry

|4| Shut off definitively |5| Not yet authorized |6| Still in constr.

Other .... ☐

* The grey section not to be filled

Figure 2. Incinerator Plant Survey Form (Part 1)

TECHNICAL CHARACTERISTICS OF THE PLANT

Kiln n° [   ]

Type of kiln:

[1] Grate   [2] Steps

[3] Fluidized bed   [4] Hearth   [5] Rotary kiln

[6] Pyrolitic   [7] Platforms

Other .... [|]

Built by .... [|]

Model .... [|]

Secondary combustion chamber: [1] Yes   [2] No

In accordance with law standards: [1] Yes   [2] No

If yes, indicate category   [1] (A: T>950°C)   [2] (B: T>1200°C)

Nominal potentiality [| | |] , [ ] (t/h-q/h)* [| | | | | | |] t/year

Effective potentiality [| | |] , [ ] (t/h-q/h)* [| | | | | | |] t/year

Working hours per day (weekly average) [|]

Working days per week   [ ]

Geometric height of stack [| |] meters

Flow of effluent [| | | | | | | |] Nm³/h   [1] Humid [2] Dry

Oxygen in effluent [|] , [ ] %

Temperature of emission fumes [| | |] °C

Smoke abatement system: [01] None

[02] Cyclone   [03] Wet   [04] Electrofilter

[05] Fabric filter   [06] Static chamber

Other .... [|]

Scrubber liquid: [1] Water   [2] Alkaline solution

Energy recovery

If yes specify:

[1] Hot water   [2] Electric energy   [3] Vapour

Other .... [ ]

*Circle unit used

Figure 3. Incinerator Plant Survey Form (Part 2)

649

DESTINATION OF SLAGS, ASHES, WASTEWATERS, AND SLUDGES

Slags

| 1 | Storage in plant | 2 | Uncontrolled landfill

| 3 | Controlled landfill

Prevailing type if controlled landfill utilized:

| 1 | I | 2 | IIA | 3 | IIB | 4 | IIC | 5 | III

| 1 | Reuse   *Type*.. ☐☐

Other .... ☐☐

Ashes

| 1 | Storage in plant | 2 | Uncontrolled landfill

| 3 | Controlled landfill

Prevailing type if controlled landfill utilized:

| 1 | I | 2 | IIA | 3 | IIB | 4 | IIC | 5 | III

| 1 | Reuse   *Type*.. ☐☐

Other .... ☐☐

Sludges

| 1 | Storage in plant | 2 | Uncontrolled landfill

| 3 | Controlled landfill

Prevailing type if controlled landfill utilized:

| 1 | I | 2 | IIA | 3 | IIB | 4 | IIC | 5 | III

| 1 | Reuse   *Type*.. ☐☐

Other .... ☐☐

Wastewaters

| 1 | Storage tank | 2 | Treatment

Final destination of wastewater:

| 1 | River | 2 | Lake | 3 | Sea | 4 | Sewage

| 5 | Soil   Other .... ☐

Note ....
....
....
....

Date

☐☐ ☐☐☐ ☐☐   ....
                                                  Compilers signature

Figure 4. Incinerator Plant Survey Form (Part 3)

Figure 5. Waste Incineration Plants in Italy (1987)

important micropollutants.   Many of them had already been involved in the
determination and standardization of procedures with international organizations   such   as   the   Bureau Comunitaire de Reference (BCR), supported by
the Economic European Community.   During the first year of work, the
Committee has prompted a draft of methods for sampling pollutants from
emission stacks and for the analysis of PAHs, PCDDs, and PCDFs.

The method proposed for sampling allows isokinetic collection of dust
and vapors.   Dust is collected on fiberglass; vapors are retained by two
trapping systems placed in series. The first one is a cryogenic apparatus
filled with a solvent, and traps PCDDs and PCDFs. Vapors escaping the
first trap are collected on graphitized carbon black.   After solvent
extraction, PCDDs and PCDFs are separated from the rest of the sample
according to a procedure developed by Buser et al.   (1978).

Two methods were proposed for the analysis. The first one (screening method) may be used to find whether PCDDs and PCDFs are present in the mixture at levels that require further investigation. The method is based on high-resolution gas chromatography (hrGC) combined with electron capture detection (ECD). The GC column allows separation of PCDDs and PCDFs according to the number of chlorine atoms in benzene rings.

The second method (reference method) allows unambiguous determination of PCDDs and PCDFs by utilizing high resolution gas chromatography combined with mass spectrometry (MS) employed in the selected ion monitoring (MID) mode. Two GC columns need be used during this step. One permits quantification of the total amounts of PCDDs and PCDFs characterized by the same number of chlorine atoms (homologue-specific determination). The isomer groups considered are tetra-, penta-, hexa-, hepta-, and octaCDDs and CDFs.

The other column is used when evaluation of specific isomers - such as those included in the so-called "dirty dozen" (see Chapter 4.1) - is requested. In both cases, quantification must be accomplished by using isotopically marked congeners as internal standards. An example of the GC-MS traces obtained from the analysis of specific congeners present in flyash from a municipal solid waste incinerator is shown in Figure 6. In defining the method, emphasis has been given to harmonization of techniques and materials. At the same time an interlaboratory comparison has been proposed to test and validate the procedure, once formal acceptance of the draft be made by the UNICHIM Committee.

2.1.5   Present Gaps and Future Needs

Currently, the main gaps in the field of environmental protection are associated with the lack of limits established by regional authorities regarding the release of PCDDs and PCDFs into the environment.

In Italy to date, only the Lombardy Region together with the Province of Bolzano, have established limits, but they apply only to the amount of PCDDs and PCDFs emitted from waste incineration stacks. In order to provide help to decision-makers, a National Toxicologic Committee has been established at the Istituto Superiore di Sanita which has developed suitable procedures for risk assessment evaluation based on a multimedia approach (that is, considering air, water, soil, and food as man's exposure routes to PCDDs and PCDFs).

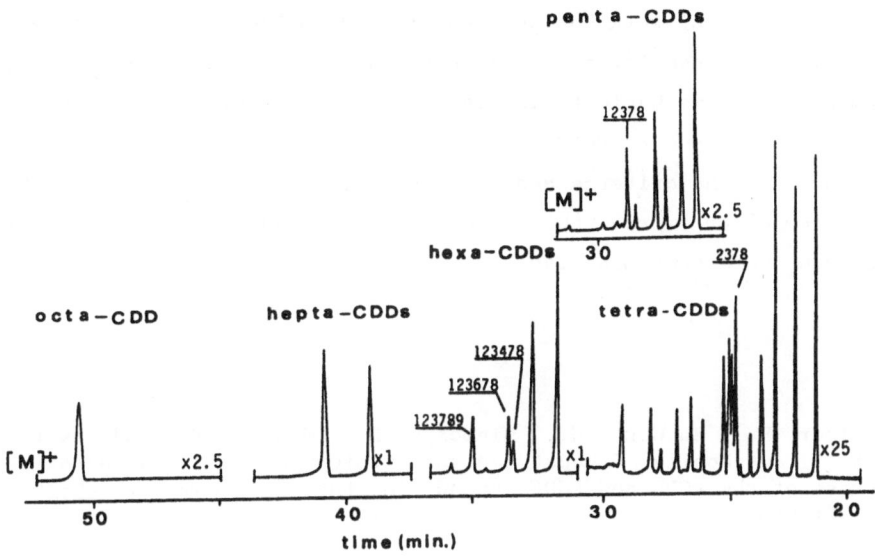

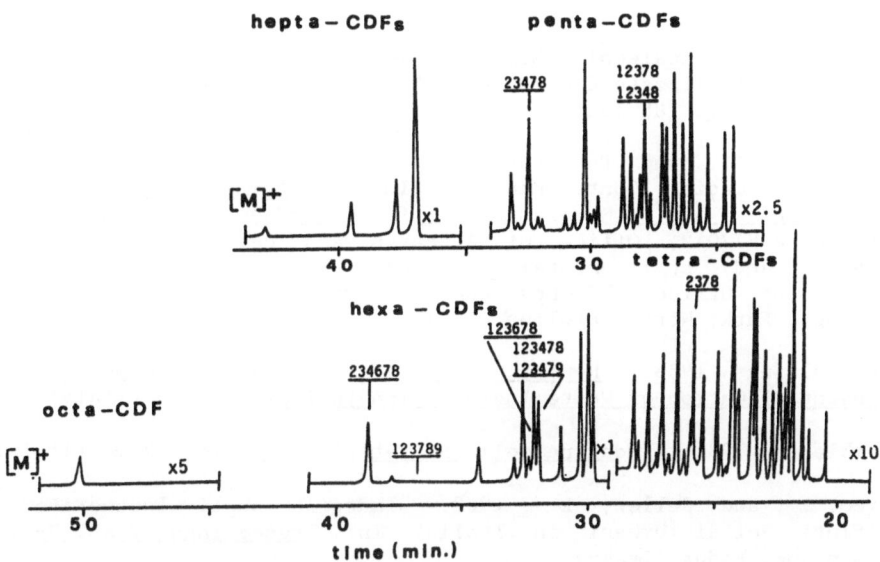

Figure 6.   PCDD and PCDF Recordings From a Municipal Solid Waste
Incinerator Flyash Extract.

The availability of this instrument will hopefully help regional authorities in deciding upon the limits to be fixed to restrain the release of PCDDs and PCDFs into the environment.

2.1.6   Conclusions

A great deal of work has been carried out by research groups, governmental bodies, and committees at a national level thus giving way to development of some basic tools regarding the assessment of risks arising from contamination due to PCDDs and PCDFs. However, only the extensive application of the criteria and methods proposed will make it possible to gain experience and improve the present knowledge on the important issue of PCDDs and PCDFs contamination.

REFERENCES

Ballschmiter, K., Buchert, H., Niemczyk, R., Munder, A., and Sweron, M. 1986. Automobile Exhaust versus Municipal-Waste Incineration as Sources of the PCDD and PCDF Found in the Environment. Chemosphere 15, 901-915.

Buser, H.R., Bosshardt, H.P., and Rappe, C. 1978. Identification of Polychlorinated Dibenzo-p-dioxin Isomers Found in Flyash. Chemosphere 2, 165-172.

Ciccioli, P., Brancaleoni, E., and Cecinato, A. 1986. Report to the "Progetto Finalizzato Energetica II" Tema: Ambiente e Salute. CNR publ. AES vol., 188-191.

di Domenico, A., and Pocchiari, F. 1987. Italian Strategy for Dealing with Multimedia Contaminants. In: Multimedia Approaches to Assessment and Management of Hazardous Air Contaminants, EPA/600/8-87/012, Office of Research and Development, Office of Health and Environmental Assessment, Environmental Criteria and Assessment Office, US Environmental Protection Agency, Research Triangle Park, North Carolina (USA).

Eco Consulting 1986. Report to the Economic European Community on the Present Situation of Waste Incineration in Italy. Milan (Italy).

Federambiente 1984. Indagine della Segreteria Generale. Rome (Italy).

Giannico, L., and Seller, L. 1982. Indagine sugli Inceneritori dei Rifiuti Solidi Urbani in Italia. In: Proceedings of SEP Pollution Symposium, Padua (Italy).

Merli, F., Viviano, G., Carrieri, M.P., and Taggi, F. 1988. Impianti di Incenerimento: Censimento, Sorveglianza ed Istituzione di una Banca Dati a Livello Nazionale. Acqua Aria (in press).

Viviano, G., and Ziemacki, G. 1987. L'Incenerimento dei Rifiuti Solidi Urbani in Italia - Valutazione delle Emissioni. Inquinamento 4, 59-63.

2.2   Contingency Planning and Emergency Response at MONTEDIPE

Luigi Corigliano

MONTEDIPE
20124 Milan, Italy

2.2.1   Abstract

The development of plans to cope with unexpected hazardous events has been a common practice in MONTEDIPE Industry. There are different patterns of emergency, some affecting a small area or a single plant, others involving several plants and eventually sites outside of the factory. MONTEDIPE has developed different types of emergency actions, each studied for a special set of events. MONTEDIPE has also generated, on behalf of FEDERCHIMICA, a data base of more than thousand substances commonly handled in chemical plants. The data base has been named SIET (Information Service for Emergency of Transports), and it is available on IBM personal computers (XT type). The system provides easy access to external users such as fire brigades and other emergency response agencies.

In June 1982 EEC issued Directive No. 82/501 on the major accident hazards of certain industrial activities. In Italy, with respect to risk of fire and explosion the Home Office has implemented the following decrees: DPR of 27 July 1982, DM of 16 November 1983, and DM of 2 August 1984. The decrees contain provisions only for new industrial installations and for significant changes to existing plants.

The Circular No. 16 of the Home Office, promulgated on 10 June 1986, provided guidelines on how to prepare a specific Notification Dossier (Safety Case). The development of a thorough Notification Dossier for both new and existing plants was deferred to a specific law for the general implementation of the EEC directive.

The Ministry of Health, with the ruling of 21 February 1985, required a census of all industrial activities involving major risks and falling in the field of application of the Directive. The ruling of 24 September 1986, issued later by the same office, required all industries to present Notification Dossiers for those activities included in Article No. 5 of the Directive prior to the 30 September 1987 deadline. Dossiers should be prepared following the guidelines elaborated by ISPESL (Superior Institute for Prevention and Safety at Work) and similar to those issued by the Home Office.

2.2.2   Procedures

On this basis MONTEDIPE, which is the petrochemical and polymers Company of MONTEDISON Group, carried out the following actions.

(1)   Elaboration of Notification Dossiers regarding new industrial installations, and/or significant changes to existing plants, following the guidelines issued by the Home Office. Three safety cases for new industrial installations were favourably received and considered to be feasible by the competent Italian authority.

(2)   Preparation of Notification Dossiers regarding all plants falling in the field of application of Article No. 5 of the Directive as required by the Ministry of Health. Dossiers should be presented to the competent authority during the period 30 September 1987 through 1 January 1989. MONTEDIPE, as a member of a Working Group within the Federation of Chemical Industry, participated in the elaboration of a Notification scheme. This scheme, regarding the operating storage of liquid chlorine and all chloroalkali plant units involving liquefied chlorine, represents a useful guide for the preparation of notification of all interested plants.

(3)   Updating of existing emergency plans in harmonization with the indications provided by a MONTEDISON Task Force, and their integration with territorial civil protection plans.

(4)   Cooperation with public authority (charged for elaborating civil protection plans) in the identification of possible areas of interest outside the factory. MONTEDIPE has so developed computerized programs for assessment of accident consequences. These programmes are based on models elaborated and validated by international institutes.

(5)   Optimization of an emergency service prearranging departmental selective communication warning systems from the centre of operations to the control rooms located in the various plants. This allows shut-down and evacuation of the plants affected by the event without having to spread the alarm and the general emotional upset, and without involving other departments where it is not needed. A continuous alarm signal (sirens) would however be heard over the entire factory.

(6)   Availability of a computerized system in the emergency centre. This offers the possibility to select the control room that must be alerted, supply information about the event, and issue the appropriate instructions on how to operate. The operator is provided with information on:

- the topical causes of emergency;
- the corresponding procedure to adopt;
- the area that will be typically affected.

(7) Arrangement of different kinds of emergency actions, each studied for a special set of events. The first response level is the plant emergency, referred to weak leakage or small accidents; the second is a multiple plant emergency, that can involve a partial area of the factory; and the third is a general emergency of the entire factory. Obviously the three levels are interconnected, in such a way as to allow the gradual transition from the first to the last response level in the case of anomalous evolution of a small accident.

(8) Setting of an Information Service for Emergency of Transports (SIET) on behalf of FEDERCHIMICA: a data base of more than a thousand substances is available to cover many events. Public authority, such as the Fire Brigade, may call SIET at any time: there, an expert is available to retrieve and provide information from the computerized data base.

(9) Training of a specialized team of workers suitably equipped for emergency cases both inhouse and outside the factory. These workers may operate also on request of public authority and/or other factories and have at their disposal special equipment such as:

- hydraulic pump ensuring safe operation even when flammable mixtures are present;
- a flare that can be assembled on the site of the accident to eliminate any flammable gases;
- portable gas detectors to check the concentration of several substances into the environment;
- portable oxygen analyzer;
- kit for repairing leaks;
- several types of manifolds and joints;
- neutralizing products which inhibit the evaporation of toxic and harmful substances;
- provision of sets of emergency protection equipment, such as: self-contained breathing apparatus, gas mask with universal and specific filters, chemical safety goggles, acid- and gasproof clothings, face shield;
- lifting device for particular cases of emergency.

2.3 Contingency Planning: An Expert System (ChEM) for Describing the Level of Response Needed

Silvia Cerlesi

Regione Lombardia
c/o Dipartimento di Fisica Nucleare e Teorica
Universit di Pavia
27100 Pavia, Italy

Flavio Argentesi

Joint Research Center
Commission of the European Communities
21027 Ispra (Varese), Italy

Wander Tumiatti

SEA Marconi Technologies
10146 Turin, Italy

G. Umberto Fortunati

Via Vincenzo Monti 29, 20123 Milan, Italy

2.3.1 Abstract

Chemical Emergency Manager (ChEM) is an expert system that has the function of aiding decision makers during chemical accidents. ChEM has the capability to estimate the level of the chemical threat by using data input by the user. At present, the system consists of four modules: (1) identification of the threat, (2) evaluation of the threat, (3) steps for mitigation, and (4) documentation and recovery of costs. The system is being developed for use by local authorities, industry, civil protection, and defense officials.

2.3.2 Introduction

It is generally agreed upon that expert systems will have an important role in the management of chemical accidents. In fact, as rapid decision-making is essential to minimize damages and risks, a sound management of this type of accidents requires the prompt availability of experts - or their expertise - from different backgrounds.

The ChEM expert system is being developed at the Joint Research Center of the Commission of European Communities at Ispra, Italy (Argentesi et al., 1987). It is intended to be an efficient tool in the chemical emergency management of a potentially wide spectrum of chemical accidents and their consequences: in fact, chemical accidents often involve environmental, social, and possibly political aspects under highly stressed conditions.

As chemical accidents are often characterized by fast-dynamic processes (Johnson and Jordan, 1983), great care must be taken to provide ChEM with information at an adequate rate so that the system response may be effective.

2.3.3   The ChEM Expert System

ChEM, whose underlying logic is rooted in the development of Artificial Intelligence (AI), was originally focused on accidents involving only chloroaromatics and, specifically, PCBs (Tumiatti and Nobile, 1984).

ChEM is virtually divided into different development stages or "modules":

(1)   identification of the threat;
(2)   analysis and evaluation of the threat;
(3)   response and neutralization of the threat;
(4)   resolution of the threat (post-emergency);

of which only the first one has been so far fully implemented. The first module is basically concerned with the identification of threat ("threat identification module") and its immediate effects, and contains a:

- glossary, of the terms used in the system;
- tutorial, as a user's guide to the expert system;
- handbook, which includes technical data on the chemicals and systems most frequently involved.

As said, ChEM is implemented in separate modules, each focused on a rather homogeneous and self-consistent topic. Module integration is acknowledged to be a difficult task and is going to be achieved on the basis of a computational model that guarantees cooperation between knowledge sources working independently.

2.3.4   Threat Identification Module

This module has been developed on the expert system shell SAVOIR (Cox et al., 1984)). SAVOIR is written in Pascal, may run on a wide variety of machines, and can link a knowledge base with external programs for numerical computation and data input. ChEM is integrated with non-symbolic components for mathematical simulations, such as the release and diffusion of a chemical into the soil and other media (for an example, see Ratti et al., 1986).

2.3.5   The Knowledge Base

The knowledge base is the component of the system associated with various (logical, mathematical, physical, etc.) models, including the expert knowledge. It exhibits the following two-level structure (Figure 1).

(1) A higher level where the semantics of the problem is defined (threat levels, intermediate semantic variables, and connection variables with lower levels).

(2) A lower level where the technical knowledge of the systems involved (equipment, process, storage, transport, and disposal) is represented.

Connection between levels is obtained through a set of "metric variables" anyone of which may be associated with a weighting value: this leads to a symbolic and qualitative characterization of the accident. Weighting values are inferred from the reported information on the accident on the basis of provided ranking scales.

The knowledge base validation has been achieved by verifying past accident histories in different dynamic phases. The knowledge base has been extended with case studies and simulations by using the ID3 algorithm (rule generation through inductive methods).

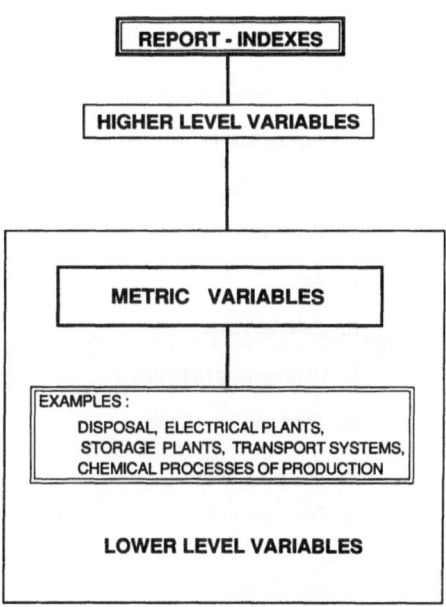

Figure 1.   Two-Level Knowledge Base Structure

2.3.6   The System Consultation

Consultation is subdivided into two phases.

(1)   In the first phase, ChEM gathers information from the external source (operator) via questions generated according to a pre-existing system menu. Operator's answers are entered as the aforecited metric variables. Graphical facilities are available to facilitate dialogue.

(2)   In the second phase, ChEM output is an estimate of the accident threat level to be used in planning counteractions and in their management.

ChEM output is obtained via the association network implemented in the knowledge base. The output report may furnish: a general information on the accident, status and dynamics of the accident, preliminary suggestions for urgent countermeasures, and values of main variables.

2.3.7   Improvements and Developments

ChEM is subjected to continuous revision to provide the "best" support in decision-making.

Activation of rule automatic-generation algorithm relies on maximum availability of information on past accidents. A method to characterize an accident through the propagation routes of chemical(s) involved is now being developed (Table 1). Figure 2 exhibits in flow-chart form as accident notification format has been arranged. Detailed information (e.g., Tables 2-4) must be provided to allow evaluation of elements in each flow-chart block. The present work is evolving partly from pertinent work carried out in 1982 (WHO, 1982).

Table 1.   Exposure and Propagation Routes

A. Ground Water

B. Surface Water

C. Indoor Air

D. Air

E. Solid Surfaces

F. Underground Networks

G. Food and Food Chains

H. Fire and Explosion

I. Direct Contact

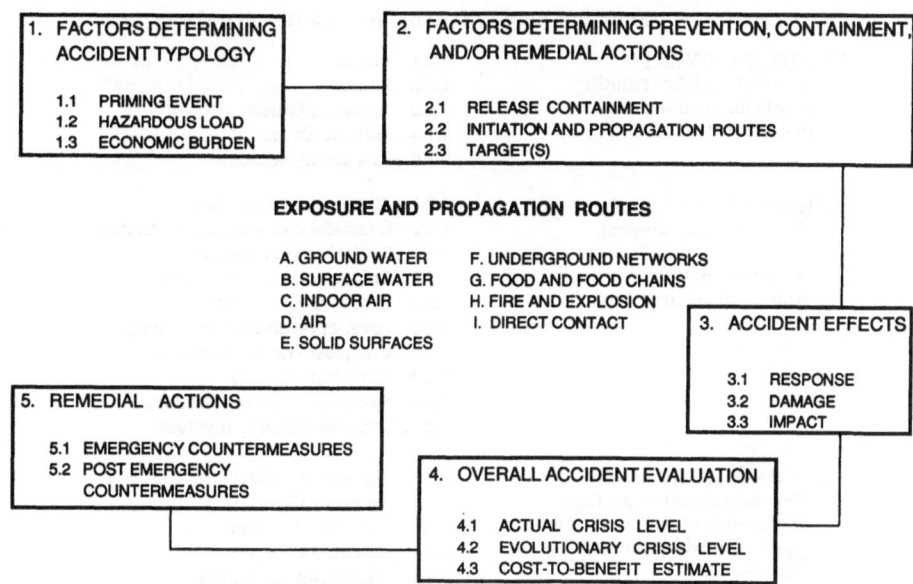

Figure 2.   Flow Chart of Accident Notification

While the threat identification module is now under extensive testing and subjected to revision, the following additional facilities are in the design phase:

(1) map-based interfaces: a system capable of visualizing the graphical information at different map complexity levels (Argentesi and Avouris, 1987);

(2) data base access: facilities for the consultation of chemical data bases relevant to the problem.

Future developments will focus on the identification of counteractions and remedial actions for a reduction of hazards and damages, and for a fast recovery of life conditions - if possible.

2.3.8   Conclusions

At present ChEM is capable of dealing with the common types of accidents involving PCB's. Adaptations of the knowledge base, which will take into account the time-dependent character of the threat, are in progress. ChEM is designed for:

(1) local authorities;
(2) large industrial facilities;
(3) civil protection; and
(4) defense.

Table 2.   Factors Determining Accident Typology

1.1 PRIMING EVENT
A complex of forerunning
events likely to act as a
primer of the accident

1.1.1	Primer
1.1.2	Source
1.1.3	Apparent Cause
1.1.4	Intrinsic Cause
1.1.5	Release Dynamics

1.2 HAZARDOUS LOAD
The chemical, physical,
and toxicological features
without which the accident
would not occur

1.2.1	Chemical Identification
1.2.2	Chemical and Physical Properties
1.2.3	Toxicological Properties
1.2.4	Ecotoxicological Properties
1.2.5	Hazard Classification
1.2.6	Level of Hazardous Impurities
1.2.7	Chemical Transformations
1.2.8	Chemical Quantity
1.2.9	Released Chemical Quantity
1.2.10	Effective Quantity Involved

1.3 ECONOMIC BURDEN
The overall cost to produce
and/or upkeep a certain product/
equipment with accident risk
minimization

1.3.1	Replacement Cost
1.3.2	Recovery Cost
1.3.3	Utilization Indicator
1.3.4	Expected Lifespan
1.3.5	Breakdown Frequency
1.3.6	Deterioration Processes
1.3.7	Maintenance
1.3.8	Servicing
1.3.9	Disposal

Table 3.   Factors Determining Prevention, Containment, and/or Remedial
Actions with Reference to Main Human Exposure Routes

2.1 RELEASE CONTAINMENT
The available natural and/or man-made
barriers apt to prevent and/or minimize
a chemical release to migration routes.
These barriers concern :

2.1.1	Ground Water
2.1.2	Surface Water
2.1.3	Indoor Air
2.1.4	Solid Surface
2.1.5	Underground Networks
2.1.6	Food and Food Chains
2.1.7	Fire and Explosion

**2.2 INITIATION AND PROPAGATION
ROUTE FEATURES**
The characteristic elements of a route
where irregularity is detected

2.2.1	Accident Site
2.2.2	Direct Evidence
2.2.3	Physical State
2.2.4	Level of Hazardous Concentration
2.2.5	Meteorological Parameters
2.2.6	Area Accessibility
2.2.7	Area Features

2.3 TARGET(S)
Something potentially damaged by the
accident. A group of main features is :

2.3.1	Land Use
2.3.2	Source-Population Distance
2.3.3	Source-Groundwater Distance
2.3.4	People Exposed through Indoor Air
2.3.5	People Exposed through Air
2.3.6	People Exposed through Surface Water
2.3.7	People Exposed through Groundwater

Table 4. Accident Effects

3.1 RESPONSE
The available resources to detect
and notify the Authority of the
emergency, and manage the crisis
afterwards

3.1.1 Detection and Notification
3.1.2 Contingency Plan(s)
3.1.3 "First-Aid" Response
3.1.4 Disposal
3.1.5 Timing

3.2 DAMAGE
The overall negative effects on the
accident site, propagation route(s),
and target(s) in terms of financial
costs and health problems

3.2.1 Direct Evidence
3.2.2 Physical Damage to Persons
3.2.3 Psycological Pressure
3.2.4 Adverse Effects on the Environment
3.2.5 Damage on Food Products
3.2.6 Damage to Vital Services

3.3 IMPACT
The overall effects in terms of
perturbation of ecological equili_
bria and human health and lifestyle

3.3.1 On Aquatic Life
3.3.2 On Terrestrial Life
3.3.3 On Aerial Fauna
3.3.4 On Human Communities

REFERENCES

Argentesi, F., and Avouris, N.M. 1987. A Map Multilevel System as a Man-Machine Interface for Emergencies Management". Presented at the Conference on Artificial Intelligence and Other Innovative Computer Applications in the Nuclear Industry, Snowbird, Utah (US), September 1987.

Argentesi, F., Bollini, L., Facchetti, S., Nobile, G., Tumiatti, W., Belli G., Ratti, S., Cerlesi, S., Fortunati, G.U., and La Porta, V. 1987. ChEM: An Expert System for the Management of Chemical Accidents Involving Halogenated Aromatic Compounds. Proceedings of the World Conference on Chemical Accidents, pp. 227-230, Rome (Italy), July 1987.

Cox, P.R., Broughton, R.K., and Rubinstein, M.G. 1984. The SAVOIR Expert System Package - User's Manual Version 1. ISI Limited, Redhill, Surrey (UK).

Johnson, C.K., and Jordan, S.R. 1983. Emergency Management of Inland Oil and Hazardous Chemical Spills: A Case Study in Knowledge Engineering. In: Building Expert Systems, pp. 349-398, Addison-Wesley, Massachusetts (US).

Ratti, S.P., Belli, G., Lanza, A., Cerlesi, S., and Fortunati, G.U. 1986. The Seveso Dioxin Episode: Time Evolution Properties and Conversion Factors between Different Analytical Methods. Chemosphere 15, 1549-1556.

Tumiatti, W., and Nobile, G. 1984. Emergency Procedures in Polychlorobiphenyls Contamination - Emergency Intervention - Procedures and Methods. Presented at the Fourth International Environmental and Safety Conference, London (UK), March 1984.

WHO 1982. Guideline Document on Rehabilitation Following Chemical Accidents. Revision of the International Workshop organized by the World Health Organization and Istituto Superiore di Sanita, Rome (Italy), November 1982.

3.   IMMEDIATE EMERGENCY RESPONSE

3.1   Emergency Health Care

Paolo Bruzzi

Servizio di Epidemiologia Clinica
Istituto Nazionale per la Ricerca sul Cancro
16132 Genova, Italy

3.1.1   Abstract

The health care measures following the occurrence of a chemical emergency can be divided into three steps: (1) prevention of (further) exposure, (2) identification and follow-up of exposed individuals, and (3) diagnosis and treatment. Each of these three procedures rely on discrete steps, but the effectiveness of the overall plan is dependent on the success and accuracy of the others. Major technical, organizational, and political problems are involved with the implementation of such measures during the emergency period.

3.1.2   Introduction

Emergency health care actions to cope with chemical accidents are clearly distinguished from the epidemiological studies: while the former try to minimize the damage caused by accidents, epidemiological studies are undertaken in order to assess accident consequences. Indeed, the aims and the methodology of the two approaches are different. Nevertheless, the feasibility, the scientific validity, and the results of the epidemiological studies are greatly influenced by the activities aimed at protecting the health of the population at risk. On the contrary, only in few instances can the results of epidemiological studies be used to overcome inadequacies of the health care program.

Prevention and/or reduction of the health consequences of chemical accidents rely on three types of measures:

(1) prevention of (further) exposure;

(2) identification and follow-up of those who were (are) presumably exposed to the toxic chemical(s), regardless of the signs or symptoms of exposure that they may or may not show;

(3) diagnosis and treatment of the pathological effects of the chemical.

Major technical, organizational, and political problems are involved in the implementation of these protective or remedial measures during the emergency period, which is critical in determining their overall effectiveness.

3.1.3   Prevention of (Further) Exposure

This is the most important issue in the emergency response to an accident involving release of dioxin in the environment. It relies mainly on a risk assessment process which should be based on analytical data accurately describing the distribution and the level of environmental pollution. However, these data are not available for some time after the accident takes place and, when available, they are usually incomplete and of questionable reliability.

As a consequence, risk assessment in the emergency phase must be based on incomplete, or indirect evidence. The latter includes information from various sources, which should be carefully searched and scrutinized. Indirect information which may help assessing the risk for the potentially exposed population includes:

(1) toxic effects (death) among farm or wild animals;

(2) presence of specific markers of exposure in humans (the only specific effect of dioxin in man is chloracne);

(3) presence of clusters of individuals showing a specific acute consequences of exposure to dioxins or related compounds:

- eye, respiratory, skin, and gastrointestinal irritation;
- headache;
- malaise.

No laboratory test, at the moment, is sufficiently specific to allow its use in risk assessment. Detection of TCDD in human fluids or tissues, though technically feasible, involves ethical, political, and logistic problems which hamper its potential use in the emergency period. Measurements of TCDD levels in the fluids (e.g., milk) or tissues of animals can be used to refine the risk assessment process and to justify the implementation of specific protective measures.

While each individual piece of information has to be evaluated in the assessment of the risk in specific communities, it is much more difficult to obtain an overall picture of the distribution of pollution over the entire area involved. Nevertheless, even in the presence of insufficient environmental data, some modeling procedure can be attempted which takes into account route of pollution, duration of the polluting process, location of the polluting source, quantity of dioxins released in the environment, wind direction and speed, etc. The resulting estimates of pollution distribution must be continuously updated as more data become available, and may prove very useful in establishing the general lines of the emergency response (allocation of resources, implementation of protective measures, etc.).

The protective measures which must be considered in the emergency period include:

(1) general hygienic advices (e.g., abstaining from consumption of local fruits, vegetables, and meat from local animals, and abstaining from working and playing outside);

(2) strict regulations (e.g., closure of shops and plants, requisition of food, etc.);

(3) evacuation of polluted areas.

Selective evacuation of children, who are likely to follow less strictly the hygienic advices, may be justified. Selective evacuation of other groups of individuals, such as pregnant women or patients with liver disease, though possibly useful from a psychological viewpoint, has no strong scientific basis.

3.1.4   Identification and Follow-up of Individuals Presumably Exposed to Dioxin

This can be based on the following items.

(1) General or personal information such as place of residence, occupation, presence in particular places in critical periods, consumption of specific foods. Much of this information can be obtained only by means of a personal interview, which is not always feasible when the number of potentially exposed individuals is large.

(2) Signs or symptoms of dioxin exposure: these include chloracne and a specific symptoms. Alterations in a number of laboratory tests, which include liver function tests, blood lipids, and others, have been inconsistently found to be associated with dioxin exposure. Whether an individual can be considered as exposed solely on the basis of the recent occurrence of alterations in one or more of these tests remains an unresolved issue. Beyond its obvious relevance to epidemiologic studies, identification of exposed individuals is of little use in emergency

health care, once adequate measures have been issued to prevent further exposure.

Acute effects of dioxin exposure disappear soon after exposure cessation, and none of the chronic conditions which have been associated with dioxin exposure appears to benefit from earlier diagnosis and treatment. Individuals with definite or presumed dioxin exposure should be reassured about their short- and long-term health risk, and might possibly be advised to reduce their alcohol consumption. Fertile women should be advised to postpone any planned pregnancy, in the light of the teratogenic effect of dioxin in animals and of the uncertain evidence concerning its effects on pregnancy outcomes in humans. Women who were in the first trimester of pregnancy at the time of exposure should receive adequate and balanced information, along with psychological and obstetrical assistance.

3.1.5   Diagnosis and Treatment of Pathological Effects Due to Dioxin Exposure

In the aftermath of a chemical accident involving exposure of the general population, all new symptoms and diseases - as well as exacerbations (either real or supposed ones) of preexisting symptoms - tend to be attributed to the chemical. As a consequence, those in charge of implementing the health care program must be prepared to cope with a large number of anxious individuals presenting various complaints or expecting to be thoroughly examined in order to detect or to rule out health consequences of exposure.

Examination and, when appropriate, reassurance of these individuals should be included among the priorities of the emergency health care plan. Thus, local medical services should be supplied with (1) extra personnel including specialists in dermatology, internal medicine, pediatrics, obstetrics, and gynecology, and (2) equipment for laboratory tests. In addition, facilities should be identified where this temporary overload of clinical activities can be carried out.

The use of media to provide adequate and correct information to the general population is recommended, although its improper use (falsely reassuring information) may discredit the political and scientific establishment, and cause more anxiety than it prevents among the general population.

From a clinical viewpoint, little can be done for those who show signs of true or suspected dioxin intoxication: acute effects tend to subside spontaneously within few days or weeks. Chloracne has a chronic behaviour which seems not to be affected by the treatments used for acne vulgaris, except perhaps, retinoic acid derivatives. No effective treatment is available for the other affections which have been associated, most with inconsistent evidence. to dioxin exposure.

3.1.6   Adverse Pregnancy Outcomes

In the light of the well known teratogenic and embryotoxic effects of TCDD in laboratory animals, pregnancy outcomes in the Seveso area have been extensively studied after the ICMESA accident. Rates of spontaneous abortions were retrospectively studied through careful search in the medical records, but neither the completeness of the ascertainment, nor the reliability of the data have been thoroughly evaluated. Indeed, the presence of a bias arising from selective reporting of voluntary abortions as "spontaneous" in the most polluted areas has been suggested, but was never confirmed. A signiicant excess in the rate of spontaneous abortions started around and soon after the accident has been observed in the most polluted areas, but its association with TCDD exposure has not been looked at on an individual basis.

The occurrence of birth defects was investigated by means of a Registry set up in 1978 to cover the whole population (220,000) of the 11 towns involved with TCDD contamination. Malformations were actively searched among all newborn children of resident mothers by clinical examination and search in the medical records. Comparisons were made between birth defect rates per 1,000 births (livebirths + stillbirths) in areas at different average level of pollution. The areas were defined according to the official Zones (A, B, R, and background), as well as according to chloracne rates, rates of unspecific skin lesions, and rates of animal deaths.

The results indicate that the Seveso accident did not cause any major change in the pattern or in the frequency of congenital anomalies among newborn children in the polluted areas. They do not provide evidence in favour of, or against, TCDD teratogenicity, nor do they permit to rule out that TCDD pollution has caused, under specific exposure circumstances, congenital anomalies in a limited number of children.

3.2   <u>The Emergency in Seveso - Measures to Limit Human Exposure</u>

G. Umberto Fortunati

Via Vincenzo Monti 29, 20123 Milan, Italy

Vito La Porta

Regione Lombardia
20100 Milan, Italy

3.2.1   <u>Abstract</u>

The emergency phase of the Seveso accident was marked by great psychological and social tensions together with uncertainty on the hazardousness
of the materials released during the accident.   These facts reflected
heavily on the precautions and countermeasures adopted thereafter. This
reevaluation of the Seveso experiece focuses on the following aspects:
(1) toxicologic limits: a cautious and pragmatic approach; (2) practical
limits in the topsoil layer according to an acceptable-risk-to-use criterion;   (3) monitoring of the territory and risk area delimitation: Zones A,
B, and R; administrative-political problems in drawing the map: property
boundaries and municipality borders; (4) temporary limitations in the use
of the territory; and (5) whether a cost/benefit approach is possible.

3.2.2.   <u>Introduction</u>

Since the appearance of animal deaths in a large portion of Seveso
territory, every attempt was made to reduce human exposure to TCDD acting
on:

(1)   the environment,
(2)   the behaviour of the population.

When the reactor safety valve ruptured, a toxic mixture burst through
the pipe into the open air. The cloud rose approximately 50 m, then sub-
sided and bent down towards earth because of the south-east wind. Sixteen
days after the toxic release the authorities evacuated all the inhabitants
(739 persons) living in an area extending about 2 km south-east from the

ICMESA factory. Zone A area was approximately 110 ha. The inhabitants of surrounding Zones B and R were subjected to a number of hygienic regulations. Zone B (270 ha) was the extension of Zone A along TCDD diffusion pathway, and Zone R (1,430 ha) a large buffer zone surrounding both Zones A and B. For the definition of the three main Zones, and related Subzones, see G.U. Fortunati "A Brief History of Risk Assessment and Management after the Seveso Accident", and A. di Domenico "Introductory Comment on Seveso (Italy) Industrial Accident", both in this chapter on Management of Accidents.

3.2.3 Practical Limits

The measures adopted in the three zones were based on the following practical limits fixed on both toxicologic data and analytical limitations. The limits were included in a Regional Law. This fact put an end to the debate on the inherent risk of assuming TCDD whose toxicology was then not yet well known:

- evacuation limit: Zone A, 400 ppt;
- nonagricultural use of soil: Zone B, 40 ppt;
- agricultural use of soil: Zone R, 6 ppt;
- background level: outside Zone R, 1-2 ppt.

In some cases as in the area east of ICMESA and at Zone R extreme south, the delimitation appeared to be barely sufficient, while on the west side it was unnecessarily extended. In addition, as the "contamination map" had been set with the same Regional Law, no changes could be brought about after the end of 1976.

3.2.4 Measures to Limit Human Exposure Acting on the Behavior of the Population

(1) Evacuation of Zone A. TCDD was diffused in the area finely and irregularly, and could not be confined. Therefore, a most drastic precaution was taken during the emergency phase, i.e., to evacuate approximately 160 houses in Zone A.

(2) Residents of Zone B. The following precautions were recommended to Zone B residents:

- intensification of personal hygiene;
- no animal breeding or vegetable planting;
- daily relocation of children up to 12 years and of pregnant women;
- abstain from procreation;
- minimize dust level in the air: the vehicles had to limit their speed to 30 km/h;
- careful emptying of vacuum cleaners.

(3) Residents of Zone R. Same precautions as for Zone B residents for what concerned personal hygiene and use of agricultural soil.

Pets had to be fed with food originating from areas not contaminated by TCDD.

(4) Measures common to the three zones. To avoid assumption of TCDD, all locally bred animals (mostly rabbits and chickens) in the three zones were slaughtered and their bodies transferred and buried in Zone A. At the same time hunting was forbidden and this prohibition was maintained for about 8 years. As an extreme precaution, even honey was collected and disposed of in Zone A.

(5) Cleanup workers and chemists. It was assumed that all this personnel was aware of the inherent risk. Each one had to use special (Tyvek) suits, full protection mask with carbon filter, rubber gloves and boots, and it was compulsory to take a shower after every 4-hour shift. Specific precautions had to be taken for dismantling the TCP reactor vessel and contaminated equipment of the ICMESA department where the accident had originated. A special decontamination unit was built alongside the department and Mururoa-type frog suits were adopted: compressed air was fed from a clean external intake point. All protective material was disposed of after each shift.

3.2.5 Measures to Limit Human Exposure Acting on the Environment

(1) Defoliation and agronomic activities. Defoliation was decided on to remove the contamination deposited on grass and leaves. Before cutting, the green parts were sprayed with vinyl acetate in aqueous solution to fix TCDD. Precautions for the population during such operations were graduated according to home density in the area.

(2) Reclamation of homes in Subzones A_6 and A_7. Subzones A_6 and A_7 on both sides of Isonzo Street at Seveso were reclaimed through the cooperation of the local authorities, Istituto Superiore di Sanita (Rome), and the Swiss Corporation Hoffmann-LaRoche, mother-Company of ICMESA. Cleanup was carried out by four teams, each one working 4 h/day in the contaminated area. Details concerning this complex procedure are reported in other papers.

(3) Schools. Schools were thoroughly cleaned and floors washed daily. Children were a risk group and schools were checked to verify that contamination was not transported in by students through their shoes. This monthly procedure was discontinued only after over a year - i.e., when TCDD levels were consistently found to be <10 ng/m^2 (analytical threshold and "action level").

3.2.6 Conclusive Considerations

All limitations in the use of the affected territory - A_1-A_5, A_6-A_7 (Zone A most southern part), A_8, B, and R - were gradually removed when the acceptable contamination levels were achieved through painstaking reclamation work. All losses and cleanup costs were borne by Hoffmann-LaRoche. A group of agronomists of Seveso Special Office (organization and purpose of which is detailed in other papers) was entrusted with the evaluation of losses due to ceased farming activities, such as raising crops or livestock, taking as a basis the activity level of 1975.

A formal cost/benefit estimate for the Seveso case is not feasible. Due to the uncertainty of estimating when TCDD chronic effects will cease, and the difficulty in placing a monetary value to human health, one can only state the following.

- The measures taken were sufficient to prevent (as of August 1987) any diffused chronic adverse health effect to the population. The acute effects were: chloracne, few miscarriages within one year after the event, and minor neurological disorders in some risk groups.

- The money spent by both the Italian authorities and the Swiss Company (300 million Swiss francs) was worthwhile being spent.

Should a similar case occur again, the accumulated experience would allow to reduce the amount of expenditures. Reclamation time would also likely be reduced by 50 percent. Although accident type and severity are different, the information provided by the Seveso case may be useful in reducing the number of mistakes in the emergency phase and in having the same positive results in a much shorter time span.

3.3   Individual Protection in the Seveso Cleanup

G. Umberto Fortunati

Via Vincenzo Monti 29, 20123 Milan, Italy

3.3.1   Abstract

The Seveso accident is reported upon in short.   Estimates of TCDD
amount released as well as some of the compound's physical and chemical
characteristics are given.   Cleanup operations were carried out between
1978 and 1984 under Lombardy Regional Government Special Office for
Seveso's management.   Personnel were protected with masks, boots, and complete coveralls.   Different levels of protection for areas at different
risk levels were adopted in reclamation operations.   The efficiency of
protective suits is assessed on the basis of the minor accidents and/or
discomfort occurred during four years.   In general, the external high
temperature was the single major factor of physical discomfort.

3.3.2   Introduction

On July 10, 1976, an accidental release in the atmosphere - deriving
from the alkaline hydrolysis of 1,2,4,5-tetrachlorobenzene to 2,4,5-tri-
chlorophenol - contaminated an area of 1,810 ha in the Milan area.   The
trichlorophenol-producing plant belonged to ICMESA (Industrie Chimiche
Meda Societ Azionaria), an Italian company of the Givaudan-Hoffmann-
LaRoche multinational corporation.

If the direct cause of the release is clear, the primary cause of the
accident is still debated (Cardillo and Girelli, 1980; Salomon, 1982).
Which were the chemical reactions that developed during the accident?
What compounds were generated and how much of them? What were the qualitative and quantitative features of the contaminating release discharged
from the reactor? These are all legitimate questions that even today are
difficult to have thorough and clear answers.

Among the various possible chemical reactions, the one involving formation of tetrachlorodibenzo-p-dioxin (TCDD) - by condensation of two molecules of sodium 2,4,5-trichlorophenate - is that of greatest concern due to the high toxicity, environmental persistence, and stability of the compound (Esposito et al., 1980; Hay, 1982). Only some bacterial exotoxins, for example, have a minimum lethal dosage lower than TCDD.

In the technical literature there are numerous estimates as to the quantity of dioxin generated during the accident and the fraction of such quantity released into the environment. The latter has been estimated anywhere from 250 g to 130 kg (Cattabeni et al., 1978; Esposito et al., 1980; Marshall, 1980; Hay, 1982).

3.3.3 Main Physical and Chemical Characteristics of TCDD

TCDD is a solid substance melting at approximately 305°C. It is thermally stable, practically insoluble in water, scantly soluble in alcohols, more soluble in hydrocarbons and, especially, in chlorinated solvents (for example, 1,400 ppm in o-dichlorobenzene). TCDD has a very low vapor pressure at room temperature, and a very limited mobility in soil: mobility is lower for higher contents of soil organic component (Schroy et al., 1984; Fregman and Schroy, 1984).

3.3.4 Removal and Disposal of Contaminated Material

Bearing in mind the characteristics of TCDD, the most common procedure for decontamination of areas, land, and in general any environment contaminated with dioxin, consists of removing the contaminated material with all necessary precautions to avoid further dispersion of toxic substance(s) and harm people. The contaminated material is then placed in controlled basins, in abandoned coal or salt mines, etc., in concrete roadbeds of highways, in containers to be kept in controlled storehouses, or in blocks of concrete to be sunk in the ocean.

However, it is possible to affirm that depositing in a suitable basin, built at an adequate site with appropriate means, and kept under control, offers an almost complete guarantee for low contamination material disposal even when the quantity involved is relevant.

In this report, the individual protection utilized for the personnel employed in decontamination activities are examined in essential terms in relation to the methods adopted for decontamination of extensive areas as well as for disposal of large quantities of material having a low degree of dioxin contamination.

3.3.5   Choice of Protection

(1) Generalities.   The choice of individual protective equipment to
enable operation in a dangerous and contaminated environment was encountered at different levels, mainly for protection from radioactive nuclides
or dangerous chemical substances.   The Harwell Hazardous Waste Service
(UK) has carried out an analysis of potential risks involved, and classified the total protections as well as the masks for the respiratory
apparatus.   Division of types of protection into classes allows immediate
identification of everything is necessary to meet different levels of
risk.

The exposure time in the contaminated areas is determined on the basis
of environmental conditions and physical effort that the work involved
implies.

The ENEA's classification (1984) consists of five classes in accordance with the degree of exposure and risk of breathing toxic substances,
from weak (Classes I and II), to high (Classes III and IV), to very high
(Class V).   The first two levels of risk require a facial mask with filter, while the third level requires a mask with self-contained breathing
apparatus.   The fourth and fifth levels require coveralls with air input
by means of a flexible pipe, the air being let in from beyond the contaminated area.   For emergency operations, the self-contained breathing apparatus is equipped with limited duration (15 min) respirators.

(2) Risk Areas.   In the Seveso TCDD case, the following three zones
were identified and associated with different risk levels:

- Zone A: high contamination (>50 $\mu g/m^2$), evacuated;
- Zone B: medium contamination ($5\text{-}50$ $\mu g/m^2$), severe limitations;
- Zone R: low contamination ($>0.75\text{-}5$ $\mu g/m^2$), less severe limitations and shorter duration.

(3) Protections for dismantling trichlorophenol production equipment.
An especially high risk area was located inside the department where trichlorophenol production had been carried out and where the accident had
taken place.   In previous experiences in other countries, there had been
chloracne cases and serious intoxication by TCDD due to insufficient
protection of workers dismantling the contaminated plants.   Therefore, in
view of the special type of operations required within the department,
special aerated suits were adopted.   These were individual protective

Mururoa-type suits made of transparent fire-proof PVC, and equipped with air-proof zipper, gloves, and welded overshoes. Suit aeration was ensured by a distributor and plastic pipes with holes placed along arms and legs and around the neck. Suit air outlet was equipped with two valves located on the back; these valves ensured complete sealing even in case of temporary interruption of compressed air or sudden depression of the suit. The hood was made of PVC with the front of methacrylate for excellent transparency. In case of suit tearing, the air outlet prevented the operator from being contaminated.

The compressed air for suit ventilation (300 l/min) was supplied by a double system: an Asterint compressor with air filter and a storage reservoir, combined with a battery of air cylinders at 150 atm. In case of compressor failure and air pressure drop in the pipes below 6 atm, the compressed air from the cylinders was let into the air circuit automatically, thus avoiding any interruption of the air supply. The operator was guaranteed 3 minutes of air previously stored in the suit, until the air distribution system was switched on.

Suits were equipped with an alarm device in case the air flow rate fell below 150 l/min. Air was kept at a temperature of 21°C by means of a water-heat exchanger to ensure a comfortable microclimatic condition inside the suit. Due to suit characteristics and risks that work implied, work shifts were limited to 2 hrs each with a maximum of 4 hrs per day for a normal two-shift working day. Before being removed, suits were washed with water and detergent in order to avoid worker contamination during undressing. After every shift within the trichlorophenol department, suits were collected in special containers, placed in a safe area, and later properly disposed of.

(4) Protective equipment for reclamation of large suburban areas. Because of TCDD toxic characteristics and expected work procedures, Seveso Special Office established the following protective equipment for operating in Zones A and B areas:

- cotton and rubber gloves; for heavy work, leather gloves;
- complete coveralls with hood, made of Tyvek;
- full face masks equipped with dust-proof filters (e.g., active carbon);
- rubber boots;
- covershoes made of Tyvek or polyethylene.

Since TCDD grips strongly to soil particles, the protection chosen was in fact suitable to avoid the operator's contact with dust, dust being the contaminating substance vehicle. The protections chosen were proportioned to contamination levels of the various areas and to the type of activity involved. However, for practical purposes and also in order to standardize purchases and stocks, the choice has been practically limited to Tyvek coveralls having different consistencies and colors, reserving the white-colored coveralls for all activities inside the fenced area, and the green- or blue-colored coveralls for operations in Zones B and R.

For what concerns masks, the nose-mouth type apparatus integrated with glasses was soon abandoned for the full-face type to unify stocks. In fact, the first mask could be easily removed while the second type was practically unremovable, thus offering a guarantee of protection to less careful operators.

Soil was kept constantly moist by means of spraying with hydrants either fixed to the ground or moved around by personnel. As an additional precaution, when wind speed as indicated by the decontamination unit's anemometer exceeded the 2-m/s limit, the technicians responsible for control operations stopped all activity inside the contaminated area.

Every worker was subjected to a sanitary checkup by the Desio Hospital Occupational Medicine Department prior to being accepted for reclamation operations. The sanitary checkup was carried out to establish the total absence of any actual or past pathologies. On the basis of such checkups, an average of one worker out of three examined, was considered suitable (Lombardy Region Bollettino Ufficiale, No. 2, January 17, 1977).

3.3.6   Accidents or Discomforts During Cleanup with Total Protection

During the four-year survey while cleanup operations were under way in high contamination areas (Zone A), no serious accidents were recorded. Only during the ICMESA factory evacuation a tractor driver of a building company fell off his vehicle when it overturned experiencing a brain shock with consequent eyesight damage. At least partly, this accident may be attributed to the reduced sight range the driver had due to the facial mask. In the area where operations were directly controlled by regional technicians, only minor accidents and discomforts were recorded.

Table 1 shows the statistical data referring to the number of people involved in Zone A operations: 45,000 man-shifts in four years. In order to make a correct evaluation of results exhibited, it is important to remember the types of work performed:

- removal of the vegetation;
- demolition of civil and industrial buildings;
- soil decortication and removal of contaminated soil layers by means of machines, under continuous water spraying to keep dust down;

Table 1. Infirmary Intervention[a] for Minor Accidents or
Discomfort in the Decontamination Team of Zone A[b].

Cause of accident or discomfort	No. of cases	% on man-shifts	Notes
Protection defects:			
- gloves	12	0.027	
- coveralls	9	0.018	
- masks	7	0.015	
	Subtotal:	0.060	
Not perfect physical conditions or other:			
- nausea	11	0.024	45%[c]
- headaches	7	0.015	71%[c]
- other	41	0.091	
	Subtotal:	0.130	
Totals:	87	0.190	

[a] Source: USRL-PO1, Desio Hospital Occupational Medicine Service, Desio (Milan), April 1985.
[b] Total number of work-attendances inside Zone A: 45,000 man-shifts in four years (June 1980-July 1984).
[c] In the three summer months (June-July-August).

- transportation of contaminated material with trucks to temporary storage places;
- building of water-proof basins followed by dumping of removed material from contaminated areas into basins;
- covering and sealing of basins, and washing up of means and tools used for work (final stage).

In prevalence, soil removal and demolition operations were carried out.

Work shifts lasted four hours, based on the Desio Hospital Occupational Medicine Department instructions. Shifts had been fixed of four hours only, bearing in mind the breathing difficulties through the mask filter as well as the physical discomforts produced by the individual protection, rather than the risk of contact with TCDD. However, in the trichlorophenol department operations were performed on the basis of two 2-h shifts.

In the light of the above, the rate of 0.19 percent (less than two minor accidents or discomforts every thousand man-shifts) appears to be really satisfactory. Moreover, less than 1/3 of infirmary interventions recorded should eventually be attributed to defective protection (foreseen for dust but not for every type of heavy activities performed during cleanup). Of this fraction, less than 1/3 (i.e., a total of approximately

10 percent) was due to coveralls which could not resist abrasion or the corrosive action of caustic products. In addition, if not worn correctly, coveralls did not protect the skin from insect punctures, stings, or thorn pricks.

Some cases of nausea and headache may have been due to the use of mask whose clasps could have been too tight at the back of the head. However, the higher rate of accidents registered in June, July, and August, compared to the rate of the other months with cooler temperatures, shows that the main factor in determining such uneasiness was climate, followed by worker's physical conditions at the time of entering the area of operations. The type of facial protection chosen could not be associated with any real inconveniences.

3.3.7 Conclusion

Looking back at the Seveso experience, one can affirm that the choice of the equipment proved to be effective in protecting the health of workers involved in cleanup activities. The lack of any apparent pathologies among the operators in comparison with the control group proves the above statement. In order to solve the Seveso case, 150,000 Tyvek personal protective suits and about 1,200 PVC Mururoa-type equipment were used. In conclusion, a few aspects of the reclamation work planning were crucial in safeguarding the health of the operators:

- accurate health control before acceptance and during actual exposure (every 3 months);
- training of selected workers;
- disposal of all the protective coveralls after being worn once; masks and boots were used several times and subject to accurate cleaning after every usage.

REFERENCES

Cardillo, P., and Girelli, A. 1980. Aspetti Chimico-Fisici dell'Evento di Seveso. Chimica* e Industria September 1980, 651-655.

Cattabeni, F., Cavallaro, A., and Galli, G., Eds. 1978. Dioxin - Toxicological and Chemical Aspects. Halsted Press, New York, New York (USA).

Esposito, M.P., Tiernan, T.O., and Dryden, F.E. 1980. Dioxin. U.S. EPA-600/280-197, Cincinnati, Ohio (USA).

Fregman, R.A., and Schroy, J.M. 1984. Environmental Mobility of Dioxins. Monsanto Co., St. Louis, Missouri (USA), April 15-17.

Hay, A. 1982. The Chemical Scythe. Plenum, New York, New York (USA).

Marshall, V.C. 1980. Seveso - An Analysis of the Official Report. The Chemical Engineer 1980, 499-516.

Salomon, C.M. 1982. Untersuchung der thermischen Stabilitat von Natrium 2,4,5-Trichlorophenolat enthaltenden Gemischten durch Differentialthermoanalyse. Chimia* 36, 133-139.

Schroy, J.M., Hileman, F.E., and Cheng, S.C. 1984. The Uniqueness of Dioxins Physical/Chemical Characteristics. Monsanto Co., St. Louis, Missouri (USA), April 15-17.

4.   LONG-TERM REHABILITATION

4.1   Design Issues in the Long-Term Surveillance of the Affected
        Population

Pietro Alberto Bertazzi, Carlo Zocchetti,
Laura Radice, and Angela Pesatori

Istituto di Medicina del Lavoro
Universita degli Studi
20122 Milan, Italy

4.1.1 Abstract

The first crucial point in long-term surveillance is to identify and enumerate the affected population. Exposure information should include the type and amount of chemicals released, their environmental fate, routes, and carriers of exposure. Early effects are relevant not only per se, but also as indicators of exposure to the chemicals. Long-term studies require a trade-off between efficiency and power. The whole population or selected samples (e.g., high-exposure groups, susceptible groups) might constitute the study group. Internal comparison is advisable to control for several possible confounders. A cohort approach should be sought since differential migration from the disaster area can be anticipated.

4.1.2. Long-Term Surveillance

This paper addresses the issue of long-term rehabilitation of populations affected by accidental exposure to dioxin (TCDD), with special emphasis placed on post-emergency medical surveillance and late effects ascertainment, and is based mainly upon the experience gained after the Seveso accident.

Understandably, after an unexpected accident, decisions have to be made rapidly and under great pressure. Attention is therefore focused on acute and immediately evident effects. There are, however, other decisions to be made rapidly, and they concern the preliminary planning of

long-term studies. Delayed effects (teratogenic, mutagenic, and carcinogenic) will occur, for their very nature, after a rather long time span;
nonetheless, the feasibility of similar studies depends heavily on steps
taken in the immediate emergency response phase. Therefore, people with
specific expertise in epidemiology should be part of the response team
from the very beginning.

When considering long-term effects, one has to bear in mind that the
affected population is not only comprised of those who suffer from immediate acute illness, but of all those who have been potentially exposed to
the agent. They include all persons living in the affected area at the
time of the accident, and persons who have been moving into the area
since, and for as long as a potential for exposure remained.

The exposed population should be identified and enumerated. The first
point requires a delimitation of the contaminated area. The second one
requires the collection of basic data for each individual resident in the
area. The nonresidents present at the time of the accident are thus lost,
unless they refer to existing reference centres. The basic data needed
are those necessary to trace people in time and space. A good example is
provided by the experience acquired in tracing occupational cohorts
(Figures 1 and 2; from Halperin, 1986). This information should be collected and checked according to a standardized protocol and stored in an
ad hoc registry. Problems of confidentiality may arise.

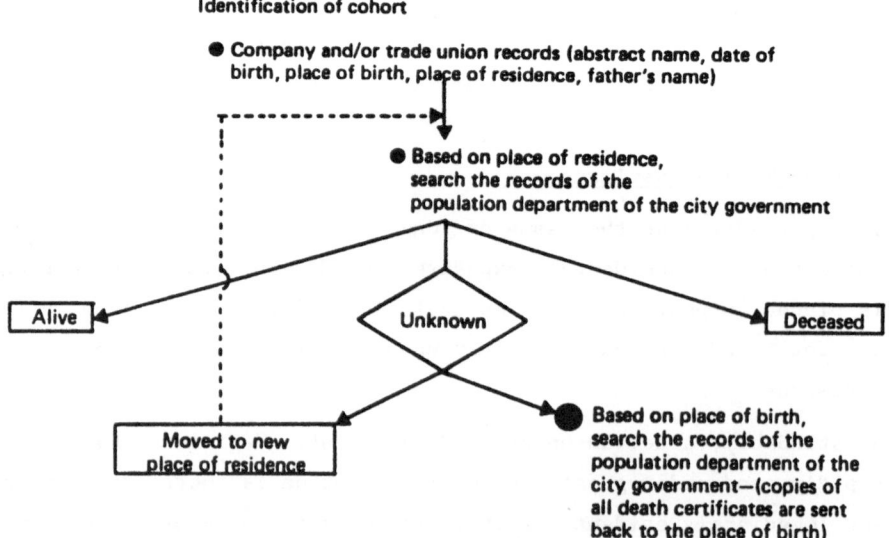

Figure 1. Vital Status Follow-Up in Italy

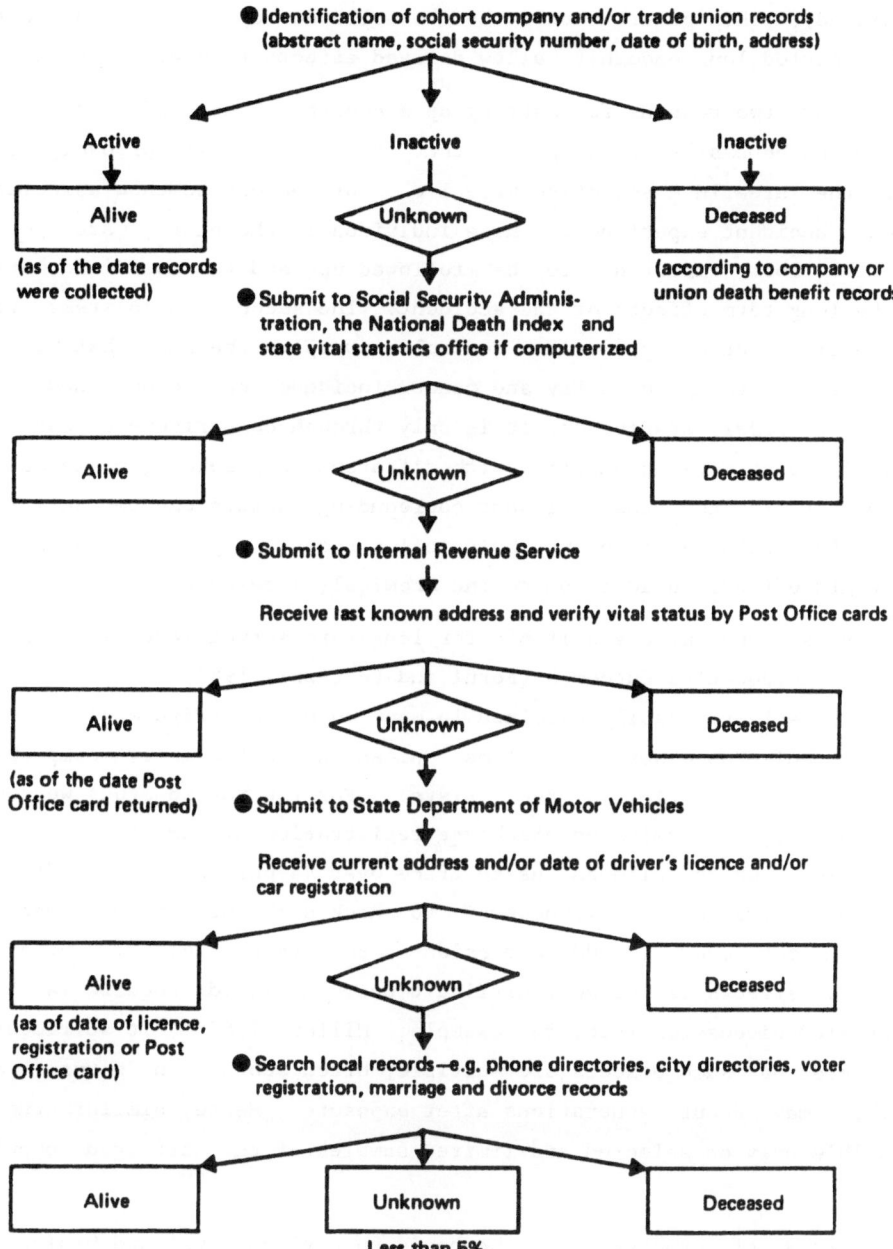

Figure 2. Vital Status Follow-Up in the United States

Such a registry has a peculiar feature in that it does not cover the population living at any time or at a single point in time in a given area, but deals with the cohort of people who have been living in the area in a specified period of time (the time during which a potential for exposure existed). This cohort should then be followed up in a prospective way for a period long enough to allow delayed effects to become manifest.

There are two reasons for setting up a cohort as the study population. First, in can be easily anticipated that residents will probably move outside the area in a selective way, i.e., for reasons somehow associated with their accident experience. These individuals, therefore, are potentially the most important to be followed up, and the most informative about the long-term effects of the accident. The second reason stems from the need of selecting appropriate controls. Despite the fact that comparison figures regarding mortality and cancer incidence can be obtained from national or local statistics, it is only through comparisons across sub-cohorts of the affected population with different exposure (in terms of level and/or length) that relevant confounding factors can be taken into account. For instance, major psychological stressors may exert their adverse health effect, in addition to the chemical(s) released.

Biological end points suitable for long-term surveillance studies are cancer and reproductive outcomes (Forni and Bertazzi, 1987). Cancer mortality can be fairly easily examined in many countries (Figure 3). Cancer incidence studies pose severe problems, unless a sufficiently comprehensive cancer registration system exists. For the Seveso study we could rely on the hospital admission/discharge registration system of the Lombardy Region, which allowed us to trace over 95 percent of the affected population. It was necessary, however, to check back the original medical information for over 20,000 admission cases (Figure 4). The study of reproductive effects is the most difficult one, and would require a long and detailed discussion (see, for example: Miller, 1983). It suffices to mention here that such effects are multiple, often subtle in nature, and that they may occur generations after exposure. Hence, similar studies are feasible only on selected and limited samples of the affected population.

Indicators of exposure may be based on environmental and biological measurements. The absence of an epidemiological approach in the early phase of response to the Seveso accident caused the waste of many opportunities for collecting biological specimens of different body fluids and tissues which could have been stored and analyzed later. Early effects, either specific (e.g., chloracne) or non specific (e.g., skin lesions,

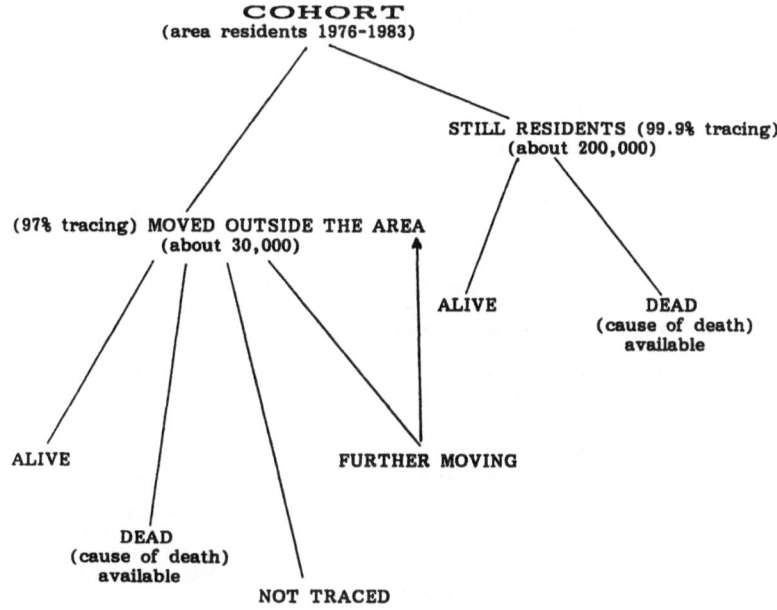

Figure 3. Follow-Up for Mortality in Italy

mucosa irritation, respiratory acute symptoms), should be viewed and treated not only as "effects", but also as important indicators of exposure.

Environmental measurements should be carefully planned and performed under controlled conditions. In Seveso it was possible to analyze thousands of measurements in order to identify contours of TCDD contamination in the affected area. As regards exposure, the basic information in the Cohort Registry should be sufficient to locate in time and place every cohort member with respect, at least, to the distance from the emission source, and, whenever possible, to the contours identified as having a varying degree of exposure. The use of questionnaires and interviews is advisable on selected samples of the population for specific research purposes.

The study conducted aims at quantifying the mortality and incidence from specified diseases in the affected population and comparing them with the corresponding experience of nonexposed populations. Provided that subcohorts with different exposure can be identified, the internal comparison will be most useful to control for several possible confounders, and to test for possible exposure-response relationships. Once the existence and the amount of given risks have been ascertained, the association of such risks with the relevant exposure can be tested by means of case-control studies. Cases and controls are drawn from the same study base

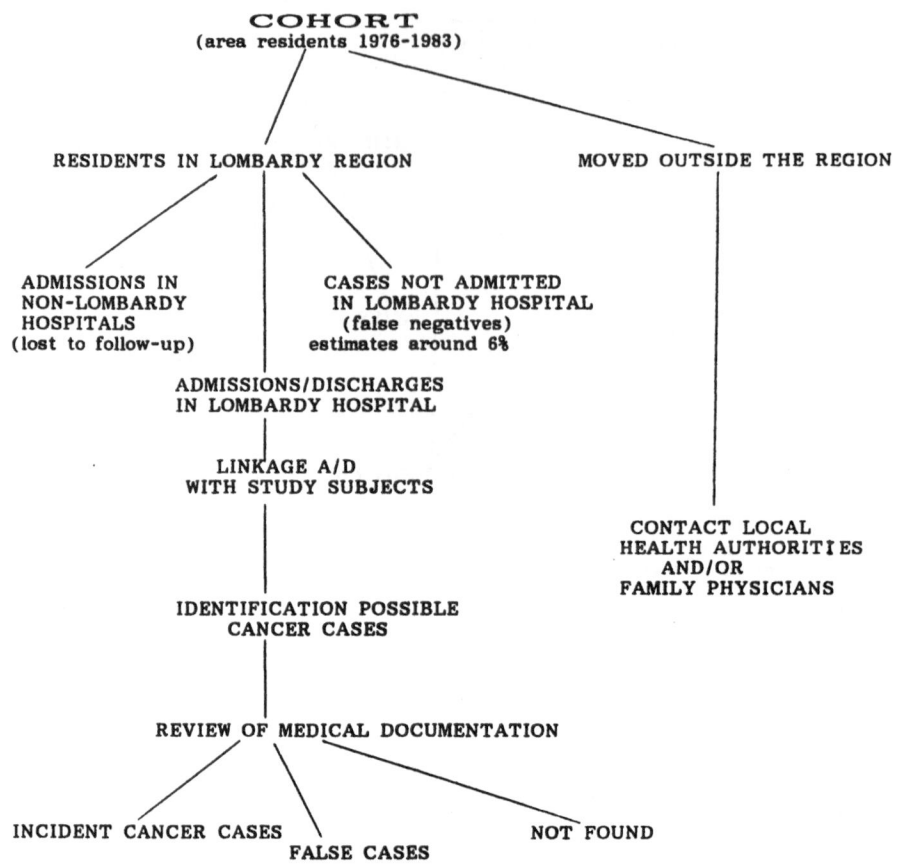

Figure 4.   Follow-Up for Cancer Incidence in Lombardy Region

(the cohort).   Similar studies are most efficient and suitable to study the specificity of the association between the chemical exposure and the noted effect.

In summary, the major point we have tried to make here is that, without an active mechanism of follow-up set up at the time when the accident occurred, it is almost impossible to conduct valid studies concerning the long-term effects of the accidental exposure.

REFERENCES

Halperin, W. 1986: The Cohort Study. In: _Epidemiology of Occupational Health_, pp. 149-180, M. Karvonen and M.I. Mikheev, Eds., World Health Organization Regional Publications, European Series No. 20.

Forni, A., and Bertazzi, P.A. 1987: Epidemiology in Protection and Prevention against Environmental Mutagens/Carcinogens. _Mutation Research 181_, 289-297.

Miller, J.R. 1983: Perspective in Mutation Epidemiology - General Principles and Considerations. _Mutation Research 114_, 425-447.

4.2   Surveillance in TCDD Cleanup Workers

Italo Ghezzi, Pasquale Cannatelli,
Paolo Mocarelli, and Franco Sicurello

Servizio di Medicina del Lavoro
Ospedale di Desio
20033 Desio (Milan), Italy

Giorgio Assennato

Istituto di Medicina del Lavoro
Universita degli Studi di Bari
Policlinico
70124 Bari, Italy

Franco Merlo

Istituto Nazionale dei Tumori
16132 Genova, Italy

Pietro Brambilla

Servizio di Patologia Clinica
Ospedale di Desio
20033 Desio (Milan), Italy

4.2.1   Abstract

A surveillance program for cleanup workers in the TCDD subareas A_1-A_5 in Seveso (Italy) was developed in order to verify the effectiveness of the safety measures taken during reclamation. The design was based on a prospective controlled study, with multiple comparisons of the cleanup and the reference groups. Clinical signs and symptoms, and biochemical parameters were measured before, during, and after the period of potential exposure, in both groups. The major study results are summarized: during the follow-up no case of overt clinical disease occurred which could be attributed to TCDD (i.e., chloracne, porphyria cutanea tarda, etc.). In conclusion, it is possible to assume that the safety measures taken during the cleanup operations were effective, reducing TCDD absorption to almost zero.

When cleanup in subareas A_1-A_5 (see, for instance, di Domenico, this report) was planned, a complex set of procedures was established in order to minimize the risk of exposure in cleanup workers because the occurrence of TCDD-related diseases among cleanup workers had been shown in several previous episodes. Along with a stringent safety and industrial hygiene program, a medical monitoring program was set up in order to verify the effectiveness of protective procedures used. The design was a two-year-long prospective controlled study, based on multiple comparisons of the cleanup and of the reference groups. The outcomes (mainly clinical signs and symptoms, biochemical parameters) were measured before and during the period of potential exposures in both groups.

The clean-up started in May, 1980 and was completed in 1984. The major study results are summarized hereafter.

(1) During the follow-up period no case of overt clinical disease which could be attributed to TCDD occurred (i.e., chloracne, porphyria cutanea tarda, peripheral neuropathy, liver disease).

(2) Five workers were considered no longer job-eligible at the job-fitness evaluations: one presented abnormal values of some liver enzymes at two examinations, returning to normal subsequently. The man had increased his weight from 95 to 108 kg in two months. However, it cannot be ruled out that it was a transient effect due to TCDD exposure.

(3) No clear-cut group difference was detected between the cleanup and the reference group for the laboratory tests (Table 1). In particular, the measure of urinary porphyrins which is considered to be a quite sensitive indicator of TCDD exposure did not show any remarkable difference. No significant difference was detected even considering the data as binary variables (normal-abnormal). The same conclusion is achieved when 13 variables are taken into account simultaneously by discriminant analysis.

(4) The analysis of individual trends for six variables in the cleanup workers, at the job-fitness evaluation, shows that some changes occurred only for triglycerides. However, it is attributable to biological variability or lack of analytical precision rather than to TCDD exposure.

In conclusion, it is possible to assume that the safety measures adopted during cleanup operations in subareas A_1-A_5, together with the compliance of workers, were effective in reducing the absorption of TCDD to almost zero. Therefore, the safety, industrial hygiene, and medical surveillance programs could be applied in future similar occasions. However, whether or not long-term effects of TCDD are also prevented, is still in question. The study stresses the importance of having a reference group for comparison, and baseline levels of the variables measured.

692

Table 1. Medical Monitoring Program

EXAMS		I C	I R	II C	II R	III C	III R	IV C	IV R	V C	V R
PORPHYRINS	NO. SUBJECTS	32	36	36	31	31	26	32	29	29	23
	MEAN	88.8	98.6	70.8	70.4	90.9	80.7	78.8	83.9	67.2	69.3
	SD	44.5	45	24.3	26.8	43.9	56.8	47.8	47.2	27	30.7
AST	NO. SUBJECTS	36	36	36	31	33	29	33	29	29	23
	MEAN	16.4	15.1	14.2	16	18.1	16.1	17.7*	14.4*	15.9	15.2
	SD	3.7	3.7	4.4	14	6.6	4.2	7.5	3.6	6.7	3.8
ALT	NO. SUBJECTS	36	36	36	31	33	29	33	29	29	23
	MEAN	18.6	18	16.1	18.5	21.8	19.9	18.8	19.7	19	20.3
	SD	5.5	5	6.1	11	13.3	9.5	6.1	7.3	7.1	10.1
AP	NO. SUBJECTS	36	36	36	31	33	29	33	29	29	23
	MEAN	28.4	33.6	29.7	31.4	30.3	30.4	30.8	30.1	28.1	28.1
	SD	8.6	6.5	7	8.9	7.5	5.2	7.2	7.7	6.6	6.3
GGT	NO. SUBJECTS	36	36	36	31	33	29	33	29	29	23
	MEAN	20.5	21.7	20.3	20.1	24.3	22.1	18.3	16.1	22.3	21.2
	SD	12.2	8.9	8.7	9.3	12.7	10.7	9.5	5.5	13.5	10.4
TOTAL BILIRUBIN	NO. SUBJECTS	35	36	36	31	33	29	33	29	29	23
	MEAN	.7	.6	.6	.5	.6	.6	.5	.4	.6	.6
	SD	.1	.2	.2	.1	.2	.2	.2	.2	.2	.2
U-ALA	NO. SUBJECTS	35	36	36	31	31	26	32	28	29	23
	MEAN	3.4	3.5	3.8	3.7	2.9	3.2	3.3	3.5	3.6	3.5
	SD	1.2	1.5	1.3	1.2	.7	1.3	1.4	1.3	1.3	1.6
CHOLESTEROL	NO. SUBJECTS	35	36	36	31	33	29	33	29	29	23
	MEAN	190.5	191.9	192.3	197.3	194.9	190.9	197.6	197.2	198.6	206.3
	SD	45.7	34.8	37.4	39.8	46.2	39.3	48.2	30.8	48.2	36.1
TRIGLICERIDES	NO. SUBJECTS	35	36	36	31	33	29	33	29	29	23
	MEAN	113.7	125.7	125.2	142.9	133.5	115.8	106.4	133.6	132.5	137.9
	SD	48.6	55.6	52.7	106.5	113.5	36.6	55.9	71.4	76.6	129.2

* = significant between-group difference

C = CLEAN-UP   R = REFERENCE GROUP

4.2.2 Methodological Limitations of the Present Study

Studies such as the present one are to be evaluated with particular caution and using criteria different from those used for etiological studies. It should be remembered that the aim was not to determine the effects of TCDD exposure, for which the study design would be inadequate, but to evaluate the effectiveness of the safety measures taken in the cleanup of a TCDD-polluted area, by means of a medical surveillance program. Even so, shortcomings are present in this study, some of which are reported below.

4.2.3 Type of Monitoring

Ideally, biological monitoring would be desirable; however, in such a situation, the only available method of biological monitoring would be fat biopsy, a procedure considered unacceptable by the Seveso Health Committee. Indirect monitoring is always a second choice, especially if the sensitivity and specificity of the tests chosen are unknown. Therefore, a wide set of laboratory tests was evaluated, aiming at identifying early toxic effects of TCDD. The specificity remains a major problem and is partially dealt with by exploring all other potential causes of abnormal findings.

4.2.4 Sample Size

It is possible to determine an a priori sample size large enough to minimize the type II statistical error of the study. In any event, awareness of such a shortcoming is essential when trying to generalize the study results.

4.2.5 Long-Term Effects

Because of the short follow-up period (two years), it is impossible to rule out any potential long-term TCDD-related effect (such as tumors and heart diseases).

4.2.6 Incompleteness of Biochemical Indicators

Other useful laboratory tests which should have been carried out in the study include cholesterol fractions (HDL, LDL, and VLDL), urinary d-glucaric acid as a marker of enzyme induction, and porphyrin patterns (which are more sensitive than just the total urinary concentration).

4.2.7 Suggestions for Future Studies

Hopefully, no disaster such as the Seveso one will happen in the future. However, if a similar situation occurs again, some actions are recommended:

(1) biological monitoring should be implemented by fat biopsy in order to detect the degree of exposure, if any;

(2) useless waste of resources should be avoided; e.g., laboratory tests for which there is no evidence of TCDD effect should be replaced by fewer, more valid tests, so reducing the number of comparisons; and

(3) strong emphasis should be given to the role played by the administrative staff, which is vital to keep the response rate as high as possible.

4.3   <u>Strategies for Rehabilitating Affected Areas and Buildings</u>

Vito La Porta

Regione Lombardia
20100 Milan, Italy

G. Umberto Fortunati

Via Vincenzo Monti 29, 20123 Milan, Italy

4.3.1   <u>Abstract</u>

Strategies for rehabilitation of areas affected by chemical accidents are selected taking into consideration a number of factors such as: (1) public opinion pressure, (2) socio-economic needs, (3) technical-economic reasons, and (4) organizational restraints. After setting acceptable limits for contaminant(s) (i.e., TCDD), priorities are defined for each type of construction. Different cleanup methods are chosen for each type of surface/material of the contaminated construction. The methodologies to check the effectiveness of reclamations carried out are also described. At the end of all the cleanup activities, the rehabilitation is marked by restitution of soil to its original use, inhabitants' return to their houses, and company's reimbursement of damages to the population.

The information exchange meetings organized within the frame of NATO/CCMS Pilot Study "Dioxin Problems" are an occasion to reevaluate, under different points of view, the experience made in Seveso by the authors. The accident is well-known as it has been already described by the same and other authors in meetings and reports.

To solve all sanitary, social, economic, and organizational problems, the Lombardy Regional Government issued a special law and set up an office in Seveso to act on the area affected by the toxic compound. The law provided for the following Task Groups:

Task Group 1: control of the contamination and reclamation activity;
Task Group 2: sanitary control for the population and veterinary assistance;
Task Group 3: social assistance, schools: it also provided new dwellings for the evacuated population;
Task Group 4: reconstruction of homes and public buildings not reclaimable;
Task Group 5: economic aids to artisans, industries and other productive facilities which were damaged by the toxic event.

The tasks within the five programs were aimed at normalization of life in the territory hit, taking into consideration both time and cost factors. For the reclamation program (Task Group 1) the options adopted had to take into consideration the following factors.

(1) Public opinion: the accident had a dramatic psychological impact on the population. Public opinion, in fact, influenced positively and negatively some of the solutions adopted. For example, an incinerator to dispose of TCDD-contaminated material could not be installed in Seveso because the local population feared smoke fallout. It was also feared that other industrial wastes might be destroyed as well.

(2) Socio-economic restraints: the area was interested by several types of urban and agricultural activities. Therefore, the choices on contamination control and cleanup priorities had to be made according to the use of territory and the social and economic importance of the area.

(3) Techno-economic restrainsts: in relation to soil contamination level and that of the other materials, cleanup methodologies adopted were diversified taking into account the type of surface to be reclaimed.

As an example, hereafter given in detail are the procedures adopted in reclaiming Subzones A_6 and A_7, characterized by a relatively low level of contamination: Zone A was subdivided into eight subareas classified according to contamination levels. The most southern area (A_6, A_7) had relevant importance because 70 percent of the population evacuated had resided there. The two subareas covered 31.8 ha and the minimum distance from the contamination source was 1,200 m. These subareas as well as the whole Zone A were analytically checked taking samples on the basis of an orthogonal reference grid made of 50-m-side squares. Soil contamination level varied between 270 $\mu g/m^2$ (2,160 ppt), and 21.7 $\mu g/m^2$ (173 ppt) with means for Subzones A_6 and A_7 of 30 and 15 $\mu g/m^2$, respectively.

The methodology adopted involved mechanical removal of TCDD from surfaces to reach acceptable thresholds.

4.3.2 Cleanup Methods

(1) Buildings. The following materials were not considered reclaimable: all textiles, like curtains, carpets, clothing, etc.; all furniture with textile upholstery; all food products; household appliances; all miscellaneous material the value of which was estimated to be inferior to reclamation cost. The above material was classified, listed, and removed starting from attics, and thereafter living quarters down to the basement; it was then transferred to Subzone A_5. The following procedures were adopted.

- Interiors were vacuum-cleaned and then washed thoroughly with water and detergent. At the end of each operation, cleansing liquids were sucked and collected in containers to be later transferred to nearby Subzone A_5.

- Furniture was first cleaned and then washed with disposable cloths impregnated with solvents. These activities started from the highest floors, in each flat starting from the farthest rooms back to the entrance. The entry hall was left at the end.

- The vacuum cleaners used were all equipped with absolute filters. In all rooms where results were not satisfactory reclamation operations were repeated.

(2) External surfaces. All external surfaces and fences were washed with high pressure water (60 bar). Vegetation was cut, ground, and transferred to Subzone A_5. A 25-30-cm soil layer was removed manually or using excavators whenever possible. With each machine there was a group of workers to collect the material dispersed during excavation. Soil scarification was stopped when the residual contamination at the bottom of excavation was $<5 \ \mu g/m^2$. Trucks were not allowed to run over reclaimed areas but used special routes. Removed soil was cautiously loaded on the trucks to avoid spilling along the roads. In every loading and unloading area, water was sprayed to keep dust level in air as low as possible.

4.3.3 Analytical Checks

Building interiors and exteriors were checked by means of wipe and/or scrape tests; the latter involved removing of a thin layer of plaster. Test points were selected taking into account the position of doors and windows - which had been exposed to the toxic cloud. Soil core samples were taken in gardens. After extraction, all samples were analyzed by GC-MS. Approximately 700 analyses were carried out on the 87 buildings and surrounding gardens. All were below acceptable level. Agricultural areas were checked in 56 points by utilizing the same orthogonal grid: again results were satisfactory, i.e., below the target.

4.3.4 Environmental Reconditioning

Buildings, gardens, and agricultural areas were finally reconditioned and checked again randomly.

4.3.5 Conclusion

At the end of all the described operations as indicated by the favorable outcomes of analytical checks, the Health Authority gave clearance for reentry of the evacuated population. For additional safety, another test was carried out a year later, by executing four wipe tests in each flat. Again analytical results did not reveal any trace of TCDD.

4.4 Decontamination of PCDDs and PCDFs In Situ by Solar and UV Radiation

Arnoldo Liberti, Domenico Brocco

Istituto Inquinamento Atmosferico, CNR
00016 Monterotondo Stazione (Rome), Italy

4.4.1 Abstract

The use of UV radiation has been found to be one of the most effective procedures to achieve the decay of dioxins and related compounds by replacing a ring chlorine with hydrogen. The reaction occurs in solution and its efficiency depends on light intensity at a specific wavelength. For TCDD the maximum light absorption occurs at 307 nm, so that a high pressure mercury lamp emitting in the range 280-320 nm can be used for photodegradation. Better results could be obtained using an excimer laser, i.e., an excited dimer laser with XeCl having an emission at exactly 308 nm. This device has been successfully used to obtain polychlorobiphenyl mineralization. This is obtained by spraying matrices with a suitable solvent (1:1 xylene-ethyl oleate) acting as a hydrogen donor. Soil contaminated with dioxin as well as areas where fallout of atmospheric dust containing dioxin(s) occurred can be decontaminated by means of this procedure.

4.4.2 Introduction

Decontamination of PCDD-polluted areas and, namely, of those close to sources whose emissions contain dioxins as urban wastes incinerators, is still an open problem and slight progress has been so far realized. Since the Seveso accident, based on the early investigations by Crosby et al. (1971), extensive work has been carried out both in the laboratory and field to demonstrate solar and UV photodecomposition of soil as well as any surfaces polluted by PCDDs (Allegrini et al., 1977; Liberti et al., 1978; Bertoni et al., 1978).

For environmental protection, public administrators should get advantage of the possibilities offered by radiation. There is a common agreement that dioxins are ubiquitous and traces of these compounds are released into the environment by a variety of sources. However, in spite of the abundance of these sources, very seldom accumulation occurs in open spaces. This seems to indicate that release into the environment is counterbalanced by removal probably occurring to a large extent through photodecomposition.

4.4.3   Photodecomposition of TCDD

The basis for photodecomposition arises from the examination of the UV spectrum of TCDD which exhibits an absorption maximum at approximately 307 nm (Figure 1). It means that using a radiation source such as a mercury lamp, whose emission spectrum covers the range 280-320 nm, experimental conditions to achieve photodecomposition may be realized. As the sun spectrum contains a certain percentage of such radiation, solar light may be used just as well for this purpose. There is however another factor which has to be accounted for, and this depends upon the mechanism of photodecomposition. Even though the photolytic mechanism has not yet fully established it seems to occur through loss of chlorine atoms by dissociation of excited molecules to free radical, or through nucleophilic displacement. Either mechanism is plausible and the occurrence of the decay process is affected by reaction medium and local environment conditions.

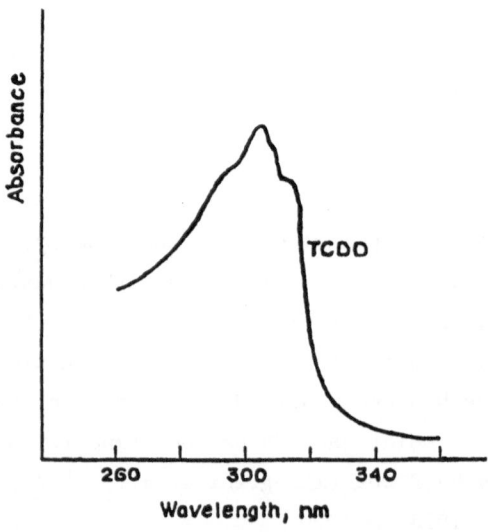

Figure 1.   Ultraviolet Absorption Spectrum of TCDD

Photodegradation of TCDD occurs only in the presence of a system acting as a hydrogen donor such as alcohols, ethers, hydrocarbons, and natural products as waxes. Such donors could be represented by the waxy cuticles of some green leaves as well as by traces of oily or aromatic solvents. However, when hydrogen donors are not directly available they must be supplied: this may be achieved by spraying the contaminated materials or any surface with a suitable solution. A liquid layer is then formed in which dioxin can dissolve. Though it is believed that a large variety of mixtures might be used, quite satisfactory results have been obtained by using a spray solution containing 1:1 xylene and ethyl oleate. Both are solvents for dioxin. Xylene is used to render the mixture less viscous: as it vaporizes, a thin film is formed consisting of ethyl oleate, transparent to UV radiations and acting as an efficient hydrogen donor.

As the primary photodecomposition product has been proved to be trichlorodibenzodioxin, reaction shown in Figure 2 can represent the first photolytic process. The free radicals or newly formed compounds are photolyzed at a higher rate than TCDD. The photolytic action can be followed by gas chromatography until no further chromatographic peaks are observed.

In the photodegradation of dioxins two factors need to be considered: environmental conditions and radiation energy. The nature of the matrix upon which TCDD is deposited plays an important role in the photolytic reaction. On silica gel and aluminum plates decay occurs to some extent, whereas no reaction takes place on glass and ceramic tiles. However, there is no limitation to have all systems added with an activating solution as the one reported on.

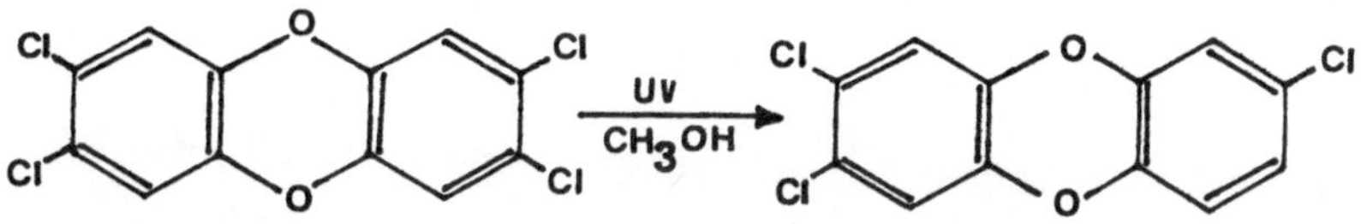

Figure 2. TCDD First Photolytic Process in Hydrogen-Donor Media

A further support to PCDD photodecomposition comes from a recent investigation on the use of excimer lasers for the safe destruction of dangerous materials. It was shown that polychlorobiphenyls (PCBs) dissolved in proper solvents (n-hexane, methanol, or ethanol) have absorption bands which resonate with the excimer laser radiation in the range 190-310 nm. Solutions of PCBs in these solvents were exposed to a KrF laser beam: by following the process through gas chromatography, new peaks were observed at the beginning of irradiation for the formation of chlorobenzenes. These disappeared for further exposure to the laser beam (Figure 3).

The use of a XeCl excimer laser, whose radiation occurs at 308 nm, seems to be the most convenient mean to achieve photodecomposition of TCDD. Though photodecomposition has not been extensively used in decontamination processes, there is experimental evidence of its efficiency in determining the decay of dioxins. It seems that in any accidents leading to release of dioxins, or in any industrial activities leading to production of PCDD traces, the spray of a solution acting as a hydrogen donor and the possibility of building up a thin oil layer where dioxins can dissolve is the best practice to favour their photodecomposition. Photodecomposition may be most efficient if the spraying of this solution is

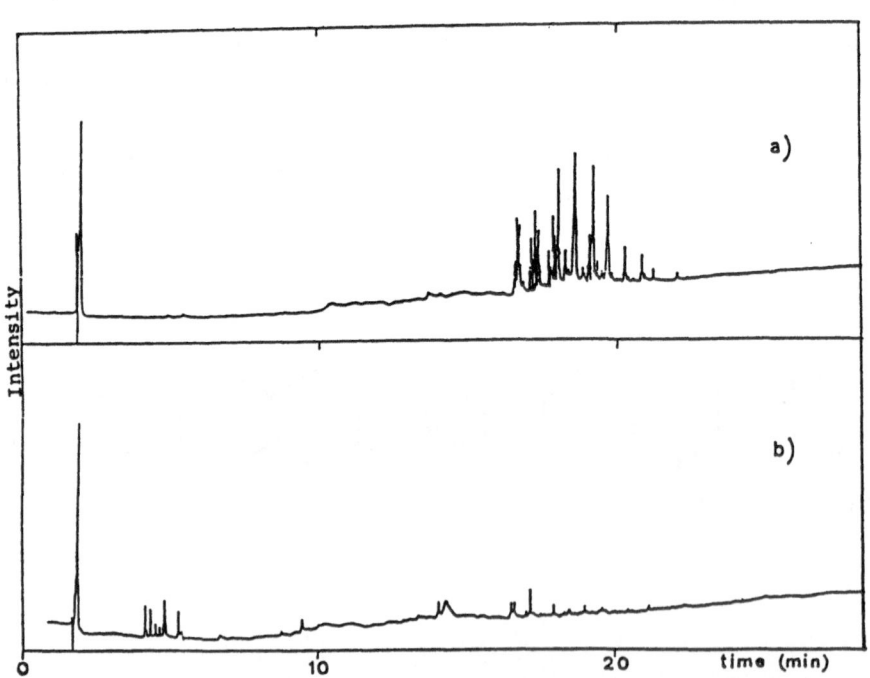

Figure 3. Gas Chromatograms of an Aroclor 1254 Solution in n-hexane:
(a) Before Irradiation; (b) After Irradiation with Laser KrF
(40mJ) for 1 hr at 2 Hz.

carried out in the summer time to take advantage of the more intensive solar radiation and of the favourable temperature.

4.4.4   Conclusions

As a result of the investigations carried out on photodegradation of dioxins and polychlorobiphenyls, the following points have been established.

TCDD and polychlorobiphenyls, and most likely all chlorodioxins and chlorodibenzofurans, readily undergo photolysis in the presence of hydrogen donors including alcohols, ethers, hydrocarbons, and natural products such as waxes. Considering all the natural removal mechanisms, photolysis appears to be the most significant degradation process. It is difficult to assess it in a quantitative fashion, as in natural systems PCDD distribution is not readily evaluated. Efficiency is however very high when PCDDs are on the surface and directly exposed to UV radiation.

Three major requirements for PCDD and PCDF photolysis have to be accounted to achieve their decay:

- dissolution in a light-transmitting film,
- presence of an organic hydrogen donor,
- ultraviolet light of appropriate wavelength.

The primary photolytic degradation pathway involves dechlorination to less chlorinated congeners. The final products of this sequence of reactions are unknown. The higher chlorinated congeners such as OCDD are less reactive than the lower chlorinated compounds.

REFERENCES

Allegrini, I., Bertoni, G., Brocco, D., Liberti, A., and Possanzini, M. 1977. Decontaminazione mediante Radiazioni Ultraviolette da Inquinamento da 2,3,7,8-Tetraclorodibenzodiossina. La Chimica e l'Industria 59, 541-544.

Bertoni, G., Brocco, D., Di Palo, V., Liberti, A., Possanzini, M., and Bruner, F. 1978. Gas Chromatographic Determination of 2,3,7,8-Tetrachlorodibenzodioxin in the Experimental Decontamination of Seveso Soil by Ultraviolet Radiation. Analytical Chemistry 50, 732-735.

Crosby, D.G., Wong, A.S., Plimmer, J.R., and Woolson, E.A. 1971. Photodecomposition of Chlorinated Dibenzo-p-dioxins. Science 173, 748-749.

Liberti, A., Brocco, D., Allegrini, I., Cecinato, A., and Possanzini, M. 1978. Solar and UV Photodecomposition of 2,3,7,8-Tetrachlorodibenzo-p-dioxin in the Environment. The Science of the Total Environment 10, 97-104.

5.   OTHER CONTRIBUTIONS

5.1   Accidental Release of 2,3,7,8-Tetrachlorodibenzo-p-Dioxin (TCDD)
          at Seveso, Italy: Estimation of Total TCDD Released

          Alessandro di Domenico, Vittorio Silano,
          Giuseppe Viviano, and Giovanni Zapponi

          Istituto Superiore di Sanita
          00161 Rome
          Italy

5.1.1   Introduction

    The following is a condensed translation of the recently-released
inquiry by experts designated by the magistrate entrusted with judiciary
proceedings on the Seveso accident. Some relevant data reported in the
Acts of the Parliamentary Commission for the Seveso accident are also
given.

5.1.2   TCDD Synthesis

    TCDD can be made up in the laboratory by condensing two molecules of
2,4,5-trichlorophenate at above 180°C, optimum reaction temperatures
being between 250 and 300°C. It is known that if the reaction occurs
in a vacuum so as to facilitate TCDD sublimation and prevent its further
reacting or decomposing, yields >30% may be had. If, on the other hand,
TCDD removal is not favored, yields are much lower, i.e., in the region
of 0.5 to 1%.

    As TCDD is formed by the thermal condensation of trichlorophenate,
secondary reactions leading to TCDD formation may occur during the production of 2,4,5-trichlorophenol (TCP) (see section 5.1.3). In theory, TCDD
cannot be formed below 153°C. At 180°C less than 1 mg is formed
per kg of TCP. If heating at 230-260°C is continued for 2 hours,
about 1.6 g of TCDD may be obtained per kg of TCP.

5.1.3   2,4,5-Trichlorophenol Production at the ICMESA Plant

TCP production at the ICMESA plant was started in 1969-1970 and brought up to full scale levels in the following years with a big production hike in 1974-1975 (Parliamentary Commission, 1978). Annual TCP production figures are summarized in Table 1. TCP production at the ICMESA plant was based essentially on the procedure set forth in a Givaudan patent (US Patent 2,509,245, filed on March 20, 1947) that foresees partially hydrolyzing 1,2,4,5-tetrachlorobenzene to sodium trichlorophenate with sodium hydroxide, and transforming the trichlorophenate to TCP by acidifying it with hydrochloric acid.

Table 1. 2,4,5-Trichlorophenol Production at ICMESA Plant, 1970-1976

1970:	6,361 kg
1971:	33,000 kg
1972:	40,350 kg
1973:	No production owing to lack of orders
1974:	38,400 kg
1975:	105,346 kg
1976:	142,820 kg (sold up to July 9, 1976)

Prior to the accident the tetrachlorobenzene hydrolysis reaction used to occur inside a 10 m^3 reactor in Sector B of the ICMESA plant (Inquiry, 1979). The following substances were introduced at the start of every reaction cycle: tetrachlorobenzene (~2,000 kg), sodium hydroxide (~1,040 kg), xylene (~600 kg) and ethylene glycol (~3,280 kg).

Controlling the process temperature is important because an exothermic reaction may occur when sodium hydroxide is mixed with ethylene glycol. This reaction, attributable to the decomposition of sodium ethoxylate ($NaOH_2C-CH_2OH$), may start at ~230°C and progress rapidly and uncontrollably with a temperature increase to ~410°C and release of large quantities of fumes (Milnes, 1971; Jirasek, et al., 1973; May, 1973; Jirasek, 1974; Carter, 1975).

During the reaction, batch temperatures were kept at 140-150°C; the xylene that distilled with the reaction water was recycled after its water content had been removed (a). At the end of the reaction, xylene and the nearly 1,500 kg of glycol that had not condensed into diethylene glycol or had not reacted with the sodium hydroxide were distilled (b). Once distillation was complete, 3,000 kg of water were introduced into the reactor and the mass so-formed transferred to a second reactor where acidification and all other necessary operations to separate out the unrefined trichlorophenol took place (c). The trichlorophenol was distilled to give ~1,300 kg of commercial product (d). For each of these four phases (a-d), 8, 6, 1, and 8 hours were required, respectively. A full production cycle could therefore be completed in 24 hours with three work shifts.

The process adopted at the ICMESA plant incorporated a few variations on the original Givaudan process (Parliamentary Commission, 1978). The Givaudan patent foresees that the solvent be distilled after the trichlorophenate has been acidified into trichlorophenol. At the ICMESA plant the order was inverted. Without such an inversion, prolonged heating of sodium ethoxylate could have been avoided as solvent distillation would not have occurred in an alkaline environment. Added to this, the process adopted at the ICMESA plant entailed a gradual removal of the solvent which necessarily led to a progressive reduction in the heat buffer. Lastly, there was a modification in initial tetrachlorobenzene, sodium hydroxide and ethylene glycol molar ratios. In the Givaudan patent said ratios are approximately 1:2:11.5 whereas the ICMESA ratios were 1:3:5.5. These variations entailed a substantial reduction in costs on one hand and an increase in TCDD formation and glycol exothermic reactions risks on the other (Parliamentary Commission, 1978).

At about 5 a.m. on July 10, 1976, workers at the ICMESA plant began shutting down the production cycle for the weekend. At that point only 500 of the 1,500 kg of glycol had been distilled (Table 2). Interrupting the cycle before distillation had been terminated and wash down procedures started was not part of the normal routine, which was usually completed prior to the beginning of the weekly rest period (Parliamentary Commission, 1978).

Table 2. ICMESA Reactor Estimated Batch Contents
at Start of Shut-Down on July 10, 1976

Sodium trichlorophenate plus other hydrolysis products[a]	2,000 kg
Ethylene glycol (still to be distilled)	1,000 kg
Diethylene glycol[b]	1,300 kg
Sodium hydroxide[c]	360 kg
Sodium chloride	540 kg
	5,200 kg

[a] ~85% as trichlorophenate.
[b] Diethylene glycol + polyethylene glycols + alcoholates.
[c] A portion was salified with hydrolysis products and another portion was contained in the glycols.

At about 1 p.m. on July 10, 1976, an exothermic reaction due to unknown causes had raised the in-reactor temperature and pressure beyond limits thereby causing a safety device tared to approximately 3 atmospheres to blow out (Inquiry, 1979).

Accident consequences, and in particular those that befell the plant surroundings, could have been avoided, or at least restrained, if the plant had been equipped with a collecting and abating system for the substances which may - as in fact did happen - possibly be released. Similar safety devices had already been installed some years before 1976 in other trichlorophenol-producing industrial plants such as those operated by the Dow Chemical and Coalite companies (Rawls and O'Sullivan, 1976).

The ICMESA company violated Article 3, Sub-section 1 of Presidential Decree 322 dated April 15, 1971, the contents of which prescribe that all industrial plants forming part of an industrial complex and which may contribute to atmospheric pollution, must be equipped with abatement systems in conformity with the regulations therein reported (Parliamentary Commission, 1978).

At the moment the safety device gave way, the major volatile compound inside the reactor was ethylene glycol (b.p. 197°C, vapor pressure 3 atmospheres at ~240°C). It cannot be established exactly at what temperature the valve blew out, because the reactor did not contain pure ethylene glycol, but a multi-component mixture and some water vapor deriving from glycol condensation. However, it can be assumed that this temperature was above 200°C and close to 240°C. After an almost instantaneous pressure drop inside the reactor when the valve gave way, the glycol present probably boiled violently thereby causing the batch temperature to drop thanks to the elimination of latent heat of vaporization (to reduce the temperature by 1°C, 10 kg of glycol had to evaporate). Glycol distillation gradually normalized reaching equilibrium conditions when the temperature inside the reactor had dropped to ~200°C. Once the glycol had been exhausted, the temperature most probably rose to a new plateau where it steadied until the ~1,300 kg of diethylene glycol (b.p. 245°C) had been eliminated. After the diethylene glycol had been exhausted, reactor temperatures probably went up even further to ~350°C at the bottom of the reactor and 450°C in the batch upper layer thus causing extensive mineralization of residual organic substances. Emission gradually dropped during this phase until it ceased altogether.

Almost certainly the violent blow-out following the valve's giving way caused the escaping vapors and entrained particles to leave the reactor at a speed of some hundreds of meters per second. Smaller particles may have been carried out by the glycol vapor during its violent initial boiling phase.

During the glycol evaporation stage following its initial boiling in over-heated conditions and during diethylene glycol evaporation and resi-due mineralization, gaseous compounds were expelled from the reactor. This could only have entrained fumes or sub-micron sized particles (Inquiry, 1979). Approximately 2,300 kg of matter remained inside the reactor, of which ~72% was sodium chloride and the rest was decomposed organic residues (Inquiry, 1979).

It may be estimated that altogether nearly 2,900 kg of organic matter left the reactor. Consequently, apart from the 2,300 kg of glycol (Table 1), at least 600 kg of trichlorophenate were expelled.

Owing to reactor ambient temperature, part of the sodium trichloro-phenate was already being transformed into TCDD when the valve gave way. Part of the TCDD inside the reactor at that moment was dispersed along with the larger particles entrained by the fluid burst. Mixed with these larger particles, the TCDD settled to the ground in accordance with pre-vailing wind directions at the moment the valve blew out (di Domenico, et al., 1980). Furthermore, favorable conditions for TCDD entrainment in a vapor stream existed while the 1,000 kg of glycol and 1,300 kg of diethylene glycol were evaporating. It seems likely that TCDD emission also continued during the final mineralization phase.

Analytical data concerning TCDD levels inside the reactor show that TCDD was subject to considerable volatilization. In fact, sampling from the reactor up to the super-structures shows that dioxin levels increase progressively in proximity of the blow-out aperture (Inquiry, 1979). As compared to the 100 ppm found in the 2,300 kg of matter that remained inside the reactor, 0.18% of that taken from the stirrer vanes was TCDD, 0.85% of that in the condensate barrel, 9.5% of that in the final bleed valve condenser while 12 and 20% levels were detected in the deposits on the inner walls of the bleed pipe. The TCDD on the internal walls could have only come from the final mineralization stages because the >100 m/s velocity of particles expelled during the initial pressure release would not only have prevented the gas-entrained particles from depositing on the bleed pipe lining but would have even favored wall cleansing by the "sand blast" effect of the escaping particles. During glycol evaporation, the internal walls were washed by the condensate that fell back into the reactor thereby taking with it any TCDD that may have been deposited. Consequently, the TCDD contents found in the deposits could only have been formed after volatilization and the washing down by the wall-cleansing condensate had terminated (Inquiry, 1979).

As stated above, the progressive removal of the TCDD being formed may enhance production yields. During the accident, removal may have been favored by the evaporation of the 2,300 kg of solvent present. Although the considerable amount of sodium hydroxide (Table 2) may have reduced reaction yields, outbound TCDD dispersion of more than 10% of the initial trichlorophenate content cannot be excluded (Inquiry, 1979). However, this figure appears to be improbable. Event dynamics, batch figures, and subsequent analyses carried out on reactor super-structures suggest dioxin yields up to 3% of the total trichlorophenate as being possible. If we assume that 2,000 kg of Na-TCP were present inside the reactor at the time the blow-out occurred, a 1% TCDD formation yield would have produced ~13 kg of TCDD.

REFERENCES

Acts of the Parliamentary Commission for the Seveso accident. Atti Parlamentari VII legislatura - Doc. XXIII n.6 1978.

Carter, C.D., et al. 1975. Science 188:738.

di Domenico, A., Silano, V., Viviano, G., Zapponi, G. 1980. Accidental release of 2,3,7,8-tetrachlorodibenzo-p-dioxin (TCDD) at Seveso (Italy): II. TCDD distribution in the soil surface layer, Istisan 1980/3.

Gribble, W.G. 1974. Chemistry 47:15.

IARC Monographs. 1977. The evaluation of the carcinogenic risk of chemicals to man. 15:54.

Inquiry of the experts designated by the magistrate entrusted with judiciary proceedings on the Seveso accident. 1979.

Jirasek, L., et al. 1973. Cesk. Dermatol. 48:306.

Jirasek, L. 1974. Cesk. Dermatol. 49:145.

Jirasek, L., et al. 1974. Hantzart 27:328.

Langer, H., et al. 1973. Environ. Health Perspect. 5:3-7.

May, G. 1973. Br. J. Ind. Med. 30:276.

Milnes, M.H. 1971. Nature 232:395.

Rawls, R.L., O'Sullivan, D.A. 1976. Chemical and Engineering News 54:27.

5.2 Needs for Skilled Personnel in Emergencies Following Industrial Accidents

Luigi Noé

ENEA
00198 Rome, Italy

5.2.1 Abstract

Management of the emergency phase of industrial accidents is acknowledged as the most critical one in terms of having the capability to (1) provide immediate and effective response to the accident effects and (2) plan the most appropriate future actions. The importance of producing good records (films, video tapes, photographs) of social behavior and reactions in the stressed situation - as well as that of recording all the technical operations performed - is emphasized. Lastly, the Seveso experience indicates that there is a strong need for skilled personnel to handle events adequately.

Because of the increasing amount of chemicals being produced, used, transported, and stored around the world, an increase in chemical accident risks is inevitable. The organization of the necessary actions following a serious industrial accident is deemed essential and delicate. It is clear that prior investigations of the work required in case of an accident can facilitate this initial phase. On the other hand, it may be said that some decisions made in the emergency period itself condition the development of subsequent work.

By definition, accidents are unexpected events. Furthermore, no two chemical accidents are identical. Consequently, emergency and remedial action management call for specific information and expertise. In particular, the toxicological, physical, and chemical properties of the material released, as well as its environmental behaviour and effects must be known at once. Suitable techniques and know-how should therefore be implemented immediately so as to contain and possibly eliminate short-term and long-term accident consequences, monitor contaminated areas and the population at risk, and assess remedial action efficiency.

In the nuclear field, codes and rules provide exactly the nuclear emergency plan. However, in spite of the differences, the nuclear field experience and policies may supply useful hints for collection of the basic data to be later utilized in the initial emergency phase of a chemical accident. It should be added that even if the nuclear field is a valid reference example, nevertheless management of a nuclear emergency has to be handled by a body, or group of experts, different from that destined to manage an industrial accident - given the different nature and object.

The capability to handle an emergency following an industrial chemical accident requires experience of a wide range of possible accident types with a great number of eventual polluting substances to be identified and evaluated for both health and environmental impact. It is necessary that the Civil Protection have access to a data base containing all the information concerning the relevant toxic substances, among which: physical and chemical characteristics, measurement methods, toxicity levels and propagation models in various environmental and biological chaines, and eventual processes for disposal.

A nation-wide system for the quick search of scientific and technical information, and its relative retrieval and analysis should be implemented so as to provide data essential for dimensioning the problem and immediate starting of a detailed planning. Indeed, organizational aspects should be studied in advance so that responsibilities may be designated, tasks assigned, and operations coordinated to ensure skilled personnel and resources availability and establish prompt and efficient information exchanges with national and international institutions, agencies, authorities, and the public.

The Chernobyl accident has shown the usefulness of atmospheric propagation models for radioactive pollutants, under various meteorological conditions. Also in some possible industrial accidents a rapid appreciation of the effect of metereological conditions is fundamental. This fact was shown at Seveso, where the prevaling atmospheric conditions strongly influenced the event. To rapidly locate this type of information and define the initial interventions, a data base can be of great help. This data base may be used in the development of an approach strategy (e.g., the need for epidemiologic studies).

In addition, it could be useful to produce video tapes and similar recordings to have in the end records of requirements and behavioral reactions at various stages (local health units, population, etc).

As for a nuclear emergency, it is extremely useful to have a list of the hospitals capable of treating injuries of various degree. This information is particularly important in accidents with injuries that require highly specialistic treatments (e.g., serious burns).

Lastly, the so-called psychological risk based on prevention has to be kept in mind; a conditioning factor in this is the quality of the official information and the level of responsibility at which this information is gathered and then disseminated.

The availability of known elements is a necessary step to reduce the critical moments of the emergency, but even more important is the choice of the personnel that must deal with it. The following observations are made from the Seveso experience.

(1) Due to the resultant potential hazards to human health and the environment, it would seem advisable for a country to set up an emergency response and advisory service capable of providing essential assistance for chemical accident occurrence.

(2) An adequate emergency organization should be predisposed on the basis of contributions from existing institutions and services, and aimed at guaranteeing the immediate mobilization of experts, technicians, resources, equipment, and information systems. Such an organization might contemplate all relevant public health organizations, and officers should be familiar with an essential set of general safety rules, to be followed immediately after the accident takes place.

(3) A complete list of national and international experts capable of providing contributions in specific fields should be drawn up in connection with those activities and studies required in most emergencies (public health surveillance, toxicological care, environmental monitoring, chemical and biological testing, and so on).

(4) According to the extension and seriousness of the accident, a proper number of experts should be selected from the list available to form a group for management of the emergency, reclamation of the land, and health assistance.

(5) The choice of personnel is an important item. Personnel should be preferably chosen locally, and complete the operative structure. They should have adequate professional qualifications, and be fully committed to their job.

5.3   Liability and Compensation in the Seveso Technological Disaster

Giuseppe Anzani

Via Isonzo 11, 22100 Como, Italy

5.3.1   Abstract

The Seveso industrial accident is the starting point to explore the difficult and delicate relationships between liability and compensation of damages. By and large, the issue of compensation of damages is a secondary problem, that is a problem arising as a result of a law violated because of an illegal act - law's primary precept and goal being "neminem ledere", i.e., do not damage another human being. At Seveso, or in general in similar cases, the complex human sacrifice is difficult to measure and almost never actually compensable. In strict legal terms, "moral damage" to the population will remain beyond monetary relief; which places justness not in reparation of wrong, but in its prevention.

5.3.2   Introduction

Technological progress requires the ever more extensive use of chemical components, many of which are highly toxic (note 1), and risky processes: in terms of control of the production method (security), or because of formation of undesirable byproducts, or because of discharge of wastes. Such risks, some of which are not even well known, represent a menace to human health and the environment. Environmental contamination in turn causes a deterioration in the quality of life. This may happen even without unforeseen disasters: in creeping and continuous fashion, thus masking the overall damage because of the absence of apparent individual damage. And because no one takes the initiative to complain privately to the courts, this ends up by being "normalized" and sustained, even for long periods of time, since the individual does not realize the difference between his condition and that of other people.

A demand for compensation usually occurs when an unforeseen disaster strikes individuals, life, health, property; the goal of the dispute is then to acknowledge that the damage is not the true situation (degraded) preceding the disaster, and recognize the ideal and "proper" situation that corresponds to full respect of the rights of the subject. The overall damage is not a differential damage, but a total damage.

One may wonder about the acceptability of those phenomena that were tacitly sustained or even unknown before the accident. It is not clear if the principles of common law concerning compensation are sufficient to solve the problem, or if special laws must be provided. Or finally, if the maximum force of law is that which avoids such disasters as much as possible with severe measures of obligatory safety.

5.3.3   The Principle of "Neminem Ledere" and Compensation of Damage

In effect, compensation of damage is a secondary problem for the law. Not in the sense that it is less important, but in the sense that it comes second: its function is not satisfied by a primary law, but reparatory to a law that has been violated by means of an illegal act and where the prior situation is almost never exactly restorable.

This fundamental principle is kept in mind at the forefront of all legal considerations concerning a comparison of collective interests at stake between the requirements of economy and the requirements of protection of health and the environment: the rule "he who damages pays" is not quite equivalent to the primary precept "neminem ledere", which is the original standard. It is asked: is it not utopian to raise the question of technological activity totally immune from damage and risk? I answer that any modern legal system, now more informed than in the past of the hidden risks of certain enterprises, must strive toward this.

Although the zero index is not possible in absolute terms, restriction of rights that arises from a calculated hazard because of permission of hazardous activity (legal, but subject to proper caution) should be done by accounting for the comparative risks and benefits derived. It is the pioneering aspect of research and technological progress that should not be suppressed; but at the same time there is a requirement not to sacrifice to it values and interests that are superior. Determination of the point of equilibrium (or safety) is an area where science can give its answer and guide the law in order to form the content of the law.

5.3.4   Private Subjective Right and "Diffuse Interests"

Compensation of damage is a principle that generally operates within the scope of individual private rights. But often the prejudice stemming from industrial initiative is collective and indivisable in nature. This is obvious in the case of creeping environmental pollution. It has been accepted for centuries that the air and water, for example, are not "res nullius" but "res omnium communes"; agreement has still not been reached in terms of the fact that legal interest, which makes fruition of certain benefits from the earth intangible, does not derive from private property (from a portion of the ground, the corresponding column of air above it, water running nearby, or what is concealed underground), but from life and from human citizenship (note 2). Thus, even those who possess nothing, or who live far from the specific site where environmental pollution occurs experience the same wrong as soon as they also suffer by merely seeing the quality of life deteriorate, either from the objective side or from the subjective and psychological side that includes emotional stress and fear. It is therefore necessary to understand that the intent of legal protection does not concern the environment per se, but rather man in relation to the environment.

In Italy preventive protection of this "diffuse" right is still rarely entertained in private lawsuits on the occasion of their violation. Intervention of public authority predominates through laws and administrative acts for authorization, control, and repression of illegal activities in terms of hazardous or unhealthy industrial operations. A Ministry for the Environment is a recent institution (note 3). In addition to suitable laws issued in this area (note 4), the health reform instituted with Law No. 833 of 1978 establishes with respect to contamination the final limits or standards of acceptability that cannot be surpassed. Validity extends to the entire national territory; for each environmental resource it includes any source of contamination, incorporates the structures for prevention and control, and establishes the unified reference value in defense of health.

No one can ignore the importance of the scientific response to determine the qualitative and quantitative standards. This is crucial for the same legal problem of defining conduct: the standard, in fact, represents the boundary between that which is legal and that which is illegal. Without the need to ascertain and measure if damage has occurred in a specific case, in the event of violation administrative and criminal sanctions apply. Control therefore belongs to public authority, like the reaction to illegal acts.

If damage has also occurred, the right to compensation is safe; however, it is deferred to the criminal process: petitions for compensation are suspended until the criminal process has reached its final result.

5.3.5   Standards and the Right to Health

We can therefore ask if industrial initiative that is respectful of standards, staying within the confines of what is legal under the criminal and administrative code, is protected from private claims of any type, when some damage has hypothetically occurred. The traditional negative answer, based on the principle that indemnifiable damage is only unjust damage, whereas conduct conforming to the law or permitted by the law cannot be defined as unjust, today is subject to criticism by scholars. The fundamental principle of the Italian compensation system is contained in Article No. 2043 of the Civil Code which states: "Any fraudulent or culpable act that causes unjust damage to another obliges the one who committed the act to compensate the damage". This is an open formulation that does not discriminate a priori harmful acts that lead to compensation and harmful acts that leave the losses at the victim's expense. Unjust is a predicate of the event, not of the conduct; it is therefore not ruled out that an event caused by one exercising his rights can be considered unjust. In Italian legal practice, however, private compensation claims in this direction are still almost unknown.

Private litigation concerning violations of the right to health are also in the shadows in Italy, when this does not involve specific injuries or diseases. This probably depends on the fact that protection of health and all its components, thus including environmental healthiness, is a task assumed in Italy by the State in accordance with the constitutional principle contained in Article No. 32 of the Constitution. What about situations where certain unhealthy activities which are defined as legal by the public administration are incorrectly determined within the limits and standards? Some scholars in Italy maintain that the right to health, having primary and constitutionally protected relevance, must supersede standards derived from administrative acts. In other words, not only the public administration, but even the ordinary legislature can suppress it with legal acts and standards contrary to the constitutional precept. It is therefore obvious that serious and complete scientific investigation appears decisive, even in the dispute phase, as a criterion for establishing the legitimacy of permissive administrative acts and even ordinary laws.

5.3.6   Evolution of the Concept of Damage

The concept of a claim for private compensation becomes manifest when a sudden and massive damage occurs, the consequence of an unforeseen technological disaster that visibly affects the person (life, health) and his property (economic loss, interruption of earnings). Responsibility and compensation follow the general principles of common law. To accentuate the obligation of the perpetrator of an action that causes damage in cases in which the form of incident remains unknown or is difficult to interpret (as is frequently the case) and considerable effort is expended in searching for the efficient causes of the accident (note 5), is the principle of presumed responsibility. One who carries out a hazardous activity is obliged to provide compensation, unless he offers positive proof of having taken "all suitable measures to avoid the damage". It is not sufficient to demonstrate not having committed any violation of legal standards or common prudence. It is necessary to prove that every effort or suitable measure was taken to prevent the incident.

According to common law, compensation includes "the loss suffered and earnings lost"; where possible, one must restore the prior situation; indemnifiable damage is that derived from the event according to a connection of "causal regularity". Italian law classifies damage into two categories: property damage and nonproperty damage. Property damage is easier to grasp and is a fully elaborated concept in doctrine and jurisprudence. Moral damage (which is only involved when the conduct of the responsible person represents the offense) is understood as "pecunia doloris", i.e., indemnification of physical and moral suffering as a result of the wrong suffered. But between these two concepts there is a vast area of prejudice and suffering in which both aspects participate jointly; "damage to health" can be included here.

Traditionally this has been considered exclusively because of its effects on property. For a long time attention has focused on the concept of "illness" as an impediment to continuation of one's ordinary occupation. Thus, to measure damage we resort to the amount of lost income due to the impossibility of working during the period of temporary disability. Similarly, for permanent injuries we use the criterion of percentage reduction of work capacity and earning capacity. In sum, that which is restored in traditional civil processes in cases of damage to health is the lost income because of the disability suffered. For a long time there has been no intuition that good "health" is a subjective right in itself and that, when attacked, it merits relief, leaving income out of consideration. This principle was finally affirmed for the first time by the Supreme Court of Appeals with a judgment in 1979, and consolidated on August 20, 1984.

In fact, at the end of its text the Court affirms that biological damage is autonomous "since it affects the value of man in all his concrete dimensions, which are not exhausted in the mere capacity to produce wealth, but are related to the sum of the natural functions inherent to the subject in the environment in which life goes on and have not only economic relevance, but also biological, social, cultural, and esthetic relevance".

It should be emphasized that since the judgment of 1979 the Supreme Court has related the right to health to the right to "healthiness of the environment"; this means that he who makes the environment unhealthy commits a wrong thereby.

5.3.7   The Problem of Equitable Settlement

What remains imprecise in this important trend is the extent of compensation. Translated into money, how much is health worth? The Court invites judges to make settlements "according to equity". But equity is a very elastic concept: different judges use different measurements and usually the only uniformity is that the awarded sums are singularly meager, in comparison with that practiced in other countries, notably the United States. For example, in 1982 the Court of Pisa awarded 500,000 lire for each percentage point of permanent disability and 20,000 lire for each day of temporary disability (today the values are more and more increased). Other judges usually adopt as parameter triple the social pension (that is, a very low level), which has a textual reference in a law dedicated to compulsory insurance for damage resulting from traffic accidents.

Is it possible to predict that the next category of indemnifiable damage will be that of "the right to quality of life", against that which disturbs, degrades, causes suffering, even by means of psychological stress and fear? Nothing opposes this conceptually. But it involves still uncertain elaborations; however, it is not sufficient, as we shall see, to exhaust the subject of the complex damage produced by the technological tragedy that occurred in Seveso in 1976.

5.3.8   Contamination with TCDD

As is known, on July 10, 1976, a toxic cloud containing TCDD was released from the ICMESA plant and contaminated an area of about 1,800 hectares, densely populated, causing as immediate and visible result damage
to vegetation, deaths of farm animals, and skin diseases among the infan-
tile population.   The contaminated zone closest to the facility was evacuated and fenced off: 212 families (732 persons) were compelled to move,
abandoning and losing all their property.   In the more external adjacent
zone (about 4,700 persons) the population remained, but was subject to
rules of life that were severely restrictive in comparison with normal
conditions.   Slaughtering of animals (80,000 head),   prohibition against
animal breeding, ban on cultivation and harvesting of any types of plants,
temporary removal of babies and pregnant women, stoppage of all building
activity,   closure of 121 craftsmen's shops and 10 industrial enterprises.
In a third even more external zone (31,800 persons) lesser restrictions
were imposed, banning consumption and sale of vegetables and raising of
farm animals with slaughtering of all those existing.

Even these first protective measures give an impression of the vast-
ness of the trauma produced within the community within the contaminated
territory.   We can attempt to outline a scheme of the items that comprise
the overall damage produced by the Seveso incident, as follows.

(1)   Personal injuries because of subjects who became temporarily or
       permanently disabled.

(2)   Damage to health: both for subjects who exhibited immediate
       skin reactions of greater or lesser severity, or others in whom
       a certain pathology later occurred.   Considering the vastness
       of the contaminated territory and the population, the different
       individual responses and incomplete understanding of the effects
       of TCDD,   this area is still open and not measurable with preci-
       sion.

(3)   Damage to private property: destroyed houses,   abandoned and
       unrecoverable land, destroyed furniture due to contamination.

(4)   Monetary disbursements necessitated by the search for other
       housing, purchase of furnishings, moving.

(5)   Loss of earnings: loss of work; closure of artisan and indus-
       trial activities;   inhibition of productive activity.

(6)   Stoppage of development projects already in progress.

(7)   Moral damage.

(8)   Damage to the overall economy due to loss of image (products
       from Seveso boycotted).

(9)   Social damage: a partially uprooted human community.

(10) Disbursements related to intervention of the State and the Region with formation of a special office assigned to carry out five operational programs.

For the items listed from (1) to (7) the problem of compensation presents no particular interpretive difficulty with common criteria. It is less easy to define "social damage". In Italian law such damage can be tested according to the traditional doctrine only at the cost of breaking it down into a fine powder of individual prejudices which replicate for each subject the hardship and suffering stemming from the hardship inflicted to the collective social fabric. Nothing prevents the entire civilian community affected by the disaster from being considered a victim of this type of damage; naturally it is necessary to find the regularity and legitimization suitable for a subjective decision of compensation. It would not be unreasonable to suggest that this should be the territorial public enterprise (for example, the commune). In other words, it is specifically seen that to repair the consequences of the disaster it is necessary to make provisions for social reassimilation and collaboration.

The last damage item poses a legal problem. Theoretically, if a private citizen must abandon his property because it is contaminated and receives in exchange another property of equal value at another location (or the equivalent market price), one might think that the damage has been totally compensated. However, if the contaminated land is reclaimed, it might happen that the cost of this enterprise proves greater than the sale value of that which is recovered. The problem then arises if this cost should be indemnified by the responsible party. In my opinion the affirmative answer that must be given can obviously not make reference to a private scheme of the usual type. In addition, at Seveso recovery of the badly contaminated area by means of reclamation does not entail restitution to the original owners, but to others. The fundamental concept in ordinary cases in which saleable property is destroyed or lost (and replaced or indemnified), is that a territory cannot be cancelled from reality: its reappropriation by a human community, wherever possible, assumes the form of an autonomous right exercised in collective form by means of State authority.

Equally delicate is the problem of the cost necessary for health monitoring, intended to last a long time. In this activity the institutional duty of the State and of local bodies, on the one hand, and the requirement to superimpose this on economic obligations determined by the illegal behavior of the perpetrators of the damage are intertwined. A good criterion could be compensation of "differential" damage. The legitimate subject is the public body directly.

5.3.9   Civil Responsibility

Concerning the problem of civil responsibility for the Seveso incident, it can be said that almost nothing was controversial. One month after the disaster the Chairman of the Board of LaRoche Co. declared in a press conference that LaRoche would compensate all damage. The company later confirmed this. At the same time collaboration began between the Lombardy Region and the Givaudan Co. for the reclamation operations and later for indemnification to private individuals. These events, extended in time with identical agreements, permitted avoidance of judicial treatment of legal problems concerning the civil responsibility of various subjects from the ICMESA Co. (whose limited resources would not have been able to cover even a minimal part of the damage) and notably of the parent companies, Givaudan and LaRoche.

In addition, satisfaction of compensatory interests was anticipated. In fact, any civil disputes that might have been undertaken would have had to be suspended until conclusion of criminal proceedings.

The latter took place in the first instance before the Court of Monza, ending with the judgment of September 24, 1983; in the second instance before the Court of Appeals of Milan with a judgment on May 14, 1985, filed on July 10, 1985. The criminal judge asserted the responsibility of some managers of the ICMESA plant.

5.3.10   Methods of Compensatory Intervention

Restoration of the environmental damage caused by the Seveso disaster was characterized by massive intervention of public authority. Two State laws and a law of the Lombardy Region took emergency measures for the population and created programs for decontamination in order to achieve "restoration of normal conditions of life". With respect to the assumption of an abandoned and neutralized territory, or to the different assumption of a "diverse" program of complex territorial order, the choice to restore the prior situation wherever possible responded to the human requirements of the population, to their expectations, and to their desires. This explains why Seveso did not become a dead city, but rather a new town.

However, little by little that which achieved the objective of "recovery and restoration" mitigated the task of "compensation": in the sense that the recovery put into action a substantial form of reparation "in specific form" of the collective prejudice. The parallel line of assistance, traditional in the case of a calamity (provision of housing for evacuated persons, activation of assistance centers, distribution of money, subsidies to industries and craftsmen, relief contributions), can be defined as an operation of "anticipated compensation" according to a legislative mode that was already tested during the Vajont disaster (when a landslide that fell into an artificial basin caused the subsidence of a dam and violent flooding of the valley). According to this contrivance the State assumes the burden for the most urgent compensatory measures and becomes surrogate in the right of credit toward those responsible up to competition of the economic burden sustained.

The effect obtained at Seveso from this strategy was that of a substantial lessening of "fragmentary" litigation and synthesis of claims for compensation, many of which were assumed and satisfied by public authority with the criterion of homogeneity of response. The Special Office of Seveso finished by becoming the reference point of a large concourse of human, economic, psychological, and social interests: the melting pot of petitions, wherever aroused, wherever diminished, wherever channeled, each supported the other in solidarity which ultimately proved fruitful despite storms and interruptions.

But there is another "mediatory" aspect of the functions developed in the events of Seveso by the public authorities: the compensatory availability of Givaudan, at least for some damage items that were not difficult to evaluate and where immediate evidence permitted acceleration of the mechanisms for compensation without judicial contention. Technicians from the Region in cooperation with technicians appointed by private individuals and technicians from ICMESA carried out an enormous number of estimates of damaged property sources which formed a basis for development of contracts for compensation, avoiding litigation or reducing it in marginal cases.

The problem of economic compensation on the part of public authorities for costs sustained and for compensation of damage directly assumed was finally developed extrajudicially by means of a transaction approved by the State government and the regional government; this, although it did not cover all the financial obligations calculated to that point (about 121.7 billion lire) did assume the major burden (by means of total obligations valued at 103.5 billion lire) and permitted the authorities to overcome the legal problem of responsibility of the parent companies.

Private litigation was not eliminated, but reduced to reasonable proportion in the judicial sense. Private litigation resulting in lawsuits permitted partial satisfaction of the claims of 7,000 persons with an economic burden of 70 million francs. Civil litigation for unresolved cases occurred in only 20 cases: 7 of these have been settled and 13 are pending; none has reached the phase of judicial decision.

5.3.11 Conclusions

We are far from the day when it may be said that all problems related to the Seveso accident have been resolved and restored. On the one hand, the health risks in the future are still not fully understood at the current state of scientific knowledge. On the other hand, the complex human sacrifice is difficult to measure and almost never actually compensable. In strict legal terms the "moral damage" to the population will remain beyond monetary relief. This prompts us to review the entire problem from the beginning, focusing on the point of origin, which places the "just" not in reparation of wrong, but in its prevention: in "neminem ledere". No good compensatory system is capable of making peace with the law if it is not accompanied by the obligation to prevent with ever stricter security any possible new disasters.

NOTES

(1) For example: "a semiconductor firm may use over 2,500 chemicals, many of them highly toxic, in the manufacture of chips and other computer parts" (Baram, M., 1985: Chemical Industry Hazards - Liability, Insurance, and the Role of Risk Analysis).

(2) Article No. 844 of the Civil Law is dedicated to the indirect environment protection, in the respect of property: "The owner of an estate cannot prevent the lettings of smoke or of heat, noises, shakes, and such propagations coming from the estate of his neighbour, if they are not over the normal tolerance, even with regards to the place conditions. In applaying this rule, the judge must adapt the production requirement to the reason of propriety and minding the priority of a potential use".

(3) Law No. 657, December 14, 1974; Law No. 5, January 29, 1975; DPR No. 805, December 3, 1975.

(4) Here are the most recent acts:

(a) Water pollution:
- Law No. 875, December 19, 1975;
- Law No. 126, April 16, 1976;
- Law No. 319, May 10, 1976.

(b) Pollution of seawater:
- Law No. 662, September 29, 1980.

(c) Hydrocarbon contamination:
- Law No. 94, January 14, 1970;
- Law No. 341, June 5, 1974;
- Law No. 185, April 6, 1975;
- Law No. 875, December 19, 1975.

(d) Atmospheric pollution:
- Law No. 615, July 13, 1966;
- DPR No. 323, February 22, 1971;
- DPR No. 322, April 15, 1971;
- DPR No. 400, June 8, 1982.

(5) In a dynamic meaning, the exact cause of the accident at Seveso has not yet been clarified. During the Court of Monza trial, after having mentioned various eventualities - which were all judged improbable and anyway not proven, the college of experts came to the conclusion that any valuable hypotheses could be set forth on the event dynamics. From a letter of Monsanto Company to Givaudan, we know that Monsanto was not able to draw a definitive conclusion on the cause of the particular reaction occurred at Seveso. Similar doubts were expressed on the occasion of other accidents involving dioxin formation.

5.4   A Brief History of Risk Assessment and Management After the Seveso Accident

G. Umberto Fortunati

Via Vincenzo Monti 29, 20123 Milan, Italy

5.4.1   Introduction

Procedures to define tolerable limits for TCDD during Seveso early emergency period in July-August, 1976, are confronted to later and more complete procedures based on experimental studies in the weakly contaminated Zone B (chronic exposure). Zone B had some 6,000 residents. A comment on the continuously evolving attitude towards acceptance of risk is presented.

Public authorities and many Italian scientific organizations, especially the ones in the Lombardy Region, were mobilized shortly after the TCDD-containing cloud was released into the environment from the ICMESA plant on July 10, 1976. Among the committees that were created during the emergency, one may mention the following:

- Consultive Committee for Reclaiming;
- Consultive Committee for the Chemical and Statistical Analysis of the Territory;
- Research Committee;
- Epidemiology Committee.

The task of fixing acceptable limits, i.e., of performing the first risk assessment for TCDD in the Seveso area soil, was carried out by technicians and scientists of various committees, with the constant presence of the regional authorities and the determinant cooperation of the Istituto Superiore di Sanita (Rome). The basic criteria were:

- NOEL (No-Observed-Effect-Level) for the most sensitive animal, i.e., the guinea pig: 1 ng/day x kg bw;
- safety factor: 1,000;
- specific safeguard of the inhabitants most exposed to risk of contaminated soil ingestion: children;

- limit dose found: 1 pg/day x kg bw, equal to approximately 3 g of soil ingested by a 20-kg child or 7.5 g by a 70-kg adult at a mean contamination level of 7 pg TCDD/g of soil (equivalent to 750 ng TCDD/m^2, limit to be reached through decontamination).

From the above criteria, the acceptable amounts for the various matrixes were assessed: land, inside of housing, equipment, and others listed in the text of the regional law of intervention. Definition of risk areas was thus carried out taking into account only TCDD levels found in soil through assays of topsoil core samples having a 7-cm diameter and a 7-cm height. From then onwards, and for all operations relative to reclaiming and analytical rechecking, reference was made to the "surface density" unit $\mu g/m^2$, always intending a 7-cm-deep square with a 1-m-long side. Conversion from $\mu g/m^2$ to the most commonly accepted ppt (10^{-12} g/g, weight of contaminating agent by weight of soil) was obtained by multiplying the concentration value in $\mu g/m^2$ by a factor of 8, and therefore:

- agricultural land (limit by   <0.75 $\mu g/m^2$ 6 ppt
  regional law)
- nonagricultural land (limit by   >5 $\mu g/m^2$ 40 ppt
  regional law)
- limit for evacuating population   >50 $\mu g/m^2$ 400 ppt

There is a remarkable difference between the regional limit in Lombardy for evacuation, especially considering the economic and social consequences deriving from such action, and the "limit of concern" - but not yet of intervention - indicated by the Sanitary Control Authority of the United States, Center for Disease Control, on the basis of a different risk assessment. Such limit is 1 ppb = 1,000 ppt, that is 2.5 times higher than that adopted in Seveso for evacuation.

Times Beach (Missouri), where most of the roads which had been sprayed with TCDD-contaminated oils had been covered with a layer of asphalt, exhibited contamination levels above 300 ppb. Times Beach was gradually evacuated after the river had flooded and some analyses had shown the extension and intensity of contamination.

In 1982 and in 1984-1985, during Seveso reclaiming, Zone B was rechecked. This area had original contamination levels between 5 and 50 $\mu g/m^2$. Confirmation of contamination pattern unevenness was obtained, as though the cloud had "bounced" on the land various times, leaving toxic prints less and less marked every time. Among such contaminated spots, the cloud left lower levels of TCDD. Therefore, a risk assessment was carried out which was more detailed than the first one during the emergency period. The exercise was performed for the approximately 6,000 people residing in Zone B.

Reference may now be made to the document elaborated by the participants to the subcommittee of the Cimmino Committee for Seveso. This Committee was organized by Lombardy Region's Seveso Special Office and by the Istituto Superiore di Sanita (Rome). In the Milan meeting of March 26-30, 1984, it was estimated that a daily TCDD ingestion of 0.28 pg/kg bw increased the cancer risk by 1/100,000, if the effects observed on the female rat liver were taken as a reference. In fact the female rat liver appeared to be an organ very sensitive to TCDD. In addition, it was assumed that a linear relationship existed between cancer risk and TCDD exposure. So, 0.028 pg/day x kg bw reduced the risk to 1/1,000,000, while 2.8 pg/day x kg bw increased it to 1/10,000, and so on.

Zone B contamination was schematically defined as follows: for 80 percent of its extension: 1.5 $\mu g/m^2$, i.e., 12 ppt of TCDD in the soil. The remaining 20 percent with contamination 10 times higher. The routes of exposure considered were:

- ingestion and contact with soil;
- consumption of epigeal and hypogeal vegetables grown in vegetable gardens;
- consumption of zootechnical products (chickens and rabbits) from that area.

On the basis of a few hypotheses, only partially verified, and assuming that total risk is given by the sum of single exposure risks, Table 1 was obtained. Owing to the above results the subcommittee of the Cimmino Committee for Seveso recommended a more detailed analysis of various parameters, especially those that previously had not been thoroughly verified, such as: (a) the relationship between TCDD levels in soil and vegetables, and (b) the feeding habits of the population (including the consumption of eggs and milk, not considered in Table 1) in order to reach a more reliable evaluation. Such detailed analysis was to be carried out also by creating cultivated plots and animal farms for experimental studies in Zone B. At the same time, TCDD dilution/degradation was to be helped through agronomical interventions.

The U.S. EPA has defined a method of risk assessment for TCDD-contaminated soil where five routes of intake are considered, that is:

- dust inhalation,
- cutaneous absorption,
- ingestion of soil,
- ingestion of food (such as veal meat and dairy products),
- fish consumption.

Table 1. Zone B Risk Assessment as Carried Out in March 1984 for Families Having a Vegetable Garden and Courtyard Animals[a]

Exposure routes	Quantity ingested (pg/day x kg bw)	Extra cancer risk per million exposed	Comments
Soil	0.007	0.25	Equivalent to 3/10,000.
Vegetables	5.7	200	This dose is associated to a
Meat	2.4	100	cancer risk of 1/3,300.
TOTAL EXPOSURE	8.107	300.25	

[a] It was assumed that: mean TCDD level in soil, 1.5 $\mu g/m^2$ = 12 ppt; average individual weight, 70 kg; Length of Exposure, 70 y.

The interest of this publication is in the discussion on the coefficients for evaluation of the amounts absorbed through each route. For a quick estimate of quantities ingested, and therefore of cancer risk, the use of nomograms is suggested.

For Zone B, using the simpler method of nomograms, the values shown in Table 2 were obtained. They express the number of extra tumours every million inhabitants exposed. Mean soil contamination level was considered to be, as in the previous case, 1.5 $\mu g/m^2$ = 12 ppt.

Table 2. Extra Cancer Incidence Over a Million People as Reevaluated for Zone B Inhabitants According to U.S. EPA, 1984.

Exposure route	Risk estimate
Inhalation	Dust in air (inhalable): 2 mg/m^3. No. of tumors: 0.4/1.0 million
Cutaneous absorption	No. of tumors: $10 \cdot 10^{-5}$/1.0 million, negligible
Soil ingestion	No. of tumors: 0.1/1.0 million. This is the only hypothesis of risk evaluated in 1976
Meat and dairy products	No. of tumors: 100/1.0 million
Fish consumption	In Zone B there are no surface waters to fish: thus, the risk of contaminated fish consumption is <u>zero</u>

By adding all single risks the total risk is obtained, as follows: total number of extra tumors = 100.5/1.0 million inhabitants

The residing population was slightly less than 6,000 people, only partly exposed to the risk of feeding with contaminated products. Thus, such number has sometimes led to the conclusion that the extra cancer frequency (10^{-4}) could not be observed and, in any case, was not statistically quantifiable. However, in view of this additional risk, Seveso Special Office, supported by Hoffmann-LaRoche Corporation, decided to proceed to a series of operations for diluting and even replacing the contaminated soil in vegetable gardens and agricultural areas. The aim was to reduce further the risk deriving from feeding on vegetables and farm animals from the contaminated area.

We owe Prof. Schlatter (Swiss Federal Institute of Toxicology), the following presentation of 2,3,7,8-TCDD risk assessment.

"Having examined the ADI (Acceptable Daily Intake, 1-10 pg/day x kg bw), here are the results of the latest researches:

- the amount accumulated in the human body is known;
- the 12 congeners having toxic relevance are known;
- the main origin of TCDD in man is known, i.e., the intake of animal fat;
- TCDD half-life period in the human body is experienced to be 5-6 years, 80 times longer than that in laboratory animals.

The population of an industrial area hit by fumes and ashes from municipal waste incinerators is estimated to assume 16 pg/day x kg bw of total dioxins, the amount being expressed in TCDD-equivalent units (TCDD toxic potency = 1). In the light of the above, the proposed ADI limit is exceeded and it is necessary to reexamine the whole problem starting from health effects of populations exposed to low contamination levels for a long period of time (chronic exposure).

The only clear and specific symptoms met in Seveso are associated with chloracne which has appeared, however, only for many ng/day x kg bw exposures".

Because of the above, Prof. Schlatter's conclusion was that the exposure of an industrial area population to this class of compounds (as evaluated on the basis of the amounts present in the adipose tissue) is at least 100 times lower than the critical limit at which chloracne appears. Even though the proposed ADI was widely exceeded by reality, the final aim must not be forgotten and the intake of dioxins must be reduced below the fixed limit of 1-10 pg/day x kg bw.

The most important conclusion of the many costly studies carried out on the spreading and toxicity of dioxins is as follows.

It is true that the exposure at Seveso has not - at least until now and apart from chloracne - determined any measurable health damages of the exposed population. However, the large number of toxic compounds of the same family, or of similar families, in the environment strongly advises us to at least reduce their release substantially. In fact, 25 percent of deaths is caused by malignant tumours and a large part of these casualties is generally attributed to the environmental chemicals: herefrom the need to eliminate them as quickly as possible and start in particular from those, like dioxins, which have no utility. This aim will not be easy to reach. In the meantime the improvement of risk assessment methods is necessary and useful for a more reliable estimate of the environmental risk associated with toxic chemicals.

5.4.2 The Evolving Attitude of Risk Acceptance

"Faced with events like the ones that occurred at Seveso, we must have the courage of humility. For the progress of science, humanity must be prepared to accept even such risks" (Attorney General C. Cecere).

This motivation was given by the Attorney-General when requesting discharge for the two ICMESA executives during the debate at the Supreme Court of Justice in Rome (May 1986), and shows once more how the judgment of facts and responsibilities is liable to change quickly and radically within very few years. It is advisable to remember that no operations, experiments, or research which could be useful to the "progress of science" were under way at the ICMESA plant in 1976. It may be stressed that about 20 similar accidents had occurred in plants for the production of trichlorophenol, even though the consequences were luckily confined within the limits of the plants themselves. The experience of others had not been taken into account. In spite of the Attorney General's pleading, the guilt of the two executives was definitely confirmed.

A similar difference in evaluation was seen some time later - after Chernobyl, May 5, 1986 - between the Lombardy Region and its Swiss neighbour Canton Ticino, both faced with the risk of exposure to radionuclides. On the Italian side, the authorities banned all forage for one month and decided to kill all rabbits fed with fresh vegetables grown in the same areas. On the Swiss side, even though with some embarassment, no actions were taken. Radioactivity has no custom boundaries and the rain had dropped an equal amount of particles coming from the USSR on both sides of the Italian-Swiss boundary. The difference in evaluation could be due, according to some people, to political considerations because Swiss authorities did not wish to raise additional difficulties for the building of nuclear power plants: in fact, such projects had already encountered the resistance of the population of the areas involved as well as that of the ecological societies.

Together with the judgement of responsibilities, the attitude to risk acceptance also changes: this attitude depends on social, cultural and economic factors which may vary widely according to time and space, and it is refined by the daily experience which all of us live through, directly or indirectly, and by the frequency and gravity of the accepted risk.

5.5 The Role of Statistical Analysis in the Management of Chemical Accidents

Sergio P. Ratti

Dipartimento di Fisica Nucleare e Teorica
Universita di Pavia, and
Istituto Nazionale di Fisica Nucleare
Sezione di Pavia
27100 Pavia, Italy

5.5.1 Abstract

Based on the Seveso experience, this paper deals with the statistical analysis and expertise required in assisting the recovery strategies to be defined after a chemical accident. The following topics are specifically focused upon due to their relevance: mapping strategies, comparison of incoherent data sets, search for the best phenomenon descriptors, optimization of analytical procedures, and fractal geometry.

5.5.2 Introduction

The role of statistical analysis and the statistical expertise required in assisting the recovery strategies following a chemical accident are obvious to most experts. Therefore, in this report, we shall focus only on those four points which turned out to be of particular relevance for the statistical assistance given by the group of the University of Pavia to the management of the Seveso accident (see A. di Domenico "Introductory Comment on the Seveso (Italy) Industrial Accident", and G.U. Fortunati "A Brief History of Risk Assessment and Management after the Seveso Accident", this publication; Figure 1a). The points are:

(1) the assistance in defining mapping strategies for a geometric description of the phenomenon, as well as the rationale guiding the related decisions;

(2) the problems introduced by (a) the evolutionary improvements of analytical procedures, (b) the time evolution of the phenomenon, and (c) the comparison of data obtained using different analytical procedures;

(3) the search for the best descriptive variables and functions, and for the best mathematical approaches;

(4) the optimization of analytical procedures in view of the specific problems to be approached.

Finally, mostly related to Item 3 above, we will mention briefly a fifth point, i.e., the very new technique of fractal geometry which appears to be of immense value as a future tool for the description of very erratic processes.

5.5.3   Mapping, Mapping Strategies, and a First Geometrical Description

The mention of some relevant figures related to the Seveso accident is important to understand the rationale used to orient the different sampling campaigns. These were requested by local authorities in order to provide the population involved with some fair information and the Government with reasonably reliable data on the phenomenon and its consequences. Immediately after the accident, a Government committee ordered a thorough mapping of the entire region affected by the contamination. Soil specimens were 7-cm thick and 7-cm in diameter. The mapping was organized to follow a regular grid made of 50-m-side squares in the most contaminated zone, and a regular grid of 150-m-side squares in the rest of the territory (approximately 2000 hectares; Figure 1b). Lacking any a priori information on the contaminant distribution, the approach selected for mapping was adequate.

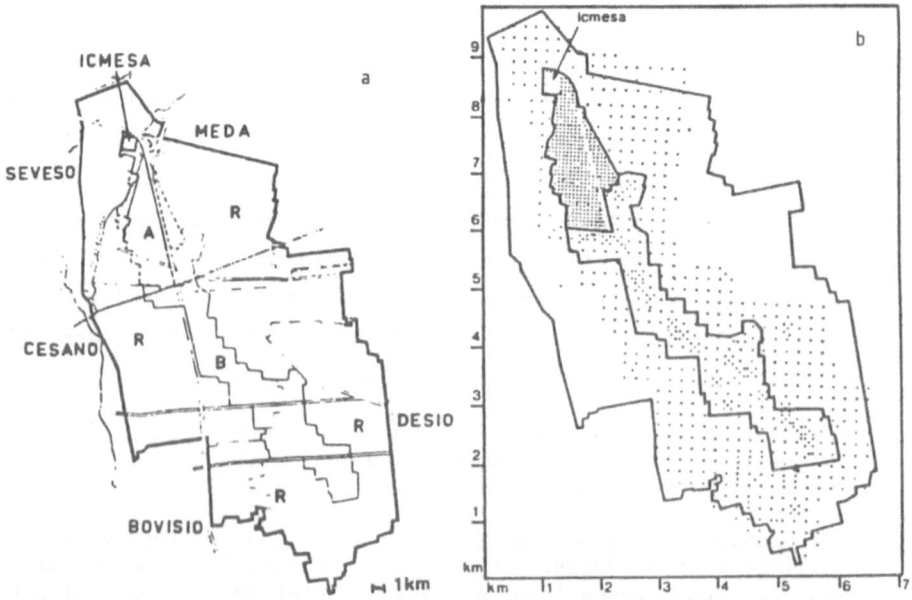

Figure 1a,b. Seveso Accident Location and Sampling Sites

According to the contamination figures, the territory was classified into three distinct zones (what follows is a classification at a certain stage during the Seveso history and is purely indicative):

ZONE A, $S_{tot}(A)$ = 0.873 million square meters, heavily contaminated;

ZONE B, $S_{tot}(B)$ = 2.690 million square meters, having intermediate contamination levels;

ZONE R, $S_{tot}(R)$ = 14.304 million square meters, exhibiting a lower level of contamination;

for a total S_{tot} = 17.867 million square meters.

During the years 1976-1977, 2013 topsoil specimens were collected. As each covered an area s_{an} = $\pi \cdot r^2$ with r = 3.5 cm, all these specimens together provided a total analyzed area S_{an} = 7.65 m^2 (7.65×10^{-6} million square meters). Limiting our attention to the most contaminated area (Zone A), $S_{an}(A)$ = 2.87 m^2. Thus, the overall statistical indices of the sampling campaigns were:

$R(A)$ = $S_{an}(A)/S_{tot}(A)$ = 3.29×10^{-6};

R = S_{an}/S_{tot} = 4.28×10^{-7}.

These are indeed very small numbers. However, such situation is most common in environmental analysis. In a new mapping campaign taken place over the years 1980-1981, although aiming at specific objectives the ratios were:

$R(A)$ = 4.74×10^{-6};

$R(B + R)$ = 4.56×10^{-7};

R = 6.65×10^{-7}.

This means that - as in most environmental monitoring cases - the "representative (statistical) sample" was a very minor part of the population to be described; therefore, the analyzed specimen was a weak statistical representation of the total phenomenon. On the other hand, to perform over two thousand analyses, it took several laboratories about two years; in addition, the analyses were - and would be - costly and paid at the expense of the community.

Statistically, therefore, on one hand the information had to be taken carefully as it might have been easy to reach misleading conclusions. On the other hand each single datum had to be considered precious and could not be disregarded.

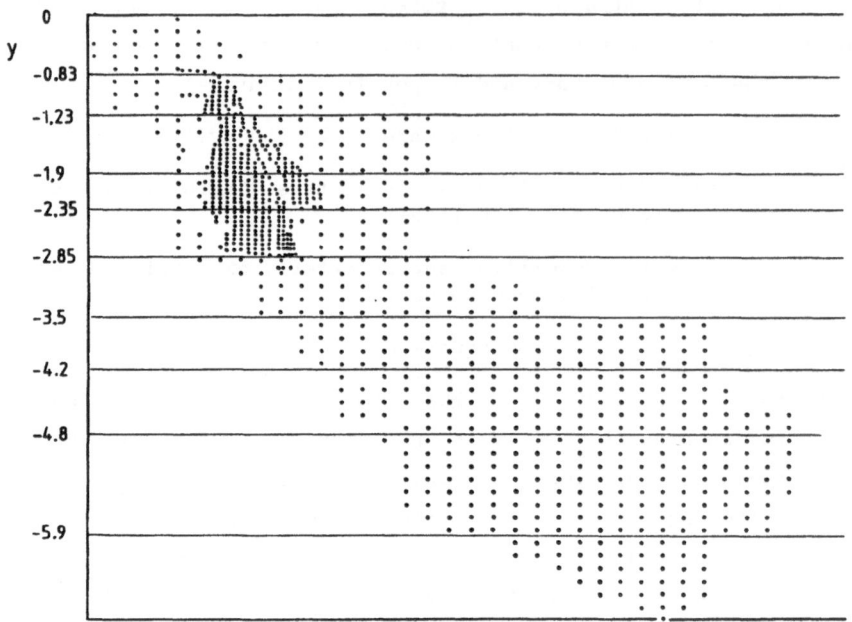

Figure 1c.   Sampling Site Coordinates (y-axis) in 1976-1977 Campaigns

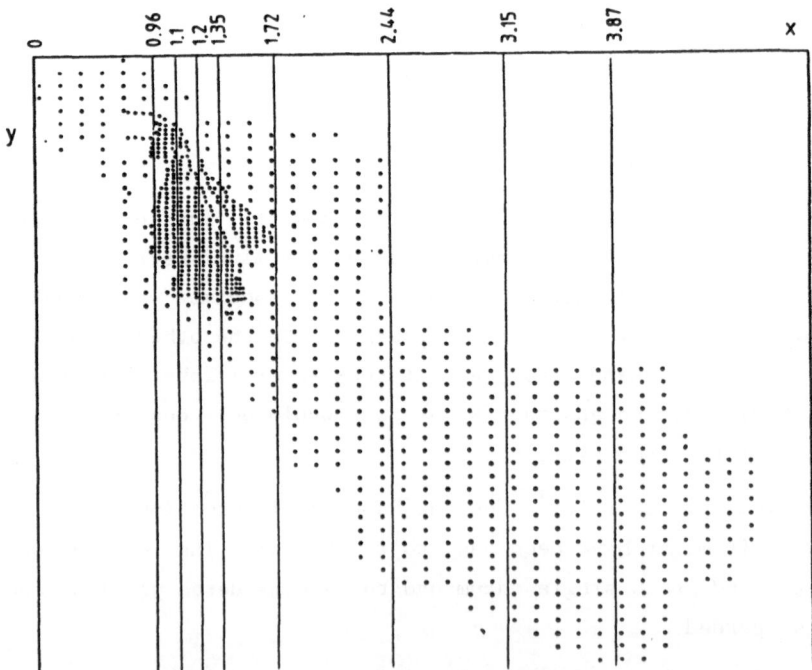

Figure 1d.   Sampling Site Coordinates (x-axis) in 1976-1977 Campaigns

740

By contrast with the above, the local communities needed answers to questions and were not willing to wait forever. Therefore, the necessity of finding strategies able to provide a satisfactory information, equally valuable, using less money as well as less time.

The strategy suggested for mapping Zones A + B + R again in 1978-1979 was guided by the perspective of obtaining additional information on the local fluctuations in topsoil using a limited number of specimens. The new approach was called "the octet method" and originated from the following considerations.

Given the initial measurements ordered by the Government committee, the whole area was sliced into large strips to determine the location of maximum contamination in the geometric variables x and y. Strips were of different dimensions in order to allow the collection of approximately an equal number of measurements per strip (Figures 1c,d). The total y-range was cut into nine intervals; the total x-range was cut into eight intervals. For each interval in x(y) all data were fitted to a contaminant Gaussian distribution in y(x). It seemed reasonable to assume that the contaminant(s) ejected from the safety valve propagated as a cone swinging according to the changing direction of wind streams at the moment of the accident. Indeed, Gaussian distributions gave an adequate description of the geometric distribution of TCDD. This procedure provided two sets of measurements (Figure 2a) that could be used to fit a regression line of the type:

$$y = a + b \ (1 - e^{\alpha \ \bullet \ x})^{\beta}.$$

The fit gave:

$$y = 1.1 + 4.47 \ (1.0 + e^{-0.5x})^{6.9},$$

with a Chi-square = 6.8 for 14 degrees of freedom. The recommendation for 1978-1979 mapping campaign was then to obtain 50 measurements only, evenly distributed along 10 perpendiculars to the regression line (Figure 2b), each measurement made of eight topsoil specimens taken over an overall 2.5-meter range and regularly spaced out (2.5/7 meters from each other). The ideogram of each octet (i.e., the sum of the eight Gaussian distributions, each associated with the analytical value of a specimen) gave the information on the local fluctuations; a typical example is reported in Figure 3.

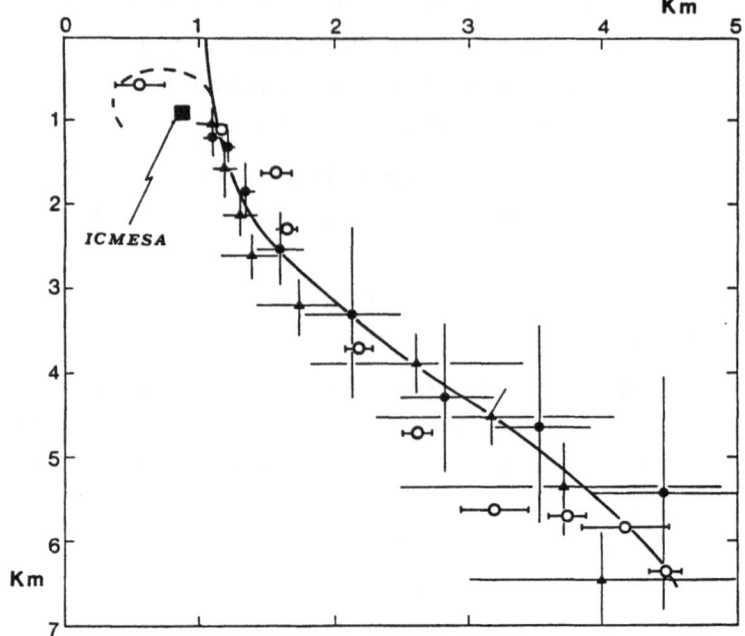

Figure 2a. Determination of Maximum Contamination Line on the Basis of
1977 Campaign. Data Points are: (O) Cuts in the
x-Coordinate; (A) Cuts in the y-Coordinate. The Solid Line
is the Best Fit Curve.

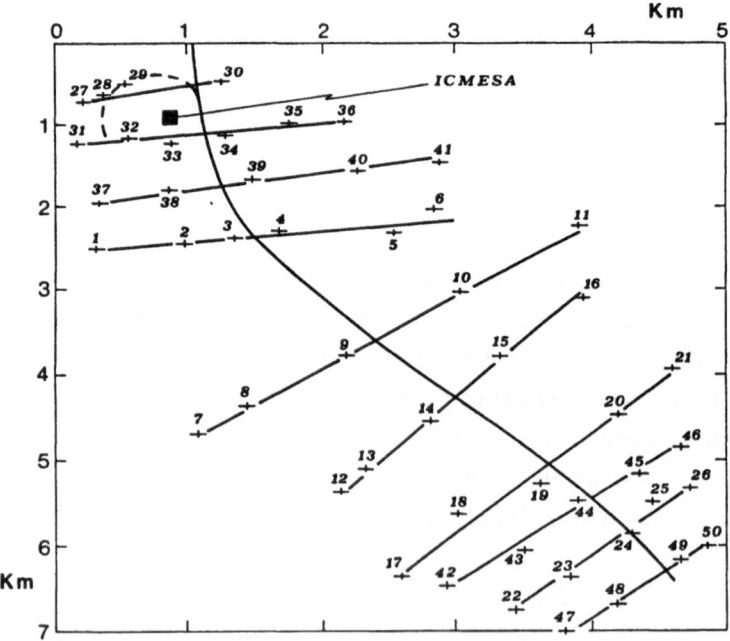

Figure 2b. Locations of Measurement Octets During 1979 Campaign. The
Solid Curve is the Same as in Figure 2a.

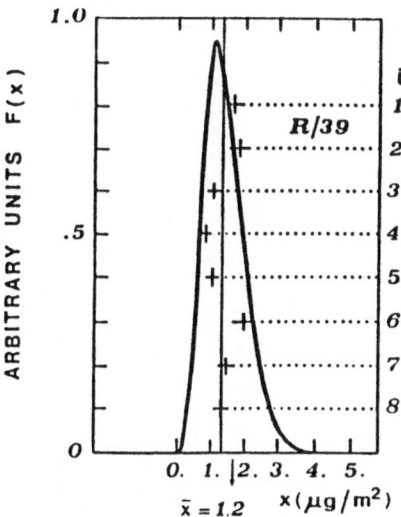

Figure 3.   Eight Measured Values of a Well-Calibrated Octet, and the
Corresponding Ideogram.

The mapping covered the whole area with 400 specimens instead of 1260,
at 1/3 of the cost.

Analysis of the 400 data to find the position of points with maximun
contamination turned out to confirm the 1976 data. The saving in time and
costs was evident and relevant. No doubt, the statistical manipulation of
the information was of great help to sound decision-making.

Next was an overall geometric description of TCDD distribution on the
territory.   The importance of this point relies on the possibility of
estimating the contamination level in any given location within certain
geometric boundaries.   Not irrelevant at all, here the problem arises of
checking whether or not the boundaries determined mostly on a geopolitical
basis are scientifically adequate, or have to be changed, and if or how
much contaminant(s) may be expected to have spread outside the defined
limits.   Mathematical tools can be different according to the amount of
data available and the problem will be definitely approached in general
terms in Section 4. Using Tchebitchev approximants and a proper technique
to enrich the data sample without introducing biases (Belli et al., 1988),
it was possible to describe analytically TCDD distribution in Zone A
(Figure 4).

A proper analysis of the first data allows to save money and time in
subsequent sampling campaigns of large areas.

The advantage of possessing an analytical description of the contaminant distribution is that of having full knowledge of the distribution

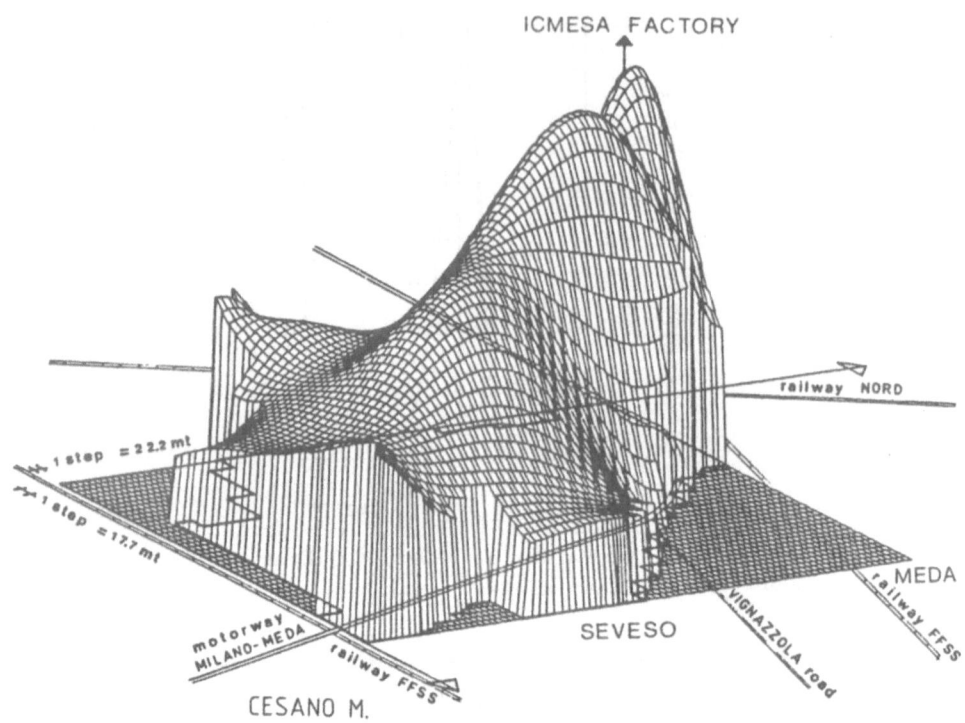

Figure 4. Graphic Display of Function Fitted to 431 Samples

function. Furthermore, given any two-dimensional trajectories within the geometric boundaries of the description, the path integrals of deposited TCDD - or another chemical - may be readily calculated. This can be of particular interest and usefulness in epidemiological studies. In addition, the total amount of material deposited within the considered boundaries may also be readily estimated. The approach limitation resides in the fact that the data base needed has to be relatively abundant. In fact, the application of the Tchebitchev approximant method to the whole Seveso area was a failure.

5.5.4 Time Evolution of a Chemical Analysis and Recovery of Biased Findings

The attempt to apply an analytical description to the entire Seveso area brought up the problem of homogenizing the data coming from different campaigns, and obtained by different laboratories in different periods eventually several years apart. It also brought up the question to recover any possible data and correct them for possible biases and systematic errors. This in order to use the whole available analytical information. On the other hand, it is well known that hunting for biases and systematic errors is among the most difficult tasks of experimental sciences.

Under the pressure of local communities and popular committees, re-measurements are often requested "only were previous measurements showed the presence of high contamination", or only in regions defined by geo-political reasons. At any rate, a large effort is often limited to very circumscribed areas. It has been shown (Ratti et al., 1986) that all these data can be recovered by means of average corrections and renormali-zations.

Although there is no general approach suggested by the statistical science, the procedure followed in the Seveso case is of a somewhat broader interest. The strategy used to compare two different methods (Method I and Method II, one of which, say Method I, may be assumed to be unbiased or, rather, affected by a minimum bias) and normalize one against the other is summarized in Figure 5. The relevant feature is that of knowing the relationship between the two data samples as obtained by Method 1 and Method 2. For example, one could think of two data samples as coming from two mapping campaigns: the first a systematic one, the second limited to an area with heavier contamination. Moreover, the two campaigns were undertaken in two different periods, a few years apart. As the campaigns were carried out in different years, one could assume Method II to be associated with a chemical procedure improved with respect to the previous one, Method I.

One must have four data samples, two treated by one method and two by the other. Each pair of data samples are related by the same bias.

In such case (Figure 5), one can apply the known bias to the unbiased sample, then extract a known biased sample, make a proper comparison, and estimate the correction factors which convert the biased sample into an unbiased one. Such correction factors may later be used to recover the real biased sample and make it homogeneous with the unbiased one. As a matter of principle, at this stage the comparison between Method I and Method II can be performed indifferently using as a reference the biased or the unbiased sample, provided samples have been made homogeneous.

It is unfortunately common knowledge that in the scientific experimen-tations there are many unknown biases, and that even the original sample - assumed to be unbiased - is not in reality bias-free. Nonetheless any corrections eliminating a bias improve the quality of the data and there-fore the quality of the conclusions that can be reached.

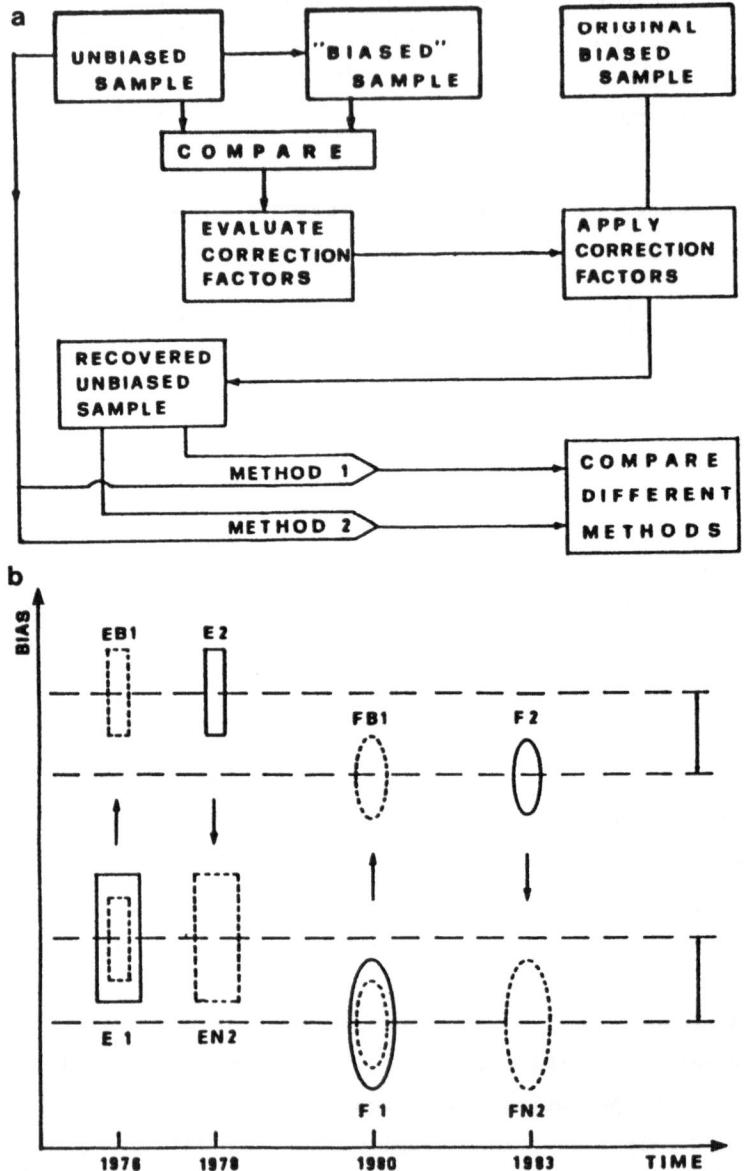

Figure 5.   Flow Chart for Recovery of Biased Samples and Comparison of
Different Analytical Procedures

To illustrate the logic of the bias correction procedure, we will assume (Figure 5) that we are dealing with two methods used in different years and affected by different biases. In a qualitative plot of bias vs. time, we may imagine to have two samples (E1, E2) analyzed by Method I at time t_1, and two samples (F1, F2) analyzed by Method II at time t_2. Samples 1 and 2 have the same relative bias or systematic error (typically, remeasurements only in the most contaminated areas). Sample F1 has some unknown bias - even if smaller than the unknown bias of E1 obtained by a less reliable method - and therefore none of the representative points lie at zero level. From samples E1 and F1 examined at t_1 with Method I and at t_2 with Method II, respectively, one extracts samples EB1 and FB1 by applying the known relative bias; estimated correction factor(s) are applied to the real biased samples E2 and F2 to obtain the new corrected unbiased samples EN2 and FN2. At such point, comparison of the two methods may be done at either the biased or unbiased level.

The results obtained from the described statistical manipulation of data allow to recover a good part of biased samples. This increases the overall statistical information on the contamination phenomenon; it also permits to check and correct time-dependent effects and normalize different data as if analyzed at the same time. All this may be of great support to epidemiological studies (Merlo et al., 1984; Bertazzi et al., 1986).

It is worth mentioning that the above may be done - on average and on a pure statistical basis - without entering into a detailed investigation (Cerlesi et al., 1987, 1988) on the technical implications of the methods to be compared: that is, it may be accomplished by any independent Agency or Institution willing to provide a complete and exhaustive study of a given phenomenon.

5.5.5 <u>Search For the Best Descriptive Variables and Analytical Functions</u>

After the recovery of all possible biased samples and correction of systematic errors and for the time evolution of analytical procedures, the problem arose of providing a geometrical description of contaminant distribution over a large area (approximately 2000 hectares). This mostly in connection with possible epidemiological studies on mortality and cancer incidence (Merlo et al., 1984; Bertazzi et al., 1986) by level and length of exposure.

In order to obtain such description one needs to select the most useful geometric variables as well as convenient mathematical functions suitable for a parametrization based on a limited number of free parameters. The use of Tchebitchev polynomials as approximant functions (Section 2; Belli et al., 1988) is not adequate in the absence of a relatively abundant data base. It may also turn out to be impossible to use each single measurement - if not in a new fractal approach as suggested in Section 6. Moreover, often it will be possible to use mean interpolated values relying upon the knowledge that the phenomenon is log-normal in nature (Belli et al., 1982a; Ratti et al., 1986), as well as all the structural information collected during the routine investigations.

In the specific case of Seveso the relationship $z = \ln(x)$ was defined, where x was the amount of TCDD in topsoil measured in $\mu g/m^2$. It turned out (Belli et al., 1982a; Ratti et al., 1987a) that z was distributed according to a log-normal distribution around the average value <z> with standard deviation σ_z:

(I) $\quad F(z) = (\sigma_z \bullet \sqrt{2\pi})^{-1} \exp [-(z - <z>)^2/2\sigma_z.$

First of all, the distance r from the source and the azimuthal angle θ were chosen as geometrical variables for an easier parametrization (Figure 6a). The problem was then to find a reasonable empirical description of <z> as a function of the geometric variables selected: $<z> = \phi(r,\theta)$. The value <z> was parametrized as:

(II) $\quad <z> = g(r) \bullet f(r,\theta)$, where:

(III) $\quad g(r) = A \exp(-B \bullet r^2)$, and

(IV) $\quad f(r,\theta) = \exp [-(\theta - \theta*)^2/2\sigma_\theta^2] - (C + D \bullet r)$, with

(V) $\quad \sigma_\theta = E \exp(-F r) + G$,

while the value of $_z$ was parametrized simply as:

(VI) $\quad \sigma_z = [\exp(P \bullet <z>)]/[Q + R \bullet \exp(P <z>)].$

Thus, 10 parameters were sufficient to describe the geometric distribution over the entire Seveso territory. As an example the distribution of $z = \ln(TCDD)$ is shown for the two 1980-1981 campaigns (Figure 6b; Ratti et al., 1987a). The fitted values of the 10 parameters as well as the measured and calculated average values are given in Table 1 for 1976-1977 and 1980-1981 campaigns. In Figure 6b the poor reproduction of the sharp initial peak and that of the valley are intrinsically due to the smoothing process utilized in the averaging procedure. In spite of that, the shape parameters were reproduced within approximately 10 percent and, performing

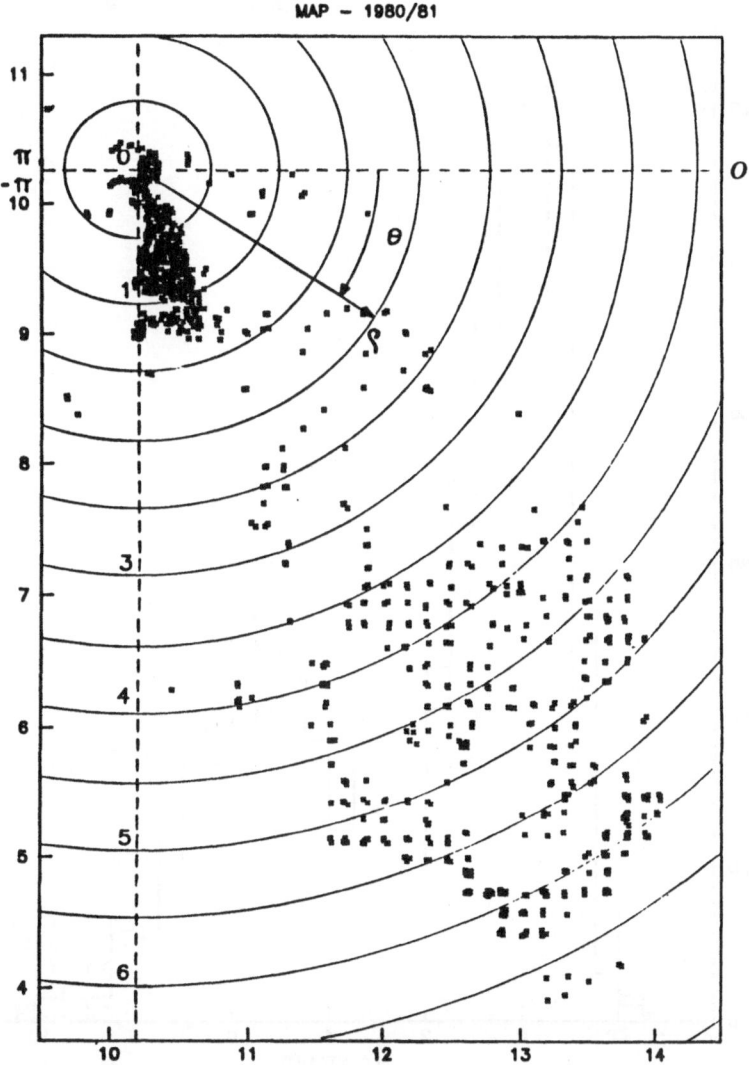

MAP — 1980/81

Figure 6a.   Choice of Cylindrical Coordinates for the 1980-81 Campaigns

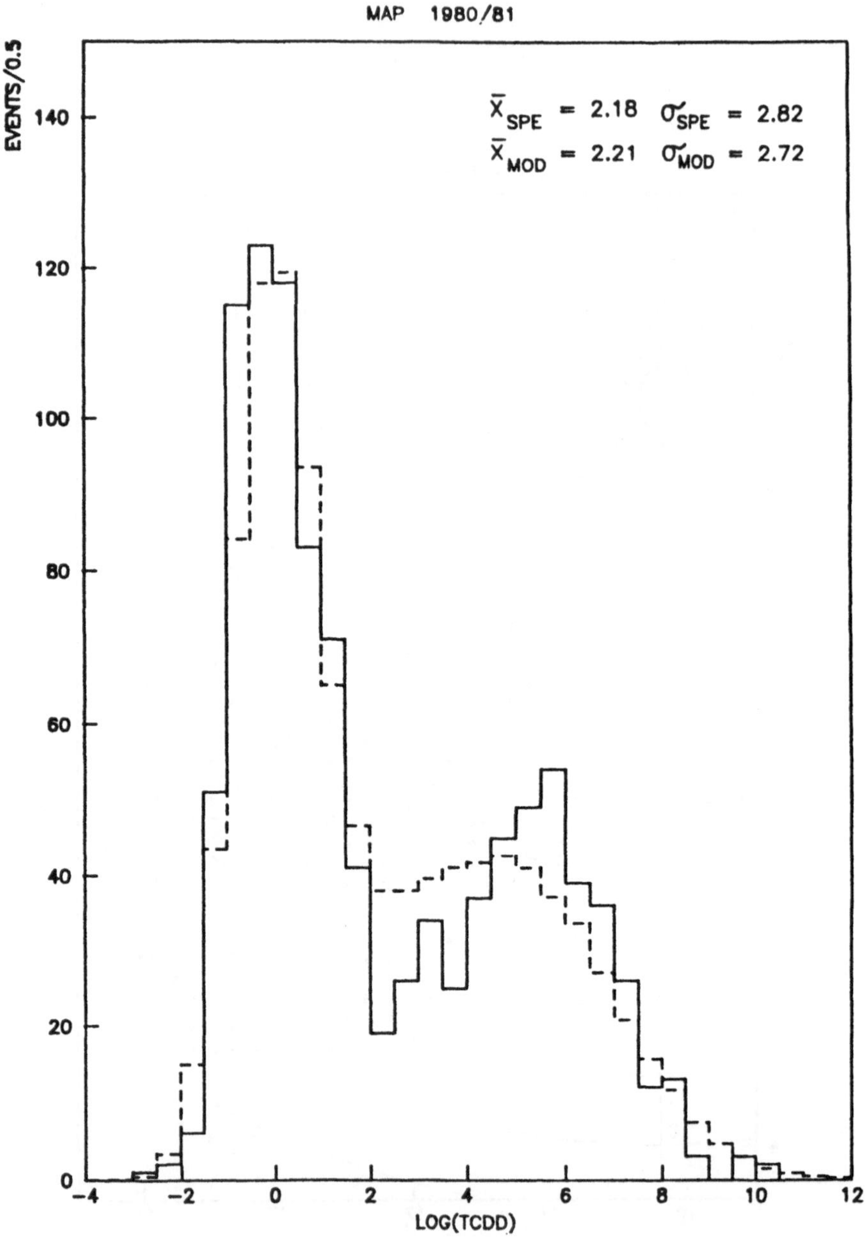

Figure 6b. Comparison of Real Data (Full-Line Histogram) Against Model
Data (Dashed-Line Histogram)

750

a fit to all data, measurements were reproduced by the model with a Chi-square probability greater than 50 percent.

Table 1. Values of Fitted Parameters and Comparison Between Experimental and Model Values for $<z>$ and σ_z.

Parameters	"A"	"B"	"C"	"D"	"E"	"F"	"G"	"P"	"Q"	"R"	Values of $<z>$ and σ_z	
											exp.	model
1980/81	6.40	0.09	-0.12	0.15	1.18	7.49	0.16	0.42	0.76	0.49	$<z>=2.18$	$<z>=2.21$
	±.33	±.01	±.05	±.02	±.32	±.85	±.01	±.16	±.10	±.09	$\sigma_z=2.82$	$\sigma_z=2.72$
1976/77	4.09	0.13	-0.53	0.24	0.00	0.00	0.16	0.65	0.47	0.61	$<z>=1.85$	$<z>=2.00$
	±.46	±.04	±.07	±.03	±.00	±.00	±.01	±.15	±.10	±.03	$\sigma_z=1.84$	$\sigma_z=1.72$

By statistical interpolation of the available information, it was possible to obtain an empirical description of TCDD distribution over the whole territory around Seveso. This was valuable in view of epidemiological studies. In fact, at least in theory, individuals living in locations characterized by coordinates (r, θ), could be easily associated with the corresponding mean TCDD environmental level used as an estimator of their potential exposure to the contaminant. As the random walk processes associated with the mobility of the population was, and generally is not negligible, there was no need of a very high accuracy in evaluating the local estimator. Nonetheless, the entire exposure history of each individual could be taken into account to a satisfactory degree.

5.5.6   Selection of the Analytical Procedures According to Specific Needs

Our statistical studies on contamination phenomena were in general not limited to describing the geographic distribution (namely, of TCDD) on the ground or underground (Belli et al., 1982b), but also aimed at improving the nature itself of the investigations (Ratti et al., 1987a). At to the Seveso accident, its nature provided very useful tools of general validity (Ratti et al., 1987b).

In the case of release of fumes or rather light materials from an exhaust pipe into the atmosphere, the extension of the central limit of statistics assured that the distribution of the quantity of material detected in/on the soil surface layer was of log-normal type (Cramr, 1955).

That is, if x is the deposited quantity, $z = \ln(x)$ is Gaussian-distributed and the relationship

(VII) $\sigma = k\, z_{av}$

between the standard deviation σ and the average value z_{av} approximately holds. This observation (Figure 7) was performed in 1982 (Belli et al., 1983) but its theoretical justification was obtained only in 1987 (Ratti et al., 1987b). In theory, when z_{av} goes to zero, should go to zero, or vice versa. Now in practice, when σ goes to zero, z_{av} does not go to zero. Indeed z is a measured value of the theoretical variable and therefore is affected by systematic errors: in particular any measuring device or analytical procedure cannot detect beyond a given detection limit which is characteristic of the given procedure/apparatus. Thus, when σ goes to zero, the zero of the measuring device goes to zero. That is, when σ goes to zero, z_{av} reaches its detection limit.

This important fact may be used in a very general way (Ratti et al., 1987b), for instance to elaborate a method effective in determining a proper analytical procedure to be utilized in particular cases. Procedures can be optimized minimizing costs and measuring time, without loosing the needed accuracy for given circumstances. The method is sketched in Figure 8.

From Figure 7, two measurements (with their standard deviations) of quantities distributed artificially according to a log-normal curve, allow determination of the detection limit of the procedure employed. Obviously a procedure without error would be represented by the x axis in Figure 7, i.e., zero error for any measured quantity. In a cost-benefit investigation, the procedure can then be varied to minimize particular parameters, such as cost or personnel time, to compromise properly with the requested accuracy for a particular problem (Figure 8b).

The statistical investigation described above brought about very useful results and provided a method to tune an analytical procedure to particular circumstances (for instance, an emergency situation). As the method has a general validity, it may be employed to choose the procedure of use to approach a given problem. Finally, it can help in providing the analytical laboratories with a strategy of preparedness in case of an emergency.

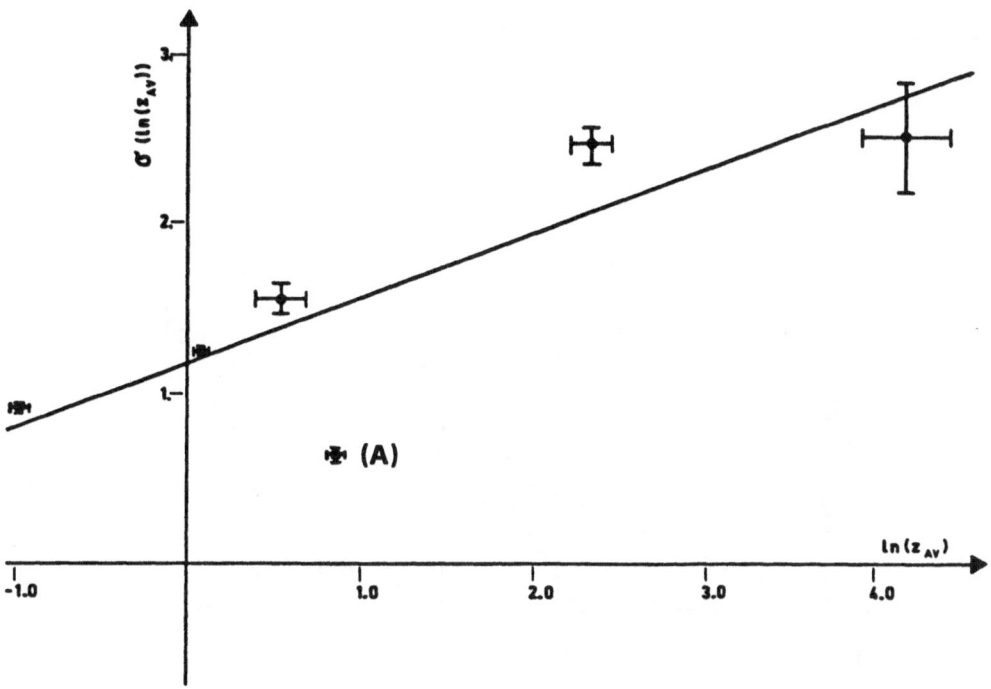

Figure 7. Correlation Among Five Independent Sampling Campaigns

5.5.7   Fractal Geometry and Random Fractals in Chemical Accidents

Fractal geometry and fractal objects (Mandelbrot, 1977) are a new technique only in environmental sciences. This technique has been applied to meteorologic problems (Lovejoy, 1981; Lovejoy and Mandelbrot, 1985), and in perspective it appears to be the most powerful tool to describe the distribution of polluting agents in the atmosphere and consequently on the ground. Although incomplete, Mandelbrot (1977) provides a rigorous mathematical treatment of fractals - from the latin "fractus", a graphic representation made of broken lines or geometric shapes having very irregular and fragmented contours that remain such on any given scale. In addition, random fractals can account mathematically for particularly erratic fluctuations, larger than allowed by the most common Gaussian distribution.

Very fragmented and irregular contours are typical of pollutions consequent a chemical accident. Similarly, large erratic fluctuations are typical of pollution situations following a chemical accident. These two simple facts allow to predict that fractals will probably be the algorithm of the future in the environmental science of chemical accidents.

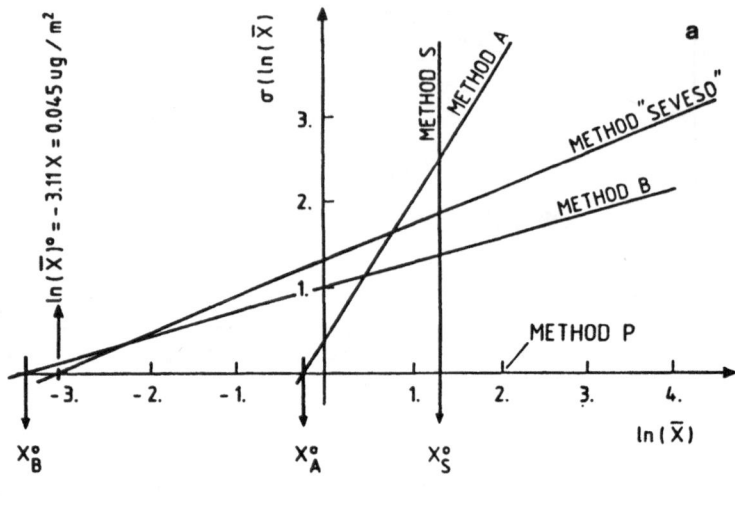

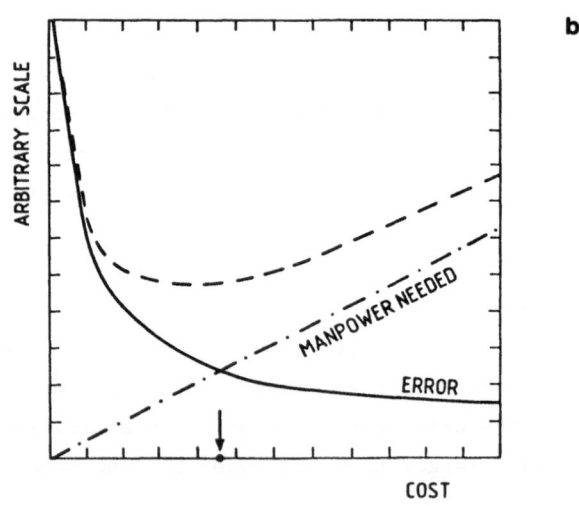

Figure 8a,b.   Sketches to Compare Different Analytical Procedures and
Cost-Benefit Optimization

To illustrate concisely the concept of fractals we will recall that in Euclidean geometry, for any given planar figure, the perimeter P is proportional to the square root of the area A. That is:

(VIII) $P = k \sqrt{A}$, or

(IX) $P = k \cdot A^{\alpha}$,

with $= 0.5$. For sake of illustration, in Table 2 some regular geometric figures have been collected together with their perimeter-area quantitative relationships.

Now, let us suppose that we have a set of small circles connected by two-way lines (bottom part of Table 2), and suppose we want to calculate the total perimeter - which is the sum of the perimeters of individual circles plus the lengths of all interconnecting lines. At the limit, let us suppose we have a set of points occupying completely a given area, and that we wish to measure the perimeter, in the sense of considering every point as a circle having a vanishing radius. In such case, the perimeter does coincide with the area. Therefore, the more fractured is the perimeter of a surface, the larger is the value of parameter in Formula IX linking the perimeter to its area.

In regular fractals such as those shown in Figure 9 this depends upon the number n of straight segments into which a straight line connecting two points can be broken down. $D = \log(n)$ is called the fractal dimension. This dimension can be applied to any straight line connecting two points, no matter how short such line is. Figure 9 is self-explanatory.

Random fractals have been applied to meteorology (Lovejoy, 1981; Lovejoy and Mandelbrot, 1985), for instance to describe the dynamic evolution of rain in a given zone and estimate the relationship between zone and contour of clouds. In Figure 10 the relationship between area (km^2) and perimeter (km) of both clouds and rain areas are shown to follow Formula IX on as many as six orders of magnitude, i.e., from areas in the order of $100 \ m^2$ to areas in the order of millions of km^2.

Based on the "fractal sum of pulses", the model is stochastic in nature. However, while in normal stochastic models, the expected values of pulse durations and amplitudes are finite, in fractal models the total sum of the elementary pulses has to approach infinity when the time interval allowed for the pulse to take place tends to zero. In the spirit of the fractal idea (Figure 9), the phenomenon has to be valid on any scale,

Table 2. The Discontinuous Nature of Geometry

	Fundamental Lenght	Perimeter	Area	Area/ Perimeter	
	L	$P = K\ A$	A	A/P	$A\ =\ f(L)$
○	radius R	$2\ \pi\ R$	$\pi\ R^2$	0.5 R	3.14 R^2
△	side L	3 L	0.433 L^2	0.144 L	0.433 L^2
▢	side L	4 L	1.0 L^2	0.24 L	1.0 L^2
⬠	side L	5 L	1.7105 L^2	0.342 L	1.711 L^2
⬡	side L	6 L	2.598 L^2	0.433 L	2.598 L^2
⬡	side L	7 L	3.634 L^2	0.519 L	3.634 L^2
○	side L	8 L	4.828 L^2	0.603 L	4.828 L^2
○	side L	9 L	6.182 L^2	0.687 L	6.182 L^2
○	side L	10 L	7.694 L^2	0.769 L	7.694 L^2
○	side L	11 L	9.366 L^2	0.859 L	9.366 L^2
○	side L	12 L	11.196 L^2	0.933 L	11.196 L^2

$$P = K\ A^\alpha$$

PEANO CURVE (year $\approx$1890) $P = A$

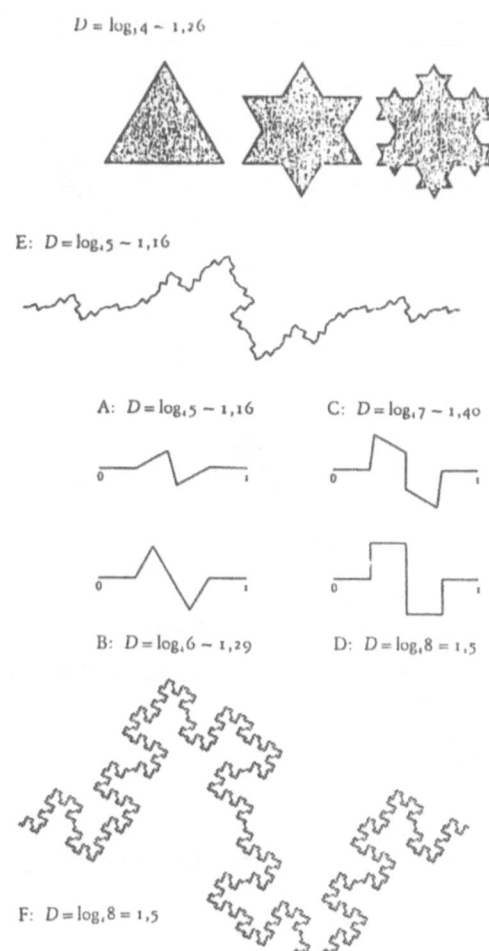

$D = \log_3 4 \sim 1{,}26$

E: $D = \log_6 5 \sim 1{,}16$

A: $D = \log_4 5 \sim 1{,}16$ C: $D = \log_4 7 \sim 1{,}40$

B: $D = \log_4 6 \sim 1{,}29$ D: $D = \log_4 8 = 1{,}5$

F: $D = \log_4 8 = 1{,}5$

Figure 9. Examples of Regular Fractals

as is the case exhibited in Figure 10. Mathematically, in order to guarantee these features, it is sufficient to assume that:

1.   the probability that an evolutionary   duration r' exceeds
      r be:

      $P \bullet r(r'>r) = 1/r;$

2.   the relative intensity of the elementary pulse Δr is a random
      contribution - either negative or positive - to the total sum.
      Positive and negative signs are equally probable and the intensi-
      ty of the signal is related to the occurrence probability by the
      equation:

      $\Delta r \sim \pm\ r^{1/\alpha},$

      where α indicates the dimension of the fractal phenomenon.

    In our laboratory, these models are now being applied to describe the
radioactive distributions and precipitations after the Chernobyl accident

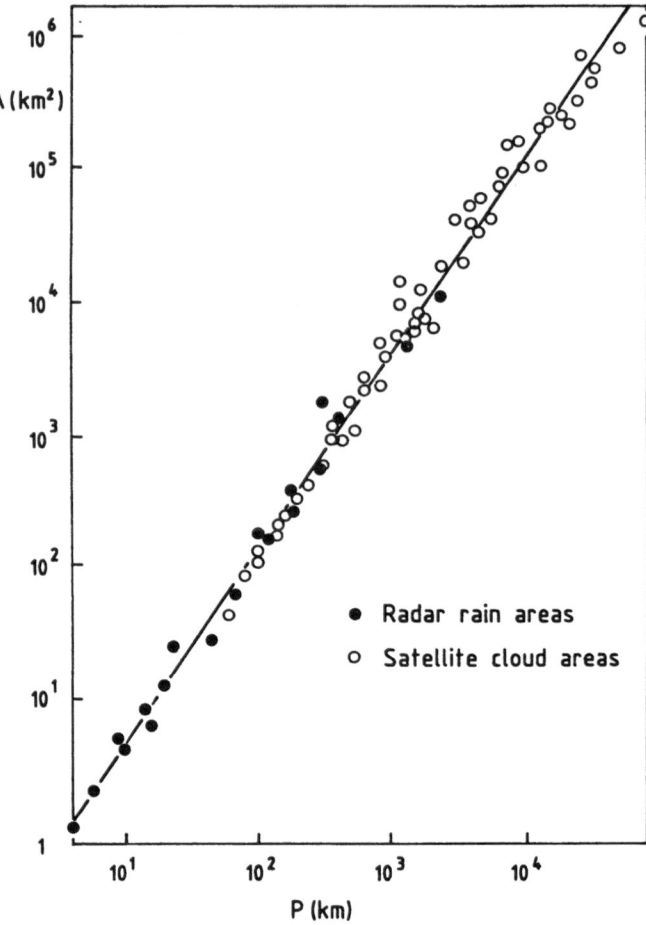

Figure 10. Correlation Plot Between Observed Areas and Observed
Perimeters for Meteorologic Events

(Graziani and Maineri, 1988) and for a global description of TCDD deposition on the ground as due to the Seveso accident (Salvadori, in preparation).

5.5.8   Acknowledgments

The author is grateful to: the members of the Pavia group, G. Belli, G. Bressi, and S. Cerlesi, for the collaboration given and the constant personal interaction on any detail of the work; Drs. A. di Domenico and S. Facchetti for numerous discussions; the student G.F. Salvadori for his "fractal" enthusiasm; and finally Mr. G. Bonaschi for his very careful drawings and technical assistance.

REFERENCES

Belli, G., Bressi, G., Calligarich, E., Cerlesi, S., and Ratti, S. 1982a. Geometrical Distribution of TCDD on the Surface Layer around the ICMESA: An Analytical Description of the Main Features and the Different Approaches in the Different Mapping Procedures. In: <u>Chlorinated Dioxins and Related Compounds - Impact on the Environment</u>, pp. 155-171, Hutzinger, O., Frey, R.W., Merian, E., and Pocchiari, F., Eds., Pergamon Press, Oxford.

Belli, G., Bressi, G., Calligarich, E., Cerlesi, S., Ratti, S. 1982b. Analysis of the TCDD Distribution as Function of the Underground Depth for Data Taken in 1977 and 1979 in Zone A at Seveso (Italy). In: <u>Chlorinated Dioxins and Related Compounds - Impact on the Environment</u>, pp. 137-153, Hutzinger, O., Frey, R.W., Merian, E., and Pocchiari, F., Eds., Pergamon Press, Oxford.

Belli, G., Bressi, G., Cerlesi, S., and Ratti, S. 1983. The Chemical Accident at Seveso (Italy): Statistical Analysis in Regions of Low Contamination. <u>Chemosphere 12</u>, 517-521.

Belli, G., Cerlesi, S., Milani, E., and Ratti, S. 1988. Statistical Interpolation Model for the Description of Ground Pollution Due to the TCDD Produced in the 1976 Chemical Accident at Seveso in the Heavily Contaminated Zone A. <u>Toxicol. Environ. Chem.</u>, in press.

Bertazzi, P.A., Ratti, S.P., Cerlesi, S., Bressi, G., and Zocchetti, C. 1986. Assessment of Exposure to TCDD in the Seveso Area for Epidemiological Research". In: Proceedings of the 50th International Symposium on Epidemiology in Occupational Health, Los Angeles, California, September 9-11, 1986. In press.

Cerlesi, S., di Domenico, A., and Ratti, S.P. 1987. Recovery Yields of Early Analytical Procedures to Detect 2,3,7,8-Tetrachlorodibenzo-p-dioxin (TCDD) in Soil Samples after the Industrial Accident of July 10, 1976, at Seveso, Italy. Presented at the Seventh International Symposium on Chlorinated Dioxins and Related Compounds "Dioxin '87", Las Vegas, Nevada, October 4-9, 1987.

Cerlesi, S., di Domenico, A., and Ratti, S.P. 1988. 2,3,7,8-Tetrachloro-dibenzo-p-Dioxin (TCDD) Persistence in the Seveso (Milan, Italy) Soil". Presented at the Eighth International Symposium on Chlorinated Dioxins and Related Compounds "Dioxin '88", Umea, Sweden, August 21-26, 1988.

Cramer, H. 1955. <u>Mathematical Methods of Statistics</u>, Princeton University Press, Princeton, New Jersey.

Graziani, G., and Maineri, M. 1988. Application of a Fractal Model to the Precipitation Field in an E.C. Country for a Day after the Chernobyl Accident. JCR-Ispra Technical Note No. 188.

Lovejoy, S. 1981. The Statistical Characterization of Rain Areas in Terms of Fractals. In: <u>Proceedings of the Toronto Conference on Radar Meteorology</u>, pp. 476-483.

Lovejoy, S., and Mandelbrot, B. 1985. Fractal Properties of Rain and a Fractal Model. <u>Tellus 37A</u>, 210-232.

Mandelbrot, B. 1977. _Fractals: Form, Chance and Dimension,_ W.H. Freeman and Co., S. Francisco, California.

Merlo, F., Cerlesi, S, Ghioldi, R., Stagnaro, E., and Ratti, S.P. 1984. Geographical Distribution of Some Indicators of TCDD Pollution in the Seveso Area. Report DFNT/RL 84/21.

Ratti, S.P., Belli, G., Lanza, A., Cerlesi, S., Fortunati, U. 1986. The Seveso Dioxin Episode - Time Evolution Properties and Conversion Factors Between Different Analytical Methods. _Chemosphere 15,_ 1549-1556.

Ratti, S.P., Belli, G., Bertazzi, P.A., Bressi, G., Cerlesi, S., Panetsos, F. 1987a. TCDD Distribution on All Territory around Seveso - Its Use in Epidemiology and a Hint into Dynamic Models. _Chemosphere 16,_ 1765-1773.

Ratti, S.P., Belli, G., and Cerlesi, S. 1987b. Mathematical Approach to Data Analysis in Environmental Science - The Lecture of Seveso. Presented at the Seventh International Symposium on Chlorinated Dioxins and Related Compounds "Dioxin '87", Las Vegas, Nevada, October 4-9, 1987.

5.6   A Summary Comment on the Seveso Experience

A. Essam Radwan

Department of Civil Engineering
Arizona State University
Tempe, Arizona 85287, USA

The Seveso incident brought to our attention how vulnerable our society is to unexpected catastrophes resulting from technological innovations. The incident was a typical low probability, high consequence event. Such events occur once a decade, but when they happen they become the focus of public attention around the globe and tend to make public officials reevaluate current policies related to emergency response and contingency plans. It is imperative that our world see more technological advances in all fields; with that, more and more hazardous substances will be produced, stored, handled, and disposed of. Furthermore, no matter how advanced industrial innovations become, the human being will always be in the loop and human error is inevitable. It is our responsibility to make the public aware of the high price we might pay to cope with technological disasters.

This report contains an interesting collection of papers that address technological, legal, and public policy issues of the Seveso incident.

More specifically, the report entails information on PCDDs, where to find them and how to assess their risk on public health. There are numerous potential sources for both PCDDs and PCDFs. Some are mobile sources (motor vehicles) and others are stationary sources (power plants, industrial and chemical plants). It has been reported that these sources produce small amounts of dioxins. Data is being collected to identify locations in Italy where the presence of both PCDDs and PCDFs could cause a health threat to the public.

Chemical industries have identified the urgency of developing contingency plans and emergency response plans in the case of unexpected hazardous events. MONTEDIPE is a good example of one of those plants where a comprehensive data base on hazardous substances was developed. This data base contains thousands of chemicals that are handled by this facility, and it provides emergency plans on how to reduce the risk resulting from incidents involving these chemicals. The data base operates on a microcomputer. It is perceived as good information service for fire brigades and other emergency response agencies.

The use of computer power to store large amounts of information and process it in short time has resulted in tremendous advances in the field of data processing. Artificial intelligence has broadened its domain from the traditional high technology defense-related applications to civilian applications. Expert systems, as a special application of Artificial Intelligence, have been applied extensively to engineering, medicine, agriculture, and public policy. The described application of an expert system to a chemical emergency in this report is to aid decision makers during and after chemical accidents. Once the system is fully developed and tested by local authorities, industry, and defense officials, it can be viewed as an excellent decision support system and may be exported to other localities in Europe and other parts of the world.

The subject of response to a chemical emergency in general and the Seveso incident in particular is successfully addressed in three papers.

The three steps identified for health care measures following the occurrence of a chemical emergency are: prevention of further exposure, identification and follow-up of exposed individuals, and diagnosis and treatment. The success of implementing these three steps depends primarily on the effectiveness of the health care plan and the success of overcoming technical, political, and organizational problems.

As far as emergency response actions to the Seveso incident are concerned, appropriate precautions were taken to minimize the risk of exposure to dioxin for pregnant women, children, and animals. No conclusive evidence was found afterwards that the incident caused major changes in the pattern or frequency of congenital anomalies among newborn children in the polluted areas. The accumulated experience with emergency health care activities and countermeasures during the few years following the incident has certainly made emergency response officials and decision makers aware of the magnitude of the problem. Furthermore, it shed some light on how costly an emergency contingency plan can be, and how this cost can be significantly reduced if a similar incident should occur in the future.

The long-term surveillance of affected population and cleanup workers of affected areas and contaminated buildings has been thoroughly addressed in this report.

It was concluded that without an active mechanism of tracing the affected population immediately after the incident occurs, it is impossible to conduct successful studies to evaluate the long-term effect of accidental exposure. Studies have concluded that the safety measures taken during the cleanup activities were effective in reducing dioxin absorption. As to the evacuated area, the health authority concluded that the evacuated population could be allowed to reenter the area without having any risks to their health due to the effectiveness of reclamation operations.

The utilization of computer networks is considered an excellent vehicle for research groups in the field of technological disasters to exchange information. Networks tie a great number of countries worldwide, and permit users to perform activities such as: electronic mail, requesting urgent information, sending and receiving list of references, and uploading and downloading files that contain technical papers or reports.

The need for skilled personnel following industrial accidents has been acknowledged as a critical issue. Without trained and experienced personnel, developing immediate and effective response and planning future actions may be an impossible task.

The liability and compensation of damages for technological disasters is a sensitive subject. No matter how much the financial compensation could be for the affected population, the "moral damage" and the complex human sacrifice are almost never compensable.

Two subjects could have used more attention in this report. The first is risk assessment, and the second is the transportation of hazardous substances.

The subject of risk assessment has been addressed in the last paper with reference to live animals and soil contamination. It was pointed out that although the exposure at Seveso did not cause measurable health damages of the exposed population, it is advisable to monitor a large number of toxic compounds and try to reduce their release into the environment as much as possible.

The transportation of hazardous substances is emerging as an international concern for hazardous material manufacturing, hazardous waste generators, and local and national regulatory agencies. Concern arises from the risk associated with handling such substances outside the industrial site and the possibility of an incident occurring which may result in a chemical release, thereby resulting in a possible health threat to the public or contamination of agricultural land and water sources.

Several countries have established formal regulations for mitigating hazardous substance disasters (United States, Canada, Czechoslovakia, Denmark, France, to mention only a few). Despite efforts made by these countries to promote safety, disasters have occurred. A few countries, including Bulgaria, Canada, Federal Republic of Germany, France, Rumania, the United Kingdom, and the United States, have laws and regulations governing the manufacture, handling, and transportation of hazardous substances.

Accident histories around the world reveal that relatively few accidents involving hazardous materials in transport occurred during the last three decades. The records of these accidents are not detailed enough to assist in planning accident mitigation efforts. Furthermore, statistics indicate that the frequency of accidents resulting from transporting hazardous substances is highest for road transport.

Results of a recent international mail survey confirmed the presence of hazardous substances in transportation systems. Flammable substances and compressed gases top the list of common categories of hazardous materials. Countries that responded to the survey reported that they do not have a record-keeping system for hazardous substances; however, there are no designated departments to keep accident statistics. All countries reported that they have some kind of disaster response systems, but the responsible units are not equipped for handling disasters. Furthermore, inadequate control and enforcement of national and international regulations related to safe packaging was a consensus among the respondents.

Certain common problems do exist on an international level. These problems are basically defined by the lack of common standards and practices related to packaging and transporting hazardous substances, and the inadequacy of reliable accident data collection systems.

INDEX